Oceanography
An Introduction

FOURTH EDITION

Oceanography

An Introduction

Dale E. Ingmanson
William J. Wallace
San Diego State University

Wadsworth Publishing Company
Belmont, California
A Division of Wadsworth, Inc.

Oceanography Editor *Jack Carey*
Special Projects Editor *Mary Arbogast*
Editorial Assistant *Susan Belmessieri*
Production Editor *Leland Moss*
Designer *Carolyn Deacy*
Print Buyer *Barbara Britton*
Art Editor *Toni Haskell*
Copy Editor *Caroline Arakelian*
Photo Researcher *Stuart Kenter*
Technical Illustrator *Tasa Graphics Arts, Inc.*
Compositor *Graphic Typesetting Service*
Cover *Carolyn Deacy*
Cover Photograph *William Garnett*

Printed in the United States of America *49*
2 3 4 5 6 7 8 9 10—93 92 91 90 89

Library of Congress Cataloging-in-Publication Data

Ingmanson, Dale E.
Oceanography : an introduction / Dale E. Ingmanson, William J. Wallace, David H. Milne.—4th ed.
p. cm.
Bibliography: p.
Includes index.
ISBN 0-534-09552-6
1. Oceanography. I. Wallace, William J. II. Milne, David. III. Title.
GC16.I53 1989
551.46—dc19 *00-17213*
CIP

To all those who make their living respectfully on, in, or about the ocean.

Contents

FIVE

Plate Tectonics / 54

SIX

Margins of the Continents / 68

SEVEN

The Chemical Properties of Water / 86

EIGHT

The Physical Properties of Seawater / 102

NINE

Climate and the Ocean / 122

TEN

Ocean Circulation / 150

ELEVEN

Waves / 184

TWELVE

Tides / 208

THIRTEEN

Coastal Processes and Estuaries / 232

FOURTEEN

Life in the Sea: Bacteria, Protists, Plants, and Invertebrates / 272

FIFTEEN

Life in the Sea: Vertebrates / 312

SIXTEEN

The Distribution and Abundance of Life in the Sea / 342

Preface

The earth has often been viewed as a possession to be subdued and exploited. The ocean that makes up about 70% of the earth has not been easily subdued or exploited despite heroic attempts. As a result, people still refer to the ocean as a frontier or as a vast resource that will enable us to solve such problems as food shortage or a shortage of valuable minerals. One reason for writing *Oceanography: An Introduction* was to explore the myths and realities according to our present understandings of the ocean.

In our opinion textbooks should present the guiding philosophy of the authors very early on. Our book was the first general oceanography text to take serious issue with environmental topics. When one considers the recent (March, 1986) statement by the Foundation for Oceanographic Research that there has been more marine pollution in the past ten years than in the previous 400 years combined, it is obvious that people must be made aware of the problems before they can become concerned. But real awareness is based on understanding. So the primary aim of this book is to present the fundamental geological, chemical, physical, and biological marine processes necessary to understand the ocean environment. We hope this understanding and awareness will help to generate the desire to help in the preservation of this planet's last frontier.

We have been on, in, and around the ocean all of our lives. From a purely pragmatic viewpoint this quote by Kenneth Grahame is worth a mention: "There is nothing—absolutely nothing—half so much worth doing as simply messing about in boats . . . or with boats . . . In or out of 'em, it doesn't matter." This would include, of course, scientific activity as well.

This fourth edition of *Oceanography* is substantially rewritten and updated. An extensive survey of faculty who have read or adopted the text and of students in our classes at SDSU has been used to help guide us to make substantial changes. The sequence of chapters remains the same as the last edition with discussions of the chemical and physical properties of sea water split into two chapters.

Dr. David Milne, marine biologist from Evergreen State College, has completely rewritten the three chapters dealing with marine biology. Highly knowledgeable about current concepts in marine biology, Dr. Milne is an experienced teacher of undergraduate students and a capable writer.

The chapters dealing with marine geology have been substantially rewritten and updated with the suggestions made by Dr. Tjeerd van Andel of Stanford University.

This text remains the only introductory oceanography text with a chapter on polar oceanography, an area receiving increasing scientific attention.

The chapters on marine resources, ocean technology, pollution, and management have been extensively updated with more than fifty new or revised tables and figures. There are also new sections on marine transportation, marine recreation, and the exclusive economic zone.

The book emphasizes visual materials: There are over four hundred line drawings and photographs, more than three hundred of which are original. Many illustrations have been added to this edition. Illustrations are essential to an oceanography text because some students have never experienced the ocean in person, while many others have seen it only from the shoreline or from the deck of a boat, usually in only a few locales. If properly executed, illustrations can be good substitutes for direct observation.

This edition continues an extensive glossary based on the *Glossary of Oceanographic Terms* published by the U.S. Naval Oceanographic Office and other specialized sources. Key terms are set in boldface type for easy identification. Also, coastal maps of the USA show coastal configuration, locations, and the 30- and 100-fathom contours.

To give the book another dimension, we display at the beginning of each chapter an original photograph along with quotations from the works of Carroll, Melville, Conrad, Whitman, Arnold, and others. We hope these quotations and photographs will convey the variety of human experience with the ocean in a stimulating way. As Alice said, "What is the use of a book without pictures?" (Lewis Carroll).

Extensive Manuscript Review

We would especially like to thank the many reviewers who contributed their time and ideas to all four editions.

Edward Aguado / *San Diego State University*
David Alt / *University of Montana*
Franz E. Anderson / *University of New Hampshire*
Scott M. Ascher / *Meramec Community College*
William Balamuth / *University of California, Berkeley*
Charles Breitsprecher / *American River College*
Harold L. Burstyn
Frank Carsey / *JPL*
Bert L. Conrey / *California State University, Long Beach*
Herbert Curl / *Oregon State University*
Christopher Dewees / *University of California, Davis*
Clive Dorman / *San Diego State University*
Michael Dowler / *San Diego State University*
Charles L. Drake / *Dartmouth College*
Robert Eberhardt / *San Diego Community College*
Donald Eidemiller / *San Diego State University*
William P. Elliot / *Oregon State University*
Theodore L. Esslinger / *North Dakota State University, Fargo*
Reinhard Flick / *University of California, San Diego*
Tom S. Garrison / *Orange Coast College*
Michael T. Ghiselin / *California Academy of Sciences*
William Glen / *College of San Mateo*
Ralph Gram / *California State University, Hayward*
Gary B. Griggs / *University of California, Santa Cruz*
Howard R. Hetzel / *Illinois State University, Normal*
Tom Hopkins / *Brookhaven National Labs*
Donald E. Keith / *Tarleton State University*
Robert C. King / *San Jose City College*
Phyllis Kingsbury / *Drake University*
Donald Klim / *Leeward Community College*
Ronald Kong / *American River College*
Eugene Kozloff / *University of Washington*
Gerald Kuhn / *Scripps Institution of Oceanography*
Rivian S. Lande / *Long Beach City College*
Norma Lang / *University of California, Davis*
Lawrence H. Larsen / *University of Washington*
Ruth LeBow / *University of California, Los Angeles*
Donald Lovejoy / *Palm Beach Atlantic College*
D. McGeary / *California State University, Sacramento*
Joseph W. MacQuade, Jr. / *North Shore Community College*
J. Robert Moore / *University of Wisconsin*
Steve Murray / *California State University, Fullerton*
Steve Neshyba / *Oregon State University*
Dennis N. Nielsen / *Winona State College*
Jerry O'Donnell / *Del Mar College*
B. L. Oostdam / *Millersville University*
Jan A. Pechenik / *Tufts University*
Charles F. Phleger / *San Diego State University*
Nan Pickett / *University of Wisconsin–Eau Claire*
K. M. Pohopien / *Mount San Antonio College*
Barry G. Quinn / *Westminster College*
Robert Riffinburgh / *Naval Ocean Systems Center*
Stuart A. Ross / *University of Southern California*
John Schopp / *San Diego State University*
Thomas W. Spence / *Office of Naval Research*
James J. Sullivan / *University of California, San Diego*
Mia Tegner / *Scripps Institution of Oceanography*
Stan Ulanski / *James Madison University*
Tjeerd Van Andel / *University of Cambridge, UK*
Ellen Weaver / *San Jose State University*
Edgar Werner / *Inter-American University*
Kelly Williams / *University of Dayton*

Oceanography
An Introduction

The sea does not reward those who are too anxious, too greedy, or too impatient. To dig for treasures shows not only impatience and greed, but lack of faith. Patience, patience, patience, is what the sea teaches. Patience and faith. One should lie empty, open, choiceless as a beach—waiting for a gift from the sea.

Anne Morrow Lindbergh

ONE

The Ocean in Perspective

You hear it long before you see it—a muffled roar, thunder in the distance. This is no ominous rumbling of some passing storm. It is the compelling, rhythmic sound of the world's ocean, its waters surging against the land.

You round a bend in the road or step through a break in the coastal forest or near the edge of a cliff, and you see it for the first time. Before you is an expanse of water so dynamic, so vast, so primal, that the turmoil of cities is forgotten. At that moment, your world expands. That moment, in Anne Morrow Lindbergh's phrase, is a gift from the sea.

Some people who experience that moment go on to become oceanographers of one sort or another. They may turn to chemical oceanography—the study of the distribution of chemical substances in the ocean water and the reactions that take place between them—or to physical oceanography—the study of the transmission of light, sound, and kinetic energy through the ocean, the distribution of temperatures, and air-sea interactions. They may become specialists in biological oceanography—the study of interactions of marine organisms with one another and with their environment. Or these people may become interested in geological oceanography—the study of the origin and physical characteristics of the ocean basins and the processes that have shaped them.

Whatever their special interest, oceanographers have all glimpsed a magnificent natural force, and they are all in their own ways responding to its challenge. Oceanography is more than a profession; it is a special way of viewing one of the great features of nature.

THE WORD *OCEANOGRAPHY*

Sir John Murray commented in his book *The Oceans,* published in 1910, on the word *oceanography:*

> The term *Thalassography* has been used, largely in the United States, to express the science which treats of the ocean. The term *Oceanography* is, however, likely to prevail. The Greeks appear to have used the word *Thalassa* almost exclusively for the Mediterranean, whereas the almost mythical "oceanus" of the ancients corresponds to the ocean basins of the modern geographer. In recent times I believe the word *Oceanography* was introduced by myself about 1880, but I find from Murray's English Dictionary that the word *océanographie* was used in French in 1584, but did not then survive.

The German word *Ozeanographie,* now largely replaced by *Meereskunde,* was used somewhat earlier than the English version.

1.1 Japanese woodcut prints (by Masanobu Kano) showing early nori cultivation techniques: preparing and planting brush as a foundation for Porphyra *spores (Kode, 1877).*

The suffix *-graphy* suggests drawing, describing, or reporting, as in *biography* and *geography.* The suffix *-logy* refers to a science or a branch of knowledge. Surely the study of the ocean has progressed beyond a pure description, and *oceanology* would be a more accurate term than *oceanography.* Still, *oceanography* retains its currency, and we shall use it throughout this book.

The term *hydrography* is sometimes used incorrectly as a synonym for *oceanography.* Hydrography deals primarily with the charting of coastlines, bottom topography, currents, and tides for practical use in ocean navigation. Oceanography is a more comprehensive discipline that uses chemical, physical, biological, and geological principles in its study of the ocean at large.

EMERGENCE OF THE SCIENCE OF OCEANOGRAPHY

Artists, poets, philosophers, admirals, and merchants have long had a passionate relationship with the ocean, and people of all ranks have harvested fish, shellfish, and seaweed from the sea for thousands of years (Figure 1.1). Seafarers have sailed the world in search of fortune and far horizons. Since the time of Homer, Plato, and Aristotle, and probably long before, poets and philosophers have reflected on the sea. Indeed, in Plato's model of the world system, which comprised earth, air, fire, and water, the water was the ocean (Figure 1.2). Aristotle devised a system for classifying living creatures and perceived the connection between marine fossils and living organisms.

However, oceanography developed late as a science. The study of the ocean began in earnest with the voyage of HMS *Challenger* (Figure 1.3). In December 1872, under the direction of Wyville Thomson of Britain, the *Challenger* embarked on the first major oceanographic expedition in history—an expedition that lasted almost three and a half years. During that time, the *Challenger* covered 68,890 nautical miles (Figure 1.4). (A nautical mile equals one minute of latitude, or 1/21,600, of a great circle of the earth; this is equal to 1,852 m [6,076.1 ft].) It was the first steamship ever to reach the Antarctic ice barrier and the first to cross the Antarctic Circle.

The wealth of information gathered during that expedition prompted J. Y. Buchanan, the chemist on the expedition, to comment:

> The history of the *Challenger* expedition is well known to all students of oceanography, which, as a special science, dates its birth from

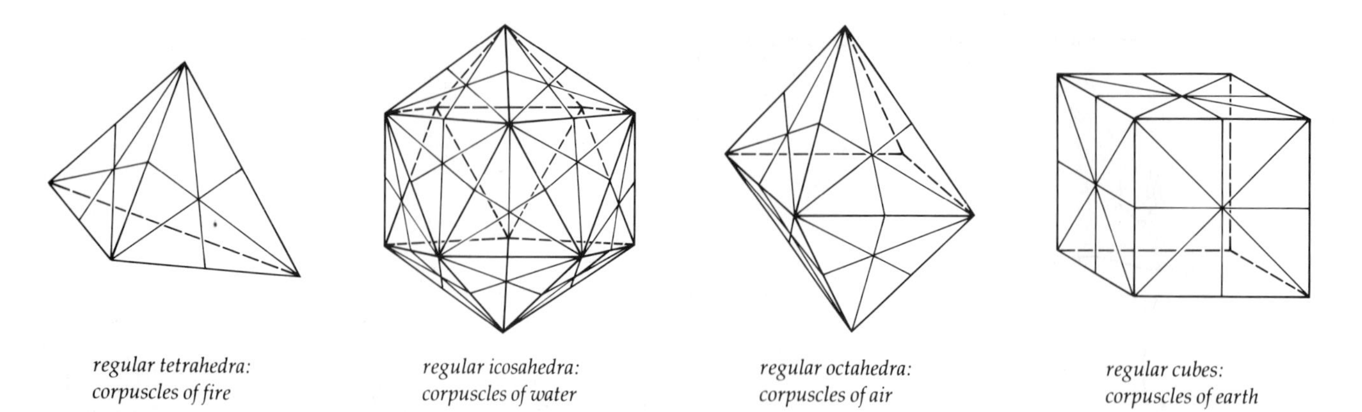

1.2 According to Plato, "First of all, take what we now call 'water.' When this is solidified we see, as we think *. . . , that it becomes earth and stones while, alternatively, that this very same thing (water), when dissolved and dispersed, becomes wind and air, and then fire when the air burns up; and that fire, conversely, when condensed and extinguished, goes over into the form of air; and that air again, when congesting and thickening up, becomes cloud and mist; and that from these, when they are further compacted, comes flowing water, and then from water come earth and stones once again, thus,* as it seems *. . . , they pass on to one another in a cycle the process of generation" (Gregory Vlastos,* Plato's Universe, *1975).*

1.3 HMS Challenger *at St. Paul's Rocks.*

that expedition. It must be remembered that when the *Challenger* expedition was planned and fitted out, the science of oceanography did not exist. . . . The work of the expedition proper began when the ship sailed from Teneriffe, and the first official station of the expedition was made to the westward of that island on 15th February 1873. It was not only the first official station of the expedition, but it was the most remarkable.

Everything that came up in the dredge was new, the relation between the result of the preliminary sounding and that of the following dredging was new; and, further, from the picturesque point of view, it was the most striking haul of the dredge or trawl which was made during the whole voyage.

Consequently, it may be taken that the Science of Oceanography was born at Sea, in Lat. 25°45′ N, Long. 20°11′ W, on 15th Feb-

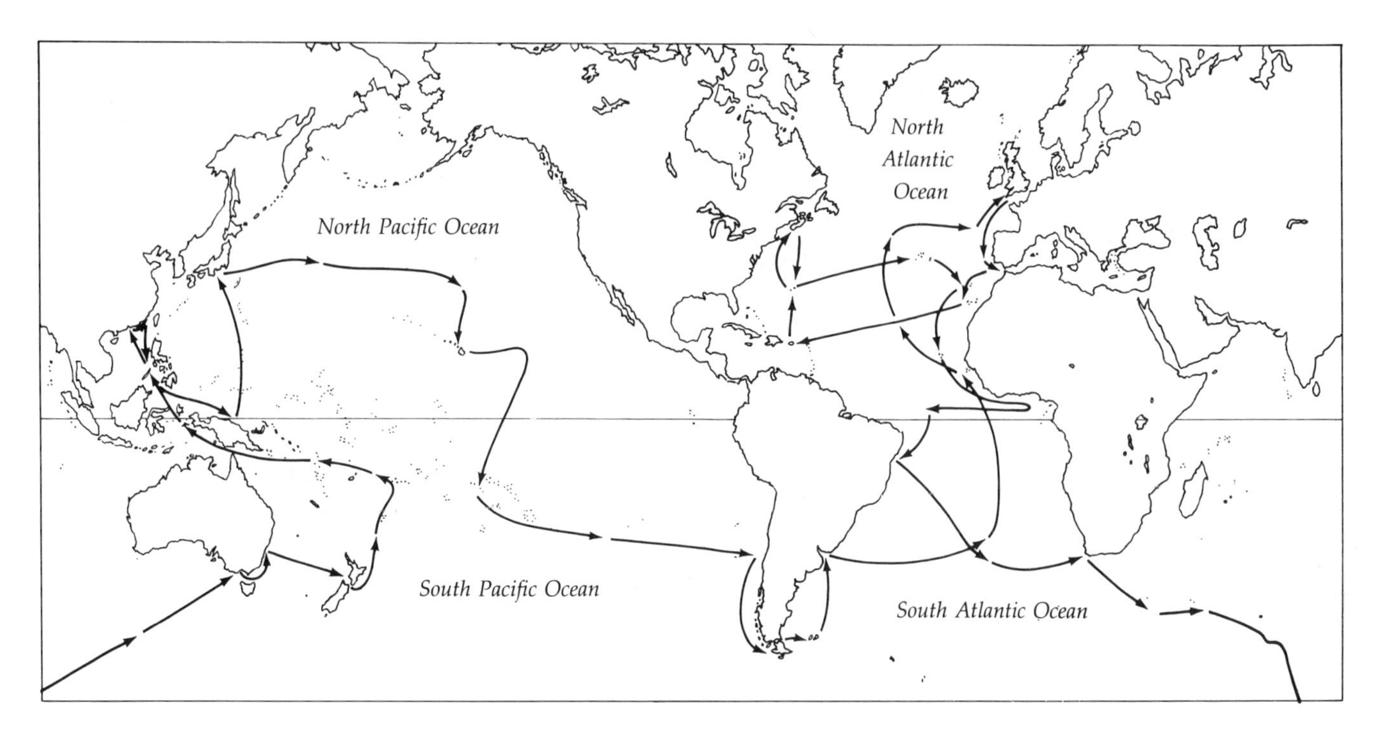

1.4 *Route of HMS* Challenger, *1872–1876.*

ruary 1873. . . . When the *Challenger* sailed from Portsmouth in December 1872, there was no word in the Dictionary for the Department of Geography in which she was to work, and when she returned to Portsmouth in May 1876, there was a heavy amount of work at the credit of the account of this department, and it had to have a name. It received the name Oceanography. (Buchanan, *Accounts Rendered of Work Done and Things Seen,* Cambridge: Cambridge University, 1919, pp. xii–xiii.)

The information gathered during the *Challenger*'s voyage was significant for every branch of the natural sciences. The scientists aboard collected and described 715 genera and 4,417 species of previously unknown organisms. They made soundings in the Mariana Trench to a depth of almost 9,000 m, proving that the ocean was much deeper than had previously been believed (Figure 1.5), and they made magnetic compass readings that enabled the *Challenger* to fix its location far more accurately than had ever been done before. The final report, titled *The Challenger Report,* published from 1880 to 1895, fills fifty large volumes. This historic voyage was responsible for the emergence of descriptive oceanography as a unique, interdisciplinary science.

As word of the successful *Challenger* expedition spread, other nations mounted their own expeditions. By the end of the nineteenth century, the science of oceanography had become firmly established.

1.5 *Dredging and sounding equipment on board HMS* Challenger.

1.6 *The chemical laboratory aboard HMS* Challenger.

CHEMISTRY AND THE SEA

Although Aristotle had recognized the saltiness of the sea and had speculated about its cause, there was little interest in the chemistry of the sea during the Middle Ages and the Renaissance. The first important work on the chemical characteristics of the sea appeared in 1674, when the English chemist and natural philosopher Robert Boyle published *Observations and Experiments about the Saltness of the Sea*. This pioneering work established Boyle as the founder of the science now known as chemical oceanography. It was the first definitive treatment of the physical and chemical characteristics of the sea, including parameters such as temperature, depth, and salt content. Boyle carried on his work at a time when his fellow scientists showed little interest in the subject.

Almost a hundred years later, in 1772, the French chemist Antoine Lavoisier published the first analysis of seawater as part of a paper on mineral waters. Twelve years later the Swedish chemist Torbern Olaf Bergman, also in a study of natural waters, included a list of the substances he had identified in seawater. Neither chemist was particularly interested in seawater itself; they simply regarded it as the most concentrated solution of salt water occurring in nature and studied it for that reason.

In the early nineteenth century many chemists turned their attention to the composition of seawater and tried to develop analytical methods that would produce consistent results. A lengthy paper by Georg Forchhammer, published in 1865, was a milestone in the history of chemical theories of seawater. Forchhammer introduced the term *salinity* and identified twenty-seven elements in seawater. His principal conclusion was that, although the salinity of the open ocean varied, the proportions of various salt constituents were the same everywhere. He introduced the coefficient of chlorine as a means of rapidly determining salinity, a method still used today. Other analytical works published by John Murray (not the later Sir John Murray of *The Challenger Report*'s fame) and Alexander Marcet in these early years attracted little attention at the time, although those works helped lay the foundations of chemical oceanography.

The publication in 1884 of William Dittmar's report on the seventy-seven water samples collected on the *Challenger* expedition marked the most extensive analysis of the chemistry of seawater undertaken until that time—and for some time to come (Figures 1.6 and 1.7). Previously, most scientists had concentrated on the components and salinity of seawater, although some of them were interested in the gases dissolved in it. There was little agreement among their findings, however.

Dittmar accepted Forchhammer's principle of the constancy of seawater constituents and used the coefficient of chlorine to determine salinity, although he suggested a different value for it. Partly because of the strong impact made by *The Challenger Report*, Dittmar's extension of Forchhammer's work won widespread acceptance in the scientific community. There was now little question about the major constituents of seawater or about the constancy of their proportions. With

1.7 Emptying one of the various types of water sample bottles after it has been brought back on deck of HMS Challenger.

this firm foundation, chemical oceanography emerged as a distinct discipline within the new science of oceanography.

During the last years of the nineteenth century, the discipline took on its modern form. Chemists began to investigate a great many aspects of the ocean water. Winkler, for example, published a test for dissolved oxygen in 1889. In 1892, Natterer analyzed dissolved organic matter, carbon dioxide, oxygen, and nutrients and related them to biological productivity. Svante Arrhenius' theory of electrolytic dissociation (1887) enabled oceanographers to "visualize" the constituents of seawater in a manner consistent with their analytical results.

Although nineteenth-century chemists learned a great deal about salinity, they still had difficulty defining the total salt content of seawater. That problem was solved by Martin Knudsen, Carl Forch, and S. P. L. Sörensen, who in 1902 produced a gravimetric definition of salinity with a procedure based on chloride content. Their work standardized procedures for determining salinity.

PHYSICS AND THE SEA

Waves, tides, and currents profoundly affect the physical, chemical, and biological characteristics of the ocean. Tides and currents were the first to receive serious scientific interest, perhaps because physicists associated them with such general forces as gravity and the motion of the earth.

In *Mathematical Principles of Natural Philosophy* (often known as the *Principia,* 1687), Isaac Newton presented a mathematical description of gravitational force that allowed, for the first time, a precise explanation of the tides. He also suggested, more than a hundred years before field measurements confirmed it, that the equatorial diameter of the earth is greater than its polar diameter. His work marked the final phase of a revolution that gave physicists the mathematics they needed to explain many natural processes.

Between 1687 and 1715, Edmund Halley, another great natural philosopher (best known for the comet that bears his name), published four influential articles dealing with the sea. In the first article, "An Estimate of the Quantity of Vapour Raised out of the Sea by the Warmth of the Sun," Halley used evaporation experiments to calculate the amount of water rising from the Mediterranean. He then calculated the amount of water that the Mediterranean receives from its nine major rivers, whose flow he equated roughly to the observed flow rate of the Thames. He demonstrated that more than three times more vapor left the Mediterranean than it received each day. Halley was the first to calculate how much water the sea lost by evaporation and how much it gained from river runoff.

If we consider Robert Boyle to be the father of chemical oceanography, we must consider Edmund Halley to be the father of physical oceanography. Strictly speaking, however, neither was an oceanographer. The ocean itself was of little interest to them except as it entered into their other studies. Boyle probably never even went to sea, and Halley made only short ocean voyages.

The scientific study of ocean waves began early in the nineteenth century. The first theoretical work was done in 1802 by Franz von Gerstner, who attempted to describe deep-water surface waves. In 1839 George Green showed that surface waves over deep water move at speeds proportional to their length, and by 1842 Georges Aimé had performed experiments to detect the depth to which wave action could be felt. Probably the most influential early work was done by Sir George Airy, who in 1845 noted that the passage of waves through water was oscillatory and that the water particles or molecules moving in elliptical orbits became progressively flatter with depth. His application of the sine-wave equation to ocean waves has remained in use ever since. In 1876 George Stokes pointed out that the velocity of a group of deep-water waves is half the velocity of the individual waves.

All these attempts to understand ocean waves were scattered efforts rather than parts of an organized, integrated study. It was World War II that revealed how little scientists really knew about the behavior of waves at sea. A thousand U.S. Marines were killed and two thousand wounded during the landing on Tarawa in the Pacific, largely through ignorance of surf conditions and the near-shore sea bottom profile. The first definitive study of waves at sea, *Wind, Waves and Swell: A Basic Theory for Forecasting,* was published in 1943 by Harald Sverdrup and Walter Munk of the

Scripps Institution of Oceanography. They reported that wave size depended both on wind speed and on duration as well as the area over which the wind was blowing, and they backed up their theory with observations. Soon they published a second book, *Breakers and Surf,* in which they treated shallow-water waves. The modern study of waves began with these influential works.

Ocean currents are another subject of lively interest to physical oceanographers. Early voyagers must have had some knowledge of ocean currents, although they left no scientific record of what they knew. As sea trade spread around the globe, ship captains kept extensive records of the currents and weather they encountered on their voyages. They kept their information secret, however, to deprive rivals of its economic and military benefits.

Georges Fournier (in 1667) and Athanasius Kircher (in 1678), both Jesuits, described the patterns of ocean currents. Kircher, a bold and imaginative observer, identified three major movements of the sea: the tides caused by the moon, the general movement of the sea from east to west in the tropics, and its movement from north and south. He noted that when the equatorial currents impinged on the eastern coasts of the continents, they were deflected north and south, creating a gyratory pattern that moved clockwise in the Northern Hemisphere and counterclockwise in the Southern. Kircher identified the evaporation and precipitation of water under the influence of the sun as the driving force behind these movements, although he suggested that winds might be a contributing cause.

Benjamin Franklin began to study ocean currents as early as 1769–70 and was later instrumental in publishing a chart of the Gulf Stream. He advised that if mariners used on-board thermometers to keep their ships within the Gulf Stream on their way to Europe and out of it on their way back, they would speed their trans-Atlantic crossings. Franklin's work was purely descriptive, rather than explanatory.

The first scientific study of currents was made by William Ferrel, who in 1856 described the effects of the earth's rotation on wind-generated currents and later derived the equation that relates the barometric-pressure gradient to wind velocity.

All the pilot charts issued by the Hydrographic Office of the U.S. Navy Department bear this legend: "Founded upon the researches made and the data collected by Lieutenant M. F. Maury, U.S. Navy." Between 1848 and 1855, Maury issued a series of papers called "Sailing Directions," and in 1855 he published a book titled *The Physical Geography of the Sea.* The book went through eight editions in the United States, growing from 274 to 474 pages; the last edition appeared in 1861. Maury dispelled many myths and errors about the sea that were generally accepted at the time. His book achieved widespread popularity and heightened public interest in the sea.

During the latter part of the nineteenth century, two major theories of ocean currents attracted the attention of many physicists. One theory held that surface winds were the propelling force behind currents. The other held that currents were the result of the cumulative effects of heating, precipitation, cooling, and evaporation.

Two methods of determining currents were commonly used at this time. Direct measurement, which had long been in use, had begun to produce more reliable results as techniques improved and new equipment, such as current meters, was developed. The second method was newer and more theoretical. Although it was based largely on direct measurements, it aimed to describe oceanic circulation by means of computations and theoretical assumptions. Pioneered by James Croll and Henrik Mohn, this method was best explained in a paper on hydrodynamics by Vilhelm Bjerknes in 1898 and in a paper on the effect of the earth's rotation on wind-driven currents by Vagn Walfrid Ekman in 1902. The complete geostrophic formula was published by J. W. Sandström and B. Helland-Hansen in 1909.

The use of submarines in the twentieth century increased dramatically the pool of information available to physical oceanographers. Even this advance and others, however, did not eliminate the need to measure currents over long periods of time. Ships supplied data from which patterns could be pieced together, but data for a detailed Gulf Stream study, for example, sometimes took several years to gather, during which time any short-term variations probably would have been missed.

The recent development of infrared satellite photography has made it possible to identify currents readily and accurately by noting relative differences in surface temperatures in adjacent waters at the same time as well as daily changes in local temperatures. Satellite data even permit the detailed study of eddies within the Gulf Stream.

BIOLOGY AND THE SEA

Biological oceanography—or marine biology, as it is also called—is a relatively new science, as is biology itself. Biology evolved not from medical studies but from natural history. The study of marine creatures goes back at least to the time of Aristotle's comments on dolphins and his brilliant treatise on a species of catfish, dating to around 350 B.C. Natural history did not come into its own, however, until a keen interest in terrestrial organisms arose during the second half of the eighteenth century. By the early nineteenth cen-

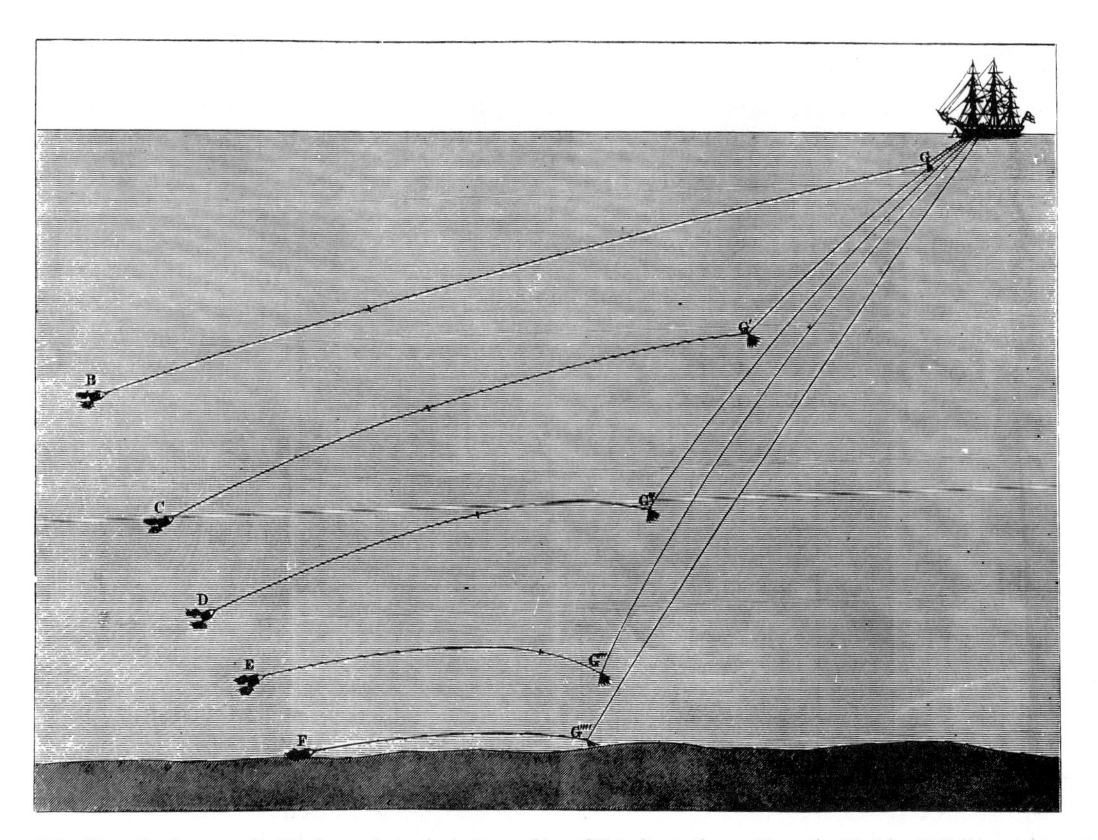

1.8 How the deep-sea dredge descends to the bottom. G' to G'''' indicate the position of a 68.2-kg (150-lb) weight as it slides down the rope.

tury, natural historians in Europe had waded through tide pools and salt marshes and had embarked on a variety of small craft to study the shallow waters along the coast.

They believed that the sea itself consisted of two layers: a thin surface layer and a deep, dark, cold abyss. They considered seawater so compressible and its density at the bottom so great that not even a lead sounding weight would survive at those depths. Clearly, there could be no life at the bottom of the sea. Anyway, they reasoned, since there was no circulation of water in the ocean deep and consequently no renewable food supply, no animals could possibly live there.

Edward Forbes proved them wrong. In 1841, in an effort to determine the upper limit of the lifeless (or azoic) zone, Forbes brought up specimens of living organisms from 421 m (230 fathoms) (1 fathom = 6 ft). Later he brought up specimens from 549 m (300 fathoms). Forbes died when he was 39, and so he never participated in the deep-sea expeditions that his work stimulated and encouraged.

The British Antarctic Expedition of 1839–43, led by Sir James Ross, dredged shellfish and worms from depths of 732 m (400 fathoms). In 1858 HMS *Bulldog* brought seastars from the bottom of the North Atlantic, from a depth of 2,305 m (1,260 fathoms). In 1860

1.9 Bringing in the deep-sea dredge aboard HMS Challenger.

the raising of a failed Mediterranean submarine cable that had been laid in 1857 revealed animals encrusted on the cable and sounded the death knell for the azoic theory.

The publication of Charles Darwin's theory of evolution in 1859, contained in his *Origin of the Species*, suggested that life in the ocean changed more slowly than life on land and therefore might shed light on earlier forms of life. Biologists then began to draw samples from greater and greater depths. Sir Wyville Thomson and W. B. Carpenter reported, for example, that they found animals at whatever depth they dredged. In 1869 they brought up an abundance of sponges and foraminifera from a depth of 4,458 m (2,437 fathoms) (Figures 1.8–1.10).

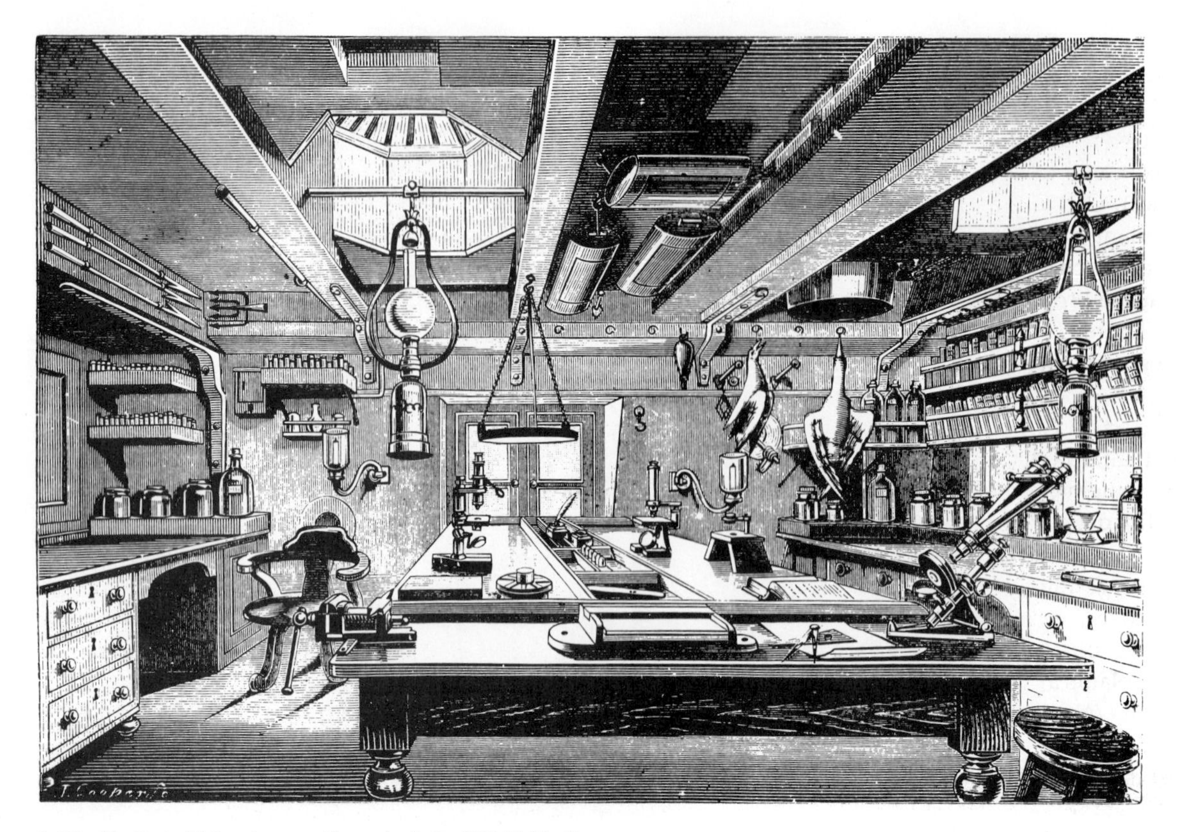

1.10 Zoological laboratory on the main deck of HMS Challenger.

As the century progressed, knowledge of deep-sea animals grew rapidly. One question remained unanswered, however: What was the food source for these animals? There was abundant life at the surface and at the bottom, but there seemed to be a relative absence of fauna at intermediate depths.

Thomson had suggested that the food source for the multicelled bottom-dwelling animals was a "protozoan" mass that included foraminifera, sponges, and unicells. Carpenter had shown through the analysis of seawater samples that the sea contained abundant organic matter in solution. Thomson concluded that this organic matter must be the source of nutrients for his protozoan mass because he did not believe that the population of organisms at intermediate levels could provide an adequate food source. In 1869, however, J. Gwen Jeffreys, on the basis of evidence that diatom shells had been found in deep sediments, suggested that both living and dead animals reach the bottom and serve as the food source for deep-dwelling creatures.

The first microscopic examination of marine phytoplankton (*phyto-* = plant; *plankton* = free-floating organism) was made in 1676 by Anton van Leeuwenhoek. Over the next 200 years, only sporadic observations were made, as when O. F. Mueller first used finely woven netting to sieve organisms out of seawater in 1777. Such organisms were regarded as curiosities of no great significance.

John Murray's studies during the *Challenger* expedition proved that plankton are abundant and widespread at the surface. Because of this finding, Jeffreys' views became accepted in subsequent years. The naturalist Alexander Agassiz conducted studies in the Caribbean and the Gulf of Mexico in 1880 that supported the view that organisms did indeed sink to the bottom. The role of phytoplankton in the food chain of the Arctic was set forth by G. O. Sars during the Norwegian North Atlantic Expedition of 1876–78. Until Viktor Hensen undertook the first systematic studies of plankton (a term that he coined) by making quantitative estimates of their populations and growth rate, information consisted of vague generalizations. The techniques and apparatus developed by Hensen for his plankton expedition of 1889 served as a model for years to come.

Earlier studies (such as by W. B. Carpenter in 1856) had shown that phytoplankton need sunlight, water, carbonic acid, and ammonia to carry out photosynthesis, and Paul Regnard had established in 1891 that photosynthesis does not occur deeper than 30 m (98 ft). Then Viktor Hensen in 1887 and Karl Brandt in 1899 showed that phytoplankton abundance was linked to the distribution of the nutrient nitrate. By 1900 it was clear that phytoplankton abundance was not simply a response to light and temperature but to nutrients as well.

Having been led to the study of deep-water plankton from the taxonomy of siphonophores (a type of colonial cnidarian), Carl Chun discovered a diverse, sometimes abundant midwater population of plankton that included such organisms as radiolarians, medusae, euphausids, copepods, siphonophores, and ostracods (see Chapter 14). In a series of deep-water investigations beginning in 1886, Chun established that deep-water animals, both benthic (bottom dwelling) and planktonic, feed on vertically migrating midwater plankton, called intermediate fauna, as well as on sinking organic remains. Chun's findings answered the question of the deep-water food source.

From about 1900 on, research into the sea's biology began to diverge into two distinct trends. One group of researchers, composed predominantly of naturalists, set out to collect, identify, and name all species of marine plants and animals and to study the total environment. Another group, intrigued and motivated by a decline in the number of fish in the North Sea, tried to answer the question, Why do fish populations fluctuate?

Despite the efforts of the latter group, little was learned about the causes of fluctuations in fish populations. By 1960, following impressive advances in fishing technology and increasing awareness of overfishing, especially in the North Atlantic, more attention was paid to the management of the fisheries than to the causes of fluctuating fish populations. While oceanographers in the Scandinavian countries and elsewhere continued to research the patterns of ocean currents with the hope that they would learn more about fish populations, such was not the case. In fact, pure science as practiced by Hensen, for example, contributed much more to marine biology than to the management of the fisheries. In the late 1950s and early 1960s, however, the fisheries management group began to publish an increasing number of papers on the marine environment, reflecting a growing awareness, for example, of the effects of pollutants on the sea as a whole and on fish in particular. This clearly indicated that the fisheries group had only begun to see fish as part of the overall ecology of the sea.

Today, the effect of overfishing on the sea's ecology still remains poorly understood. Neither is much known about how the larger marine predators are affected by their environment or how they compete with one another. Even though the environmental factors that affect the top terrestrial predators are well understood, that knowledge is of little help in understanding the relationships between marine predators and their environment. Terrestrial carnivores have relatively few offspring, and they survive with parental protection even in a harsh environment. In contrast, predatory fish lay large numbers of eggs, have virtually no parental care, and only a few survive to become adults. Since fish depend on their environment for survival, we might expect the commercial fisheries of the world to sponsor rigorous studies of the marine environment. Biologists have been urging them to do so since the end of the nineteenth century.

The divergent paths of marine research are now beginning to converge under the influence of such powerful modern tools as satellite imagery, which is used to determine chlorophyll concentrations in surface waters. Biological oceanography *in toto* seems to be moving toward an ecological approach in its study of the ocean environment.

The recent discovery of new species of large mussels and tube worms around deep-sea hot springs has stimulated research on deep-ocean life. These organisms seem to rely exclusively on dissolved nutrients for support. Biological oceanography continues to revise its theories in response to new information about life in the ocean.

GEOLOGICAL OCEANOGRAPHY

Prior to the 1960s, little was known about the ocean floor. Eighteenth-century theorists had conjectured about the effects of the Old Testament flood, but not until the end of that century did scientists undertake a serious investigation of the geology of the ocean.

In 1785, two papers were published that had a profound effect on the history of geology: one by Abraham G. Werner and the other by James Hutton. Although the ideas contained in these papers had been known previously, their publication triggered a lively controversy. Hutton and Werner, regarded as the fathers of modern geology, held opposing views. Werner's view, known at the time as the "Neptunian" view, was that the earth was relatively static and solid. Werner suggested that the earth's crust had been formed by the precipitation of solid matter from a receding ocean. Hutton's view, known as the "Plutonian" view, held that the earth was dynamic, in constant change. He believed that the crust had crystallized from molten material from within the earth. The controversy, then, was much deeper than a simple disagreement over the origin of basalt and granite or the deposition of sediment from the sea. Since the sediments could be observed only where they were exposed on land, the role of the sea in the formation of the sediments was in question.

Ancient philosophers believed that the ocean was concave, or bowl-like. If that were true, the deepest part of the ocean would be in the middle. That belief prevailed for a long time because mariners trying to

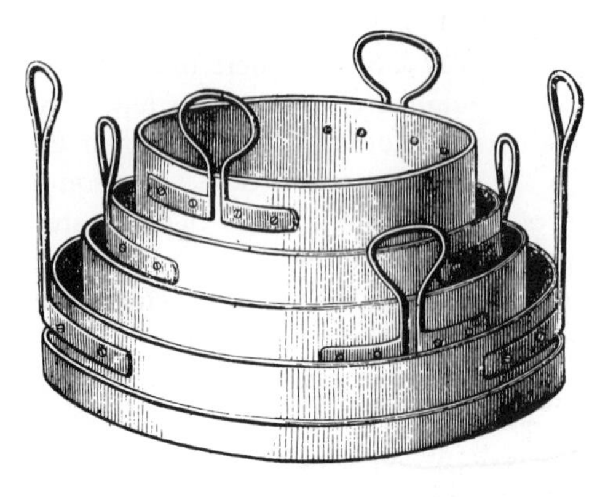

1.11 *The sieves used aboard HMS* Challenger *and their method of employ. Scientists dredge sediment from the seafloor and bring it on board to sieve and analyze.*

plumb the depth of the open ocean lacked sufficient line to reach the bottom. For example, Magellan failed in his attempts to sound the bottom of the Pacific during his voyages in 1521. Sir James Ross made soundings of the Weddell Sea in the 3,659 m (2,000-fathom) range but failed to complete a sounding at 7,317 m (4,000 fathoms) in 1843. HMS *Challenger* took 370 soundings at 362 stations approximately 324 km (200 mi) apart, with the deepest sounding at 8,186 m (4,475 fathoms). The *Challenger* used a steam-driven winch (a device that had been in use since the early 1850s) and a recently invented sounding machine that read the length of payed-out line. Before 1850, soundings had been made at random. The laying of the trans-Atlantic telegraph cable led to bottom surveys by H. O. Berryman in 1853 and 1856. From the data assembled from those surveys, Maury constructed the first bathymetric (depth measurement) chart of the North Atlantic.

As time passed, researchers began to assemble information about the composition of the ocean floor as well as its topography. As early as the eighteenth century, British hydrographers had studied bottom samples that had become embedded in the soft lead of the sounding weights. Later sounding weights had a hollow space in the bottom to allow the collection of sediments. These cavities were sometimes lined with tallow to improve sediment adherence. Over the years, as dredging techniques improved, more and more samples of mud and rocks were brought up from the ocean bottom. The German Atlantic Expedition of 1925–27 used a coring device capable of taking cylindrical samples 1–1.2 m (3–4 ft) long. In 1935, Charles Piggott fashioned an improved device that produced cores 3 m (9.8 ft) long. In 1945 a piston corer, invented by Borje Kullenberg, produced cores up to 21.3 m (70 ft) in length.

By the time of the *Challenger* expedition, the scientific community had recognized that a dynamic relationship existed between the ocean and the continents. Data assembled by the *Challenger* suggested that deep-sea deposits had formed slowly over a long period of time (Figure 1.11). Wyville Thomson and John Murray believed that the ocean basins were permanent features, unchanged since early geological times. In 1889, Osmond Fisher calculated that the suboceanic crust was denser than the continental crust and thus had sunk more deeply into the substratum, which was then assumed to be liquid. Fisher also suggested that convection currents in the earth's crust ascend beneath the oceans and descend beneath the continents, causing the continental crust to rise and buckle into mountains.

These suggestions did not meet with universal agreement. Early in the nineteenth century, James Hutton and Charles Lyell, British geologists, proposed that sediments might shift from one location to another and that both the seafloor and the land could rise and fall. They could not, however, conceive of any force that could transport the earth's crust laterally for great distances. They explained the presence of tropical fossils in rocks in England as the result of climatic change rather than what later was referred to as continental drift. Most geologists of the nineteenth century accepted this view. By 1900, the theory that the ocean basin and the continents were permanently fixed had won general but not universal acceptance. Some geologists, however, favored a more dynamic view.

In the 1920s, a new tool became available that enabled oceanographers to gain accurate knowledge of ocean depths. This was the echo sounder (originally called the fathometer), which had been invented during World War I as a means of locating submarines.

1.12 The Deep Sea Drilling Project (DSDP) vessel Glomar Challenger *drilled and cored for ocean sediment in all the oceans of the world. The Scripps Institution of Oceanography was the managing institution for DSDP. (Scripps Institution of Oceanography)*

After World War II, military and commercial vessels alike recorded bottom profiles on strip charts along their routes of travel. As more and more data accumulated, it became obvious that the ocean floor was highly irregular, with mountain chains and ridges even greater than those found in the North Atlantic by the *Challenger*. The deepest regions of the ocean, it became clear, were not at its center but close to its edges, that is, closer to the continents.

In January 1912, Alfred Wegener presented a paper on lateral continental displacement and shortly thereafter published a book titled *The Origin of Continents and Oceans*. The notion that the continents had at one time been joined and had subsequently moved apart was not new—F. B. Taylor and H. B. Baker had arrived at a similar conclusion independently. Furthermore, as early as 1807 Alexander von Humboldt had noted the seeming fit between the opposing coasts of Africa and South America. Nevertheless, the publication of Wegener's views led to a heated controversy. Most of his contemporaries rejected his theory, objecting that no known mechanism was capable of moving continents and that it was difficult to conceive of immense continental crustal blocks moving over oceanic rock. But Wegener had set the stage for the emergence of modern theories of geological oceanography. He proposed convection in the earth's mantle as the mechanism for moving continents, an idea which has become accepted in the form of seafloor spreading proposed in the early 1960s by Harry Hess and R. S. Dietz.

Following World War II, offshore oil and gas exploration provided a wealth of new information about the ocean and the ocean floor near the continents. In the 1960s, increased federal funding of deep-sea geological research culminated in the Deep Sea Drilling Project, for which a ship, the *Glomar Challenger*, was specially designed (Figure 1.12). As a result of this project, many assumptions on which geological oceanography had been based have been tested and reappraised.

Table 1.1 *Milestones in the Study of the World Ocean*

Year	Event
1674	Robert Boyle publishes *Observations and Experiments on the Saltness of the Sea,* in which he proves that evaporation methods alone cannot reproduce salt content values for seawater. Included is a description of the first use of silver nitrate as a test reagant for seawater, as well as comments on the sea's temperature and pressure.
1687	Isaac Newton publishes the *Principia,* which describes the three principles of mechanics and the principle of universal gravitation, as well as an explanation for the tides based on gravity.
1725	Luigi Marsigli writes the first book dealing entirely with the sea, *Histoire physique de la mer.*
1740	Leonhard Euler discusses a tangential force generating the ocean tides, then shows that the moon's attractive force has maximum effect on water directly below it.
1769–70	Benjamin Franklin is instrumental in developing the first published chart of the Gulf Stream. He suggests that mail packets stay in the Gulf Stream going to Europe and avoid it when returning. Franklin measures sea surface temperature during Atlantic crossings in 1775, 1776, and 1785.
1772	The French chemist Antoine Lavoisier publishes one of the first analyses of seawater.
1768–80	Three important voyages take place led by James Cook. Extensive data on currents, depths, tides, and temperatures are collected.
1814	The narratives of the travels of Alexander von Humboldt are published. They contain sea temperature measurements from around the globe at surface and deep waters, cold water temperature at depth in the tropics, notes on ocean circulation, and examples of what is now called upwelling.
1818–19	John Ross leads Arctic explorations in HMS *Isabella* and HMS *Alexander;* temperature measurements are made to a depth of 2,000 m.
1831–36	HMS *Beagle* embarks on a five-year voyage to chart the South American coastline. The journey helps to shape the ideas of the ship's naturalist, Charles Darwin, about coral reef and atoll formation and the process of biological evolution.
1835	Gaspard de Coriolis publishes his initial paper on changes in an object's horizontal motion across the earth's surface.
1845	George Airy noted that the water particles or molecules moved only in an oscillatory manner as an ocean wave passed through them.

POSTSCRIPT

The foregoing discussion is an all-too-brief sketch of the history of oceanography and of the contributions of various disciplines to the study of the sea (Table 1.1). We have mentioned only a few of the people who played a role in the emergence of the science, but we have tried to suggest that the history of oceanography, like the history of any science, is a history of ideas.

The highest goal of science is to explain the nature of the world. Its goal is not to promote the production of material goods or even to increase the life span of human beings. Rather, it is to uplift the human mind and spirit by providing an explanation of the world. It was toward that goal that Newton, Thomson, and Dittmar were striving.

Today, the fate of oceanography as a science seems to be closely tied to the pursuit of wealth and military power. Fortunately, however, that pursuit justifies the expenditure of as much as $50,000 a day just to operate a single oceanographic research ship. It also fosters the development of sophisticated satellite technology that contributes significantly to our knowledge of the ocean. The prospects for oceanographers in the United States and around the world are bright, especially in areas related to the development of marine resources (such as fishery management, shipping, and ocean engineering) and in research related to military needs (such as acoustics and physical oceanography).

Over the years the sea has attracted many scientists curious enough to ask questions and diligent enough to seek answers through hard work and painstaking research. The sea does not give up its secrets easily.

The era of great expeditions on ships like the *Challenger* is nearing its end. Satellite studies of surface waters and robotic vehicles to study the deep sea are opening a new era of research. Tomorrow's oceanographers may never go to sea!

FURTHER READINGS

Bailey, H. S. "The Voyage of the Challenger." *Scientific American,* 1953, *188*(5), 88–94.

Buchanan, J. Y. *Accountants Rendered of Work Done and Things Seen.* Cambridge: Cambridge University Press, 1919.

1854	Edward Forbes publishes "Distribution of Marine Life," which describes his concept of an azoic deep-sea zone.
1855	Matthew Maury publishes the first of eight American editions of *The Physical Geography of the Sea,* which becomes a popular and influential book.
1868–70	Wyville Thomson and William Carpenter make trial cruises on HMS *Lightning* and HMS *Porcupine,* which are preludes to the voyage of the *Challenger.*
1872–76	HMS *Challenger* expedition, led by Thomson, makes the first extensive, purely scientific cruise to study ocean chemistry, physics, geology, and biology.
1877–80	The American naturalist Alexander Agassiz takes part in three voyages of the U.S. Coast and Geodetic Survey Ship *Blake;* he takes 355 deep-sea stations and introduces wire rather than heavy rope for dredge tows following its use for deep-sea sounding on the USS *Tuscarora.*
1884–1901	The USS *Albatross* is built especially for oceanographic research. Scientists on board include Agassiz, W. Dall, D. Jordan, J. Murray, C. Gilbert, G. Goode, M. Rathbun, and R. Rathbun.
1899–1900	The Dutch survey vessel HMS *Siboga* explores the Indian Ocean and what is now coastal Indonesia.
1912	The German geophysicist Alfred Wegener presents his theory of continental drift.
1925	Oriented toward physical oceanography, work on the German *Meteor* during an Atlantic expedition heralds a new era in oceanology; using modern equipment, the most accurate and systematic data to this point are taken.
1942	Harald Sverdrup, Richard Fleming, and Martin Johnson publish the oceanographic classic, *The Oceans.*
1947–48	Hans Pettersson leads a major Swedish deep-sea expedition in the Atlantic, Pacific, and Indian oceans and the Red and Mediterranean seas.
1950–52	The Danish *Galathea* expedition explores the world ocean.
1960–61	Publication of the seafloor spreading hypothesis by Harry Hess and Robert Dietz.
1968–83	The drill ship *Glomar Challenger,* put to sea in 1968, begins an extensive series of operations to explore the ocean bottom and to test the newly emerging concept of global plate tectonics.
1985–present	Ocean Margin Drilling Program

Challenger Expedition Centenary. Proceedings of the Royal Society of Edinburgh, sec. 3, vols. 72–73. Edinburgh, 1972.

Davies, G. L. *The Earth in Decay: A History of British Geomorphology, 1578 to 1878.* Amsterdam: Elsevier, 1969.

Deacon, M. *Scientists and the Sea, 1650–1900.* London: Academic Press, 1971.

Kode, Elichi. "Illustrations of Japanese Manufacturers," vols. 1–4. Japanese Ministry of Home Affairs, 1877.

Maury, M. F. *The Physical Geography of the Sea.* New York: Harper, 1855.

Raitt, H., and B. Moulton. *Scripps Institution of Oceanography: The First Fifty Years.* Los Angeles: Ward Ritchie, 1967.

Sarton, G. *A History of Science.* Cambridge: Harvard University Press, 1959.

Schlee, S. *The Edge of an Unfamiliar World: A History of Oceanography.* New York: Dutton, 1973.

———. *On Almost Any Wind.* Ithaca, N.Y.: Cornell University Press, 1978.

Schneer, C. J., ed. *Toward a History of Geology.* Cambridge: MIT Press, 1969.

Sears, M., and D. Merriman, eds. *Oceanography and the Past.* New York: Springer-Verlag, 1980.

Shor, E. N. *Scripps Institution of Oceanography: Probing the Oceans.* San Diego: Tofua Press, 1978.

Thomson, C. W., and J. Murray, eds. *Report on the Scientific Results of the Voyage of H.M.S.* Challenger *during the Years 1873–1876.* 50 vols. London: HMSO, 1880–95.

Vlastos, G. *Plato's Universe.* Seattle: University of Washington Press, 1975.

Wallace, W. J. *The Development of the Chlorinity/Salinity Concept in Oceanography.* Amsterdam: Elsevier, 1974.

Jonathan Fisher

In the beginning God created the Heavens and the Earth. The Earth was without form and void, and darkness was upon the face of the deep; and the Spirit of God was moving over the face of the waters.

Genesis 1:1–2

TWO

The Origin of the Earth, Ocean, and Life

I. *The Origin of the Universe*
II. *The Origin of the Sun*
III. *The Origin of the Planets*
IV. *The Origin of the Ocean and Atmosphere*
V. *The Origin of Life*

All through recorded history—and possibly long before—the ocean has been a recurrent theme in religious accounts, folklore, and scientific investigations. Geologists believe that the ocean covered the face of the earth for about 200 million years between 3.9 and 4.1 billion years ago, and according to recent theories, life originated in geothermal springs deep in the ocean. Archeologists confirm that human societies have long had close ties to the ocean and its shores—ties that persist to this very day. Of the world's twenty largest cities, eighteen have direct access to the ocean. We swim in the ocean, fish in it, sail on it, dump waste into it, and probe it to discover its origins and the processes that govern its surface waters and its depths.

The ocean contains 1,360,000,000 km^3 (3×10^8 mi^3) of water and covers more than 70% of the earth's surface. This vast ocean is unique in the solar system; only on Earth is there a large body of water. Water does exist on other planets, but it is either locked in ice or suspended as vapor in thick, hot atmospheres, prevented from condensing and falling to the surface below.

Why is the earth unique in this respect? For a possible answer, we shall review current ideas about the origin of the universe, the earth, and the ocean.

THE ORIGIN OF THE UNIVERSE

When did the universe originate? Three approaches to this question suggest that the universe came into existence between 10 and 20 billion years ago (Table 2.1). These three approaches are: (1) nuclear chronology (based on rates of formation and relative amounts of the elements uranium, thorium, osmium, plutonium, and rhenium), (2) studies of the age of the oldest stars, and (3) measurements of the rate at which the universe has expanded.

According to the model most widely accepted by astronomers, the universe originated in a great explosion, the so-called "Big Bang." This model is consistent with observations first made in 1929 that distant galaxies are receding from the earth at velocities proportional to their distance from Earth. In 1948 George Gamow predicted that astronomers would one day detect background microwave radiation left over from the Big Bang. In 1965, A. A. Penzias and R. W. Wilson proved Gamow right when they detected such radiation, and subsequent measurements provided further confirmation.

Other theoretical models have been proposed to explain the origin of the universe, but they have proved deficient when tested against observations and physical measurements.

Table 2.1 *Chronological History of the Origin of the Universe, Earth, and Life*

Event	Time Before Present
Origin	20 billion years
Particle creation	20 billion years
Universe becomes matter-dominated	20 billion years
Galaxy formation begins	18–19 billion years
Galaxy clustering begins	17 billion years
Our protogalaxy collapses	16 billion years
First stars form	15.9 billion years
Our parent interstellar cloud forms	4.8 billion years
Protosolar nebula collapses	4.7 billion years
Planets form; rock solidifies	4.6 billion years
Intense cratering of planets	4.3 billion years
Oldest terrestrial rocks form	3.9 billion years
Microscopic life forms	3.8 billion years
Oxygen-rich atmosphere develops	2 billion years
Macroscopic fossils appear	600 million years
Early land plants appear	450 million years
Fish appear	400 million years
Ferns appear	300 million years
Conifers appear	250 million years
Reptiles appear	200 million years
Dinosaurs appear	150 million years
First mammals appear	50 million years
Homo sapiens appears	2 million years

THE ORIGIN OF THE SUN

Although we shall never know all the details of how the sun formed, many astronomers accept the gravitational collapse theory (Figure 2.1). This theory is also called the nebular theory and the dust-cloud theory. According to this theory all stars, including the sun, are formed in the same way, and planets sometimes emerge as a natural by-product of their formation.

Interstellar space contains vast amounts of gases, 99% of which consist of hydrogen and helium atoms. These gases frequently accumulate into more or less coherent clouds, or nebulae. (The singular *nebula* is Latin for "mist" or "cloud.") One such nebula is believed to have collapsed in response to gravity to form our solar system. Its initial mass was probably slightly greater than the present mass of our sun (approximately 2×10^{33} g, or 2×10^{27} tons).

As the nebula contracted, its rate of rotation increased and the nebula began to flatten as a result. It continued to contract until most of the matter had coalesced into a central mass, which ultimately became the sun. A small portion of the nebula survived as a flat disc spinning around the central mass, and it was from the matter contained in that disc that the planets eventually formed.

As the protosun (*proto-* from the Greek for "first," "foremost," or "earliest form of") continued to contract, its internal temperature rose to several million degrees Kelvin. The immense internal pressure that developed due to particle collisions eventually halted further gravitational contraction, and the sun stabilized. Nuclear fusion, which occurs at such extreme temperatures, released sufficient energy to maintain the temperature and pressure at constant levels, thus stabilizing the sun at essentially the size it is now. This process of formation, from nebula to stable star, probably required several tens of millions of years and occurred some 4.6 billion years ago.

THE ORIGIN OF THE PLANETS

While the protosun was undergoing the final stages of contraction, the flat disc of gases, solids, and liquids spinning around it was forming into planets. Again, the details of the process are uncertain, but we can sketch a plausible scenario.

The planets are believed to have grown through a steady process of accretion in which dust particles, molecules, and atoms at first joined together to form larger bodies, which in turn coalesced into larger and larger bodies. In time, through collision and gravitational attraction, these bodies developed into what we call planets.

In the region of the disc closest to the still-contracting protosun, the temperature was considerably higher (up to perhaps 2,000° K) than it was farther out in the disc. This was due to gravitational acceleration and collisions. Consequently, the bodies that formed in this inner region were composed mainly of the stablest molecules, forming the so-called rocky planets. Only in the more distant regions of the disc, where the temperature was very low, did the gases and less stable molecules form into solids and thus contribute to the formation of protoplanets and eventually the giant planets (see Table 2.2).

Because of the high temperatures, all regions of the disc must have consisted largely of hydrogen and helium initially, with the heavier elements making up only 1% or 2% of the total. The planets closest to the sun (Mercury to Mars) contain only very small amounts of hydrogen and helium, which are highly volatile

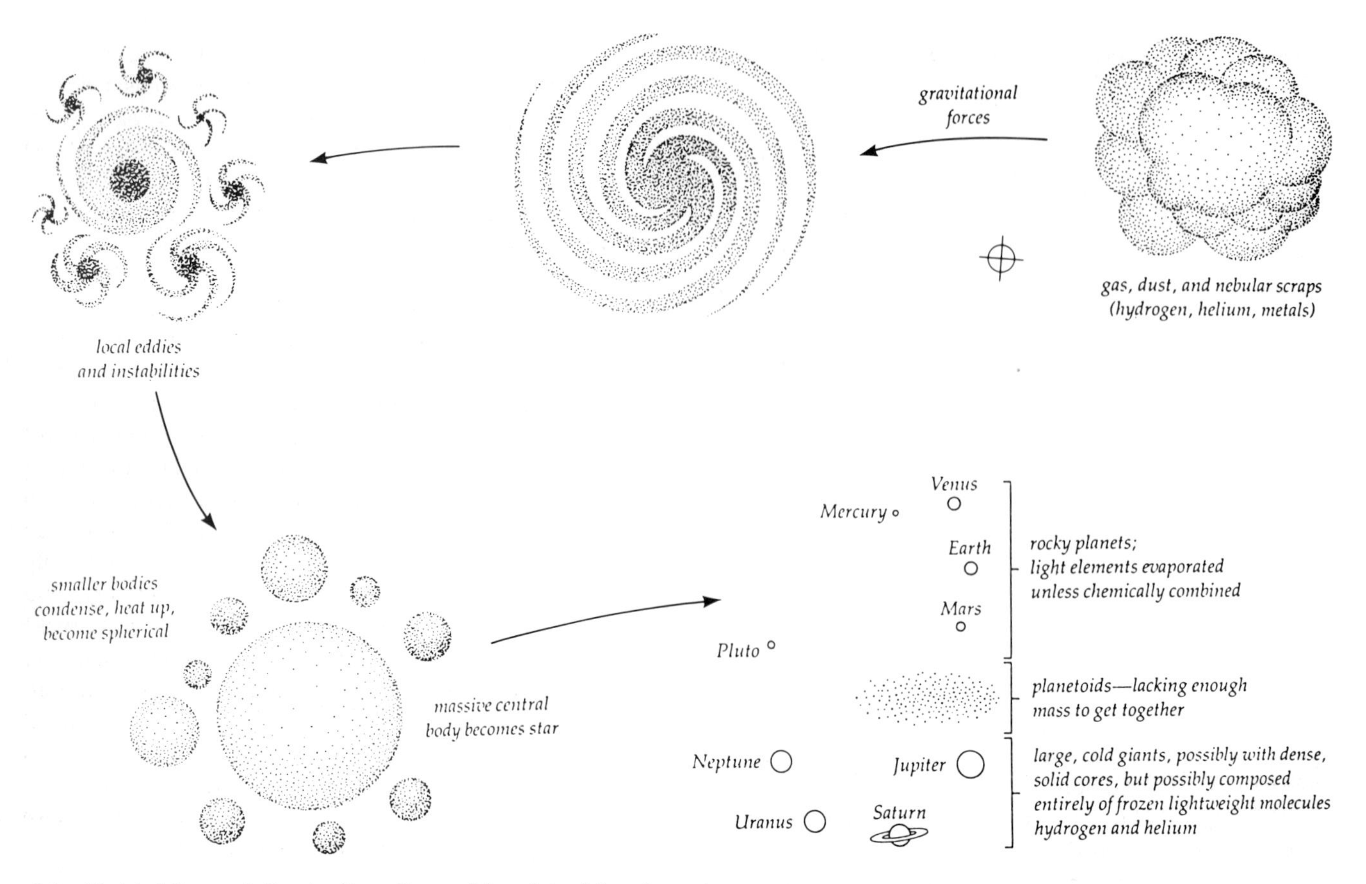

2.1 *Model of the gravitational collapse theory of the origin of the solar system.*

elements. Out in what were the colder reaches of the disc, hydrogen and helium account for a major portion of the mass of the planets (Jupiter to Neptune).

The evidence indicating that the planets were formed from a disc spinning about the sun is compelling. The orbits of the planets lie in roughly the same plane, and they revolve around the sun in the same direction and in virtually circular orbits. It seems likely that these highly regular orbital characteristics were established during the collapse of the nebula, before the planets formed.

The third planet from the evolving sun was the earth. This planet had a violent early history, marked by countless collisions with meteors. As it grew in mass, its temperature increased as a result of the energy released by impacts with meteors and the decay of radioactive elements within the planet. Although its temperature never rose to the level needed to initiate nuclear reactions, it did rise high enough to melt the interior. When this happened heavier elements, such as iron and nickel, were differentiated from lighter elements, such as carbon, and light minerals, such as quartz. The heavier elements formed the earth's core, and the lighter materials formed the mantle and crust.

The lightest gases, hydrogen and helium, were too light to be held by the earth's gravitational field. In fact, in these very early stages of the earth's history, the gravitational field was probably not strong enough to hold any gases at all. Since the heavier, chemically inert gases (neon, argon, and xenon) are less abundant on the earth than on other planets, scientists infer that the earth lost its early atmosphere to space.

THE ORIGIN OF THE OCEAN AND ATMOSPHERE

Where did the water now contained in the earth's oceans and atmosphere come from? The answer lies in the assumption that volcanoes were abundant early in the earth's history and that impacts by meteors caused gases to escape from the earth's surface.

Volcanic gases consist mainly of water vapor, nitrogen gas, and carbon dioxide. On the early earth, what would have happened to those gases? If the surface temperature of the early earth were about the same as it is now, the water vapor would have condensed to liquid water and the nitrogen gas and carbon dioxide would have formed the atmosphere.

Would the condensation of water vapor into liquid water have been sufficient to form the oceans? At the present rate of volcanism, the earth would have to be

Table 2.2 *Masses and Densities of the Planets*

Planet	Mass Relative to Earth	Mean Density (g/cm^3)
Inner Planets		
Mercury	0.06	5.4
Venus	0.82	5.2
Earth	1.00	5.5
Mars	0.11	3.9
Outer Planets		
Jupiter	318	1.3
Saturn	95	0.7
Uranus	15	1.3
Neptune	17	1.5
Pluto	0.002	0.8

three times as old as we believe it to be (4.5 billion years) for condensation to have produced the oceans as they exist today. Of course, the rate of volcanism may have been considerably greater in the past than it is today, in which case condensation of the water vapor produced by volcanoes might have been sufficient to create the present-day oceans.

Water vapor may also have been released when the impact of meteors raised the surface temperature of the early earth high enough to melt the outer layers. If the composition of those layers were similar to that of meteorites, which contain about 0.5% water, melting would have released large amounts of water vapor. As time passed, the frequency of impacts would have declined, since the meteors near the earth would have collided with it early in its history. The earth would have subsequently cooled, and the water vapor would have condensed, contributing to the formation of the ocean. Volcanic activity has probably continued to increase the volume of water in the ocean.

Scientific opinion now favors the idea that the water in our present oceans came predominantly from water vapor in the early earth's mantle, which condensed to form liquid water once the earth had cooled sufficiently. Calculations based on several assumptions show that the earth's mantle, which is thought to contain about 0.5% water, would have had to lose only about one-fifteenth of its suspected water content (via volcanism) to account for the present ocean. New water is thought to be released from the mantle at a yearly rate of approximately 0.1 km^3.

A new theory began forming in 1981 when the first attempt was made to prepare a global map of the earth's aurora—the reradiated ultraviolet images received from a high altitude (14,500 mi or 23,345 km). The polar-orbiting satellite *Dynamics Explorer 1* (DE 1) showed puzzling dark spots. In the next four years 2,000 hours of observing time identified over 30,000 spots. Initially these were thought to be the equivalent of static caused by telemetry interference or perhaps deteriorating sensors of computer errors. But retesting and cross-checking over this time period confirmed the fact that these spots did exist and that they were absorbing ultraviolet radiation over a large part of that band of the light spectrum. Of the common materials in the universe, only one substance does this—water. Calculations suggest that the spots could have been caused by 300 metric tons of ice entering the atmosphere every hour and vaporizing to form a cloud with an initial diameter of 30 mi (48 km) within 300 mi (483 km) of the earth's surface and eventually mixing with the atmosphere, precipitating, and becoming part of the ocean. Assuming that these "spots" occurred at the same rate through time, the calculations also show that they could supply enough water to compose the present ocean water's mass of 1.4×10^{27} grams. Although this seemingly radical theory of the ocean's origin has yet to gain acceptance by the scientific community as a whole, more and more data seem to offer additional support.

THE ORIGIN OF LIFE

What is the origin of life? This question has intrigued people throughout recorded history and has elicited sharply varying hypotheses. Many of the first explanations were mythical or supernatural. According to this view, God or gods created life. People who believe in the literal interpretation of the Bible adhere to this view.

However, it is important to distinguish between scientific explanations and theological explanations. Scientific explanations begin with data and observations, which constitute the raw materials of science. Models or theories are then devised to explain these data and observations. As more and more data and observations are assembled, the models and theories are elaborated and revised. This is a continuous process, although a model may prove valid and useful for hundreds of years. Theological explanations start with a model or a theory. These constitute the truth of religions. Data and observations are then cited selectively to confirm the theory. New observations and data are either accepted or rejected as valid, depending on whether they support the theory. The theory cannot be revised, since it is by definition true.

One of the most enduring theories about the origin of life was spontaneous generation. First proposed by Aristotle, it commanded almost universal accep-

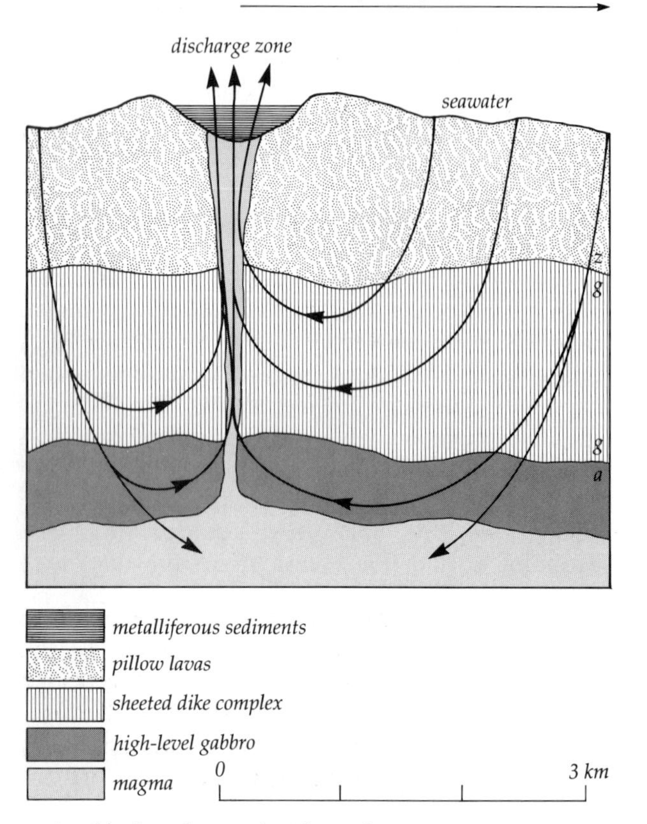

2.2 Hypothetical model of a seafloor hot spring as a reactor for synthesizing living organisms or their immediate precursors. Seawater percolates down through fractures in the seafloor, gets heated, and emerges through other fractures.

tance in the Western world until the nineteenth century. This theory holds that if conditions are right, life emerges spontaneously from nonliving matter. People found evidence of the validity of this theory when they saw maggots crawling out of decaying meat and frogs emerging from mud. Then, in the mid-nineteenth century, Louis Pasteur demonstrated that no microbes would grow in a nutrient-rich broth so long as the broth had been boiled long enough and was properly protected from the atmosphere. So much for the theory of spontaneous generation.

For any scientific theory to be accepted as valid, it must satisfy three requirements: First, it must be presented in abstract terms that are logical. Second, it must be tested by experiments conducted under rigorously controlled conditions. Third, its validity must be confirmed by observations of the natural environment.

In 1923, A. I. Oparin proposed the theory of the chemical evolution of life. That theory, which the scientific community now accepts as valid, states that life originated in a natural environment when inorganic chemicals combined to produce organic chemicals, which in turn produced a living cell.

In 1953, Stanley Miller reported experiments in which he used a hypothetical reducing atmosphere (chemicals combined with hydrogen rather than oxygen; for example, CH_4, not CO_2)—the atmosphere that Oparin had theorized would be required. Miller's atmosphere contained the inorganic molecules methane, ammonia, water vapor, and hydrogen. After periodic electric discharges through that atmosphere, amino acids, the building blocks of proteins, were observed as well as hydrogen cyanide (HCN) and formaldehyde (a simple carbohydrate, H_2CO). Since then, extensive research has confirmed that the complex organic compounds essential to life can be synthesized from inorganic substances.

When scientists refer to the theory of chemical evolution, they mean the series of chemical processes that occurred in the atmosphere, lakes, and ocean of the primitive earth and that resulted in the appearance of living organisms. The possible energy sources could have been lightning, ultraviolet rays from the sun, shock waves from earthquakes and meteor impacts, and heat from molten rock. Organic compounds would have accumulated in the lakes or ocean. Higher concentrations could have been achieved by evaporation of water in tide pools or mud flats, by freezing, and by adsorption on clay or mineral surfaces.

Until about twenty years ago, scientists were tentative in their acceptance of the theory of chemical evolution because they could make no observations in the natural world to substantiate it. However, two recent discoveries have brought them closer to outright acceptance. One was the discovery of such organic molecules as carbohydrates and amino acids in interstellar space and on meteorites. Since the earth is thought to have had the same origin as meteorites (indeed, they are of the same age), carbohydrates and amino acids were probably on the early planet as well. These are significant, of course, because sugars are carbohydrates and amino acids link together to form proteins.

The second discovery was the presence of a form of cyanide and of amino acids in a hot spring on the seafloor. For many years researchers were convinced that reactive molecules like cyanide and formaldehyde must have formed in the early earth's atmosphere and then entered into solution in the ocean. Subsequently, these researchers believed that those substances became concentrated in tide pools and began to react with one another. Current geological observations do not support these assumptions, however.

Recent research suggests that the earth's early atmosphere contained free oxygen but not much methane or ammonia. Researchers at NASA report that ultraviolet radiation reaching the earth from the young sun was 100,000 times greater in energy than it is now. Such radiation would have caused carbon dioxide and some water vapor to break down, releasing oxygen gas. It would also have caused organic compounds to break down. In addition, geological evidence of abundant iron oxide in the oldest rocks (3.8 billion years) confirms that the early atmosphere contained free oxygen.

The only natural environment on earth containing methane and ammonia that have not been produced by decaying organic material is the system of hot springs found at the seafloor (Figure 2.2). This is important because Oparin's theory assumes an oxygen-free environment consisting of methane, ammonia, hydrogen, and water. In addition, Miller and others have demonstrated in the laboratory that introducing energy into a mixture of these compounds will produce hydrogen cyanide, formaldehyde, and amino acids. As a result, scientists are now exploring the possibility that the reactive molecules were (and are) produced along the ocean floor in areas of geothermal activity or in space.

The first hot springs were discovered in the 1950s in the Red Sea. Since then many more have been found in the Atlantic, Indian, and Pacific oceans. Whatever their location, they are all associated with the oceanic ridges that form a 75,000-km (45,000-mi) system that splits the earth's crust into a series of polygons. Such hot springs could have been a prevalent feature of the early earth.

Scientists seeking to validate the theory of the chemical evolution of life are keenly interested in these newly discovered hot springs. These geothermal systems (deep-sea hot springs through which water seeps along fractures) exhibit all the varied conditions and provide all the compounds required (Table 2.3) for the synthesis of molecules essential to cell division and the inheritance of traits.

Can we be certain that geothermal springs were active along the oceanic ridges 4 billion years ago? Oceanic ridges are the sites of active crust formation, an activity that was surely occurring 4 billion years ago, and ocean water was certainly present. It seems likely that water circulated down into fractures, became hot, leached chemicals from the surrounding rock, and concentrated those chemicals, just as it does today. It is also likely that the water protected any newly formed organic compounds from decomposing, just as the atmosphere protects life today.

The ocean makes the earth unique. The ocean exists as liquid water because the earth is the proper distance from the sun to allow the appropriate energy to be absorbed. The proper combination of gases is present in the atmosphere to maintain the ocean's temperature between −2°C and 100°C. Also, the earth is sufficiently inactive geologically to maintain an appropriate heat flow to avoid either freezing or vaporizing the water. The interaction between the ocean and the formative processes of the earth's crust probably resulted in the origin of life or at least in the flourishing of life here. The origins of the earth, the ocean, and life itself may be intertwined in a complex fashion.

Table 2.3 *Requirements for Chemical Evolution*

1. Reducing environment
2. Presence of cyanide or derivative
3. Presence of water
4. Presence of carbon dioxide
5. Presence of hydrogen sulfide
6. Presence of nitrogen
7. Presence of inorganic phosphate
8. Known agents for catalyzing the synthesis of organic molecules
   a. Clays c. Zinc ions
   b. Lead ions d. Iron ions
9. Alternating wet and dry conditions for synthesizing organic molecules*
10. Temperature ranging from 4°C to 1,000°C

Note: Abiotic amino acids may also be necessary for chemical evolution.
*Deep-sea hot springs flow for a while, then dry out. Also, earthquakes can shift the spring pipelines.

FURTHER READINGS

Abell, G. O. *Realm of the Universe.* New York: Saunders, 1984.

Apfel, N. H., and J. A. Hynek. *Architecture of the Universe.* Menlo Park, Calif.: Benjamin/Cummings, 1979.

Cairns-Smith, A. G. *Genetic Takeover and the Mineral Origins of Life.* Cambridge: Cambridge University Press, 1984.

Ingmanson, D. E., and M. J. Dowler. "Chemical Evolution and Plate Tectonics." In *Origins of Life,* edited by Y. Wolman. Boston: Reidel, 1981.

Orgel, L. E. *Origins of Life.* New York: John Wiley, 1973.

Seeds, Michael A. *Foundations of Astronomy,* 2d ed. Belmont, Calif.: Wadsworth, 1988.

Silk, J. *The Big Bang.* San Francisco: W. H. Freeman, 1980.

Strahler, A. N. *The Earth Sciences.* 2d ed. New York: Harper & Row, 1971.

Wetherill, G. W. "The Foundation of the Earth from Planetesimals." *Scientific American,* June 1981, 162–174.

Young, L. B. *The Blue Planet.* New York: New American Library, 1983.

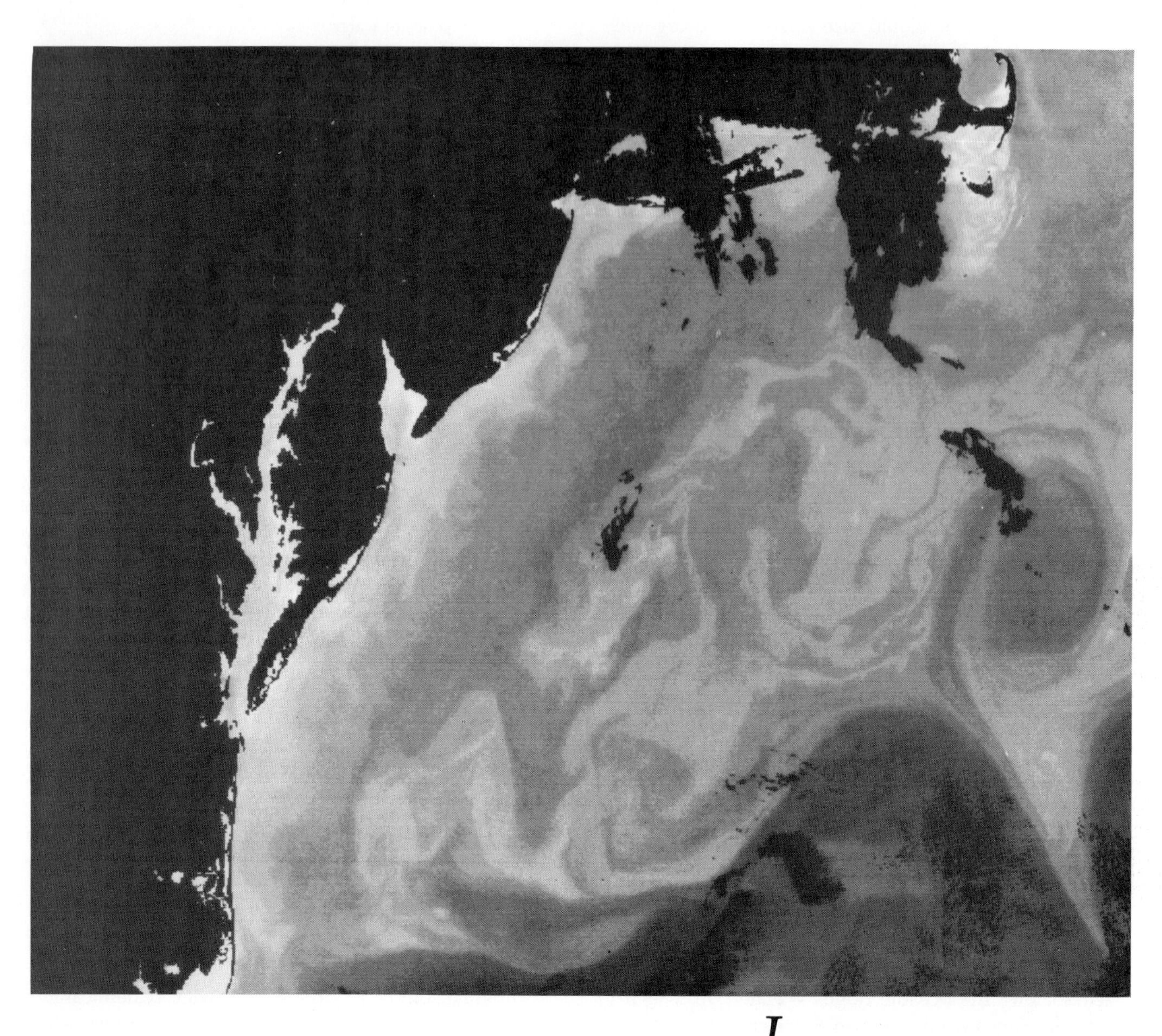

It's not been done, the sea, not yet been done,
From the inside, by one who really knows;
I'd give up all if I could be the one,
But art comes dear the way the money goes.
So I have come to sea, and I suppose
Three years will teach me all I want to learn
And make enough to keep me till I earn.

John Masefield

THREE

Obtaining Information About the Ocean Basins

I. *Research Ships*
II. *Methods of Sampling Sediments*
  A. *Depth Recording*
  B. *Dredging*
  C. *Coring*
III. *Properties of the Ocean Crust*
  A. *Heat Conduction*
  B. *Rock Magnetism*
  C. *Gravity Anomalies*
  D. *Seismic Profiling*

How do oceanographers obtain information about the ocean floor? Geologists working on land can hike across mountains and plains and easily collect samples along the way, or they can analyze cores taken at sites where oil or water wells are being drilled. But to obtain samples of rocks and sediments from the ocean floor, oceanographers must resort to a complex system of investigation conducted from ships that cost as much as $50,000 a day to operate. Also, oceanographers study more than just sediment and rock samples. They study patterns in earthquake activity, heat conduction through the earth's crust, rock magnetism, and variations in the earth's gravitational field. These patterns enable them to propose models that, for example, explain why earthquakes occur where they do and why deep trenches or volcanoes form.

In this chapter, we shall describe how oceanographers locate the position of their ship, collect bottom samples, and measure depth, heat conduction, rock magnetism, and gravity anomalies and how they undertake seismic profiling. We shall discuss the origin of rock magnetism in Chapter 5.

RESEARCH SHIPS

Before 1876, most of what was known about the physical and geological aspects of the ocean was summarized in one book, *The Physical Geography of the Sea*, by Matthew Maury. Prior to World War II, so little had been learned that, again, most of what was known appeared in a single volume, *The Oceans*, by H. Sverdrup, M. Johnson, and R. Fleming. By 1950, there were still few reliable maps of the seafloor beyond coastal waters, and our knowledge of seafloor features was minimal.

How could this scientific knowledge gap be closed? Clearly, scientists had to go to sea on ships. Since 1950, some 1,000 research ships have been built, and about 720 are registered today. The U.S.A. has 113 research ships longer than 25 m (80 ft), 43 of which are operated by governmental agencies, 25 by universities (Table 3.1), and 45 by private industry. The U.S. government defines an oceanographic research vessel as:

> a vessel which the Secretary of the department in which the Coast Guard is operating finds is being employed exclusively in instruction in oceanography or limnology, or both, or exclusively in oceanographic research, including, but not limited to, such studies pertaining to the sea as seismic, gravity meter, and magnetic exploration and other marine geophysical or geological surveys, atmospheric research, and biological research.

Table 3.1 *University Fleet—1982 (more than 80 feet LOA)*

Ship's Name	Length (ft)	Built/Converted	Crew/Scientists	Owner	Operating Laboratory
Large Ships					
Melville	245	1970	22/26	U.S. Navy	Scripps
Knorr	245	1969	24/25	U.S. Navy	W.H.O.I.
Atlantis II	210	1963	24/25	W.H.O.I.*	W.H.O.I.
T. G. Thompson	208	1965	22/19	U.S. Navy	U. Washington
T. Washington	208	1965	19/23	U.S. Navy	Scripps
Conrad	208	1963	25/18	U.S. Navy	Lamont-Doherty (Columbia U.)
Intermediate Ships					
Oceanus	177	1975	12/12	N.S.F.	W.H.O.I.
Wecoma	177	1975	12/16	N.S.F.	Oregon State U.
Endeavor	177	1976	12/16	N.S.F.	U. Rhode Island
Gyre	174	1973	11/18	U.S. Navy	Texas A. & M.
Columbus Iselin	170	1972	12/13	U. Miami*	U. Miami
New Horizon	170	1978	12/13	S.I.O.	Scripps
Fred H. Moore	165	1967/1978	9/20	U. Texas	U. Texas
Kana Keoki	156	1967	12/16	U. Hawaii	U. Hawaii
Small Ships					
Cape Florida	135	1981	9/10	N.S.F.	U. Miami
Cape Hatteras	135	1981	9/12	N.S.F.	Duke U.
Alpha Helix	133	1965	12/12	N.S.F.	U. Alaska
Ida Green	130	1965/1972	7/12	U. Texas	U. Texas
Cape Henlopen	122	1975	6/12	U. Delaware	U. Delaware
Velero IV	110	1948	11/12	U.S.C.	U. Southern California
Ridgely Warfield	106	1967	8/10	J.H.U.*	Johns Hopkins U.
E. B. Scripps	95	1965	5/8	S.I.O.	Scripps
Cayuse	80	1968	7/8	N.S.F.	Moss Landing Marine Lab
Longhorn	80	1971	5/10	U. Texas	U. Texas
Laurentian	80	1974	5/8	U. Michigan	U. Michigan

*Although title is held by the operator, ship construction was funded by the National Science Foundation, which holds a conditional lien on the title.
Abbreviations: Woods Hole Oceanographic Institution (W.H.O.I.); Scripps Institution of Oceanography (S.I.O.); National Science Foundation (N.S.F.); and Oregon State University (O.S.U.).
Source: After R. Dinsmore, *Oceans*, Spring 1982, p. 9.

The period from 1872 to 1987 has been an era in which research vessels have been the principal instruments by which scientists obtain ocean data. Without their specialized equipment, progress in oceanography would have been impossible (Figure 3.1). Some examples of this equipment include shipboard laboratories, stern-mounted A-frames for towing and lifting equipment, side thrusters to maintain ship position, and drilling and coring equipment. It may be that robotic systems and satellites will play increasing roles as oceanographic instruments, but ships will continue to be fundamental to oceanography in the foreseeable future.

Among the major advances in ocean technology, none has had such far-reaching effect as the design and implementation of shipboard computers. There have been special problems associated with this application: (1) limited space on ships, (2) maintenance of proper power output and temperature control, (3) secure mountings to compensate for ship movement, and (4) insulation from moisture. To a large extent, these problems have been overcome and computers

3.1 Research vessel Melville *was built under U.S. Navy contract for Scripps Institution of Oceanography, University of California, San Diego. The 245-foot, 2,075-ton, steel-hulled ship is capable of cruising 10 knots with a range of 9,000 miles. The blue-and-gray vessel carries out research in biological, geological, physical, and chemical oceanography. The ship's unusual propulsion system gives it extraordinary maneuverability for handling scientific investigations at sea. It uses vertically mounted, multi-bladed cycloidal propellers, one near the bow and one near the stern. (Scripps Institution of Oceanography, University of California, San Diego)*

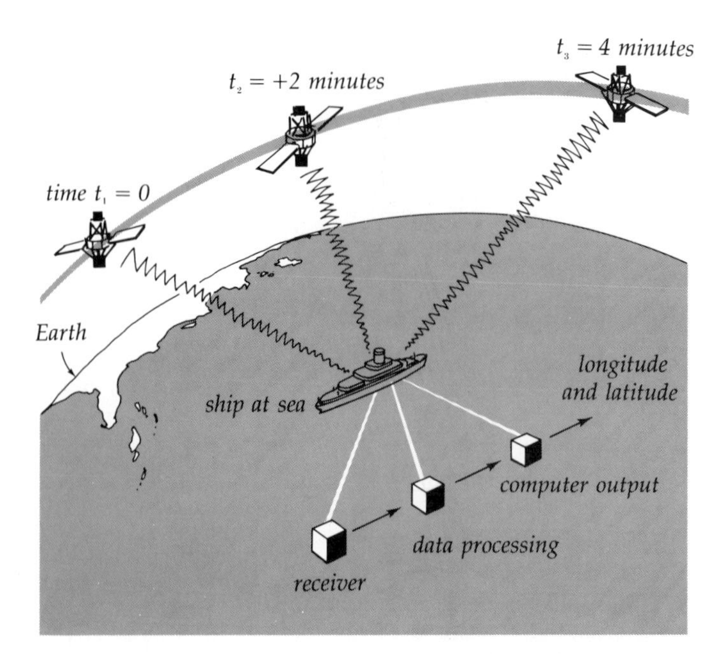

3.2 Satellite radio navigation system. The shipboard computer receives a radio output from the satellite and determines the ship's position relative to the satellite. The computer is also fed the precise orbit of the satellite and the most accurate possible time from a satellite control computer on land. From this, the position of the ship is determined.

are now used widely. The two most significant contributions are that they allow the feeding of raw data directly into a program for rapid preservation and analysis, and they introduce rapid and accurate positioning when coupled with satellite transmissions. Ships with computers on board can use a satellite radio navigation system (Figure 3.2). The system requires an accurate chronometer. The shipboard computer is fed information on the orbit of the satellite by a land-based control computer via radio. The shipboard computer then records the radio output of the satellite as it passes over and calculates the relative position of the ship with respect to the satellite. Then the latitude and longitude of the ship are computed by extrapolation. Greater accuracy is achieved by calculating the change in relative positions of the satellite and ship in succeeding time intervals. This system has increased navigation accuracy at least tenfold over previous systems.

METHODS OF SAMPLING SEDIMENTS

Depth Recording

The first accurate data concerning the ocean bottom came from depth recordings made during the *Challenger*'s explorations of the Atlantic, Pacific, Indian, and Antarctic oceans. Oceanographers on these voyages used steam-driven winches, designed by the physicist Lord Kelvin, and piano wire. Paying out more than 6,000 m (19,680 ft) of wire and reeling it back often took as long as twelve hours.

Today, data are gathered by means of sonar graphs of the ocean bottom (see Figure 3.3). First, a sound known as a *ping* is transmitted toward the bottom. The bottom reflects the ping, and a special microphone on the ship's hull receives the reflected sound. A timer records the time span between transmission and reception. The depth is calculated by multiplying the elapsed time by the speed of sound through water and then dividing that quantity by 2. All this is done automatically and is displayed on paper strips. Sometimes the ping is reflected by schools of fish, by the surfaces of water masses of different densities, or by loosely packed mud and ooze. Generally, however, such distortions produce recognizable patterns on the sonar graph.

Where many ships have passed, the topography of the seafloor (better known as the **bathymetry**) is well determined, but the picture is incomplete where few ships have passed. William Haxby of the Lamont-Doherty Geological Observatory has assembled all such ship soundings and, using computer imagery techniques, has produced maps of sea level and ocean depth.

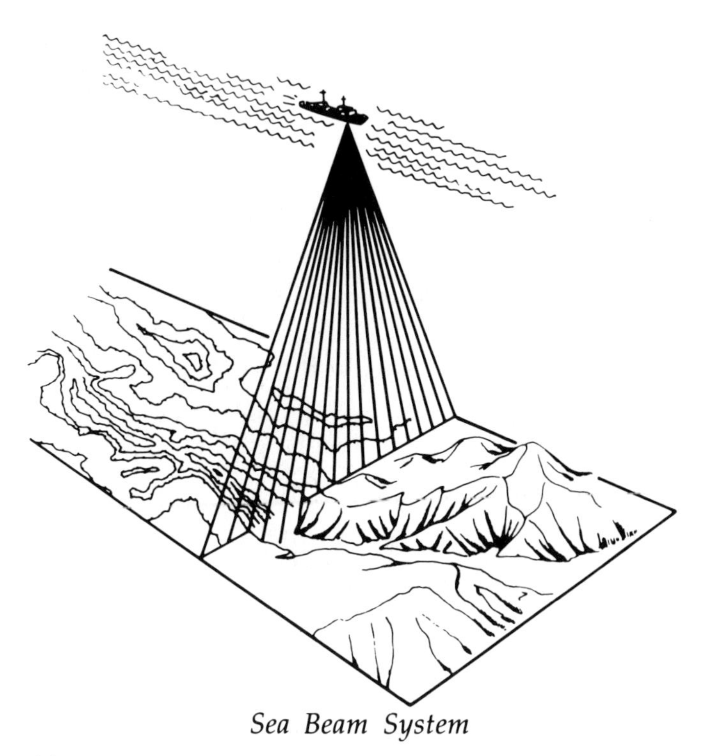

(a)

(b)

3.3 *(a) The newly installed Sea Beam multibeam echo-sounding system aboard the research vessel* Thomas Washington *has been routinely conducting detailed bathymetric surveys in the Pacific. This modern sonar system measures the water depth at sixteen locations beneath the ship out to a swath width equal to about three-quarters the water depth. From this depth data, bathymetric contour maps are automatically created aboard ship. (b) Printout of the ocean floor produced by the Sea Beam. The circular contours of the map show a small string of volcanoes with heights up to 210 m (690 ft) and crater depths as low as 48 m (158 ft). This particular map was made at a depth of 3,421 m (11,220 ft), approximately 2,900 km (1,800 mi) east of Tahiti. (Peter F. Lonsdale, Scripps Institution of Oceanography, University of California, San Diego)*

In addition to changes in sea level due to ocean waves, tides, and currents, the sea surface has larger-scale hills, valleys, and bulges associated with variations in gravity from place to place on the surface of the earth. Over large concentrations of mass such as the mid-ocean ridges, gravity is greater than over deep-ocean trenches. Thus, the "pull" of gravity over a ridge will be stronger and consequently will "pile up" water above it, so that the sea surface will rise as much as five meters (16 ft). Over a trench, however, gravity is less and the sea surface may be depressed by as much as 60 m (192 ft). Thus, depth can be inferred by sea level.

The map of sea level was obtained from the altimeter that flew aboard the three-month *Seasat* mission in 1978. Just as an echo sounder acoustically measures water depth, the altimeter electromagnetically measures the height of the satellite above the sea surface. Given knowledge of the position of the satellite (obtained by accurate tracking), sea level was measured by *Seasat* with 5-cm (2-in) precision and a 65-cm (26-in) accuracy.

This accuracy will not only improve our knowledge of the solid earth, but will also enable resolution of sea-level variations due to the effects of ocean tides and currents. A global view of ocean circulation will be possible for the first time.

Dredging

From time immemorial, fishermen have been dragging nets across the shallow floors of the sea to catch crabs and fish. Present-day scientists have adopted that ancient practice to bring samples of organisms and bottom sediments up from the deep ocean waters. They use a dredging net made of metal chain or nylon mesh with one end held open by a rigid metal frame, as shown in Figure 3.4(a). The frame is secured to the ship by a line. The dredge is lowered and dragged along the bottom and then hoisted by an electric winch. Unfortunately, small samples pass through the mesh, and fragile specimens are often broken by the action. Moreover, it is impossible to determine precise loca-

3.4 *Sampling the ocean bottom. (a) Marine geologists from Scripps Institution of Oceanography bring up a lightweight shell dredge during the research ship* Spencer F. Baird *Vermilion Sea expedition to the Gulf of California in 1959. (b) A Scripps marine geologist starts a grab sampler on its way to the ocean floor. (c) A gravity corer. (Courtesy Scripps Institution of Oceanography, University of California, San Diego)*

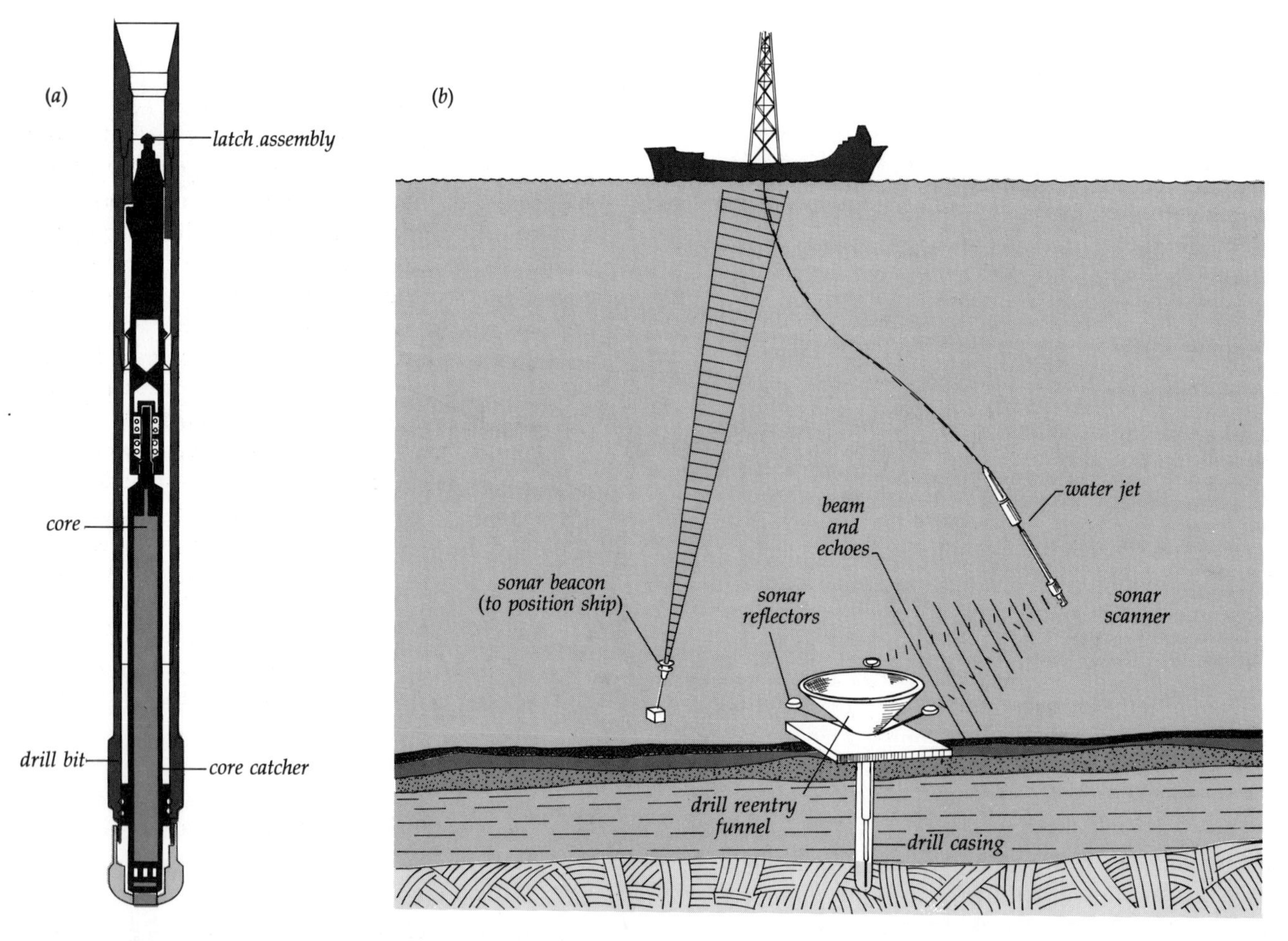

3.5 *(a) Drilling pipe and core barrel. (b) Reentering a drill hole in the deep-ocean floor. A newly developed mechanism guides the drill bit into a previously drilled hole. The reentry funnel is attached to the drill string when it is first lowered to the bottom, and it remains on the ocean floor when the string is withdrawn. At the time reentry is attempted, the drill string is relowered with a sonar scanner on the bit assembly that emits sound signals. These signals are echoed back from three reflectors spaced around the funnel. Position information is relayed to the ship, and the water jet is used to steer the bit directly over the funnel.*

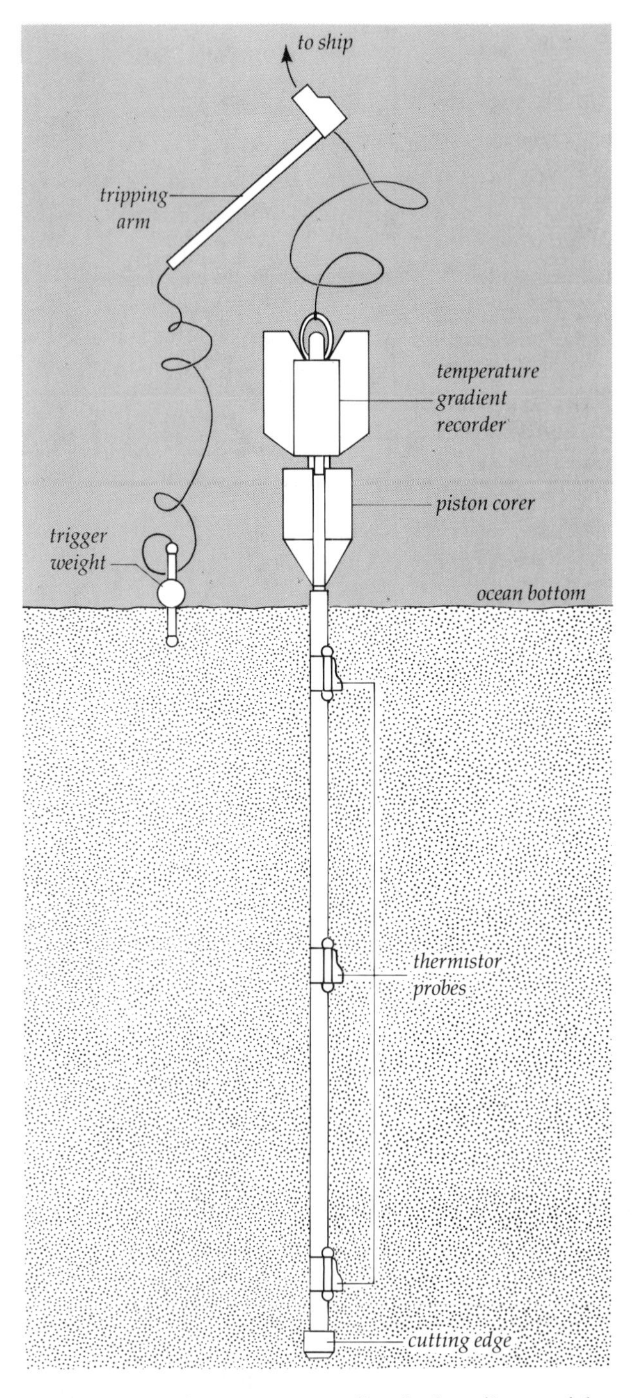

3.6 To measure the temperature gradient in the sediments of the seafloor, heat probes (temperature-sensitive resistors) are strapped to a piston corer. The temperature gradients, combined with measurements of thermal conductivity of the cores that are made aboard ship, determine the amount of heat flowing out from the seafloor. The amounts are very low.

tions and interrelationships among samples by examining the contents of the dredge.

A more satisfactory research device is the grab sampler, shown in Figure 3.4(b). A grab sampler has spring-loaded trap doors held open by clamps. When

3.7 A heat probe is mounted on the steel frame to measure heat flow through seafloor sediments. (Eric Abramson)

the sampler hits bottom, the force of the impact releases the clamps and the trap doors snap shut around a sample of sediment, thus including small organisms from a single location.

Coring

Various coring devices have been designed for research into the origin and history of the ocean basins and their sediments. The gravity corer (Figure 3.4[c]), for example, is a hollow pipe about 8 cm in diameter with a sharp cutting end. The pipe is sometimes equipped with fins to keep it vertical as it falls. To guide the pipe in its descent, a pilot weight is first sent down. When the weight hits bottom, a messenger releases the pipe, which plunges into the sediment. The corer is then hauled back to the surface. Gravity corers produce sample cores up to a few meters in length.

Another device, the piston corer, is capable of producing cores up to 12 m (39 ft) long. Longer than the gravity corer, it contains a piston to prevent compaction of the sediment under the impact of the corer.

A third technique of coring used in deep-sea research is platform drilling, modeled on oil-drilling apparatus. The research vessel *Glomar Challenger,* which was designed for platform drilling, has recovered cores from holes more than 1,100 m (3,608 ft) deep (Figure 3.5).

PROPERTIES OF THE OCEAN CRUST

Heat Conduction

Heat flow through the earth's crust is greater in some regions of the ocean basins than in others. The heat results mainly from radioactive decay of elements in the earth. Temperature-sensitive devices called **heat probes** are used to measure heat flow in ocean-bottom sediments (Figures 3.6 and 3.7). In a heat probe, a

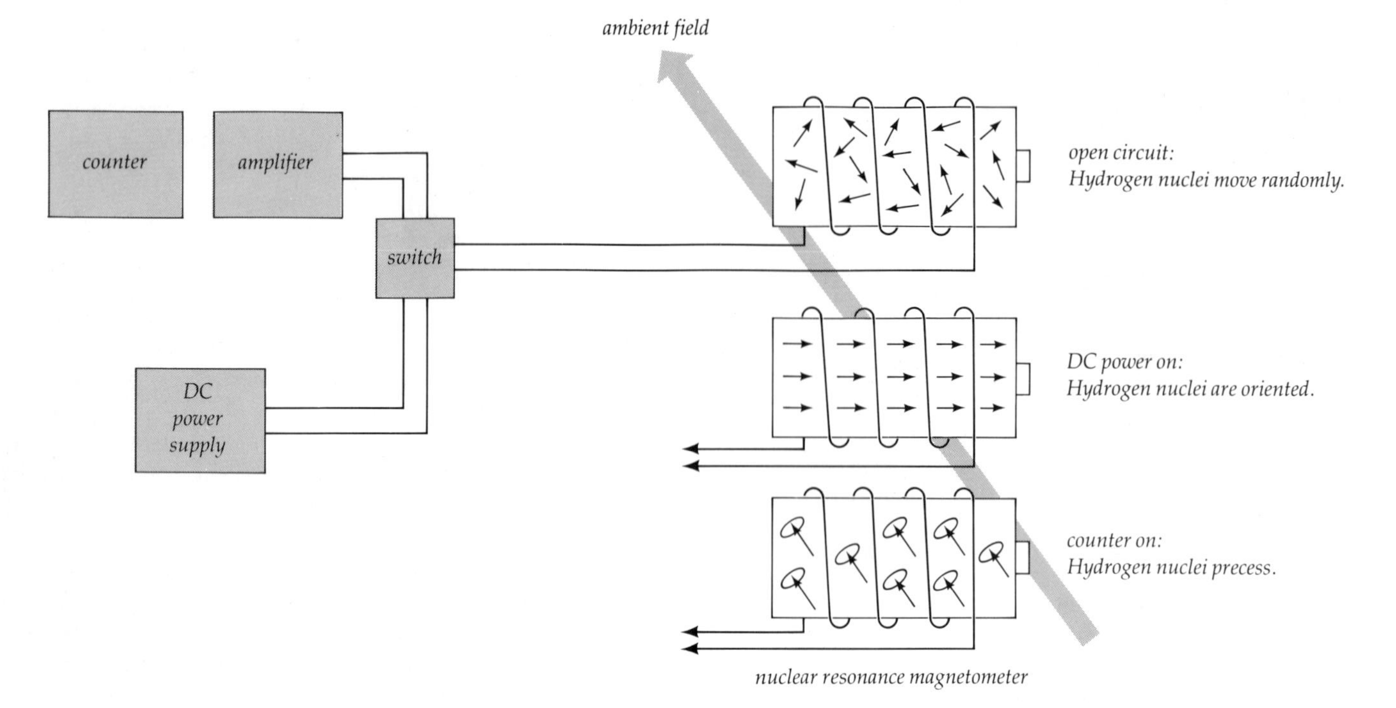

3.8 Most magnetometers now used at sea take advantage of the fact that hydrogen nuclei have magnetic properties. The detector consists of a coil of wire wrapped around a bottle of water. When a strong direct electrical current is applied to the coil, a magnetic field is created, and the nuclei align themselves along this field. When the current is switched off, the nuclei try to orient themselves in the direction of the earth's magnetic field, but because they are spinning, they behave like tiny gyroscopes and wobble rather than flopping over immediately. The combined effect of these nuclei wobbling produces a weak alternating current, the frequency of which is directly related to the absolute strength of the earth's magnetic field.

metal conductor called a thermistor transmits known quantities of electric current. The conductor's resistance to the amount of electricity flowing through it is inversely proportional to its temperature. Its resistance at different temperatures is measured before the field studies are made. Thermistors are mounted on a probe, and the probe is inserted into ocean-bottom sediments. When the thermistors achieve temperature equilibrium with the surroundings, the heat flow of the sediments can be determined, provided the conductivity of the sediments is known. The conductivity is determined by inserting a probe consisting of two electrodes (positive and negative), attached to a meter and a power source, into the core taken at the same time.

Rock Magnetism

Earthquake or seismic data suggest that the inner core of the earth and the mantle are solid. As you will discover in Chapter 5, the patterns of rock magnetism provide a quantitative measure of crustal movements—in both direction and rate of motion!

To measure the intensity and direction of ancient poles of magnetism in the ocean basin, a **magnetometer** is towed beneath the surface of the sea (Figure 3.8). The heart of a magnetometer is a water-filled cylinder fitted inside a copper coil. When a direct current is sent through the coil, protons (hydrogen nuclei) in the liquid become aligned with the induced magnetic field. When the current is shut off, the coil briefly continues to carry a weak current because of the wobbling of aligned protons about the earth's magnetic field. This movement is analogous to the wobbling of a spinning top as it slows. This weak current is amplified and measured with a digital counter. When the magnetometer passes over crystalline rocks containing different magnetic properties, differences in the strength and direction of the field can be measured.

Gravity Anomalies

The earth is not perfectly round, and neither is its mass distributed uniformly. As a result, the acceleration of any free-falling body varies slightly from place to place on the earth's surface from the constant of

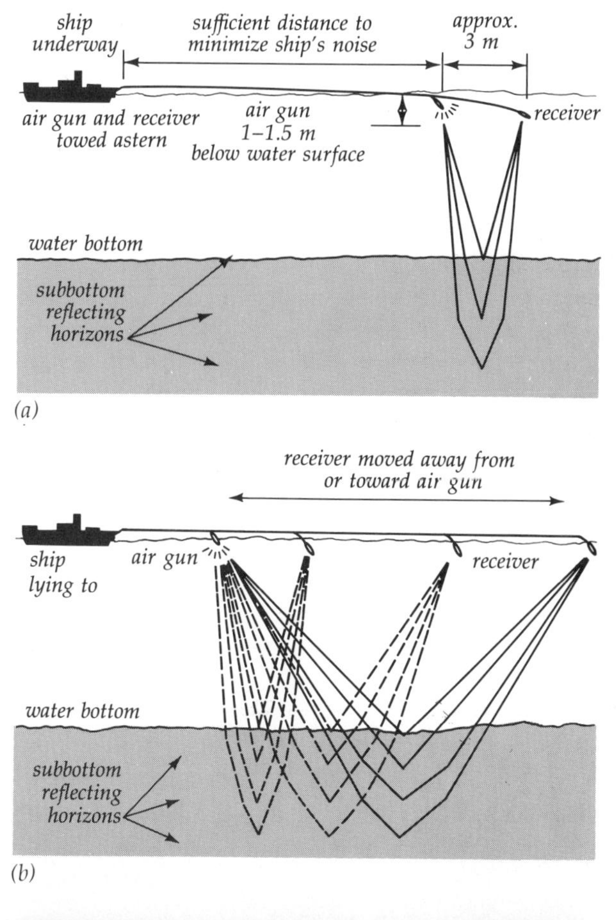

3.9 (a, b) Seismic profiling of subbottom sediments. (c) Seismic airgun. (Rodger Larson)

9.8 m/s^2 (34 ft/s^2). Instruments called **gravimeters** measure these slight variations. One type of gravimeter measures the period of a pendulum. Most modern instruments measure changes in the length of a spring that supports a mass.

The difference between the measured value of the acceleration of gravity at a given place and the predicted value for a theoretical earth model is called a **gravity anomaly.** Because a gravity anomaly indicates the earth's density beneath a given location, gravity anomalies can be used to make inferences about the rocks beneath the ocean bottom.

Seismic Profiling

Seismic profiling was developed by oceanographers in the 1950s. Its principal use is to determine the orientation of rock and sediment layers below the seafloor, their thickness, how they tilt, and how well consolidated they are.

To create a seismic profile, seismic air guns produce shock waves that travel through the water and the rocks below (Figure 3.9). Hydrophones trailed behind the ship pick up the shock waves after they are reflected and refracted by ocean sediments. Sound waves move through sand at one speed, through limestone at another, and through the basalt of the oceanic crust at still another. Broad average velocities for different rocks have been determined experimentally and are published in tables. The data not only reveal what types of rocks are present but also the general structure of the rocks. Structural features such as tilted layers, faults, and buried salt domes are revealed.

FURTHER READINGS

Drake, C. L., and others. *Oceanography.* New York: Holt, Rinehart & Winston, 1978.

Instrumentation Society of America. *Proceedings of the National ISA Marine Sciences Symposia.* New York: Plenum Press (annual).

Kennett, J. P. *Marine Geology.* Englewood Cliffs, N.J.: Prentice-Hall, 1982.

Marine Technology Society. *Proceedings of the MTS and IEEE Ocean Engineering Society.* Washington, D.C.: Marine Technology Society (annual).

Ross, D. A. *Introduction to Oceanography.* Englewood Cliffs, N.J.: Prentice-Hall, 1982.

Sea Technology. Arlington, Va.: Compass Publications (monthly journal).

The face of places, and their forms decay;
And that is solid earth that once was sea;
Seas, in their turn, retreating from the shore,
Make solid land, what ocean was before.

Ovid

FOUR

Ocean Basins and Sediments

Rising majestically above the remote ranges of the Himalayas is Mount Everest—8,850 m (29,028 ft) above sea level and the highest point on earth. What, you might ask, does this have to do with the ocean? Buried within the flanks of this windswept peak are marine fossils—remains of tiny organisms that lived many millions of years ago beneath the surface of the sea. How did those organisms get there? As you will see, it is impossible to study the forces that shape the ocean basins without studying the forces that shape the land as well.

In this chapter we shall get an overview of marine geology: What is known about ocean basins, oceanic ridges, abyssal plains, coral reefs, island arcs, trenches, seamounts, and island chains. We shall also review how research data on the ocean basins are obtained. Along the way we shall describe such catastrophic events as the volcanic eruptions and earthquakes that jar the earth's crust. Throughout the chapter we shall be setting the stage for one of the most sweeping geological concepts of our time: the theory of global plate tectonics. This theory explains how all these seemingly unrelated geological features and events—from marine fossils buried in mountains to rumblings of the crust beneath the sea—are related in a fundamental way.

If Mount Everest, towering though it is, could somehow be dropped down to the deepest point of theocean, its peak would reach to no more than 2,000 m (about 6,560 ft) below sea level (see Figure 4.1). By contrast, Mauna Kea volcano on the island of Hawaii stands about 4,180 m (13,710 ft) above sea level and extends more than 5,000 m (16,400 ft) below sea level to the ocean floor. Thus, Mauna Kea is the highest peak in the world. In the last 30 years, oceanographers have learned a great deal about the relatively inaccessible regions of the ocean bottom. Today, working through colleges and universities, research institutions, and navies throughout the world, they are carrying on research and compiling data at an ever-accelerating pace.

WHAT IS AN OCEAN BASIN?

An **ocean basin** is defined as a large area of ocean floor lying at a depth of more than 2,000 m. A continent is defined as a large continuous area of land. The vague thing about these definitions is the word *large.* Why is Australia regarded as a continent and Greenland (about 2 million sq km, or 0.77 million sq mi) as an island? Why is the Caribbean said to be a sea and the Arctic an ocean? Names, it seems, are sometimes arbitrary. Yet, through common usage over the centuries, names become fixed in the language. In oceanography, the term **sea** (as in Sea of Cortez) now refers to a large, salty, relatively landlocked body of water or to a mass of water with physical and chemical characteristics that distinguish it from a nearby ocean. The International Hydrographic Bureau recognizes fifty-four seas but only six oceans: the Arctic, the North Atlantic, the South Atlantic, the North Pacific, the South Pacific, and the Indian.

Just as continents have distinct features, such as mountains and valleys, so do ocean basins. Among the features of the ocean basins that we shall consider in this chapter are oceanic ridges, abyssal plains, coral reefs, island arcs, trenches, seamounts, and island chains (Figure 4.2). The deepest region is the Challenger Deep in the Mariana Trench, which is 11 km (6.8 mi) below sea level.

Composition

The difference between ocean basins and continents extends beyond large topographic features (Figure 4.3). There are major differences in rock composition and density, sediment thickness, and related properties. The oceanic crust consists of a layer composed largely of rocks called **basalt** and **gabbro** resting on a layer of another rock called **peridotite.** These are igneous rocks—that is, they are composed of molten material that has crystallized. Basalt and gabbro differ in the size of their crystals but not in their composition. They

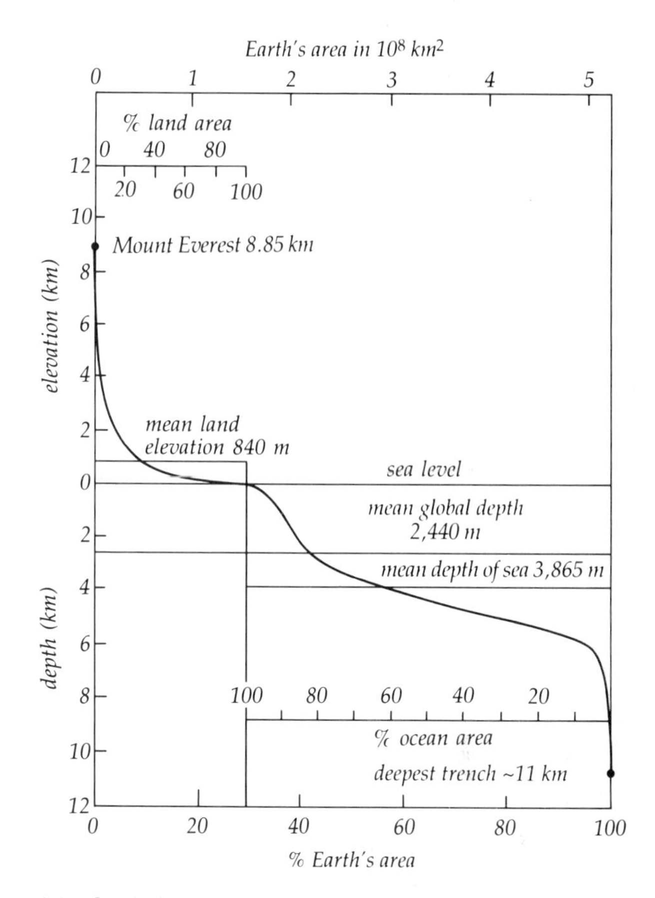

4.1 Graph showing distribution of elevations and depths on the earth. Compare the relative elevations of the deepest trench and Mount Everest.

4.2 Map of a portion of the Atlantic Ocean floor showing some major oceanic features: oceanic ridge, transverse fracture zones, canyons, seamounts, rises, alluvial (sediment) cones seaward of deltas, trenches, and abyssal plains. Depths are in meters. (From Heezen, Tharp, and Ewing)

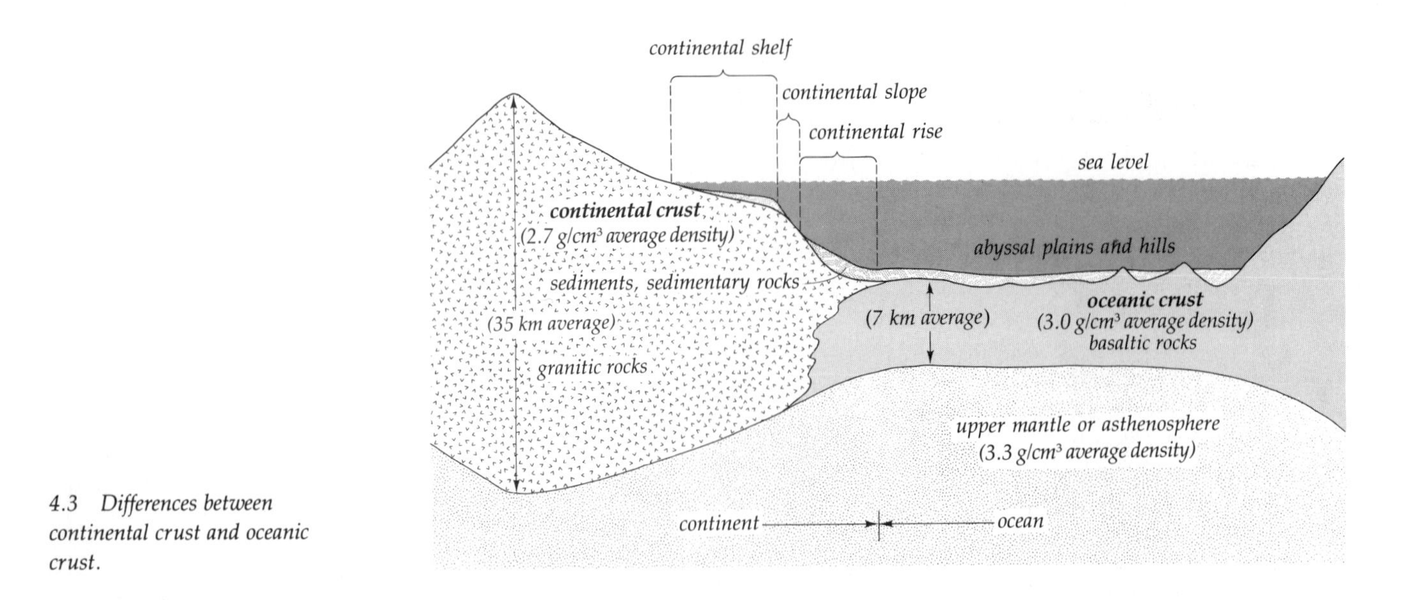

4.3 Differences between continental crust and oceanic crust.

both contain such minerals as pyroxene, olivine, and calcium-rich feldspar. Peridotite contains pyroxene and olivine but no feldspar. All are rich in iron and magnesium.

The continental crust is less homogeneous. It is made up of igneous rocks that are chiefly of granitic composition. A typical granite is an igneous rock composed of quartz, potassium, and sodium-rich feldspar, with minor amounts of iron and magnesium-containing minerals. The continental crust contains less iron, magnesium, and calcium than the oceanic crust, but more silicon, sodium, and potassium. (We shall discuss these minerals in more detail in a later part of this chapter called Characteristics of Ocean Sediments.)

Density

Because of the differences in composition between the continental crust and the oceanic crust, their densities differ. The average density of continental crust is about 2.7 g/cm^3, whereas the average density of the oceanic crust is about 3.0 g/cm^3. Isostacy accounts for the relative elevations of the two types of crust.

Isostacy

Isostacy is the concept that all large regions of the earth's crust are in balance as if they were floating on a denser underlying layer (Figure 4.4). Regions of less dense crust float higher topographically than regions of denser crust, provided they are of equal thickness. This explains why continents, which consist of relatively low-density rocks, stand topographically higher than ocean basins.

Assume that a section of the crust—a continent, for example—is floating in equilibrium with its density. If weight is added to or removed from the continent, the equilibrium will be disturbed. Assume that Canada is floating in isostatic equilibrium. If a thick cover of ice were to blanket the area, as actually happened during the ice ages, the additional weight would make the land sink. Once the ice was removed, Canada would rise again. The water contained in great glacial ice caps ultimately comes from the oceans. Consequently, as glaciers form, the volume of water in the oceans and the weight of water on the ocean floor decrease, and the ocean floor rises. During the ice ages, Canada was depressed from 500 to 700 m (1,640 to 2,296 ft) below its present level under a cover of ice between 2 and 3 km (1.2 and 1.9 mi) thick, and the ocean floor rose about 40 m (131 ft). Once the ice had melted and the water had returned to the ocean, the ocean floor was depressed once more.

FEATURES OF THE SEAFLOOR

Oceanic Ridges

For more than two centuries, geologists have recognized that islands such as those that make up Iceland are the peaks of underwater ridges. But not until 1928, when the German oceanographer L. Kober first began charting **oceanic ridges** throughout the world, did the extent of the vast system of which Iceland is a part begin to be understood. And not until the work of Harry H. Hess, Maurice Ewing, Bruce Heezen, H. Menard, and others from 1950 to 1962 did the details emerge.

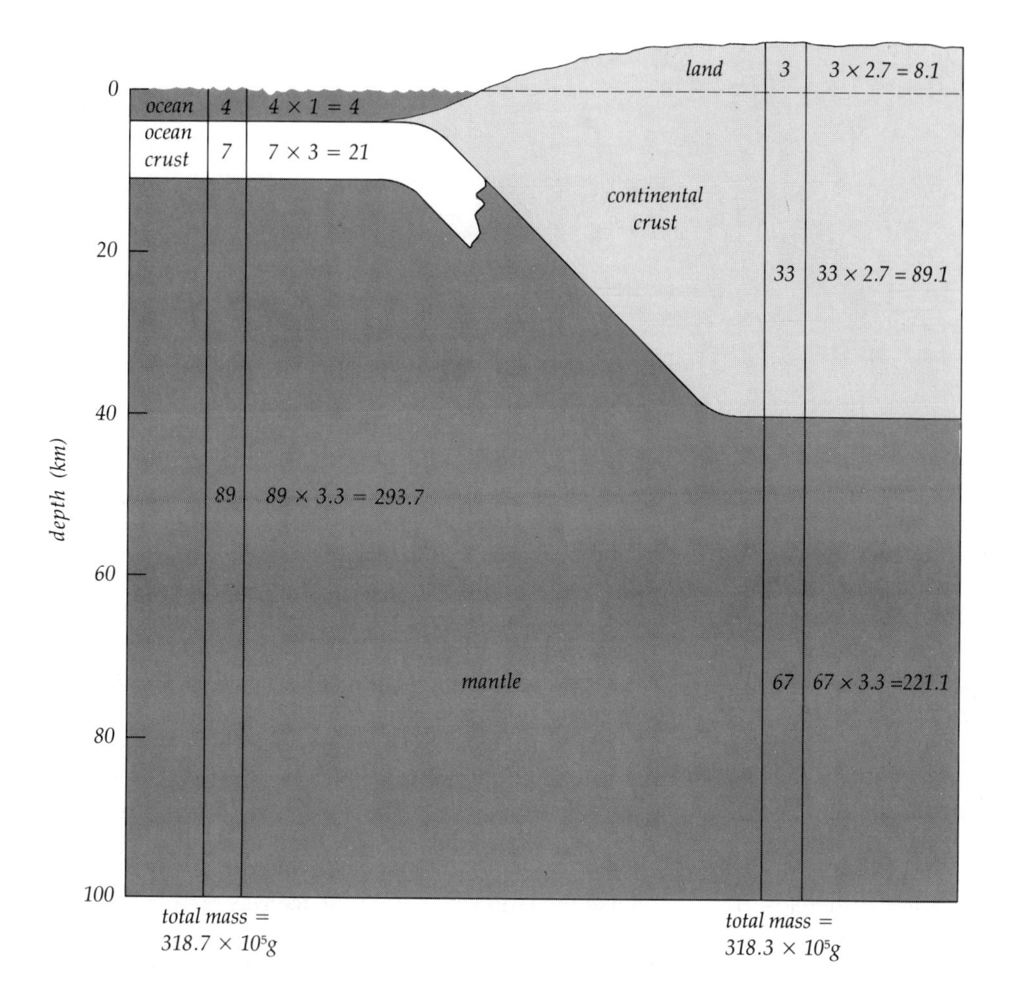

4.4 Isostacy is the concept of crust balance in the earth. Differences in density of layers of the earth are compensated by differences in volume of the layers. This diagram demonstrates that two columns 100 km × 1 cm × 1 cm are in approximate balance. If the ocean is 4 km deep, the ocean crust is 7 km deep, and the upper mantle is 89 km deep, then it balances a column consisting of 43 km of continental crust, 7 km of ocean crust, and 50 km of upper mantle. Note: Density = Mass ÷ Volume.

The Mid-Atlantic Ridge runs the length of the entire Atlantic Ocean, from 55°S to 70°N (Figure 4.5). In the South Atlantic, the ridge swings east to the Indian Ocean, where it becomes the Mid-Indian Ridge, which in turn branches at 25°S. One branch goes northward to the Red Sea; the other goes southeastward between Australia and Antarctica. There it becomes the Pacific-Antarctic Ridge. The system continues up through the eastern Pacific Ocean, where it is called the East Pacific Rise, and on to the Gulf of California. There it disappears, only to reappear west of Oregon! This extensive system of oceanic ridges is a major topographical feature of the earth's crust.

Iceland and its neighboring islands are almost entirely volcanic in origin, and active volcanoes are still found there (Figure 4.6). One of the most graphic accounts of volcanic activity along oceanic ridges is Sigurdur Thorarinsson's article "Surtsey: Iceland's New Island," which was published in June 1966 in the *American Scandinavian Review.* His description follows:

> Although eruptions in Iceland have often caused astonishment in other countries, probably none of them has attracted so much attention in so many places as that which, in November 1963, gave birth to a new island off the south coast of Iceland. This eruption has been in progress ever since, and there is no sign of its ending. The island that appeared on November 15, 1963, continued to grow until the middle of May 1965, at which time it had attained an area of one square mile, or more than half the size of Central Park in New York City. Its height was then 568 feet above sea level. The Icelandic place-name committee named it Surtsey after the black giant Surtr, the leader of the fire giants who fought against the gods at Ragnarok.
>
> Surtsey was not the only island born during this eruption. When volcanic activity ceased in Surtsey, it broke out in another place in the sea a little ENE of that island, and during the spring and summer of 1965 a new island arose, popularly called Syrtlingur (Little Surtr—it never received an official name).
>
> By the middle of September 1965, Syrtlingur had reached a height of 230 feet above sea level and an area of 35 acres. Yet this sizable

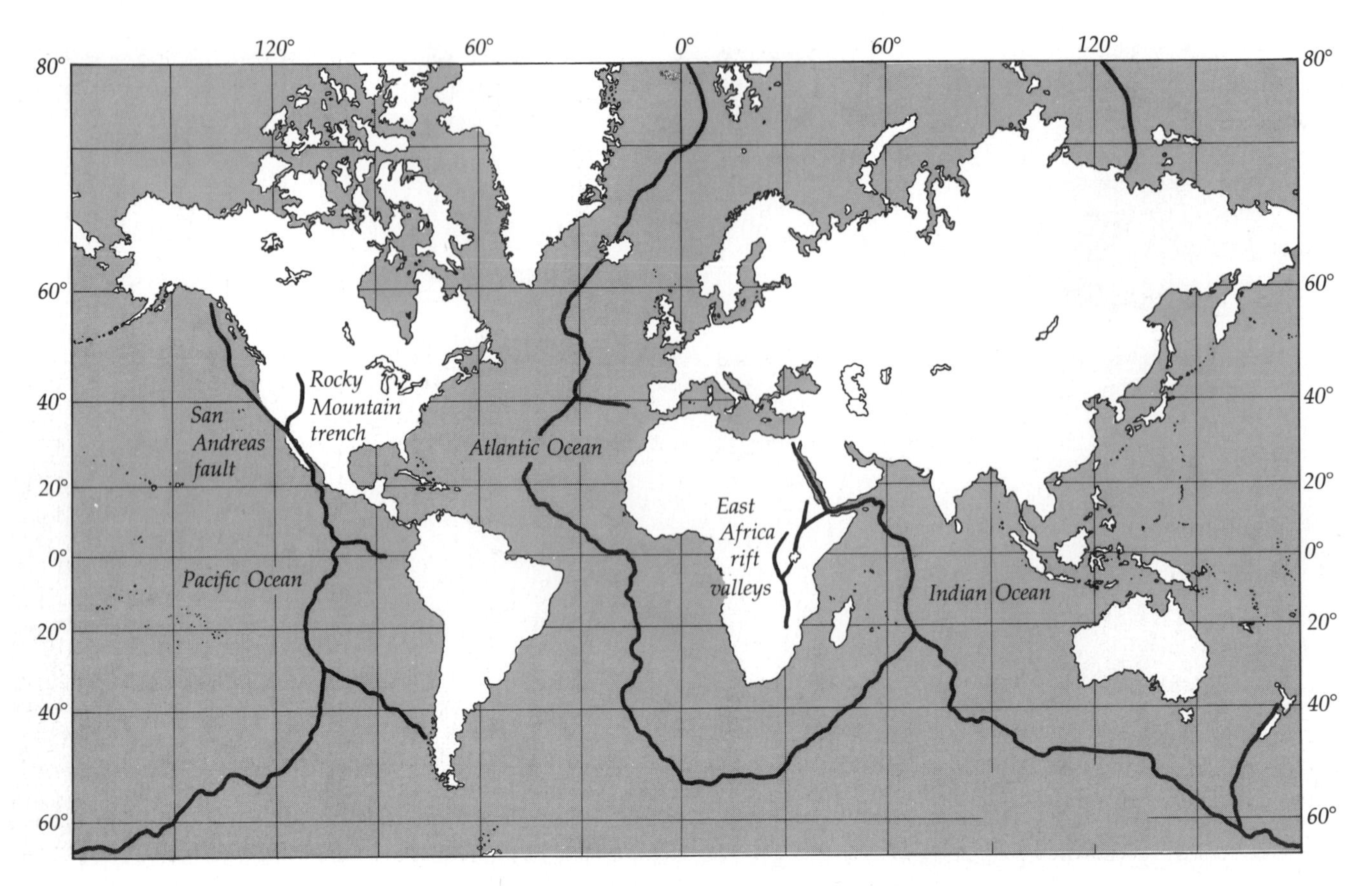

4.5 Oceanic ridges. (Adapted from Surtsey, *by Sigurdur Thorarinsson. Copyright © 1966 by Almenna Bókafélagið. Reprinted by permission of The Viking Press, Inc.)*

island was completely swept away by the breakers during a single week of storms and high seas in late October 1965. On December 28 still another island was born, this time on the opposite side (SW) of Surtsey, half a mile offshore. This island has since continued to disappear and reappear.

The system of oceanic ridges is characterized by rifting, frequent shallow earthquakes (usually shallower than 20 to 30 km, or 12 to 19 mi) and active volcanism. When the crust is pulled apart, it fractures and lava rises in the fissures (cracks in the crust). Because cooling prevents the lava from rising all the way to the top of the fissure, it forms a floor in the crack below the adjacent seafloor. As the crust continues to be pulled apart, new fissures form, fill with lava, cool, and crack. The older, now cooler portions are pushed up to form the crests of the ridge (Figure 4.7). This action produces a rift valley and fractures in the earth's crust whose angles approach 90° to the ridge. The volcanic rocks that form the ridge are basalt. Photographs and direct observations can reveal hot springs and recent lava extruded in rift valleys. Sediment is rarely present on the igneous rock that forms the ridge. When it is present, it is always less than 100 m (328 ft) thick.

A ridge system is cross-hatched by extensive fractures along which the ridge axis has been offset. An example on land is the San Andreas Fault in California, which is a fracture separating the East Pacific Rise (southeast of Baja California) from the Gorda Ridge (west of Oregon).

What caused the ridge system? Why do deep trenches and island arcs ring the Pacific Ocean? Why do the zones of frequent earthquakes appear as narrow bands? How did Hawaii form? In Chapter 5, we shall see how the steady advancement of scientific knowledge has provided answers to these questions.

Abyssal Plains

Abyssal plains are broad, somewhat flat regions found in all oceans at depths ranging from 4,000 to over 5,000 m (13,120 to 16,400 ft). They are bounded by oceanic ridges or continents. Where many submarine canyons cut across the continental slope, large fan-shaped aprons of sediments transported by **turbidity currents** (a dense, gravity-driven current containing

(a)

(b)

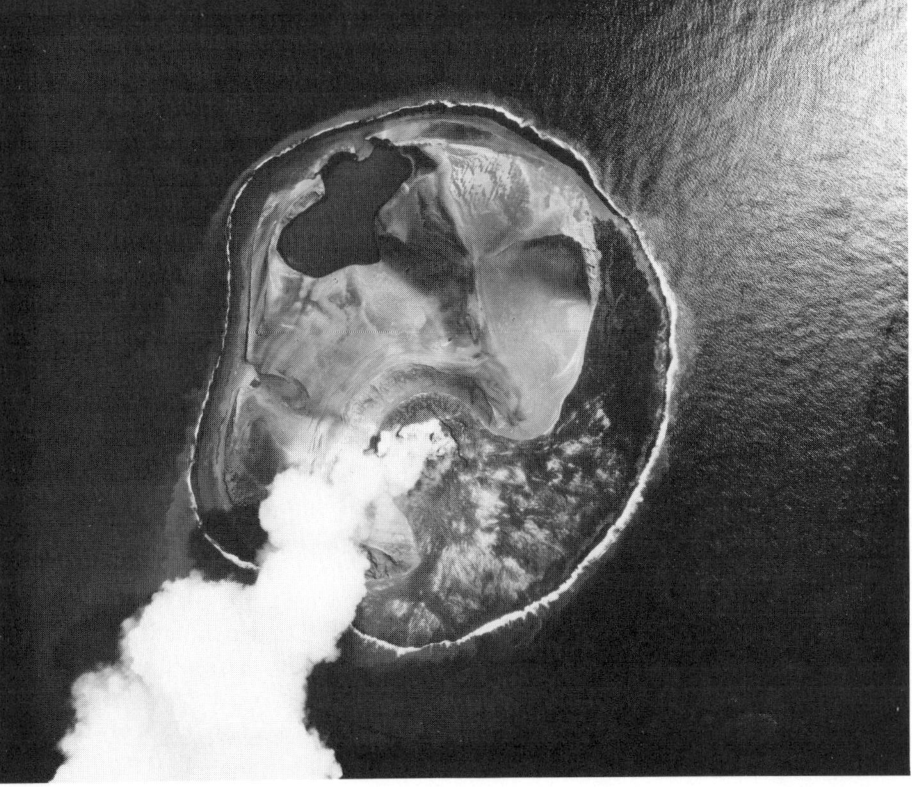

(c)

4.6 Heimaey Island is a volcanic island belonging to Iceland and located on the Mid-Atlantic Ridge. A volcanic eruption and ensuing lava flows caused extensive damage to the village while increasing the size of the island. Townspeople attempted to slow the advance and divert the flowing lava to save homes and other property. (a) Before the eruption. (b) After the eruption. Note the changes to the lower left of the harbor entrance. (Icelandic Federal Government) (c) Surtsey Island. Location is approximately 60°N by 21°W. (From Surtsey *by Sigurdur Thorarinsson. Copyright © 1966 by Almenna Bókafélagið. All rights reserved. Reprinted by permission of The Viking Press, Inc.)*

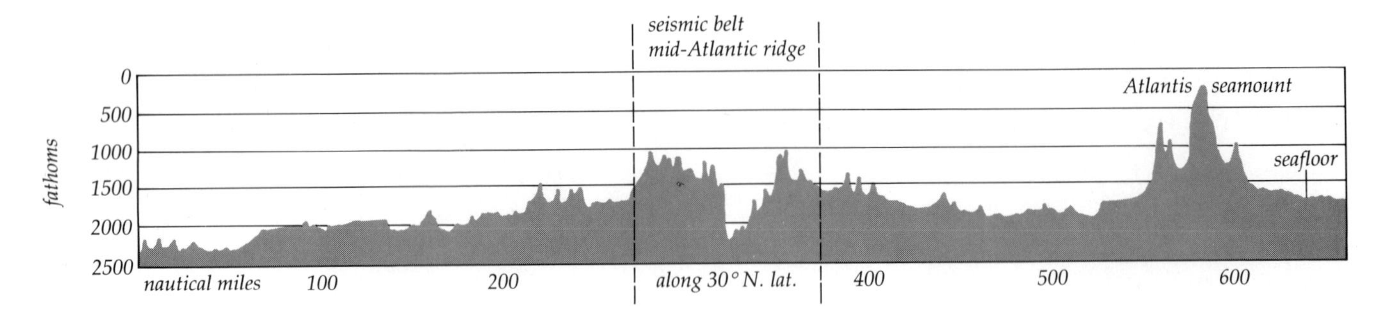

4.7 Profile of the Mid-Atlantic Ridge along 30°N latitude.

sediment in suspension) spread out onto the abyssal plain. Such regions are called submarine, abyssal, or deep-sea fans. They are particularly well developed on abyssal plains in the Indian Ocean off the Ganges and Indus rivers.

Most abyssal plains are gently sloping regions often covered with relatively unconsolidated sediments resulting from turbidity currents. Sediments on the abyssal plains tend to be much thinner than those on the continents, ordinarily ranging from a few hundred meters thick to more than 1,000 m (3,281 ft) thick.

Having formed at oceanic ridges, the basalt layer underlying ocean sediment is highly irregular. As the ocean crust moves away from the ridges, turbidity currents carry sediment from the continents, which buries the irregular basalt layer and forms the abyssal plain. In the Pacific Ocean, **trenches** lie between the continents and the abyssal plains. They trap the turbidity currents, dumping sediment into the trench. Consequently, irregular features of the basalt layer seaward of the trenches receive little sediment and stand as **abyssal hills.** The abyssal plains and abyssal hills are covered by thin **pelagic sediments** (discussed later in this chapter), which consist mainly of **biogenic oozes** with some clay. The biogenic oozes consist of the calcareous or siliceous skeletons of microscopic organisms living near the ocean surface. Some of the clay is picked up by strong near-shore currents at river mouths, held in suspension, and dropped far out to sea. Other particles are swept up by the wind from desert regions, such as the Sahara, and are released on the open ocean, where they slowly settle to the bottom. Still other particles consist of volcanic ash blown into the air and carried out over the ocean by the wind. Vast areas of the abyssal plains are covered by manganese nodules that are so closely packed in some places that they obscure the underlying sediment.

Seamounts, Coral Reefs, and Island Chains

Submerged mountains, many of which have been cored and dredged, rise above the abyssal plains. These mountains are of two general types: seamounts, which have rounded or irregular peaks, and guyots (pronounced "GEE-ohz"), which have flat tops. Both are extinct volcanoes. Although their peaks may be considerably below sea level, fossils of such shallow-water organisms as corals and clams have been dredged from their surfaces. The flat tops of the guyots appear to be the result of wave erosion that took place before the guyots were submerged.

A few chains of volcanic islands have a low incidence of earthquakes and relatively few faults. One such chain comprises the Hawaiian Islands and the Emperor Seamount Chain to the north. In this chain, volcanic activity is currently confined to the southernmost island, Hawaii, although a submerged volcano southwest of Hawaii is also active. Similar chains in the Pacific include those formed by the Line Islands, Tuamotu Archipelago, and the Chile Rise and by the Austral, Gilbert, and Marshall islands.

In tropical regions, especially in the Pacific Ocean, scattered islands fringed with **coral reefs** rise from the ocean floor. These reefs are composed largely of shells, algae, and the remains of tiny colonial animals called corals. Some coral reefs form long offshore barriers, as they do off the eastern coast of Australia. Others form crescent-shaped reefs and islands arranged in a circular pattern called **atolls.**

How did these islands and reefs form? Oceanographers have been debating that question ever since Captain Cook first described these formations during his exploration of the seas between Tahiti and New Zealand in 1768 and 1769. During the voyage of HMS *Beagle* in the 1830s, Charles Darwin described many coral islands in the Pacific and Indian oceans and conjectured about their origin (see Figure 4.8). He also observed many volcanic islands in the Pacific, some of which were active, including the Galápagos Islands, some 850 km (527 mi) west of the coast of Ecuador. As he traveled westward, Darwin noted that volcanic islands gave way to coral atolls. Could there be a relationship between the two?

Darwin described his interpretation of coral reefs and volcanic islands in his book, *The Structure and Distribution of Coral Reefs.* He summed up his thoughts this way:

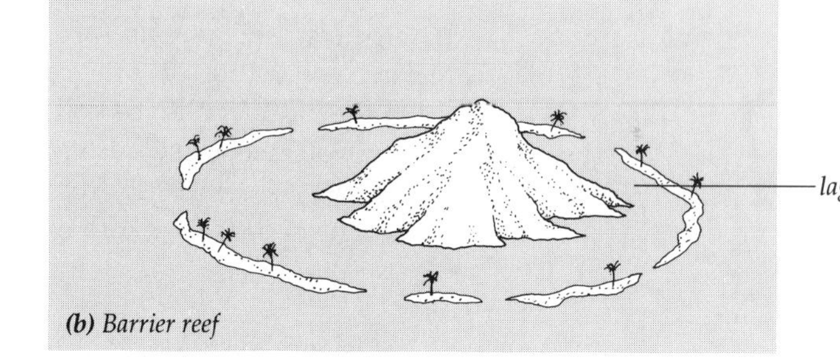

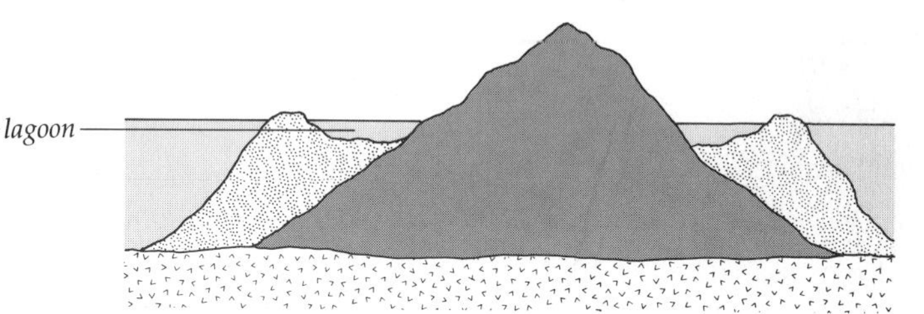

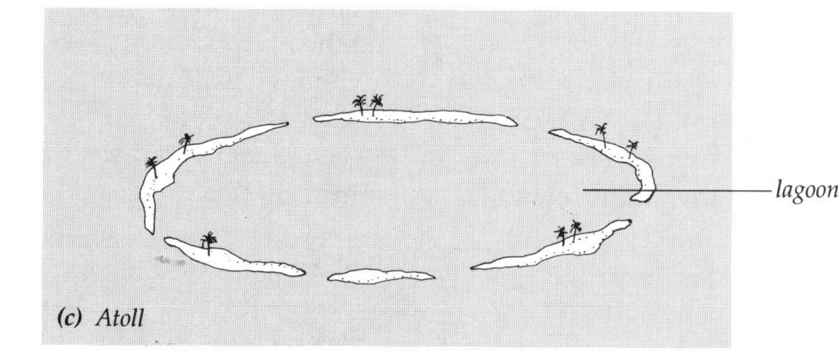

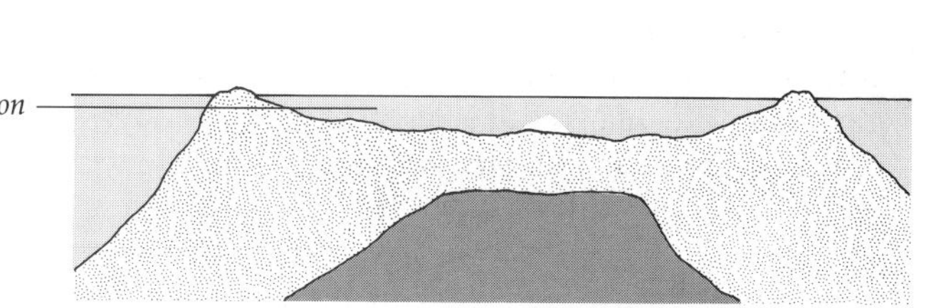

4.8 Sequence of coral atoll development according to Darwin. (a) Fringing reef. (b) Barrier reef. (c) Coral atoll. (d) Kossol Reef (Mariana Islands), showing a coral atoll in the background. (Douglas Faulkner)

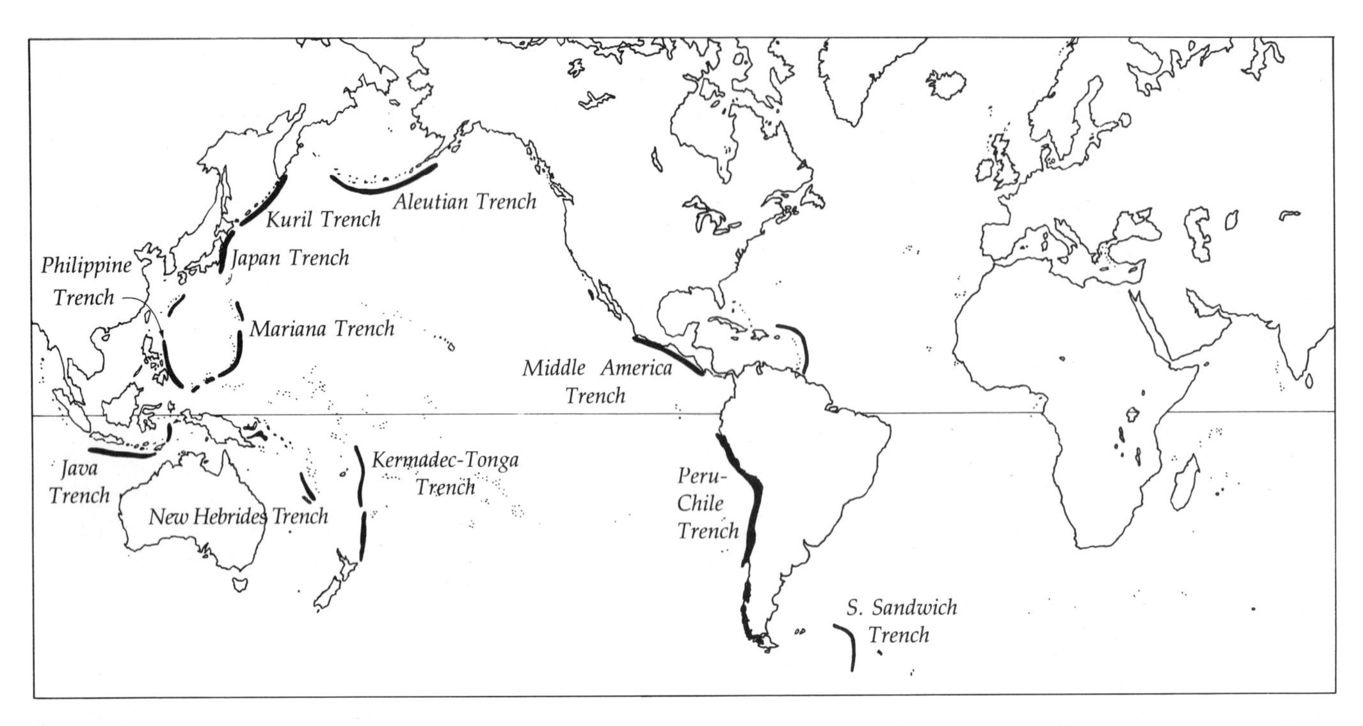

4.9 *Trench regions of the world. (After B. Gutenberg and C. F. Richter,* Seismicity of the Earth and Associated Phenomena, *1954, Princeton University Press)*

The principal kinds of coral reefs . . . were found to differ little, as far as relates to the actual surface of the reef. An atoll differs from an encircling barrier reef only in the absence of land within its central expanse; and a barrier reef differs from a fringing reef, in being placed at a much greater distance from the land with reference to the probable inclination of its submarine foundation, and in the presence of a deep-water lagoon-like space or moat within the reef. . . . When the two great types of structure, namely barrier reefs and atolls on the one hand, and fringing reefs on the other, were laid down in colours on our map, a magnificent and harmonious picture of the movements, which the crust of the earth has within a late period undergone, is presented to us. We there see vast areas rising, with volcanic matter every now and then bursting forth through the vents or fissures with which they are traversed. We see other wide spaces slowly sinking without any volcanic outbursts; and we may feel sure, that this sinking must have been immense in amount as well as in area, thus to have buried over the broad face of the ocean every one of those mountains, above which atolls now stand like monuments, marking the place of their former existence.

About a century later Harry Hess, a geologist and geophysicist at Princeton University, proposed that cores be drilled from atolls in order to test Darwin's theory. These drillings revealed that each atoll has a volcanic root buried under as much as 800 m (2,624 ft) of limestone—Darwin was basically correct. During the last few decades, a plausible answer to a more fundamental question has emerged. Why did the volcanic roots gradually sink beneath the surface of the sea? We will return to this question in Chapter 5.

Island Arcs and Trenches

Many of the world's active volcanoes are located in **island arcs** that have formed near deep trenches. The pattern typically seen is arcuate volcanic islands, a deep trench and two ridges on the seaward side of the islands, and the ocean basin. The trenches themselves closely parallel the arc of the island group; rarely are they more than 130 km (81 mi) wide, but they may be 1,500 km (930 mi) long and more than 10,000 m (32,800 ft) deep.

Most island-arc/trench systems fringe the Pacific or separate the Pacific from the Indian Ocean (Figure 4.9). Exceptions include the West Indies, where an island arc separates the Caribbean Sea from the Atlantic Ocean, and the South Sandwich Islands, off the southern tip of South America. The Aleutian, Kuril, Ryukyu, and Philippine islands, along with Japan and

Table 4.1 *Classification of Sediment Particles by Size (Wentworth Scale)*

Particle Term	Particle Diameter (mm)
Boulder	>256
Cobble	64–256
Pebble	4–64
Granule	2–4
Sand	0.062–2 (62–2,000 μm)
Silt	0.004–0.062 (4–62 μm)
Clay	<0.004 (<4 μm)

Note: 1μm is a micrometer, or one-millionth of a meter, which equals 39.4 millionths of an inch.

the Malay Archipelago, are parts of island-arc/trench systems. Since many of these islands are volcanically active, the boundary of the Pacific Ocean is sometimes called the Ring of Fire. Rarely does a year pass without a major volcanic eruption on at least one of the islands on the borders of the Pacific. Frequently the eruptions are explosive, producing fiery outpourings of molten masses and lava. Clouds of water vapor and other gases are also spewed, which cause extensive damage to vegetation and occasionally take human lives.

The characteristic volcanic rock of the islands ringing the Pacific is andesite, named after the Andes Mountains of South America. **Andesite** is closely related to basalt but consists mostly of the minerals sodium plagioclase and biotite or other iron-bearing silicate. Recall that the lava spewed by volcanoes on mid-ocean islands and on the ocean floor consists of nothing but basalt.

Island arcs and trenches are sites of frequent earthquakes as well as volcanic eruptions. The frequency is of the same order as is found on the ocean ridges. The earth movement that generates the earthquakes tends to be shallow beneath the trenches and deeper beneath the volcanic islands. (This pattern, which forms what is called a subduction, or Benioff, zone, is discussed in Chapter 5).

The Aleutian Islands trend southwest from Alaska. At the western end, a S–N cross section reveals a trench with shallow earthquakes and volcanic islands with deeper earthquakes. At the eastern end, however, two parallel island arcs have formed between the mainland and the Aleutian Trench. The arc closest to the mainland is volcanic; the one closest to the trench is composed of folded sedimentary rock. Between the two arcs is a "minitrench," some 700 m (2,310 ft) deep. By what process did it form? Again, we shall find the answer to this question in Chapter 5.

Another interesting system is the Peru-Chile Trench. This trench is similar to other trenches except that no island arc lies between it and the South American continent. Active volcanoes near the continental coast and in the Andes are made up of andesitic lavas, however, suggesting that the volcanic arc has been superimposed on the continent.

CHARACTERISTICS OF OCEAN SEDIMENTS

Oceanographers use several variables to classify ocean sediments: particle size and shape, density, color, composition, and bedding features. With this knowledge, oceanographers can establish the distribution and age of sediments and piece together the history of the ocean basins.

Particle Size

Geologists classify sediments according to the size of the constituent particles. They use the Wentworth scale (Table 4.1) to classify both land sediments and ocean sediments. In ocean sediments, sand, silt, and clay are the most common components.

The distribution of particles of various sizes gives us insight into both the origin of a particular sediment and how it was transported to its current site. For example, marine sediment that is poorly sorted, with different-sized particles randomly mixed, indicates that it was transported by turbidity currents or by sea ice. A sediment in which the particles are of uniform size indicates transport by wind or by waves. Oceanographers use standard statistical techniques to describe the distribution of particles by mean size, standard deviation, range, and so forth.

Density and Shape

The density and shape of particles provide clues to the origin and the strength of the transporting agent. **Density,** the ratio of mass to volume, is usually expressed in grams per cubic centimeter. Minerals of different compositions have different densities. The presence of particles of different densities suggests that they were transported by agents of different energy levels and that they settled to the bottom at different rates.

The shape of a particle also reveals significant information. For instance, a rounded particle and an elongate particle, both with the same composition, will settle to the bottom at different rates and be carried different distances by currents. Shape may also indicate how far a particle has been transported by a stream or an alongshore current. Abrasion over a long distance tends to round off the edges of a particle eroded from its parent rock. Uniformly rounded particles have been transported over great distances and sorted by wind or waves.

Color

Since particles of clay are so small that they cannot be identified with the naked eye, the color of a clay sample is commonly recorded as soon as it is brought up from the bottom. The color is a general indicator of the clay's composition and the chemical environment in which the clay exists.

Marine sediments may be of various colors, depending on their mineral composition. White generally indicates calcium carbonate. Black indicates most often the presence of iron or manganese sulfides. Yellow indicates iron content in the form of limonite or a hydroxide. Brownish red usually indicates the presence of iron oxide in the form of hematite. In addition, black usually indicates a reducing environment—that is, an environment that contains almost no free oxygen. Yellow, brown, and red, however, usually indicate an oxidizing environment.

Mineral Composition

The minerals that occur most commonly in ocean sediments are, in order of abundance, silicates, carbonates, oxides and hydroxides, phosphates, and sulfates.

Silicate minerals contain silicon and oxygen, with one silicon atom surrounded by four oxygen atoms. They may contain other elements as well, such as potassium, aluminum, sodium, calcium, magnesium, and iron. The most common silicates are the feldspars, amphibole, pyroxene, olivine, quartz, opal, micas, clays, and zeolites.

Carbonate minerals contain carbon and oxygen, with one carbon atom surrounded by three oxygen atoms (CO_3^{-2}). In ocean sediments, calcium or magnesium (sometimes both) is often bound to the carbonate group to form the minerals calcite, aragonite, and dolomite.

The oxides and hydroxides are minerals that contain negatively charged chemical entities consisting of oxygen alone (O^{-2}) or oxygen and hydrogen combined in a hydroxyl group (OH^{-1}). The most common of these minerals are hematite (iron oxide), goethite (hydrous iron oxide), and manganese-iron hydroxides.

Phosphate (PO_4^{-3}) minerals consist of one atom of phosphorus surrounded by four atoms of oxygen, arranged like the atoms in a silicate group. The most common phosphate mineral in marine sediments is apatite, $Ca_5(PO_4)_3F$.

The sulfate minerals are less common in ocean sediments, because the sulfate group, SO_4^{-2}, tends to dissolve in seawater. The mineral barite ($BaSO_4$) is sometimes found in deep-sea sediment.

Thickness of Sediments and Rate of Deposition

The thickness of sediments provides clues to their origin and age and even to the history of the ocean basin itself. Near the continents, sediments may measure 20 km in thickness. In regions near the mouths of large rivers, sediments may accumulate at the rate of several meters per year.

The turbidity current is a flow of sediment-laden water that occurs when an unstable mass of sediment at the top of a relatively steep slope under water is jarred loose and slides down the slope. As the slide travels down the slope, more and more water becomes mixed in the mass, increasing its fluidity. The currents reach very high velocities, enabling them to spread for vast distances across the abyssal plains. Turbidity currents are triggered by earthquakes or storms. As large turbidity currents lose energy, the sediment settles to the ocean floor, and sediment layers several meters thick may accumulate in just a few days.

In deeper waters, far beyond the reach of turbidity currents, inorganic clays carried out to sea by winds and currents, along with the shells (tests) of plankton, slowly sink to the ocean floor to form deep-sea sediments. Clays accumulate very slowly, at a rate less than 2 mm every thousand years. In regions where surface plankton are abundant, shells may accumulate at a rate of 20 mm every thousand years.

Over large regions of the seafloor, the rate of deposition has remained relatively constant for millions of years. Thus, the thickness of these sediments gives us an indication of their age.

CLASSIFICATION OF OCEAN SEDIMENTS

In 1891, Sir John Murray and A. F. Renard, in the first major study of ocean sediments, published a report on samples of sediments taken on the *Challenger* expedition (1872–76). In that report they proposed a system of classification of three major types of ocean sediments (Table 4.2) that remains useful today: **Pelagic sediments** are derived from processes occurring within the ocean and far from the continents; **terrigenous sediments** are derived from land sources, usually continents; **cosmogenic sediments** are made up of particles of extraterrestrial origin.

Table 4.2 *Classification of Marine Sediments*

Sediment Type	Components
Pelagic	Biogenic, pelagic clay, authigenic, and volcanic deposits
Terrigenous	Terrigenous muds; slump deposits; turbidites; glacial deposits
Cosmogenic	Meteorites; tektites

Pelagic Sediments

Biogenic Sediments In large regions of the ocean floor, the sediments contain abundant skeletal material. Billions of microorganisms live near the ocean's surface and in the waters throughout the ocean. Over eons of growth, reproduction, and death, the shells and other hard parts of these organisms have slowly accumulated on the ocean floor. The most common types are foraminifera, radiolaria, pteropods, diatoms, and coccolithophores (Figure 4.10).

Foraminifera, which are single-celled members of the kingdom Protista, measure up to 2 mm across. Although they are related to amoebas, they have protective shells. To ensnare food, they send out sticky projections through tiny openings, or pores, in their shells. In fact, the name *foraminifera* means "bearers of pores." These microorganisms secrete calcium carbonate to form their shells. Most of the foraminiferal shells found in the sediments of abyssal plains are the shells of plankton, which are microscopic organisms that float in the upper 100 m (328 ft) of ocean water and move with surface currents. Some plankton also live on the continental shelf, from which turbidity currents transport them down submarine canyons to the abyssal plains. They do not make up an appreciable component of sediments on the shelf because there is so much terrigenous sediment.

Radiolaria, another form of plankton, have shells composed of silica in the form of opal, a hydrous variety of quartz. The shells of radiolaria have long, fragile spines radiating from a central chamber. They are distributed on the abyssal plains in the same manner as foraminifera.

Diatoms are single-celled, photosynthetic organisms with silica hard parts and are among the most common phytoplankton in both freshwater and ocean environments. Diatoms, which are found worldwide, form the base of the oceanic food chain (see Chapter 14). The hard parts of coccolithophores, a microscopic plant, are also common in deep-sea sediments. Their hard shells are made of calcium carbonate.

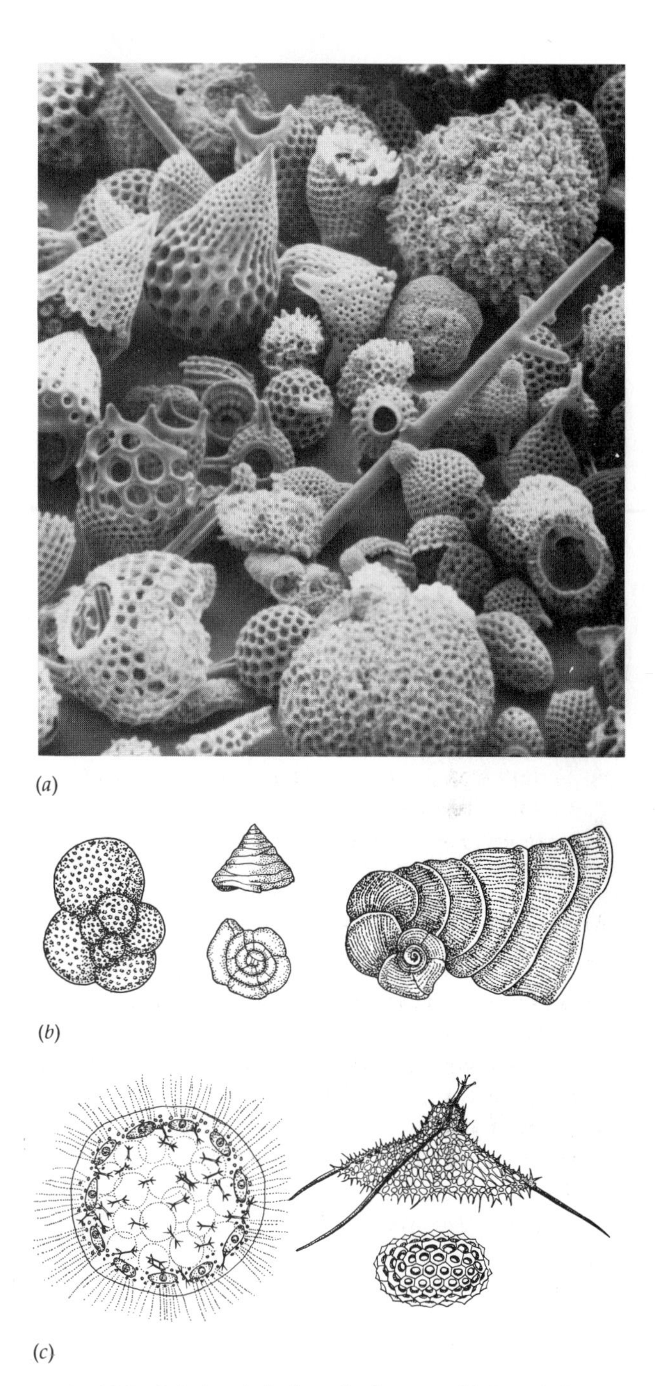

4.10 (a) Radiolarian shells from the deep sea. (b) Foraminifera. (c) Radiolarians.

Biogenous deposits such as these enable scientists to date sediments and to interpret ocean temperatures at various times in the past. By studying the layers of sediments, which contain sequences of fossils, with the oldest usually at the bottom, scientists can infer the relative ages of the layers. They also use carbon-14 dating to establish the age of sediments (see Chapter 6).

4.11 Growth rings of a manganese nodule are shown in this cross section, magnified about five times. There is no generally accepted theory that organisms are responsible for the nodules. (Scripps Institution of Oceanography, University of California, San Diego)

Pelagic Clay Pelagic clay consists of small (less than 20 microns) mineral particles. Some of the particles are carried by the wind from the land. These are usually illite (a clay mineral) or quartz. They also contain some authigenic minerals, cosmogenic iron-nickel dust, and occasional phosphatic fish teeth.

Authigenic Sediments In the deep sea, **authigenic** minerals are common. These minerals form in place from ocean water or within the sediments. Iron oxides and zeolites in pelagic clays are examples. Zeolites are hydrous silicate crystals thought to have formed when volcanic glass containing high amounts of magnesium and iron crystallized in the seawater, which altered the glass. They are especially widespread in the Pacific Ocean. Other authigenic minerals are barite in biogenic sediments, marine phosphorite deposits, manganese nodules, and metalliferous muds at submarine springs.

Manganese nodules (discussed in Chapter 18; see Figure 4.11) are widely distributed on the seafloor. These nodules contain about 65% manganese dioxide; about 30% iron oxide; less than 5% oxides of nickel, copper, and cobalt; and traces of other minerals. Manganese nodules are particularly interesting because of their potential economic value. Their principal constituents are present in only trace amounts in seawater and in marine organisms, and they are rare in other marine sediments except in hydrothermal deposits at mid-oceanic ridges. Moreover, these nodules tend to be scattered over the ocean floor rather than buried in sediments.

There are several hypotheses about the origin of manganese nodules. One is that their components originated in land rocks carried to the sea in solution, which were then slowly precipitated from the seawater. A second theory proposes that the iron, manganese, copper, and nickel they contain dissolved from marine sediments under the reducing conditions of the sediments and then migrated up to the seafloor, where they were oxidized and precipitated. A third suggests that the components of the nodules were released by metalliferous hot springs at mid-oceanic ridges and were carried by bottom currents across the seafloor, where they slowly oxidized and precipitated. Recent research tends to favor this third hypothesis.

Phosphate nodules are found on the continental shelf or on banks and plateaus of the continental margins. The nodules may become abundant where the rate of sedimentation of other particles is low. Examples are found on the southern California borderland and on the Agulhas plateau off southwest Africa.

The phosphates precipitate from seawater under reducing conditions the details of which are poorly understood. Since the nodules are found along coasts with upwelling and high biological productivity in the water, chemical oceanographers suspect that the decay process on the shelf floor creates the reducing environment that promotes phosphate precipitation. Phosphate is an important agricultural fertilizer, and the marine deposits represent a resource with increasing value as terrestrial sources are depleted (see Chapter 18).

Metalliferous deposits along the East Pacific Rise and in the Red Sea have been studied recently. These deposits are characterized by such heavy metals as iron, copper, and manganese, along with lead, cobalt, silver, nickel, zinc, and clays such as montmorillonite.

Volcanic Deposits Explosive volcanic eruptions throw great quantities of ash into the atmosphere, where winds transport them around the earth. Ash that falls onto the ocean surface and settles to the bottom degrades into siliceous mud. These deposits are found especially downwind from the explosive volcanoes associated with island arcs and the west coast of South America. Other deposits are the result of submarine lava flows that congeal on the seafloor. They are especially common at the mid-oceanic ridges and along the submerged flanks of volcanic islands, such as the island of Hawaii.

Terrigenous Sediments

Muds The terrigenous (derived from the land) muds consist of silt and clay originating on the continents. The principal minerals in these sediments are quartz, feldspars, and such clay minerals as illite, kaolinite, and chlorite. Coastal rivers carry small particles of these minerals to the ocean, where, if the currents are strong enough, they remain in suspension. The particles may then be carried far out to sea before they slowly settle to the seafloor. This suspended sediment represents a relatively small portion of seafloor sediment because most coastal currents move along the coast rather than away from it.

The largest amount of terrigenous muds are transported to the seafloor by turbidity currents. Much of this mud moves down submarine canyons and spreads out on the seafloor. Where shelves are wide (more than 10 km) or where canyons are not present, sediment has no easy route to the ocean basins. Small quantities are, nonetheless, transported seaward during storms.

Trade winds carry dust from tropical deserts to the ocean, where the dust settles slowly to become mud on the seafloor. Dust from the Sahara Desert is carried across the Atlantic Ocean to the Caribbean Sea. Storms in higher latitudes carry dust from arid regions also. Dust from Mongolia is transported to the North Pacific, and dust from Australia is carried to the Indian Ocean.

Turbidites Terrigenous sediments hundreds of meters thick accumulate on the continental shelf and the continental slope. The pore spaces of these sediments are filled with seawater. The angle of decline in these areas of the seafloor ranges from 2° to 7°, occasionally greater, and the unstable sediments tend to move if there is no natural impediment to movement. An earthquake or storm of sufficient magnitude may throw sediments into suspension, or gravity may dislodge them. When this occurs, a turbidity current forms, consisting of water, sand, silt, and clay. Because the current is more dense than seawater, it moves downslope until the slope flattens. Then the energy is lost by friction with the bottom, and the sediment settles. The largest particles settle first, and there is a gradation in the sediment produced from coarse to fine particles (graded bedding). The resulting deposit is called a **turbidite.** Sometimes even boulders show up in turbidites, as in the Gulf of Elat. Turbidites may also contain shallow-water fossils and fossils of terrestrial organisms.

Since turbidity currents last from only a few minutes to perhaps fifteen hours, it is difficult to observe the process—and dangerous as well. The most carefully documented turbidite was produced by an earthquake in 1929. The current was so powerful that it snapped and buried the trans-Atlantic submarine telephone cables. It moved down submarine canyons for some 600 km (372 mi) before settling to form a turbidite 350 m (1,148 ft) thick (Figure 4.12).

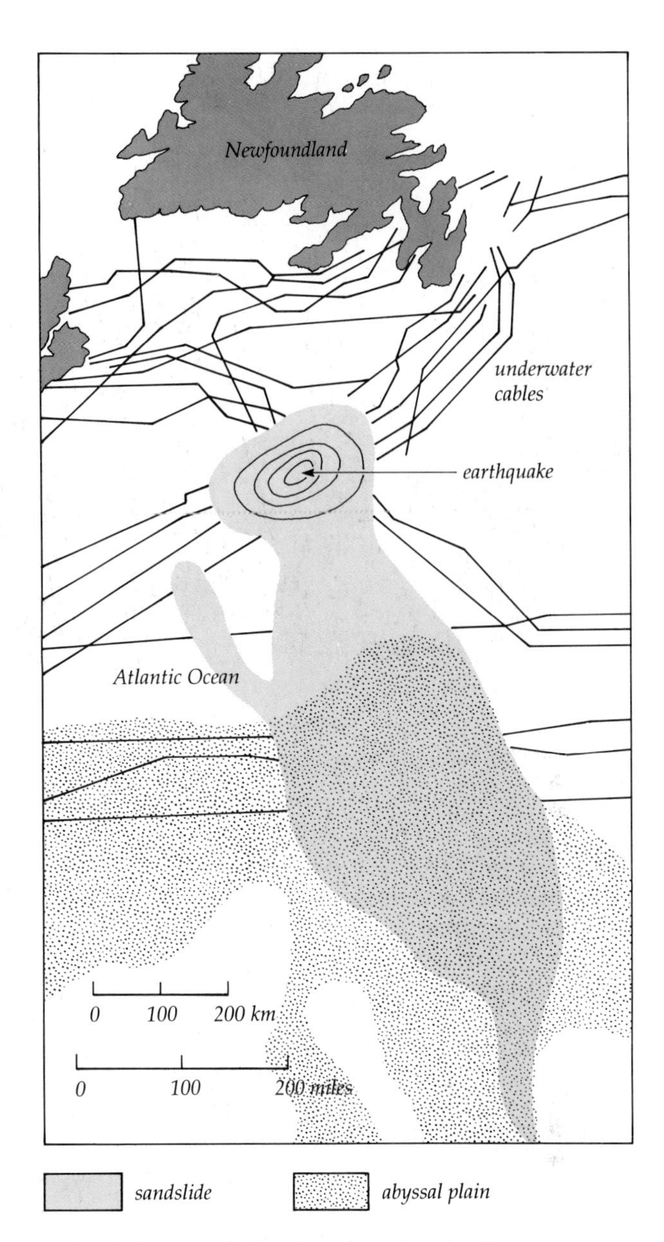

4.12 Undersea sand slides from an earthquake off Newfoundland were discovered by Bruce Heezen and Maurice Ewing to be the cause of severed trans-Atlantic telephone cables. (After Heezen and Ewing, 1952, see Further Readings)

Glacial Deposits Sediments near the Antarctic continent and Greenland consist of materials derived from melting icebergs. As glacial ice moves downslope toward the sea, the lower regions periodically melt and refreeze in response to pressure and the heat of friction. As the ice travels, it erodes and transports

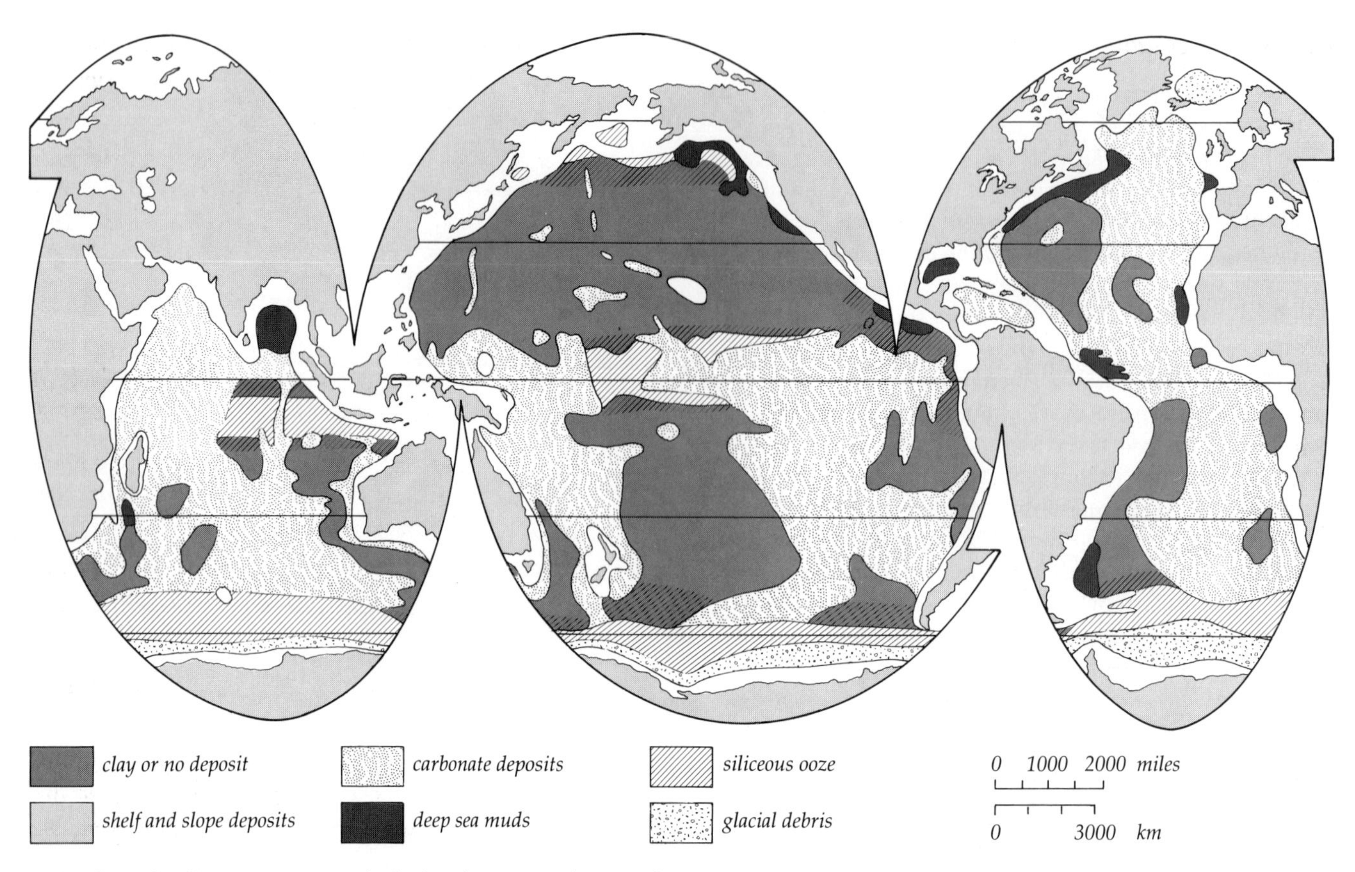

4.13 General sedimentary pattern of pelagic sediments on the ocean floor. Sediments of terrigenous origin, except for glacial deposits, are not shown. (From W. H. Berger)

sand, silt, and gravel, which become incorporated into the ice. When the glacial ice finally reaches the sea, large blocks break off as icebergs, which are carried out to sea by ocean currents. As the icebergs melt, they drop their load to the ocean floor. The resulting sediments are distinguished from turbidites by their lack of graded bedding and the presence of random cobbles and boulders.

Cosmogenic Sediments

It is estimated that about 5,000 tons of meteorites reach the earth's surface every year. Most of these extraterrestrial objects consist of small particles that fall into the ocean and become incorporated into other sediments. They can be recognized by their characteristic composition and their magnetism. Radioactive dating indicates that they are about 4.5 billion years old, clear evidence that they did not form in the ocean.

Cosmogenous components called **tektites** are sometimes found in ocean sediments. They are fairly numerous in the eastern Indian Ocean and occur in some land areas. Tektites consist of 70% to 90% silica glass (SiO_2). They are less than 1.5 mm in diameter and are typically elongate. Although they are thought to be of extraterrestrial origin, their precise origin is unknown. Some scientists suggest that they came from comets, planetoids, or the moon. However, all agree that tektites have fallen through the earth's atmosphere.

DISTRIBUTION OF MARINE SEDIMENTS

The worldwide distribution of carbonate and siliceous deposits and clay-to-silt-sized sediments follows distinct patterns (Figure 4.13). Carbonate deposits are most common in tropical regions, and siliceous deposits in the higher southern latitudes. Clays are abundant throughout the North Pacific and near the mouths of major rivers. Glacial deposits ring Antarctica. Carbonates are relatively scarce in sediments deeper than 4,000 m (13,120 ft).

As we have noted, carbonate deposits are rare in deep-sea sediments. Experiments in which calcium carbonate was suspended at various depths revealed that at about 2,000 m, the rate of dissolution increases. The fact that seawater is undersaturated in carbonates below this level accounts for the dissolving. In sediments less than 2,000 m deep, where the water is saturated or in some cases supersaturated, calcium carbonate sediment is common. Usually it is biogenic.

FURTHER READINGS

Berger, W. H. "Deep-Sea Sedimentation." In *The Geology of Continental Margins,* edited by C. A. Burk and C. D. Drake. New York: Springer-Verlag, 1974.

Blatt, H., and others. *Origin of Sedimentary Rocks.* Englewood Cliffs, N.J.: Prentice-Hall, 1972.

Cronan, D. S. "Authigenic Minerals in Deep-Sea Sediments." In *The Sea, Marine Chemistry,* edited by E. D. Goldberg, vol. 5. New York: John Wiley, 1974.

———. "Manganese Nodules and Other Ferro-Manganese Oxide Deposits." In *Treatise on Chemical Oceanography,* edited by J. P. Riley and R. Chester, vol. 5. New York: Academic Press, 1976.

Cronin, T. M. "Rapid Sea-Level and Climate Changes." *Quaternary Scientific Reviews,* 1982, *1,* 177–214.

Darwin, C. *The Structure and Distribution of Coral Reefs.* Berkeley: University of California Press, 1962.

Davies, T. A., and D. S. Gorsline. "Oceanic Sediments and Sedimentary Processes." In *Treatise on Chemical Oceanography,* edited by J. P. Riley and R. Chester, vol. 5. New York: Academic Press, 1976.

Dietrich, G., and others. *General Oceanography.* 2d ed. New York: John Wiley, 1980.

Emery, K. O. "Characteristics of Continental Shelves and Slopes." *American Association of Petroleum Geologists Bulletin,* 1965, *49*(9), 1379–84.

———. "The Atlantic Continental Margin of the U.S. during the Past 70 Million Years." In Geological Association of Canada, *Geology of the Atlantic Region,* Special Paper no. 4, November 1967, pp. 53–70.

———. "The Continental Shelves." *Scientific American,* September 1969, pp. 39–52.

Emery, K. O., and B. J. Skinner. "Mineral Deposits of the Deep-Ocean Floor." *Marine Mining,* 1977, *1,* 1–71.

Hall, J. M., and P. T. Robinson. "Deep Crustal Drilling in the North Atlantic Coast." *Science,* 1979, *204,* 573–86.

Hay, W. W. *Studies in Paleo Oceanography.* SEPM Special Publication 20. Tulsa, Okla.: Society for Economic Paleontology and Mineralogy, 1974.

Heezen, B. C., and M. Ewing. "Turbidity Currents and Submarine Slumps and the 1929 Grand Banks Earthquake." *American Journal of Science,* 1952, *250,* 849–873.

Heezen, B. C., and C. D. Hollister. *The Face of the Deep.* New York: Oxford University Press, 1971.

Heezen, B. C., and I. D. MacGregor. "The Evolution of the Pacific." *Scientific American,* May 1973, pp. 102–15.

Heezen, B. C., M. Tharp, and M. Ewing. "The Floors of the Ocean." Denver: *Geological Society of America,* Special Paper 65 (map), 1959.

Heirtzler, J. R., and W. B. Bryan. "The Floor of the Mid-Atlantic Rift." *Scientific American,* August 1975, pp. 79–90.

Hill, N. N., ed. *The Sea.* Vol. 3. New York: John Wiley, 1963.

Hsü, K. J., and H. C. Jenkins, eds. *Pelagic Sediments: On Land and Under the Sea.* Special Publication of the International Association of Sedimentologists, 1. Oxford: Blackwell Science, 1974.

Kennett, J. P. *Marine Geology.* Englewood Cliffs, N.J.: Prentice-Hall, 1981.

Lisitzin, A. P. *Sedimentation in the World Oceans.* SEPM Special Publication 17. Tulsa, Okla.: Society for Economic Paleontology and Mineralogy, 1972.

McKenzie, D. P., and J. G. Sclater. "The Evolution of the Indian Ocean." *Scientific American,* May 1973, pp. 62–74.

Menard, H. W. "The Deep Ocean Floor." *Scientific American,* September 1969, pp. 53–63.

Sheldon, R. P. "Phosphate Rock." *Scientific American,* June 1982, pp. 45–51.

Thorarinsson, S. "Surtsey: Iceland's New Island." *American Scandinavian Review,* 1966, *54*(2), 117–25.

To see a World in a Grain of Sand
And a Heaven in a Wild Flower
Hold Infinity in the palm of your hand
and Eternity in an hour.

William Blake

FIVE

Plate Tectonics

Until the early twentieth century, geologists assumed that the ocean basins and the continents of the earth were fixed in position relative to one another. Even with the emergence of geology as a branch of natural science in the eighteenth century, scientists continued to view these features as permanent. There was abundant evidence that mountains consist in part of rocks that were formed below sea level and subsequently uplifted. This required modification of the location of the shoreline, but the relative positions of the continents and ocean basins were thought to be fixed. During the nineteenth century, some geologists remarked on the apparent fit between the Atlantic coasts of Africa and South America, but they failed to recognize the possible significance of that curious coincidence. However, a bold new hypothesis challenged the beliefs of geologists in the early twentieth century.

CONTINENTAL DRIFT: THE FIRST APPROACH

In 1912, Alfred Wegener published a paper questioning the idea that the position and shape of the ocean basins and continents were fixed. He cited the fit of the African and South American coasts and pointed out that the coastlines of other land areas exhibit a similar apparent fit (Figure 5.1). He also wrote about certain rocks that appear in the mountain ranges of both Africa and South America and about plant and animal fossils common to both continents.

Wegener asked an intriguing question: Could these two great land masses have been joined together at some distant time and subsequently drifted apart? Most geologists at the time chose to ridicule the question, but Wegener had raised the issue of **continental drift**, an issue too compelling to be ignored indefinitely. Finally, during the 1950s and the 1960s, it became the central issue of geological science.

Wegener went on to suggest a sequence of events that might account for the global migration of continents. He proposed that initially a supercontinent, which he called Pangaea, had included all the present continents. Over time, Pangaea split apart, and the resulting fragments drifted to their present locations (Figure 5.2). In support of this hypothesis, Wegener again cited the similarities of rocks and fossils on different continents. He also pointed out that glacial striations (grooves produced by ice moving over rocks) on the southern continents would form a simple radiating pattern if the continents were viewed as having once constituted a single mass.

Wegener's proposal was startling. Influential geologists attacked it as preposterous, noting that Wegener was not even a geologist—he was a me-

5.1 The Gulf of Aden (center) and the southern Red Sea (left) appear to have been formed by rifting between Africa and Arabia, which began within the past 30 million years. (NASA) This apparent match of land margins is compelling evidence for continental drift.

teorologist. What force, they demanded, could have been powerful enough to pull the continents apart and send them across the face of the earth?

The controversy was laid to rest in 1928 by geologists led by T. C. Chamberlain. He, ironically, had previously written an influential paper on the "multiple working hypothesis," which advocated that scientists not confine themselves to one explanation of data until the evidence in support of one is overwhelming. Of the leading geologists of the time, only Sir Arthur Holmes, of England, and A. L. DuToit, of South Africa, continued to pursue the Wegener model in the 1930s and 1940s.

Several studies launched in the 1950s laid the groundwork for the reemergence of Wegener's model. These were studies of the seafloor by Bruce Heezen, Keith Runcorn, and Harry Hess. The studies revealed the magnetism of rocks on the seafloor, submarine topography, and the description of the ocean ridge system.

ROCK MAGNETISM

Scientists had long been aware of rock magnetism. In the seventeenth century, Sir William Gilbert, physician to Queen Elizabeth I, reported experiments he had made with the mineral magnetite. For some time explorers had been using magnetite (also known as lodestone) as an aid in navigation. Gilbert compared measurements of the earth's magnetic field they had made in various parts of the world with measurements he himself had made in England. He concluded that the earth behaves as though it were a giant magnet.

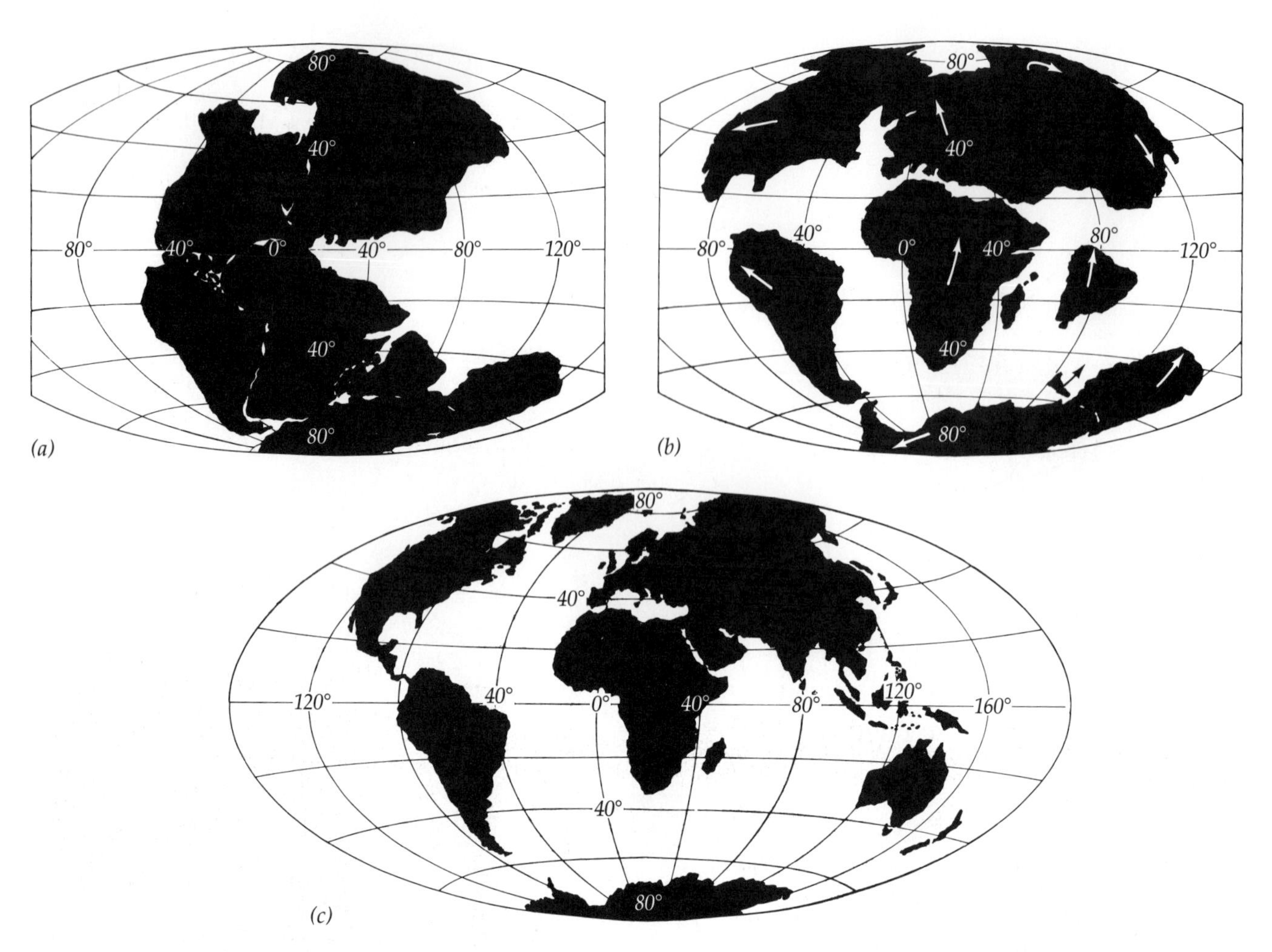

5.2 *Possible configurations and positions of the continents (a) 150 to 200 million years ago, (b) 65 to 80 million years ago, and (c) at present.*

The earth's magnetic field is analogous to the field that would be produced by a giant iron bar magnet located a few hundred kilometers from the center of the earth and tilted about 11.5° to the earth's axis (see Figure 5.3). We know now that magnetic fields are caused by moving electrical charges—in other words, by electrical current. The current that creates the earth's magnetic field is believed to be generated by heat from the earth's core and by the relative motions of the mantle, the inner core, and the outer core. (This theory is known as the *dynamo model*.)

During the nineteenth century, scientists noted that between 1576 and 1823, magnetic north as measured in London had drifted from 8° east of true north (the north geographic pole) to 24° west. They inferred that the magnetic pole wanders around the geographic pole in a cycle of about 500 years. Moreover, we know that over the last hundred years, the strength of the earth's magnetic field has declined by about 6%.

Many rocks carry a record of the earth's magnetic field of the past. As molten material (magma) starts to cool, magnetite crystals begin to form even while the temperature is still high. Not uncommonly, lava erupting from a volcano, at a temperature of about 1,000°C, already contains suspended magnetite crystals. As the magma continues to cool, the crystals align themselves with the earth's magnetic field. Thus, the rock formed when the whole magma has cooled contains magnetite crystals aligned with the earth's magnetic field as that field existed at the time of crystallization. What we have, then, is a permanent record of the magnetic field at that time (Figure 5.4).

Studies of that record in rocks of different ages reveal that the intensity of the earth's magnetic field has changed over time, that the location of the earth's magnetic poles has shifted, and that the positions of the magnetic North and South poles have reversed from time to time on about a 700,000-year frequency. Detailed maps based on these studies suggested that at particular times in the past, each continent has had a magnetic pole different from the poles of the other continents (Figure 5.5). This is clearly impossible, since the earth as a whole has only two poles. But when the continents are reassembled as Wegener proposed, the divergent magnetic patterns almost coincide! This is compelling confirmation of Wegener's proposal.

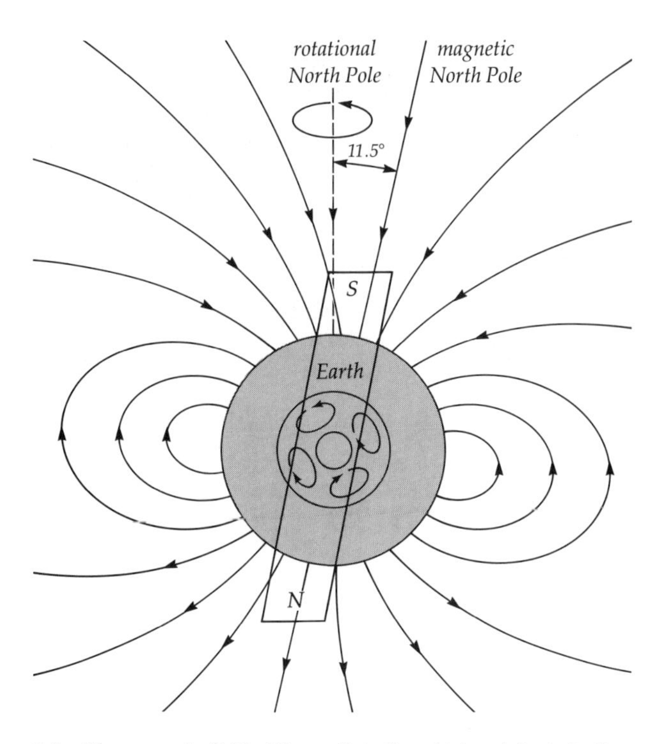

5.3 The magnetic field of the earth is thought to originate in the motion of the liquid iron in the outer core. The core motions are shown; however, nobody really knows what they are. At present the magnetic North Pole is inclined by about 11.5° to the rotational North Pole. The lines outside the earth represent lines of force, which show the direction of the field, the direction that a compass needle free to move in a vertical plane would point to in order to line up with the field. On the surface of the earth near the pole, the field is nearly vertical, whereas near the equator the field becomes horizontal. The closer the lines of force are, the stronger the field. It reaches its greatest strength at the magnetic poles and is only half as strong at the magnetic equator.

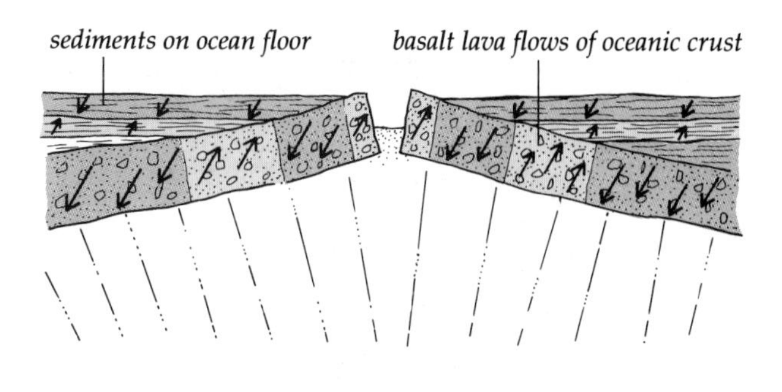

5.4 Deep-sea cores reveal exactly the same vertical sequence of magnetic reversals as appears horizontally in the pattern of magnetic anomalies in the oceanic crust. Arrows indicate the direction of magnetic pole. (D. Alt, Physical Geology. *Belmont, Calif.: Wadsworth, 1982)*

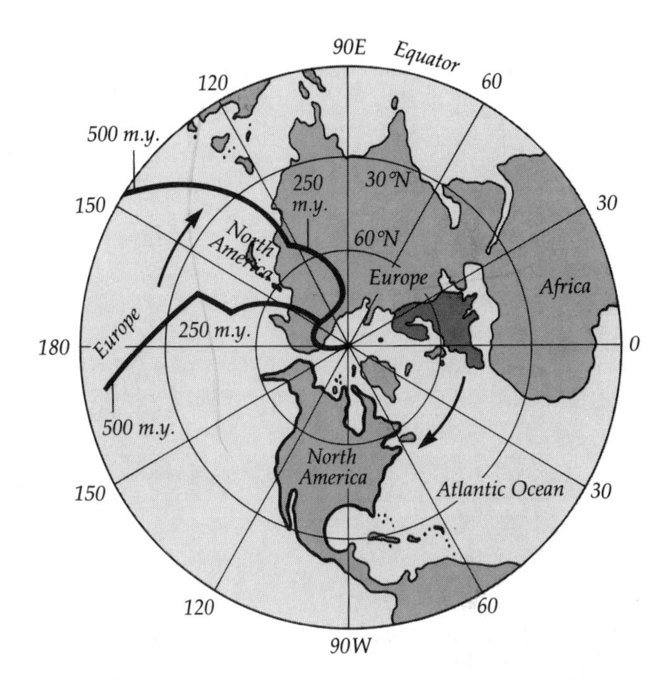

5.5 The apparent polar wandering curve for North America would very nearly fit that for Europe if it were moved east through the same angle North America would have to move to close the Atlantic Ocean. Similar relationships obtain for the polar wandering curves of other continents that reassemble to make Pangaea. (D. Alt, Physical Geology. *Belmont, Calif.: Wadsworth, 1982)*

SEAFLOOR SPREADING: THE DATA CONVERGE

While these studies of rock magnetism were being conducted, other geologists were exploring the ocean depths. They discovered a worldwide system of ocean ridges in which the ridge in the Atlantic Ocean, which had already been recognized, was connected with newly identified ridges in the Arctic, Indian, and Pacific oceans. Moreover, they found that all the ridges were characterized by rift valleys and by frequent earthquakes. Responding to this new knowledge, marine geologists dredged rock samples from the ridges and nearby seamounts. They made two findings: (1) The rocks were mainly of basaltic composition, and (2) they were relatively young at the ridges and seemed to grow progressively older at increasing distances from the ridges.

At about the same time, Vening Meinesz and J. H. F. Umbgrove revived a proposal made by Arthur Holmes in the 1930s that convection currents occur in the earth's mantle beneath the crust. These currents, which produce mass movements within the mantle, were thought to result from differences in temperature and density.

Notice that scientific knowledge advances through the accretion of discrete bits of information. Once these bits have achieved "critical mass," they can be brought together into one unified statement that helps us understand an underlying pattern or principle.

Then, around 1960, Harry Hess suggested that convection currents welling up under mid-oceanic

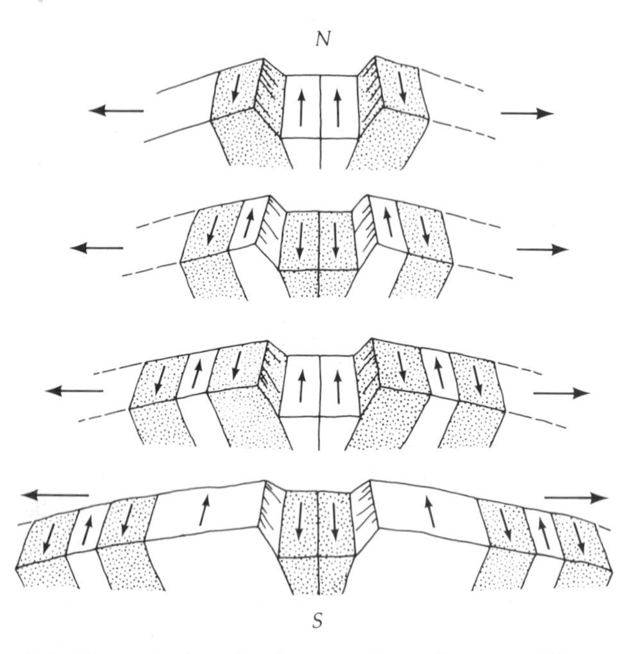

5.6 The mechanism of seafloor spreading as demonstrated by the pattern of oceanic rock magnetism. Basalt lava flows erupt into the ridge rift to make new oceanic crust, which then splits and moves down opposite sides of the ridge, carrying the record of the earth's magnetic field reversals with it. Some authors have compared the rift in the ridge crest to a recording head and the moving oceanic crust to a magnetic tape. Arrows pointing up designate present-day polarity. Arrows pointing down indicate polar reversals. Lateral arrows indicate the directions of plate movements. (D. Alt, Physical Geology, *1982)*

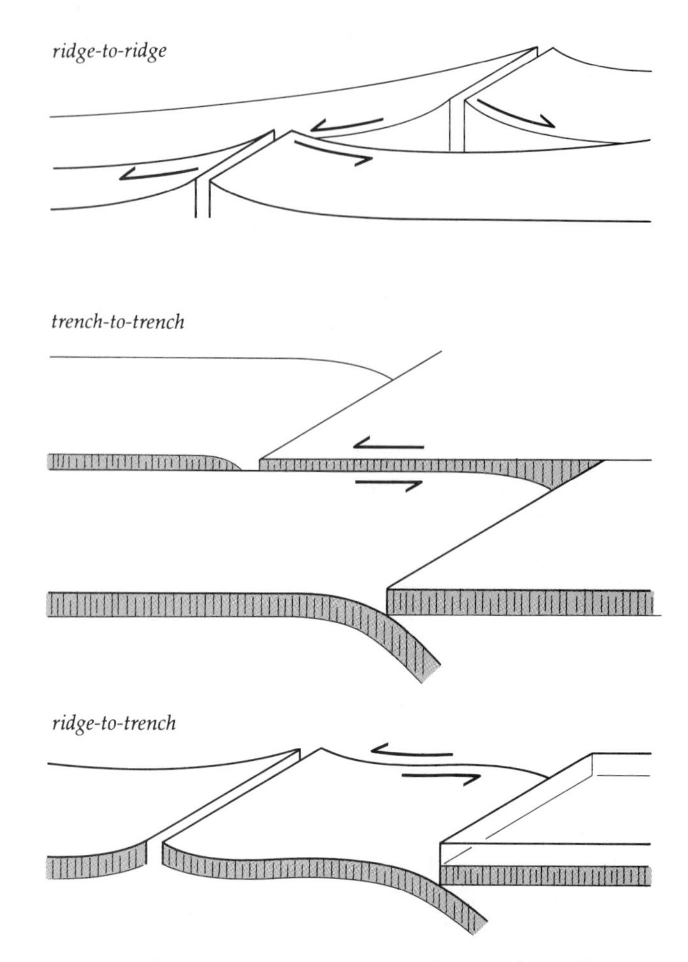

5.7 Transform boundaries can connect ridges to ridges, ridges to trenches, and trenches to trenches in a total of fourteen possible right- and left-handed combinations, of which these are a sampling. (D. Alt, Physical Geology, *1982)*

ridges might explain many observations. The process Hess envisioned was later given the name **seafloor spreading.** In this process, a current of magma (molten material) rises in the mantle under a mid-oceanic ridge, heats the overlying crust, and causes it to expand. The crust then stretches and fractures, and magma rises, producing rift valleys, lava flows, and volcanoes on the seafloor. The rising convection current in the mantle beneath the crust is deflected sideways away from the ridge. As it moves along below the crust, the current cools, becomes less dense, and eventually sinks. As it sinks, the current drags the crust down, forming trenches. Meanwhile, the overlying crust, which has been fractured at the ridges, glides toward the trench. As the process continues, new magma seeps up along fresh fractures at the ridge crest, forming new crust. At the trenches, the oldest seafloor crust is dragged down into the mantle to be recycled. In this one model, oceanic ridges, rift valleys, ridge flanks, trenches, island arcs, and all associated seismicity, magnetism, and volcanism appeared to have been explained.

The concept of seafloor spreading provided a plausible explanation for the known differences in age between the continents (3.8 billion years) and the ocean basins (less than 200 million years). This, too, lent credibility to Wegener's ideas. Soon geologists began looking for more evidence to check his ideas against the new model of seafloor spreading.

The knowledge that rocks containing magnetite provide a record of polar reversals and changing intensities of the earth's magnetic field was now used to test the soundness of the seafloor-spreading model. If the model was correct, there should be a pattern of magnetic field anomalies in the rocks paralleling a seismic ridge. Moreover, that pattern should correspond to the distance from the axis of the ridge. No one had observed such a pattern until 1963, when F. J. Vine and D. H. Mathews published evidence of a pattern of magnetic field anomalies that was consistent with predictions based on seafloor spreading. They observed a mirror image of the pattern on both sides of a ridge, with similar anomalies equidistant from the axis (Figure 5.6).

Equally convincing evidence for seafloor spreading was supplied by J. T. Wilson in his explanation of transform faults and seismic motions observed along oceanic ridges. Faults occur at right angles to and offset the axes of ridges and rift valleys (Figure 5.7). If the rift valley lying north of such a fault is offset to the left of the rift valley lying south of the fault, one would expect that the north side of the fault is moving to the left relative to the south side, which is moving to the right. Seismic studies reveal that the opposite is true at the offset of the ridges. The north side moves to the right and the south side to the left along the portion of the fault lying between the ridge axis. Wilson explained that this can only occur if the seafloor is spreading in opposite directions perpendicular to the ridge axis.

PLATE TECTONICS: A UNIFYING CONCEPT

Geologists tend to be observers and experimenters rather than theorists. Despite the mounting evidence supporting the theory of seafloor spreading, many geologists continued to resist the new concept. Do the continents move horizontally like the seafloor? they asked. Is the earth expanding? How long has the process been going on? Why are there no trenches at the edges of the Atlantic Ocean?

The seafloor-spreading model did not permit the continents to move laterally. Clearly, a more inclusive model was needed to account for the apparent separation of South America and Africa and other similar observations. Serveral scientists independently came up with modifications in 1967 and 1968. The scientists noted that world maps of earthquake frequency and volcanic activity reveal narrow bands of high activity (Figure 5.8). They suggested that these bands represent the edges of rigid crustal plates. When plates interact with one another, earthquakes and volcanic activity occur. They further noted that the new plate boundaries do not always coincide with the boundaries of contintents and ocean basins; some plates have both, others have one. This suggests that the plates might be thicker than oceanic crust or even continental crust. The depth of the deepest earthquakes provided the clue. They occur no deeper than about 100 km, and those are beneath the trenches. The plates, then, must be about 100 km thick and must include the crust and upper mantle. The new term, **lithosphere,** was given to this zone. Beneath the lithosphere must be a region capable of allowing convection currents. This region was called the **asthenosphere.**

To answer the question of what happens to the new crust, they concentrated on the margins of the oceans and the continents. In some areas, like the east coasts of North and South America, there are few earthquakes, no volcanism, and no evidence of tectonic activity. However, along the west coasts of North and South America, there is an active margin with frequent earthquakes, volcanism, faults, and hot springs. This is tectonic activity even though the topographical features of these areas are very different from the oceanic ridges.

Volcanic island arcs and trenches are found along the eastern margin of Asia and in the Aleutian Islands. A deep trench exists along the western margin of South America. Earthquakes are frequent, occurring in a narrow band extending from beneath the trenches back toward the continents at increasing depth. Over the trenches, strong negative gravity anomalies are observed, indicating that the earth's crust must be descending (subducting) into the mantle in these regions.

The new crust that forms at the ridges descends at the trenches. Zones of descending crust are called **subduction**, or Benioff, **zones** (Figure 5.9). This explanation answered the question, What happens to the new crust? However, it raised a new question: What makes the new crust descend? One answer is that it may be pulled down by a descending convection cell in the mantle beneath the trench. Another possibility is that the mantle drags the plate along as it descends. Although its composition is similar to that of the upper mantle, the new crust is cooler and denser. Thus it sinks through and displaces the hotter material.

The zones of high earthquake and volcanic frequency define a number of rigid plates that are moving in relation to one another. Earthquakes, according to this view, occur only when plates collide, move away from one another, or slide past one another. It follows that the central areas of the plates should be relatively inactive tectonically, with few earthquakes or volcanoes. The east coasts of North and South America, which lie in the central area of plates whose eastern edge is at the Mid-Atlantic Ridge and whose western edge is along the west coasts of these continents, is such an inactive region. The west coasts, by contrast, are active margins because they coincide with plate boundaries.

The Pacific plate extends from the East Pacific Rise and the San Andreas Fault to the Aleutian, Japan, and Mariana trenches. How is it that active volcanoes occur on the island of Hawaii even though that island is situated in the middle of a rigid plate? Perhaps there are random "hot spots" in some places beneath the crust that erupt from time to time. As a plate moves over a hot spot, a volcano forms. In the case of Hawaii, the plate continues to move northwestward, older volcanoes become extinct and form a ridge or islands, and a new active volcano forms to the southeast over the hot spot (Figure 5.10). In fact, the volcanic rocks

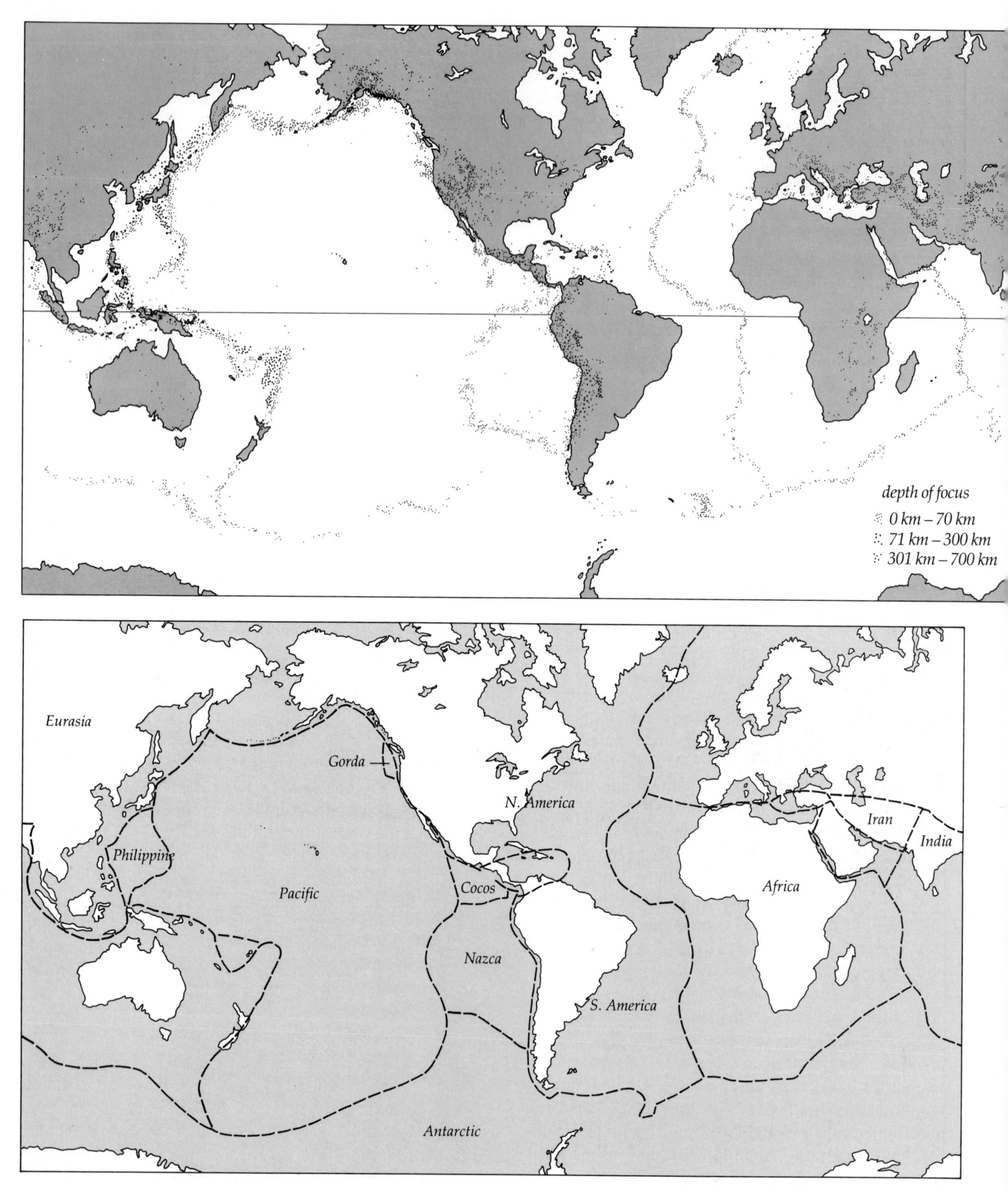

5.8 (above) Worldwide earthquake epicenter distribution, 1963–77. The dark bands on the map consist of thousands of small dots, each representing the epicenter of an earthquake of magnitude 4.5 or greater. Ten or more stations were used to determine each epicenter. (Data from National Oceanic and Atmospheric Administration)

(below) When lines are drawn along the zones of highest earthquake frequency, a new view of the earth's surface containing a few major plates emerges. Continents and ocean basins are no longer dominant in this view.

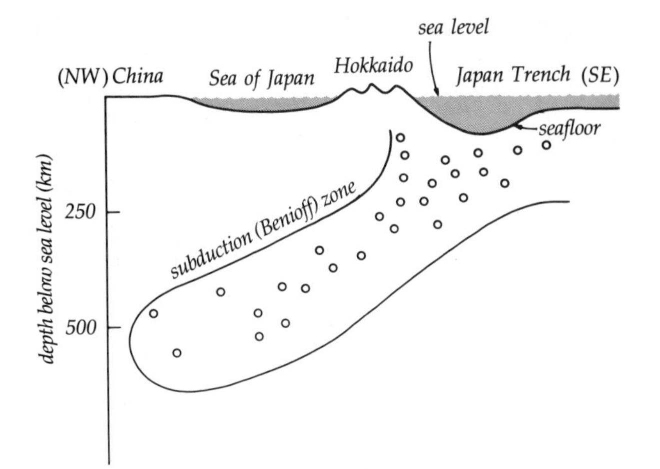

5.9 *Island arc cross section, showing earthquake foci clustered in a zone thought to be evidence of subduction of an oceanic plate. (After B. Gutenberg and C. F. Richter,* Seismicity of the Earth and Associated Phenomena, *Princeton, N.J.: Princeton University Press, 1954)*

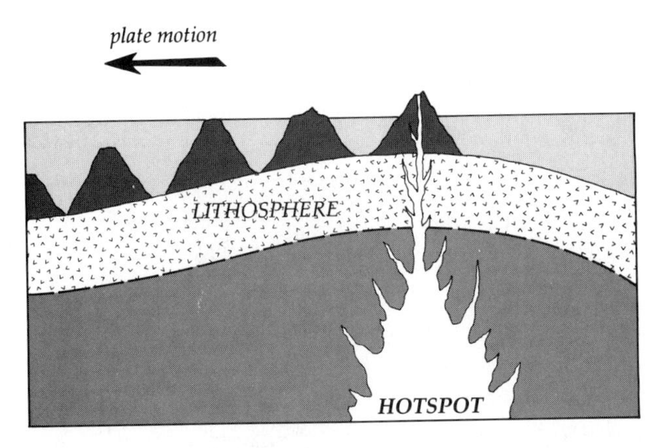

5.10 *Volcanic hotspots, such as the island of Hawaii, appear to mark places where an ascending plume of hot rock within the mantle perforates the lithosphere. A trail of extinct volcanoes with an active one at its head marks the path of the plate across the plume. These produce linear chains of volcanoes and seamounts on the seafloor notably free of seismic activity except at the hotspot. (Van Andel, 1985; see Further Readings at end of the chapter)*

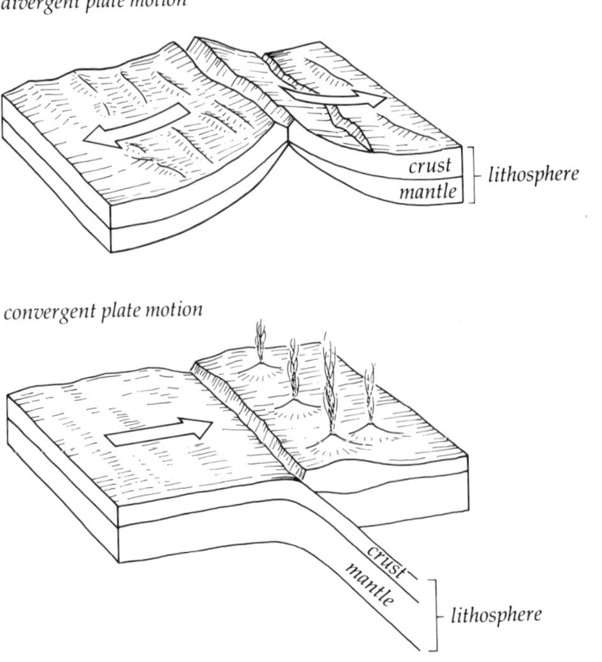

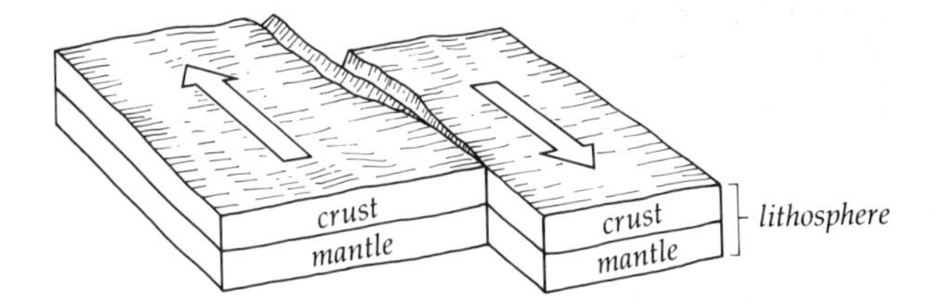

5.11 *The three possible kinds of relative plate motion. The first kind, where the plates move away from each other, is called a divergent plate boundary. The process of seafloor spreading creates new oceanic lithosphere at divergent plate boundaries. The second kind, where the plates approach each other, is called a convergent plate boundary. The process of subduction destroys lithosphere at convergent plate boundaries. The third kind, where the plates slide past each other in contact without either approaching or diverging, is called a transform fault boundary. Transform faulting is the corresponding process (see Figure 5.7).*

of the Hawaiian Islands do increase in age to the northwest.

The new crust that forms at an oceanic ridge marks the trailing edge of a plate (see Figure 5.11). Here two plates are moving in opposite directions, away from the ridge. Such a ridge is said to form the **divergent boundary** of the plates. Where plates collide, a **convergent boundary** exists. Where oceanic crust collides with continental crust (which is less dense and hence more buoyant),the oceanic crust subducts, producing a trench and volcanoes (Figure 5.12). When two plates with continental crust collide, they squeeze sediments between them to form mountains and the overthrusting of rocks. Since continents are composed of granitic rocks that are less dense than oceanic basalt, the continental crust does not descend into the mantle. When India collided with Asia, for example, the Himalaya Mountains were thrust upward. Finally, where plates are sliding by one another, a **transform boundary** is said to exist. The San Andreas Fault in California is such a boundary.

The rocks at the ridges cool slowly. The material at the lower part of the crust cools even more slowly than the mantle above, and as it cools, it contracts. This contraction accounts for the deepening of the ocean basin as the distance from the ridge increases, until

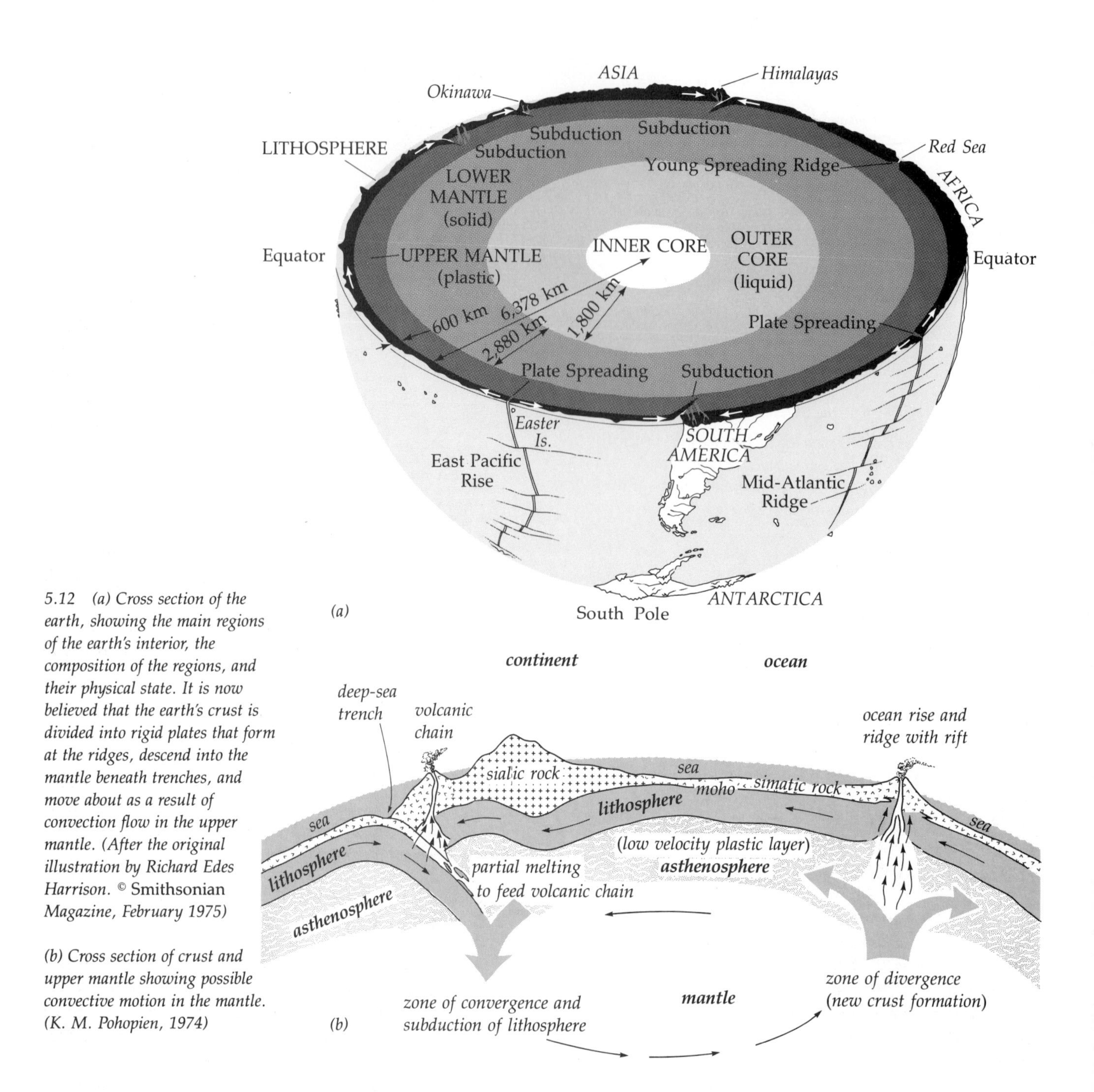

5.12 (a) Cross section of the earth, showing the main regions of the earth's interior, the composition of the regions, and their physical state. It is now believed that the earth's crust is divided into rigid plates that form at the ridges, descend into the mantle beneath trenches, and move about as a result of convection flow in the upper mantle. (After the original illustration by Richard Edes Harrison. © Smithsonian *Magazine, February 1975)*

(b) Cross section of crust and upper mantle showing possible convective motion in the mantle. (K. M. Pohopien, 1974)

the continental margins are reached. Except for trenches, which form tectonically, the deepest parts of the ocean basins lie farthest from the ridges.

The western Pacific Ocean is the oldest part of the seafloor, since it lies farthest away from any divergent boundary. Yet even the oldest rocks in the ocean basins are less than 200 million years old. The oldest rocks on the continents, by contrast, are more than 3.8 billion years old. It seems clear that the continents are relatively permanent features of the earth's crust, despite their horizontal and vertical motion. The rocks of the ocean basin seem impermanent, and most of them are destined to be recycled down into the mantle.

One possible way to produce that evidence was to examine, test, and date cores extracted from the seafloor paralleling the ridges. If the dates of rocks equidistant from the ridges proved to be the same, the model must be correct.

In 1968, the deep-sea drilling ship *Glomar Challenger* set about this mission. Cores drilled across the ridges did show that the thickness of sediment (Figure 5.13), the time span of the sediment column, and the ages of the underlying basaltic rocks increased with distance from the ridges and that the pattern was repeated on both sides of the ridges.

Analysis of these samples made it possible to calculate the rates at which the tectonic plates are moving. The rates range from about 1 cm per year along the Mid-Atlantic Ridge to more than 15 cm per year along parts of the East Pacific Rise. These rates cor-

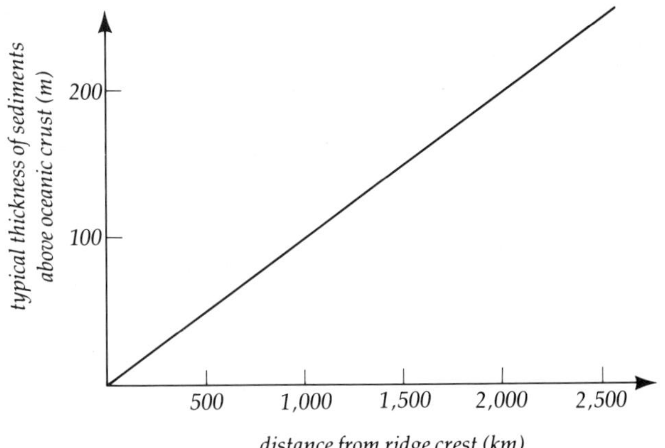

5.13 Plot illustrating how the thickness of sediments covering oceanic crust increases as a function of increasing distance from a spreading ridge system.

respond to the rates calculated on the basis of magnetic anomalies. While such rates may not seem great, they can produce a shift of 200 km (124 mi) in 10 million years along the San Andreas Fault.

THE DRIVING FORCE

Many geologists and geophysicists now accept the action of convection cells in the upper mantle as a force capable of moving the rigid crustal plates (see Figure 5.12). The main evidence is twofold. First, the absence of earthquakes within the mantle indicates that internal adjustments must occur by plastic deformation or flow. Second, there is a zone in the upper mantle where the velocity of shock waves is lower than would be expected if the material were solid and composed of a peridotite-type substance. This low-velocity zone can be explained if the material, though not molten, is capable of plastic deformation or flow under high pressure. Convection, then, may be a driving force. The evidence does not require convection currents, but it does permit them. The driving force behind the movement of tectonic plates remains uncertain, as it was in Wegener's time.

SUMMARY OF EVIDENCE

Many types of evidence support the current model of global plate tectonics, which we shall summarize here (Table 5.1 and Figure 5.14).

Earthquake Locations

The distribution of earthquake foci in the earth's crust reveals shallow, intermediate, and deep foci at the trenches and shallow foci at the ridges. The location of foci supports the idea of sinking and rising material, respectively. Shallow earthquake foci are also located along transform boundaries. This earthquake distribution is the cornerstone of plate tectonics.

Magnetic Bands

Perhaps the second most significant evidence supporting the model of plate tectonics is the matching bands of magnetic anomaly that parallel oceanic ridges. These bands show relative spreading on opposite sides of the ridge and the rate at which such spreading occurs. Rates measured so far vary between 1 cm per year at the Mid-Atlantic Ridge to 15 cm per year at the East Pacific Rise.

Sediment Age and Thickness

Sediments overlying bedrock are younger than the bedrock. Recent sediment is, of course, distributed all over the ocean basin. The thickness of the sediment and the time span it represents, determined mainly by dating microfossils, increase with distance from the axis of the oceanic ridges.

Continental Margin Configuration

Although not sufficient in itself to support the theory, it is an important fact that the continental blocks and the configuration of the edges of the continental shelves fit surprisingly well.

Bedrock Age

Cores of bedrock obtained in scattered parts of the ocean basins have yielded no rocks over 200 million years old. The oldest, about 175 million years, have been found in the western corner of the Pacific plate, a place far from the actively spreading new crust at the East Pacific Rise. The rocks in active rift valleys are less than 1 million years old.

Heat Flow

Heat-flow patterns for the ocean bottom are low in the vicinity of trenches and exceptionally high on the oceanic ridges. This tends to confirm the idea of sinking crust or deep mantle at the trenches and rising crust or shallow mantle at the ridges.

Lithologic Correlation

The relative percentages of the occurrence of minerals and the detailed mapping of fractures in rocks in east-

Table 5.1 *Summary of Plate Tectonic Interactions*

	Type of Plate Boundary		
Characteristic	Divergent	Convergent	Lateral
Plate movement	Plates move apart	Plates approach each other, one goes down into mantle	Plates slide past each other in contact
Name of process	Seafloor spreading	Subduction	Transform faulting
Topographic expression	Mid-oceanic ridges	Oceanic trenches, island arcs, volcanic mountain chains; continental collision mountain belts (nonvolcanic)	Fault ridges and troughs, oceanic fracture zones
Type of magmatism	Basaltic volcanism	Volcanism of intermediate average composition (andesite); very little activity in continental collision mountain belts	None
Source of magma	Partial melting of mantle	Partial melting of basaltic layer covering downgoing slab (second stage of double distillation) and oceanic and margin sediment (mainly responsible for high gas content and explosive nature)	None
Effects of size of plates	Increases area of plates involved	Decreases area of at least one plate	Leaves area of plates unchanged
Geological features generated	Oceanic crust and lithosphere	Continental crust and lithosphere	Faults
Earthquakes	Small, numerous, and shallow	Large, deep, and destructive	Large and destructive
Other effects	Creates linear magnetic anomalies that permit measurement of rates and history of motion; moves continents apart	Creates island arcs, volcanic mountain chains, continental collision mountain belts	Creates major strike-slip (lateral) faults
Examples	Mid-Atlantic Ridge, East Pacific Rise	Tonga, Japan Trenches and arcs (island arcs caused by oceanic-oceanic lithosphere collision), Andes (volcanic mountain chain caused by oceanic-continental lithosphere collision), Himalayas, Alps (fold mountain belts caused by continental-continental lithosphere collision)	San Andreas Fault (on land), Eltanin Fracture Zone (submarine)

ern South America closely match those of rocks in western Africa. The chance that the same set of forces produced the same fracture patterns in rocks on separate continents is slight. The ages of the rocks are also the same. Similar correlations exist between New England–Canada and the British Isles.

Terrestrial Fossil Correlation

The famous Gondwana series of sedimentary rocks in Africa contain fossils of ferns, amphibians, and other freshwater swamp organisms. The fact that a number of the same fossils have also been found in Antarctica and South America is strong evidence of a previous connection between the continents, because such organisms are incapable of migrating across vast oceans.

FURTHER READINGS

Alt, David. *Physical Geology*. Belmont, Calif.: Wadsworth, 1982.

Bullard, Sir E. "Reversals of the Earth's Magnetic Field." *Philosophical Transactions of the Royal Society of London*, Series A, 1968, *263*(1143), 481–524.

———. "The Origin of the Oceans." *Scientific American*, September 1969, pp. 66–75.

Calder, N. *The Restless Earth*. New York: Viking Press, 1973.

Cox, A. *Plate Tectonics and Geomagnetic Reversals*. San Francisco: W. H. Freeman, 1973.

Dewey, J. F. "Plate Tectonics." *Scientific American*, May 1972, pp. 56–72.

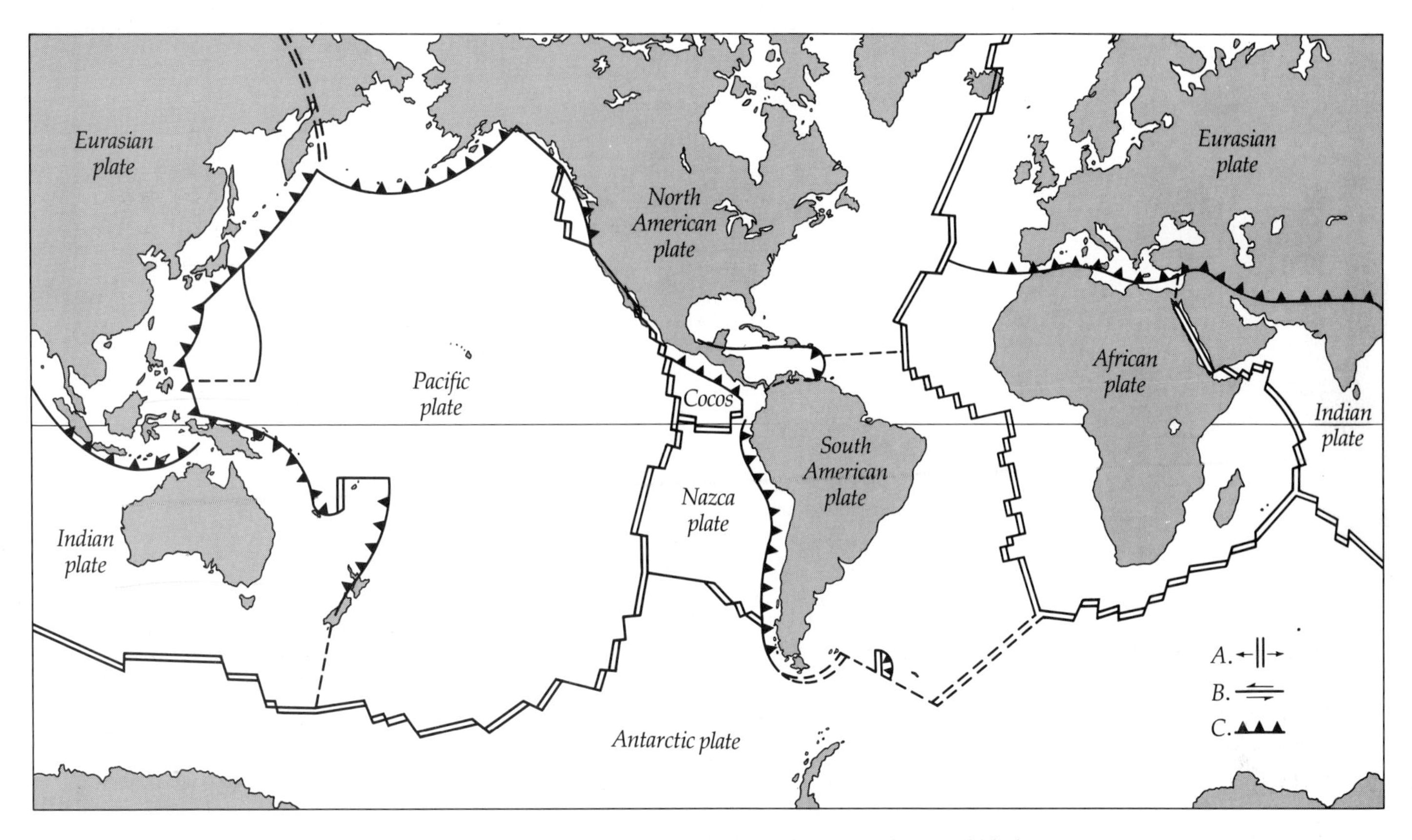

5.14 Major plates of the world. Plate boundaries are marked by heavy lines. Mid-oceanic ridges at which the plates move apart are represented by double lines. Transform fault boundaries are shown by single lines. Trenches and other subduction zones are marked by lines with teeth on one side. The teeth point down on the descending slab. Dashed lines are used where the exact location or nature of the boundary is uncertain.

Dietz, R. S., and J. C. Holden. "The Breakup of Pangaea." *Scientific American*, October 1970, pp. 30–41.

Dott, R., and R. Batten. *Evolution of the Earth.* New York: McGraw-Hill, 1976.

Heirtzler, J. R. "Sea Floor Spreading." *Scientific American*, December 1968, pp. 60–70.

Hess, H. H. "History of Ocean Basins." In *Petrologic Studies: A Volume in Honor of A. F. Buddington.* A.E.J. Engel and others, eds. Boulder, Colo.: Geological Society of America, 1962, pp. 599–620.

Holmes, A. *Principles of Physical Geology.* Ontario: Nelson and Sons, 1965.

Hurley, P. M. "The Confirmation of Continental Drift." *Scientific American*, April 1968, pp. 52–64.

Isacks, B., and others. "Seismology and the New Global Plate Tectonics." *Journal of Geophysical Research*, 1968, *73*(18), 5855–99.

Kennett, J. P. *Marine Geology.* Englewood Cliffs, N.J.: Prentice-Hall, 1981.

Le Pichon, X. "Sea-Floor Spreading and Continental Drift." *Journal of Geophysical Research*, 1968, *73*(12), 3661–97.

Morgan, W. J. "Rises, Trenches, Great Faults, and Crustal Blocks." *Journal of Geophysical Research*, 1968, *73*(6), 1959–82.

Phinney, R. A. *The History of the Earth's Crust.* Princeton, N.J.: Princeton University Press, 1968.

Rudman, A. J. "The Role of Geophysics in the New Global Plate Tectonics." *Science Teacher*, 1969, *36*(7), 21–26.

Sawkins, F. J., and others. *The Evolving Earth.* New York: Macmillan, 1978.

Uyeda, S. *The New View of the Earth.* San Francisco: W. H. Freeman, 1978.

Van Andel, T. H. *New Views on an Old Planet.* Cambridge: Cambridge University Press, 1985.

Vine, F. J. "Spreading of the Ocean Floor: New Evidence." *Science*, 1966, *154*, 1405–15.

Wegener, A. "Die Entstehung der Kontinente und Ozeane" ("The Origin of Continents and Oceans"). *Sammlung Vieweg*, 23, Branunschweig, 1915.

Dennis Brokaw

The rocks beneath the water, clearest water,
Are what thoughts ? what's wrought beneath the tossing
And the wind ? I walk. I sleep. My head is almost lost.
The world rolls round the galaxy, the green sea
Agitates its glory and its mystery. Deeper than history
Is a thought's journey. The clear stones are a mystery.

John Tagliabue

SIX

Margins of the Continents

The **continental margin** includes the coast, the continental shelf, the continental slope, and the continental rise (Figure 6.1). It consists of the submerged part of the continents and constitutes a transition zone between the continents and the ocean basin.

The continental margins are currently attracting intense interest among maritime nations eager to develop marine resources. Several nations have accepted the concept of a 200-mi exclusive economic zone, according to which each maritime nation has control of the water and underlying sediments to a distance of 200 mi from its coast. Where less than 400 mi of water lie between two nations, the boundary is to be negotiated bilaterally. This concept is embodied in the UN Law of the Sea Treaty (not ratified as of this writing).

Why are the continental margins attracting so much interest? For two reasons: fish and oil. Most of the world's commercial fishing is concentrated over the continental margins, where the waters are rich in nutrients. In addition, certain areas of the continental margins contain abundant deposits of oil, gas, gravel, phosphate, coal, and other nonliving resources. Some oil and gas resources are already under development, and an active search is under way for additional resources. The continental margins, especially the slopes and the rises, contain vast resources that are still untapped although they are accessible by means of modern technology.

As natural resources on land begin to dwindle, social and political pressures are forcing industrialized nations to focus attention on the potential riches of this region. In this chapter we shall discuss the origin and physical characteristics of the continental margins. We shall deal with the features of the coast at length in Chapter 13 and with the resources of the continental margins in Chapter 18.

CLASSIFICATION OF MARGINS

Continental margins are of three main types (see Figure 5.8). The margins along the western coasts of North and South America are characterized by frequent earthquakes, active volcanoes, and active faults. Such margins are termed **active margins**. The eastern margins of those continents are characterized by relatively few earthquakes, active volcanoes, and active faults. Such margins are termed **passive margins.** A third type of margin is called a **translation margin.** This is a margin characterized by faults with large horizontal offsets and shallow-focus earthquakes. They differ from active margins by having no subduction zone. The coast of southern California is an example (Figure 6.2).

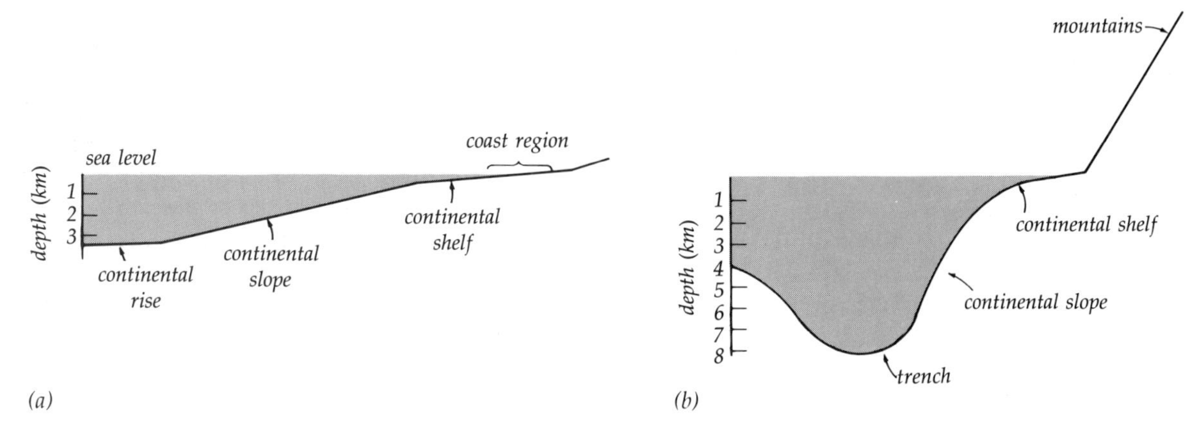

6.1 *Continental margins. Typical profile for continental margins on (a) the Atlantic, and (b) the Pacific Ocean. Note that in the Pacific a trench replaces the continental rise found in the Atlantic.*

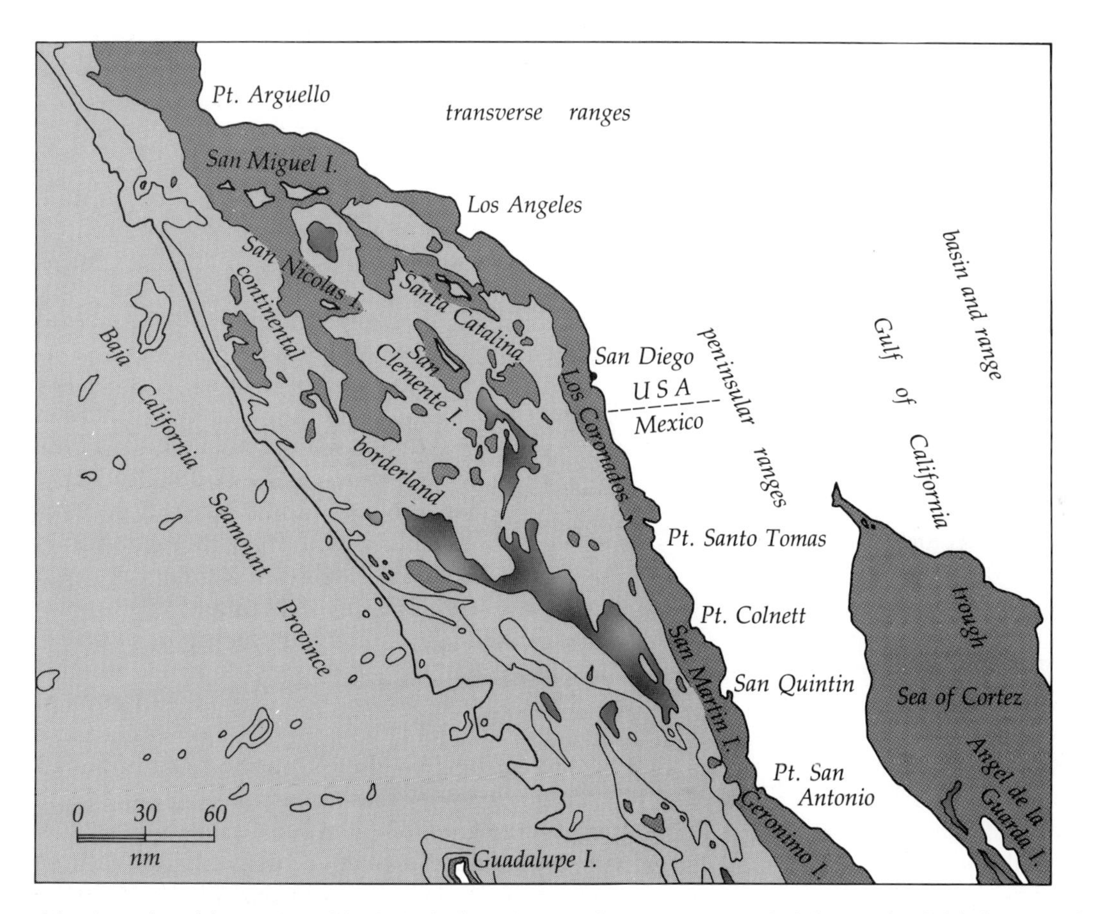

6.2 *A portion of the continental borderland off southern California. Basins are shaded. For a detailed look at a portion of the borderland, see Figures 6.12 and 6.13. (From D. J. Moore, GSA Special Paper 107)*

Differences in tectonic activity produce differences in rock structure and sedimentation along the margins. Alternating narrow banks and troughs, faults, and relatively narrow shelves are typical of translation margins. Deep trenches and volcanic island arcs are common along active margins.

Passive margins typically have wide continental shelves (such as the eastern coast of the United States), broad deltas (such as the Mississippi River where it enters the Gulf of Mexico), or widely distributed coral reefs (such as northeastern Australia) (Figure 6.3). They show no active faulting or volcanoes, and their con-

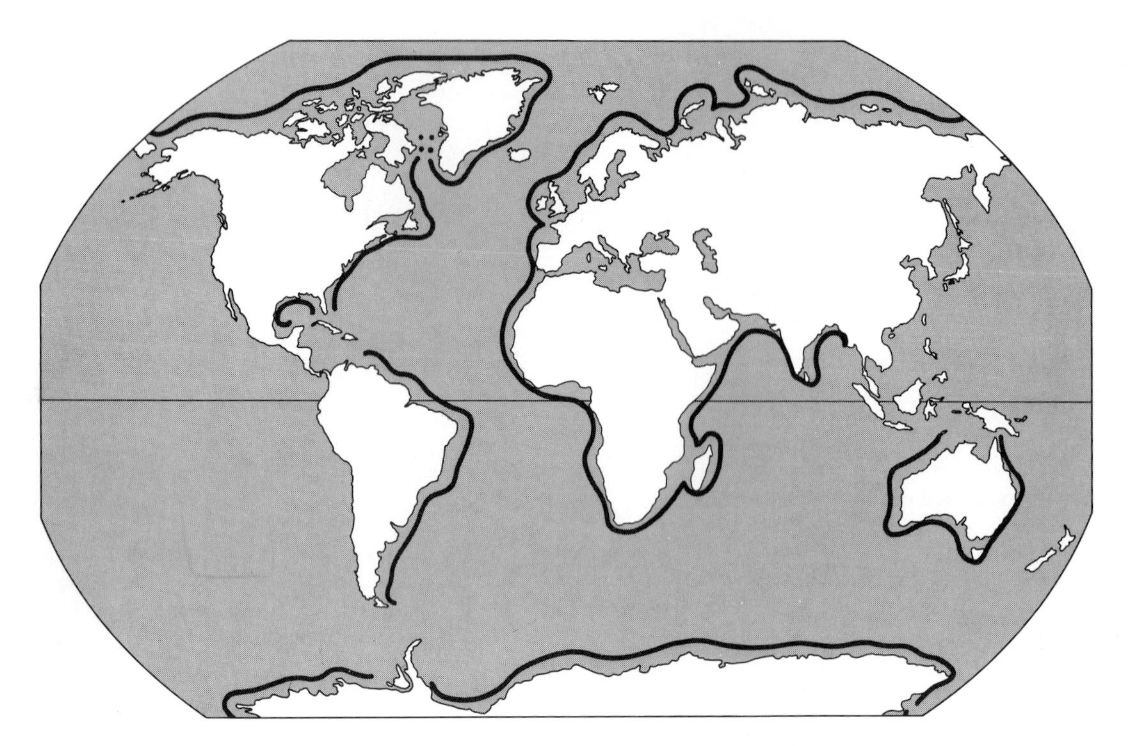

6.3 World map showing present-day location of passive continental margins.

tinental shelves range from about 20 km (12 mi) to more than 200 km (124 mi) in width. What accounts for these differences?

The concept of **plate tectonics** explains these differences. Active margins occur at convergent plate boundaries where the edge of a continental plate comes into contact with another plate. The result of this collision is a subduction zone, which produces volcanic island arcs and trenches. In addition, when sediments are caught between the converging plates, such a collision can cause mountain building (Figure 6.4). A translation margin is explained by the intersection of a transform fault system with the margin. In southern California, the margin is cut by a series of lateral, parallel faults trending northwest to southeast. This is considered to be a transform fault system offsetting the East Pacific Rise south of the Sea of Cortez and the Gordo Ridge west of northern California and Oregon. Off southern California, the margin consists of a series of blocks that indicate uplift (Catalina Island) and subsidence (San Diego trough).

Passive margins occur at some distance away from the edge of a tectonic plate that is diverging from another plate (Figure 6.5). At the boundaries of the divergent plates, an ever-widening ocean stands between the continents. As the plates move farther and farther away from the spreading center (marked by an oceanic ridge), sediment is deposited on the seafloor adjacent to the coast. At the same time, the crust cools, contracts, and sinks. Accumulating sediment along the continental margin produces a broad continental shelf (Figure 6.5).

How does a passive margin form? This is a key question because its answer provides the mechanism for continental drift and for the accumulation of the major oil and gas resources of the world. Most of the passive margins recognized today began their histories when the supercontinent, Pangaea, began to split apart about 200 million years ago. First, rifting occurred (Figure 6.6). This is the process of stretching that produces mid-oceanic ridges and divergence of plates. The East African rift valley may be a current example of this process in action. This marks the birth of an ocean basin.

Second, the continental fragments spread apart producing a gap over new crust, which then fills with sea water (Figure 6.7). The Red Sea is an example of this youthful ocean basin. Sediment is eroded from the adjacent land masses and is deposited in the basin. New oceanic crust continues to form and the basin widens. In some cases, the basin may become isolated from the ocean, allowing the seawater to evaporate and producing great salt deposits. Then, further movement of a transform fault will allow seawater to inundate the basin again. As the crust between the new basin and the land cools, it contracts and subsides. This produces a deep basin between the land and the mid-oceanic ridge that will eventually become an abyssal plain. Coral reefs may be built along the margin of the land fragment.

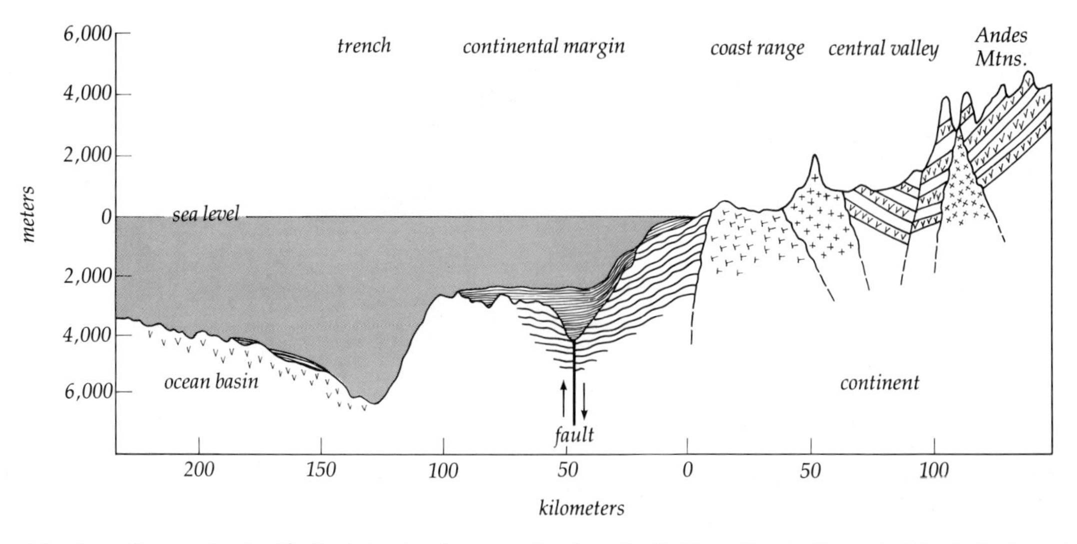

6.4 An active margin. An idealized structural cross section from the Pacific seafloor to the crest of the Andes in central Chile. Note that there is an active fault at the continental margin and an offshore trench indicating subduction. (From D. Scholl and others, "Peru-Chile Trench Sediments and Sea Floor Spreading," Bulletin of the Geological Society of America, *1970,* 81, *1339–60)*

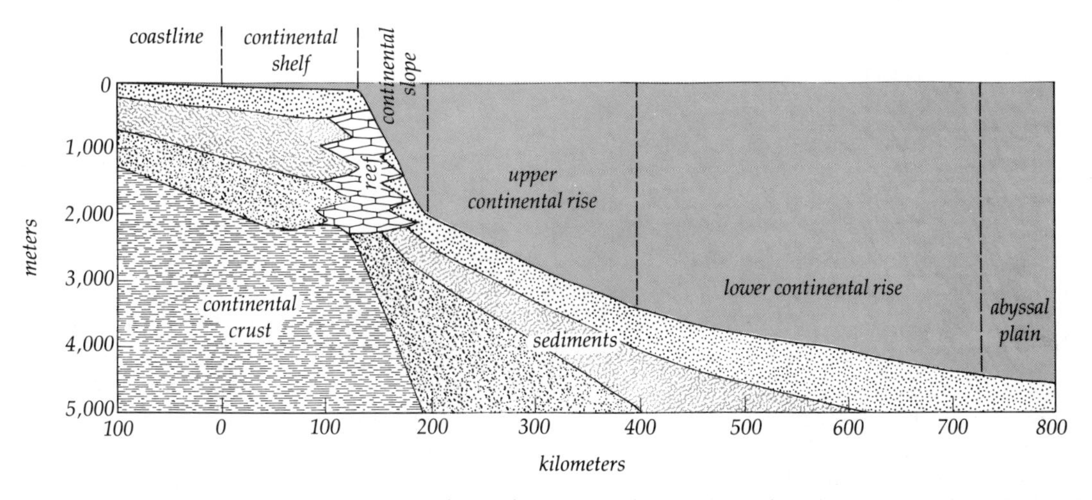

6.5 A passive margin. Stratigraphic section of the sedimentary rocks underlying the Atlantic coastal plain, shelf, and continental slope in the vicinity of Cape Hatteras, North Carolina. Note the thickening of layers oceanward and the extreme vertical exaggeration.

Third, terrigenous sediment continues to be deposited at the margin of the land. Normally, some type of sediment dam develops at the outer margin that prevents or delays sediment from spilling out onto the seafloor. This dam may be caused by a coral reef or by the formation of salt domes (see Chapter 18). How other dams form is an area of active research now, and it is poorly understood. The land mass moves so far from the ridge that the terrigenous sediment is no longer transported all the way to the ridge. These older margins are also characterized by submarine canyons that are the avenues for the terrigenous sediment to reach the seafloor.

Oil and gas are thought to form when abundant organic material settles to an oxygen depleted seafloor where rapid burial prevents complete decay (see Chapter 18). This is initiated early in the history of a passive margin when circulation in the basin is limited.

CONTINENTAL SHELVES

Continental shelves are the submerged areas of continents (Figure 6.8). Although most continental shelves are located on the boundary between a continent and an ocean, some are nearly landlocked, such as Cana-

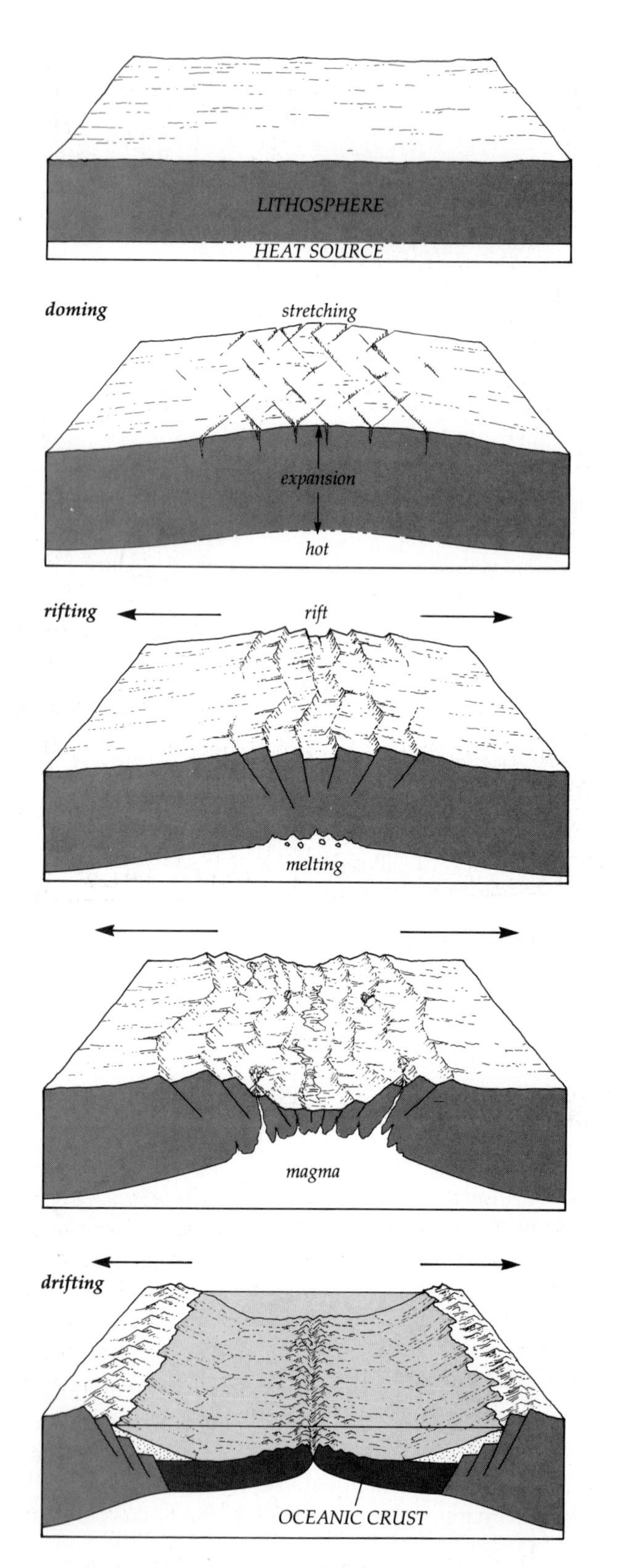

6.6 *The breakup of a continent probably begins with local heating beneath the lithosphere. This produces by expansion a surface bulge, which cracks to form a rift valley, accompanied by much volcanism. Eventually, though not always, the stationary phase of rifting is followed by drift, oceanic crust is intruded, and the sea invades the rift valley to form an embryonic ocean basin. (T. Van Andel,* New Views on an Old Planet, *1985)*

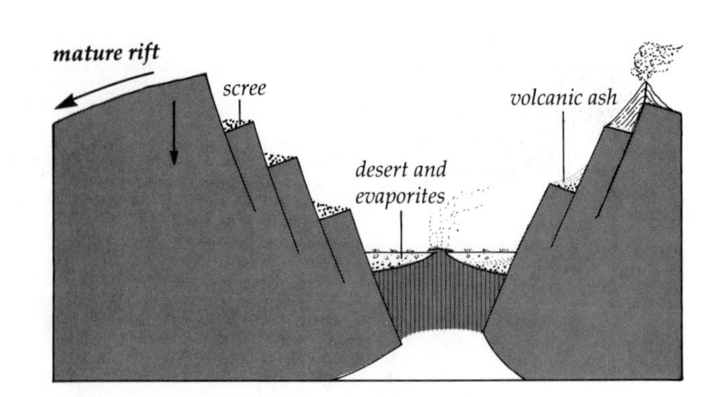

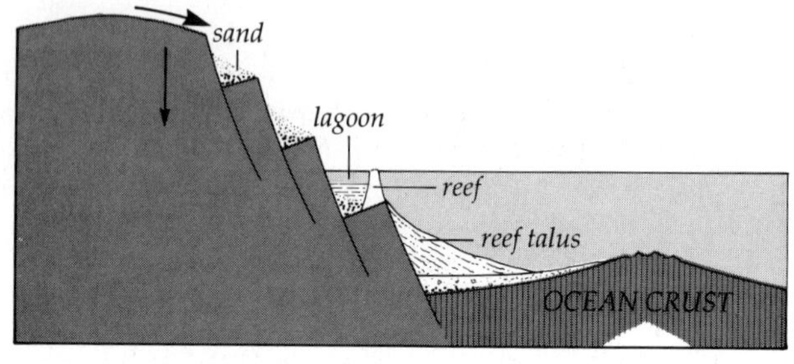

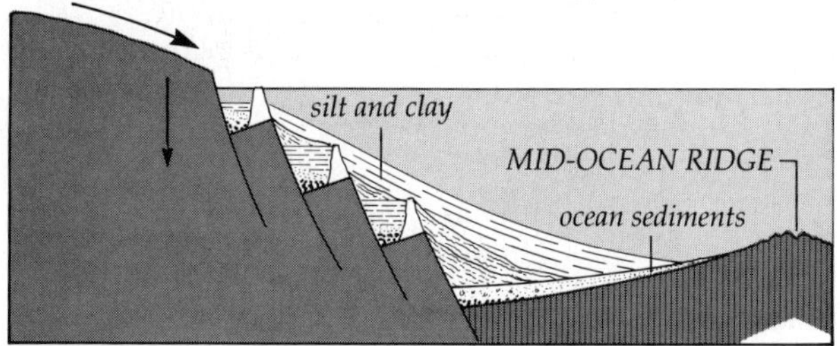

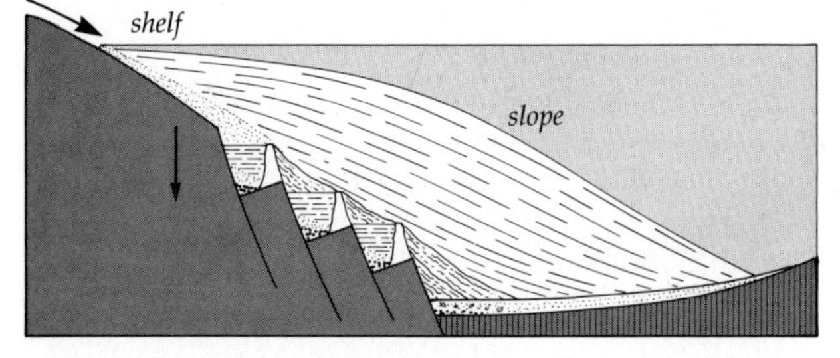

6.7 *A mature continental rift is a shallow basin near sea level where evaporites form, together with volcanic sands and material washed down from the slopes. With the onset of drifting, the rift widens, and the margins cool and subside. The sea invades, coral reefs grow, organic matter accumulates in the sediments, and beaches form along the shores. Ultimately, the margins sink enough to reverse the direction of continental drainage, and large quantities of silt and clay bury the reefs and the black organic shales, creating the continental shelves of a passive margin. (T. Van Andel,* New Views on an Old Planet, *1985)*

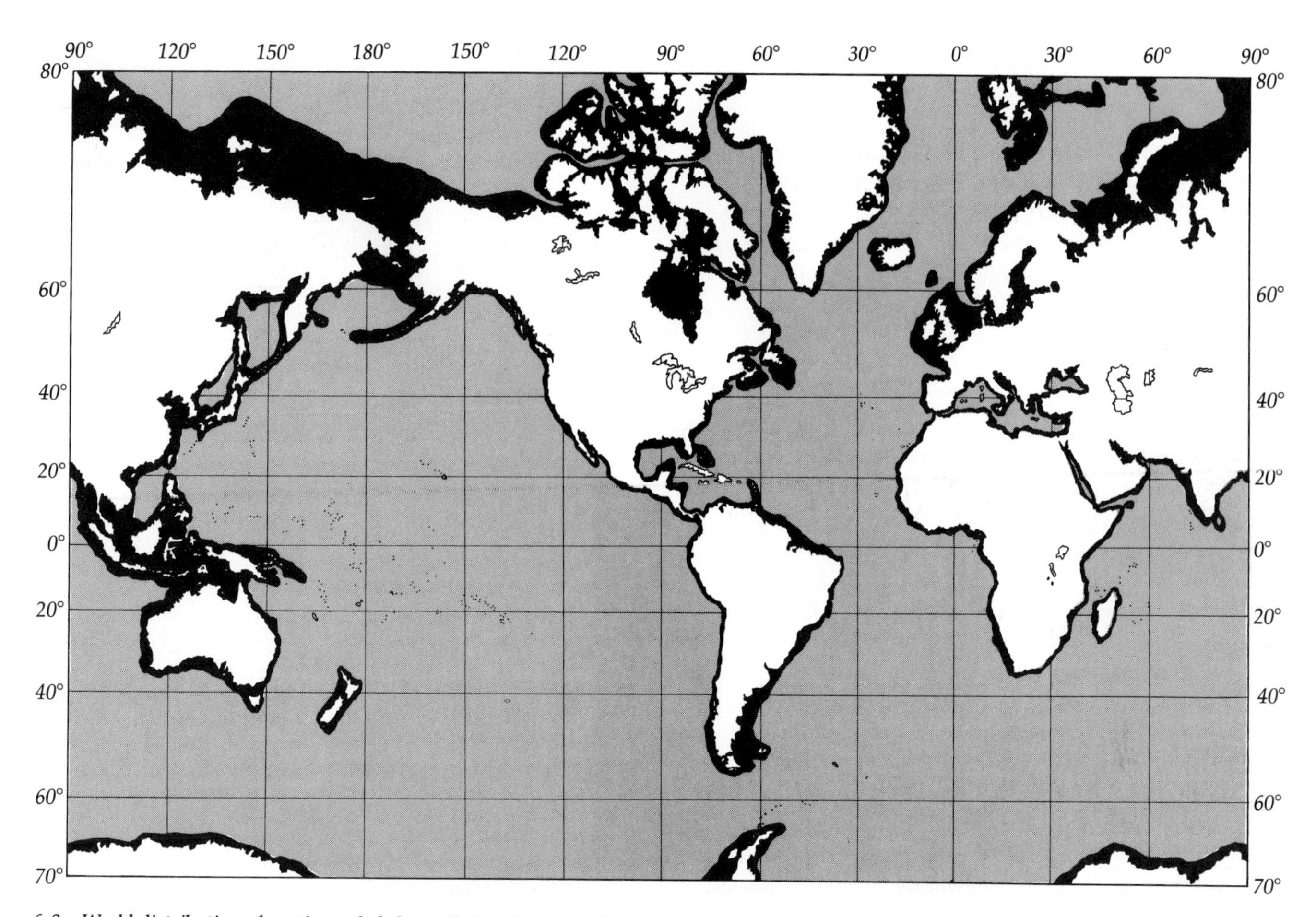

6.8 World distribution of continental shelves. Shelves in the northern latitudes have been exaggerated because of the projection. Note that the continental shelves vary greatly in width. Great Britain is part of the continent of Europe, and Alaska connects to Siberia.

da's Hudson Bay. Continental shelves form 7.5% of the total area of the seafloor. They may be as wide as 1,500 km (930 mi), but their average width is between 50 and 100 km (31 and 62 mi).

The outer edge of the continental shelf is typically about 130 m (426 ft) deep, the depth at which the slope of the seafloor increases dramatically. From the coast, the continental shelf drops off about 1 m per 1,000 m, a rate that increases to 10 m per 1,000 m on the continental slope. The inclination of the shelf is about 0°07′. There are many exceptions to this average rate, however. For example, glacially eroded shelves generally extend to a depth of nearly 500 m (1,640 ft) before giving way to the continental slope, and many coral shelves reach a depth of only 20 m (66 ft).

We can attribute most of the detailed features of continental shelves to seven processes: glaciation, sea-level changes, waves and currents, sedimentation of terrigenous material, deposition of carbonates, faulting, and volcanism.

Glaciation

During the past two million years, glaciers have constantly covered most of Antarctica and Greenland. During that time, North America, Europe, and Asia have been periodically blanketed by glaciers thick enough and large enough to be called ice sheets. Between these episodes of massive **glaciation**, interglacial thaws occurred due to relatively mild climate, similar to the climate that exists today. Researchers have compiled a complex history of glaciation, identifying many advances and retreats of the ice sheets (Figure 6.9). The extent of glaciation has varied with regional variations in climate and topography.

The climatic fluctuations shown in Figure 6.9, producing warm-and-cold epochs are recent geological events. In fact, during most of the earth's history, the temperature contrast between the equator and the poles has been much less than it is today. The polar climates in the past were more like the present climates of the mid-latitudes (45° to 60°). The reason why, at times, the earth passes into an ice age is not fully understood. Continental drift, ocean circulation, and solar energy output are all possible factors.

Once the last ice age began, why did periodic swings of glacial and interglacial periods occur? The most widely accepted theory was suggested by M. Milankovitch. He attributed the changes in climate to changes in the earth's position and orbital path. First, the orbit of the earth is elliptical. This means that the distance from the earth to the sun varies during the year. The present variation is from 91.5 million miles to 94.5 million miles. The ratio of these distances is called the eccentricity of the orbit. The orbit stretches and contracts by about 2% during a cycle of about 100,000 years (Figure 6.10a). This means that the distance from the earth to the sun not only changes during the year but also from year to year. Second, the axis of the earth is tilted at 23.8° from the plane of the earth's orbit around the sun. That plane is called the ecliptic. This tilt produces the change of seasons for the northern and southern hemispheres. The tilt, called obliquity, varies from 21.8° to 24.2° during a cycle of 41,000 years (Figure 6.10b). Increasing the tilt increases the amount of sunlight reaching the poles. Third, the earth's axis wobbles in a way similar to a spinning top that is slowing down. The wobble is called precession and has a period of 21,000 years. Precession means that the present North Star, Polaris, was not always the pole star (Figure 6.10c). The combination of the effects of this stretch (eccentricity), wobble (precession), and roll (obliquity) may cause the glacial and interglacial periods.

What can we expect in the future? The earth is still in an ice age. The warm period probably peaked about 5,000 to 6,000 years ago. The climate is in a general cooling trend. When will continental ice sheets descend again? No one knows. Our present situation is approaching the ice-growth condition depicted in Figure 6.11. The obliquity must increase beyond 24° to reach this condition. That should be at least 1,000 years from now.

The alternating periods of freezing and thawing undoubtedly had a significant effect on the ocean, the atmosphere, marine organisms, and the continents themselves. When the sea level stood 100 to 125 m (328 to 410 ft) below its current level, large parts of what is now the continental shelf were exposed. Fossil mammoth teeth discovered in dredgings of the continental shelf off New England provide convincing evidence of such exposure. In addition, samples of bottom sediments give evidence of submerged peat bogs rich in plant fossils and pollen grains.

The shells of many marine microorganisms such as foraminifera are composed of calcium carbonate ($CaCO_3$). Researchers have used their fossils to date ocean sediments and to make inferences about ocean water temperatures at various periods of glaciation. By studying fossils that consist of calcium carbonate, they have pieced together a history of the ice ages. A brief digression here will suggest the ingenuity with which scientists have applied modern chemistry to oceanographic and geological problems.

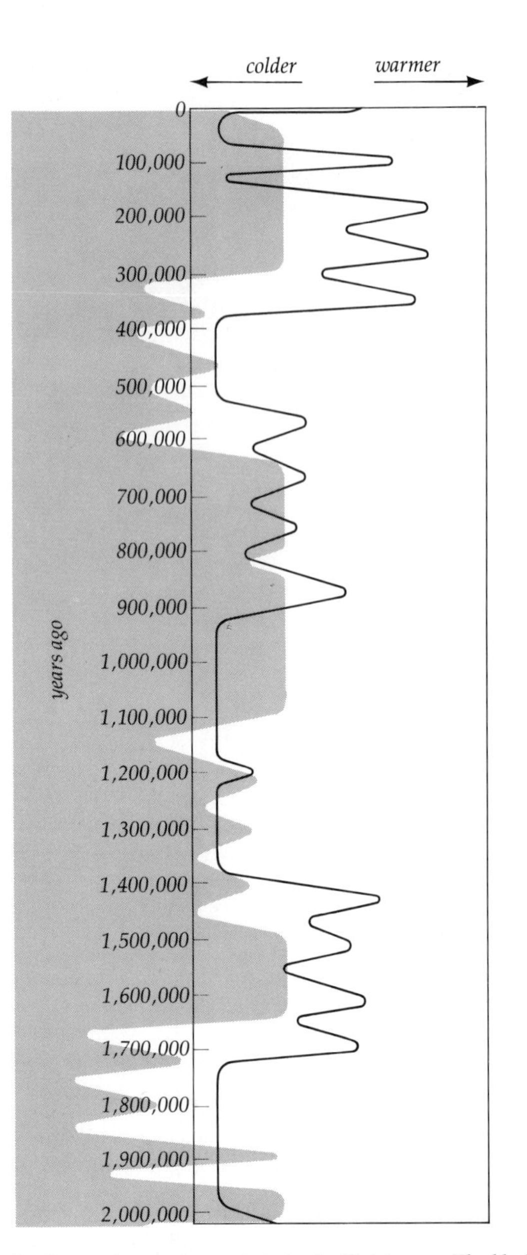

6.9 Cooling and warming periods in the Pleistocene. The black line is based on the relative abundance of cold-water and warm-water species of foraminifera. The grey-shadow part is based on oxygen isotope ratios found in fossil foraminifera tests (shells).

Carbon atoms may occur as isotopes with an atomic mass of 12, 13, or 14. The carbon 12 (^{12}C) form is by far the most common. The ^{14}C form is especially interesting, however, because it is radioactive. ^{14}C is produced when a cosmic ray neutron is captured by a nitrogen atom (^{14}N) in the atmosphere:

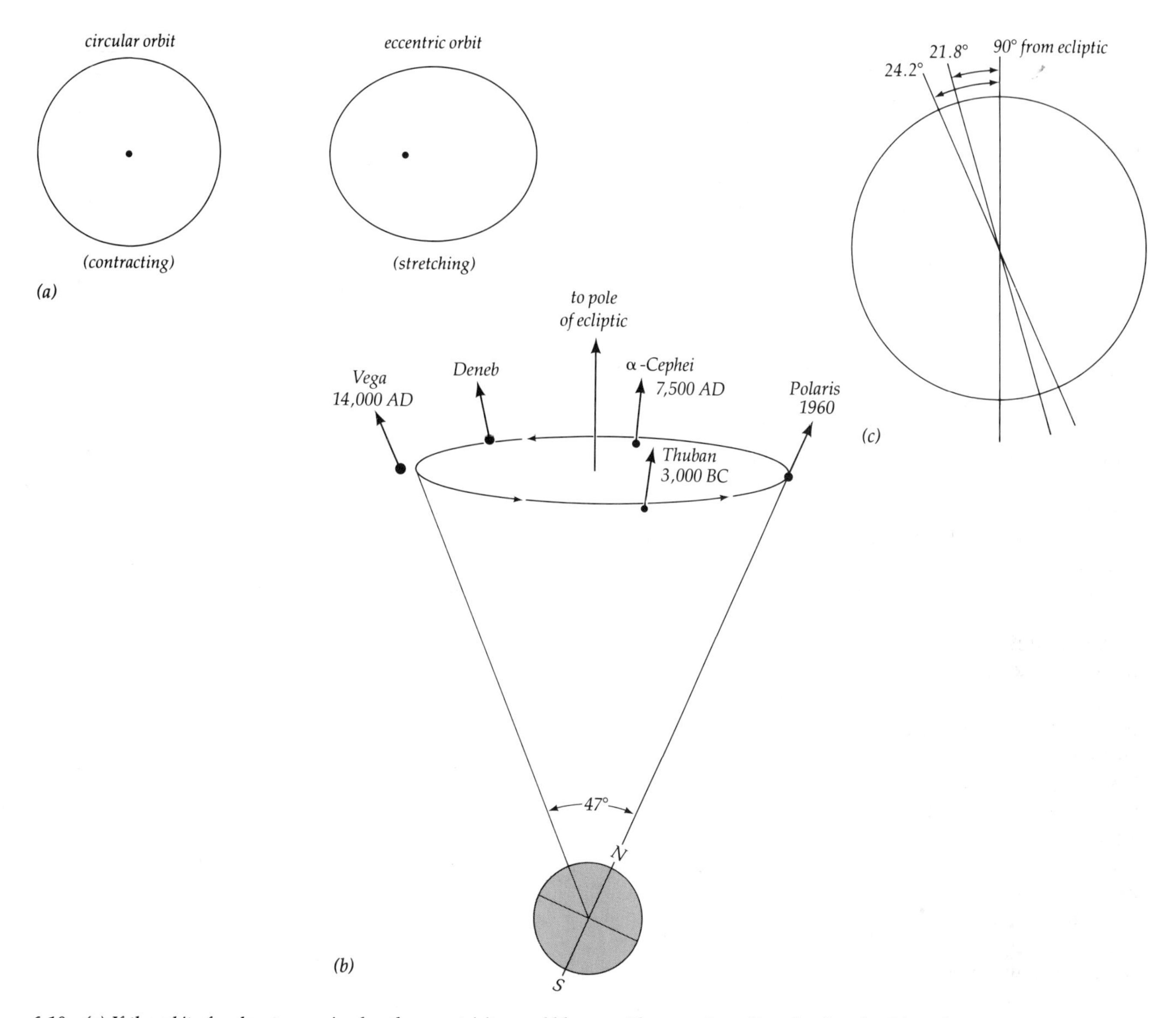

6.10 (a) If the orbit of a planet were circular, the eccentricity would be zero. The more the radius of a planet's orbit varies during the year, the greater its eccentricity. The eccentricity of the earth's orbit varies slightly, with a 100,000-year cycle. We can think of this variation as stretching and contracting. (b) The obliquity, or tilt, of the earth's axis varies from 21.8° to 24.2° from the vertical. It is now at 23.8° and is decreasing. The period of a complete cycle is 40,000 years. We can think of these variations as a rolling movement. (c) As the axis of the earth rolls, it also wobbles like a top that is slowing down. This wobble, which is called precession, has a period of 21,000 years.

$$^{14}N + {}^{1}n \rightarrow {}^{14}C + {}^{1}H.$$

The "extra" neutron in the ^{14}C atom eventually transforms itself to a proton and releases an electron. This is called beta decay and results in the formation of ^{14}N. The half-life of ^{14}C is 5,730 years, which means that half of the ^{14}C is transformed into ^{14}N in that time.

In the atmosphere, ^{14}C combines with oxygen to form radioactive carbon dioxide (CO_2). Carbon dioxide is used by plants in photosynthesis and becomes incorporated into the organic molecules of plants. Since the rate at which ^{14}C is produced equals its rate of decay, the amount of ^{14}C remains constant in the environment, and the $^{14}C/^{12}C$ ratio is constant. After a plant dies, however, it no longer takes up CO_2 and the ^{14}C decays, causing the $^{14}C/^{12}C$ ratio to decrease with time. Thus, the $^{14}C/^{12}C$ ratio in a plant fossil can be used to date that fossil.

Oxygen atoms also occur as isotopes with atomic masses of 16, 17, and 18. Oxygen 16 (^{16}O) accounts for more than 99% of the oxygen, while ^{18}O is next in abundance. When calcium carbonate forms, most of the oxygen used is ^{16}O. The extent to which ^{18}O is included depends on temperature and pressure. At

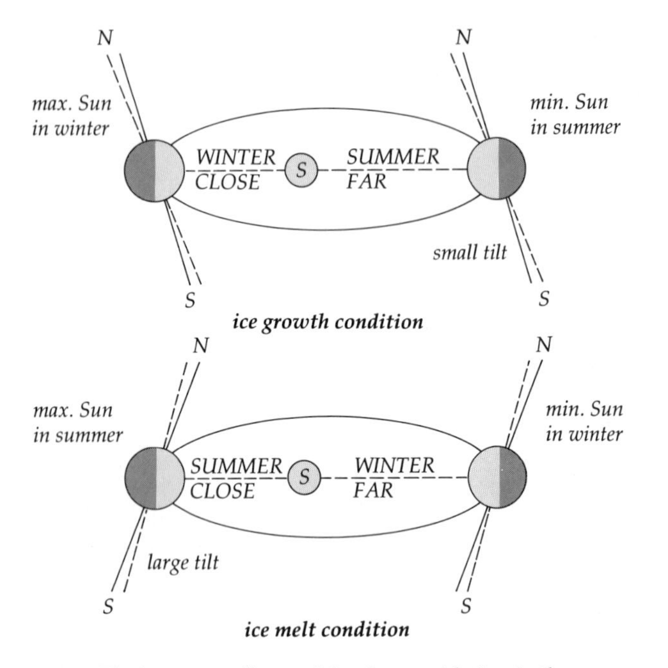

6.11 The impact on climate of the change with time in the orbital behavior of the earth is illustrated here for two extreme conditions. The first is a state conducive to the growth of icecaps, the second, to their melting. What takes place in the Southern Hemisphere is the opposite of what is shown here.

lower temperatures, the amount of ^{18}O used increases slightly. Also, ^{16}O has a higher vapor pressure than ^{18}O, so when evaporation takes place at the air-sea interface, a higher proportion of ^{16}O is removed than of ^{18}O. Since cold water evaporates more slowly than warm water under the same atmospheric conditions, and since more ^{18}O is bonded in the water molecule in cold water than in warm water, the $^{16}O/^{18}O$ ratio in calcium carbonate shells can be used to infer ocean-water temperatures and ice accumulation on land.

Because the accumulation of ice on land and the lowering of the temperature of seawater both increase the ^{18}O content of seawater, the relative contributions of the two effects are difficult to determine. To resolve this problem, a comparison is made between the fossils of surface plankton, which are affected by climate change, and the fossils of bottom-dwelling (benthic) organisms, which are unaffected by atmospheric climate.

These two uses of isotopes have provided detailed evidence of the history of the ice ages. During the past two million years, surface-water temperatures in northern temperate latitudes have varied about 4°C in the Pacific and about 7°C in the Atlantic.

During the advances and retreats of glaciers, parts of the continental shelf at high latitudes were eroded by the scouring effect of rocks and boulders embedded in the ice. In some areas, glaciers deposited tremendous amounts of sediment; in other areas, sea-level changes corresponding to ice movements led to wave erosion. Chapter 13 discusses the consequences these processes have had on the coastlines.

Sea-Level Changes

Continental glaciation has been a major cause of sea-level fluctuations, creating a rise and fall of ±130 m (±425 ft) from present-day sea level. Sea-level changes due to glaciation have dominated recent geological history, but significant changes in the past were the result of other processes. Periodic changes of as much as several hundred meters have taken place over millions of years, the causes of which are not known.

During periods when the sea level was low, the continents were nearly 25% larger than they are now. Rivers flowed across what are now continental shelves and dumped sediment directly onto the ocean floor. At other times, such as the Cretaceous period (see Appendix VIII) when the sea level was high, up to 40% of the continents were flooded. Shallow seas were predominant, and the deep sea was far from land.

Waves and Currents

Various forces have influenced the features of the continental shelf over time. Features long submerged have been revealed by changes in sea level. Waves, rivers, and faults have had the most pronounced effect on the features of the continental shelf. To a lesser extent, tides, **seismic sea waves** (long-period sea waves produced by a submarine earthquake or a volcanic eruption), **internal waves** (waves that occur within a fluid at an interface between layers of different densities), and **storm surges** (rises in sea level that accompany intense low-pressure storms) also affect the physical features of the shelf.

In some areas, tidal currents flowing between land masses sweep the shelf clean and prevent unconsolidated debris from accumulating. Similarly, turbidity currents flowing through canyons carry sediment from the shelf out to the deep-sea floor. These currents flow downslope until they lose their energy in the deep sea.

Both tides and seismic sea waves affect the ocean water from top to bottom, even in the deep sea. The extent of their effect on the continental shelf is not known, although we know that sea waves can transport large amounts of unconsolidated sediment long distances in a short period. Storm surges are thought to have the same effect.

Sedimentation

As erosion steadily reduces the elevation of continents to sea level, most of the eroded material is carried off by running water and deposited as sediment on the

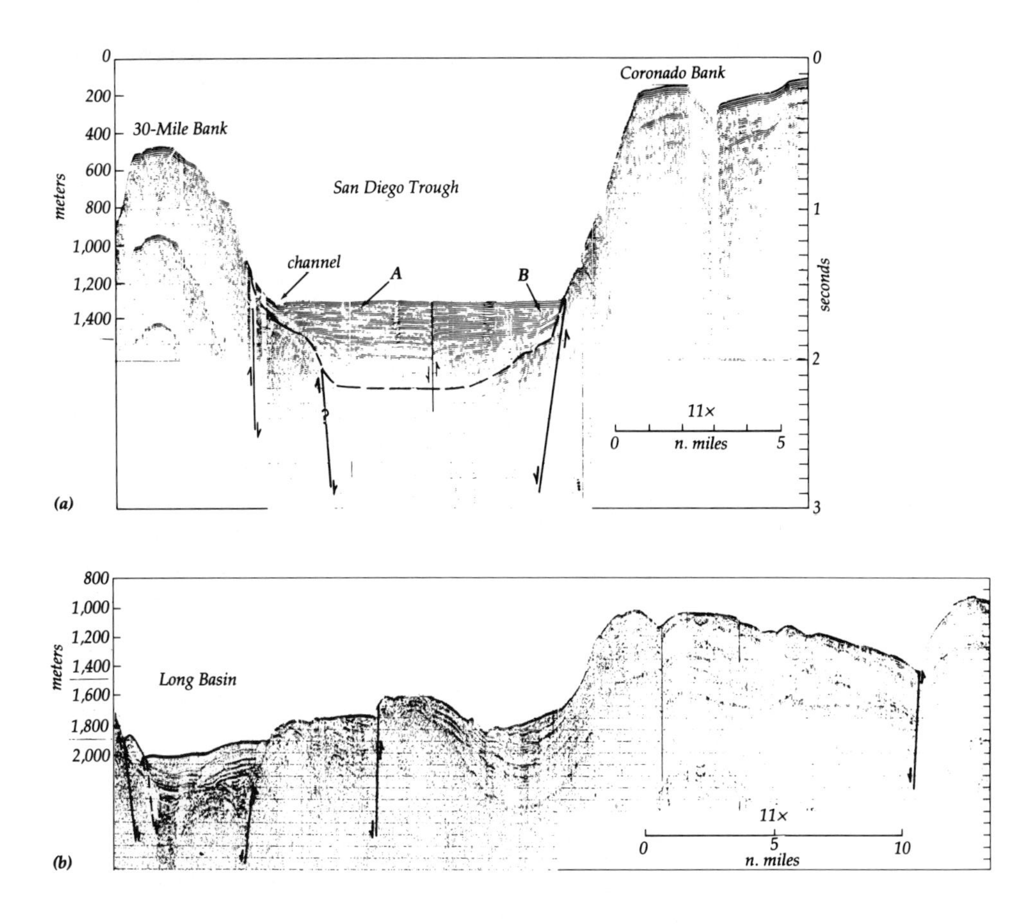

6.12 Faulting and folding off southern California. (a) A profile across the San Diego Trough shows faulting and distributary channels. A *and* B *are buried channels. (b) A profile across the central borderland off southern California. Note that the folds are largely transformed to faults here. (From D. G. Moore, courtesy of the Geological Society of America and Moore)*

continental shelf. All the world's major rivers—the Mississippi, Nile, Indus, Ganges, Irrawaddy, and many more—have deltas that extend out over the continental shelf. Herodotus, who lived around 450 B.C., made this comment on the Nile delta: "The nature of the land of Egypt is such that when a ship is approaching it and is yet one day's sail from the shore, if a man try sounding, he will bring up mud even at a depth of 11 fathoms [20 m]."

Some deltas, such as the Mississippi delta, extend many kilometers beyond the shoreline and contain an abundance of quartz and clay. The quartz, which is a principal ingredient of continental rocks, is relatively insoluble and is deposited directly onto the shelf. Feldspar, another major ingredient of continental rocks, alters chemically to clay in soil. In the process, ions such as sodium are dissolved in the water of the rivers carrying the eroded material. The solid clay is transported along with the quartz to the shelf. Oil and various minerals also tend to be concentrated in the great river deltas.

Carbonate Deposits

Although carbonate deposits are not common on the continental shelf bordering the United States, they do appear off Texas and Florida and in many other regions of the world, especially near tropical volcanic islands and atolls and Australia. Such regions are characterized by warm, shallow, sunlit waters, which encourage the rapid reproduction of marine life, including those forms that secrete calcium carbonate. Carbonate deposition must have been much more common in the past than it is today. Great, thick carbonate deposits underlie Florida and are found in the mountains of the western United States, the Alps, and the Himalayas. In Florida, some deposits below the land surface and on the continental shelf are more than 3,000 m (9,840 ft) thick.

Faulting and Volcanism

Since continental shelves are the surface expression of the transition zone between continents and ocean basins, it is not surprising that they exhibit some instability. As we have seen, the map of earthquakes reveals that many continental margins are unstable. Stresses within the earth fracture rocks and cause movement along the resulting fractures that is known as **faulting** (Figure 6.12). We find evidence of faulting on every continental shelf explored to date. Indeed, seismic activity and faulting are occurring on many continental shelves and borderlands at the present time. The region off southern California is called a borderland.

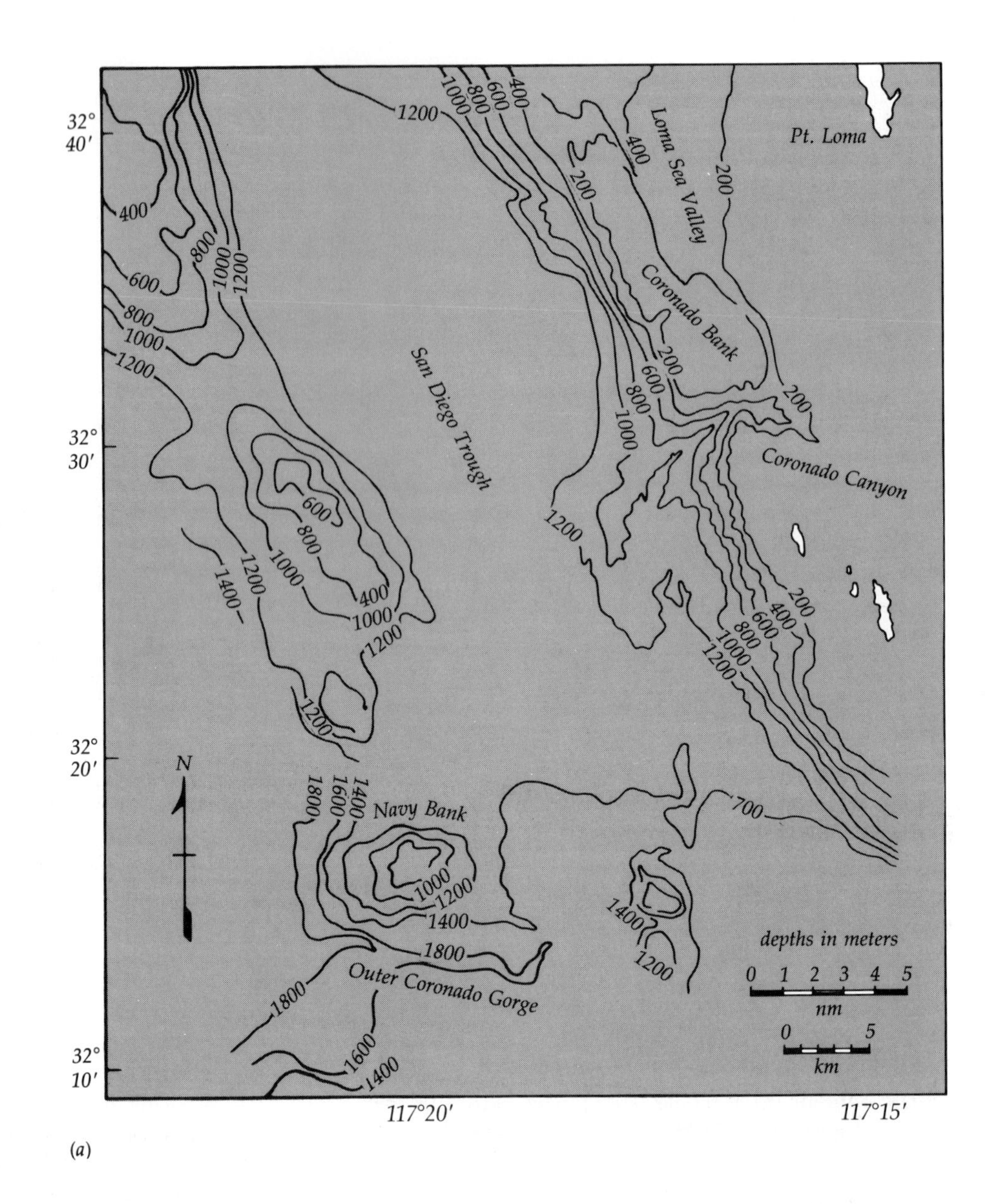

6.13 *Submarine canyons in (a) the southern California borderland and (b) off Scripps Institution of Oceanography. Depths are given in meters. (From Francis P. Shepard,* Submarine Geology, *2d ed., New York: Harper & Row, 1963) In (c), divers inspect the wall of a submarine canyon. (Official U.S. Navy photograph)*

This is a region containing continental rocks so intersected by faults that there is a narrow shelf followed seaward by a series of troughs and banks before the slope is reached some 165 km (100 mi) offshore. During the great Alaskan earthquake of 1964, segments of Kodiak Island dropped more than 2 m (7 ft). The displacement transported a supermarket down to the foreshore, where it was partially submerged at each high tide! Volcanoes, too, are numerous along active continental margins, especially around the Ring of Fire bordering the Pacific Ocean and along the European margin of the Mediterranean.

CONTINENTAL SLOPES

The **continental slope** begins where the gentle slope of the continental shelf gives way to a zone of sharply increasing steepness. Its incline gradually decreases until it meets the continental rise, the angle averaging about 4°. Continental slopes make up about 8.5% of the total area of the ocean floor and about 40% of the earth's sediment.

In the Pacific trench system of the Pacific Ocean, the continental slope reaches depths of between 5 and 10 km (3 to 6 mi). In other oceans, the slope seldom extends to that depth. Unconsolidated sediments on the continental slopes tend to be finer-grained than those on the continental shelf, and they contain fewer fossils. Consolidated and crystalline rock outcroppings along ridges and canyons are common on the slopes off northeastern Australia.

The term *continental slope* is somewhat misleading because it does not suggest the diverse topography of these regions. Submarine landslides, earth movements, and erosion have produced a wealth of distinctive features, and the sediments of the slopes have been extensively folded and faulted. Perhaps the most

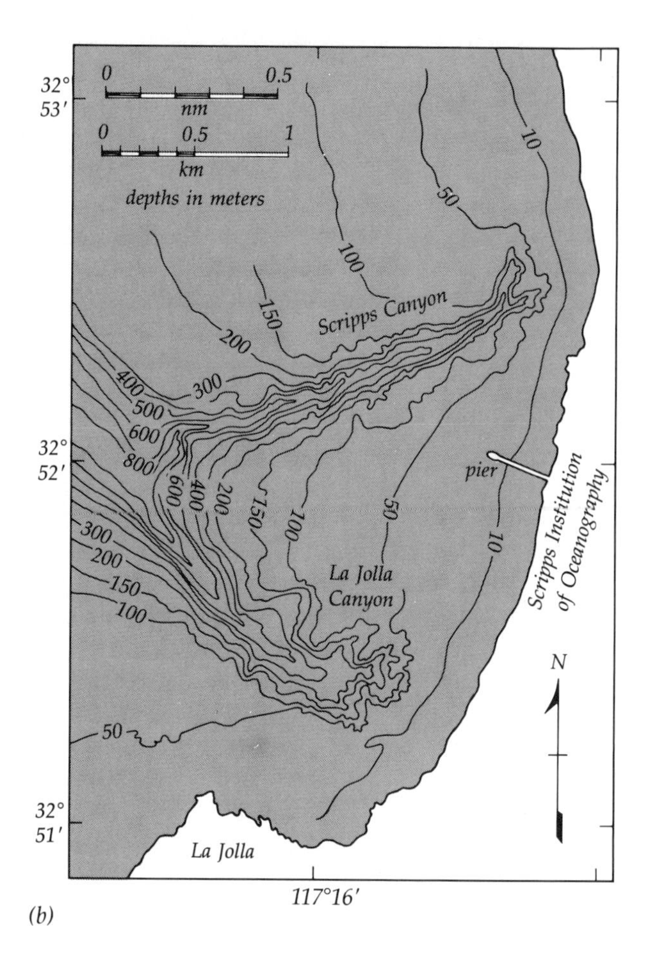

(b)

(c)

impressive features are the submarine canyons that have been cut into the surface of the slopes.

Submarine Canyons

Since World War II, the U.S. Navy, the U.S. Coast and Geodetic Survey (now called the National Ocean Survey), and various oceanographic and research institutions have been conducting intensive studies of the continental margins because of their military importance and their commercial potential. Those studies have revealed great submarine canyons on almost all the continental shelves investigated so far. Some extend many kilometers from shore and can be traced to depths of over 1,800 m (5,904 ft). Among the best known are those off California, New York, Colombia, southern France, the Mediterranean shore of North Africa, and the Congo.

On land, canyons are created mainly by erosion and faulting. Marine geologists assume that those same processes have produced the canyons of the continental slopes. Their evolutionary history differs greatly from region to region, however.

It seems likely that the submarine canyons were first eroded into the continental shelf by rivers or glaciers when the shelf was exposed during the Pleistocene (see Appendix VIII). A drop in sea level of 125 m (410 ft) over two million years would have allowed considerable erosion of the upper parts of the canyons. Moreover, sediment moving down from the shoreline would have abraded the walls and floor of the developing canyons. Cameras submerged in these canyons have confirmed that this process is continuing at the present time. True, the process is slow and small-scale, hardly seeming capable of gouging a canyon 500 m (1,640 ft) deep. But over many thousands of years, even such slight abrasion is significant. Finally, the millions of tons of sediment swept through the canyons by powerful turbidity currents would have caused extensive erosion in a short period. The presence of thick, graded layers of clay, silt, and sand at the base of the continental slopes is consistent with that fact.

One of the best-known submarine canyons is Scripps–La Jolla Canyon off La Jolla, California, just north of San Diego (Figure 6.13). (Before this canyon

(*a*) (*b*)

6.14 (a) Sediment fan at the mouth of San Lucas Canyon, off the tip of Baja California, Mexico. (b) Sand falling down over a steep slope in this canyon. (Official U.S. Navy photographs)

was discovered, the Scripps Institution of Oceanography had been built on the shore between the two heads of the canyon.) The shallow end of Scripps Canyon, the northern branch, is about 1 km north of Scripps Institution and extends almost to the shoreline of Black's Beach. This configuration produces a steady succession of well-formed waves, providing some of the best surfing in the continental United States. Nearshore, where the canyon is relatively narrow, it slopes steeply to a depth of 20 to 30 m (66 to 98 ft). Its walls are covered with many organisms, including sea anemones, rock scallops, and abalone. Sediment brought southward by alongshore currents accumulates at the head of the canyon and eventually slides down through it.

The head of the southern branch, La Jolla Canyon, is about 200 m (656 ft) offshore and 1.5 km (0.9 mi) south of Scripps Institution. Close to shore the canyon exhibits eroded bedrock ledges and mud, but farther offshore it resembles Scripps Canyon. About 1,500 m (4,920 ft) offshore the two canyons converge, reaching a depth of about 250 m (820 ft). The canyon meanders westward across the continental shelf another 12 to 15 km (7 to 9 mi) and then swings south to empty into the San Diego Trough at a depth of 800 m (2,624 ft). A broad, fan-shaped delta spreading out from the base of the canyon into the trough contains sediments and shells from the nearshore that are reworked by burrowing deep-sea organisms.

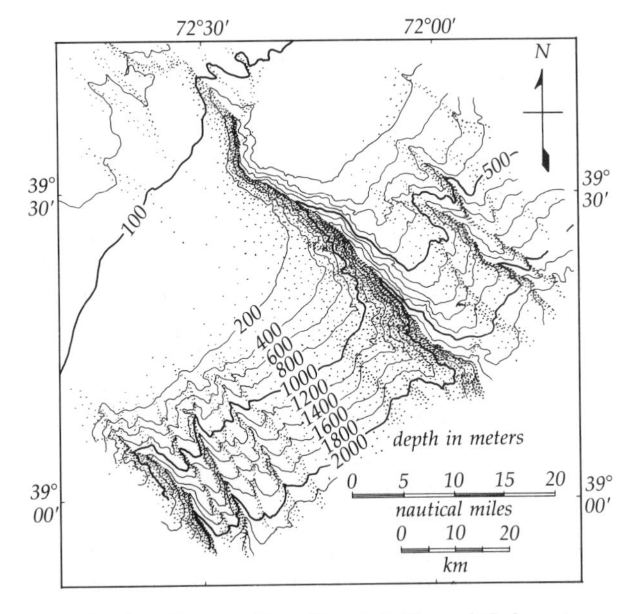

6.15 Hudson Canyon. (From Francis P. Shepard, Submarine Geology, *2d ed., New York: Harper & Row, 1963)*

San Lucas Canyon, located off the southern tip of Baja California (Figure 6.14), is similar to the Scripps–La Jolla Canyon complex. Granite outcrops occur in its deeper reaches.

The Hudson River has been draining the area of eastern New York and Vermont for several million years.

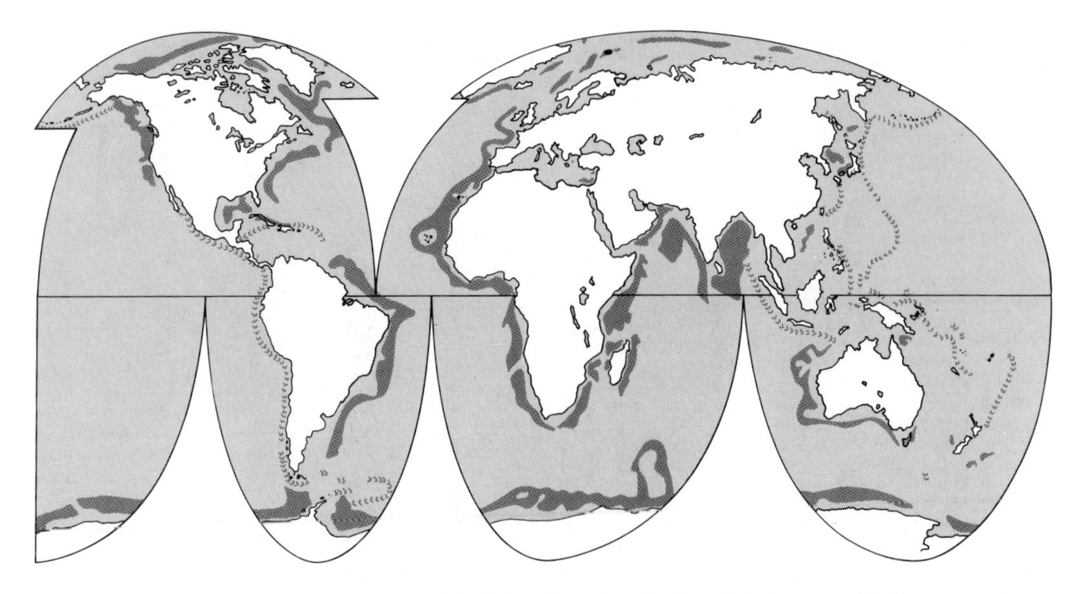

6.16 World distribution of continental rises (shaded) and trenches (hash marks). Compare this figure with the tectonic diagram in Figure 5.8a to see why continental rises are rarely well developed around the Pacific. (From K. O. Emery, "Continental Rises and Oil Potential," Oil and Gas Journal, *1969, 19(67), 231–43)*

Its straight valley follows a fault-and-fracture system that extends from Canada into the continental shelf south of New York (Figure 6.15). As it continues across the shelf, it deepens at a relatively constant rate offshore to a depth of about 100 m (328 ft), at which point it steepens and widens. At 20 km (12 mi) offshore, it is 200 m (656 ft) below sea level and forms a spectacular submarine canyon. At 50 km (31 mi) offshore, the canyon is 3,000 m (9,840 ft) deep and terminates at the floor of the Atlantic Ocean.

CONTINENTAL RISES

Continental rises, which constitute the final transition zone between the continents and the ocean basin (Figure 6.16), make up about 3% of the seafloor and about 30% of the earth's sediment. In most regions, the **continental rise** consists of unconsolidated sediments of mud, silt, and sand that have been carried down the continental shelf and slope by turbidity currents, submarine slides, and other processes. Turbidity currents are responsible for the large, fan-shaped deposits that have been laid down at the base of submarine canyons. These deposits are marked by channels, levees, and low hills. Graded bedding is common in these sediments, as indicated by the gradual decrease in particle size from the bottom to the top of cores extracted for analysis.

In the Pacific Ocean, continental rises occur only along the edge of the Antarctic continent and the northeastern Pacific, off the coast of North America (Figure 6.17). In other oceans, rises are more common, ranging from depths of about 3 to 4 km (2 to 2.5 mi). The pattern of the continental rises is clearly related to plate tectonic movements. On active margins, the sediments have been altered and carried to the mantle by subduction. On passive margins, the sediments have been preserved and extend outward to the ocean floor.

FURTHER READINGS

Baldwin, E. M. *Geology of Oregon*. Dubuque, Ia.: Kendall/Hunt, 1981.

Bascom, W. *Waves and Beaches*. Garden City, N.Y.: Doubleday, 1964.

Bird, E. C. F. *Coasts: An Introduction to Systematic Geomorphology*. Vol. 4. Cambridge, Mass.: MIT Press, 1969.

Donn, W. L., and D. M. Shaw. "Model of Climate Evolution Based on Continental Drift and Polar Wandering." *GSA Bulletin*, 1977, *88*(3), 390–96.

Emery, K. O. "The Continental Shelves." *Scientific American*, March 1969, pp. 107–22.

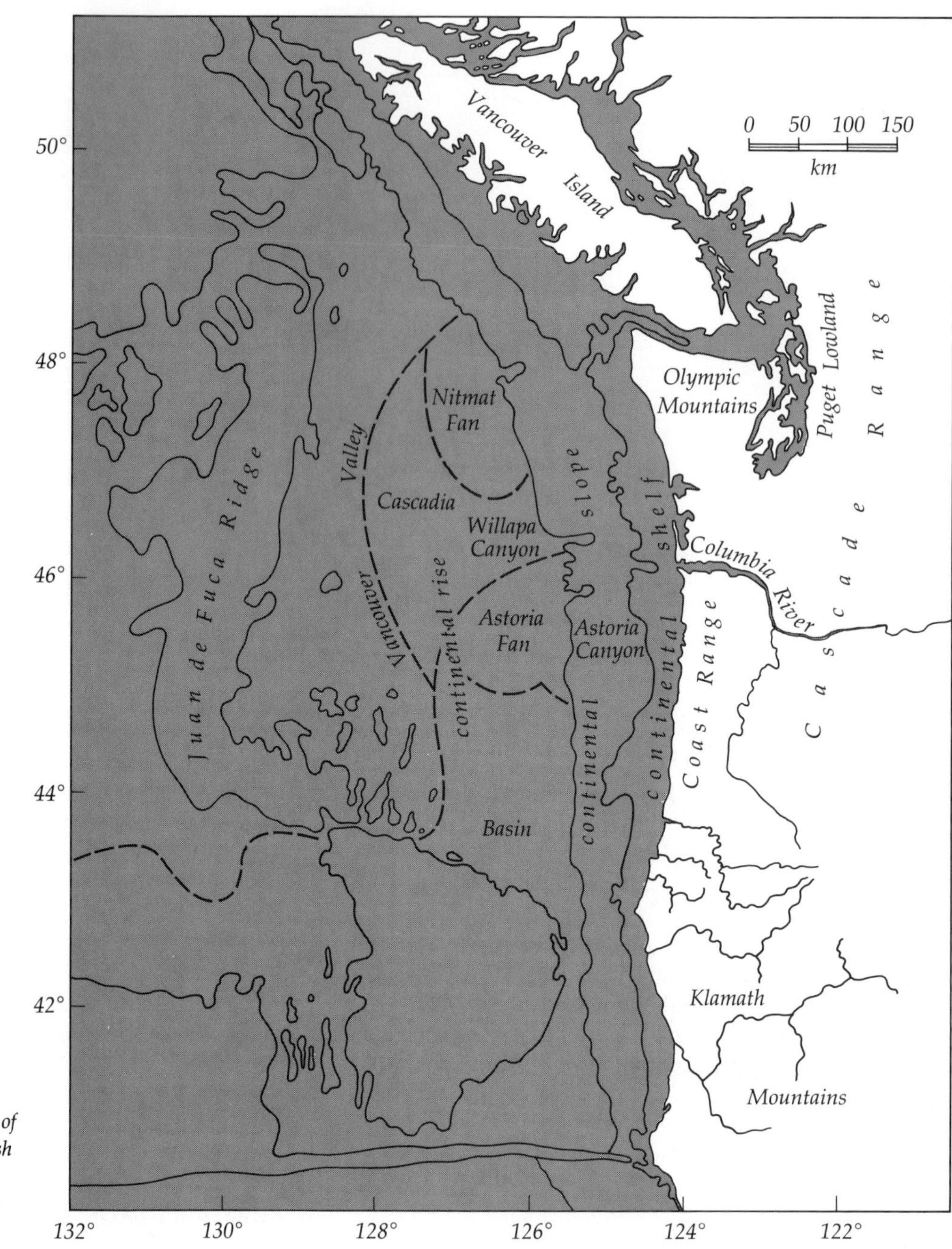

6.17 A portion of the continental margin off the coast of Oregon, Washington, and British Columbia. (After B. McKee, Cascadia, *New York: McGraw-Hill, 1972)*

Hays, J. D., J. Imbrie, and N. J. Shackleton. "Variations in the Earth's Orbit: Pacemaker of the Ice Ages." *Science*, 1976, *194*, 1121–32.

Imbrie, J., and K. P. Imbrie. *Ice Ages: Solving the Mystery.* Short Hills, N.J.: Enslow Publications, 1979.

Inman, D. L., and R. A. Bagnold. "Littoral Processes." In *The Sea*, edited by M. N. Hill, vol. 3. New York: Interscience, 1963.

Inman, D. L., and C. E. Nordstrom. "On the Tectonic and Morphologic Classification of Coasts." *Journal of Geology*, 1971, *79*(1), 1–21.

Kennett, J. P. *Marine Geology.* Englewood Cliffs, N.J.: Prentice-Hall, 1982.

King, C. A. M. *Beaches and Coasts*. London: Edward Arnold, 1972.

Komar, P. D. *Beach Processes and Sedimentation*. Englewood Cliffs, N.J.: Prentice-Hall, 1976.

Lauff, G. H., ed. *Estuaries*. Publication 83. Washington, D.C.: American Association for the Advancement of Science, 1967.

Ocean Science Board. *Continental Margins*. Washington, D.C.: National Academy of Sciences, 1979.

Ricketts, E. F., and J. Calvin. *Between Pacific Tides*. Stanford, Calif.: Stanford University Press, 1968.

Shepard, F. P., and R. F. Dill. *Submarine Canyons and Other Sea Valleys*. New York: Wiley, 1966.

Shepard, F. P., and H. R. Wanless. *Our Changing Coastlines*. New York: McGraw-Hill, 1971.

Teal, J., and M. Teal. *Life and Death of a Salt Marsh*. Boston: Little, Brown, 1961.

Van Andel, T. H. *New Views on an Old Planet*. Cambridge, Eng.: Cambridge University Press, 1985.

Waters, J. F. *Exploring New England Shores*. Lexington, Mass.: Stone Wall Press, 1974.

Jonathan Fisher

I am the daughter of Earth and Water,
And the nursling of the sky;
I pass through the pores of the oceans and shores;
I change, but I cannot die.

Percy Bysshe Shelley

SEVEN

The Chemical Properties of Water

Biologists have long noted a close resemblance between the plasma of human blood and seawater. The fluids have similar densities. They contain similar percentages of dissolved salts. Both are very slightly basic; in fact, the basicity (alkalinity) of both fluids is almost identical. Moreover, seawater closely resembles the aqueous humor of the human eye. In a sense, we look at the world through ocean water. Life may have evolved on this planet in a number of watery environments: the ocean, fresh water, or geothermally heated springs. There is some evidence that supports each possibility. The similarities between blood and seawater give some support to seawater.

Certainly when the first animals invaded the land, they already had to be adapted in ways that would allow them to carry some of the sea with them. Every body part, every organ system, had to continue to be bathed constantly in the water-salt solution in which it had first evolved. Internal circulatory systems for bathing all cells and body parts would be a clear survival advantage, as would having systems for retaining and regulating the volume of water and salts in a body that was no longer able to draw them from the sea. Thus, our circulating blood and our highly efficient kidneys perpetuate the illusion that we have escaped from the sea, when in fact we carry our ancient saltwater heritage with us.

PROPERTIES OF WATER

To appreciate the properties of seawater, we must first begin with the characteristics of water itself. Water is one of the most remarkable substances in the universe. It is one of the substances most transparent to light, and it has the greatest heat capacity of all liquids except liquid ammonia. It conducts heat better than any other common liquid, and no other liquid has greater surface tension. Table 8.2 (p. 112) lists a few more of its remarkable properties.

What is the physical basis of these properties? For the answer, we must understand the bonds holding each water molecule together:

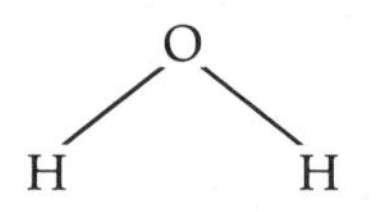

The two hydrogen atoms and the lone oxygen atom are held together by shared electrons. Roughly speaking, the electrical charge of these electrons is the "glue" between the atoms. The oxygen atom, however, has a greater number of protons and thus exerts more of a pull on the electrons than the hydrogen atoms do. In other words, on the average the oxygen atom gets a bigger share of the electrons. As a result, the water molecule is *polar*—it has more positive charge at the hydrogen end and more negative charge at the oxygen end:

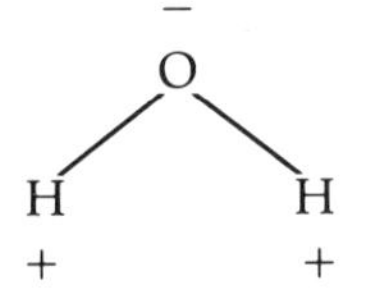

Interestingly, from time to time the electron sharing between the oxygen atom and the hydrogen atoms becomes especially unequal. In fact, a small percentage of water molecules typically breaks apart into two charged parts, OH^- and H^+, which is why distilled water conducts an electric current weakly. These symbols tell us that the OH^- part is a **negatively charged ion** (it has taken on an extra electron), and the H^+ part is a **positively charged ion** (it has given up an electron completely). Even so, since the parts are oppositely charged, they are attracted to each other and rejoin almost at once. The ease with which water molecules break apart into ions and then form new bonds is one reason why water is a liquid, not, as expected, a gas, at normal temperatures.

What we call **pH** is simply a way of describing the relative amounts of OH^- and H^+ ions in any given solution. A substance that has a low pH has an excess amount of H^+ ions and is an **acid**. A substance that has a high pH has an excess of OH^- ions and is a **base**. When equal amounts of both kinds of ions are present, the substance is said to be **neutral,** and its pH will be 7.0.

Another interesting fact about water molecules is how they interact with one another and with other substances. When water molecules are near each other, they "share" their hydrogen atoms:

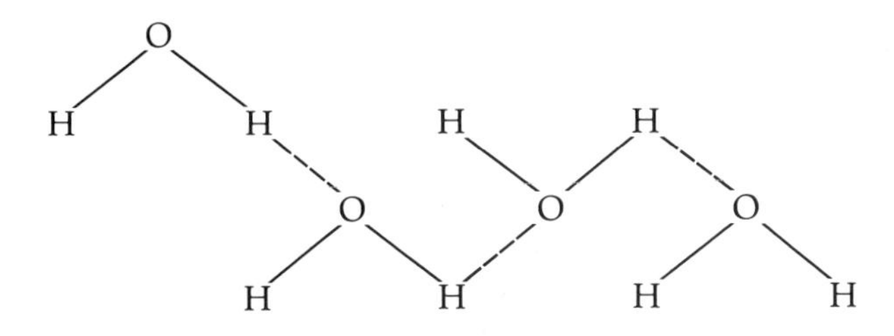

These so-called **hydrogen bonds** between the oxygen atom of one molecule and a hydrogen atom of another are the reason why water freezes and boils at much higher temperatures than other molecules of similar size. The hydrogen bonds are just strong enough to make water molecules act like larger molecules having stronger bonds. The ease with which water molecules form hydrogen bonds is also the source of water's most remarkable property: its ability to dissolve other substances. When something dissolves in water, its molecules become hydrogen-bonded with the water molecules surrounding it.

Table 7.1 *Constituents of Seawater*

Substance	Symbol	‰ Seawater	% Total Weight of Salt
Chloride	Cl^-	18.980	55.04
Sodium	Na^+	10.556	30.61
Sulfate	$SO_4{}^{-2}$	2.649	7.68
Magnesium	Mg^{+2}	1.272	3.69
Calcium	Ca^{+2}	0.400	1.16
Potassium	K^+	0.380	1.10
Bicarbonate	$HCO_3{}^-$	0.140	0.41
Bromide	Br^-	0.065	0.19
Boric acid	H_3BO_3	0.026	0.07
Strontium	Sr^{+2}	0.013	0.04
Fluoride	F^-	0.001	0.00
Total		34.482‰	99.99%

Even glass is slightly soluble in water. If water is sealed in a soft glass bottle for a few years, the water will become quite cloudy as some of the glass dissolves. Nevertheless, the rate at which some substances dissolve is very slow. Oil, for example, has an extremely low solubility in water or seawater. It happens to be a *saturated* molecule, which means that almost all of its atoms have a full share of electrons and have little tendency to bond with the atoms of other molecules. In fact, they find other molecules repulsive, which is why oil tends to separate from water. Each oil molecule gets as far away from the water molecules as it can—and that is also why the sea takes a long time to rid itself of an oil spill.

PROPERTIES OF SEAWATER

Of all the surface water in the world, 97.2% is in the ocean. Only 2.15% is locked in icecaps, snow, and glaciers; only 0.65% is found in rivers, groundwater, and lakes; and only 0.001% is found in the atmosphere. Although plants and animals on land depend on fresh water, the fact remains that most of the water available is the salty water in the sea.

Each kilogram of seawater contains about 35 g of dissolved salts; alternatively, we may say that the salt content of seawater is 35 parts per thousand, or 35‰. This expression of salt content of a solution is called its **salinity**. The salt in seawater is not the same as regular table salt (sodium chloride, NaCl). When table salt is dissolved in water, it breaks up into ions:

$$NaCl_{(s)} \xrightarrow{\text{water}} Na^+_{(AQ)} + Cl^-_{(AQ)}$$

In this case, the solution is neutral, for there are equal amounts of positive and negative ions. The ions that make up the salt content of seawater are shown in Table 7.1. (The presence of these various ions is the

Table 7.2 *Effects of Environmental Changes on Seawater*

Property	Effect on Property of Increasing: Salinity	Temperature	Pressure
Absorption of light	No effect	No effect	No effect
Boiling point	Increases	—	Increases
Density	Increases	Decreases	Increases
Electrical conductivity	Increases	Increases	Decreases
Freezing point	Decreases	—	—
Latent heat of vaporization	No change	Decreases	—
Osmotic pressure	Increases	Increases	Increases
Refractive index	Increases	Decreases	Increases
Sound velocity	Increases	Increases	Increases
Specific heat capacity	Decreases	Increases	Decreases
Surface tension	Increases	Decreases	—
Temperature of maximum density	Decreases	—	—
Thermal conductivity	Decreases	Increases	Increases
Thermal expansion	Increases	Increases	Increases
Vapor pressure	Decreases	Increases	—
Viscosity	Increases	Decreases	No change

reason why seawater is an excellent conductor of electricity, whereas pure water is not. The ions readily conduct electrical current. The conductivity increases directly as a function of salinity.) As Table 7.1 shows, the chloride ion (chlorine) is the most abundant substance next to water itself.

The salt content of seawater has some effect on its properties (Table 7.2). Pure water freezes at 0°C, but since the ocean's salt content acts somewhat like an antifreeze, seawater freezes at a lower temperature. As you might expect, the higher the salinity, the lower the freezing point (Table 7.3). Salt content also affects the ability of water to foam. What we call foam in water is simply the coalescing of myriad tiny bubbles. Unlike fresh water, which allows the bubbles to come together, the salt in seawater makes them bounce off one another. As a result, true foam generally does not form in the sea unless aided by the action of some other agent (Figure 7.1).

As Table 7.1 indicates, the bulk of the dissolved inorganic components in seawater consists of only eleven ions. The remaining 0.1% are trace constituents, which include those elements necessary for life—that is, nutrients such as nitrogen and phosphorus. Also present in seawater are a number of dissolved gases and dissolved and suspended organic substances. These are not reckoned as part of the salinity, since salinity is a measure of dissolved inorganic constituents only.

Table 7.3 *Freezing Point of Seawater as a Function of Salinity*

Salinity (‰)	Freezing Point (°C)
0	0
10	−0.53
20	−1.08
30	−1.63
35	−1.91

Origin of Salinity

We may consider seawater as a system made up of two components: water and dissolved substances. The origin of the water in the ocean was discussed in Chapter 2. Now let's look at the origin of the sea's saltiness.

When we compare the dissolved ion concentration of seawater (Table 7.1) with that of river water (Table 7.4), we notice a distinct difference in the order of constituents. In seawater, chloride is the most abundant; in river water the most abundant ion is bicarbonate, with chloride only the sixth. The saline composition of large contained bodies of water such as the Great Salt Lake, the Dead Sea, and the Caspian Sea is not at all like the composition of seawater. The salt content of the Dead Sea (240‰) is made up of 60% magnesium chloride and almost 20% sodium chloride.

Thus, it is obvious that seawater cannot be explained as merely concentrated river water. Even with many turnovers of the hydrologic cycle (see Chapter 9), it is clear that the relative percentages are all wrong. The explanation of the chemistry of seawater must be more complicated. As long as there has been liquid water on this planet, that water has gone through the hydrologic cycle, and the rain moving

7.1 Seawater does not foam. Air bubbles from surf and propellers find their way into seawater, giving the appearance of foam, but it is short-lived; the bubbles do not coalesce. Here, foam is caused by exudate (in this case alginic acid) from large amounts of kelp. (W. J. Wallace)

over the same rocks through time has dissolved the same materials and carried them to the sea. So the ocean must always have had some salt content.

It might seem that the ninety or so naturally occurring elements that compose the lithosphere (Table 7.5) should also occur in seawater, but note that those elements in shaded blocks do not appear in seawater. The presence of an element in rock obviously does not guarantee its presence in the sea, for two reasons: First there is the solubility of the rock's compounds. Most elements do not exist in the elemental form in nature; therefore, as compounds they must be soluble in water to find their way via rivers to the sea. Even with this in mind, it is still a surprise to find only eleven major ions. Therefore the reason must have to do with the complex chemistry taking place in seawater. Water is an ideal medium for chemical reactions. As a result the eleven major ions that make up over 99.99% of the sea's salinity are simply those that have remained behind in the seawater after the complex and not yet fully understood formation of detrital matter, sediments, and organism shells.

Each of the constituents in seawater has its own complicated chemistry. (Potassium, for example, is removed from the sea in two ways: by absorption into river-supplied clay minerals and by reaction with submarine basalts that liberate calcium.) Present-day salinity is the sum total of all ions present. Each individual ionic concentration is determined by that ion's unique chemistry, and all of the processes involved have been going on over a vast period of time.

The chloride ion is the key to understanding present ocean salinity. It is also the easiest to measure accurately. Its **residence time**, the average length of time

Table 7.4 *Constituents in River Water*

Constituent	Symbol	% Total Weight of Constituents
Bicarbonate	HCO_3^{-1}	35.15
Calcium	Ca^{+2}	20.39
Sulfate	SO_4^{-2}	12.14
Silicon dioxide	SiO_2	11.67
Sodium	Na^+	5.79
Chloride	Cl^-	5.68
Magnesium	Mg^{+2}	3.41
Oxides	$(Fe, Al)_2O_3$	2.75
Potassium	K^+	2.12
Nitrate	NO_3^-	0.90
Total		100.00%

the ion remains in the sea, is arbitrarily given as infinity, since no specific mechanism removes it from the sea. The chloride content of seawater cannot be accounted for by the chloride content in the terrestrial environment—no other place in nature has as proportionally high a chloride content as the ocean. Various gases are emitted by volcanoes, including hydrogen chloride (HCl), which is very water soluble. The HCl gas emitted by volcanoes dissolves in rain, ionizes, and eventually makes its way into the ocean as chloride ions. Since HCl is heavier than H_2O, it would have begun to remain on a cooling planet prior to the water.

Again, residence time denotes the average length of time a particle of an element remains in seawater. It is defined as:

$$\text{residence time} = \frac{\text{amount of given element in ocean}}{\text{amount of given element added per year}}$$

Table 7.6 contains the residence times of several elements. Elements like aluminum, which have a short residence time, are more reactive in the ocean. We can also speak of the residence time of water itself in the sea. Because of continous evaporation, this is fairly short, only about 54,000 years.

In short, ocean salinity over time is generally uniform. Ion inputs and exhausts have achieved a geochemical balance that allows the amount of salt in the ocean to change proportionally over time with the ocean's volume.

Determination of Salinity

Of the inorganic matter dissolved in seawater, 99% consists of chloride, sodium, magnesium, sulfur (as sulfate), calcium, and potassium. You might think that

Table 7.5 *Periodic Table of the Elements (Shaded elements are not present in seawater.)*

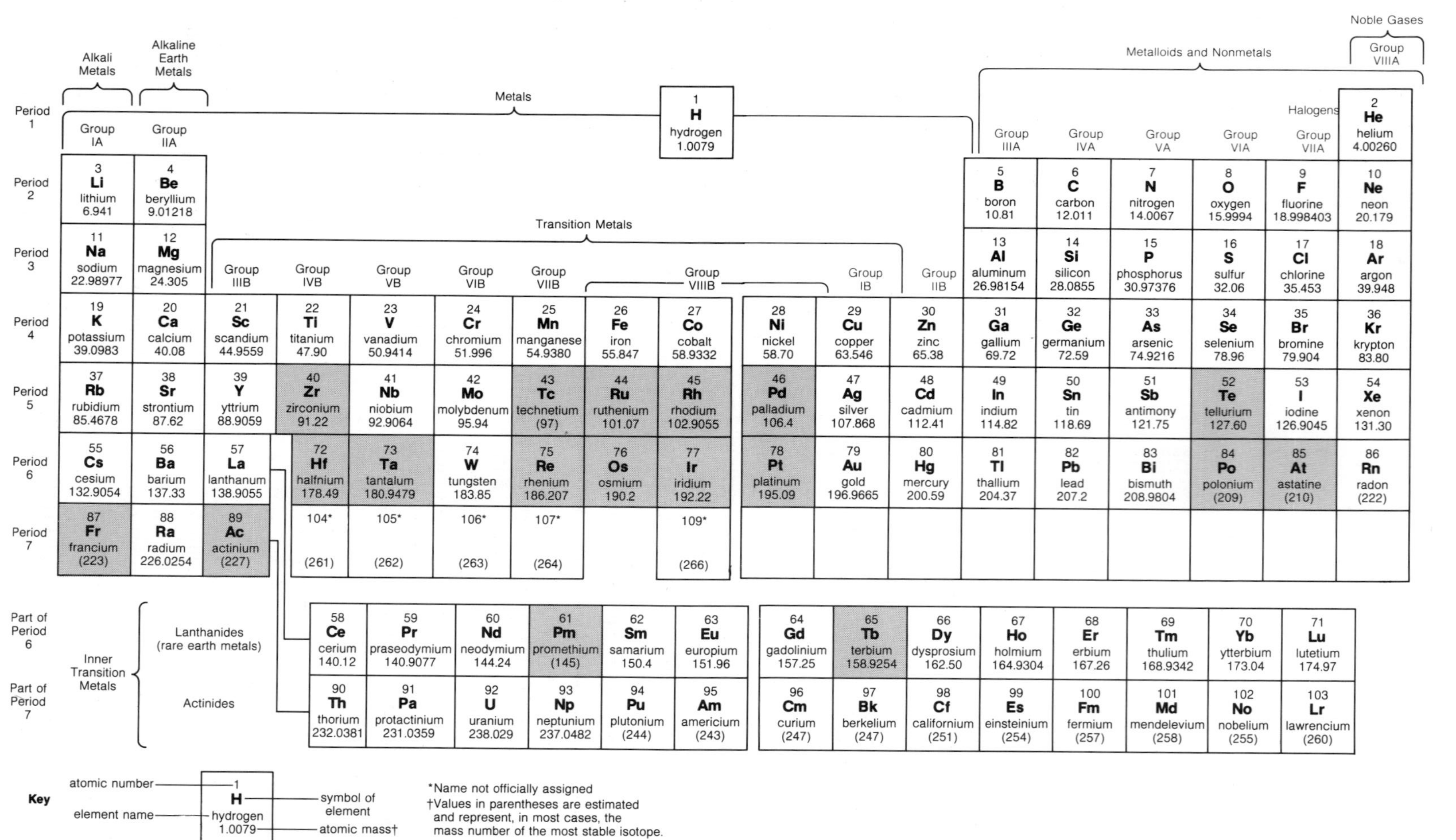

	Group IA (Alkali Metals)	Group IIA (Alkaline Earth Metals)	Group IIIB	Group IVB	Group VB	Group VIB	Group VIIB	Group VIIIB	Group VIIIB	Group VIIIB	Group IB	Group IIB	Group IIIA	Group IVA	Group VA	Group VIA	Group VIIA (Halogens)	Group VIIIA (Noble Gases)
Period 1	1 **H** hydrogen 1.0079																	2 **He** helium 4.00260
Period 2	3 **Li** lithium 6.941	4 **Be** beryllium 9.01218											5 **B** boron 10.81	6 **C** carbon 12.011	7 **N** nitrogen 14.0067	8 **O** oxygen 15.9994	9 **F** fluorine 18.998403	10 **Ne** neon 20.179
Period 3	11 **Na** sodium 22.98977	12 **Mg** magnesium 24.305											13 **Al** aluminum 26.98154	14 **Si** silicon 28.0855	15 **P** phosphorus 30.97376	16 **S** sulfur 32.06	17 **Cl** chlorine 35.453	18 **Ar** argon 39.948
Period 4	19 **K** potassium 39.0983	20 **Ca** calcium 40.08	21 **Sc** scandium 44.9559	22 **Ti** titanium 47.90	23 **V** vanadium 50.9414	24 **Cr** chromium 51.996	25 **Mn** manganese 54.9380	26 **Fe** iron 55.847	27 **Co** cobalt 58.9332	28 **Ni** nickel 58.70	29 **Cu** copper 63.546	30 **Zn** zinc 65.38	31 **Ga** gallium 69.72	32 **Ge** germanium 72.59	33 **As** arsenic 74.9216	34 **Se** selenium 78.96	35 **Br** bromine 79.904	36 **Kr** krypton 83.80
Period 5	37 **Rb** rubidium 85.4678	38 **Sr** strontium 87.62	39 **Y** yttrium 88.9059	40 **Zr** zirconium 91.22	41 **Nb** niobium 92.9064	42 **Mo** molybdenum 95.94	43 **Tc** technetium (97)	44 **Ru** ruthenium 101.07	45 **Rh** rhodium 102.9055	46 **Pd** palladium 106.4	47 **Ag** silver 107.868	48 **Cd** cadmium 112.41	49 **In** indium 114.82	50 **Sn** tin 118.69	51 **Sb** antimony 121.75	52 **Te** tellurium 127.60	53 **I** iodine 126.9045	54 **Xe** xenon 131.30
Period 6	55 **Cs** cesium 132.9054	56 **Ba** barium 137.33	57 **La** lanthanum 138.9055	72 **Hf** halfnium 178.49	73 **Ta** tantalum 180.9479	74 **W** tungsten 183.85	75 **Re** rhenium 186.207	76 **Os** osmium 190.2	77 **Ir** iridium 192.22	78 **Pt** platinum 195.09	79 **Au** gold 196.9665	80 **Hg** mercury 200.59	81 **Tl** thallium 204.37	82 **Pb** lead 207.2	83 **Bi** bismuth 208.9804	84 **Po** polonium (209)	85 **At** astatine (210)	86 **Rn** radon (222)
Period 7	87 **Fr** francium (223)	88 **Ra** radium 226.0254	89 **Ac** actinium (227)	104* (261)	105* (262)	106* (263)	107* (264)		109* (266)									

Inner Transition Metals

Part of Period 6 — Lanthanides (rare earth metals)	58 **Ce** cerium 140.12	59 **Pr** praseodymium 140.9077	60 **Nd** neodymium 144.24	61 **Pm** promethium (145)	62 **Sm** samarium 150.4	63 **Eu** europium 151.96	64 **Gd** gadolinium 157.25	65 **Tb** terbium 158.9254	66 **Dy** dysprosium 162.50	67 **Ho** holmium 164.9304	68 **Er** erbium 167.26	69 **Tm** thulium 168.9342	70 **Yb** ytterbium 173.04	71 **Lu** lutetium 174.97
Part of Period 7 — Actinides	90 **Th** thorium 232.0381	91 **Pa** protactinium 231.0359	92 **U** uranium 238.029	93 **Np** neptunium 237.0482	94 **Pu** plutonium (244)	95 **Am** americium (243)	96 **Cm** curium (247)	97 **Bk** berkelium (247)	98 **Cf** californium (251)	99 **Es** einsteinium (254)	100 **Fm** fermium (257)	101 **Md** mendelevium (258)	102 **No** nobelium (255)	103 **Lr** lawrencium (260)

*Name not officially assigned

†Values in parentheses are estimated and represent, in most cases, the mass number of the most stable isotope.

Table 7.6 *Residence Times of Various Constituents*

Constituent	Symbol	Residence Time (years)
Chloride	Cl^-	∞
Sodium	Na^+	260,000,000
Magnesium	Mg^{+2}	45,000,000
Potassium	K^+	11,000,000
Sulphate	SO_4^{-2}	11,000,000
Calcium	Ca^{+2}	8,000,000
Silicon	Si	8,000
Iron	Fe	140
Aluminum	Al	100

determining the salt content of a seawater sample would be relatively easy by using advanced methods of modern chemistry. In fact, of the arbitrary branches into which oceanography is divided, the chemical study of the sea is probably the oldest. Yet the problem of determining total salinity, let alone separating and identifying individual constituents, has always been a difficult one, even though eleven substances make up 99.9% of the total salt content.

If you dissolve 10 g of sodium chloride and 10 g of potassium bromide in 1 L of pure water and then evaporate the solution to dryness, you will end up with 20 g of solids—but not all of the residue would be the two original salts! Depending on such factors as the evaporation rate, you might find trace amounts of sodium bromide and potassium chloride. Once dissolved, these two salts exist as ions, and as they dry, they randomly combine with other ions to form different salts. You can thus see why, with a total of eleven kinds of ions, the problem becomes truly complex. Moreover, not even the total solid residue obtained is the same each time. Substances such as magnesium chloride (which gives seawater its slightly bitter taste) may partially break down from the heat used to evaporate the sample yielding different compounds.

In 1902, the International Council for the Exploration of the Sea recommended that salinity be calculated on the basis of actual determinations of chloride content by titration. (In **titration**, an exact quantity of some substance that is known to react in a definite proportion with the constituent being analyzed is added to a measured sample that contains the constituent.) The council suggested an equation that has been used by oceanographers until recently:

$$\text{salinity} = (1.805 \times \text{chlorinity}) + 0.030$$

The problem with this equation, however, is that the definition of salinity based on chloride determinations is not the same as the actual salt content. A kilogram of seawater must contain a certain amount of dissolved salts, but the amounts of salt described by the equation and by the titration method are slightly different. What are the reasons for the discrepancy? First, having a constant of 0.030 in the original equation implies that the salinity of distilled water is 0.030, but obviously its actual salinity must be zero. Second, the underlying assumption is that the ratios of the ions in seawater are constant, but this is now known to be only an approximation.

Consequently, a new definition, again based on chloride ion determination, has been proposed:

$$\text{salinity} = 1.80655 \times \text{chlorinity}$$

Although the difference in values between the two equations is small, it is large enough to be important to oceanographers. Yet, even though this equation eliminates the troublesome constant 0.030, it still does not give us an exact calculation of the salt content.

More recently, yet another method of measuring salinity has been developed. As mentioned previously, seawater conducts electricity, and its degree of conductance is a function of the number of ions present in a sample. Modern conductivity meters, or **salinometers** as they are generally called, are used to determine the salinity of seawater. They are as accurate as chloride titrations measured in the laboratory (an accuracy to about 20 parts per million), but their advantage is that the sample need not be taken out of the water for testing but can be run in situ. Salinity is determined in situ by lowering the sensing coil of the salinometer into the water. Salinometers function to depths of more than 6,000 m (19,680 ft).

The concept of constant ionic ratios of the major constituents is useful in oceanography. Formerly, the fact that salinity as defined by oceanographers was not the same as the total salt content made little difference. In 1902 it was more important to agree on a particular definition (even though the definition chosen did not coincide with the real physical properties of seawater) than to have an exact, accurate measure of total salts. In all probability, the term *salinity* will remain in use but most salinity determinations will be based on electrical conductivity.

So what have all the analyses told us about the overall salt content of the sea? The open oceans are fairly well mixed, so salinity there is reasonably consistent (Figure 7.2). Small but measurable differences occur in various parts of the ocean, and there are slight seasonal variations. Evaporation, freezing, and dissolving of salt will add salt or remove water; hence these processes increase salinity. In contrast, rain and snow, runoff from rivers, melting ice, and precipitation out of solution will remove salt or add water and thus decrease salinity. An inflowing or outflowing water current and subsequent mixing may do either. Of these, evaporation and precipitation are the most important.

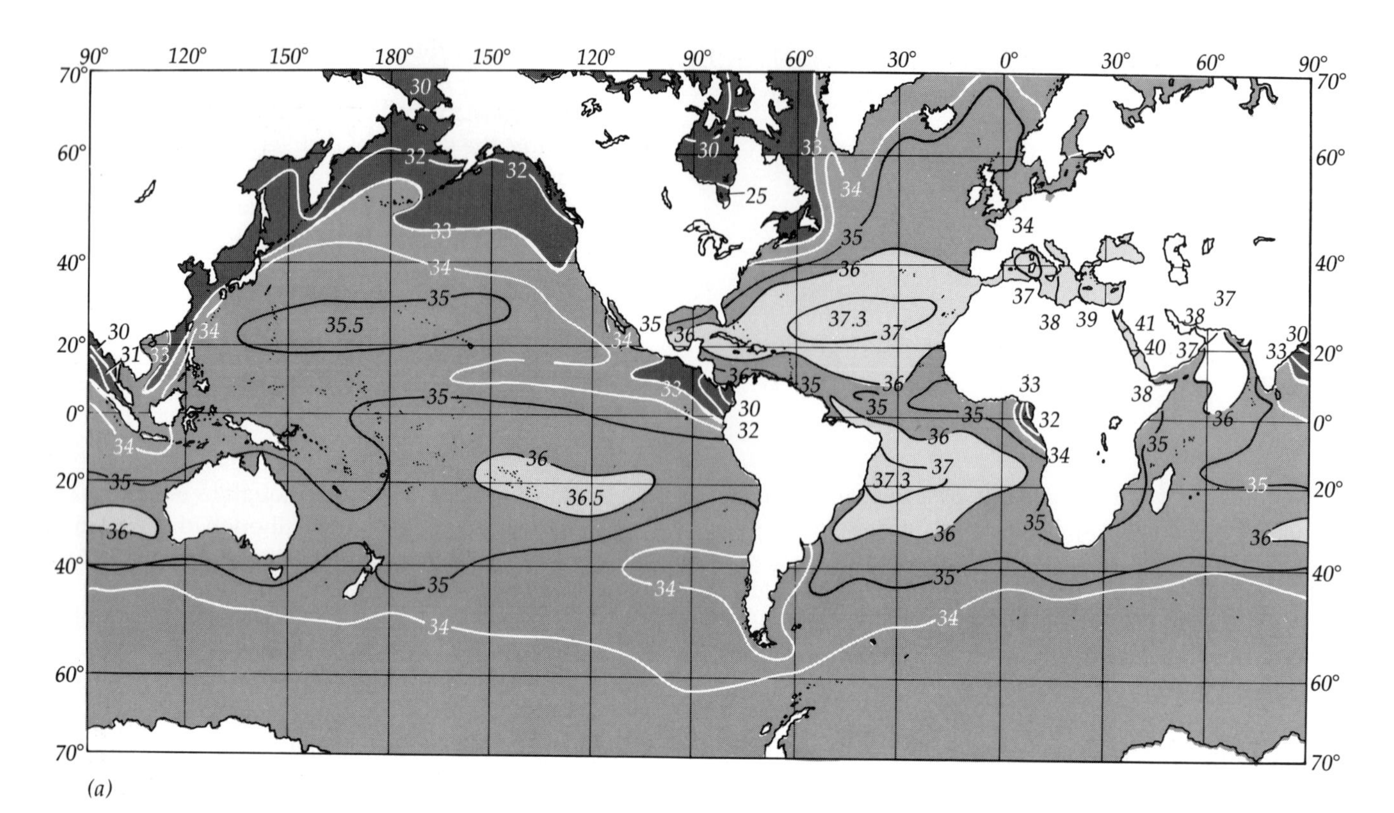

7.2 *(a) Ocean-surface salinities (‰) in August. (b) Typical salinity profiles for the open ocean. Solid lines indicate low and middle latitudes. Dashed lines indicate high latitudes. (c) overleaf.*

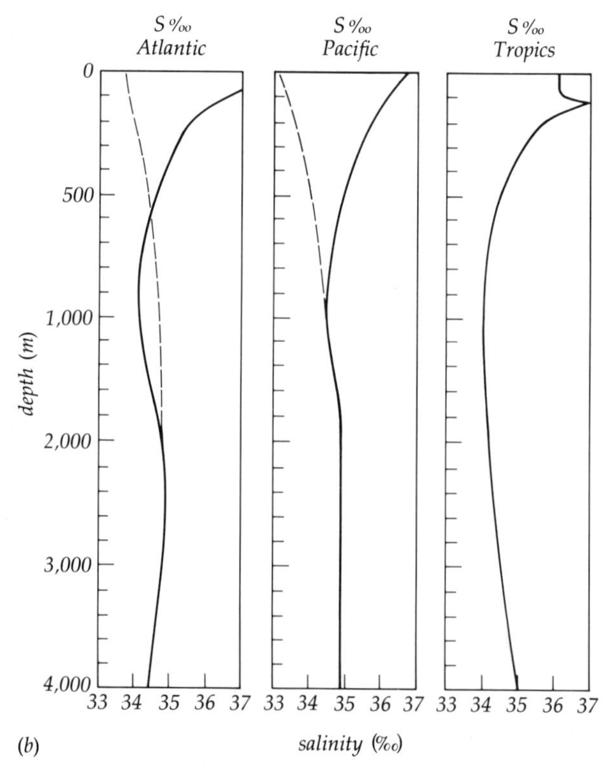

Regions of the highest salinity occur where evaporation is greater than precipitation. Conversely, regions of lowest salinity occur where precipitation exceeds evaporation. In the open ocean, these factors are balanced, and variations are not great (Table 7.7 and Figure 7.2c).

In almost all parts of the ocean, salinity lies between 34‰ and 37‰. The salinity of the Red Sea, which has little incoming fresh water and high evaporation, is 40‰ to 42‰. In regions such as the Gulf of Bothnia and the waters off Finland, salinity may be as low as 5‰. During the International Geophysical Year (1958–59), local pockets of deep, salty water were found in the Red Sea; below 2,000 m (6,560 ft), salinities were greater than 250‰. That is very close to the saturation point for salt in water. The saltiest open ocean is in the subtropical North Atlantic, where salinity is 37.5‰ at the surface. The surface water of the Pacific is less salty than that of the Atlantic because it is less subject to drying winds from neighboring continents. The average salinity for the Pacific Ocean is 34.62‰; for the Atlantic, 34.90‰; and for the Indian, 34.76‰.

Only in high latitudes does the effect of ice formation on salinity become important. Salt water under forming ice has a higher salinity, for salt tends to separate out of the water as freezing occurs. This helps to account for the fact that waters near the poles, particularly near the Antarctic, are more saline than might be expected. Since the salinity of adjacent water decreases when the ice melts, the net effect of the freezing and melting of seawater probably averages out to zero over the year.

Table 7.7 *Average Values of Salinity (S), Evaporation (E), and Precipitation (P), and the Difference (E − P) for Seawater*

Latitude	Atlantic Ocean S (‰)	Atlantic Ocean E (cm/yr)	Atlantic Ocean P (cm/yr)	Atlantic Ocean E − P (cm/yr)	Indian Ocean S (‰)	Indian Ocean E (cm/yr)	Indian Ocean P (cm/yr)	Indian Ocean E − P (cm/yr)
40°N	35.80	94	76	18				
35	36.46	107	64	43				
30	36.79	121	54	67				
25	36.87	140	42	98				
20	36.47	149	40	110	(35.05)*	(125)	(74)	(51)
15	35.92	145	62	83	(35.07)	(125)	(73)	(52)
10	35.62	132	101	31	(34.92)	(125)	(88)	(37)
5	34.98	105	144	−39	(34.82)	(125)	(107)	(18)
0	35.67	116	96	20	35.14	125	131	−6
5°S	35.77	141	42	99	34.93	121	167	−46
10	36.45	143	22	121	34.57	99	156	−57
15	36.79	138	19	119	34.75	121	83	38
20	36.54	132	30	102	35.15	143	59	84
25	36.20	124	40	84	35.45	145	46	99
30	35.72	116	45	71	35.89	134	58	76
35	35.35	99	55	44	35.60	121	60	61
40	34.65	81	72	9	35.10	83	73	10
45	34.19	64	73	−9	34.25	64	79	−15
50	33.94	43	72	−29	33.87	43	79	−36

Latitude	Pacific Ocean S (‰)	Pacific Ocean E (cm/yr)	Pacific Ocean P (cm/yr)	Pacific Ocean E − P (cm/yr)	All Oceans S (‰)	All Oceans E (cm/yr)	All Oceans P (cm/yr)	All Oceans E − P (cm/yr)
40°N	33.64	94	93	1	34.54	94	93	1
35	34.10	106	79	27	35.05	106	79	27
30	34.77	116	65	51	35.56	120	65	55
25	35.00	127	55	72	35.79	129	55	74
20	34.88	130	62	68	35.44	133	65	68
15	34.67	128	82	46	35.09	130	82	48
10	34.29	123	127	−4	34.72	129	127	2
5	34.29	102	(177)	(−75)	34.54	110	177	−67
0	34.85	116	98	18	35.08	119	102	17
5°S	35.11	131	91	40	35.20	124	91	33
10	35.38	131	96	35	35.34	130	96	34
15	35.57	125	85	40	35.54	134	85	49
20	35.70	121	70	51	35.69	134	70	64
25	35.62	116	61	55	35.69	124	62	62
30	35.40	110	64	46	35.62	111	64	47
35	35.00	97	64	33	35.32	99	64	35
40	34.61	81	84	−3	34.79	81	84	−3
45	34.32	64	85	−21	34.14	64	85	−21
50	34.16	43	84	−41	33.99	43	84	−41

* Parentheses indicate figures based on limited data.
From H. U. Sverdrup, Martin W. Johnson, and Richard H. Fleming, *The Oceans: Their Physics, Chemistry, and General Biology*, © 1942, renewed 1970. By permission of Prentice-Hall, Inc., Englewood Cliffs, N.J.

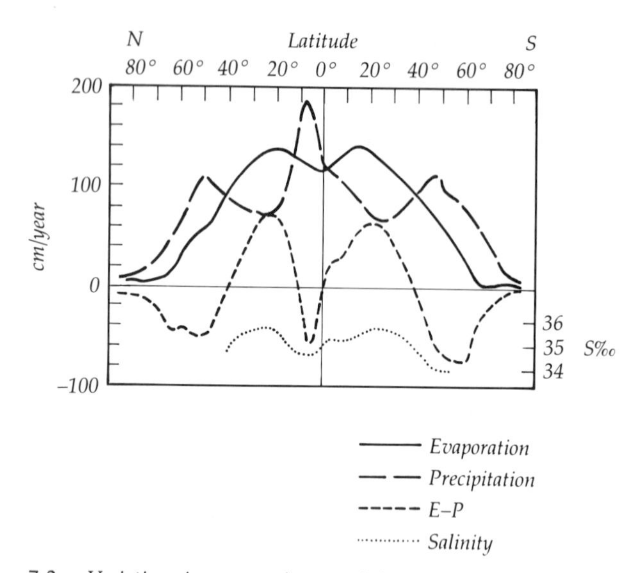

7.2c Variations in evaporation, precipitation, and salinity as a function of latitude.

Most of what we have said so far applies to the salinity of surface waters. Of the processes described for changes in salinity, only one—mixing—can change the salinity of subsurface waters. The vertical distribution of salinity in seawater is complex. Salinity does not merely increase with depth. In equatorial, tropical, and subtropical regions, there is a minimum salinity zone between 600 and 1,000 m (1,968 and 3,280 ft). Beyond this, salinity increases to a depth of about 2,000 m (6,560 ft), below which it decreases very slightly in the Atlantic and Pacific.

For all practical purposes, the salinity of a large volume of ocean water well below the surface changes very little, if at all, with time. Salinity, like temperature and therefore density, is said to be a **conservative property**, because once it is away from the surface it is conserved. Since only processes such as mixing or diffusion can change the values of these properties away from the surface, these values can be used to identify and trace the paths of specific waters in the deep ocean. Because oxygen, carbon dioxide, nitrate, phosphate, and silicate can be produced or consumed away from the surface by life processes, their respective concentrations are not conserved with depth.

Seawater Sampling Methods

How do oceanographers collect seawater samples? It's fairly simple to throw a bucket over the side of a ship so as to avoid the hot water effluent of the engine or bilge pumps and then haul in some water. Sampling water from below the surface is more of a problem. Today, most subsurface-water samples are taken in a Nansen bottle (Figure 7.3), which is really a tube (usually brass) with valves on both ends. It is lowered

(a)

(b)

7.3 (a) A marine technician attaches a Nansen bottle to the hydrographic wire. Activated by sliding weights, the bottles on a line flip over and collect a seawater sample at a specified depth. Thermometers to the side record temperature. (Scripps Institution of Oceanography) (b) A Nansen bottle. Notice the externally attached tubes for housing the protected and unprotected thermometers. (W. J. Wallace)

vertically attached to a cable called a hydrowire. Both valves are open, so that water passes through the bottle as it descends. When the desired depth is reached, a collarlike brass weight (called a **messenger**) is dropped along the wire. The messenger trips the Nansen bottle, which reverses or flips over, closing its valves. A number of Nansen bottles are lowered at the same time on the cable.

Usually samples are taken at the following depth intervals for a 1-km cast: surface, 50, 100, 130, 190, 220, 300, 350, 400, 450, 500, 600, 650, 700, 800, 900, and 1,000 m. For deeper casts, samples are taken at

500-m intervals. As the cable is payed out, it is temporarily stopped by the winch operator at the desired intervals (in reverse, since the deepest bottles are going down first). To study a specific depth, Nansen bottles may be clustered around the depth in question at intervals other than the standard ones just listed.

Once the bottles reach the desired depth, a certain amount of time is allowed for them to reach temperature equilibrium with the surrounding water. Then a messenger is released at the surface. As the messenger trips a given Nansen bottle, it releases a subsequent messenger, which drops to the next Nansen bottle; thus each bottle is tripped in sequence. The operator must be careful not to release a messenger prematurely while he or she is fixing each bottle to the hydrowire as it goes over the side; to do so would trigger all the bottles already on the line. When all the bottles in the series have been tripped, the entire hydrowire with all its bottles, referred to as a **hydrocast**, is brought up. At this point, tests may be run at once in an adjacent "wet" laboratory (which literally it very often is), or the water samples may be stored in containers and cooled or frozen for later study.

The hydrowire is a basic tool for collecting ocean data. Oceanographers attach all kinds of instruments to it. Today's hydrowire is a twisted or braided steel cable that varies in size, depending on the equipment to be tied to it. Historically, piano wire was used, and before that, hemp rope. The time necessary for a 5,000-m hydrocast ranges from five to six hours. The hydrowire itself presents many problems. Care must be taken to keep it away from the ship's propellers. Sometimes bottles don't trip, and the bottles and the wire must be checked regularly for corrosion. In deep sampling, there is always the chance that the wire might tangle. Then, too, the ship is not a steady platform on which to work, the bottles and the wire together are very heavy, and the working environment is often cold and wet.

However, the biggest problem with hydrowire is its inability to support heavy loads. A 4-mm (0.16-in) wire has a rated breaking point of about 478 kg (1,052 lb). The weight attached to make the cable sink weighs about 45 kg (99 lb), and the weight of 5,000 m (16,400 ft) of the wire itself makes the total weight more than 295 kg (649 lb). Commonly the wire is loaded only to about one-third of its maximum strength, or about 160 kg (353 lb). In short, about 180 kg (397 lb) are left for instruments. Because a Nansen bottle weighs about 3.4 kg (7.5 lb) in water, only fifty bottles can be used for a 5,000-m cast. In most hydrocasts, especially in coastal waters, a smaller number of bottles are used per cast.

Because of the high cost of operating a research vessel at sea, including the salaries of the crew, a representative cost for a seawater sample is about $21 a fifth (750 mL). And this figure does not include the salaries for the scientific crew.

Table 7.8 *Concentration of Minor Elements in Seawater*

Element Symbol	Concentration (mg/L)
Sr	8.0
B	4.6
Si	3.0
Li	.17
Rb	.12
P	.07
I	.06
N	.05
Ba	.03
Fe	.01
Zn	.01
Al	.01
Mo	.01
Cu	.003
As	.003
Ni	.002
Mn	.002

Engineers are constantly trying to improve water-sampling techniques. One of the most promising procedures is the use of submersible pumps located just below the surface. They would enable oceanographers to take much larger samples without the headaches associated with sampling bottles, and it would also allow them to collect samples while the ship is under way.

Other measuring methods are being considered to cut down on the high cost of operating staffed research vessels. Among these are automatic buoys to be moored off the coast. Designed to withstand large storms, they would constantly take and transmit data on salinity, pressure, wind velocity, and temperature. Also, aircraft have been used for some time to collect surface data such as temperature, and a number of offshore towers have been built on the continental shelf. The use of satellites for collecting surface data is especially promising. With the growing interest in scuba diving, we can anticipate further development of underwater laboratories and research submersibles (Chapter 18).

MINOR CONSTITUENTS

Table 7.8 indicates a number of important minor seawater constituents. Fortunately, these are generally nonconserved, since they are commonly involved in biological processes and their changes in concentration allow them to serve as indicators to trace-water

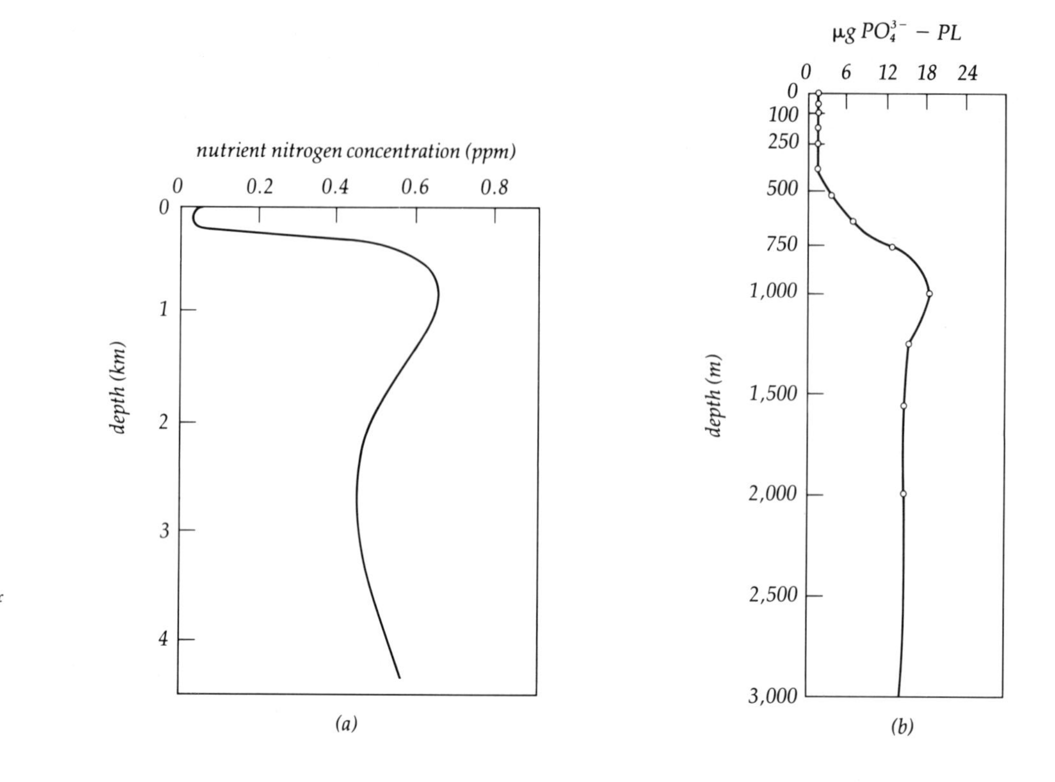

7.4 (a) Typical concentration of dissolved nitrogen (NO_3^-, NO_2^-, and NH_4^+) at varying depths. (b) Typical concentration of phosphate at varying depths.

movement (Chapter 10). There would seem to be about thirty elements (in their various compounds, ions, and so on) that are necessary for life in the sea. Those constituents that are biologically involved may be grouped into three categories: the biomaterial itself, composed of carbon, nitrogen, oxygen and phosphorus in their organic and inorganic forms; body fluids very similar to seawater in composition; and skeletal structure composed of calcium and magnesium carbonates, phosphates, and silicates. Most biomaterial elements, expecially nitrogen and phosphorus, are commonly called nutrients. The nitrogen and phosphorus exist primarily in the form nitrate (NO_3^-) and phosphate (PO_4^{-3}). Their vertical distribution varies markedly as a function of their biological activity. Figure 7.4 shows typical vertical nitrogen and phosphorus distribution.

Table 7.9 *Solubility of Various Gases in Surface Seawater at 0°C*

Gas	mL/L
N_2	14.0
O_2	8.0
CO_2	0.47
Ar	0.36

DISSOLVED GASES

The gases that compose the air are soluble in seawater, but they are not found in the same proportions in the sea. Table 7.9 gives their maximum solubility. Of the gases that dissolve in the sea, oxygen and carbon dioxide are the most important.

Oxygen

Oxygen is an abundant gas in seawater (Table 7.9) because it is fairly soluble. There are between 1 and 8 mL of oxygen in each liter of seawater, with an average of 6 mL. Oxygen enters the water from the atmosphere or is released as a by-product of photosynthesis. Both processes are essentially surface phenomena. At levels below the surface, the oxygen content of water decreases steadily, because it is used by organisms at all depths and is used in the decomposition of plant and animal remains.

The vertical profile of oxygen content (Figure 7.5) shows a high concentration of oxygen in the upper 10 to 20 m (33 to 66 ft). We would expect its concentration to be at a minimum at depth, where the amount of light available to plants decreases. An oxygen minimum does exist, usually between 700 and 1,000 m (2,300 to 3,280 ft) below the surface of the open ocean, but there is considerable variability. There are two explanations of the oxygen minimum, neither entirely satisfactory. One suggests that biological detritus collects in this zone as a result of increased density and that it consumes the oxygen present as it decays. The other attributes the oxygen minimum to the absence of any

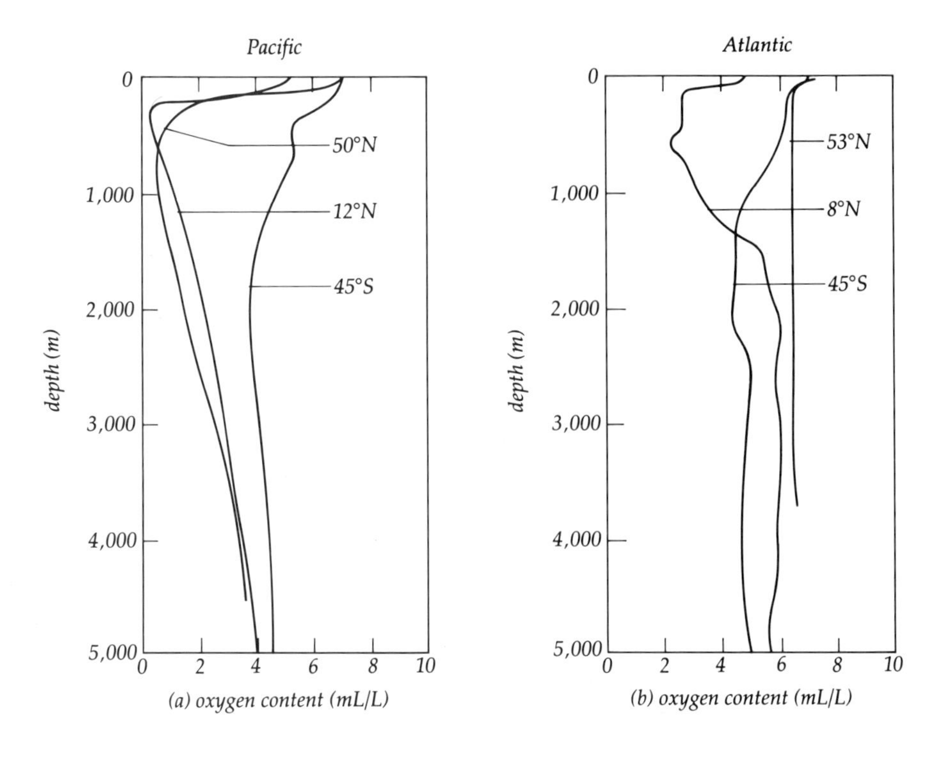

7.5 *Change in the amount of dissolved oxygen with depth (a) in the eastern tropical Pacific and (b) in the equatorial North Atlantic.*

Table 7.10 *Solubility of Oxygen in Seawater at Various Temperatures*

Temperature (°C)	mL/L
0	8.0
10	6.3
20	5.2
30	4.3

inflow of mixing waters to replace the oxygen that is used up.

Because oxygen is more soluble in cold water than in warm, more oxygen is dissolved in surface waters at higher latitudes. Seawater at 0°C may contain 50% more oxygen than seawater at 20°C (Table 7.10). Because deep-water circulation in the Pacific is more sluggish than it is in the Atlantic, the waters there contain less dissolved oxygen.

Carbon Dioxide

Water does more than just provide a medium in which life can exist. Its physical and chemical stability protects organisms from sudden change, and it provides physical support for organisms from diatoms to whales.

The sea is slightly basic and is buffered against changes in its basicity. Pure distilled water is neither acidic nor basic. It is neutral, with a pH of 7.0. As the pH moves from 6.99 to 1, water grows more acidic; as the pH moves from 7.01 to 14, water becomes more basic. The sea has various buffering systems that resist changes in pH even when an acid or a base enters it. The major buffering agent in the sea is carbon dioxide, which reacts with water to form carbonic acid:

$$CO_2 + H_2O \rightleftarrows \underset{\text{carbonic acid}}{H_2CO_3}$$

Carbonic acid is a weak acid, which means that in solution it dissociates only partially into ions. The dissociation of carbonic acid may be written as follows:

$$H_2CO_3 \rightleftarrows HCO_3^- + H^+ \rightleftarrows CO_3^{-2} + 2H^+$$

If we add acid to the solution, which will increase the number of H^+ ions present, the reaction will shift. The acid is taken up, and the pH remains constant. Figure 7.6 shows the relationships of the carbon dioxide equilibria.

This buffering mechanism provides a stable setting for marine organisms. If the pH of seawater varied appreciably, many organisms would die. For example, if the pH shifted even weakly to the acid range, mollusks, whose shells are made up largely of calcium carbonate and whose growth is most rapid in a basic solution, would eventually perish.

Carbon dioxide is a by-product of respiration. Generally, more oxygen than carbon dioxide is produced in surface waters. However, in deeper waters the reverse is true. The depth at which the production

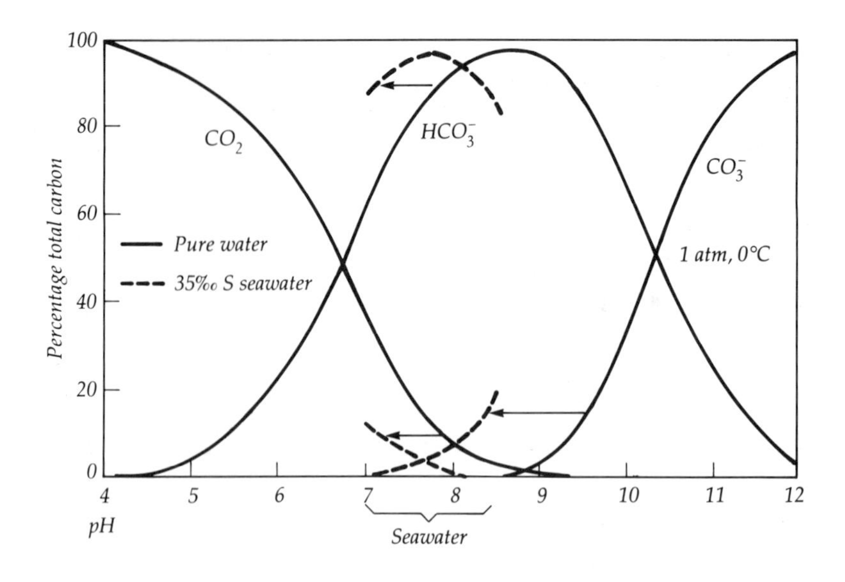

7.6 *The distribution of total inorganic carbon in the CO_2—HCO_3^{-1}—CO_3^{-2} equilibria as a function of pH at 1 atm pressure and 0°C.*

rate of carbon dioxide equals that of oxygen is the **compensation depth**. Generally, the oxygen concentration decreases beyond this depth, and the carbon dioxide concentration increases. At the compensation depth, oxygen production equals oxygen consumption. Below this depth no sustained plant population can exist. Exceptions occur when there is a slow influx of oxygen-rich waters at depths below the compensation depth. In many areas of restricted circulation, oxygen may be so depleted that anoxic conditions result (Chapter 16). The Black Sea and many fjords are examples.

Interestingly, the sea acts as a sink for the tremendous amounts of carbon dioxide released by the burning of fossil fuels. The carbon dioxide reacts with water molecules to form carbonic acid, forming a concentration gradient between the carbon dioxide in the atmosphere and that in the sea.

During certain periods in the earth's history, various factors have affected the amount of incoming sunlight. Ice caps, for instance, reflected more light during the glacial periods, but changes in the protective umbrella of the atmosphere have caused the greatest variations. The recent increase in carbon dioxide in the atmosphere is cause for concern. Some of the sunlight penetrating the atmosphere is absorbed by gases, and the energy given off is reradiated as heat energy. Carbon dioxide and water are largely responsible for this transfer. The same thing happens when an automobile is parked in the sun with the windows rolled up. The temperature inside becomes greater than that of the outside air. As the energy of the sunlight hits the interior of the car, it is absorbed and heat energy is given off. Since the closed windows prevent the heat energy from leaving at the same rate, the temperature rises. This phenomenon is referred to as the **greenhouse effect**, because a greenhouse receives and retains energy in the same manner.

The greenhouse effect is a major factor in determining the temperature of the earth's atmosphere. Because of the tremendous increase in the burning of fossil fuels, which adds carbon dioxide to the air, scientists are worried about the possibility of a large increase in the earth's temperature. In fact, variations in the carbon dioxide content of the atmosphere may explain the occurrence of past ice ages.

The amount of particulate matter in the atmosphere is also a source of alarm. As a result of the volcanic ash and dust pumped into the atmosphere in 1815 by a volcanic eruption of Mount Pambora in Sumbawa, Indonesia, there was no summer in the entire Northern Hemisphere. In Maine, for example, farmers tried to plant crops three times during the summer, only to have them all killed by frost. The tremendous amount of airborne matter reflected much more sunlight than normal. Likewise, by burning fossil fuels, we are rapidly increasing the amount of particles in the atmosphere. Every time a jet plane takes off, its engines emit dark trails of particulate carbon. Despite a marked increase in turbidity in the atmosphere—as much as 30% in some areas—so far few particulates have been unleashed into the sensitive upper atmosphere.

Although massive amounts of carbon dioxide have been released into the atmosphere since the onset of the industrial revolution, the amount in the atmosphere is actually quite small (about 0.03% by volume), and its atmospheric residence time is short, about 2 to 10 years. Where does it go? Into the ocean. The ocean now contains about sixty times as much carbon dioxide as the atmosphere. The residence time of carbon dioxide in the ocean is about 5 to 10 years, with most of it being incorporated into marine sediments. Even so, the amount of this gas in the atmosphere continues to increase steadily.

FURTHER READINGS

Broecker, W. S. *Chemical Oceanography*. New York: Harcourt Brace Jovanovich, 1974.

Davis, K. S., and J. S. Day. *Water: The Mirror of Science*. Garden City, N.Y.: Doubleday, 1961.

Gabianelli, V. J. "Water—The Fluid of Life." *Sea Frontiers*, 1970, *16*(5), 258–70.

Goldberg, E. D., and others, eds. *The Sea, Ideas and Observations on Progress in the Study of the Seas*. Vol. 6. New York: John Wiley, 1977, chaps. 17–19.

Hill, M. N. *The Sea: Composition of Sea Water*. New York: Interscience, 1963.

Horne, R. A. *Marine Chemistry*. New York: John Wiley, 1969.

Kuenen, P. H. *Realms of Water*. New York: Science Editions, 1963.

MacIntyre, F. "Why the Sea Is Salt." *Scientific American*, November 1970, pp. 104–15.

______ . "The Top Millimeter of the Ocean." *Scientific American*, May 1974, pp. 62–77.

Martin, D. F. *Marine Chemistry*. 2 vols. New York: Dekker, 1968.

Riley, J. P., and G. Skirrow. *Chemical Oceanography*. 2 vols. New York: Academic Press, 1965.

Strickland, J. D. H., and T. R. Parsons. "A Practical Handbook of Sea Water Analysis." Bulletin no. 167. Ottawa: Fisheries Research Board of Canada, 1968.

Wallace, W. J. *The Development of the Chlorinity/Salinity Concept in Oceanography*. Amsterdam: Elsevier, 1974.

Weyl, P. K. *Oceanography: An Introduction to the Marine Environment*. New York: John Wiley, 1970.

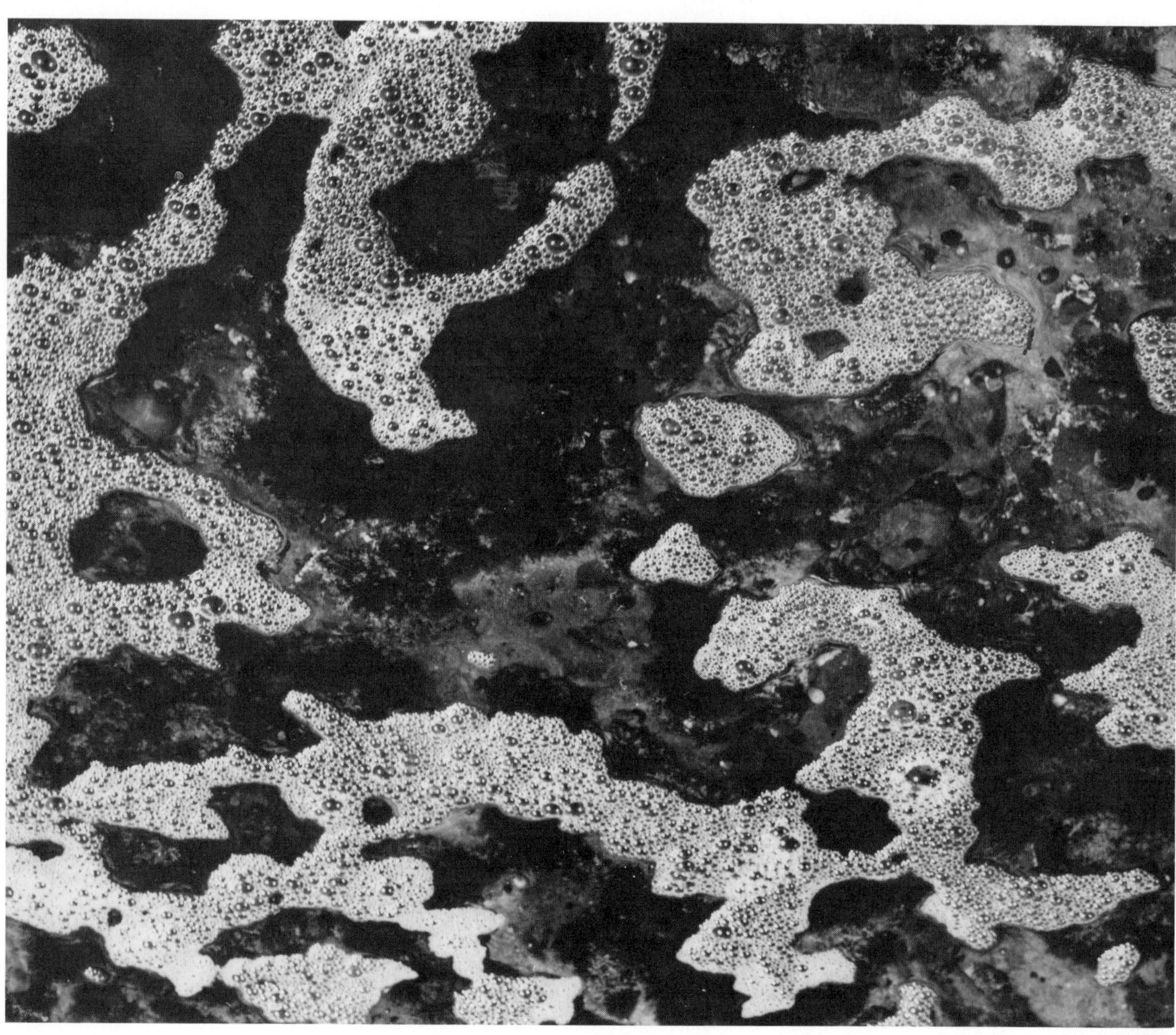

Dennis Brokaw

Everything flows; nothing remains. . . . One cannot step twice into the same river.

Heracleitus

EIGHT

The Physical Properties of Seawater

As a sensing creature, once any human has seen the ocean, he or she intrinsically knows something about it: its color, smell, transparency, and perhaps its refractive qualities. If a person has been inside the ocean, then the taste, temperature, and even density are familiar.

Of all the ocean's physical properties that oceanographers measure, temperature is the most important. One learns more about the water by knowing its temperature than by knowing any other single parameter.

TEMPERATURE DISTRIBUTION IN SEAWATER

Water has a very high specific heat, which means that it warms up and cools down very slowly. The surface waters of the ocean absorb a tremendous amount of incoming solar energy. Even when evaporation, which is a cooling process, is taken into consideration, surface waters are more effectively heated than cooled. While warmed water remains at the surface, chilled water sinks and is replaced by water from below.

The amount of solar energy incident on the earth's surface varies with the seasons, with latitude (Table 8.1), and with topography, so surface temperatures are not the same everywhere. Figure 8.1 gives the temperatures of surface waters for all the world's oceans in August and February.

Table 8.1 *Average Surface Temperature of the Oceans (°C)*

Latitude	Atlantic Ocean	Indian Ocean	Pacific Ocean
70–60°N	5.60	—	—
60–50	8.66	—	5.74
50–40	13.16	—	9.99
40–30	20.40	—	18.62
30–20	24.16	26.14	23.38
20–10	25.81	27.23	26.42
10–0	26.66	27.88	27.20
70–60°S	−1.30	−1.50	−1.30
60–50	1.76	1.63	5.00
50–40	8.68	8.67	11.16
40–30	16.90	17.00	16.98
30–20	21.20	22.53	21.53
20–10	23.16	22.85	25.11
10–0	25.18	27.41	26.01

From H. U. Sverdrup, Martin W. Johnson, and Richard H. Fleming, *The Oceans: Their Physics, Chemistry, and General Biology*, © 1942, renewed 1970. By permission of Prentice-Hall, Inc., Englewood Cliffs, N.J.

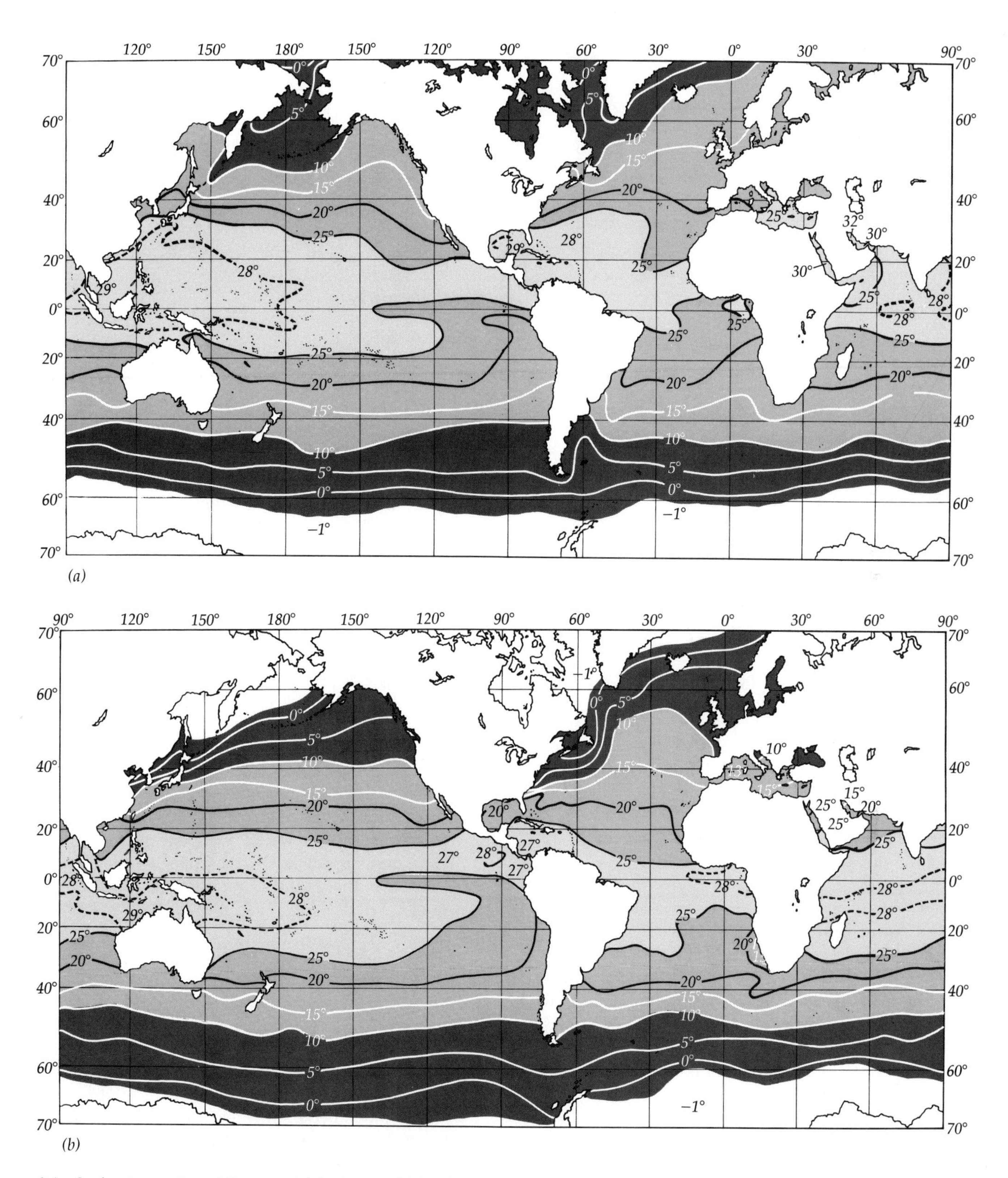

8.1 *Surface temperature of the oceans (a) in August, (b) in February.*

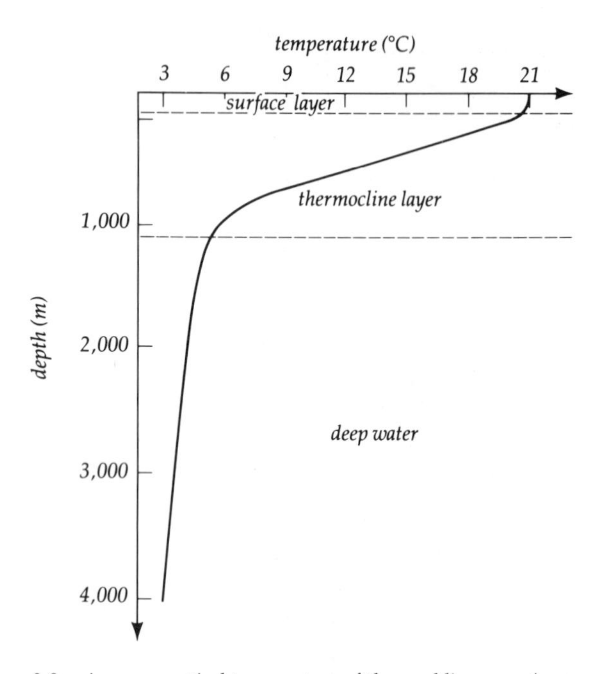

8.2 *Average vertical temperature of the world's oceans (away from the Arctic and Antarctic regions).*

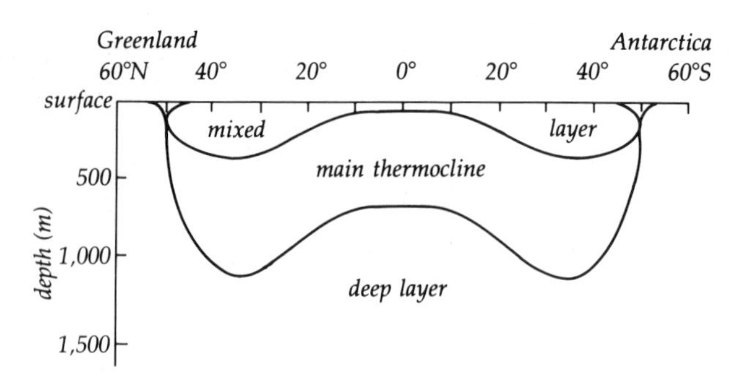

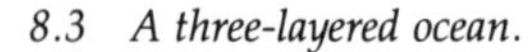
8.3 *A three-layered ocean.*

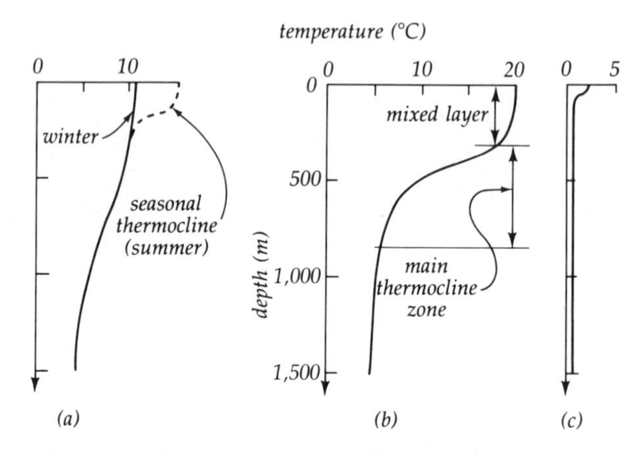

8.4 *Average temperature profiles for the open ocean at (a) middle latitudes, (b) low latitudes, and (c) high latitudes.*

A diver knows it gets colder and colder at increasing depths. This is generally true for all oceans. (One exception is the deep, salty region of the Red Sea, where temperatures approach 65°C [132.8°F] at 2,000 m [6,560 ft]. Heat flow, by conduction and convection, from a major rift zone seems to cause this variation.) As Figure 8.2 shows, there is a region where the temperature changes abruptly with depth. This region is called the **thermocline**.

The world ocean may be thought of as having three layers (Figure 8.3). In the deepest layer (below 200 to 1,000 m [656 to 3,280 ft]), the water temperature is uniformly quite cold. Although the total ocean temperature ranges from −2°C to 30°C, with depth the range narrows considerably; no other extensive region of the planet has as narrow a temperature range. A vertical temperature profile for the middle latitudes, where seasonal differences are pronounced, shows that the water warms up in summer; a temperature profile for the equator shows no change during the year. Although the regions near the poles (high latitudes) differ in actual temperature from regions on the equator, their profiles are similar in shape (Figure 8.4).

Recently it has been determined that temperature layering in the ocean is much more extensive than had been previously thought (Figure 8.5). Variations in temperature are sometimes referred to as **microlayering** (or *microstratification*), because many layers vary only a fraction of a degree from adjacent layers.

Differences in surface-water temperature and air temperature give rise to many phenomena. Fog is one; a somewhat more unusual occurrence is sea smoke, which may appear when the water temperature is greater than that of the surrounding air (Figure 8.6).

Measurements of Temperature

When a hydrocast is made, the temperature of the water column at each interval is recorded. The cast is left at depth for about ten minutes so that the thermometer can reach equilibrium with its surroundings. Each Nansen bottle carries two reversing thermometers mounted in a frame and attached to the outside of the cylinder. Because the temperature registered by a thermometer usually decreases as the thermometer descends and increases as it rises, these thermometers are designed to fix a temperature reading at the desired depth. When the messenger triggers the Nansen bottle to flip, the thermometers also flip. A reversing thermometer has a small capillary constriction so that when the bottle flips, part of the mercury column is separated, which "locks in" the thermometer reading (Figure 8.7a).

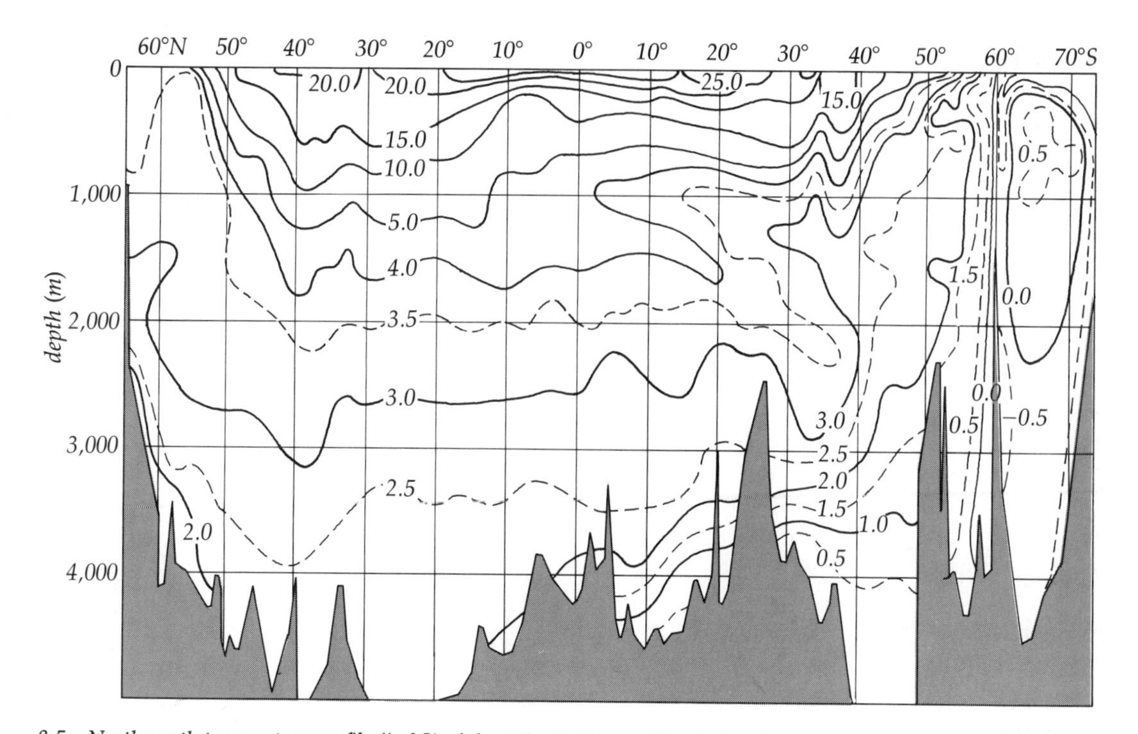

8.5 North-south temperature profile (in °C) of the Atlantic Ocean. The dark region represents a profile of the ocean floor, although the horizontal scale is compressed considerably.

8.6 Sea smoke. (Official U.S. Navy Photograph)

Of the two thermometers attached to the Nansen bottle housing, one is shielded against pressure, and the other is unprotected (Figure 8.7b). When a hydrocast is submerged, there is no way to ascertain that the cable is vertical; it might well have a kink or a bend in it, which could give a false depth recording. Thus, the true calculation of a sample depth is determined from the two thermometer readings. This is done as follows: Because the glass of the unprotected thermometer is squeezed in by the pressure of the surrounding water, its capillary size decreases. As a result, the mercury level rises, giving a higher temperature reading. By checking the difference between two different temperature readings against the known empirical relationship, the pressure to which the unprotected thermometer was subjected is calculated. From this pressure, the depth can be determined. To an oceanographer, thermometers that have proved reliable and accurate are treasured almost as friends and cared for accordingly. They are the most important part of a hydrocast.

Other methods are used to measure temperature. In the past the most common method was to use the **bathythermograph**, or **BT** (Figure 8.8). Temperature readings from the bathythermograph are not as accurate as those from a reversing thermometer. Generally, the BT registers only to a depth of about 270 m (886 ft), which is adequate for many purposes. A BT can be used while the ship is moving, although the maximum towing speed is about 19 km/h (10 knots). More recently an **expendable bathythermograph** (**XBT**) has become widely used (Figure 8.9). An XBT contains a temperature sensor and a spool of wire. After it is thrown over the side, the data are relayed electrically through the wire to the ship. The depth is determined by the device's known rate of sinking.

Still another method of determining temperature is the use of a chain of **thermistors**. A number of these electrical temperature sensors are attached to a cable.

The cable is towed, and a continuous reading of temperature throughout the depths is made (Figure 8.10). For shallow waters and ocean-surface studies, thermistors have become more common than the hydrocast, especially when they are coupled with conductivity and density sensors. Because long thermistor chains cost a great deal and are difficult to deploy, they are not used in studies of deep-sea regions.

But by far the most effective, far-reaching, and efficent method of measuring the sea surface temperature (SST) is by remote sensing (Figure 8.11). Unlike conventional temperature methods, SST maps produced by satellite imagery are for a very wide area for a short period of time. This means that SST maps for the entire ocean can be produced daily, and they can show short-term changes even over a large area, something not possible by traditional shipboard methods.

Presently there are about 4,000 ship-generated sea surface temperature reports daily. Satellites like the NOAA polar orbiters equipped with the Advanced Very High Resolution Radiometer (AVHRR) now provide between 30,000 to 70,000 SST data reports each day. The AVHRR, however, operates in the visible and near infrared channels and is therefore limited by cloud cover.

Within a very few years, if all goes according to schedule, three new satellites, the U.S. Navy's Remote Ocean Sensing Satellite (NROSS), the European Space Agency's first earth resources (ERS-1), and NASA's TOPEX, will provide a virtual data explosion in SST data, yielding about 650,000 daily data reports by 1990.

PRESSURE

Of all the characteristics of the sea, the only truly predictable one is pressure. Pressure increases with depth. For each 10 m (33 ft) further down, pressure rises by 1 atmosphere (atm). An atmosphere is the pressure (force per unit area) exerted on a surface at sea level by a column of air extending to the outer limit of the atmosphere (about 160 km [100 mi]). At sea level, the weight of such a column of air with a cross section of 6.45 cm^2 (1 in^2) is 6.67 kg (14.7 lb). It is equal in weight to a column of mercury 76 cm high, or a column of water about 10 m high. Thus, the pressure on the bottom of each container in Figure 8.12 is the same. Pressure is a function of the height of the water column on a unit area at the bottom. The unit area usually chosen in the metric system is 1 cm^2 (1.033 kg/cm^2 = 14.7 lb/in^2). From place to place at a particular depth, pressure varies slightly as a result of small variations in the density of the seawater above.

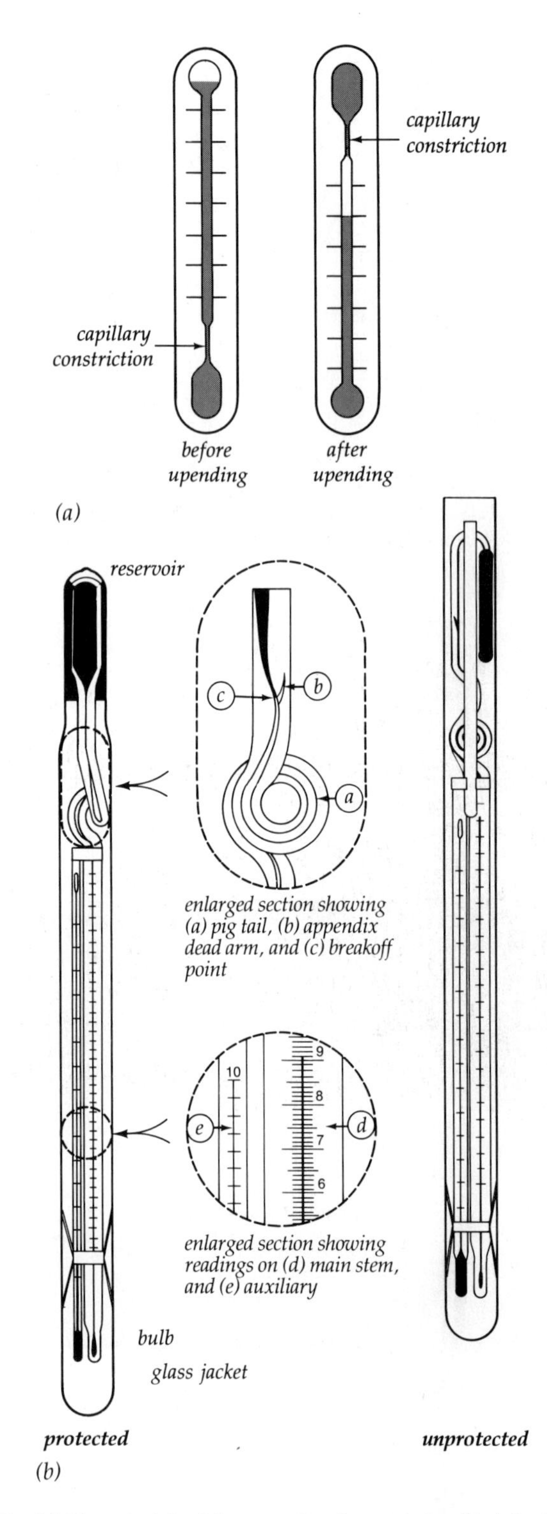

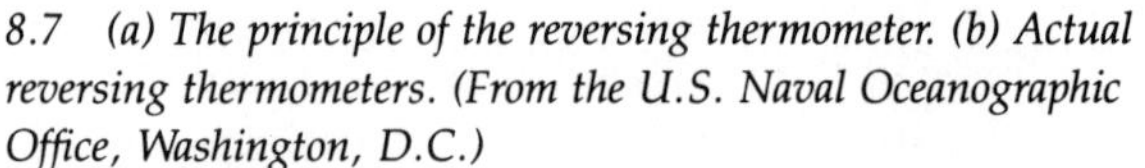
8.7 *(a) The principle of the reversing thermometer. (b) Actual reversing thermometers. (From the U.S. Naval Oceanographic Office, Washington, D.C.)*

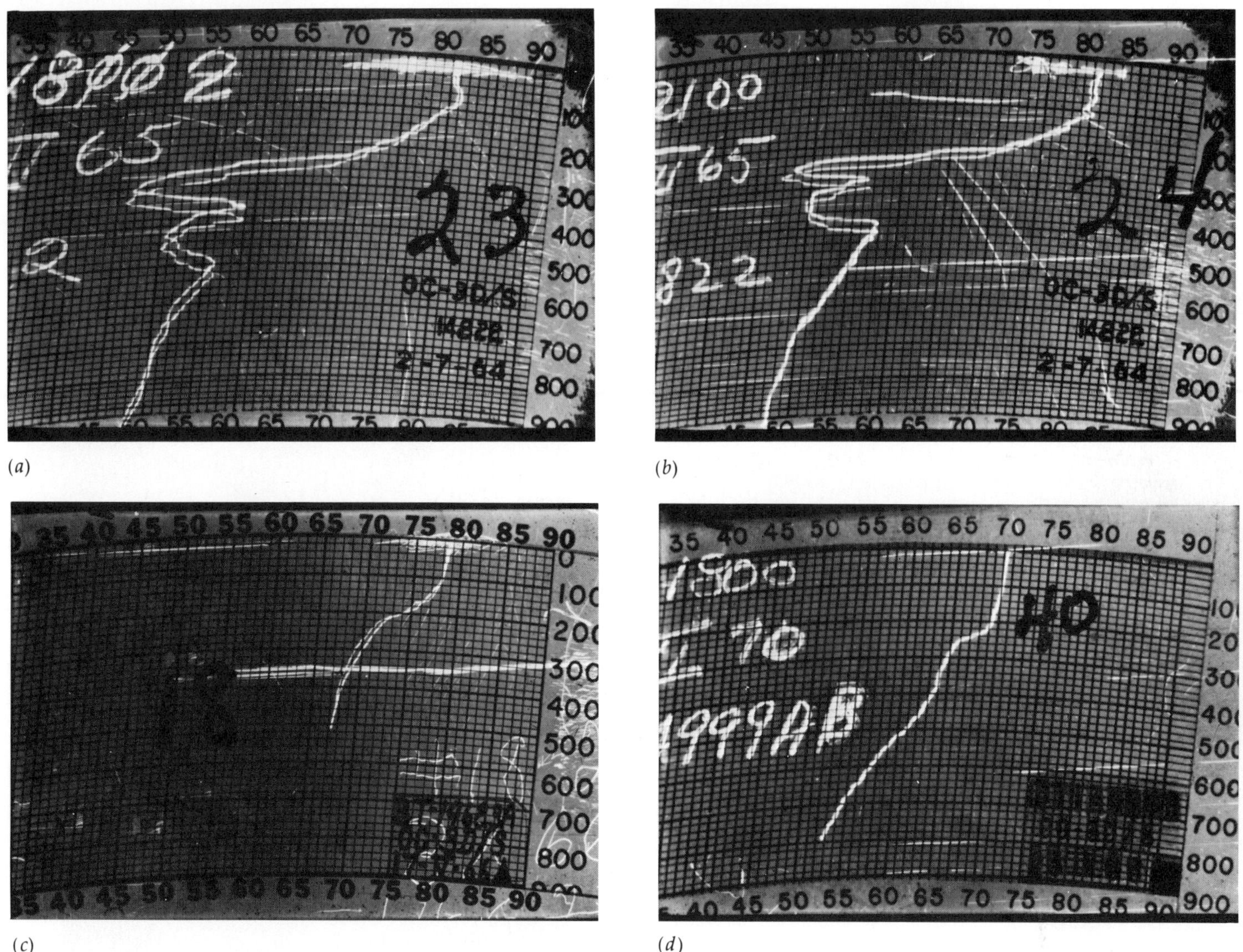

8.8 Representative BTs made on National Oceanographic Data Center cruises. (a, b) From Hudson Canyon in August 1965. (c) From the central Atlantic west of the Mid-Atlantic Ridge in October 1966. (d) From a Pacific fracture zone halfway between Honolulu and San Francisco in November 1970. Coordinates are depth (feet) and temperature (°F). (Scripps Institution of Oceanography)

Numerically, the pressure exerted at any depth below the sea's surface is a function of the weight of a column of water (with a unit cross-sectional area) from the surface to that depth. We can show this as:

$$\text{pressure} = \text{density} \times \text{depth} \times \text{acceleration due to gravity}$$

At the surface, the pressure exerted by the sea is, of course, zero, since only atmospheric pressure acts on the surface.

The deepest part of the ocean is 11 km (6.8 mi) deep. The pressure there is almost 1,000 times atmospheric pressure—1,000 kg/cm^2. (A pressure of just 94.87 kg/cm^2 would compress a block of wood to about half its volume, causing it to sink.) These great pressures are the main reason why submarines have not explored more of the deep ocean. Even so, in 1960 the U.S. Navy's research vessel *Trieste* descended 10,920 m (35,818 ft) to the bottom of the Challenger Deep in the Mariana Trench. The trip down took 4 hours, 48 minutes. But the *Trieste* is not capable of many dives; it is slow and purely research-oriented. Submarines of World War II had very shallow diving depths, seldom greater than 60 to 100 m (197 to 328 ft). Modern nuclear submarines can dive much deeper. Although scuba divers can descend to 180 m (590 ft), for all practical purposes they are limited to a working depth of about 60 m (197 ft) because of the restriction imposed by the amount of air that can be carried. A helmeted or hard-

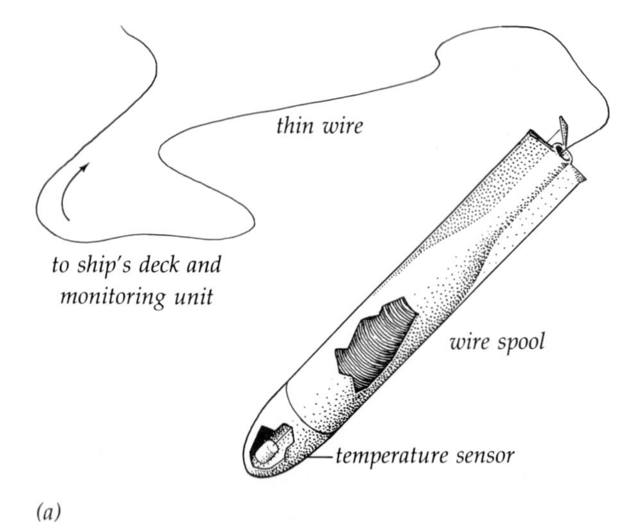

(a)

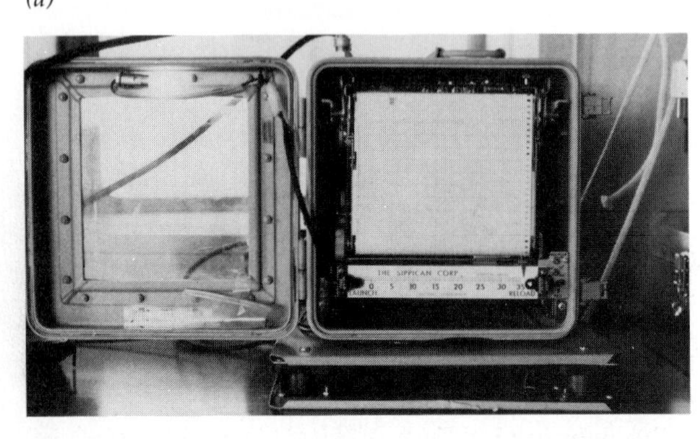

(b)

8.9 (a) An expendable bathythermograph, or XBT. XBTs are used to obtain a temperature-versus-depth profile of the top 500 m of the ocean. (b) The profile is drawn automatically on a recorder. After ninety seconds the recorder stops, the wire link to the XBT is snapped by the technician, and the XBT falls to the seafloor.

hat diver can normally work at 60 m (197 ft) for three to four hours.

Associated with deep dives is an affliction known as the **bends**. At high pressures, nitrogen dissolves in body tissue and in blood. If the rate of ascent is too rapid to allow for the normal, slow molecular dissolution of nitrogen, the gas will collect as bubbles in the blood, joints, bone marrow, and tissues, causing extreme pain. Usually the affliction is not fatal unless the dive has been very deep. In that case, if normal decompression procedures are not followed, bubbles can collect in the central nervous system tissue—the brain and spinal cord—and cause death. Should a diver fail to decompress adequately on the way up, he or she is put into an air pressure chamber in which the pressure is the same as at depth; then the pressure is slowly decreased.

In scuba diving, decompression is not necessary unless the diver goes below 10 m (33 ft), the pressure

8.10 Thermistor chain on a winch mounted on the afterdeck of U.S. Navy ship S. P. Lee. *(Official U.S. Navy photograph)*

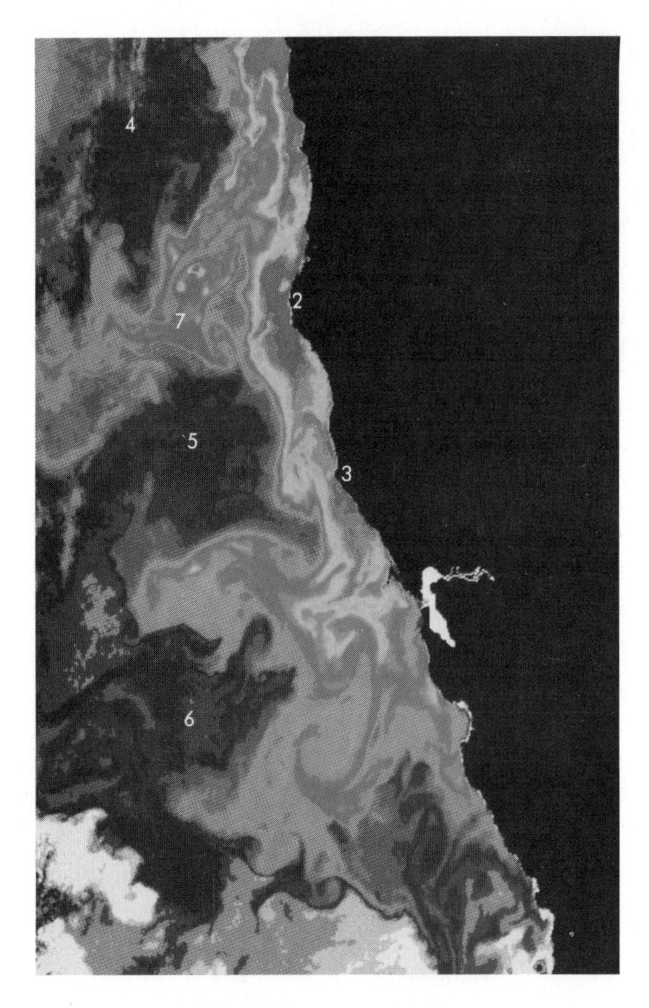

8.11 Ocean circulation patterns revealed by temperature image for 7 July 1981 showing the offshore movement of cool, productive, upwelled waters. (Thermal infrared [TIR] from NOAA)

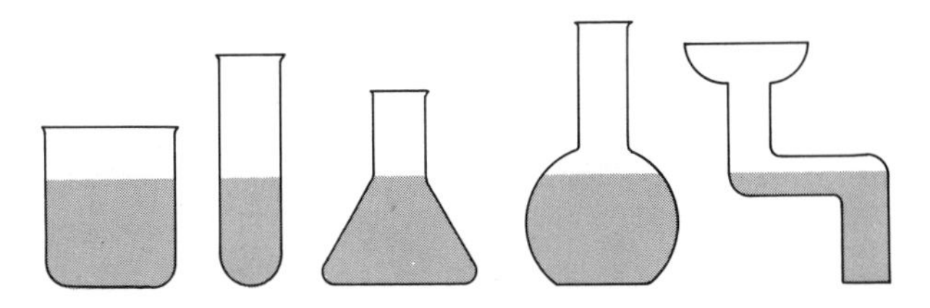

8.12 The pressure per unit of area is the same at the bottom of each container.

of 1 atm. For shallow dives a good rule to follow is never ascend faster than one's own air bubbles. In snorkling, dives are not usually deep enough to cause concern about pressure. Air of a different pressure is not taken into the system, so the bends cannot occur. Whales and other diving aquatic animals do not get the bends. Photographs taken of dolphins at depth show that their bodies are compressed to reduce the effects of internal gas pressure.

DENSITY

The terms *density* and *viscosity* are often confused. Of the substances that naturally occur as liquids, water—especially seawater—has one of the highest densities and a low viscosity. Oil, in contrast, is less dense than water yet has a higher viscosity.

Viscosity (Table 8.2) is the measure of a fluid's ability to flow. If a fluid flows easily, it is said to have a low viscosity. If it flows slowly, it has a high viscosity. This property is not a function of density but of the internal structure of the particles making up the substance and their attraction for one another. Temperature is often a factor (remember how molasses moves in January). An old pane of glass is not optically clear, as a new one is; it shows distortion. Measurement of the pane thickness at top and bottom would show that the bottom is thicker than the top. This is because glass, which is technically a supercooled liquid, has an extremely high viscosity, and because of gravity, it moves very, very slowly "downhill" over time.

Density is defined as mass per unit volume, or $D = m/v$. Density is simply a measure of the amount of matter in a particular volume. The terms *weight* and *mass* are commonly used interchangeably. Although scientists knew the difference, it was not until the advent of satellites and space travel that it really became necessary to differentiate between the two. Any object, such as a half-dollar, is made up of a certain number of atoms or molecules. The sum of these particles is its mass. The half-dollar has weight. You can feel it in your hand. If the half-dollar were orbiting 1,000 km (625 mi) above the earth's surface, though, it would be weightless. It would still have the same number of particles, so its mass would not have changed. Weight is only a measure of the attraction a gravitational field (such as the earth's) has on matter. The same half-dollar would weigh one-sixth as much on the moon as it does on earth, because the moon's gravitational field is only one-sixth that of the earth's.

The value for gravitational acceleration on the earth, normally considered to be 980 cm/s^2 (32 ft/s^2), varies from place to place on the earth's surface, but only slightly. If, say, feathers and lead are at the same place, the old question "Which weighs more, a ton of feathers or a ton of lead?" is silly. The mass is the same, but the volume occupied by the feathers would be the size of a room, and that of the lead might be roughly the size of a hassock. The density of feathers, then, is much less than that of lead.

Suppose you weigh a flask and leave the stopper sitting on the balance pan; the stopper, too, will be weighed. Now suppose you put the stopper into the flask to seal it. Is there a change in total weight? Or assume that the mass of an airplane in flight can somehow be determined from the ground. If a bird sat on the wing of the plane, would the weight of the plane increase? If a bird were sitting on a chair inside a closed, pressurized jet, would the plane weigh more or less than if the bird were flying around in the cabin? All these questions are similar. In the first example, there is no change in the mass of the bottle and stopper. By using an air pump to evacuate the flask, you could determine the density of air by weighing the absence of the air. Similarly, even if you used a scale that could function in deep water, you can't weigh water in water. Archimedes' principle states that any object in a fluid is buoyed by a force equal to the weight of the displaced fluid. In water, a volume of water is buoyed by a force equal to the weight of the displaced water; it has zero weight. Water can be weighed in air, however, because the weights are different. Technically, the weight of water in air is slightly less than the weight of water in a vacuum, because in air the water is buoyed to some extent by a force equal to the weight of the displaced air.

With the distinction between viscosity and density in mind, let's see what these characteristics mean in terms of the ocean. The high density and low viscosity of water make it easy to travel over the sea's surface. Heavily loaded ships float high in the water because of water's high density, and they move easily through the water because of its low viscosity. If the viscosity were greater than it is, the job of pushing a ship through the water would be much more expensive and difficult. Yet these two factors also produce a danger to ships and shore. Because of water's low viscosity, waves can build up to extraordinary heights and exert crushing forces on ships, piers, and coasts. If the ocean were as viscous and syrupy as molasses,

Table 8.2 *Properties of Water*

Property	Remarks	Importance to the Ocean Environment
Physical state	Only substance occurring naturally in all three phases as solid, liquid, and gas (or vapor) on the earth's surface	Transfer of heat between ocean and atmosphere
Quantity present on the earth's surface	Three times as abundant as all other substances combined	Massive abundance forms the oceans
Dissolving ability	Dissolves more substances in greater quantities than any other common liquid	Obvious in chemical, physical, and biological processes
Density: mass per unit volume Specific gravity: ratio of the density of a substance to the density of pure water at 4°C (in the mks and cgs system, the two terms density and specific gravity have the same numerical value)	Density determined by (1) temperature, (2) salinity, and (3) pressure, in that order of importance. The temperature of maximum density for pure water is 4°C. For seawater, the freezing point decreases with increasing salinity. For water with salinities higher than 24.70‰, the temperature of maximum density is below that of the initial freezing point.	Controls oceanic vertical circulation, aids in heat distribution, allows seasonal stratification, and tends to prolong lake winter freeze-up
Surface tension	Highest of all common liquids	Controls drop formation in rain and clouds; important in cell physiology
Conduction of heat	Highest of all common liquids except mercury	Important on the small scale, especially on cellular level
Heat capacity: quantity of heat required to raise the temperature of 1 g of a substance 1°C	Highest of all common solids and liquids	Prevents extreme range in the earth's temperatures; great heat moderator
Latent heat of fusion: quantity of heat gained or lost per unit mass by a substance changing from a solid to a liquid or a liquid to a solid phase without an accompanying rise in temperature	Highest of all common liquids and most solids	Thermostatic heat regulating effect due to the release of heat on freezing and absorption on melting
Latent heat of vaporization: quantity of heat gained or lost per unit mass by a substance changing from a liquid to a gas or a gas to a liquid phase without an increase in temperature	Highest of all common substances	Immense importance: a major factor in the transfer of heat in and between ocean and atmosphere
Viscosity	Relatively low for a liquid (decreases with increasing temperature)	Low internal friction allows for dampening of sea motions
Refractive index	Increases with increasing salinity and decreases with increasing temperature	Objects appear closer than in air
Transparency	Relatively great for visible light	Absorption high for infrared and ultraviolet; important in photosynthesis (with depth)
Sound transmission	Good compared with other fluids	Allows sonar and precision depth recorders to rapidly determine water depth, detect subsurface features and animals; sounds can be heard great distances underwater
Compressibility	Only slightly	Density changes only slightly with pressure/depth
Boiling and melting points	Usually high	Allows water to exist as a liquid on most of the earth

there would be virtually no waves, even at the shoreline.

The high density of water in a gravitational field is responsible for the tremendous pressure at great depths. Oceanographers express the density of seawater in metric units. By definition, pure water has a density of 1.0 g/mL at 4°C and 101.3 kPa (1 atm). Seawater, being more dense because of its dissolved solids, has a density between 1.02 and 1.03 g/mL. This is not a great variation. The density of pure water is a function of temperature and pressure, but in the oceans there is the additional effect of salinity. Temperature, salinity, and pressure determine, to a decreasing extent, respectively, the density of seawater. Even though water is not very compressible, pressure is still a factor at depths.

How is density measured? The simplest method is to throw a bucket over the side of a ship, bring up some water, weigh it, measure its volume, and calculate the density at some known temperature. Because water (like other substances) changes its volume as the temperature changes, its density also changes with temperature. To be meaningful, density must be expressed at a particular temperature. This method will work for surface waters, but what about subsurface waters? Once subsurface water is raised, its pressure and density will change even if its temperature is kept constant. Only the salinity of the sample will remain unchanged.

Several new methods show promise in determining the actual in situ density of ocean water, but the most commonly used method is the calculation of density. To do this, you must know the temperature, salinity, and depth or pressure. Tables have been constructed to aid in this determination.

As a matter of convenience, it has become customary for marine scientists to abbreviate density values. Since the actual density values of seawater always begin with 1.0 . . . and usually lie between 1.02400 and 1.03000 g/cm^3, the 1.0 . . . part has been dropped and the decimal point moved to the right three places. Thus, a density of 1.02475 becomes 24.75, which is called sigma *t* by oceanographers. The graphs in Figure 8.13 show the individual relationships of salinity and temperature to density.

Why is it necessary to determine density when it is obvious that density must generally increase with depth? Density does increase with depth, but the increase is not necessarily uniform. The heaviest or densest waters are found at the ocean bottom. The lightest or least dense waters are found at sea level. In some regions of the deep sea, the density is constant for some depth. But in other areas, called **pycnoclines**, density increases rapidly with a small change in depth. Such regions are quite stable. A water column is stable (that is, it has little tendency to mix vertically) if less dense water is above denser water and there is a sharp gradation in density. In some cases, less dense water may lie below denser water, which makes the water column unstable. When density is the same over a wide vertical region, the region has neutral stability. Two masses of water having the same density can have different temperatures and salinities. Consequently, it is not unusual to find warm waters below cold waters or less saline waters below more saline waters. But if the densities are the same, the water structure will still be stable. A region with a clearly defined pycnocline is the most stable of all (see Figure 8.14).

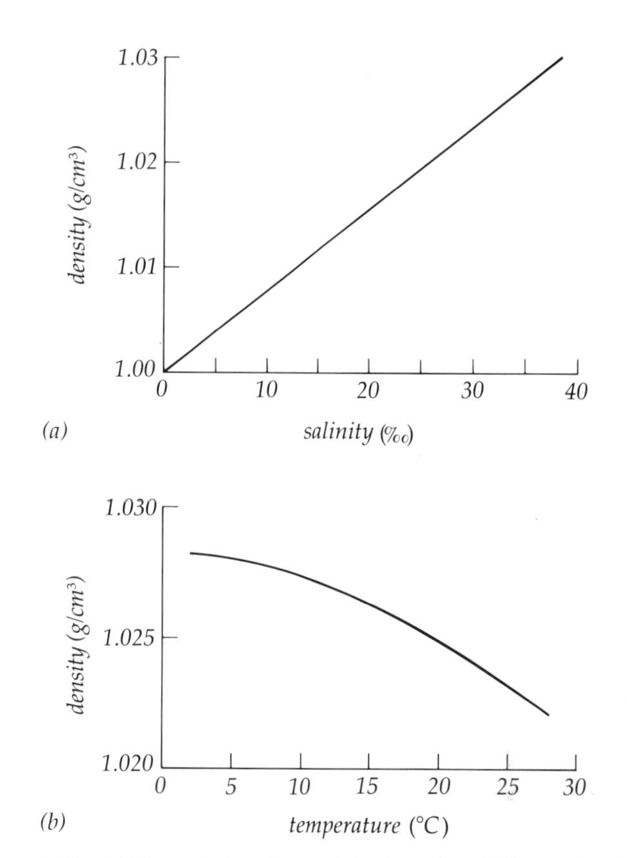

8.13 (a) The variation of water's density with salinity at a pressure of 101.3 kPa (1 atm). (b) The variation of the density of seawater (35‰) with temperature at 101.3 kPa (1 atm).

It is important to realize that density is a true property of water, whereas temperature, for example, is a *condition* or *state parameter.*

The primary reason for going to all the trouble of determining density and its stratification is to study deep currents. Because of the low viscosity of water, the ocean responds to small changes in the distribution of mass. In the deep ocean, waters sink until they reach layers of equal density and then move along these density layers or discontinuities. The study of subsurface currents is based largely on the density differences of deep waters.

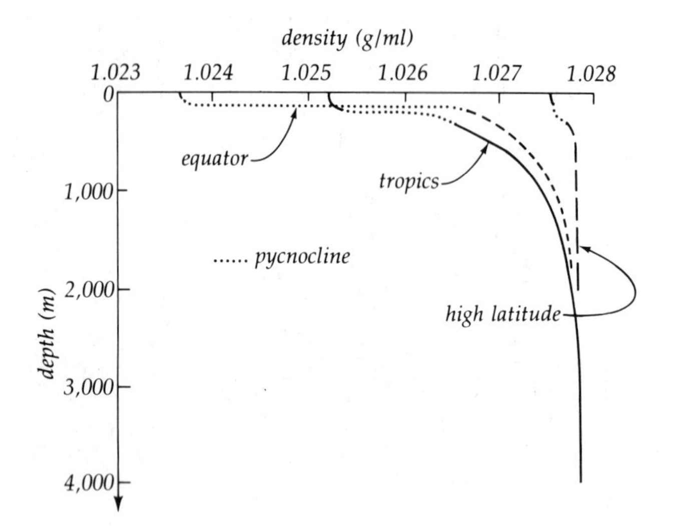

8.14 *Typical density profiles of ocean water.*

LIGHT AND COLOR

The world's oceans and seas have a variety of colors. Only about 10% of the ocean has opaque, dark blue-green waters. Clear azure blue waters are found in warm regions. The Black Sea looks dark because its bottom is covered with black sediment; the Yellow Sea off northern Siberia contains river-derived yellow mud, especially during flood season. The Red Sea and the Gulf of California (formerly called the Vermilion Sea) appear red because of the presence of blue-green algae, which have an overall reddish color despite their name. The White Sea owes its name to the fact that it is frozen solid more than 200 days of the year. The Kuroshio Current, sometimes called the Japan Current (not altogether correctly), is dark or colorless—hence the name, which in Japanese means "black stream."

Color is a function of the visible-light portion of the electromagnetic spectrum (see Figure 9.2, p. 126). The color of a given seawater surface may change as it is observed because a cloud passes overhead or the angle of the sun changes. The sea in general is blue for the same reason that the sky is blue. Blue light, which has a short wavelength (compared with red), is most easily scattered by water particles and by the microscopic matter that water contains, just as particles in the air scatter blue light. In total, the color of the sea is a function of light scattering through suspended particles, the reflection of the sky's color, and the nature of the suspended and dissolved matter in the water. All incoming light originates in the sun (Chapter 9); a small amount of light emanates from **bioluminescence,** which is also called phosphorescence.

Table 8.3 *Penetration of Light through the Sea Surface*

Depth (m)	Clearest Ocean Water	Turbid Coastal Water
0	100%	100%
1	45	18
2	39	8
10	22	0
50	5	0
100	0.5	0

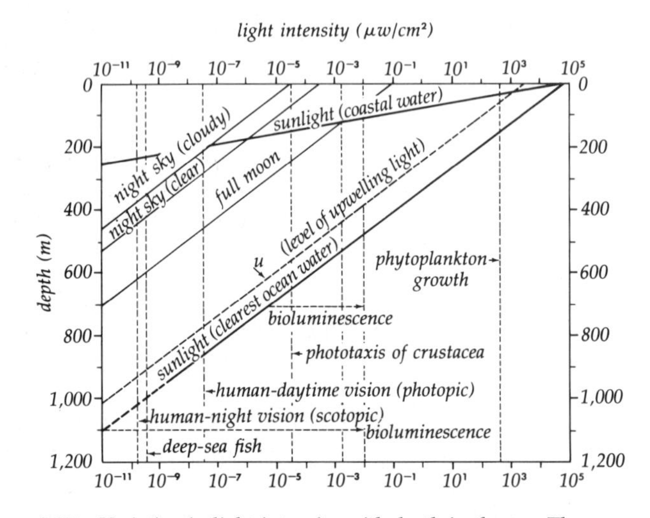

8.15 *Variation in light intensity with depth in the sea. The diagonal lines represent the penetration of sunlight and moonlight into the sea. The vertical dashed lines represent the light intensity limits for certain phenomena, such as the growth of phytoplankton (about 10^3 μw/cm²). The horizontal dashed lines denote the range for phenomena such as bioluminescence and human color vision. (After George L. Clarke)*

Light in the Sea

Water is not totally transparent to light from the sun, despite appearances to the contrary. It's just that the light intensity permeating a glass of water is so great that the amount of light absorbed is too small to notice. In the ocean, absorption is far more pronounced, and there is ultimately darkness at depth. How far down does light penetrate? Whether the water in a given region is clear or turbid has some bearing on the question.

Microscopic photosynthetic organisms are good indicators of light penetration. They can exist only in the **euphotic zone**: a vertical region extending from the ocean's surface to the depth at which only about 1% of the light still penetrates. In such extremely clear waters as the Mediterranean, the Caribbean, and especially the open ocean, the euphotic zone may be as deep as 100 to 160 m (328 to 525 ft). Maximum light

(*a*)

(*b*)

8.16 An example of refraction. (a) The dish contains a penny, but it is not visible because of the angle at which the photograph was taken. (b) Neither the dish nor the penny has been moved, but the dish has been filled with water. The refraction of light as it enters and leaves the water renders the penny visible. (W. J. Wallace)

penetration may be as low as 15 m (49 ft) near rugged coasts. However, even though coastal waters are not as clear, they are among the most productive regions on earth because of the movement of colder waters to the surface and the subsequent enrichment of the euphotic zone with nutrients (see Chapter 16).

When you descend into the ocean, the white light changes rather abruptly to blue-green. If you descend slowly, you may notice more yellow at first, which rapidly gives way to the blue-green. The red end of the spectrum is absorbed so fast by the water that it is not noticeable beneath the surface. Why, then, you might ask, can brilliant orange and red fish be seen at depths greater than a meter or so, where red light never penetrates? Such fish have pigments in their scales that absorb blue-green wavelengths of light and reemit them as wavelengths of red or orange light. The color of these fish cannot be seen at a distance underwater.

How the Sea Looks from Within

Many beginning underwater photographers have been surprised to find that below 10 m (about 33 ft), the red numbers on their cameras appear black. Of course, this is because red light is the first to be absorbed. Past 33 m (108 ft), distinguishing colors is impossible. In addition, there are no shadows; light seems to come from all around rather than from any single source. At 100 m (330 ft), visibility is limited to about arm's length. At 305 m (1,000 ft), the sea is very, very dark. Normally, light intensity decreases steadily with depth (Table 8.3 and Figure 8.15). The deeper you go, the shorter your horizontal visibility gets. In some regions, though, greater amounts of suspended particles near the surface actually enhance horizontal visibility at depth, especially if you are carrying a light. Where waters are turbid, visibility decreases vertically for awhile, then increases below the region of turbidity. It slowly decreases after that until, at about 300 m (984 ft), normal lighting barely illuminates objects only a short distance away.

Visual distortions occur underwater. When you look at a fish through the glass of an aquarium, it appears closer than it really is; the image is magnified. The effect is even greater when you look at the fish out of the corner of your eye (that is, with your peripheral vision). Similarly, when you swim underwater, objects look closer than they really are. The cause is refraction and is quantified by the **index of refraction**. For any substance, the index of refraction is the ratio of the velocity of light in a vacuum to its velocity in the substance itself. Since the speed of light in water is less than it is in air, light rays bend (are refracted) at a different angle at the interface of the two media (Figure 8.16). Light striking objects underwater makes them appear to be only three-fourths of their actual distance away. Thus, a pier piling 12 m (39.4 ft) distant would appear to be only 9 m (29.5 ft) away.

Water's refractive index is one reason why it is difficult to see through the sea surface. Also, much of the light available at the air-sea interface is reflected from the water's surface. Some fishers wear polarized sunglasses, which reduce the glare of the reflected light, to help them spot fish in the water.

If the water surface were extremely placid, what would you see if you held yourself just below the surface and looked straight up? First, you might see a reflection, because the sea surface has two sides. Light reflects off the top side, but it also reflects off the bottom side, much like a mirror. So you may see yourself, as you do in a mirror. If you were only a few meters below the surface, you might see a circle of light above you, as if you were looking up through a manhole. In that circle, you would have a 360° distorted panorama of the total outside area, much as if it were photographed through a fish-eye camera lens. You might see a boat coming across the surface, but since it would be at the edge of the circle, its image would be greatly distorted. This effect is seldom seen except in pools or small ponds where the water surface is almost mirrorlike. The panorama closes in as you descend from the surface, and it eventually vanishes.

SOUND IN THE SEA

Sound is a form of energy that is transported from particle to particle in the air. Unlike light, sound depends on some medium for its transmission. You can demonstrate this dependency by placing an electronic device such as a buzzer in a bell jar connected to a vacuum pump. The level of the buzzer's sound will remain much the same until the pressure level is low. Then the sound will diminish and finally vanish because there are no longer enough air particles to transmit the sound, even though the buzzer is still emitting mechanical energy.

A drumhead, such as the human eardrum, will vibrate when struck. As it does, it constantly strikes billions upon billions of air particles (about 99% nitrogen and oxygen molecules). These molecules in turn bounce into other molecules, which cause the sound energy to be transmitted. Small pressure variations in the air are created as the drumhead pushes into the air particles (thus increasing the density) and as it moves back (decreasing the density). The smallest sound in air that can be heard by the average human ear is about 2×10^{-10} atm of pressure. The loudest sound that the human ear can experience without pain is about 2×10^{-3} atm of pressure.

Sound Velocity in the Sea

Velocity is the rate of motion in a fixed direction through time. Because the type and number of particles in a volume may vary, sound moves with different velocities through different substances. In dry air at 20°C, the speed of sound is 344 m/s (1,128 ft/s). (Would sound move faster in wet air or dry air?) In aluminum, which is much denser than water, sound moves at 5,104 m/s. The speed of sound in water is between 1,400 and 1,550 m/s (4,590 to 5,084 ft/s). Because seawater is slightly denser than fresh water, sound travels through seawater at a greater velocity. However, this is only an approximation, and there is no precise direct relationship between the densities of substances and the speed of sound through them. The presence of certain salts, especially magnesium sulfate, affects the velocity of sound. Moreover, the constituent salts in a vertical column of seawater are not exactly the same throughout; for instance, the amount of calcium ions is slightly greater in deep water than at the surface.

The velocity of sound in seawater depends on salinity, temperature, and pressure. For water with a salinity of 34.85‰ and a temperature of 0°C, the velocity would be 1,445 m/s. Increasing the salinity by 1‰ will increase the velocity by 1.5 m/s. Raising the temperature 1°C will increase it by 4 m/s, and a 1,000-m increase in depth will increase it by about 18 m/s. Knowledge of all three of these factors is necessary in order to specify the velocity of sound at a particular depth (Figure 8.17). Since seasonal changes and latitude affect both salinity and temperature, velocity will vary accordingly (Figure 8.18). The sources of the sea's "background" sounds are given in Figure 8.18.

Sonar

Most of what we know about the topography of the ocean bottom has been gathered by **sonar** (*so*und *na*vigation *r*anging). Sonar is a collective term describing either active or passive systems. Passive sonar, a system for listening to noises in the sea, is used to determine the presence and relative direction of a sound source, be it a shrimp or a submarine. Active sonar (originally called an echo sounder) was invented by August Hayes in the early 1920s. It is what produces the pings that bring sweat to the brows of submariners in war movies.

In principle, active sonar operates much like radar except that it uses sound waves rather than radio waves. The sonar gear (sometimes called a precision depth recorder, or PDR, to signify its great accuracy although PDR is only vertical, whereas sonar can be directional) sends out a sound impulse, or a ping, which radiates in all directions. When the sound hits an object or the ocean bottom, it reflects and travels back to the vessel, where it is picked up on listening gear. Because sound travels at an average speed of 1,500 m/s, the depth is equal to one-half the time it takes for the impulse to leave and return multiplied by 1,500 m/s. In this way, the location of a submarine or a school of fish can be pinpointed. People in a submarine can actually hear the sound impulse bounce off the pressure hull of their underwater vessel.

A ship equipped with sonar can chart a continuous ocean bottom profile as it proceeds normally on

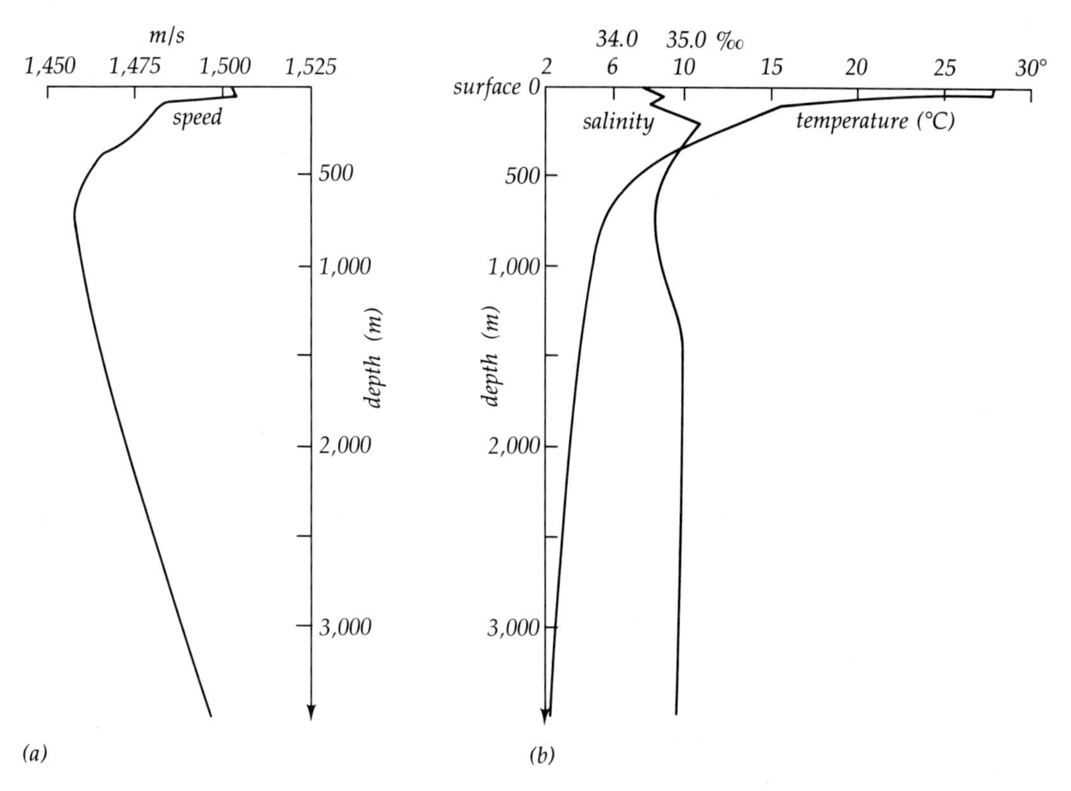

8.17 *(a) Speed of sound versus depth. (b) Salinity and temperature versus depth.*

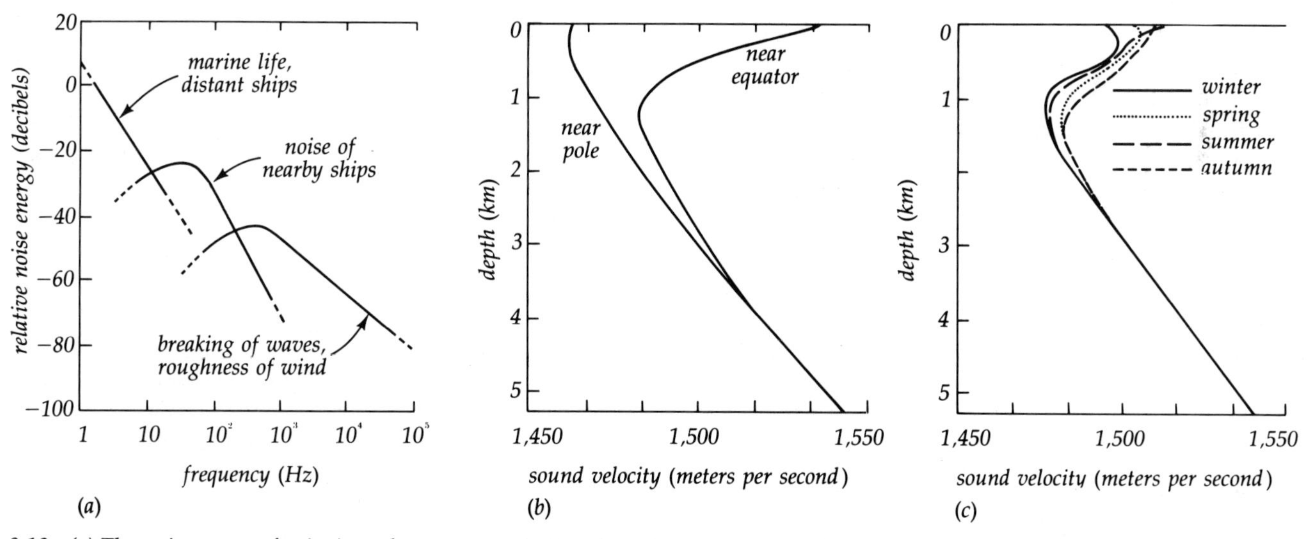

8.18 *(a) The main sources of noise in undersea communication channels may be grouped into three classes according to frequency ranges. Ships passing at a distance and marine life are the two significant factors below 10 Hz (cycles per second). Nearby shipping greatly affects the range from 10 to about 300 Hz. Above this value, the major noise sources are waves, which, of course, depend on the wind. (b) In the ocean, the speed of sound near the surface depends to a large extent on water temperature. Near the equator (near 19°N), the minimum speed is attained at a much greater depth than near the polar regions (near 61°N and S latitude). From the minimum value, however, both speeds increase almost linearly with depth, regardless of geographic location. (c) Seasonal speed fluctuations for an area in the North Atlantic Ocean are more dominant near the surface. Smaller speed changes can also occur daily and even hourly, depending on local weather. (Copyright 1969, Bell Laboratories, Inc. Reprinted by permission, Editor, Bell Laboratories* Record*)*

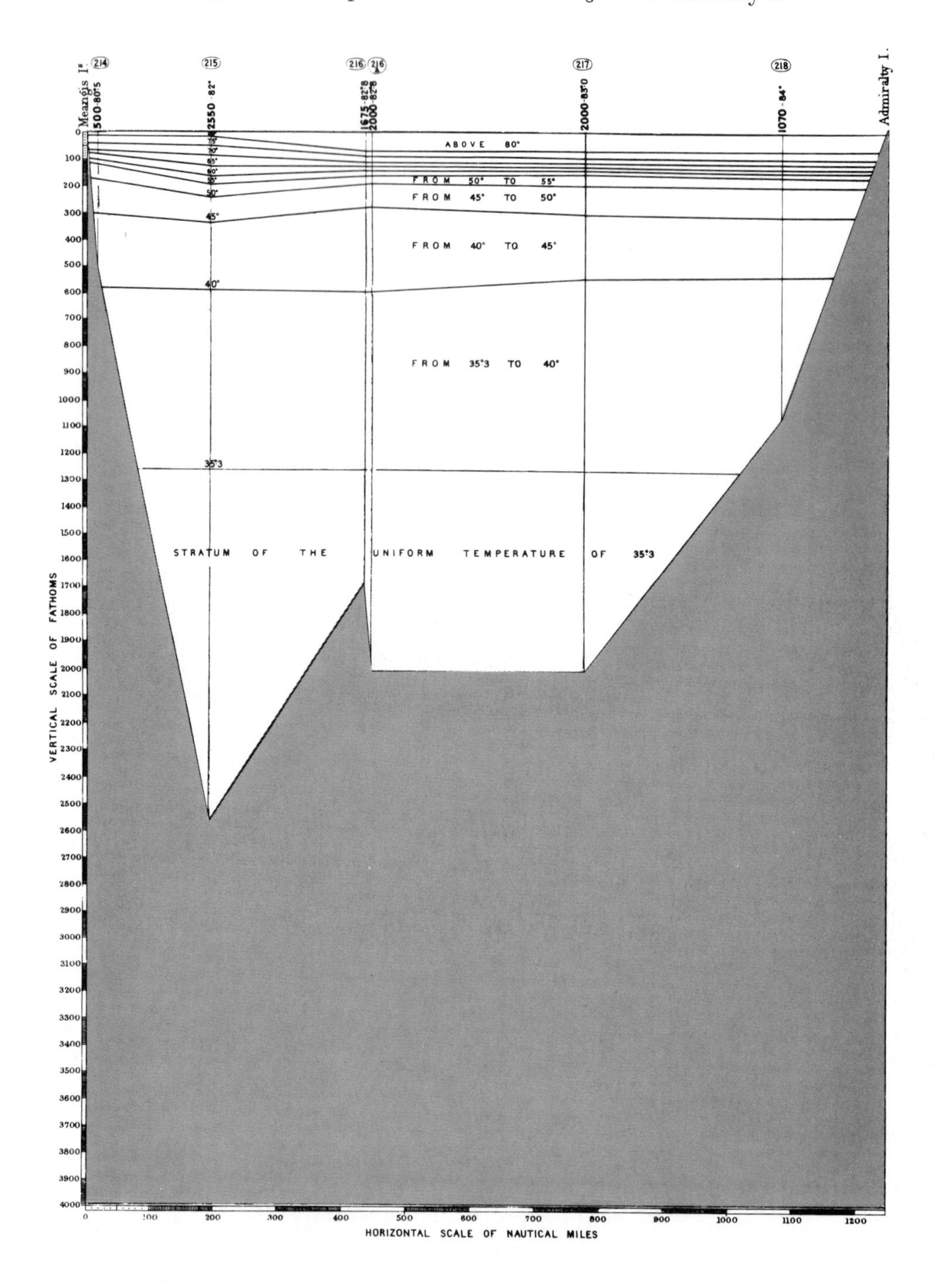

8.19 A lead-line bottom profile of a segment of the Pacific as determined during the voyage of HMS Challenger. *Because the soundings were long and tedious, they could only be made periodically. Notice that this does not give a good idea of bottom topography.*

its course. The countless profiles recorded by oceanographic, naval, and merchant vessels have made it possible to map the ocean's depth throughout much of the world. Without sonar, water depth recordings would be scarce and tedious to make. One sounding with a weighted line in water several kilometers deep takes three to four hours and requires tons of wire or wet hemp rope. Before the invention of sonar, bottom contour charts were generally misleading (Figures 8.19 and 8.20).

Shadow Zones and SOFAR Channels

One problem with sonar is that the speed of sound may differ from the calibrated speed at which the sonar gear is set, although this fact is important enough to

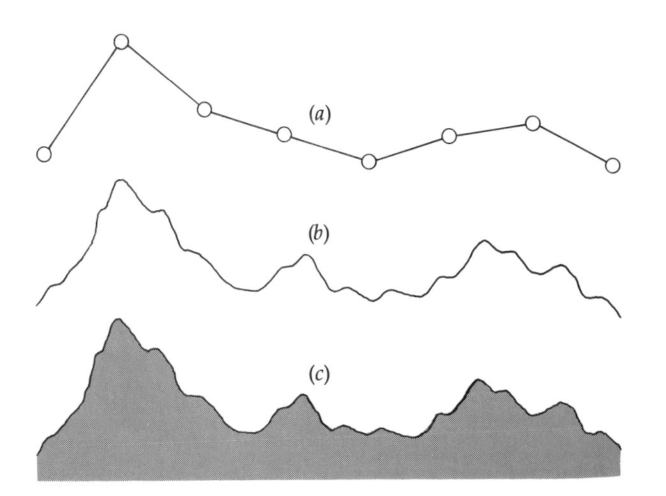

8.20 *Bottom contour. (a) Lead-line sounding, (b) sonar (PDR), and (c) actual contour.*

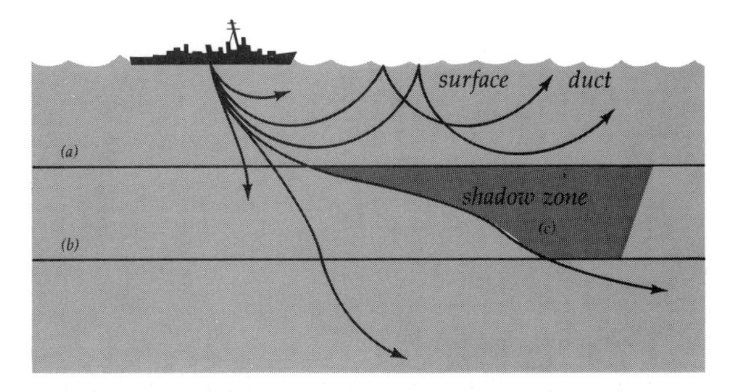

8.21 *Sound surface ducts (a) and (b) and shadow zone (c) caused by the bending and reflection of sound fronts at layers of different densities.*

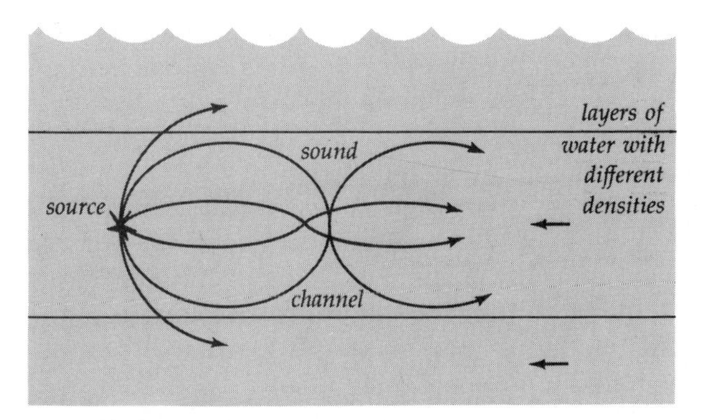

8.22 *SOFAR channel.*

require corrections in only very precise depth measurements. However, there is a problem of another sort. As sound energy moves away from its source, it decreases in intensity. This decrease, known as **propagation loss**, is the cumulative effect of energy absorption, scattering, and spreading. Some of the sound energy is absorbed by water particles and some is lost as the sound spreads out, usually in a spherical front but sometimes in a cylindrical front. The amount of energy per surface area decreases with distance from the source because the same amount of energy covers a wider front. Scattering occurs when anything in the water with different sonic properties reflects the sound energy. Fish, suspended inorganic particles, and especially bubbles of gases such as carbon dioxide can cause scattering.

Most of the world's oceans, except for the Arctic, the Antarctic (or Southern), and possibly parts of the central Pacific, have a region of depth that reflects or scatters sound. This is often called the **deep scattering layer (DSL).** Some regions may have more than one such layer. On sonar gear, the DSL may be thick enough to produce a false bottom. The DSL is a layer of living organisms, ranging from almost microscopic zooplankton to copepods, shrimp, siphonophores, fish, and squid. These organisms migrate vertically as layers in the water column, depending on light intensity. They occur at depths between 200 and 800 m (650 to 2,625 ft) during the day, with most concentrated between 310 and 460 m (1,017 to 1,509 ft); they occur close to the surface at night. A DSL is at its greatest depth at noon on a cloudless day.

Another hindrance to sonar operation is the existence of different density layers in the water column. When sound impulses strike these layers, some of the wavelengths bend and penetrate the layers, while some are reflected back (Figure 8.21). For the sonar operator, the net result is a **shadow zone**, which little or no sound penetrates. It is quite possible for submarines to hide in these regions. Hence a knowledge of temperature, salinity, and the density structure of water is crucial for submarine warfare. A submarine in, above, or below one or more of these density layers may also notice the same effect on its own sonar.

Density layers hold promise for communications. Figure 8.17 showed that sound speed is at a minimum at the bottom of a thermocline. Sound emitted or generated in this region has a strong tendency to remain in the layer; in other words, the sound has a greater tendency to bounce back into the layer than to escape it (Figure 8.22).

Most sound travels horizontally within this layer and is affected only by absorption and scattering rather than by vertical scattering. As a result, energy losses in this layer, called the **SOFAR** channel (*so*und *f*ixing *a*nd *r*anging), are extremely small. Sound remains coherent and may be picked up as far away as 25,000 km (15,500 mi)! A small explosion can normally be heard in the air for about 1 km. Underwater, the same explosion can easily be heard as far away as 160 km (99 mi), but the same sound in the SOFAR channel may be heard 20,000 km (12,400 mi) away. Because underwater telephones do not yet use the SOFAR channel, the range of such equipment is limited to short distances at best, generally much less than a mile, because the sound is not channeled and spreads out very rapidly.

How the Sea Sounds from Within

If you were at the surface within 3 m (10 ft) of a 25-horsepower outboard motor, the sound level would not be uncomfortable. But underwater the sound would be just at the level of discomfort. The sound level at about 30 m (100 ft) from a jet aircraft taking off is about the same as the sound level underwater at about 100 m (328 ft) from a dynamite explosion. The average sound level of a television set is about equal to that generated by a fairly calm sea surface.The intensity of a sound is much greater in air than in water, but because water is a much denser medium, the pressure generated by an equivalent sound is much greater. Therefore, a high-decibel (loud or high-pressure) level in air is less dangerous or damaging to the ear than the same sound in water. Furthermore, because audible sound is non-directional in water, a diver cannot tell from which direction a sound is coming.

What sounds can be heard in the ocean? To a sonar operator, the sea is full of background noises. As sound waves penetrate temperature and density caused layers, thermoclines, and pycnoclines, the operator's signals are bent, distorted, and muffled. In addition, natural events such as earthquakes and volcanism give the ocean a dull background rumble. Superimposed on this are the sounds made by the sea's many inhabitants. Some of these sounds come from within an animal's body. Fish, for instance, produce sounds as they inflate or deflate their swim bladders (a gas sac used to change depth). Often drums, sea bass, and catfish sound like foghorns. The crustaceans—shrimp, crabs, and lobsters—make audible clicks as they move their claws and snap their legs together. Sea mammals, too, make noises.

Of all the creatures in the sea, the porpoises may be the most fascinating. Perhaps it is because they are mammals; perhaps it is because in an environment as seemingly alien as the sea, they seem to be friends. Whatever the reason, they hold tremendous appeal for us. The clicking sound they make is an audio signal for a sophisticated natural sonar system that these creatures have evolved. Their whistles and squeals are apparently used to communicate with other members of the species—and perhaps, in some cases, with humans.

So much for the world of the sonar operator. What is it like when you put your head underwater? In all probability, the first thing you would hear would be yourself splashing. With scuba gear, the only thing you would hear would be the sound of your own breathing. As usual, humans make so much noise that it overrides everything else in the environment. Unless we listen, we shall never hear the sounds of the sea. Of course, this may not always be such a bad thing. Some nuclear submarines, for instance, may be 1,600 km (992 mi) away, yet at certain depths they may sound on sonar gear like an express train in one's backyard.

ICE

If you tasted a piece of ice chipped from the frozen ice cover of a harbor on the Maine coast, it would taste salty—not as salty as seawater, but salty nevertheless. If you tasted a piece of ice chipped from an iceberg well out to sea in the North Atlantic, it would have no salty taste at all. Ice with some salt content, correctly termed *sea ice,* accounts for most of the ice in the oceans. An iceberg, however, is not sea ice but glacial ice that happens to be floating in the sea. (We shall return to this subject in Chapter 17.)

Pure water freezes at 0°C. If any other substance, such as salt or clay, is mixed with the water, the freezing point is lowered. The lower freezing point is entirely a collective (or colligative) property that depends only on the total concentration of foreign particles in the water. Such particles are normally dissolved, but muddy water containing suspended silt or clay will also freeze at a slightly lower temperature than pure water. Even so, not nearly as much matter can be suspended in water as can be dissolved. Unlike most substances, water expands when it solidifies and generates tremendous pressure, perhaps as much as 2,000 kg/cm^2 (28,000 lb/in^2). Since deviations in the freezing point of water depend on the dissolved matter present, the freezing point of seawater cannot be given unless the salinity is specified (Figure 8.23).

Because water expands as it freezes, the density of ice is less than the density of liquid water. If this were not the case, icebergs would not float and lakes would freeze from the bottom up. Sea ice varies somewhat in density, but it is always less dense than seawater itself.

Fresh water has a maximum density at 4°C. As the surface-water temperature of a lake drops, its density increases until the temperature reaches that point. Then the water that is at 4°C begins to sink and warmer water rises to take its place. Extensive mixing occurs until the entire body of water is at 4°C. Only then, as the surface waters cool to 0°C, can freezing begin at the surface.

Seawater exposed to cold air also undergoes this overturning and mixing. Unless either the water is shallow or the surface waters are underlain by a region

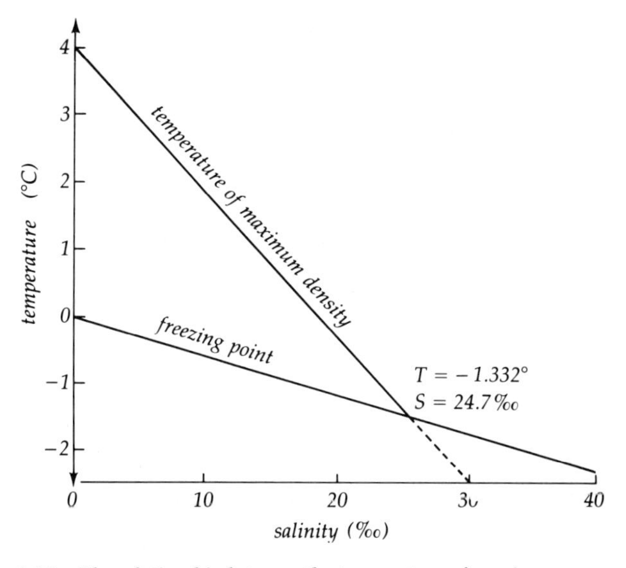

8.23 The relationship between the temperature of maximum density and the freezing point of seawater at different salinities. For salinities less than 24.7‰, when cooling occurs the temperature of maximum density is reached before freezing begins. This means that waters with salinities greater than 24.7‰ have a temperature of maximum density lower than the freezing point.

of high salinity (say, 35‰) not far from the surface (say, 150 to 250 m, or 492 to 820 ft), seawater will not freeze. Because the maximum density of seawater is below its freezing point, it may seem that seawater in shallow regions should freeze faster than lake water. Such is seldom the case, however. Lakes are in constant contact with the surrounding land, which is a poor insulator; therefore, they lose heat rapidly to the land. This means that a lake generally cools more rapidly than the sea.

FURTHER READINGS

Clarke, George L. *The Sea*, Vol. 1, edited by M. N. Hill. New York: Interscience, 1968.

Gregg, M. "Microstructure of the Ocean." *Scientific American*, February 1968, pp. 64–77.

Jerlov, N. G. *Optical Oceanography*. New York: Elsevier, 1968.

Knauss, J. A. *Introduction to Physical Oceanography*. Englewood Cliffs, N.J.: Prentice-Hall, 1978.

Morel, A. "Optical Properties of Pure Water and Pure Sea Water." In *Optical Aspects of Oceanography*, edited by N. G. Jerlov and E. S. Nielson. New York: Academic Press, 1974, pp. 1–24.

Neuman, G., and W. J. Pierson, Jr. *Principles of Physical Oceanography*. Englewood Cliffs, N.J.: Prentice-Hall, 1966.

Picard, G. L., and W. J. Emery. *Descriptive Physical Oceanography*. Oxford: Pergamon Press, 1982.

Pond, S., and G. L. Pickard. *Introductory Dynamic Oceanography*. Oxford: Pergamon Press, 1978.

Revelle, R. "Water." *Scientific American*, September 1963, pp. 93–108.

Tchernia, P. *Descriptive Regional Oceanography*. Oxford: Pergamon Press, 1980.

Urick, R. J. *Principles of Underwater Sound*. New York: McGraw-Hill, 1975.

Von Arx, W. S. *An Introduction to Physical Oceanography*. Reading, Mass.: Addison-Wesley, 1964.

Williams, J. *Optical Properties of the Sea*. Annapolis, Md.: U.S. Naval Institute, 1970.

Dennis Brokaw

In the harbor, in the island, in the Spanish Seas,
Are the tiny white houses and the orangetrees,
And day-long, night-long, the cool and pleasant breeze
Of the steady Trade Winds blowing.

There is the red wine, the nutty Spanish ale,
The shuffle of the dancers, the old salt's tale,
The squeaking fiddle, and the soughing in the sail
Of the steady Trade Winds blowing.

And o' nights there's fire-flies and the yellow moon,
And in the ghostly palm trees the sleepy tune
Of the quiet voice calling me, the long, low croon
Of the steady Trade Winds blowing.

John Masefield

NINE

Climate and the Ocean

We live at the bottom of an invisible ocean of air—the atmosphere—which is made up of a mixture of gases (Table 9.1). Although its mass is only 1/100th that of the oceans, the atmosphere weighs 5.1×10^{21}g (5.1 billion trillion grams, or over 5,600 trillion English tons). The force of gravity causes the atmosphere to exert a pressure (called atmospheric pressure) on everything immersed in it. At sea level, a column of air one inch square reaching to the outer edge of the atmosphere weighs 14.7 lb (6.6 kg) and consequently exerts a pressure of 14.7 lb/in^2 (1.04 kg/cm^2, or 1,013 mbar). Changes in atmospheric pressure exert a powerful influence on what we call weather (Figure 9.1).

Human beings are weather-sensitive creatures. Our industry, agriculture, sport, leisure, commerce, transportation, even our disposition are all affected by weather. We are comfortable at temperatures around 20.4°C (69°F) and at relative humidities of about 40–50%, and we become uncomfortable when temperatures exceed 26.1°C (79°F) and relative humidities exceed 75%. Yet the temperature and humidity at many places on the earth are well above and below these values, and the temperature at many places commonly varies by as much as 20°C within a single twenty-four-hour period.

The atmosphere and the hydrosphere are intimately related. In fact, sailors, fishers, and coastal dwellers tend to perceive the sea and the air essentially as one. Since almost 71% of the earth's surface is covered by oceans, the atmosphere and the hydrosphere are in extensive contact. The primary aim of this chapter is to describe that intimate relationship. Our discussion of the atmosphere will largely be limited to weather over the oceans and the effect of weather on the oceans.

Although weather is the primary concern of meteorologists, we cannot understand the hydrosphere without some appreciation of the meteorological influences on the oceans of the earth. For instance, in regions of low rainfall, where evaporation exceeds precipitation, the salinity of the surface waters of the ocean is

Table 9.1 *Gaseous Composition of the Atmosphere at Sea Level*

Gas	Symbol	% Total
Nitrogen	N_2	78.084
Oxygen	O_2	20.946
Argon	Ar	0.934
Carbon dioxide	CO_2	0.032[1]
Water vapor	H_2O	Up to 4

[1]This is an average figure; figure can go up to 0.1%.

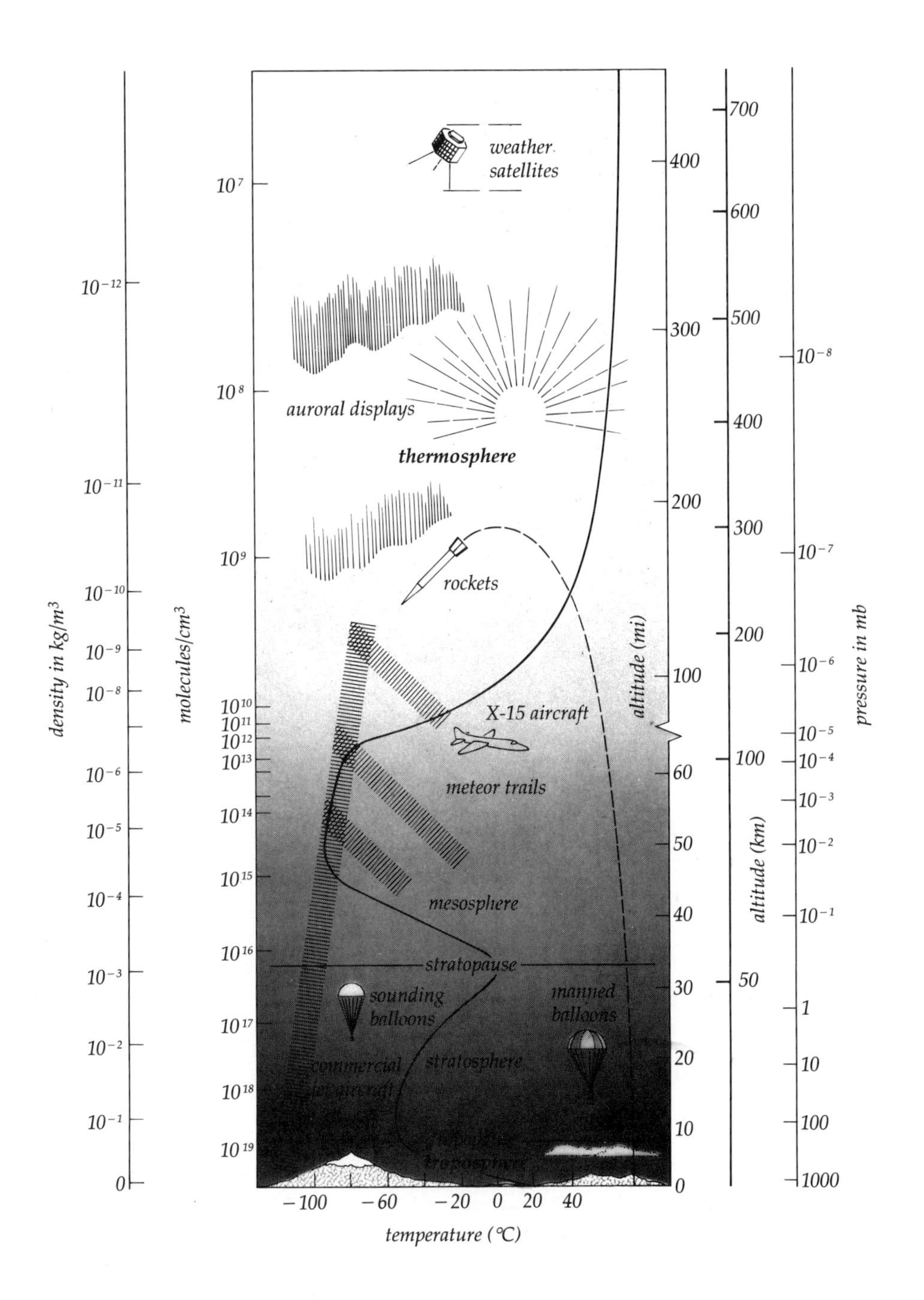

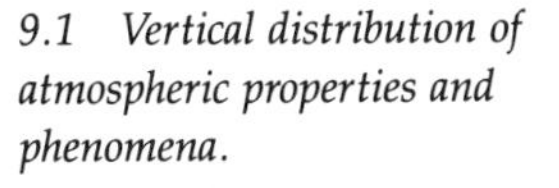

9.1 Vertical distribution of atmospheric properties and phenomena.

higher than it is in regions where precipitation exceeds evaporation. Prevailing surface winds are the major cause of ocean currents (Chapter 10), and storm winds at sea cause waves (Chapter 11). Thus, we must study some meteorology to understand the ocean.

The weather today may be the same as the weather tomorrow, but then again it may not be. And the weather today may or may not be the same as it will be on this date next year. Day to day, week to week, month to month, and even year to year, changes in meteorological conditions bring the changes called weather. When we average these short-term fluctuations over a long term, we get an overall picture of the climate of a region.

For example, if we averaged the rainfall or the temperature of a region for fifty years, we would interpret a period of unseasonal wetness or warmth as an anomaly. The longer the period on which we based our average, the more insignificant the anomaly would become. Over the long run, weather tends to repeat itself fairly reliably—a fact attested to by folk sayings like "Red sky in the morning is a sailor's fair warning; red sky at night is a sailor's delight" or by the *Old Farmer's Almanac*'s claim to being able to predict weather

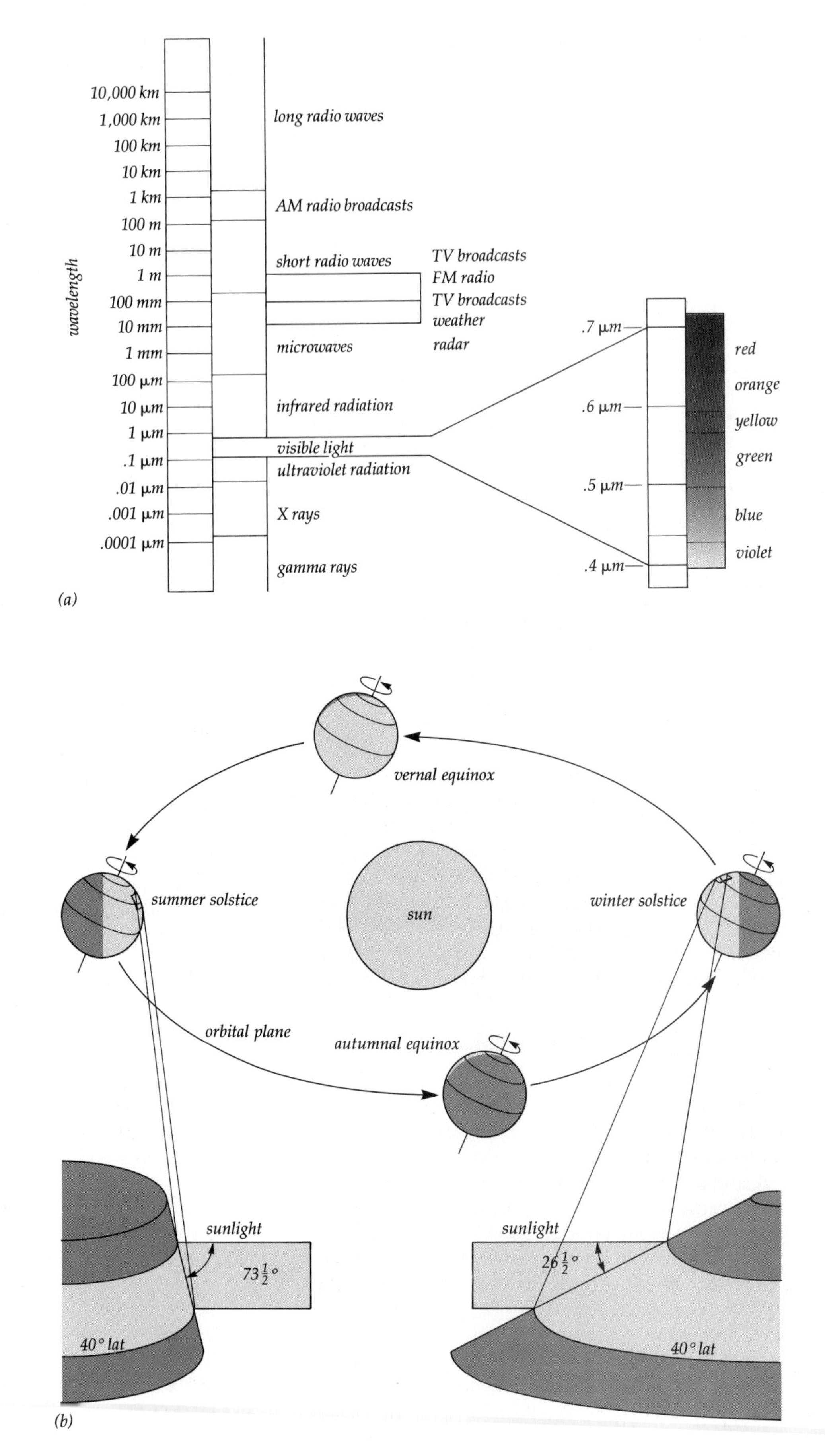

9.2 (a) The electromagnetic spectrum. (b) The summer solstice on June 21 or 22 is the time of year when the earth's axis is tipped 23½° toward the sun in the northern hemisphere. Since the sun's rays are more perpendicular to the earth's surface in the northern hemisphere, this corresponds to the time of maximum receipt of radiation. Six months later at the winter solstice on December 22 or 23, the earth's axis is tipped 23½° away from the sun in the northern hemisphere, giving colder temperatures. (The elevation of the sun above the horizon at noontime is 73½° for the summer solstice and 26½° for the winter solstice.) During the vernal and autumnal equinoxes on March 20 or 21 and September 22 or 23, the sun is directly over the equator at noon.

Table 9.2 *Wavelengths of Solar Energy Emitted*

Wavelength (nm)	Term	% Total Energy Emitted
< 1.0	X rays and γ rays	0.02
1–200	Far ultraviolet	
200–315	Middle ultraviolet	1.95
315–380	Near ultraviolet	5.32
380–720	Visible	43.50
720–1,500	Near infrared	36.80
1,500–5,600	Middle infrared	12.00
5,600–1,000,000	Far infrared	0.41
> 1,000,000	Micro-and radio waves	

a year in advance. Long-time inhabitants of a region develop an intuitive sense of the weather, although their predictions may be far off the mark when they are in another region with different weather patterns.

The driving force behind weather the world over is heat energy. The source of the world's heat energy is solar radiation except for a very small percentage (less than 0.1%) emanating from the earth's interior due to such processes as natural radioactive decay. The amount of this solar radiation varies very little from year to year, but there are short-term and long-term fluctuations that need to be explained.

Of all the energy that the sun radiates into space, only about one or two billionths strike the earth. The visible light is only one of several components of the full spectrum of solar radiation the earth receives. If the full spectrum of radiation reached the earth, all forms of life on this planet would perish. To understand why, we shall consider the components of solar energy, both in free space and at the earth's surface.

THE NATURE OF SOLAR ENERGY

Light, unlike sound waves and water waves, is electromagnetic radiation (Figure 9.2). It needs no medium for its transfer and can travel through the vacuum of outer space. In addition to light, other forms of electromagnetic energy travel from the sun through space and reach the earth's atmosphere. We perceive **infrared radiation** as heat. **Ultraviolet radiation** (often called black light) causes skin to tan. In bright, cloudless regions such as the Sahara Desert or Death Valley, ultraviolet radiation may cause sunstroke and death.

In Figure 9.2 solar radiation is broken down by wavelengths. Table 9.2 shows the percentages of the solar energy emitted at different wavelengths. Fortunately, the lethal components that reach the earth's upper atmosphere never reach the surface of the earth. Figure 9.3(a) indicates the fate of the incoming solar radiation. The atmosphere absorbs much of the harmful ultraviolet radiation in the stratosphere, and clouds reflect a large percentage of solar radiation. Most of the solar energy that reaches the earth is in the shorter, visible wavelengths, whereas most of the energy that the earth emits is in the longer, infrared wavelengths (Figure 9.3b). The solar energy received by the earth is the equivalent of 23 trillion horsepower a minute—more than human beings use in the course of a whole year.

The Heat Budget

With such a vast amount of solar energy reaching the earth's surface, why don't the oceans and the continents grow hotter and hotter? The answer is that the energy received is balanced by the energy lost or reradiated back into space. A very hot body like the sun radiates energy whose spectrum is drastically skewed to the shorter, higher-energy wavelengths. A cool body like the earth radiates energy at much longer, less energetic wavelengths, almost entirely in the infrared region.

Approximately 50% of all incoming solar energy actually reaches the earth's surface. The rest is reflected back to space by clouds and the gases of the atmosphere or is absorbed by the atmosphere. Because most of the earth is covered by water, most of that energy enters the ocean. Together, the land and the ocean radiate only about 6% of the absorbed energy directly back to space; the remaining 44% is transferred to the atmosphere by several mechanisms, and from there it is radiated back into space. The earth, like any other physical system over time, has arrived at a steady-state thermal equilibrium, called the **heat budget**. Although the atmosphere contains only about 1/1,000th as much heat as the oceans, the atmosphere and the hydrosphere serve as gigantic moderators that keep the earth's temperature relatively constant.

The ocean has its own heat budget, which may be formulated as follows:

$$Q_s = Q_b + Q_h + Q_e$$

Q_s is the shortwave radiation absorbed by the ocean as heat, Q_b is the heat given up to the atmosphere and to space as net longwave radiation, Q_h is the sensible (that is, detectable) heat leaving the sea by conduction, and Q_e is the heat lost by evaporation. Because the total heat entering the ocean must equal the heat leaving the ocean, the heat budget equation may also be written as:

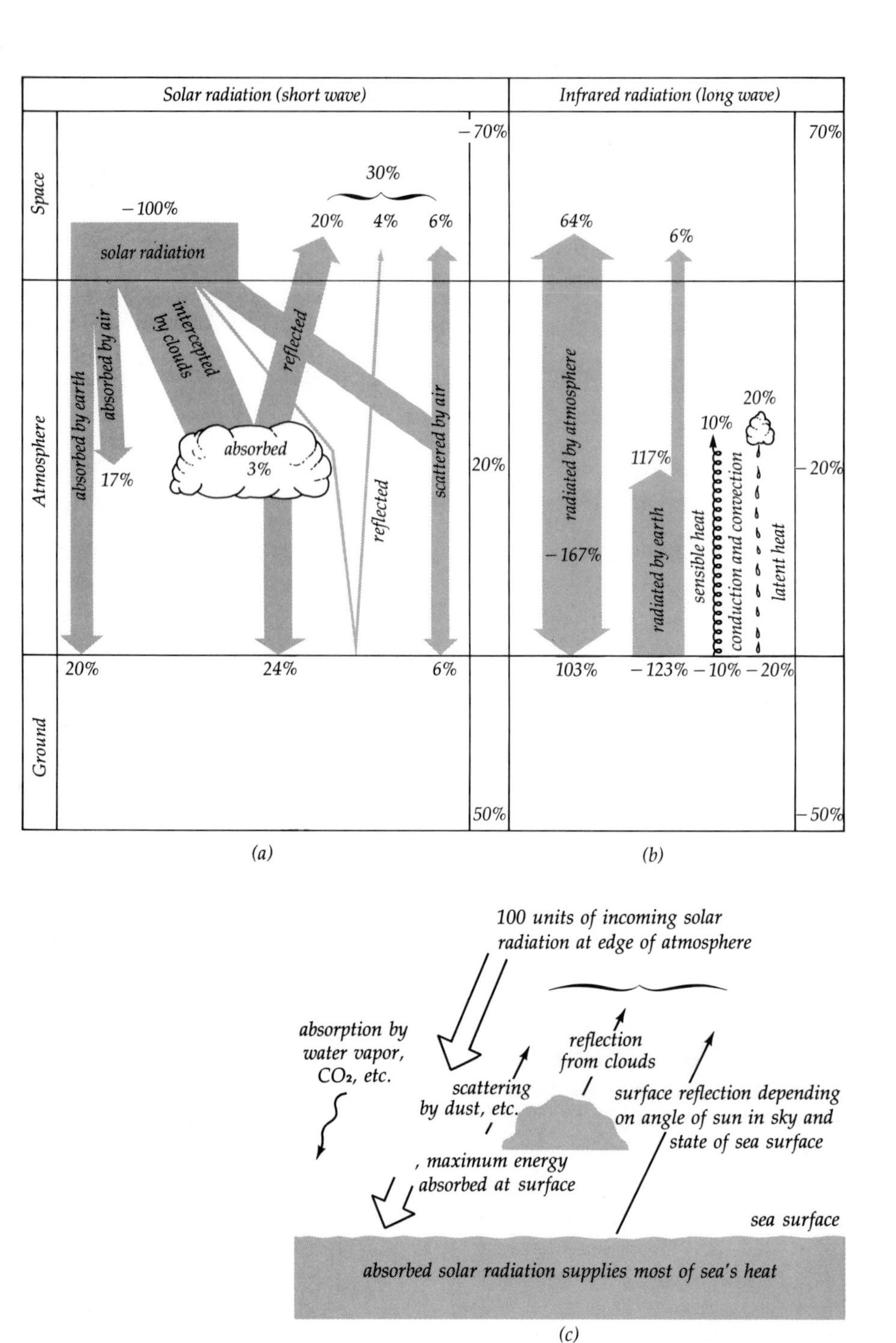

9.3 Global energy budget for (a) solar and (b) terrestrial radiation illustrating the losses and gains to space within the atmosphere and at the earth's surface, in percentages. (c) The fate of incoming solar energy over the open ocean.

$$Q_s - Q_b - Q_h - Q_e = 0$$

These equations are valid only when the values are averaged over at least a year. Cloud cover and other local phenomena may disrupt the energy exchange over shorter periods of time.

Underlying the heat budget equation is the fact that temperatures in nature tend to equalize. When warm air is in contact with cold ocean water, heat is transferred from the air to the water; when the water is warmer, heat is transferred from the water to the air. Heat is transferred from the ocean to the atmosphere through the movement of heat from particle to particle, a process known as **conduction**. Heat is transferred from the atmosphere to the ocean through the circulation of masses of particles, a process known as **convection**.

The main mechanism for the transfer of heat from the ocean to the air is the formation of water vapor. It takes a great deal of heat energy to evaporate 1 kg (2.2 lb) of water (called the latent **heat of vaporization**, which equals 540 cal/gm). The heat released through

Table 9.3 *Length of Day as a Function of Latitude*

Latitude	Winter Solstice	Vernal or Autumnal Equinox	Summer Solstice
0°	12 hr 0 min	12 hr 0 min	12 hr 0 min
10°	11 hr 25 min	12 hr 0 min	12 hr 38 min
20°	10 hr 48 min	12 hr 0 min	13 hr 12 min
30°	10 hr 4 min	12 hr 0 min	13 hr 56 min
40°	9 hr 8 min	12 hr 0 min	14 hr 52 min
50°	7 hr 42 min	12 hr 0 min	16 hr 18 min
60°	5 hr 33 min	12 hr 0 min	18 hr 27 min
70°	0	12 hr 0 min	2 months
80°	0	12 hr 0 min	4 months
90°	0	12 hr 0 min	6 months

Table 9.4 *Percentage of Incoming Solar Radiation Reflected by Various Surfaces (or Albedo)*

Surface	%
Snow	
New	80–90
Old	45–70
Clouds	50–55
Dry, light, sandy soils	35–45
Sea ice	30–40
Green fields	15–25
Sand	13–18
Concrete	17–27
Forests	
Deciduous	15–20
Evergreen	10–15
Asphalt	5–10

evaporation is subsequently released to the air when the water condenses as fog or clouds. It is estimated that latent heat of vaporization accounts for nine times as much heat transfer as sensible heat.

In regions where the temperature of the air is higher than that of the sea and where humidity is high, direct condensation produces fog at the air-sea interface and brings about heat transfer from the air to the sea. Among these fog-shrouded regions are the Grand Banks off Newfoundland and Nova Scotia and the coastal waters off northern California in summer and off southern California in spring. Some of the parameters that affect the heat budget equation are the length of the day, which varies with latitude and with season (Table 9.3), the amount of cloud cover, and the reflectance and absorbance of the land and sea surfaces. Table 9.4 gives the reflectance of a variety of surfaces.

The amount of sunlight striking the earth varies with latitude; the greatest amount strikes the tropics and subtropics. Because more energy is received than is lost at low latitudes, these regions would become very much hotter if it were not for the redistribution of heat by the atmosphere and the ocean. Air movement and water currents move vast amounts of heat to the poles, which act as heat sinks.

Because the relative angle of the earth's attitude to the sun changes during the year, the amount of solar energy striking the earth varies with distance away from the equator. This variation is responsible for the seasons. The Northern Hemisphere is 4.8 million kilometers (3 million miles) closer to the sun during its winter than during its midsummer, and yet the winters there are colder than the summers. Obviously, the changing temperatures are not a result of proximity to the sun but of changes in the earth's attitude to the sun. During the summer, when the Northern Hemisphere is tilted toward the sun, each unit of area receives more solar energy than it does in winter. The climate in equatorial regions remains much the same all year long because there is little variation in the attitude of the earth at the equator. Changes in the energy and heat reaching a given area largely determine that area's climate. Local changes such as cloudiness and wind patterns cause the daily shifts of the weather.

THE ROLE OF WATER IN WEATHER

Before we discuss the air movements that cause changes in the weather, we need to know something about the characteristics of water. The specific heat and density of water are much higher than the specific heat and density of air (see Chapter 8). In fact, the uppermost 3 m of the ocean contain the same quantity of heat as the entire atmosphere. The ocean is the great modifier of temperature, moving massive amounts of heat slowly

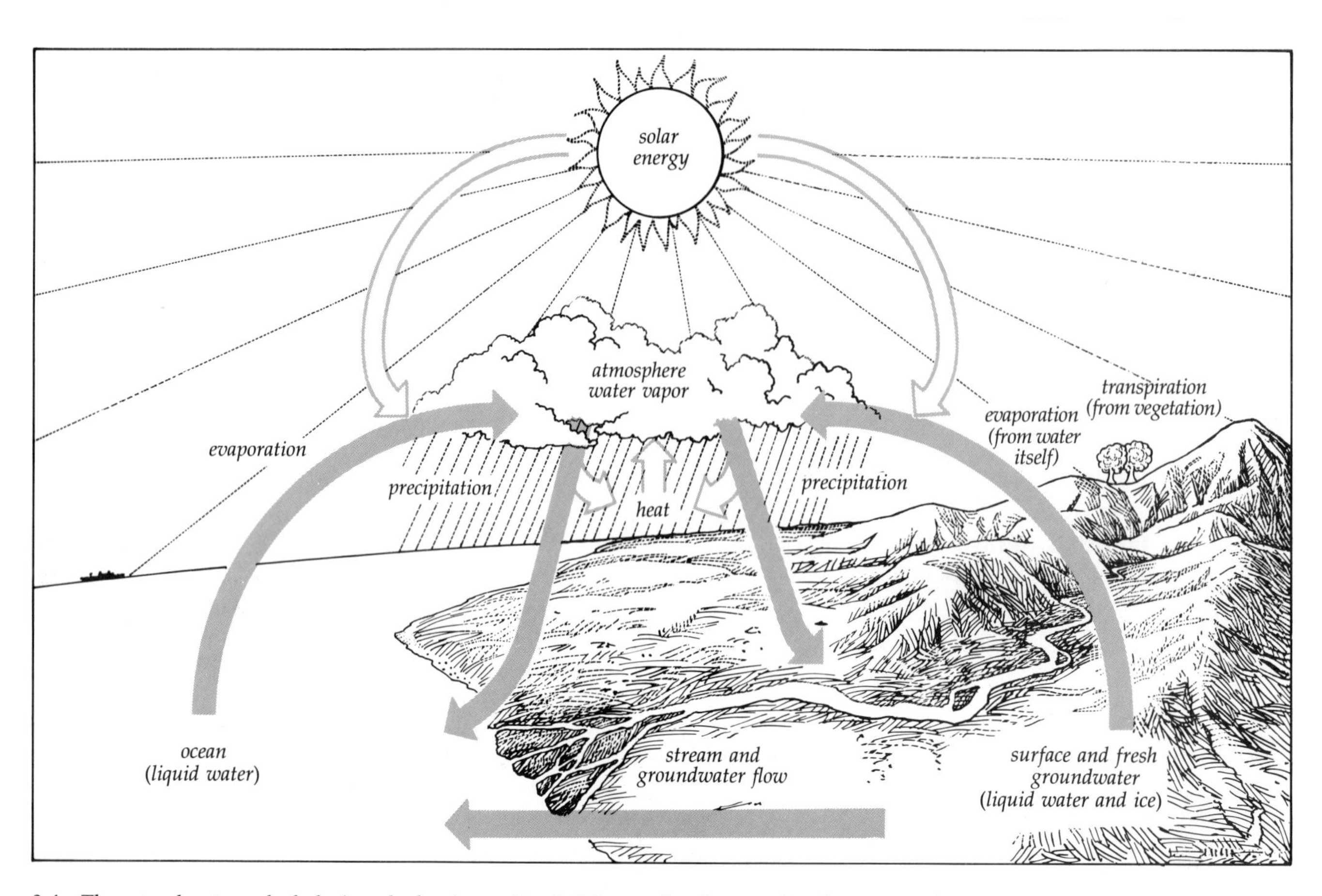

9.4 The natural water or hydrologic cycle showing cycling (solid arrows) and energy flow (open arrows).

from place to place. The atmosphere moves much smaller amounts of heat but moves it more rapidly. The net effect is that the ocean and the atmosphere share about equally the task of distributing heat from one zone to another.

Only about 0.001% of the total global surface water is contained in the atmosphere. The amount varies with latitude, decreasing as latitude increases. It also varies with altitude, again decreasing as altitude increases. The water content of the atmosphere at 3,000 m (9,840 ft) is only 18% of the atmosphere at sea level, and it varies with the seasons, being highest in summer. Even though the percentage is never high, however, atmospheric water content is very important because of the high value of water's heat of vaporization.

The ocean is a saline, aqueous solution whose motion is controlled by the sun's energy, tides, the motion of the earth, and the differing densities of water masses. All these factors interact to produce a vast hydrologic cycle (Figure 9.4). Basic to this cycle is the energy of the sun, which moves water by convection, evaporation, and precipitation. The turnover time (that is, the time it takes for water to sink and return to the surface) for water in the ocean varies from about 700 years in the Atlantic Ocean to about 1,500 years in the Pacific. Ocean water is cycled through the atmosphere and land by precipitation, groundwater seepage, and river flow about every 3,600 years.

When a gram of water evaporates, it takes 2,259 J (540 cal) of heat from its surroundings. When a gram of water condenses as rain or dew, or simply as moisture on a cold beverage can, it gives off the same 2,259 J (540 cal). When vapor condenses to form 2.5 cm (1 in) of rain, it releases as much heat to an area as three days of sunshine. Because of the moderating effect of water, areas near the ocean have less variable daily and annual temperatures than inland areas (Figures 9.5 and 9.6). The moist air that condenses into the warm winter rains west of the Cascade Mountains of the Pacific Northwest actually releases solar energy that entered the waters off Japan.

The heat energy processed by the ocean every year is staggering. The following is one method of approximating the amount: The mass of the ocean is 1.4×10^{24}g, and the residence time of a water molecule is 54,000 years; thus

$$\frac{(1.4 \times 10^{24}\text{g of } H_2O \text{ in ocean})(540\ \text{cal/g})}{(5.4 \times 10^{4}\ \text{y})} = 1.4 \times 10^{22}\ \text{cal/y}$$

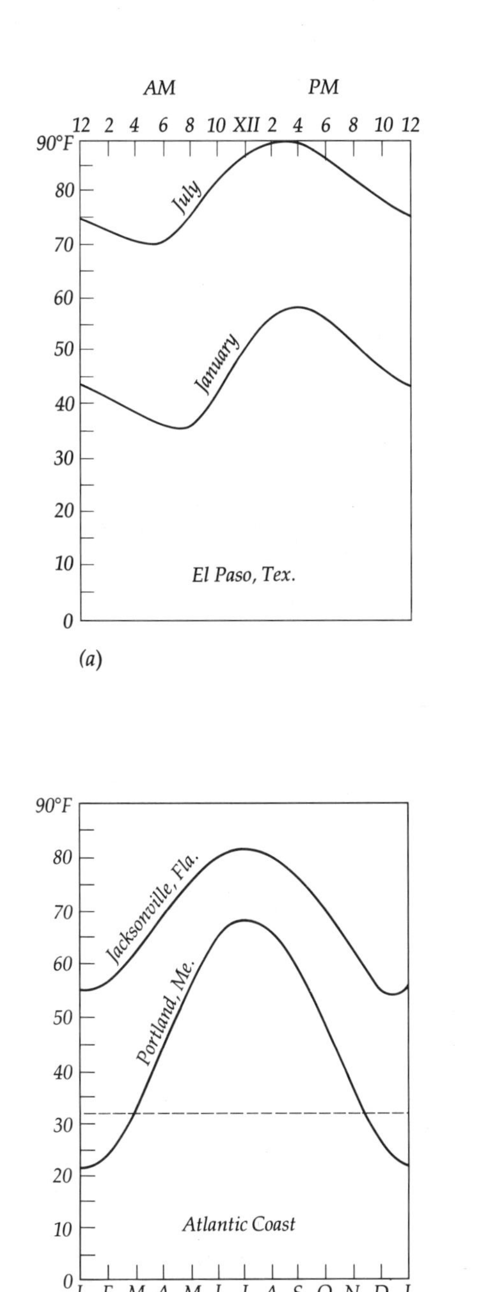

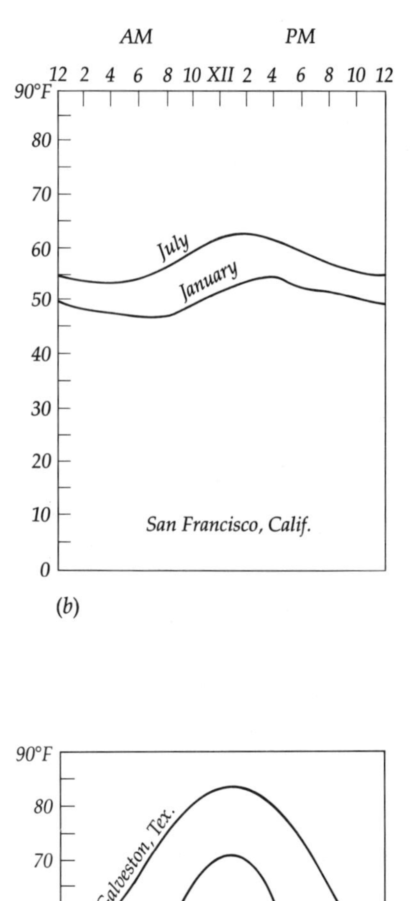

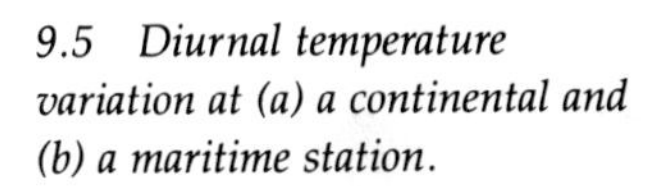

9.5 *Diurnal temperature variation at (a) a continental and (b) a maritime station.*

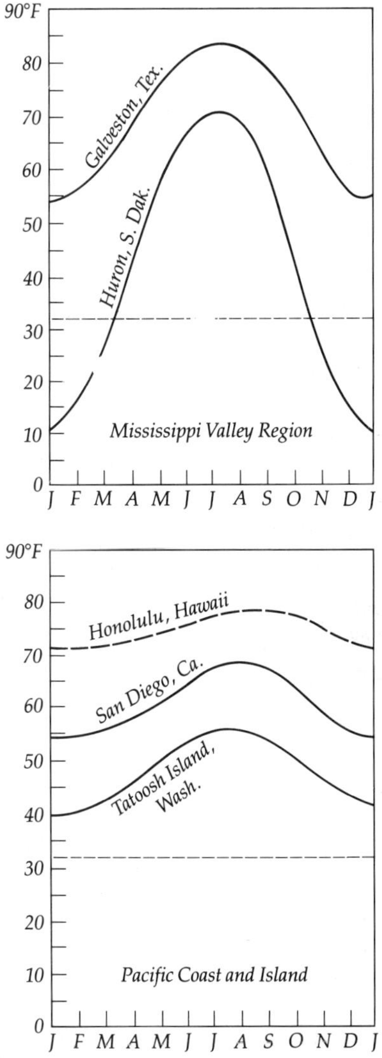

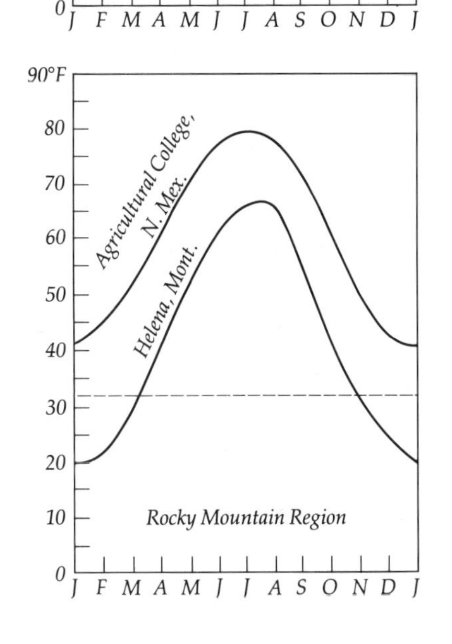

9.6 *Annual temperature variation at continental and marine stations.*

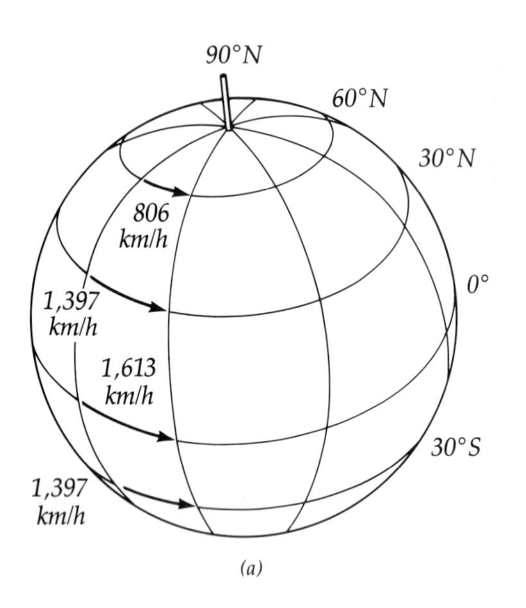

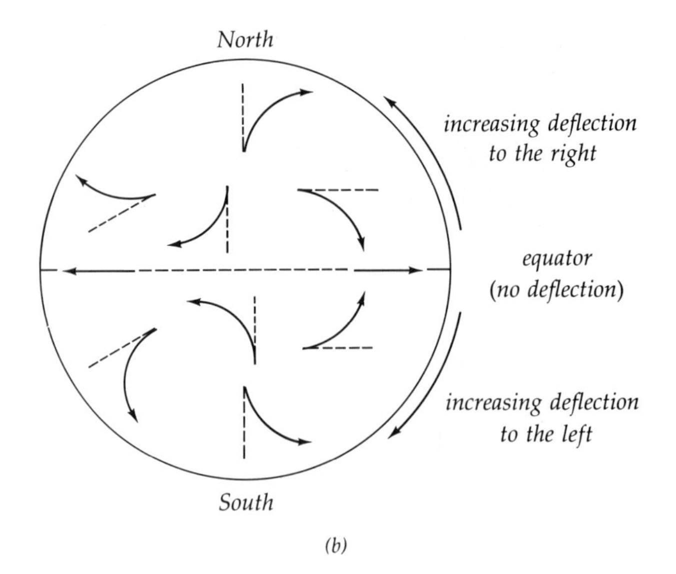

9.7 *(a) Rotational velocities of points on the globe at various latitudes. (b) The paths of moving objects are deflected by the Coriolis effect. There is no deflection at the equator; the deflection increases toward the poles. (After A. Strahler,* Physical Geography, *2d ed., New York: John Wiley, 1960) (c) A ball thrown from the North Pole will appear to follow path* A, *but the actual course is path* B. *While the ball was in flight, the earth revolved from* X *to* Y. *(d) Coriolis acceleration develops because of the earth's rotation. If two rockets are fired north and south from the central United States, in three hours the earth has rotated enough to change the north-south direction significantly compared to a fixed point in space, producing a real acceleration to the right of the path of motion as an observer looks at the departing rockets. (e) Coriolis acceleration follows right of the path of motion of an object traveling east-west because the earth is rotating; an observer on the rotating earth would see acceleration to the right of the path of motion for any east-west motion. (f) The paths of two satellites as they appear to observers on earth. (g) A turntable's direction of rotation is clockwise when viewed from above and counterclockwise when looked at from below.*

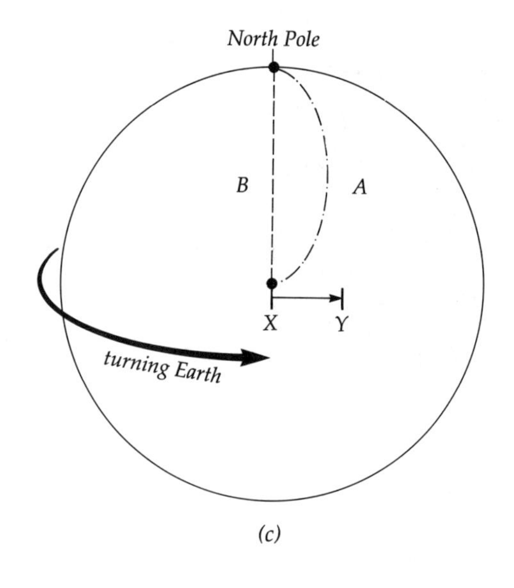

About 71% of the earth's surface is covered by the ocean, and about 88% of the solar radiation that enters the ocean causes evaporation. Clearly, evaporation and the subsequent condensation constitute the earth's major mechanism for heat exchange. The total water vapor content of the atmosphere is turned over about every twelve days.

THE CORIOLIS EFFECT

Wind is simply air that is moving relative to the earth. However, the rotation of the earth (Figure 9.7a) affects any mass that is moving across its surface. This effect, known as the **Coriolis effect**, causes winds to veer to the right, or in a counterclockwise manner, in the Northern Hemisphere and to the left, or in a counterclockwise direction, in the Southern Hemisphere (Figure 9.7b). It also causes winds to blow clockwise around high-pressure centers in the Northern Hemisphere and counterclockwise in the Southern Hemisphere; the circulation is reversed when air flows into low-pressure systems. Over short distances these deflections are usually too small to be noticed, but over large distances the deflections can be appreciable.

When a spacecraft reenters the earth's atmosphere, the movement of the earth must be taken into account if the spacecraft is to splash down at a specified site. Because the earth is spinning from west to east, the landing site moves as the spacecraft approaches. To an observer on the earth's surface, it appears as if the spacecraft were being deflected. This

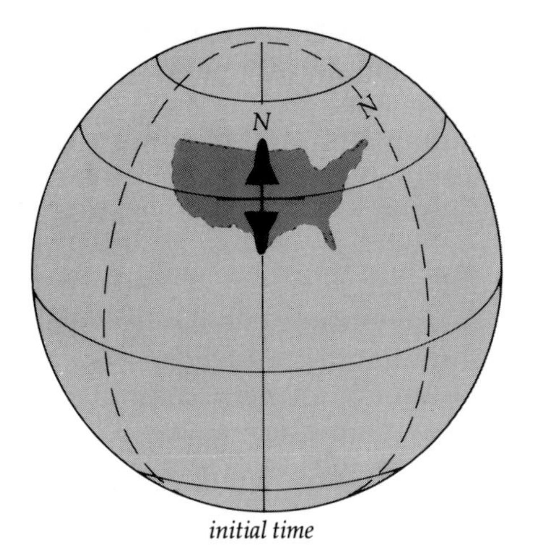
N
N
initial time

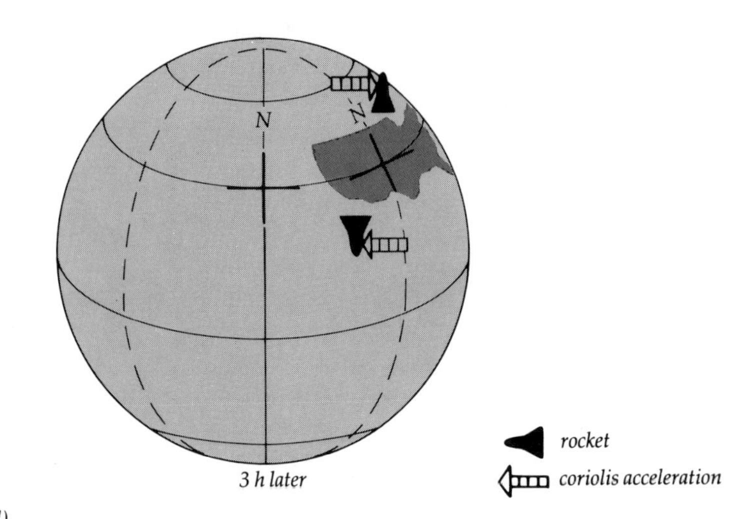
N
N
3 h later
rocket
coriolis acceleration

(d)

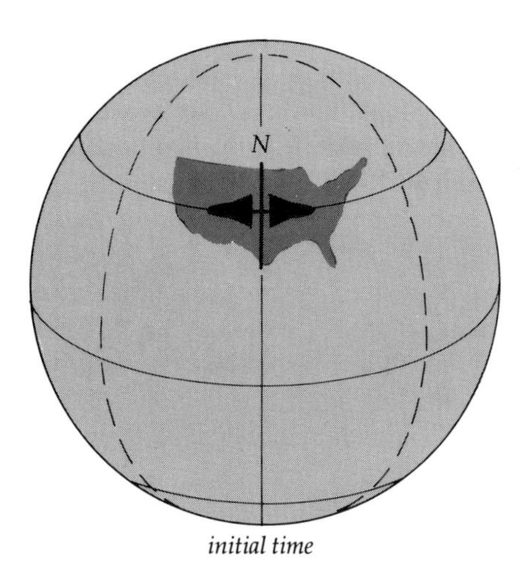
N
initial time

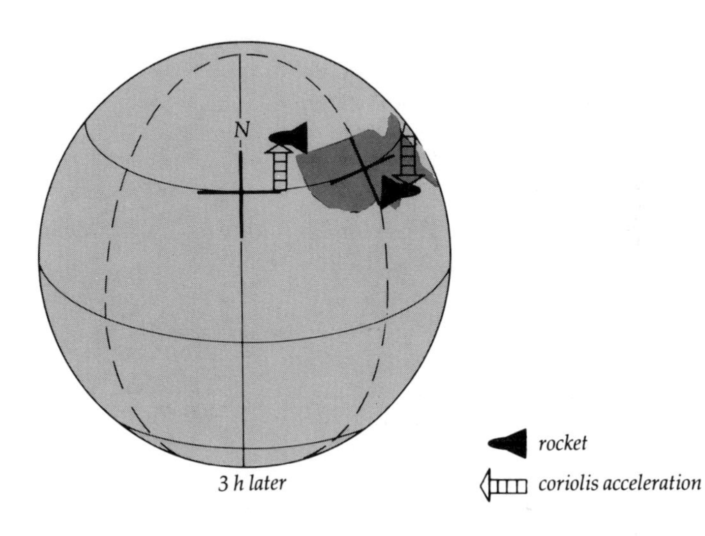
N
3 h later
rocket
coriolis acceleration

(e)

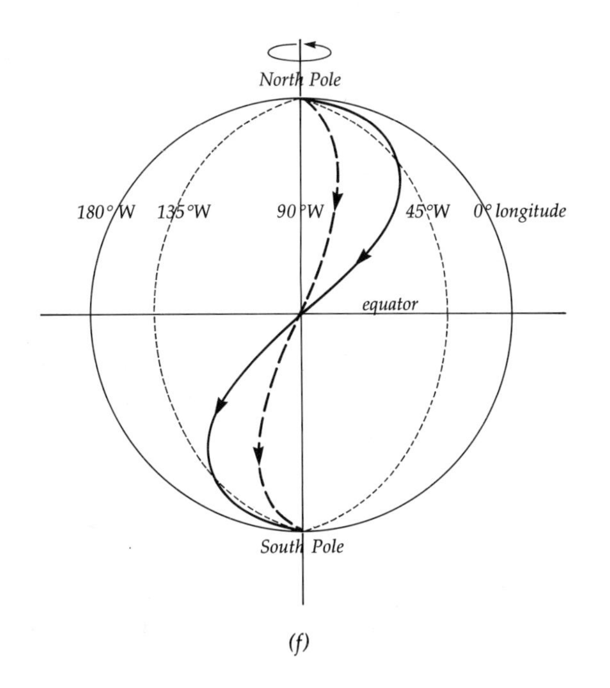
North Pole
180°W
135°W
90°W
45°W
0° longitude
equator
South Pole

(f)

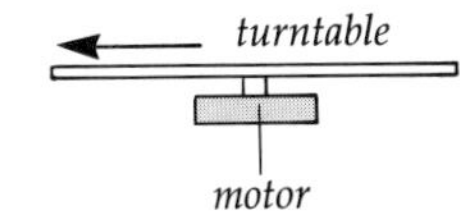
top view
turntable
motor

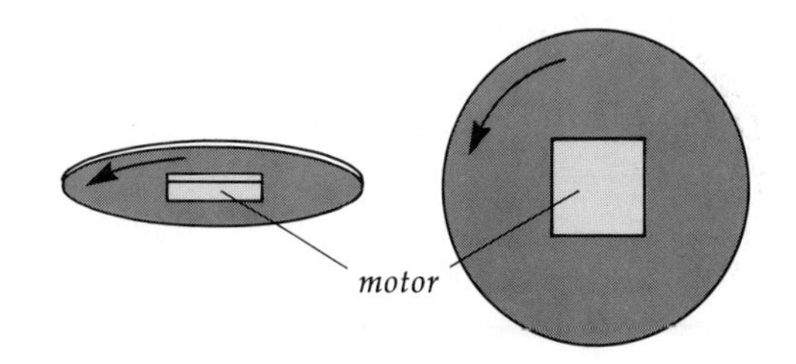
motor

(g)

apparent force is the Coriolis effect, named after the nineteenth-century French mathematician and physicist Gaspard G. Coriolis, who first described it. The Coriolis effect alters the apparent path of moving objects, although it exercises no actual force on them. You experience a similar effect when an automobile moving in a straight line at constant speed enters a sharp curve. If the curve is to the right, you respond as if a force were pushing you to the left. Because of its momentum, your body tries to keep moving in a straight line as the automobile turns.

A device called the Foucault pendulum demonstrates the idea of relative motion called the Coriolis effect. To an observer in the Northern Hemisphere, the pendulum seems to be deflecting to the right, or in a clockwise direction. Actually, however, it is the building that is moving about the pendulum. At 90° north and south latitudes, the pendulum has a period of 24 hours, and one of infinity at the equator.

Suppose you cut a straight line with a penknife across a stationary phonograph record from its center to the edge. If you try to make the same cut while the record is spinning on a turntable, the second cut will be curved, not straight, because the record was moving while you made the "straight" cut. The Coriolis effect deflects the movement of winds and currents in a similar fashion.

Any object moving across the surface of the earth experiences an apparent deflection, as shown in Figure 9.7b and 9.7c. In the Southern Hemisphere, the apparent deflection is to the left. The term *apparent* indicates that the change in direction is relative. Suppose a rocket were moving from west to east in the Northern Hemisphere (see Figure 9.7d and 9.7e). From a satellite far above the earth, the rocket would appear to be traveling in a straight path. But to an observer on the earth it would seem to be deflecting somewhat to the right. The extent of the apparent deflection is primarily a function of latitude. The maximum apparent deflection occurs at the poles, the minimum (which is zero) at the equator.

To an observer on the earth, the deflection appears to be real, although it is actually the observer who is "being deflected" (see Figure 9.7f). Because deflection is normally caused by some force, this deflection is often called the Coriolis force. If this term is used, however, it should be kept in mind that it is only an apparent or illusionary force. No force (F) or acceleration (a) is involved ($F = ma$). The Coriolis force can only affect existing motion; it cannot initiate or cause motion.

The Coriolis effect is augmented by centrifugal force. However, centrifugal "force" is not a true force either; it is only an inertial effect. Moving objects tend to move in a straight line unless some force acts on them. When the automobile turns sharply to the right, you do not actually move to the left; you are simply trying to move in a straight line. *Relative to the car,* however, you are moving to the left. Any object at rest at a given place on the earth experiences the same centrifugal force as the earth does at that particular place. The force is not constant at all places, however, because it is a function of the earth's rate of rotation from place to place, as shown in Figure 9.7a. The total velocity of a wind moving from west to east in the Northern Hemisphere is equal to the velocity of the earth's rotation *plus* the velocity of the wind. Consequently, the wind experiences a greater centrifugal acceleration than the earth does beneath it.

The deflective force is proportional to the mass and speed of the moving body and does not depend on the direction of horizontal motion. Therefore, the Coriolis deflection of a moving object is not necessarily greater at higher latitudes. The velocity of the object can vary. A light plane moving at 100 mph would take ten hours to fly from Los Angeles to Portland, Oregon. A commercial jet aircraft could do it in two hours. They would both cover the same distance, but their relative deflection would be different. The slower aircraft would experience the Coriolis effect longer and would be deflected a greater distance to its right. At any point in time, however, faster-moving objects experience a greater Coriolis effect.

In summary, the following points should be remembered when dealing with the Coriolis effect: The deflection is to the right of an object's motion in the Northern Hemisphere and to the left in the Southern Hemisphere (see Figure 9.7g). There is no horizontal deflection at the equator, and the maximum deflection is at the poles. The effect exists only when there is velocity relative to the earth's surface. It is not a true force and no work is done. Finally, the magnitude of the effect is a function of the object's horizontal velocity and the angular velocity of the earth on its axis as a function of the latitude.

For objects in vertical motion the Coriolis acceleration is greatest at the equator and least at the poles. A more detailed explanation of the Coriolis effect is given in Appendix VI.

GENERAL CIRCULATION OF THE ATMOSPHERE

The general circulation of the atmosphere is propelled by global winds and is the overriding source of weather. No matter how unpredictable the winds may seem, they behave in accordance with physical laws. The uneven heating of the earth by the sun causes differences in atmospheric pressure, and the winds are simply gravitational responses to the varying densities of the air. As the moving air begins to rotate as a result

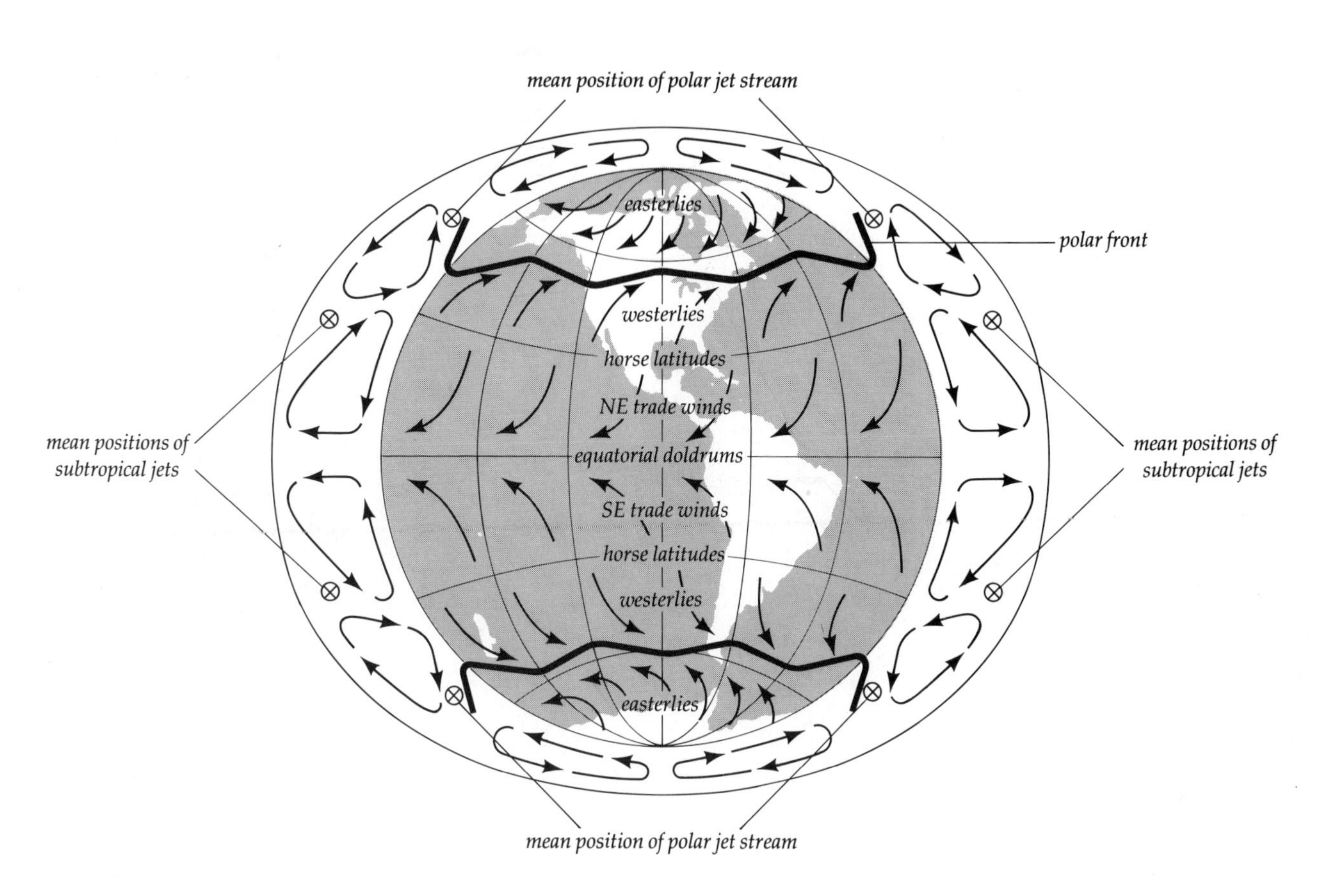

9.8 The general theoretical circulation of the atmosphere. The vertical scale is greatly exaggerated.

of the earth's rotation and the Coriolis effect, it is transformed into wind. Wind flows essentially along lines of equal pressure, called **isobars**, at a speed proportional to the pressure gradient (change in pressure per unit distance). This phenomenon allows us to calculate wind velocities by measuring atmospheric pressure at different regions. The driving force behind the winds is the difference in temperatures between the poles and the equator. The Coriolis effect imposes a zonal or beltlike pattern on the movement induced by those differences in temperature.

Figure 9.8 shows this zonal horizontal and vertical pattern of the general circulation of the atmosphere. That pattern varies with the seasons, and the pattern shown here is an average. This general circulation is constantly transporting the earth's atmosphere from equatorial regions toward the poles and back again. Within the large-scale flow are the smaller, short-term variations we call weather.

The equatorial belt between 5°N and 5°S is a region of variable winds, calms, and warm, moist air. Because there are no prevailing winds in this region, and because about a third of the time there are no winds here at all, this area has long been known as the **doldrums**. Small, violent thunderstorms and squalls are frequent, however. Because of the weak Coriolis effect here, the weather is difficult to predict. Since the surface air rises in response to solar heating, this region is marked by the surface meeting or convergence of winds. The climatic region or belt is narrower and more consistent over the ocean than over land. Over land, it may extend to 10° of latitude.

North and south of the doldrums, to about 30° to 35° of latitude, are the wide belts of the **trade winds.** The term *trade winds* comes from the days when wind-driven ships were the vehicles of commerce. As the air that rises in the equatorial region moves away from the equator, it cools and begins to descend. As it does so, its pressure increases, it is warmed by compression, and it becomes drier and cloudless. When the upper air moving north from the equator returns to the surface, it moves south (more specifically, from the northeast to the southwest because of the Coriolis effect).

Ocean currents are described according to the direction they come from: A northerly current flows to the north. But a north wind comes *from* the north. So the winds we have just mentioned are the northeast trade winds of the Northern Hemisphere. The corresponding winds in the Southern Hemisphere are the southeast trade winds. The trades are remarkably consistent both in speed and direction; they blow from only about one-sixteenth of the 360 degrees of the compass.

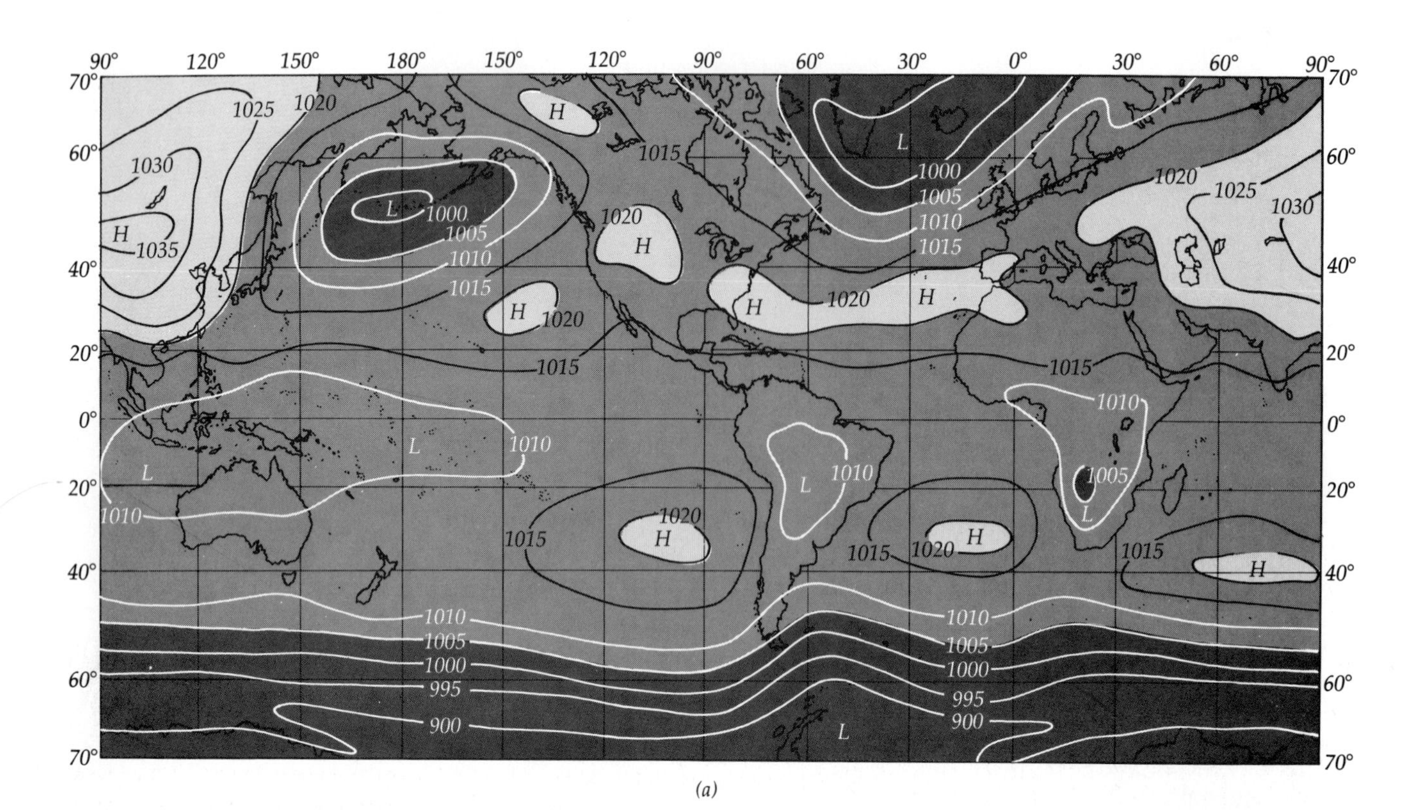

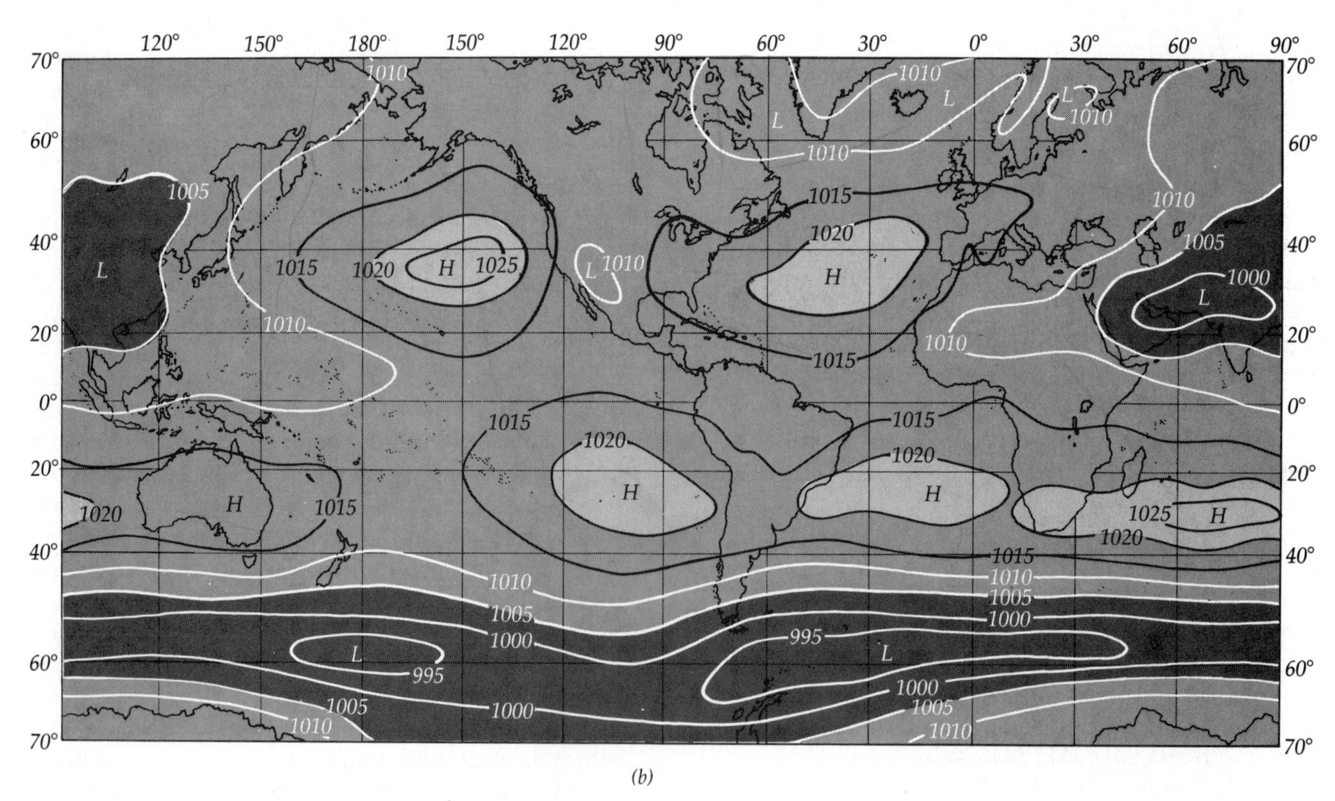

9.9 *Atmospheric pressure in millibars at mean sea level during (a) January and (b) July.*

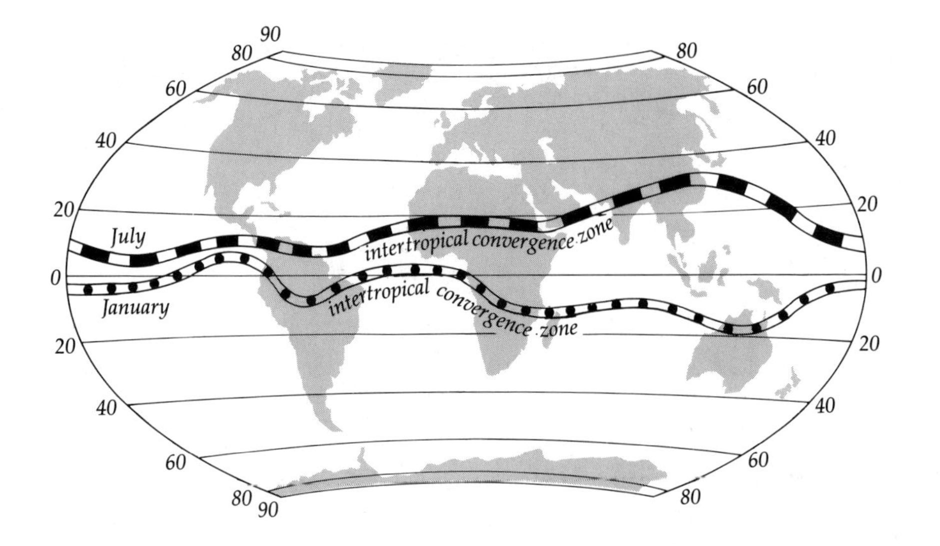

9.10 *The average location of the intertropical convergence, or equatorial trough, in January and July.*

In the atmosphere over the ocean corresponding to about 30° to 40° north and south latitude are regions of descending, dry air. In these subtropical high-pressure regions, the winds are light and variable, with calms occurring as often as a quarter of the time. The regions are marked by low rainfall and little cloudiness, especially on the eastern margins of the oceans and over the world's great deserts. In the North Atlantic, for example, the Sargasso Sea, which is the marine equivalent of a terrestrial desert, is centered in these latitudes. In Figure 7.2a, note the high surface salinity at the centers of the world's oceans, especially the North Atlantic. Here the fair weather and the relative lack of rain create a high rate of evaporation, thereby raising the surface salinity. These regions are called the **horse latitudes,** because the crews of becalmed Spanish ships short of food and water are said to have thrown their horses overboard while waiting for the wind to rise.

North and south of these latitudes, roughly between 35° and 60° north and south latitude, are well-developed belts in which descending low-level air moves poleward. Deflected by the Coriolis effect, the winds tend to blow mainly from the west, though they may blow from any direction. Widespread storms are common in these regions, though they are less intense than those in the regions of the trade winds. At latitudes between 40° and 60°, the prevailing **westerlies** encounter the polar easterlies, bringing frigid air from the Arctic and Antarctic regions. These regions are generally cloudy, with highly changeable weather and abundant rainfall, especially where the warm, moist air encounters the colder flow from the polar regions.

A comparison of global wind patterns (Figure 9.8) with the patterns of the ocean's surface currents (Figure 10.2) reveals the close relationship between the two. (We shall discuss currents in Chapter 10.) Each hemisphere has two zones of rising air: the tropics and the interface of the prevailing westerlies and the polar easterlies (usually called the polar front). Each hemisphere also has two zones of descending air—the horse latitudes and the polar regions—which are arid. While the amount of precipitation at both poles is low, the snow and ice remain there for a long time, making it seem that the poles experience heavy snowfall. We might expect these zones of rising and sinking air to form regular year-round bands of low and high pressure, as shown in Figure 9.9. Recall, however, that solar radiation varies significantly with the seasons from latitude to latitude. Moreover, atmospheric pressure over land masses is significantly different from atmospheric pressure over the ocean, even at the same latitude.

In the doldrums we find an equatorial trough formed by the convergence of the trade winds, as shown in Figure 9.10. This narrow, shifting zone, which is 480 to 960 km (300 to 600 mi) in width, is called the **intertropical convergence (ITC)** zone. It is rather persistent over the ocean, but it often disappears altogether over land areas. The weather of this region is very changeable, with sudden, severe showers lasting no more than thirty minutes or so. Immense cumulus clouds rise up more than 16 km (10 mi), releasing billions of calories each second into the upper atmosphere and triggering the mechanism that fuels the general circulation.

It was once thought that tropical weather was totally random, but satellite photos now show regular patterns. Tropical clouds form in clusters 150 to 950 km (93 to 589 mi) across. On a typical day, roughly thirty clusters lie scattered around the tropical zone, mainly over water. They all travel westward at about 12 knots, and they persist from two days to a week. From the water's surface, wet air is pumped up to 12,000 m (39,360 ft) and more. Below each cluster is an area of rain and thunderstorms. These great clusters deliver pulses of energy from the tropical ocean to the global winds for distances of up to 3,000 km (1,860 mi) north and south.

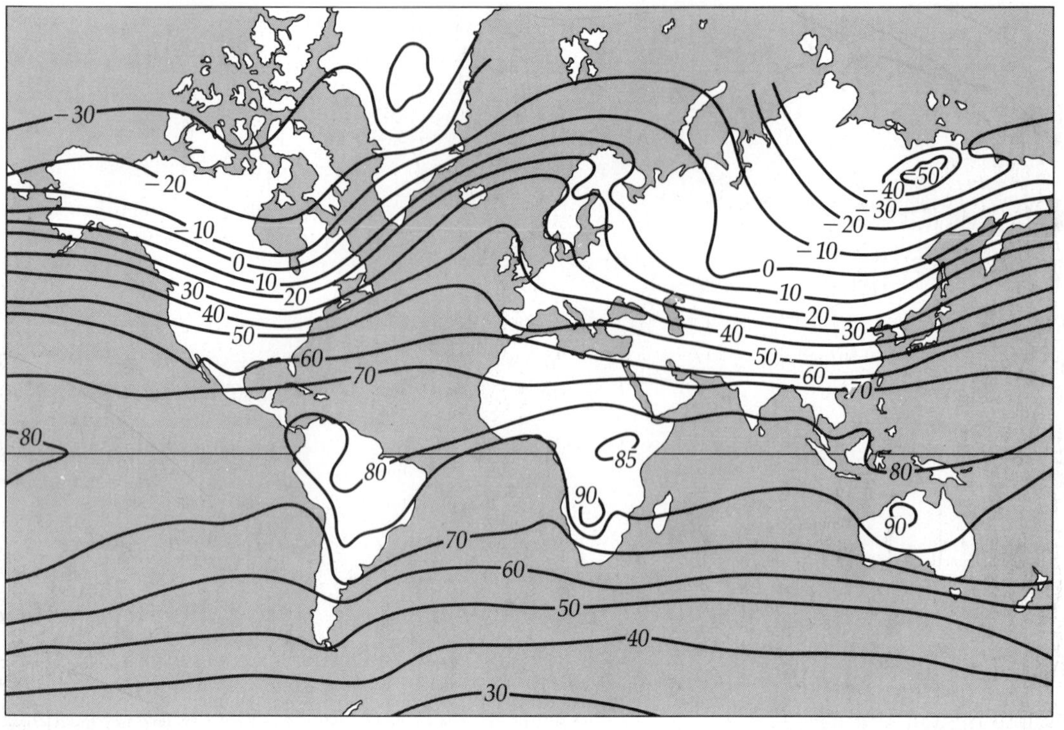

(a)

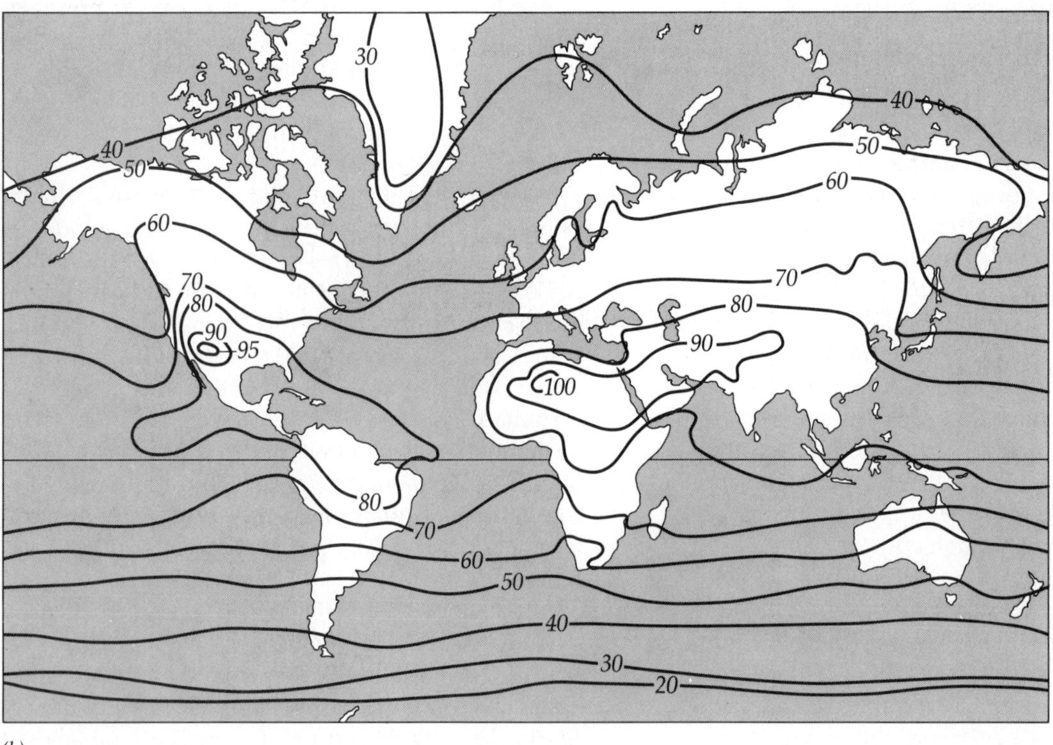

(b)

9.11 World mean surface temperature for (a) January and (b) July. (°F)

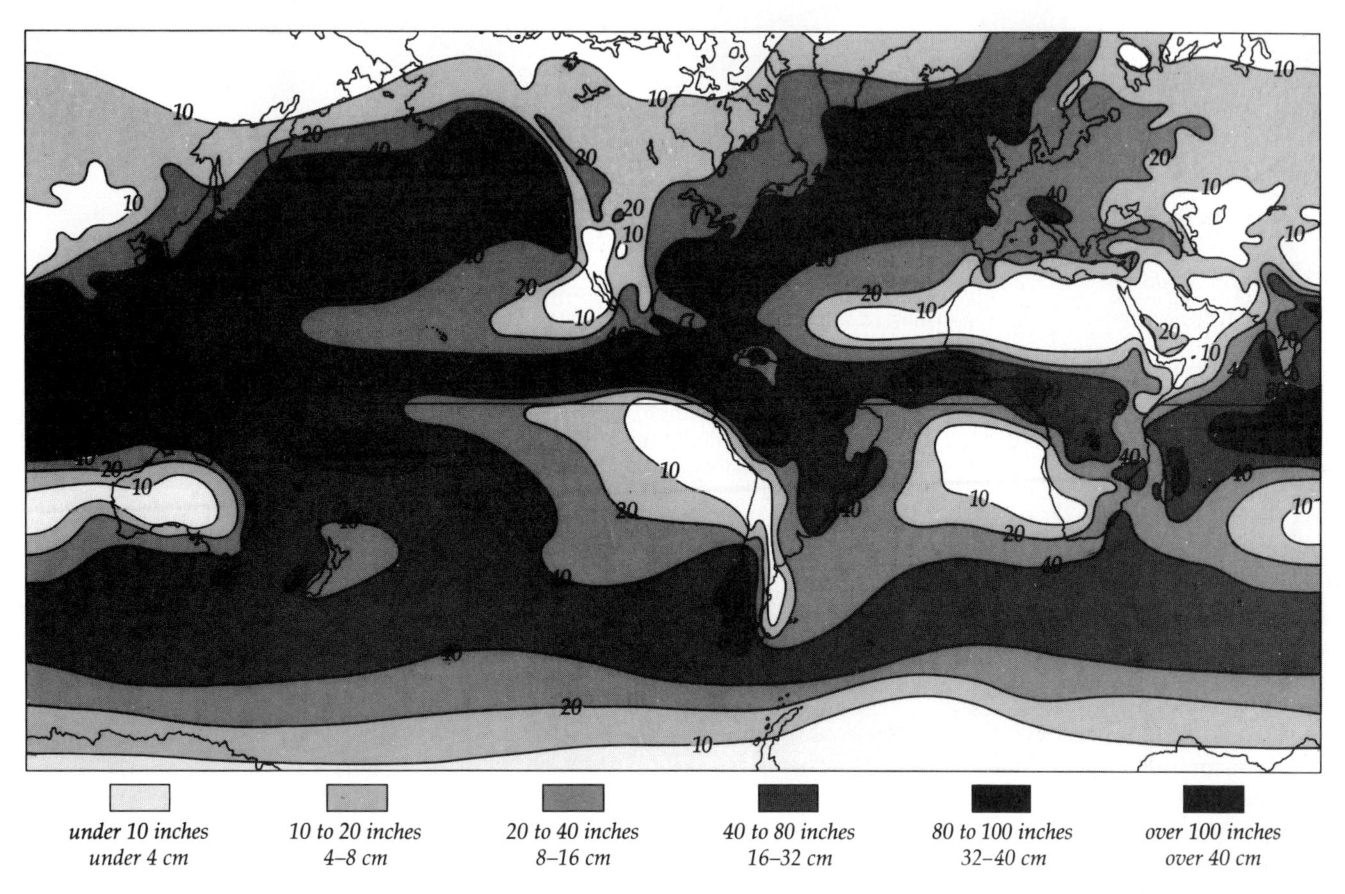

9.12 *Mean worldwide annual precipitation.*

Figure 9.9 shows the presence of subtropical high-pressure belts, known as oceanic highs. These belts are more or less permanent and are well defined in the summer months. In the winter they move about 5° nearer to the equator. They are not continuous, however; they are broken up into cells over the ocean—two in the Northern Hemisphere and three in the Southern. Air from higher altitudes accumulates to form these persistent high-pressure systems. They are the principal centers of air mass dispersion for the general circulation of the atmosphere.

Figure 9.9 also shows the presence of a large low-pressure region in the North Pacific called the Aleutian Low and another in the far North Atlantic called the Icelandic Low. These systems are formed when the warm, moist westerlies converge with the cold, polar easterlies. They are called oceanic subpolar lows. In the North Pacific and North Atlantic, the Aleutian and Icelandic lows generate a series of smaller cyclonic storms throughout the winter months. As Figure 9.9 indicates, these systems virtually disappear during the summer, especially in the North Pacific. The major storms of the mid-latitude regions originate and develop their greatest intensity in these immense low-pressure systems, bringing severe winter weather to the North American continent. The systems weaken considerably during summer, when the storms track further northward because of the warming trend in both sea and air and because of the contraction of polar sea ice.

Note that in both parts of Figure 9.9 there is a continuous band of low pressure around the Antarctic, stretching for 29,000 km (18,000 mi). An endless series of storm depressions, aptly termed the *Roaring Forties,* move through these oceanic subpolar lows. The westerlies are, on the average, appreciably stronger in the Southern Hemisphere than in the Northern.

The distribution of land within each hemisphere also influences climate. The Southern Hemisphere consists of 20% land and 80% ocean, whereas the Northern Hemisphere is 40% land and 60% ocean. Because the thermal properties of land are very different from those of the ocean, seasonal changes in the Northern Hemisphere are more pronounced. Away from the tropics, the land-sea distribution results in a climate division that considerably modifies the idealized zone boundaries into maritime and continental climatic regions. Maritime climates are characterized by small daily and seasonal temperature variations and high rainfall, whereas climates in the central continents tend to have low rainfall and large daily and seasonal temperature ranges (see Figures 9.5, 9.6, 9.11, and 9.12).

Table 9.5 *Velocities of Various Winds*

Name of Wind	Horizontal Velocity
Jet stream	Average 105 km/h (67 mph); maximum 630 km/h (392 mph)
Trade winds	Seldom in excess of 55 km/h (34 mph)
Hurricanes, Typhoons, Cyclones (strong tropical cyclonic storms)	Commonly to 220 km/h (135 mph); maximum 250–280 km/h (155–175 mph)
Weak tropical storms	64–87 km/h (40–54 mph)
Thunderstorms	Forward speed 20 km/h (12.4 mph); maximum range 65–80 km/h (40–50 mph). *Vertical velocity:* maximum at 10 km (33,000 ft), 291 km/h (181 mph); maximum at sea level, 58 km/h (36 mph)
Tornadoes	Rotational 480 km/h (300 mph); maximum 800 km/h (500 mph); average forward speed 48–64 km/h (30–40 mph); range 0–112 km/h (0–70 mph). *Vertical velocity:* maximum 322 km/h (200 mph)
Waterspouts	Forward speed 65–80 km/h (40–50 mph); rotational and vertical speed about ½ of tornadoes
Breezes	Temperate zones, maximum 12.9–19.3 km/h (8–12 mph); tropics, maximum 32.2–38.6 km/h (20–24 mph)

High-pressure and low-pressure systems vary in both magnitude and duration. Such major systems as the subtropical highs and the subpolar lows may range from a few thousand to 10,000 km (6,200 mi) in horizontal diameter and may persist for weeks or months. Cyclones (areas of lower pressure) and anticyclones (high-pressure systems) may measure from hundreds to several thousands of kilometers across and may persist for days or weeks. These systems are primarily responsible for day-to-day weather changes. They are also the main cause of surface waves in the ocean, commonly called wind waves. Thunderstorms, land-sea breezes, and tornadoes range from 1 to 100 km (0.62 to 62 mi) in width and last only for a few hours. Eddies range from a few centimeters to a kilometer or two in width and persist only for a few minutes. The winds that compose these systems differ appreciably in horizontal and vertical velocities, as shown in Table 9.5.

WEATHER

In the Northern Hemisphere, at varying altitudes up to about 10 km (6 mi), a wind blows unceasingly from west to east at high velocities, typically 60 knots but commonly over 200 knots. It moves in a zigzag path, much like a meandering river. Although the air of the upper atmosphere commonly concentrates into thin filaments 100 km (62 mi) wide and 1,900 to 3,000 m (1.2 to 1.9 mi) thick called jet streams, *the* jet stream usually refers to the stream situated on the polar side of the westerlies. Its general flow over Europe, North America, and the Soviet Union directs the path of the storms below it. As the jet stream zigzags around the globe, it carries warm air toward the North Pole and draws cold air away from it. The stream exhibits from two to six oscillations worldwide, depending on conditions at the earth's surface and the tendency of the stream to remain unbroken (Figures 9.13 and 9.14).

The jet stream is largely responsible for the persistent highs and lows. When it swerves toward the tropics, it narrows, concentrating the air and pushing it downward. In so doing, it creates a region of warm, dry, high-pressure air—a surface high marked by good weather (Figure 9.15). As it swerves toward the Arctic, it spreads out, rises, and draws air upward. In so doing, it creates a surface region of windy, wet weather called a low. Its movement influences the location of storms and the incidence of good weather in the lower air.

Associated with the jet stream are mobile surface storms (traveling depressions) that create the wet weather in the region from about 35°N to 60°N. These whirling storms are the chief source of rain in this region. They move eastward across the Atlantic from Newfoundland to Europe. They move more slowly than the jet stream, taking three to four days to make the crossing. These storms are caused by irregularities in the jet stream that produce regions of uplift 1,600 km (992 mi) in diameter. These storms commonly arise just off the east coast of North America, where changes in the path of the jet stream may cause significant changes in weather, and these storms are responsible for much of Europe's weather. The warm water off North America favors the production of these mobile depressions. The zigzag pattern that is responsible for a mild winter over Iceland may also bring tongues of

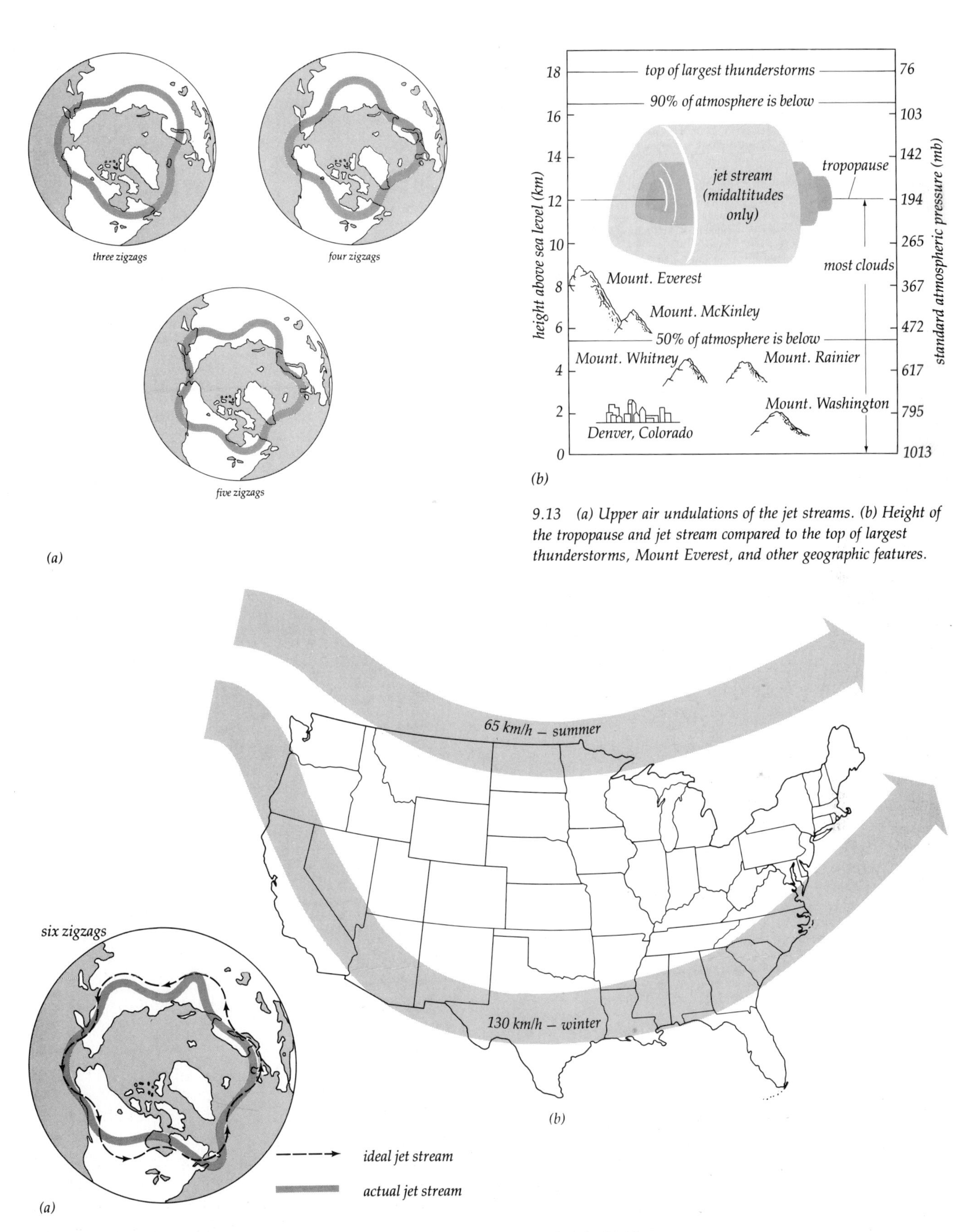

9.13 *(a) Upper air undulations of the jet streams. (b) Height of the tropopause and jet stream compared to the top of largest thunderstorms, Mount Everest, and other geographic features.*

9.14 *(a) An idealized six-zigzag pattern of the jet stream compared with an actual path. (b) The average position of the jet stream changes with the seasons; it is much farther south in the wintertime. The velocity of the jet stream also changes with seasons; the velocity is almost twice as great in the wintertime as in the summertime.*

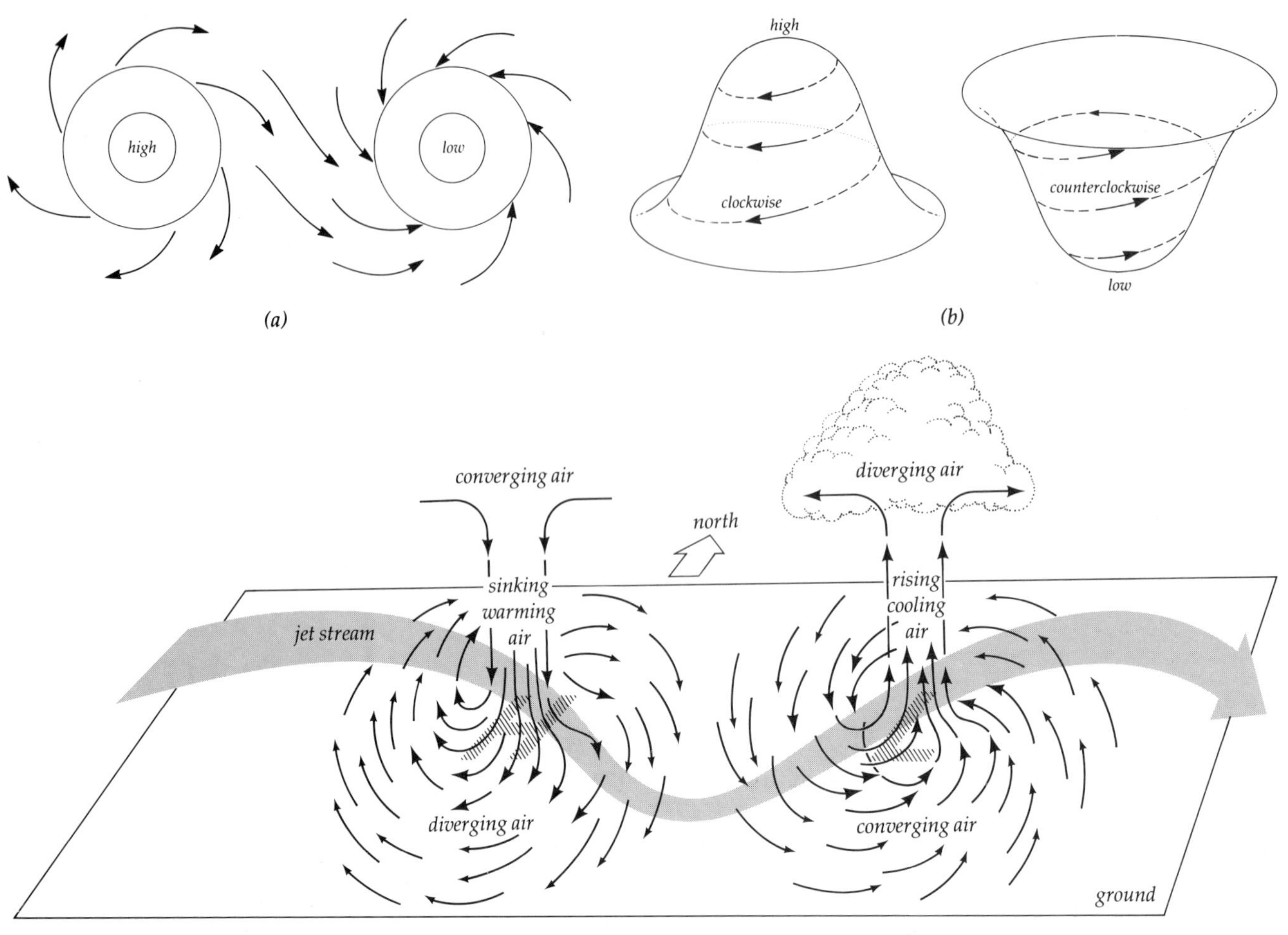

9.15 *(a) Winds flow out of highs into lows but in a circular path, not in a straight line, because the Coriolis effect causes them to deflect. (b) Motion of a parcel of air in a high and low. (c) Moving to the south, the narrowing jet stream concentrates air, driving it downward and creating a dry, warm region (that is, a high-pressure region, or a high). In swerving to the north, it spreads out, rises, and draws air upward, creating a wet, windy region (that is, a low-pressure region, or a low).*

9.16 *(a) Wave damage to the Seal Beach Pier in southern California during the severe winter storms of 1983. (Tom Heflick, U.S. Coast Guard) (b) Wind and wave damage to local vessels during the 1982–1983 winter, San Diego Harbor. (W. J. Wallace)*

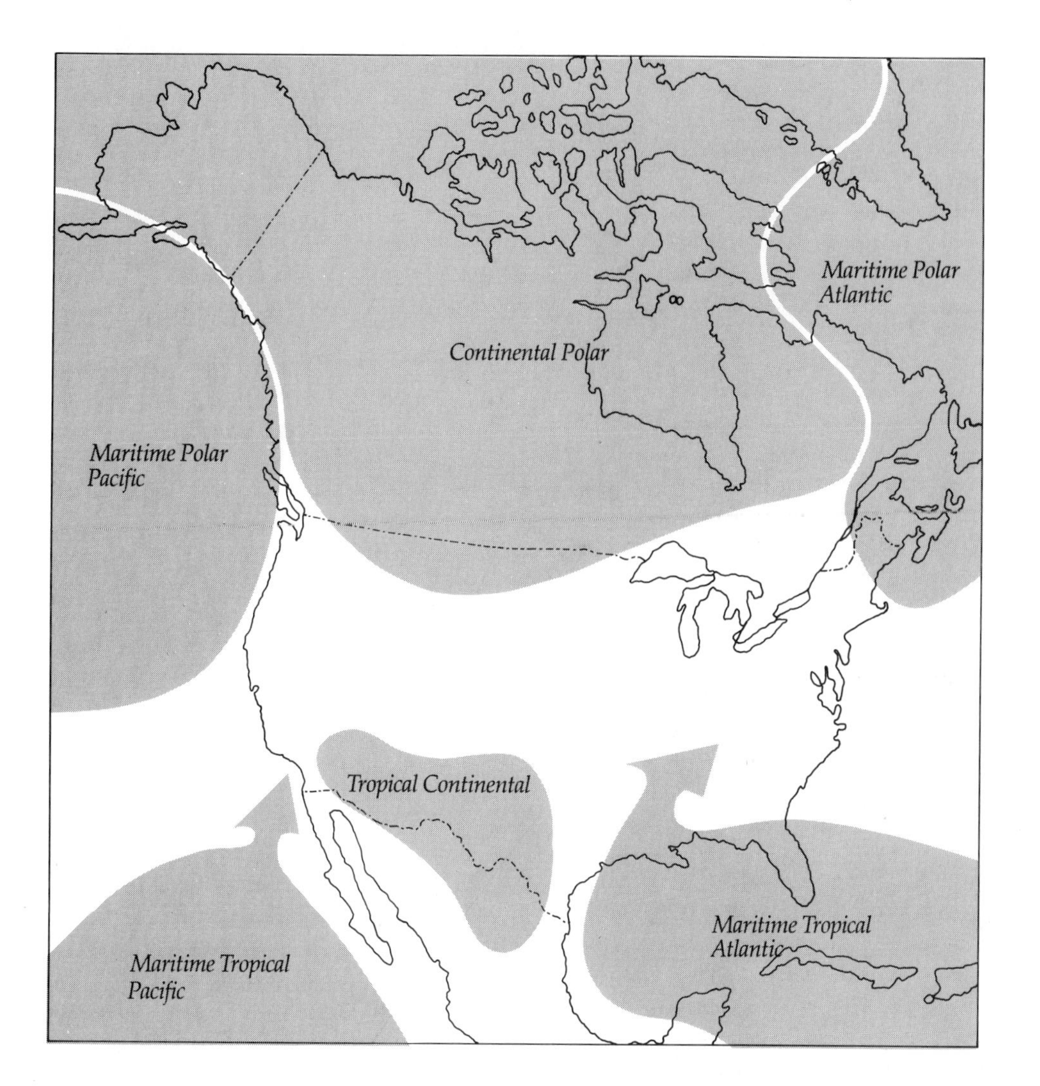

9.17 The six air masses that greatly influence weather in the Northern Hemisphere.

frigid air southward over New England and eastern Siberia as well.

The winter of 1982–83 lashed the southern California coast with a series of uncommonly violent storms (Figure 9.16). Apparently lower atmospheric pressures around the Aleutian Low and perturbations in the jet stream caused by large-scale changes in the surface temperatures of the North Pacific caused the storms to track through California.

The large masses of air that the jet stream transports are relatively homogeneous in temperature and humidity. They are classified according to their geographic source: polar or tropical and continental or maritime. North America is affected chiefly by the continental polar (cP), the maritime polar (mP), and the maritime tropical (mT) masses (Figure 9.17). The continental polar mass delivers most of the cold air, and the maritime polar mass brings gales to the Northeast and snows to the Pacific Northwest. The maritime tropical mass, heated by tropical waters, brings the United States much of its warmth.

When a cold-air mass collides with a warm-air mass, the boundary of their meeting is called a frontal zone, or simply a front. The front typically slopes upward over the colder, denser mass. Although fronts are marked by stormy weather, they are not themselves the causes of storms. They, too, are the by-product of depressions. Figure 9.18 shows a typical (winter) weather map with the fronts delineated.

TROPICAL CYCLONES

Tropical cyclones are great storms that whip the sea to a froth, generate immense waves, and do great damage to ships and coastal areas. They are a common feature of cyclonic circulation. Tropical cyclones are known by different names around the world (Table

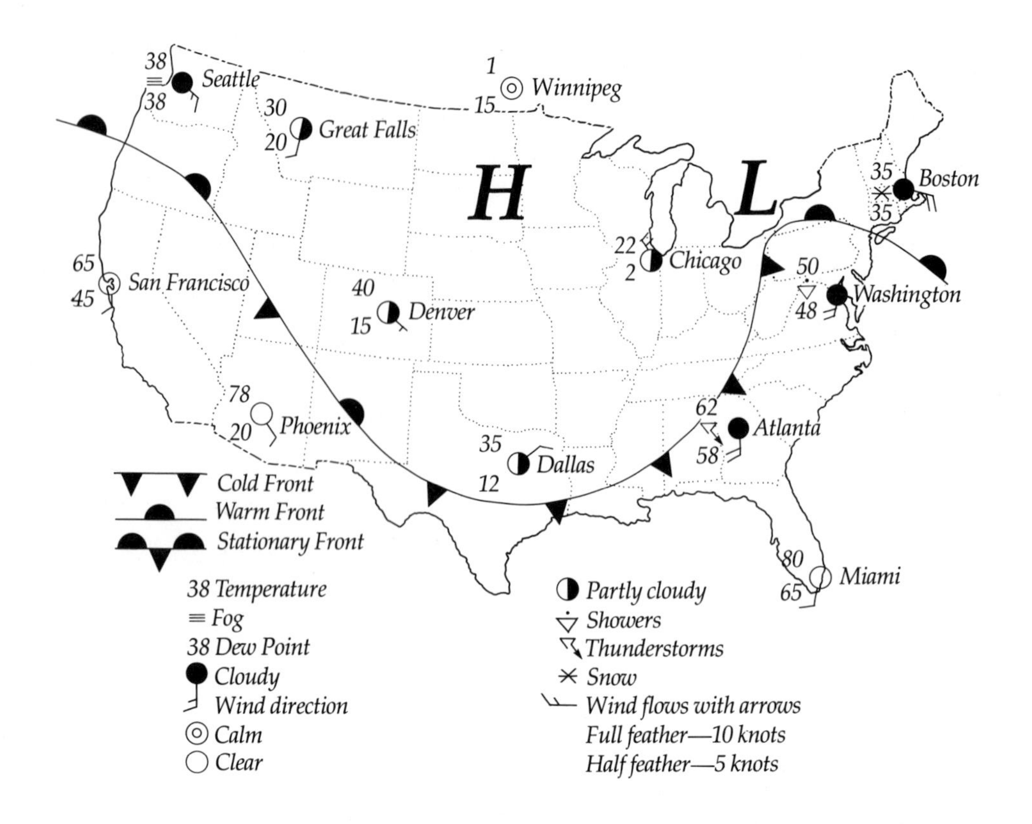

9.18 A simplified surface-weather map.

Table 9.6 *Tropical Cyclones*

Region	Local Name	Season
Arabian Sea	Cyclone	May–June, October–November
Australian region	Willy-willy	December–March
Bay of Bengal	Cyclone	May–June, September–November
North Atlantic	Hurricane	June–November
Northeast Pacific	Hurricane	June–November
Northwest Pacific and western Pacific	Typhoon	May–December primarily
Southwest Indian Ocean	Cyclone	December–April
Southwest Pacific Ocean (Philippines)	Baguio	December–March

9.6). They form over all the tropical oceans except the South Atlantic between 5° and 20° of latitude (Figure 9.19). The typhoons of the North Pacific are the most savage, and the cyclones of the Indian Ocean cause the greatest loss of life.

Tropical cyclones are formed from the scattered thunderstorms of a tropical cloud cluster above very warm water (over 26°C). They arise close to the equator, but far enough away from the Coriolis effect to produce a vigorous eddy. Figure 9.20 shows a cross section of a hurricane. Within the center, or the eye, which is about 24 km (15 mi) in diameter, the winds are light. Around the eye is a wall-like updraft. This updraft fuels the storm by creating a partial vacuum that brings hot, damp air into the eye.

After forming, tropical cyclones usually move westward, although their path is often erratic. Winds of up to 110 knots may persist over a ten-day period. They break up rapidly when they encounter land because of friction and because of the loss of their energy source. Figure 9.21 shows a satellite photograph of a hurricane. These great storms create severe flooding, called storm surges, and cause many deaths in coastal areas.

Storm Surges

The effects of a **storm surge** are similar to those of a tsunami (Chapter 11), although the phenomenon itself is quite different. A tsunami often appears during a

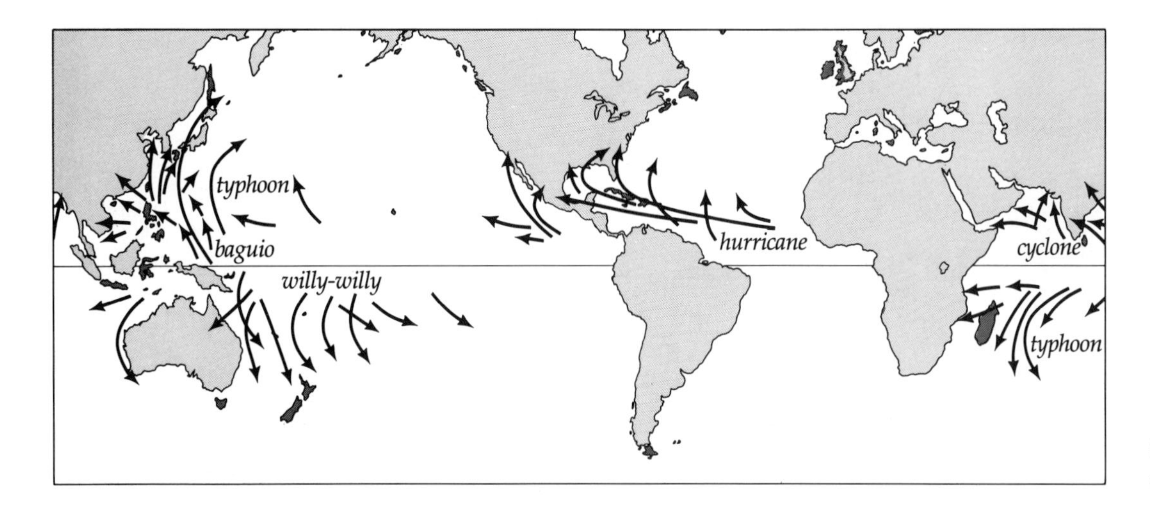

9.19 Typical tropical cyclone tracks.

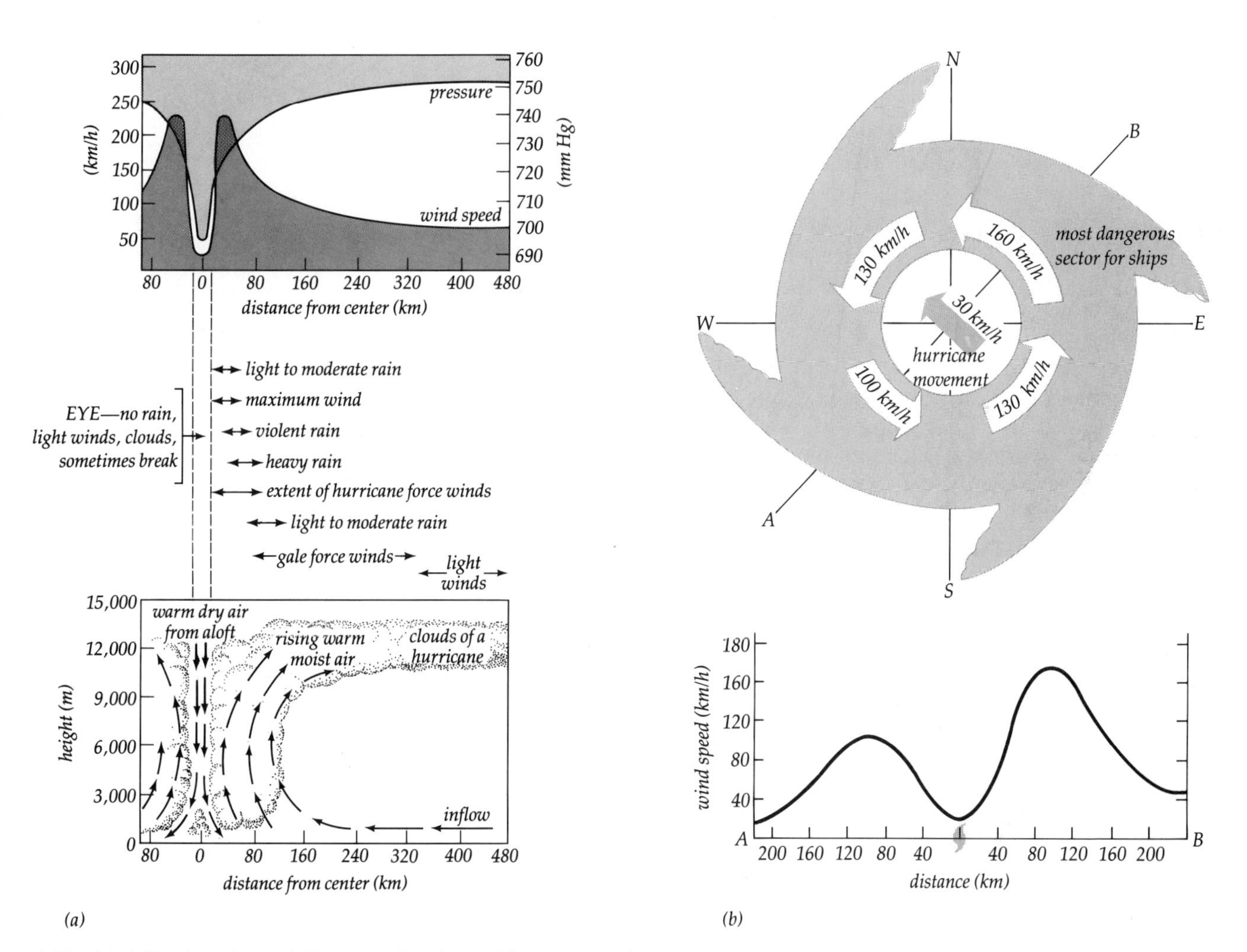

9.20 (a) A Hurricane in vertical cross section along with its pattern of pressure, rain, and wind. (b) Wind speeds around the eye of a hurricane are influenced by the forward motion of the storm. A forward motion of 30 km/h toward the northwest would add to the winds in the northeast quadrant of the storm while subtracting from the winds in the southwest quadrant. As a result, the right-hand side of the hurricane is more dangerous and destructive as indicated in the wind speeds measured in hurricanes.

9.21 *A vertical view of a hurricane from Apollo 9.*

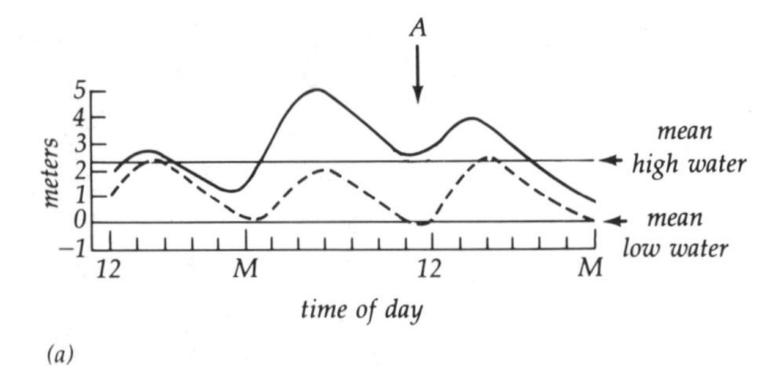

(a)

(b)

9.22 *(a) Water-level diagram for a storm surge. The solid line indicates water level during the surge. At its low level (point A, for example), the surge is higher than mean high water, even though the coming and going of the water over this 36-hour period corresponds to the times of the normal tides (the dotted line). (b) The Richelieu Apartments on U.S. Highway 90 in Pass Christian, Mississippi, before and after Hurricane Camille, August 17, 1969. Twenty-five people decided to ignore the storm warnings and hold a hurricane party to ride out the hurricane in this apartment building. The two who survived the storm were severely injured as winds greater than 200 km/h and a storm surge of 7 m struck. (Photographs by Chauncey T. Hinman)*

spell of good weather with little or no warning. In contrast, storm surges result from a combination of high tides and rising sea level during periods of bad weather. The storms that produce them are characterized by low-pressure systems, which cause a bulge in the sea surface below them. Driven as well by steady, high winds blowing for ten to twelve hours or more, vast volumes of water crash onto the shore. The flooding and surging water may persist over more than one tidal cycle (Figure 9.22), and storm waves piled atop the flood water can add to the devastation.

When a moving storm or a change in wind direction coincides with the natural period of oscillation of a basin, exceptionally high tides may occur along the coastal regions of semi-enclosed seas like the North Sea. The gale that swept down from the North Sea and piled water against the Dutch coast on February 1, 1953, was such a storm surge. Almost 1,800 people were drowned, and 3,200 km^2 (800,000 acres) of this dike-rimmed, low-lying country were flooded. The chances that all the essential conditions for such a disaster will combine at a given time and place have been calculated at once in 400 years.

The Galveston flood of 1900 was actually a storm surge. A hurricane with high winds of 193 km/h (120 mph) and storm waves pushed the normal 0.7-m (2.3-ft) tides to 5 m (16.4 ft), and low pressure raised the water level to 8 m (26 ft). More than 5,000 people died, and the city was virtually destroyed. The 1938 New England hurricane, which caused the sea's surface to rise 4.6 m (15 ft), brought devastating waves and flooding and altered many coastal features. In August 1954, Hurricane Carol battered the New England coast and caused damage of more than $40 million.

In October 1960, a storm surge near the mouth of the Ganges River picked up an American freighter and set it down 1.6 km (1 mi) inland. In November 1970, a massive bulge of water 6 m (19.7 ft) high, created by a great cyclone, struck the densely populated coast of Bangladesh. More than 500,000 people died in the worst natural catastrophe since the Yellow River flood of 1931 in China.

Unfortunately, despite improved warning systems, the potential danger of storm surges is increasing—not because they are increasing in frequency or

9.23 *A waterspout. (National Oceanic and Atmospheric Administration)*

intensity, but because the world's coastal populations are growing rapidly. There are more people in the path of storm surges.

Waterspouts

A less dramatic atmospheric disturbance is the waterspout (Figure 9.23). A waterspout is not really a spout of water; it is a weak tornadolike phenomenon, formed like tornadoes over land. Just as the tornado's funnel sucks up dirt and debris, the waterspout picks up water. It is a rapidly swirling mass of air that carries water as high as 100 m (328 ft). Unlike its land counterpart, however, it is short-lived and not particularly dangerous. Waterspouts can raise the water level by as much as 3 m (10 ft). Like tornadoes, they are generated when cold air squalls from neighboring thunderclouds collide and set a vertical core of air spinning. Table 9.7 gives the locations along the U.S. coasts of waterspout frequencies of occurrence.

Table 9.7 *Locations of Frequent Waterspouts in the United States*

Location	Waterspouts per year per 1,000 km^2
Florida Keys	43.8
Palm Beach, Fla.	3.1
Tampa Bay, Fla.	2.4
Ft. Lauderdale, Fla.	2.4
Corpus Christi, Tex.	2.3
Greater Miami, Fla.	2.2
Pensacola, Fla.	1.8
Mississippi Sound, Miss.	1.2
Port Arthur, Tex.	1.2
Galveston Bay, Tex.	0.9
Mississippi River Delta	0.6
Ft. Meyers, Fla.	0.5

Source: J. H. Golden, "An Assessment of Waterspout Frequencies along the U.S. East and Gulf Coasts," *Journal of Applied Meteorology,* 1977, *16*(3); based on reports from 1959 to 1973.

RECIPROCAL INFLUENCES OF ATMOSPHERE AND OCEAN

The most obvious effect of the atmosphere on the ocean is on the ocean's surface circulation (to be discussed in Chapter 10). Once surface currents have been set in motion by the atmosphere, they persist long after the causative conditions have ceased to exist in the rapidly changing atmosphere. Moreover, the varying temperatures and humidities of the winds affect the temperature and salinity of the ocean and influence circulation deep below the surface. In addition, storms directly affect the temperature of the surface waters and create waves capable of altering the sea level temporarily.

Conversely, the surface temperature of the ocean may affect storms. Hurricanes in the Gulf of Mexico have intensified as they have passed over a region of anomalously high surface temperature, and hurricanes in the North Pacific have diminished in intensity as they have passed over the cool waters west of Baja California. In coastal areas, the temperature of the ocean determines fog or frosting conditions and the incidence of sea ice.

Temperature shifts of only a degree or two in patches of the ocean may affect weather around the world. The winter of 1971–72 was unusually mild in the eastern United States, and the coastal waters of the western Atlantic were appreciably warmer than normal. Greenland waters, however, were exceptionally cold. The contrast between these two patches of water caused a strengthening of the jet stream and diverted it far to the north. Its oscillations brought warm, dry weather to Russia, causing the grain harvest of 1972 to fail. Other undulations brought cold, wet weather to the eastern United States and Great Britain, producing a dismal summer. A low-pressure area along the East Coast of the United States drew in Hurricane Agnes, which caused extensive flooding and loss of life and property. Agnes was the costliest storm in American history.

Patches of the ocean 100 km (62 mi) wide and 100 m (328 ft) deep have been associated with sea surface temperature (SST) anomalies of 1°C or 2°C (2° or 4°F). These slight anomalies have influenced seasonal weather in several regions. Data assembled over thirty years suggest a relationship between large-scale SST anomalies and atmospheric changes that influence weather and climate. However, interactions at the air-sea boundary are extremely complicated, and the specific mechanism by which SST anomalies influence climate are not yet fully understood.

THE FUTURE

Recent research has made it clear that weather conditions around the earth are related. The good effects that the jet stream brings to one region or country may well be associated with bad effects in another. While the weather in one hemisphere generally seems to have little to do with weather in the other, they may be interdependent.

The "little ice age" of 1431 to 1850 altered the course of European social and political history and brought about radical changes in technology. The Great Fire of London in 1666 was in large part due to the extreme summer droughts of the 1660s. Weather has affected people throughout history. It seems that the last of this century's really good summers in northwestern Europe occurred in 1949 and that the onset of really cold winters came in 1939–40.

Is our climate changing? Certainly! Why? Because it always has. In what direction will it change? Only time will tell.

FURTHER READINGS

Calder, N. *The Weather Machine*. New York: Viking Press, 1975.

Eagleman, J. R. *Meteorology: The Atmosphere in Action*. Belmont, Calif.: Wadsworth, 1985.

Emiliani, C. "Climate Cycles." *Sea Frontiers*, 1971, *17*(2), 108–20.

Harvey, J. G. *Atmosphere and Ocean: Our Fluid Environments*. New York: Crane, Russak, 1978.

Heidorn, K. C. "Land and Sea Breezes." *Sea Frontiers*, 1975, *21*(6), 340.

McIntosh, D. H., and A. S. Thom. *Essentials of Meteorology*. London: Wykeham, 1972.

Miller, A., and J. C. Thompson. *Elements of Meteorology*. Columbus, Ohio: Charles E. Merrill, 1975.

Mooney, M. J. "Waterspout vs. Marina." *Sea Frontiers*, 1978, *24*(3), 159.

"Oceans and Climate." *Oceanus*, Fall 1978, *21*(4).

Perry, A. H., and J. M. Walker. *The Ocean Atmosphere System*. New York: Longman, 1977.

Picard, G. L., and W. J. Emery. *Descriptive Physical Oceanography*. Oxford: Pergamon Press, 1982.

Robinson, J. P., Jr. "Galveston's Killer Hurricane of 1900." *Sea Frontiers*, 1981, *27*(3), 166.

Smith, F. G. W. "Planet's Powerhouse." *Sea Frontiers*, 1974, *27*(4), 195–203.

Sobey, E. "The Ocean-Climate Connection." *Sea Frontiers*, 1980, *26*(1), 25.

Stewart, R. W. "The Atmosphere and the Oceans." *Scientific American*, September 1969, pp. 76–105.

Tchernia, P. *Descriptive Regional Oceanography*. Oxford: Pergamon Press, 1980.

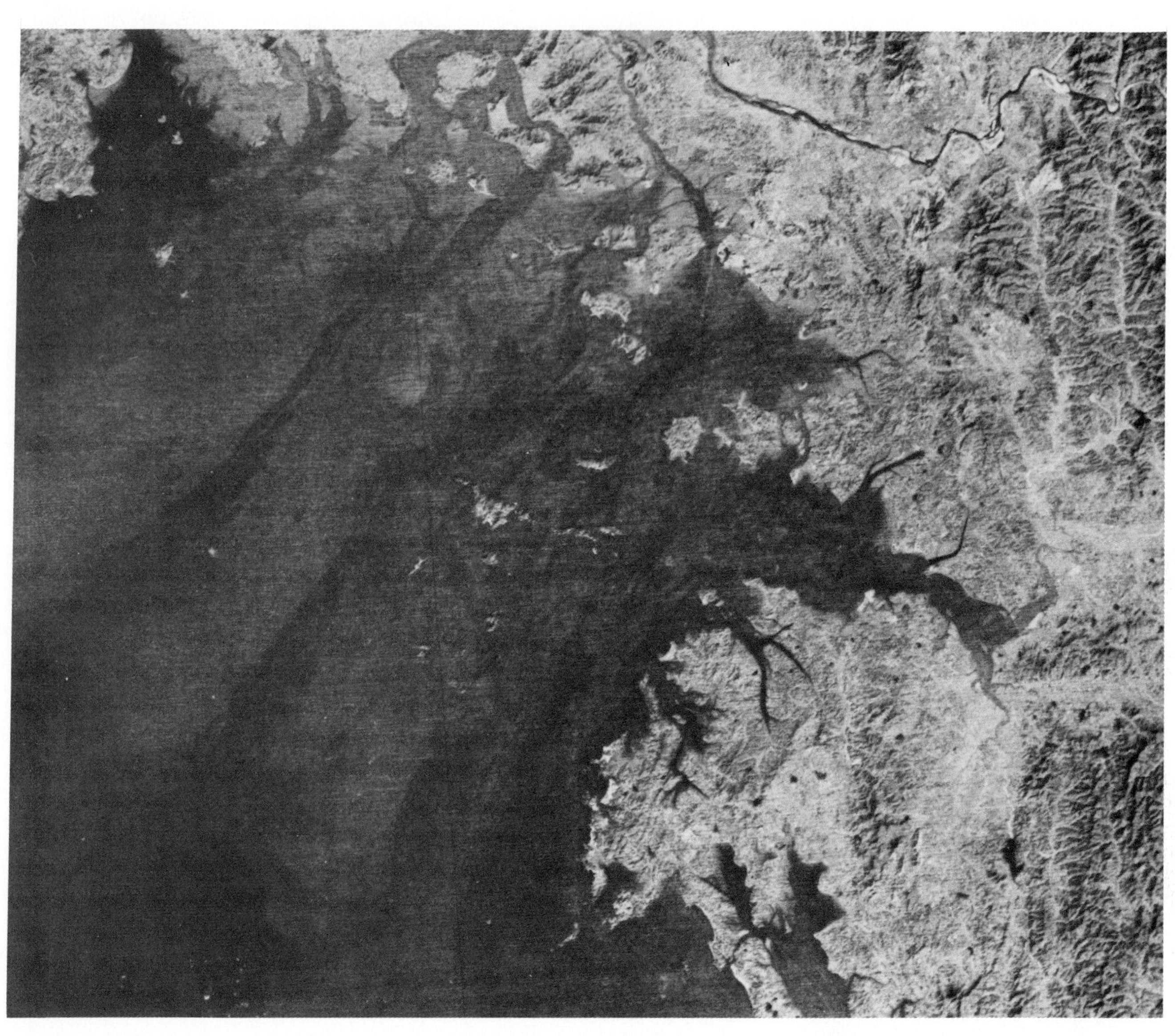

Fueled by the blazing tropic sun,
Fanned by the constant winds that blow,
The sprawling North Pacific spins
In clockwise motion, sure and slow,
Swinging past the stony headlands,
Unfettered in its timeless flow.

Herb Minshall

TEN

Ocean Circulation

Mariners long ago became familiar with the surface currents of the ocean, probably as they compared the courses they attempted with the courses they actually achieved. Commanders of the Spanish galleons of the fifteenth and sixteenth centuries took advantage of the prevailing winds and of certain currents. Ponce de León, in his pursuit of the Fountain of Youth, described the Florida Current in 1513.

In 1770 Benjamin Franklin, then deputy postmaster general, had a chart of the Gulf Stream drawn to help ships accelerate mail delivery between Europe and the colonies (Figure 10.1). He had noted that mail packets often took two weeks longer than merchant ships and whalers to make the Atlantic crossing. In his numerous journeys to Europe, Franklin measured the temperature of the water and observed its colors to help plot the course of this major current.

For convenience, we can divide the circulation of the ocean into two related parts: surface circulation and deep-ocean circulation. Since **surface circulation** is largely caused by atmospheric circulation (that is, winds), it is usually referred to as wind-driven circulation. **Deep-ocean circulation** refers to water movements caused by changes in the water's density resulting from changes in temperature and salinity.

SURFACE CIRCULATION

Unlike waves, which bring about virtually no water transport, surface currents move and mix the waters of the ocean. Their patterns are essentially the same today as they were when they were described by the Spanish in the sixteenth century. The reason is that surface currents, which occur mainly in the first several hundred meters of depth, are caused by the wind. The English natural philosopher Edmund Halley first suggested the relationship between currents and wind in 1686. Surface currents, like the wind, follow fixed patterns.

Figure 10.2 charts the ocean's surface circulation. Note that in the North Atlantic and North Pacific, the currents form large, circular, closed, clockwise (or anticyclonic) cells called **gyres**, centered about latitude 30°N. In the Southern Hemisphere, the patterns are the same, but the direction of the gyre is reversed.

The patterns shown in Figure 10.2 are **mean currents**. Seasonal shifts in wind patterns also shift the patterns of the surface currents. Nevertheless, the overall circulation patterns of the northern and southern oceans are very similar, although differences are evident in specific regions.

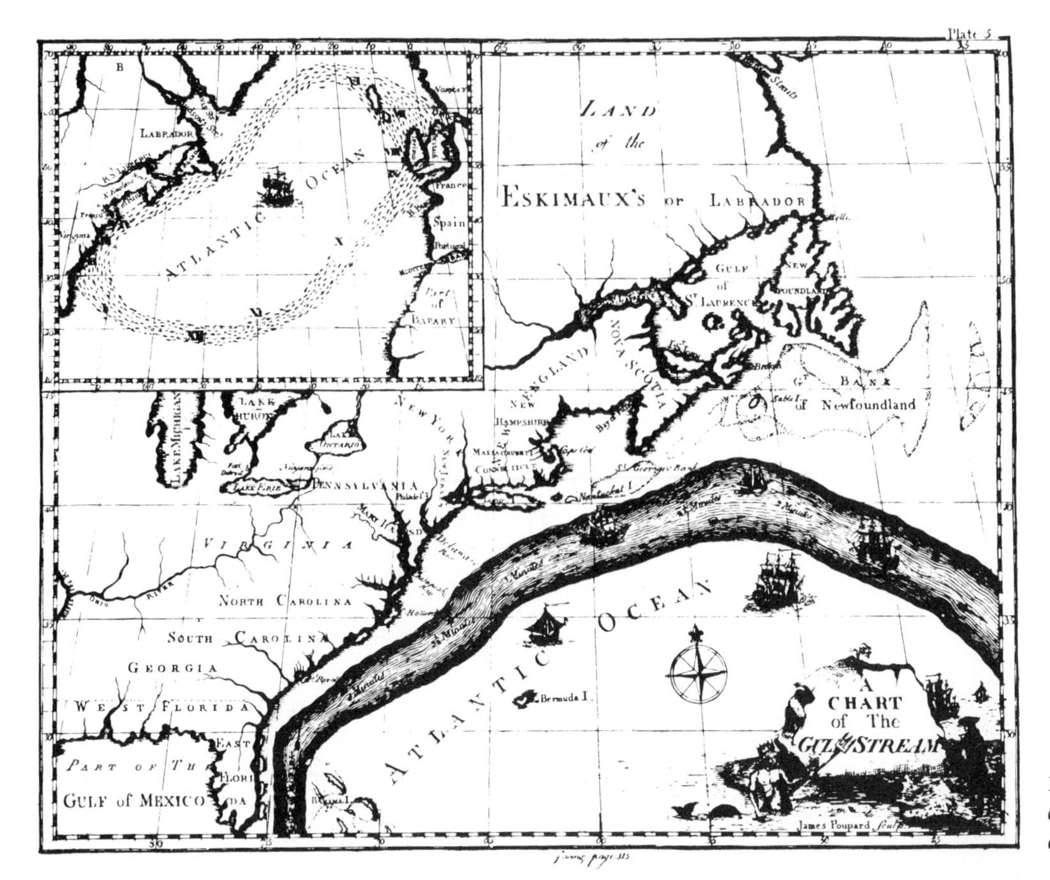

10.1 *Benjamin Franklin's chart of the Gulf Stream. (Photo courtesy U.S. Navy)*

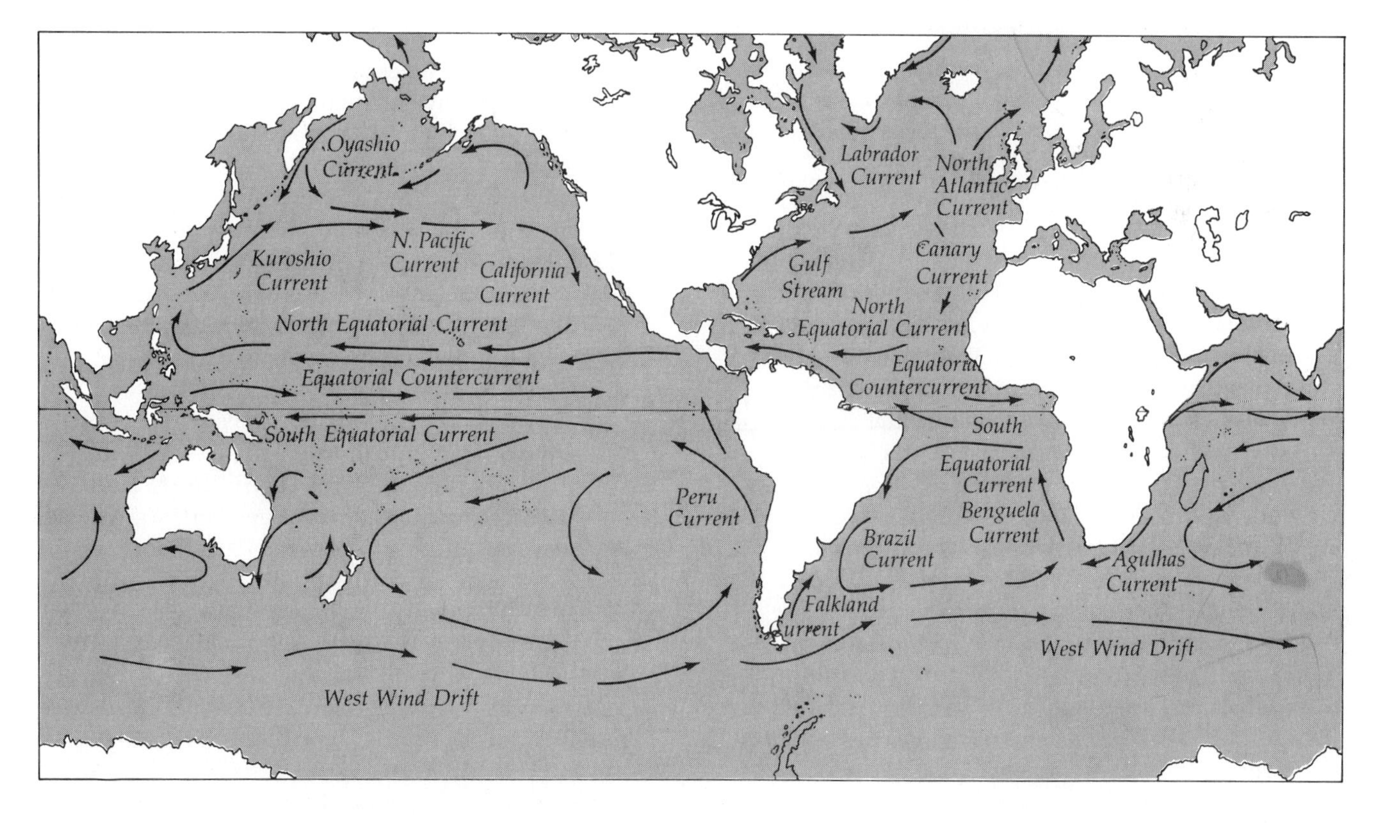

10.2 *Average ocean surface circulation.*

Some Major Ocean Currents

The currents in each of the five ocean basins form distinct gyres. Though similar from basin to basin, the gyres differ in size and in direction of flow. Let's look more closely at the surface circulation of the North Atlantic and the North Pacific. When we use the terms **surface water**, or *upper water,* in the following discussion, we are referring to the water between the surface and the **thermocline** (the depth interval at which temperature changes rapidly with further depth, usually between about 300 to 1,000 m (984 to 3,280 ft). The waters below this depth are known as deep waters.

The North Atlantic Refer to the world wind patterns in Figure 9.8 and to Figure 10.2 once again. The North Atlantic clockwise gyre consists of four segments: the North Equatorial Current, the Gulf Stream, the North Atlantic Current, and the Canary Current. Originating near the Cape Verde Islands, the **North Equatorial Current**, located between latitudes 10°N and 25°N in summer and 5°N to 25°N in winter, is driven by the northeast trade winds across the North Atlantic. There the major part flows into the Caribbean through openings between the Lesser and Greater Antilles. Some of the flow is deflected northward as the **Antilles Current**. Local winds then drive the waters west into the Gulf of Mexico through the Yucatán Channel. Following the gulf's own anticyclonic gyre, they escape into the North Atlantic through the Straits of Florida as the **Florida Current**, ultimately forming the **Gulf Stream**.

The following description of the Gulf Stream is taken from the National Oceanic and Atmospheric Administration's annual current tables:

> The region where the Gulf of Mexico narrows to form the channel between the Florida Keys and Cuba may be regarded as the head of the Gulf Stream. From this region the stream sets eastward and northward through the Straits of Florida, and after passing Little Bahama Bank it continues northward and then northeastward, following the general direction of the 100-fathom curve as far as Cape Hatteras. The flow in the Straits is frequently referred to as the Florida Current.
>
> Shortly after emerging from the Straits of Florida, the stream is joined by the Antilles Current, which flows northwesterly along the open ocean side of the West Indies before uniting with the water which has passed through the straits. Beyond Cape Hatteras, the combined current turns more and more eastward under the combined effects of the deflecting force of the earth's rotation and the eastwardly trending coastline, until the region of the Grand Banks of Newfoundland is reached.
>
> Eastward of the Grand Banks the whole surface is slowly driven eastward and northeastward by the prevailing westerly winds to the coastal waters of northwestern Europe. For distinction, this broad and variable wind-driven surface movement is sometimes referred to as the North Atlantic Drift or Gulf Stream Drift.
>
> In general, the Gulf Stream as it issues into the sea through the Straits of Florida may be characterized as a swift, highly saline current of blue water whose upper stratum is composed of warm water.
>
> On its western or inner side, the Gulf Stream is separated from the coastal waters by a zone of rapidly falling temperature, to which the term "cold wall" has been applied. It is most clearly marked north of Cape Hatteras but extends, more or less well defined, from the Straits to the Grand Banks.
>
> Throughout the whole stretch of 400 miles in the Straits of Florida, the stream flows with considerable velocity. Abreast of Havana, the average surface velocity in the axis of the stream is about 2.5 knots. As the cross-sectional area of the stream decreases, the velocity increases gradually, until abreast of Cape Florida it becomes about 3.5 knots. From this point within the narrows of the straits, the velocity along the axis gradually decreases to about 2.5 knots off Cape Hatteras, N.C. These values are for the axis of the stream where the current is at a maximum, the velocity of the stream decreasing gradually from the axis as the edges of the stream are approached. The velocity of the stream, furthermore, is subject to fluctuations brought about by variations in winds and barometric pressure.
>
> From within the straits, the axis of the Gulf Stream runs approximately parallel with the 100-fathom curve as far as Cape Hatteras. Since this stretch of coast line sweeps northward in a sharper curve than does the 100-fathom line, the stream lies at varying distances from the shore. The lateral boundaries of the current within the straits are fairly well fixed, but when the stream flows into the sea, the eastern boundary becomes somewhat vague. On the western side, the limits can be defined approximately, since the waters of the stream differ in color, temperature, salinity, and flow from the inshore coastal waters. On the east, however, the Antilles Current combines with the Gulf Stream, so that its waters here merge

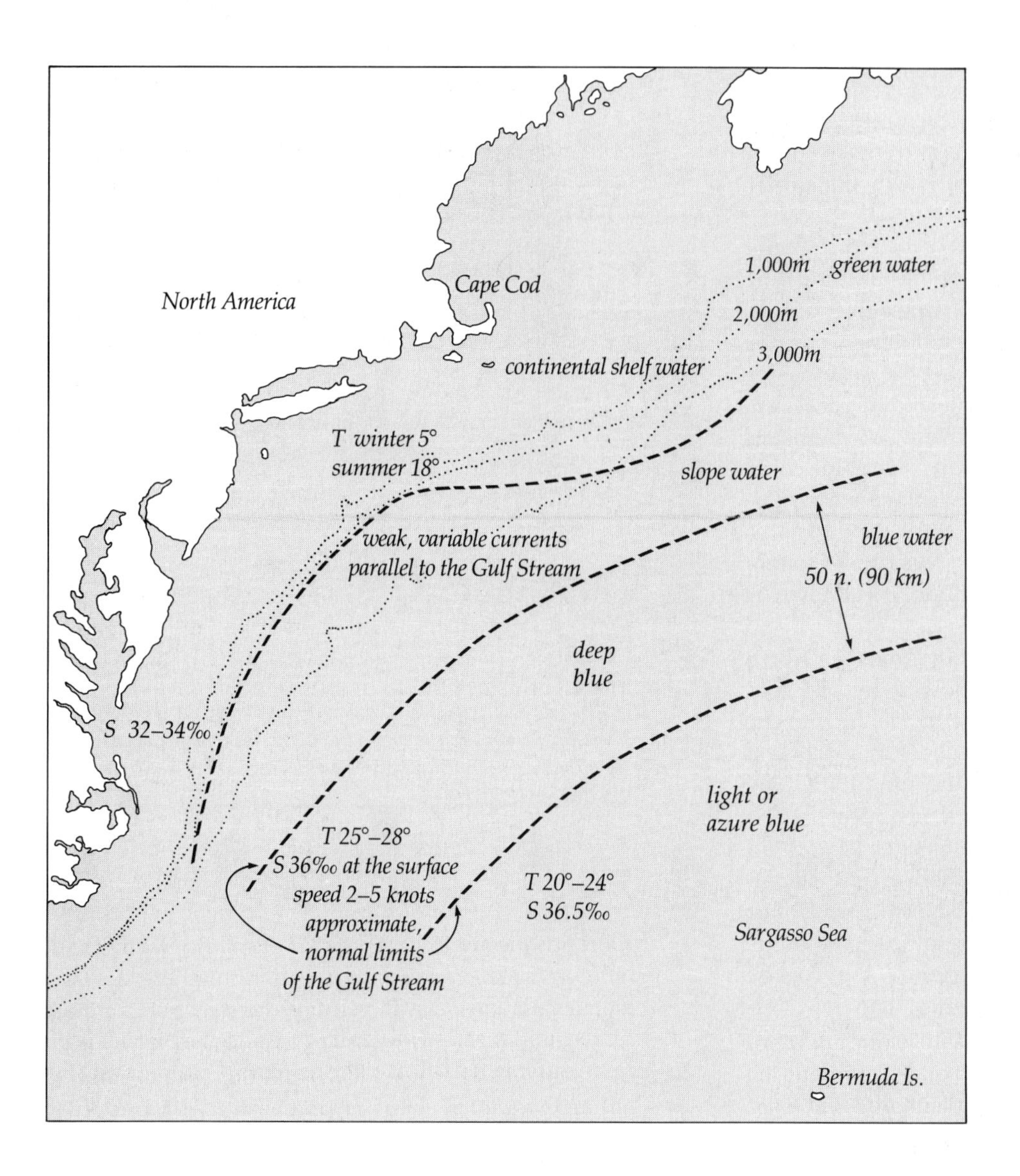

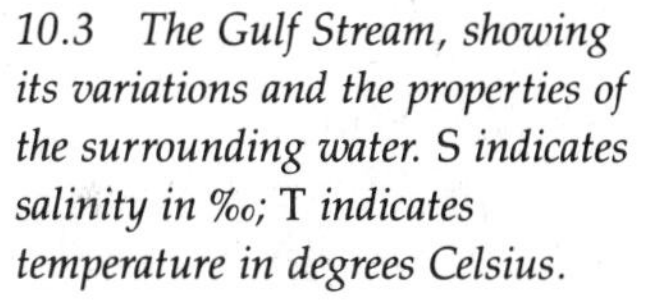

10.3 The Gulf Stream, showing its variations and the properties of the surrounding water. S *indicates salinity in ‰;* T *indicates temperature in degrees Celsius.*

gradually with the waters of the open Atlantic. Observations of the National Ocean Survey indicate that, in general, the average position of the inner edge of the Gulf Stream as far as Cape Hatteras lies inside the 50-fathom curve. The Gulf Stream, however, shifts somewhat with the seasons, and is considerably influenced by the winds which cause fluctuations in its position, direction, and velocity; consequently, any limits which are assigned refer to mean or average positions.

Figure 10.3 shows the approximate location of the Gulf Stream along the East Coast of the United States. The narrow band of colder, less saline water at the western edge of the Gulf Stream (the **"cold wall"**) is shown cross-sectionally in Figure 10.4. The deep bluish green color of the water in this band gives way to the ultramarine of the Gulf Stream itself.

East of the Grand Banks, the Gulf Stream encounters the south-flowing **Labrador Current**. Here it becomes less defined and divides into separate filaments known collectively as the **North Atlantic Current**, or drift. A major northern branch flows into the North Atlantic above 60°N and forms a complicated cyclonic gyre (Figure 10.5). The southeastern branch reaches the coasts of western Europe, sending filaments into the English Channel, and continues its flow southward as the **Canary Current** along the Iberian Peninsula. At last it returns to the region where the North Equatorial Current originates, thus closing the gyre.

The North Pacific The pattern of surface circulation in the North Pacific is similar to that of the North Atlantic. The anticyclonic (clockwise) gyre here is driven by the same wind patterns, including the northeast trades. The Pacific, however, is larger than the Atlantic and is less affected by the presence of continents. Moreover, unlike the Atlantic, the Pacific is essentially closed at its northern extremity. The currents that make up the North Pacific gyre are the North Equatorial, the Kuroshio, and the California.

The North Equatorial segment of the North Pacific gyre flows west toward the Philippines and north to about 12°N, where it divides into two segments. The northern flow moves along the Philippines, curving more and more to the northeast along the eastern edge of Taiwan and then on to the eastern coast of Japan. There it is called the **Kuroshio Current,** which means "black current," referring to its deep cobalt-blue color. (This color and the color of the Gulf Stream are characteristic of water with low biological productivity.) Leaving the Japanese coast, the Kuroshio Current moves to the east. By 140°E longitude, it is known as the **Kuroshio Extension**.

Beyond about 170°E, the flow becomes the **North Pacific Current**, which is then joined by the cold **Oyashio Current**. The combined currents flow to the western coast of North America, where they divide at about 50°N in summer (42°N in winter) and 150°E. The southern branch flows southward and appears as the cold **California Current** along Vancouver Island and the U.S. West Coast. The California Current is analogous to the Canary Current in the North Atlantic but is better defined because the driving anticyclonic winds (Figure 9.9b) are compressed against the coastal mountains of western North America. The flow continues back to the North Pacific Equatorial Current, thus closing the gyre.

As the northern extension of the North Pacific Current flows into the Gulf of Alaska, it forms a complicated cyclonic eddy. This current, which is warmer than the air above it, reaches to 60°N and is responsible for the year-round rain and fog characteristic of that region.

Flow Rate

The average surface velocity of the Atlantic North Equatorial Current is about 0.7 knots, and the mass of water it transports past any point is approximately 16×10^6 m^3/s, or 16 sv (1 sv = 1×10^6 m^3/s). The "sv" is an abbreviation for **sverdrup**, a unit of measure named for the Norwegian physical oceanographer Harald U. Sverdrup.

The Florida Current has a surface velocity of 3.5 to 5 knots and an average flow rate of 30 sv. The Gulf Stream transport is 85 sv off Cape Hatteras, increasing to 150 sv by 65°W. At a point about 480 km (298 mi) further along its course, its average velocity and transport decrease to 3 knots and 55 sv. Off western Europe its current velocity is as low as 0.2 knots, and its transport is typically 15 sv. The currents of the North Pacific show similar variations. In addition, there are seasonal variations in both velocity and transport for these currents. The North Atlantic gyre and the North Pacific gyre differ in both flow rate and mass of water transport.

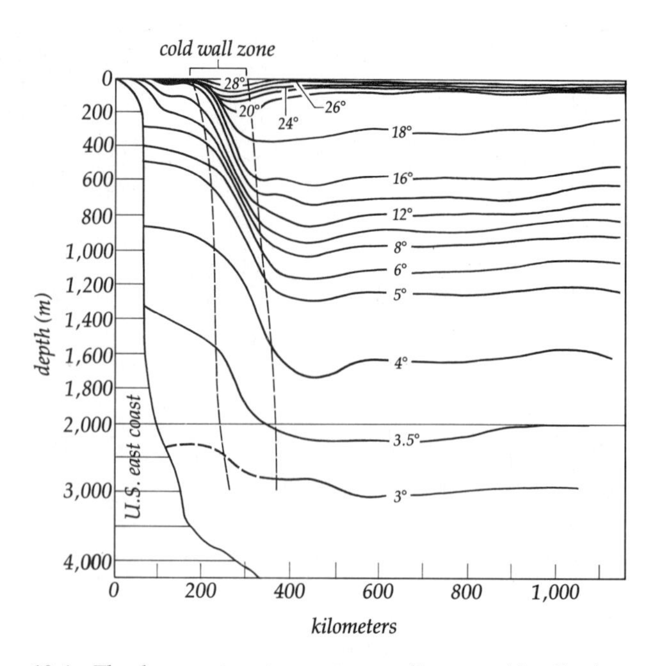

10.4 The sharp western temperature gradient, or cold wall, of the Gulf Stream, as shown by the isotherm slopes. The current is into the page. All temperatures are degrees Celsius.

The Boundary Currents

Each ocean gyre is made up of three separate currents: western boundary currents, eastern boundary currents, and equatorial currents. There are two types of equatorial currents.

Western Boundary Currents In the Northern Hemisphere, the western boundary currents move in a northerly direction. They are less than 100 km (62 mi) in width and extend to a maximum depth of 2 km (1.2 mi). Reaching a speed of several knots, they cover 100 to 150 km per day, and consequently have a large transport (25 to 75 sv). They are found on the west side of each of the five major gyres.

The Gulf Stream is a western boundary current. Once regarded as a river within the sea, the Gulf Stream, like all western boundary currents in their respective oceans, is the most pronounced current in the North Atlantic. Its name stems from an early belief that it originated in the Gulf of Mexico. The Gulf Stream

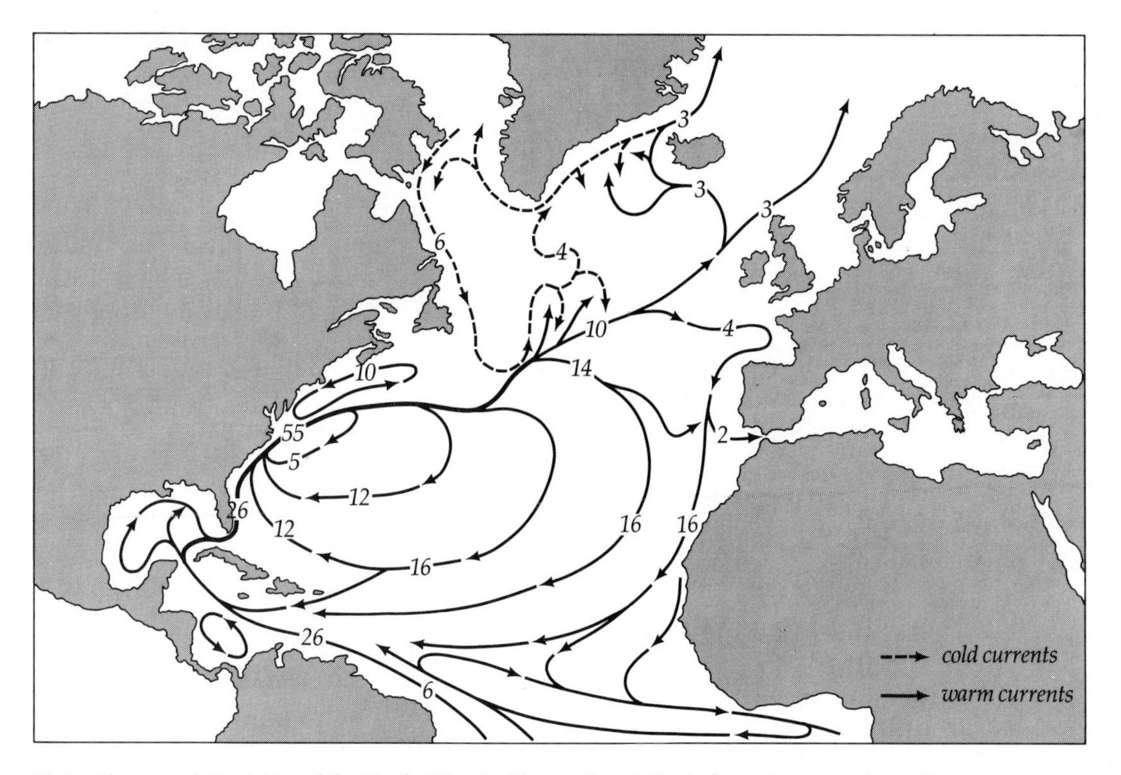

10.5 *The general circulation of the North Atlantic. The numbers indicate flow rates in sverdrups (1 sv = 1 × $10^6 m^3/s$).*

carries a tremendous amount of water: Each minute more than 4 × 10^{12} kg (4 billion tons) of water pass Miami in a stream that is 30 km (19 mi) wide and more than 300 m (984 ft) deep. The volume of flow is more than 100 times that of the Mississippi River.

Eastern Boundary Currents On the other side of the oceans are the eastern boundary currents. These currents flow toward the equator, especially at low latitudes. They are weak, poorly defined, and subject to seasonal variations. The California, Canary, and **Peru** currents are typical. They have low velocities of about 10 cm/s (0.2 knot) and are usually no more than 500 m (1,640 ft) deep. However, since they are quite wide—often more than 1,000 km (620 mi)—their transport can be as high as 15 sv, although it is usually lower.

The velocity of eastern boundary currents varies noticeably from place to place, and its flow along the coastline may be quite complex. For example, the California Current flows southward all year, bringing cold water to equatorial regions (Figure 10.6). Its weak, diffuse flow is confined to the upper 500 m (1,640 ft), and its transport tends to be less than 15 sv. Satellite photos show that its flow is not unidirectional at any given instant and that its velocity deviates sharply from its average. The pattern of the currents consists of large eddies superimposed on the wide, weak, southerly flow.

Equatorial Currents Equatorial currents, which in a sense are the oceanic counterparts of the trade winds (Chapter 9), are well established and quite permanent. They have low velocities (0.5 to 1 knot or so), and although they are fairly shallow (to a maximum of about 500 m) they are wide enough to move an appreciable volume of water. The North Equatorial Current of the Pacific has a transport of 30 sv, and the **South Equatorial Current** more than 30 sv. Those currents vary seasonally, moving slightly north in summer. At the edges of the ocean basins, the equatorial currents become more complex.

Forces Causing the Surface Currents

Winds blowing across the surface of the ocean create friction that sets the water in motion. That motion is a function of wind speed and, consequently, of the energy transferred to the ocean's surface. If there were no friction between wind and water, there would be no motion of the water. The surface-current velocity generated by the wind may be only about 3% of the wind velocity; thus, a 30-knot wind may create only a 0.9-knot current.

Surface currents may be regarded as a function of wind speed (and drag) and wind patterns. Since winds occur as banded patterns around the earth (see Figure 9.8), we might expect the ocean currents to follow

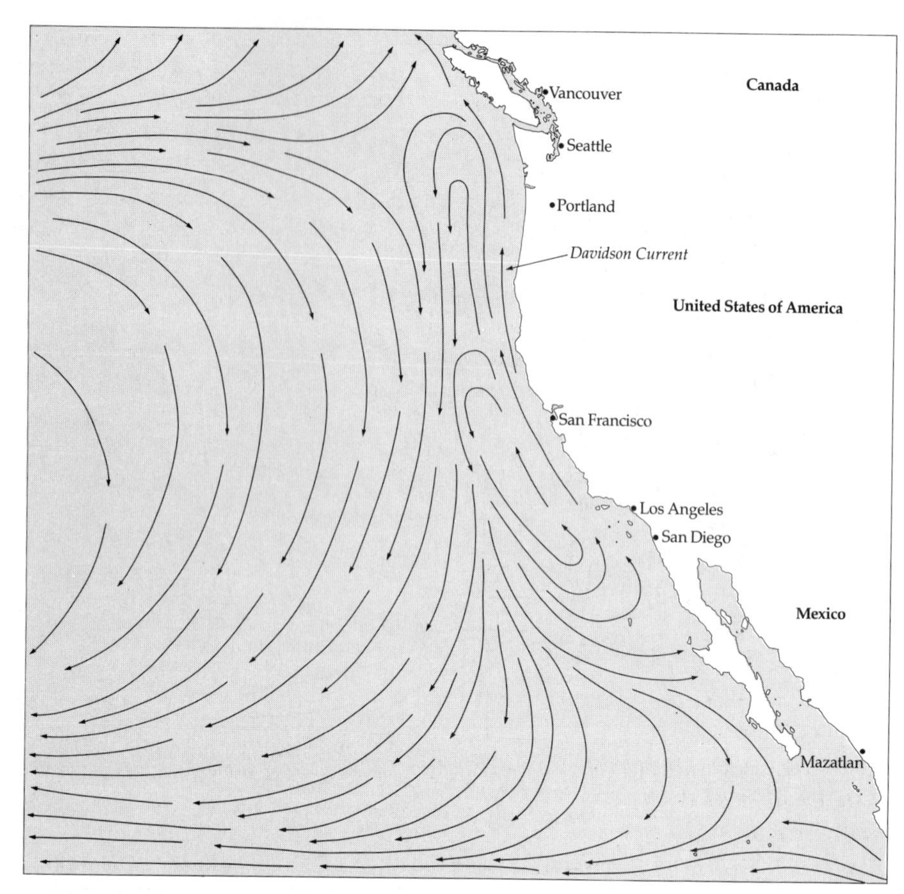

(a)

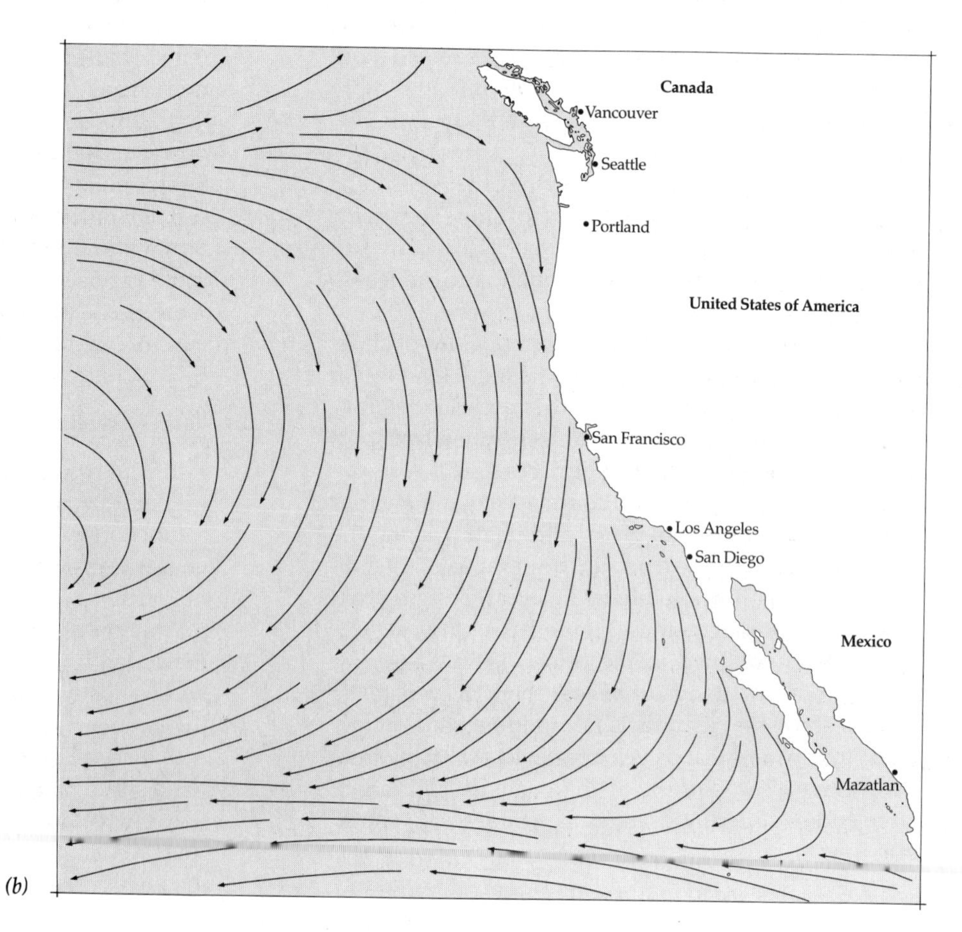

(b)

10.6 California current in (a) February and (b) August.

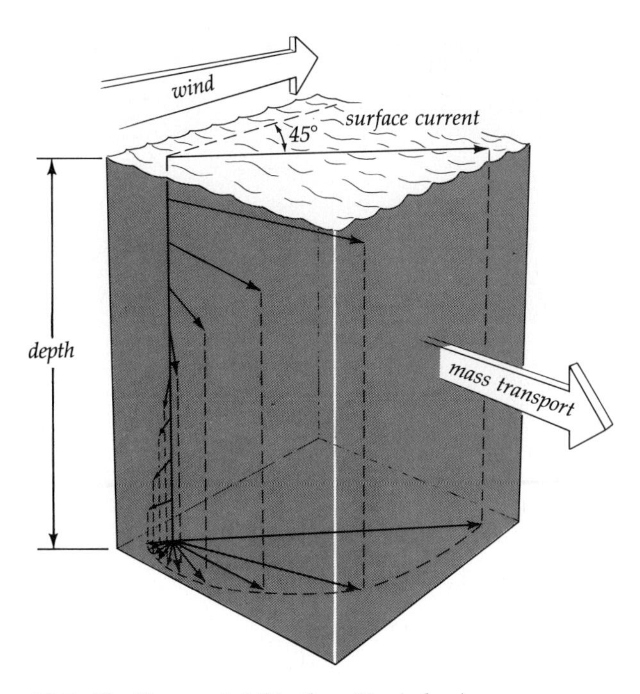

10.7 *The Ekman spiral (Northern Hemisphere).*

similar patterns. They do not, however, because continents, oceanic islands, and ridges distort the expected patterns. When we superimpose wind patterns onto the North Atlantic gyre, we see that the northeast trade winds power the North Equatorial Current and that the westerlies power the North Atlantic Current. Since these winds move in opposite directions, they exert a turning motion, or a torque, on the water, giving the entire North Atlantic a rotational motion. That motion is aided by the Coriolis effect. (As mentioned in Chapter 9, a north wind comes *from* the north, but a north current moves *to* the north. Unfortunately, this terminology leaves something to be desired.)

The Scandinavian physicist W. Walfrid Ekman explained how water responds to wind, duress, or stress in a model that has come to be known as the **Ekman spiral** (Figure 10.7). Nansen had previously noted that icebergs drift 40° to the right of the wind. Surface water moves at a 45° angle to a steadily blowing wind. Each layer of water exerts a pull by friction on the layer beneath it. Consequently, each layer is set in motion, but at a lower velocity than that of the layer above. The lower the velocity, the greater the Coriolis effect and the more the water moves to the right (in the Northern Hemisphere). At about 100 m (328 ft), the velocity of the water is about 4% of the velocity of the surface water, and its direction of movement is opposite to the direction of the surface water movement. Below this depth, which is called the **depth of frictional resistance** (or influence), wind effects are negligible.

According to the model, the net mass transport (**Ekman transport**) is directed 90° to the right of the surface wind (see Figure 10.7). Evidence of this phenomenon has recently been observed in the hydrosphere as well as in the laboratory. Aside from the tendency for sea ice to move at some angle to the right of the wind direction (Figure 10.8), there was formerly little direct evidence that the Ekman spiral exists in the open ocean.

The surface of the ocean does not match the earth's curvature exactly. Each ocean has a mound of water at its center, which corresponds to the center of its subtropical atmospheric gyre (Chapter 9). The center of the Atlantic oceanic gyre, for example, is about 150 cm (4.9 ft) higher than the western edge of the Gulf Stream. Note, however, that because of the earth's rotation, the center is offset to the west, accounting for the westward intensification of the currents. These mounded centers result from the accumulation of surface waters resulting from the Ekman transport. The mounded center of the North Atlantic is the most symmetrical and pronounced.

Density differences may also contribute to the existence of these bulges. The heights of two water columns of the same mass differ if one column has warmer and/or less saline water at its surface (above the thermocline). Since lower-density water of a given ocean mass occupies a greater volume, it rises and causes the sea surface to slope. The cold wall of the Gulf Stream and the slope of the sea surface across it illustrate an example of density distribution. In the Northern Hemisphere, even in simple wind-driven currents, denser waters, which are invariably colder, appear on the outside or to the left of the current flow (see Figure 10.5); warmer, less dense waters appear on the inside or to the right of the current flow. In the Gulf Stream, there is a well-defined temperature gradient at the western (or outside) edge along the U.S. coast. In 1922 the U.S. Coast Guard cutter *Tampa*, which was 73 m (239 ft) long, was stationed across the boundary. The water at her bow was measured at 1.1°C (34°F); at her stern, 13.3°C (56°F). Exploiting this fact, merchant vessels use temperature-sensing devices to move with the current when going up the coast and out of the current when going down the coast, thus saving time and money.

As water flows down the gradient of the mounds, it begins to deflect more and more to the right of its direction of travel (in the Northern Hemisphere) because of the Coriolis effect (Figure 10.9a). When water moves without friction in a straight line at a constant velocity, its movement is called **geostrophic flow** (Figure 10.9b). In perfect geostrophic flow, the horizontal pressure gradient (due to gravity—the water moving downhill) equals the Coriolis effect. As Figure 10.9(b) shows, the two "forces" balance each other, and the

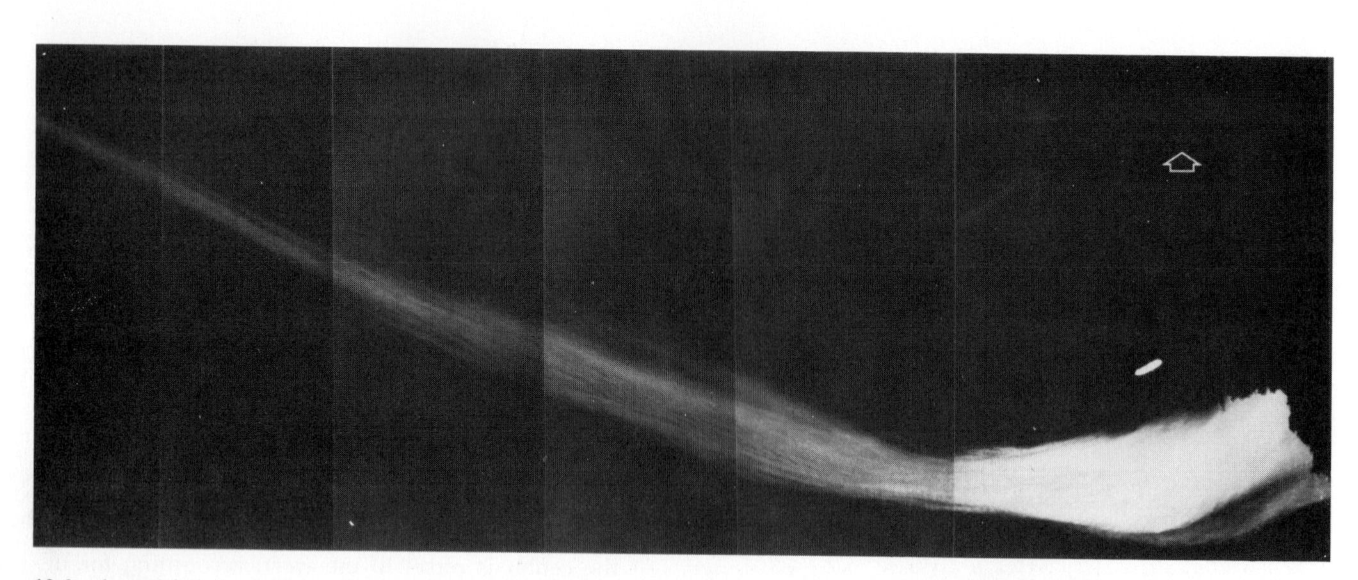

10.8 *An aerial photograph of an Ekman spiral in the ocean as dye settles in clear water. The arrow points north, the direction of mass transport. The wind is from the northwest. (Richard Linfield)*

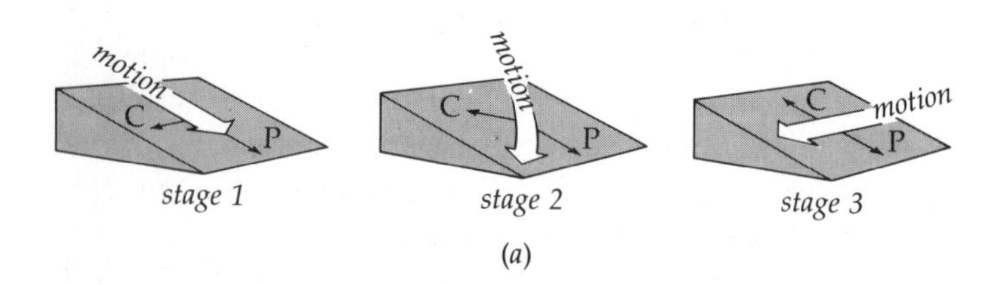

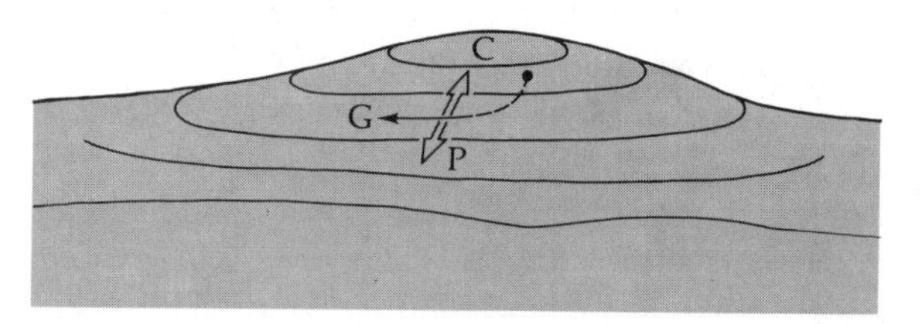

10.9 *(a) The development of geostrophic flow. Here C is the Coriolis effect and P is the pressure gradient due to gravity. (b) The resulting geostrophic flow G moves across the face of the water's slope. (c) In gradient flow around a sea-surface high (or an atmospheric high-pressure system), the pressure gradient force P and the centrifugal force C_F balance the Coriolis effect. Around an oceanic depression (or an atmospheric low-pressure region), the pressure gradient is balanced by the sum of the centrifugal force and Coriolis effect. These diagrams pertain to the Northern Hemisphere.*

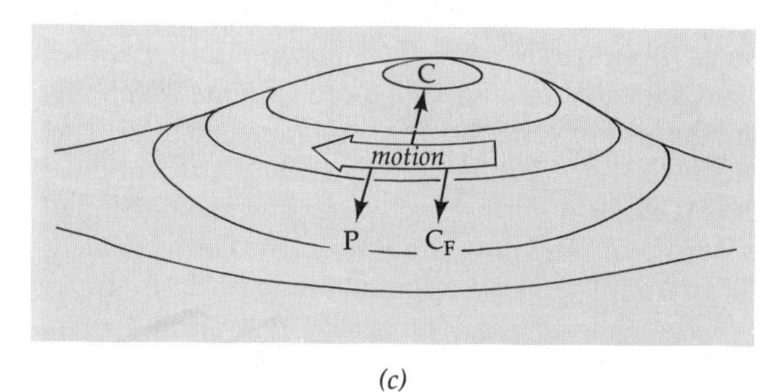

resultant flow is not downhill but across or parallel to the slope. The effects of friction, however, which are not considered in geostrophic flow, cause the resultant flow to be at some angle to the isobars or contours of the gradient surface.

Relative (not absolute) current flow can be calculated at particular locales. Well before currents were successfully measured in the open ocean, current flows were computed by these geostrophic means. Today surface slopes of less than 10^{-5} (1 m in 100,000 m) must still be inferred, since they cannot be measured directly.

When currents move in a curved path rather than in a straight line, centrifugal force must be considered. The resulting motion is called **gradient flow**. In cyclonic motion, the centrifugal force augments the Coriolis force. In anticyclonic motion, the centrifugal force augments the pressure-gradient force (Figure10.9c). Geostrophic flow is actually a special case of gradient flow in which the radius of curvature is infinite; consequently, it may be regarded as a straight line. The flow is both unaccelerated and frictionless. Since the gyres are so large, the distinction between geostrophic and gradient flow is not crucial.

Why are the western boundary currents stronger and better defined than the eastern boundary currents? Wind stress cannot be the reason because it is about equal on both sides of the ocean. Neither can it be just the piling up of water on the eastern edges of the continents by the North Equatorial Currents. A better explanation may be that the counterclockwise rotation of the earth compresses the water on the western edges of the oceans and narrows the currents there.

SUBSURFACE CURRENTS OR UNDERCURRENTS

We might assume that the volume of water transported north on the western side of an ocean would be the same as the volume of water transported south on the eastern side. Actually, however, more water flows north than south. Why the discrepancy?

In 1951, researchers with the U.S. Fish and Wildlife Service were testing a tuna-fishing technique in the westerly flowing waters of the South Equatorial Current of the central Pacific. They noted with surprise that their deeply paid-out fishing gear drifted east instead of drifting west with the prevailing surface current. The next year, Townsend Cromwell reported the existence of a **subsurface current** running from west to east almost exactly beneath the equator. This Pacific Equatorial Undercurrent—or **Cromwell Undercurrent**, as it is called—though well below the surface, is shallow and is associated with the surface currents. It usually coincides with the thermocline. Its depth varies between 50 and 300 m (164 and 984 ft), and it is 300 km (186 mi) wide and only 200 m (655 ft) thick. Its average velocity is 1.5 knots, although it can reach 3 knots. The Cromwell Undercurrent runs from New Guinea to Ecuador, well over 14,000 km (8,680 mi), and it is closer to the surface in the east than in the west. Its total water transport is 40 sv, approximately half that of the Gulf Stream.

Similar equatorial undercurrents have been identified in the Atlantic and Indian oceans. The Atlantic undercurrent was first noted in 1886 by J. Y. Buchanan of the *Challenger* expedition but was subsequently overlooked. The Atlantic and Pacific undercurrents apparently persist throughout the year, whereas the Indian Ocean undercurrent is evident only from November to March.

At the equator, the wind drives the surface currents toward the west. As they pile up against the land barriers, they create a thicker surface layer, which slopes to the east; the layer below the thermocline slopes to the west. Away from the wind-stressed surface layer, the pressure gradient causes downhill flow to the east along lines of equal density (called *isopycnals*). As a result, the undercurrent is closer to the surface at its eastern limit than at its western.

The Coriolis effect, which is zero for horizontal motion at the equator, stabilizes eastward-flowing currents by diverting them to the right (toward the equator) if they stray north and to the left (again toward the equator) if they stray south. In contrast, the Coriolis effect accentuates the meandering of westward-flowing currents.

During the International Geophysical Year, an undercurrent was discovered beneath the Gulf Stream, flowing in the opposite direction at depths of 2,000 to 3,000 m (6,560 to 9,840 ft), much deeper than the Cromwell Undercurrent. Much remains to be learned about this and other undercurrents.

OTHER SURFACE-CIRCULATION PHENOMENA

Convergence and Divergence

Look once again at Figure 9.2. Notice that the world's horizontal winds blow toward one another in some places and away from one another in others. Where they converge, the surface waters beneath them also tend to converge; where they diverge, so do the surface waters. **Convergences** and **divergences** occur in polar, tropical, and subtropical regions. The length of these oceanic convergences and divergences is at least 20,000 km (12,400 mi). Figure 17.4 (p. 390) indicates the extent of the Antarctic Convergence and the Antarctic Divergence, for example.

Where water parcels meet in a convergence, they form a slight hill, thickening the surface layer. The mixed water is usually of higher density than the surrounding water, and consequently it sinks. Local convergence is often discernible in the choppy water created by the coming together of river currents or tidal currents.

In the Antarctic Convergence, which surrounds Antarctica between 50°S and 60°S, warm waters from the middle latitudes meet the cold waters of the Antarctic. The mixed waters sink, in the manner characteristic of convergences. From north to south, the temperature may drop by 5°C over a few kilometers across the convergence. Although the Antarctic Convergence exerts little influence on surface currents, it significantly influences deep-ocean circulation.

In a divergence, the surface waters move away from one another and deeper waters move up to the surface. This action thins the surface layer and usually lowers its temperature.

Upwelling

Upwelling is a special case of divergence. The term usually refers to the welling up of colder, deeper waters along a coast, constituting half of a divergent system or cell, but the term also refers to the upward movement of waters in open-ocean divergence.

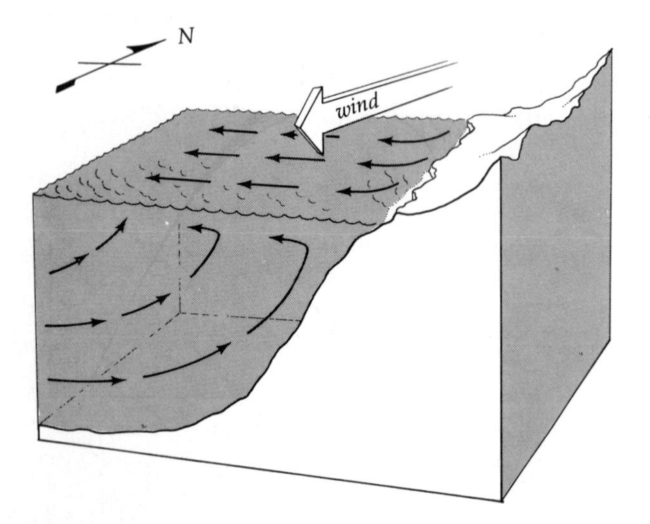

10.10 Upwelling as it occurs along the coast of Oregon during summer.

Along the Oregon coast, for example, winter winds drive the ocean waters against the land. As a result, the surface level is slightly higher, about 3 cm (1.2 in). In summer, however, the winds blow southward along the coast, moving the surface water at an angle to the right of the wind direction (Figure 10.10). Deep water then wells up to replace the surface water.

Upwelling encourages biological productivity by bringing nutrient-rich waters from a depth of about 200 to 300 m (656 to 984 ft) to the euphotic zone of the surface waters. Without this vertical transfer, the surface waters would be devoid of nutrients except where they adjoined the world's many sewage outfalls. Because of this mixing, the water off the coast of Oregon is colder in summer than in winter. Only when the wind stops for a few days and upwelling temporarily ceases does the surface water warm up enough for comfortable swimming.

Upwelling commonly occurs in the eastern regions of the oceans. In the Northern Hemisphere, the winds must blow south along the coast for upwelling to occur (which usually happens during the summer months), and in the Southern Hemisphere they must blow north. Coastal upwelling of this sort is in keeping with the Ekman spiral. In shallow waters, however, where the water depth is less than that of frictional resistance, the water moves more in the direction of the wind. About 90% of the world's finfish fisheries occur in areas of upwelling, which constitute only 3% of the total area of the oceans. Coastal upwelling occurs in only about one-thousandth of the ocean's total area, but half of the world's fish are taken in these regions of upwelling.

El Niño

The northward flow of the cold, nutrient-rich Peru Current along the western coast of South America is accompanied by coastal upwelling caused by southerly winds. The combination supplies the nitrates and phosphates that support one of the world's largest populations of marine life. That abundance in turn supports the fishing industry, and the guano droppings of millions of fish-eating birds support the huge Peruvian fertilizer industry.

Weakly each year, and roughly every seven years, the southerly winds drop off and the warm equatorial waters move in to displace the cold waters of the Peru Current. Then upwelling ceases, and the picture changes. The warm equatorial waters rapidly deplete the nutrients and cause massive plankton and fish kills. Tons of dead fish pile up on the beaches. This phenomenon is known as **El Niño** (the "Christ child") because the activity reaches its height during the Christmas season. As the oxygen content of the water is depleted, foul-smelling hydrogen sulfide and other gases are produced, blackening the paint on ships and houses and producing other discoloring effects. This action gives rise to the name **Callao Painter**. With the disappearance of anchovies and other marine organisms, thousands of birds die of starvation. El Niño has made frequent appearances over the last century, with particularly severe consequences in 1891, 1925, 1953, 1972–73, and 1982–83. It accounted for severe drops in the seabird populations in 1957 and 1965.

The Peru Current, an eastern boundary current, is both weak and slow, with a speed of between 0.2 and 0.3 knots and a transport of only 15 sv. Near the coast it is only about 200 m (656 ft) deep, increasing to 700 m (2,296 ft) offshore. El Niño, which is related to changes in the patterns of atmospheric pressure, occurs when the winds cease to blow from the east or southeast toward the west. Under normal conditions, the winds cause Ekman transport to the left (in the Southern Hemisphere), or away from the coast, with subsequent upwelling. When they cease to blow toward the west, upwelling no longer occurs. Then abnormal winds from the west move a warm current from the equatorial region south toward Peru and Ecuador, creating an area of warm water thousands of kilometers in length. The mixed layer deepens, and the deeper cold waters are isolated. The thermocline falls, and birds cannot dive to pursue fish, which have descended to the colder waters. The sun warms the surface layer still further.

It is not always simple to separate meteorological and oceanographic phenomena and the El Niño is such an example. In 1924, Sir Gilbert Walker first recognized the existence of the *Southern Oscillation*. The horse latitudes of the Pacific Ocean are characterized by high air pressure systems and low rainfall. The equatorial region of the Indian Ocean is, at the same time, characterized by low air pressure and abundant rainfall. Periodically, the situation reverses. A reversal results in the omission of monsoons over India, Southeast Asia, and Indonesia, and in the initiation of an El Niño and storms in the eastern Pacific region. This reversal has come to be known as the El Niño/Southern Oscillation (ENSO). In 1982–1983, the ENSO link was firmly established and the global weather implications were recognized.

The ENSO destabilizes the flow of energy along the equator and from the tropics poleward, thus altering weather patterns globally. The reversal of pressure gradients reverses the trade winds and equatorial currents as well. Warm water flowing from west to east causes the sea level to rise and prevents upwelling along the west coasts of North and South America.

The El Niño of 1982 began in July with the first appearance of anomalous trade winds and elevated surface temperatures. Although this event would normally have been over by late 1982, it lasted until October of 1983, making it the longest in over 100 years. By October of 1982, surface temperatures had increased by 5°C. By December, the normally westward-flowing North Equatorial Countercurrent, between 5°N and 8°N, flowed predominantly eastward. Warmer equatorial water moved into the eastern Pacific, causing a continued rise in sea level to 22 cm at the Galápagos Islands and 10 cm at San Diego. The coastal regions of Ecuador and Peru experienced a disastrous increase in rainfall and flooding, causing massive fishery and agricultural losses and grave economic hardship.

Apparently each El Niño is different to some degree, but the episode of 1982–83 was the most extreme in this century. Not only were there the aforementioned effects along the coasts of Peru but there were also far-reaching impacts elsewhere. Droughts in Australia, southern India, Hawaii, Mexico, the Philippines, and Sri Lanka, as well as floods in Bolivia, Brazil, Ecuador, and Peru were linked to this El Niño. Elevated surface temperatures (Figure 10.11) caused unusual migration-like movements of sea animals. Penguins and other sea birds moved south along the coast of South America; bluefin tuna were caught off British Columbia; salmon moved further northward; barracuda appeared off the Oregon coast; small pelagic crustaceans washed up on San Diego beaches; and many sea birds did not return to their traditional breeding grounds. In the Farallon Islands, 27 miles from San Francisco's Golden Gate Bridge, 12 species of birds usually gather annually to breed, producing about 150,000 chicks. The count for the spring of 1983 was only 15,000.

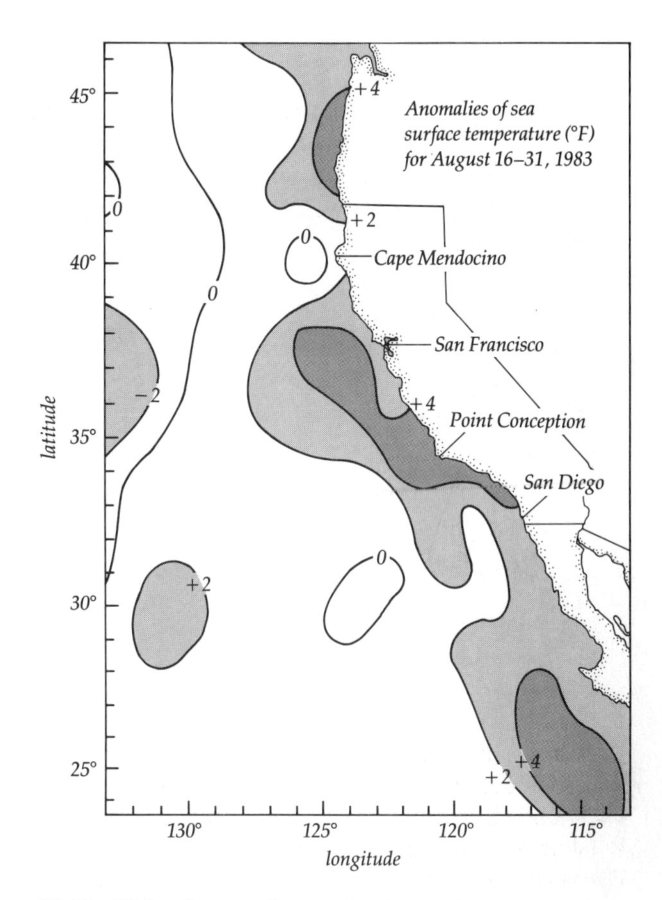

10.11 Higher-than-usual sea-surface temperatures, or temperature anomalies, along the California coast for August 16–31, 1983, during the El Niño effect. (George T. Hemingway, Scripps Institute of Oceanography)

Through the course of this El Niño a number of records were set for rainfall, drought, and temperature, as well as unseasonal and particularly devastating hurricanes in Tahiti, suppressed hurricane activity in the Atlantic, and a series of vicious storms along southern California that caused appreciable flooding and damage.

The 1982–83 ENSO, the longest and most intense episode of this century, has been well described. In piecing information together after the fact, we can outline the causal contributing factors of the El Niño as follows: In a normal year a high-pressure system exists in the eastern Pacific, and therefore the trade winds blow toward a moist, low-pressure system residing over Indonesia. These winds and the ocean currents moving westward cause water to pile up in the western Pacific, causing the sea level in Southeast Asia to rise as much as 46 cm (18 in) higher than along the west coast of South America. In October and November preceding the El Niño, this weather pattern broke down. The previous May satellite data showed a rise

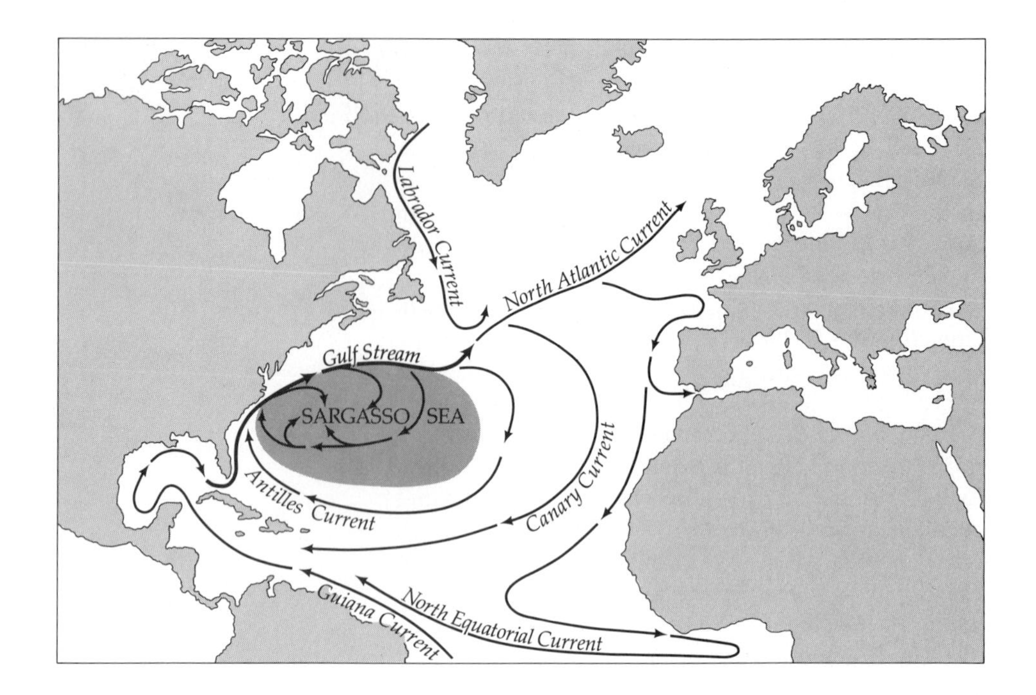

10.12 *The Sargasso Sea.*

in sea surface temperatures in the western Pacific and the Indian Ocean. A month later the trade winds began to falter. With the decline in the trade winds a region of weak subtropical easterly winds began to develop. During the normal flow of the trade winds, warm surface waters are moved westward. But this year there was a reversal of atmospheric pressures and winds, and the warm sea surface flowed to the east. Warmed rising air formed major updrafts, which are associated with abnormal storms. Compared to the usual situation 9 million square miles of the Pacific extending several thousand miles eastward became overheated and caused massive evaporation, which gave rise to storms. Subsequently there was a rise in sea surface temperatures, changes in surface currents as well as in subsurface currents, rises in sea level along much of the west coasts of North and South America, and the bringing on of the very unusual weather we call El Niño.

Since the ENSO of 1982–83, a great deal of research has been focused on the phenomenon. Programs like the multinational Tropical Ocean and Global Atmosphere (TOGA), viewed as a ten-year study, will attempt to understand the ENSO and its causes and integrate it into global weather models. This should not only lead to improved weather forecasting but also to the prediction of future ENSO events as well. Already identified is a slow atmospheric oscillation that develops in the Indian Ocean once every 40 to 50 days and has winds traveling at up to 90 kph (55 mph). It sweeps across the Pacific at up to 10° N–S latitude in an easterly direction, beginning with a speed of 30 kph (20 mph) and intensifying as it moves until it dissipates in the eastern Pacific. It may play a role in regulating the Indian monsoon and the jet-stream path as well.

The Sargasso Sea

Inside the great circle of the North Atlantic gyre is a fabled region where the world's eels go to spawn, where there are almost no surface currents, and where the water is almost stagnant. This is the once-dreaded **Sargasso Sea** (Figure 10.12). Between the Bahamas and the Azores, the surface abounds with patches of floating seaweed, sometimes aligned in rows up to 2 to 3 m (7 to 10 ft) across. Sailors as far back as Columbus feared that the seaweed might ensnare their ships or that the ships might run aground (seaweed is associated with shallow water). The small flotation bladders common to seaweed reminded Portuguese sailors of a small grape called the *sargaço*, which was indigenous to their country; hence the name Sargasso Sea.

Although the Sargasso poses no real threat to navigation, its warm, blue waters mix very little with the deeper, nutrient-rich waters that surround and underly it; consequently, the Sargasso supports less microscopic plant life than any other region of the sea. Since this phytoplankton abundance (upon which fishery stocks, and almost all other sea life depend) is so low, total productivity of the Sargasso Sea is extremely low.

Coastal Currents

In most coastal areas, especially over the continental shelves, we find varied, complex currents more or less independent of the massive currents of the oceans.

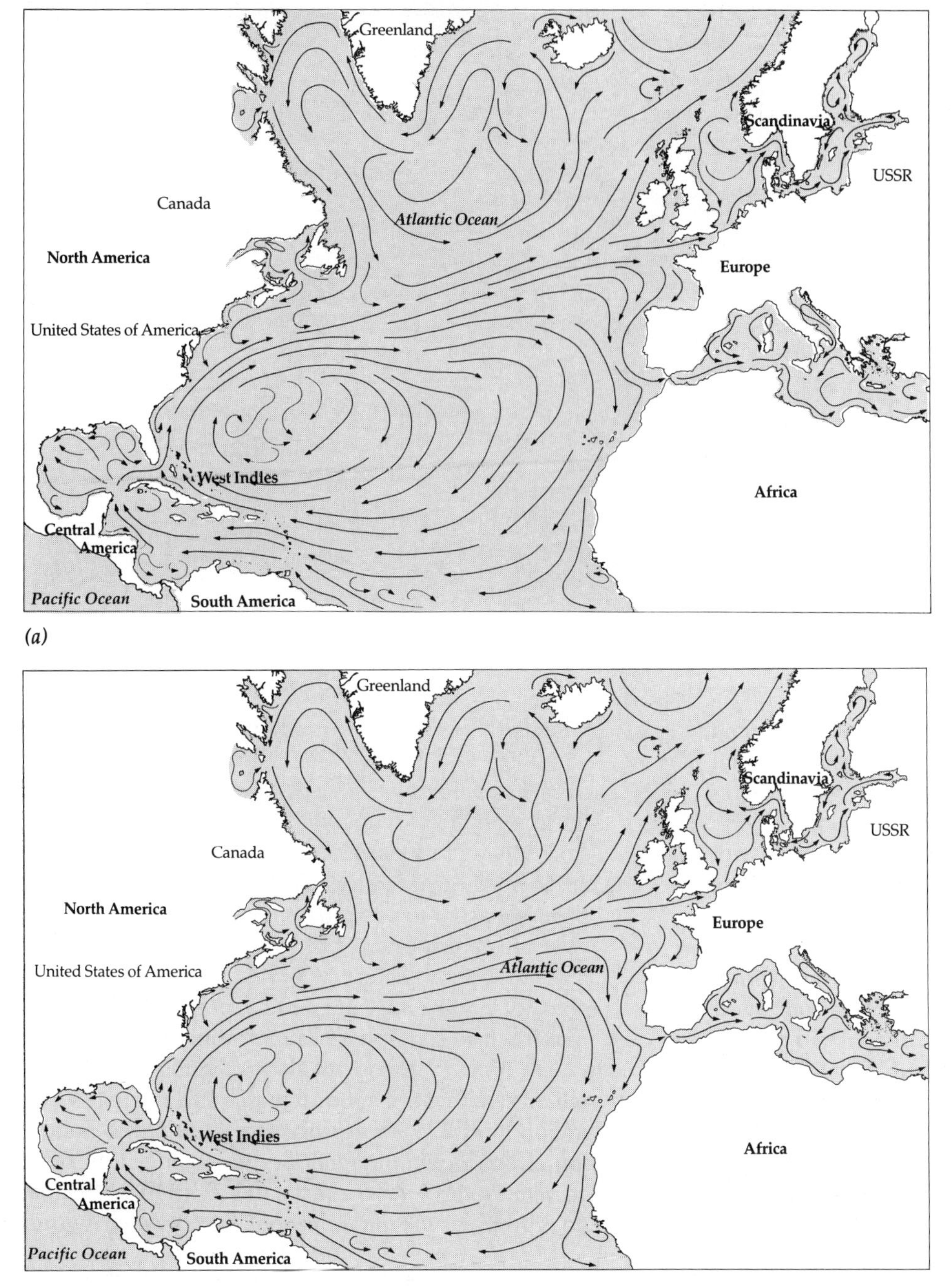

10.13 NOAA monthly pilot charts for the western North Atlantic. (a) February 1984. (b) August 1984.

These **coastal currents** are affected by local winds, tides, bottom and coastal topography, and river discharge. They vary with the seasons, along with local weather and tidal conditions, but they tend to flow parallel to the coast.

Off the U.S. Atlantic coast, the oceanic boundary of the coastal currents is well marked by the Gulf Stream system, with its heavy northerly flow, lighter color, and higher temperature. The temperature and salinity of the coastal currents change with the seasons. The surface temperature is lowest in late winter (−1°C to +1°C), when surface salinity is highest (20‰ to 33‰) because of the salty discharge from frozen rivers. From Newfoundland south to the Chesapeake Bay, the coastal current flows generally to the southwest, along the coast. Supplied in part by the cold Labrador Current, it appears as a compressed, elongated gyre.

The National Oceanic and Atmospheric Administration (NOAA) publishes monthly pilot charts for the use of mariners plying these waters. Figure 10.13 shows the NOAA's predictions for February and August 1984 for the western North Atlantic coastal regions. Note the variation in patterns. The inflow of cold, low-salinity water from the Labrador Current into a region

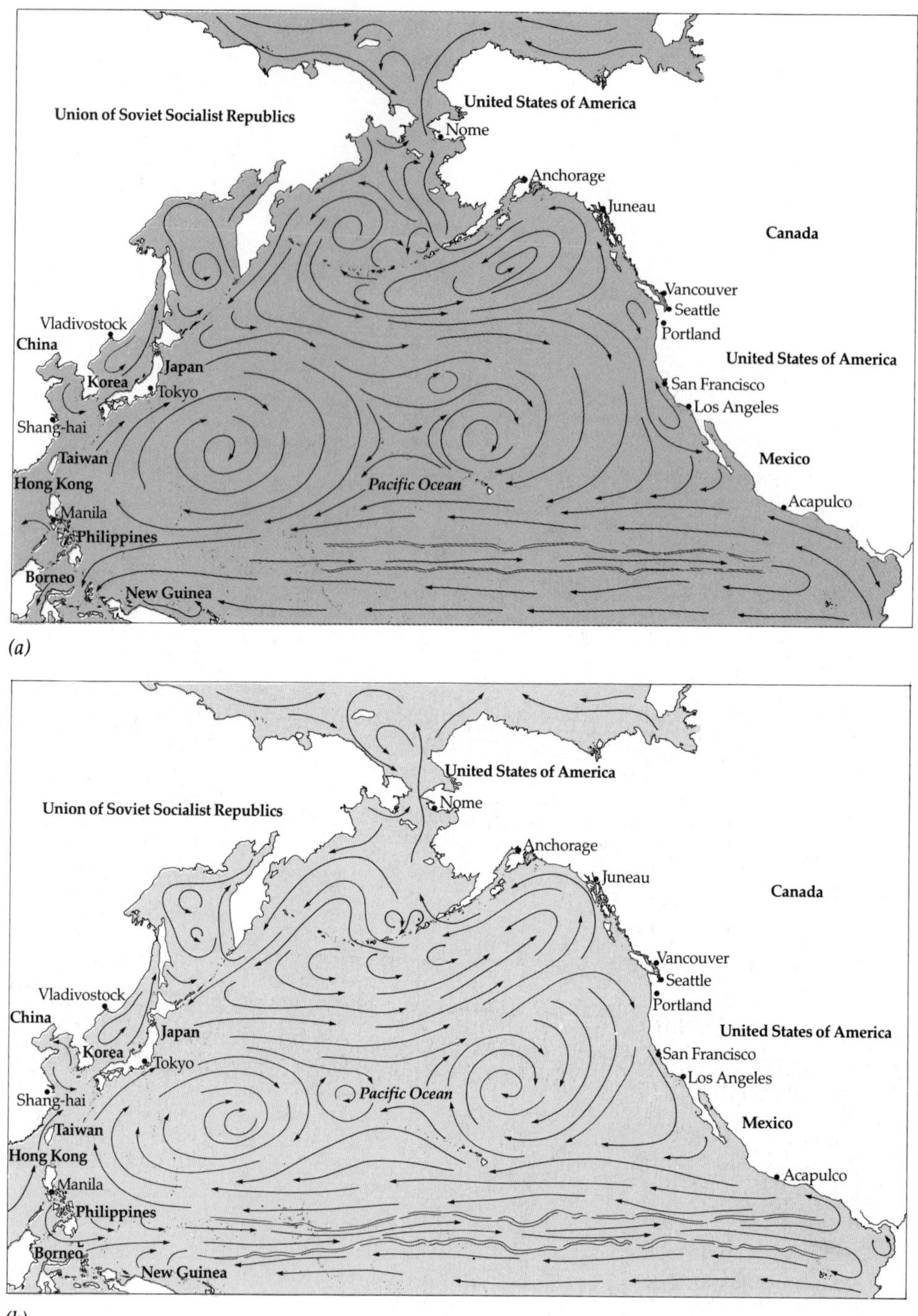

10.14 NOAA monthly pilot charts for the West Coast of the United States. (a) February 1984. (b) August 1984.

close to the warm, high-salinity Gulf Stream creates a complicated coastal regime.

Off the U.S. West Coast, the boundary between the coastal currents and the weak, diffuse California Current is not easily discerned. River runoff, however, has a marked effect on coastal currents, and so a salinity line (isohaline) of 32.5‰ is used to delineate the coastal currents from the oceanic currents. During the summer, when river discharge is low, or during periods of weak winds, the coastal currents of this region diminish and may disappear altogether. However, with the onset of winter, with its storm winds, heavy rains, and renewed river discharge, they quickly regenerate.

The coastal currents of the East Coast flow south; those of the West Coast flow north. There are significant seasonal variations in both regions, however. In the winter months from northern California to Washington's Juan de Fuca Straits, for example, winds from the southwest keep the water from river discharge near the coast, creating a seaward-sloping wedge. The clockwise deflection of the Coriolis effect combines

(a)

(b)

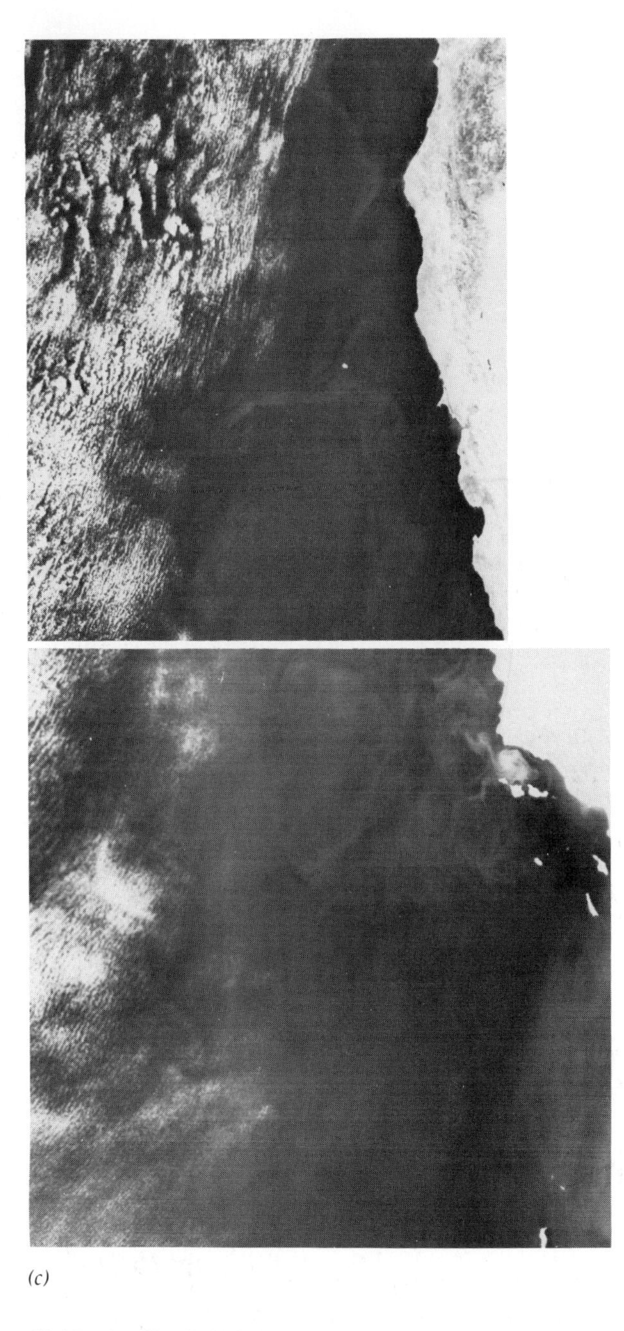

(c)

10.15 Satellite Coastal Zone Color Scanner (CZCS) images showing the wide day-to-day variability of coastal currents off the southern California coast on (a) May 12, 1979, and (b) May 13, 1979. (c) Shows the currents in the following August.

with the influence of the winds to generate a northward flow called the **Davidson Current** (Figure 10.6a). During winter and spring, the large discharge of the Columbia River moves north because of this current. During the summer and fall, however, low rainfall and low river discharge, coupled with a shift in wind direction, cause the Davidson Current to weaken and disappear. The Columbia River discharge is then carried south by the California Current and to some extent by the southerly winds that cause upwelling near the coast.

In Figure 10.14, note the absence of the northerly coastal flow in August. Since these pilot charts are intended as a general aid to mariners and are compiled from past observations, they do not show short-term and fine-scale variations. The satellite photos in Figure 10.15, however, clearly show the variability of these coastal currents.

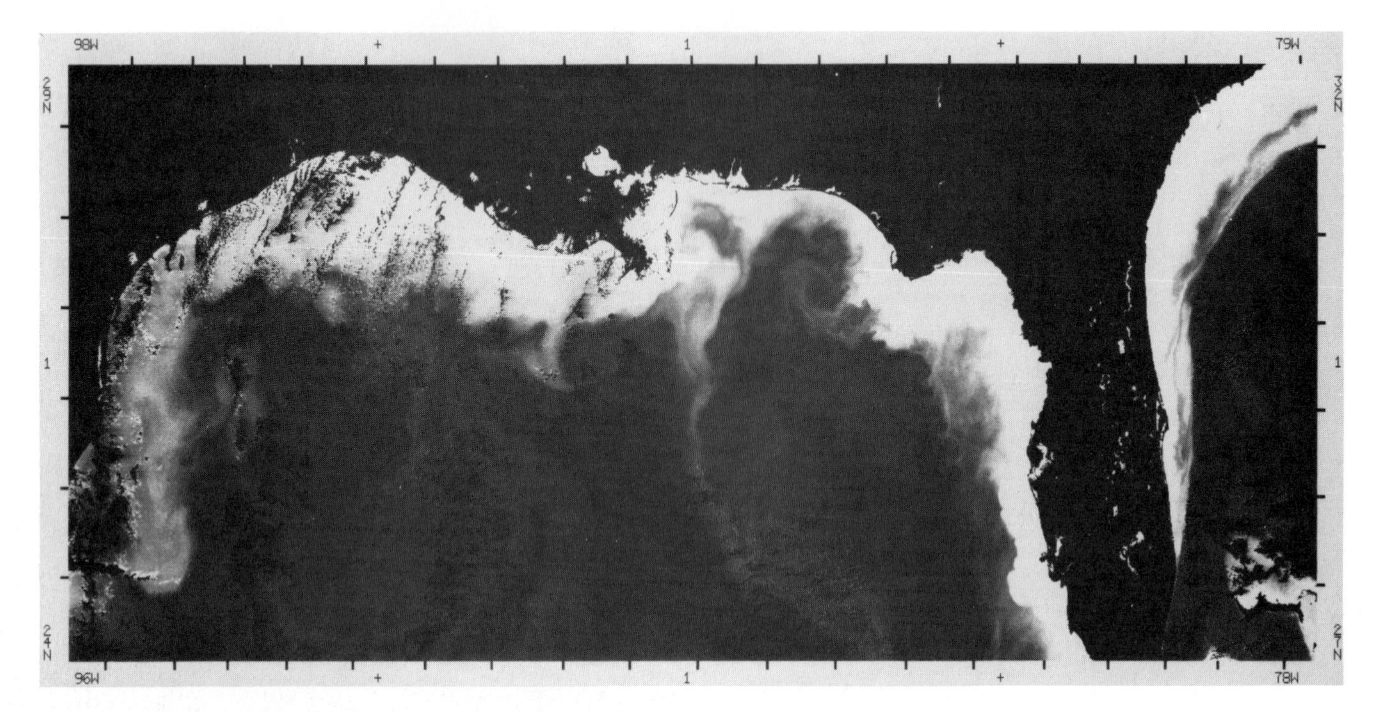

10.16 A Nimbus *7 satellite Coastal Zone Color Scanner (CZCS) image of October 8, 1979, for the Gulf of Mexico from Mexico east to Florida and beyond. The parameter actually photographed is the light penetration into the water, which is related to turbidity and gives a superlative representation of water movement. Obviously, seawater next to the continent will be the most turbid. The outflow effect of the Mississippi River can readily be seen. The dotlike mottling is due to the clouds present. (Nimbus Experimental Team)*

Beneath the coastal currents of the Pacific Northwest coast and below the pycnocline is a counterflow that runs in the opposite direction. A similar, better-defined counterflow exists off the East Coast. Within the southern California Channel Islands area, between Point Conception and San Diego, a large, variable counterclockwise eddy seems to exist all year except during March, April, and May. This eddy is accompanied by a nearshore flow to the southeast. The satellite photograph in Figure 10.16 shows the coastal current patterns off the Gulf Coast states.

Strong offshore currents pose a grave danger to mariners. One of the most treacherous regions in the North Atlantic is near Sable Island, off Nova Scotia (Figure 10.17). Here tricky currents coupled with bad weather and strong winds have wrecked many ships. Another perilous region is off Cape Hatteras. Severe, sudden storms, along with large shoals that extend well out to sea, shifting sandbars, and currents that vary in force and reverse their direction, have earned this region the name "Graveyard of the Atlantic." This phrase was first used in 1790 by Alexander Hamilton, who, as secretary of the Treasury, founded the Coast Guard and had the first lighthouse built on Cape Hatteras.

INSTRUMENTS FOR MEASURING CURRENT

Oceanographers have developed a wide array of instruments to measure currents. In addition to observing ship drift, they use drift bottles and drift cards, Woodhead seabed drifters, drogues, and parachute drogues (Figure 10.18). They also use more sophisticated gear, such as the Swallow float, free-falling devices, buoys (both fixed and free-floating) and current meters (Figure 10.19). Some of these devices transmit data directly to shore-based installations and computers.

A relatively new technique is **remote sensing**, which is used in conjunction with in situ data measurements. In this technique a variety of devices are installed in satellites, including infrared, color, and microwave radiometers, which sense emissions from the sea's surface; microwave radar altimeters; and scatterometers, which measure the sea's roughness. Satellites can now scan the surface temperature of the entire ocean system daily, reporting short-term changes (Figure 10.20) and measuring surface-current velocities (Figure 10.21). It now appears feasible to measure surface winds as well. In addition, the Coastal Zone

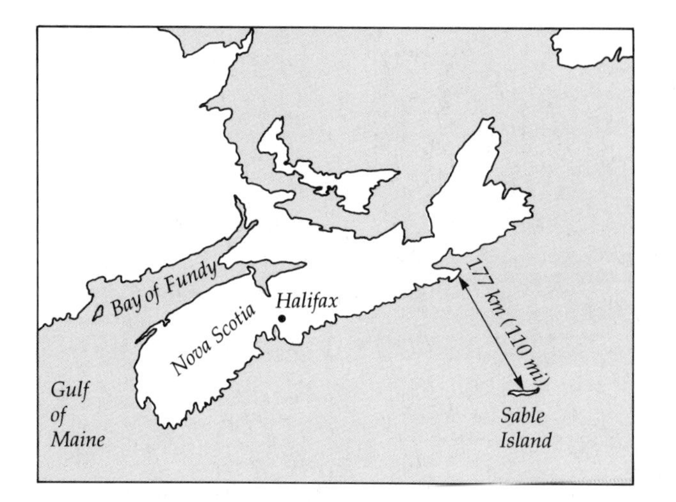

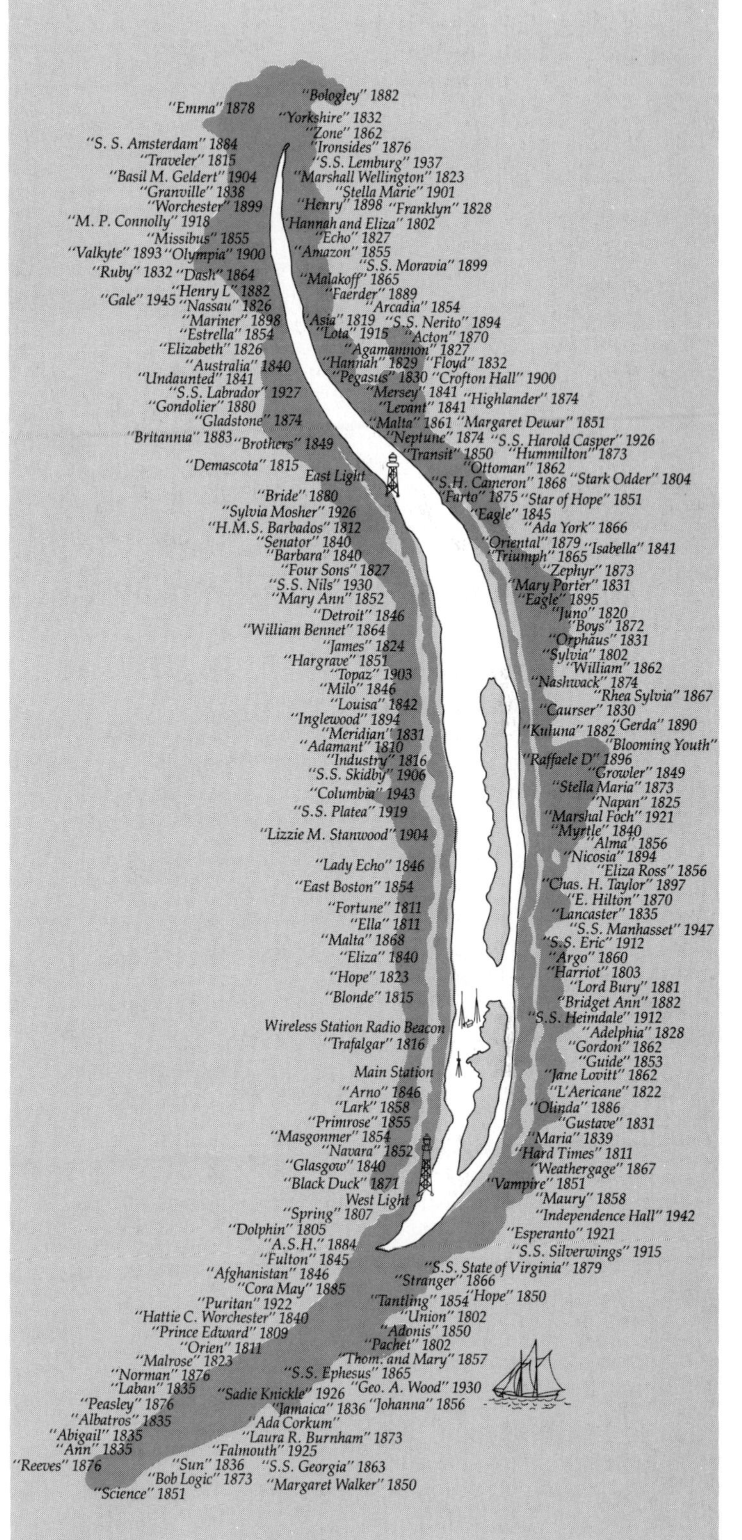

(b)

10.17 (a) Sable Island, often called the "Graveyard of the Atlantic," lies off the coast of Nova Scotia. (b) This figure lists the names and dates of known shipwrecks since 1800.

Color Scanner (as on *Nimbus 7*) can identify biologically productive areas by finding areas with high concentrations of chlorophyll (see Figure 10.23c). The findings are swiftly relayed to fishers. Satellites also monitor the type, distribution, and thickness of polar ice and snow cover (Figure 10.22).

Whereas the crew of a ship moving at only 10 or 12 knots may not notice short-range changes even in an area of a few square kilometers, satellites collect data rapidly and accurately. They can provide both large-scale, short-term reports as well as long-term, cumulative reports. Although satellite imagery is limited to changes in the surface of the ocean, those changes are caused in part by energy passing through the surface layer from greater depths. Thus, satellite imagery provides far more information than just data on surface conditions.

Another advance in the study of ocean currents is acoustic tomography, a technique similar to that used in CAT scans in medicine. In the CAT (computerized axial tomography) scan, ultrasonic sound is used to produce an image of internal body structures. In acoustic tomography, sound is generated by explosives. Sound reaches a receiver more quickly in water of identical physical properties when the water is

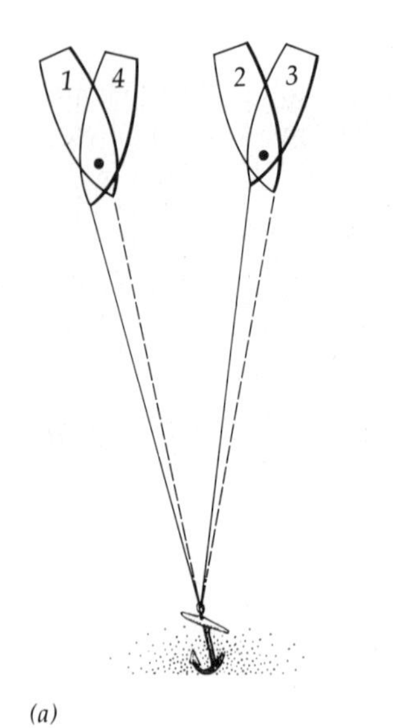

(a)

(b)

(c)

(d)

CUT ALONG DOTTED LINES BEFORE MAILING

U.M.S.G. № 1894

PLEASE REMOVE CARD, LET DRY IF WET, FILL IN INFORMATION AS REQUESTED, AND SEND BY MAIL

Where found (name of beach or place on shore, near what landmark in what town or city, on what lake)..

....

When found (date)..

Your Name (print)..

Your Home Address (print)..

....

....

Your return of this card will assist the addressee in a study of Lake Currents. The location and date of release will be sent to finder upon return of the card. THANK YOU !

(e)

10.18 Simple methods for measuring surface currents by (a) the movements of a ship or boat at anchor, (b) the degree of deflection of a weighted line, (c) ship drift, (d) drift bottles, and (e) drift cards.

moving toward the receiver than when it is flowing away. Acoustic tomography is much like seismological studies of the structure of the earth. A computer interprets the various arrival times of the sound and draws a three-dimensional "picture" of the ocean's structure and currents in a given region.

Deep currents are traced indirectly by means of chemical compounds labeled with radioactive isotopes like ^{14}C as well as by oxygen, nitrogen, and phosphorus concentrations. Since these properties are essentially determined by surface phenomena, changes in concentration with depth indicate both subsurface currents and the degree of mixing.

Over the years, the Woods Hole Oceanographic Institution, in Massachusetts, has been releasing between ten and twenty thousand bottles per year to study long-term current drift. Each bottle contains a self-addressed postcard, along with some sand to keep the bottle upright. Some of the bottles have traveled as far as Europe, and many have made a complete

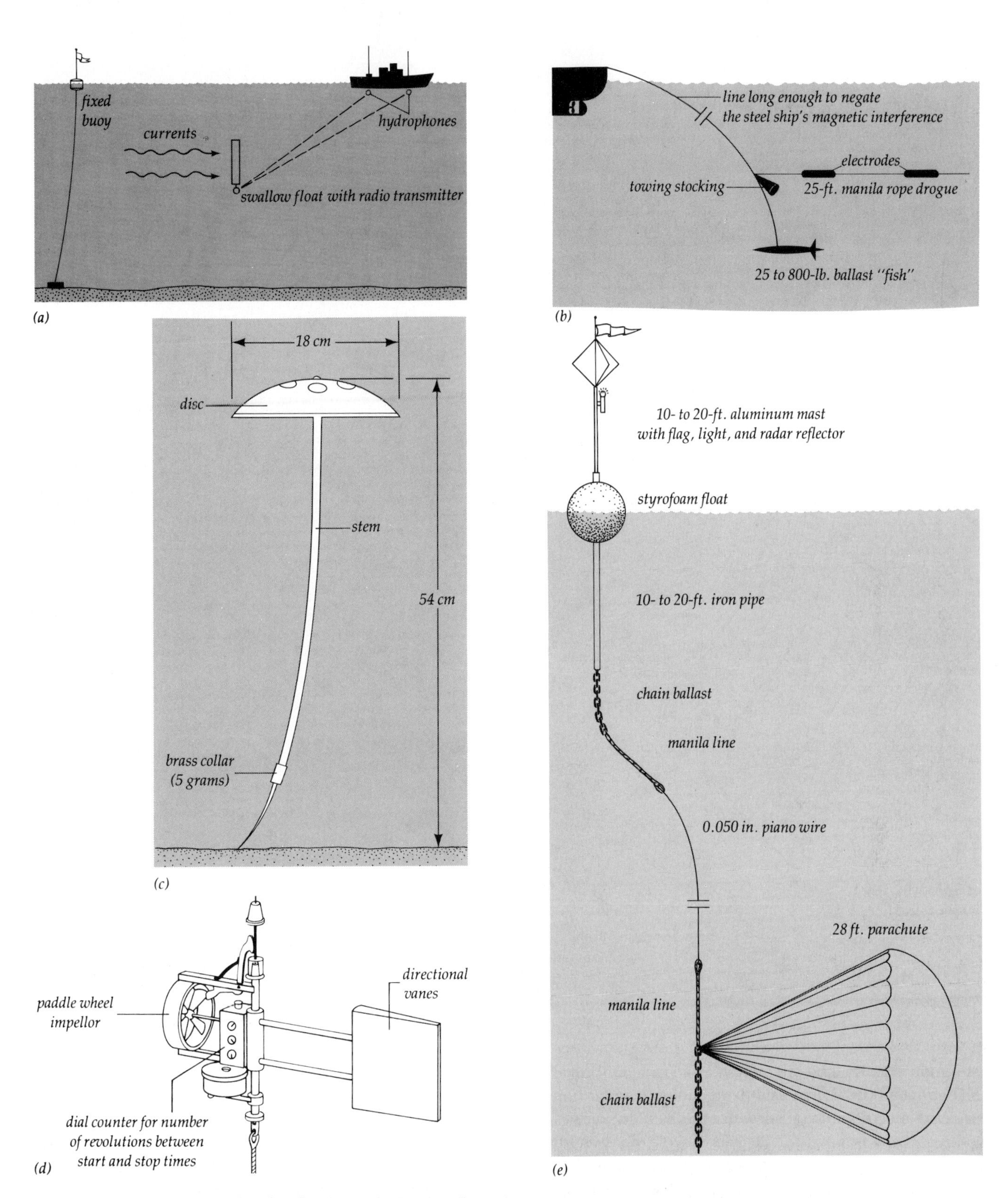

10.19 Methods to measure subsurface flow by (a) the Swallow float. The ship maintains its position relative to a moored buoy or a land station by radio or radar or a satnav fix and monitors several floats, each emitting a different sound frequency. (b) Ship-towed GEK equipment (Geomagnetic Electrokinetograph), which measures the speed as a function of the electromagnetic force the seawater generates as it moves in the earth's magnetic field. (c) The Woodhead seabed drifter is placed on the bottom in clusters, weighed down by a salt ring. When the salt dissolves, they slowly drift just above the bottom and ultimately wash ashore where finders send back the attached plastic postcards with the location. (d) Mechanical current meter such as the once very popular Ekman type. (e) Drogues or parachute drogues simply drift with the current, the parachute offering a large frontal area to catch the current.

10.20 Infrared image of the thin upper layer of the sea surface. Long Island to Nova Scotia stands out clearly, as do Cape Cod, Nantucket, Martha's Vineyard, and Chesapeake Bay. (N.O.A.A.)

circuit of the North Atlantic gyre—a distance of 8,000 to 10,000 km (4,960 to 6,200 mi). The recovery rate runs about 10%.

The longest trip was made by a bottle released in 1962 at Perth, in southwestern Australia. Five years later it turned up near Miami—a distance of 26,000 km (16,120 mi). (From the chart of currents in Figure 10.2, can you determine what route it probably traveled?)

The Gulf Stream Revisited

Traditional techniques of measuring ocean currents provided data only for a given point. To gain an idea of large-scale circulation, oceanographers had to combine many direct measurements with indirect calculations. However, this procedure posed two problems: None of the measuring devices could be left in the ocean for long periods of time, and they could not be distributed over a wide study area. As a result, variations in time and distance tended to be overlooked. Telemetering buoys, systems approaches to data collection, satellite measurements, and acoustic tomography have made it possible to generate synoptic, or large scale, short-term pictures of how ocean currents behave.

These sophisticated new techniques suggest that only a small amount of the total kinetic energy of the oceans is contained in the large circular currents, perhaps as little as 10% to 20%. The remaining energy seems to be contained in large random eddies up to 300 km (186 mi) in diameter. These eddies are analogous to storms in the atmosphere, although they may persist for months.

In some areas, the ocean currents split into separate branches or filaments, and the filaments and the main flow itself may meander somewhat, much like a river, in response to coastal configurations or bottom topography. Occasionally one of these meanders will break off and become a large loop or eddy, called a **ring** (see Figure 10.23). Such rings were first noted in association with the Gulf Stream, which can meander sideways up to 25 km (15.5 mi) in a day, with total amplitudes or widths of 15 to 45 km (9 to 28 mi). They also exist in the Kuroshio and **Antarctic Circumpolar** current systems.

The rings formed by the Gulf Stream rotate in different directions because the current meanders to both sides (Figure 10.23). Anticyclonic (clockwise) warm-water rings form between the Gulf Stream and the coast. These rings are analogous to an atmospheric high-pressure system. Cyclonic (counterclockwise) cold-water rings form on the other side of the Gulf Stream. These are analogous to an atmospheric low-pressure system. The rings range from 150 to 300 km

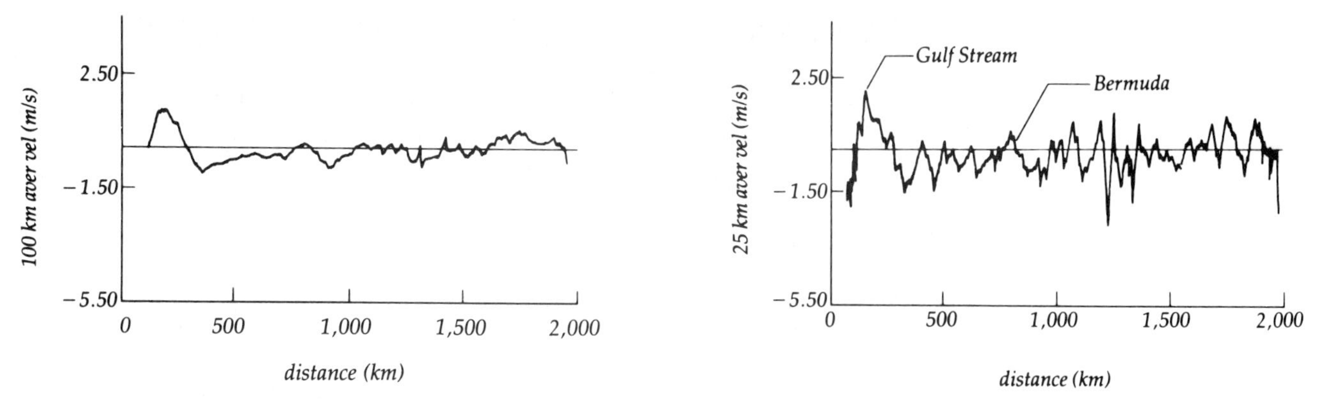

10.21 *Estimate of the surface velocity in the Gulf Stream and its immediate vicinity from a* Seasat *altimetric measurement. (From Wunsch and Gaposchkin, 1980)*

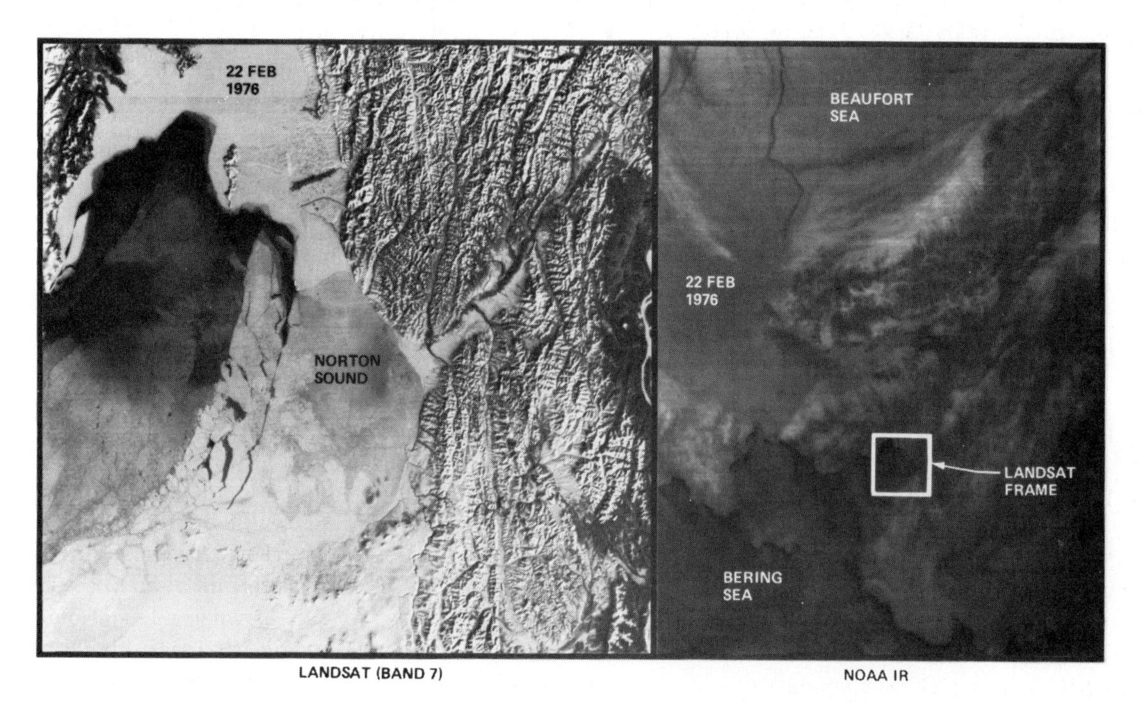

10.22 *The NOAA infrared, or thermal infrared (TIR), image at the right senses the temperature of the ice (or snow) at the surface. Thin ice surfaces are warmer than those of thicker ice. The Landsat image on the left was taken from an altitude of 950 km (589 mi), the image width is 180-by-180 km (112-by-112 mi). Band 7 refers to the spectrum range of 0.8–1.1 μm (very red), which improves images, especially when clouds are present, and gives a very good contrast between ice or snow and water. (Courtesy of Frank Carsey)*

(93 to 186 mi) in diameter and may extend as deep as 3,500 m (11,480 ft). The surface at the center of the warm rings, as measured by *Seasat* satellite, may be 75 cm (29 in) higher than the surface at their periphery; that of the cold rings may be 95 cm (37 in) lower. The Gulf Stream produces approximately five of these rings each year. They are remarkably persistent, having a mean lifetime of 4.5 months, but a few may persist for two years or more. As many as ten may exist at any given time. Like the ocean eddies, these rings transport large amounts of water, but the effect of the Gulf Stream rings and ocean eddies on the ocean as a whole remains to be determined.

Energy from Ocean Currents

By mounting tethered, buoyant underwater turbines across the flow of the largest ocean streams, such as the Gulf Stream and the western boundary currents, it may be possible to generate tremendous amounts of energy (Figure 10.24). In 1973, W. J. Moulton proposed the installation in the Gulf Stream of 242 hydro-

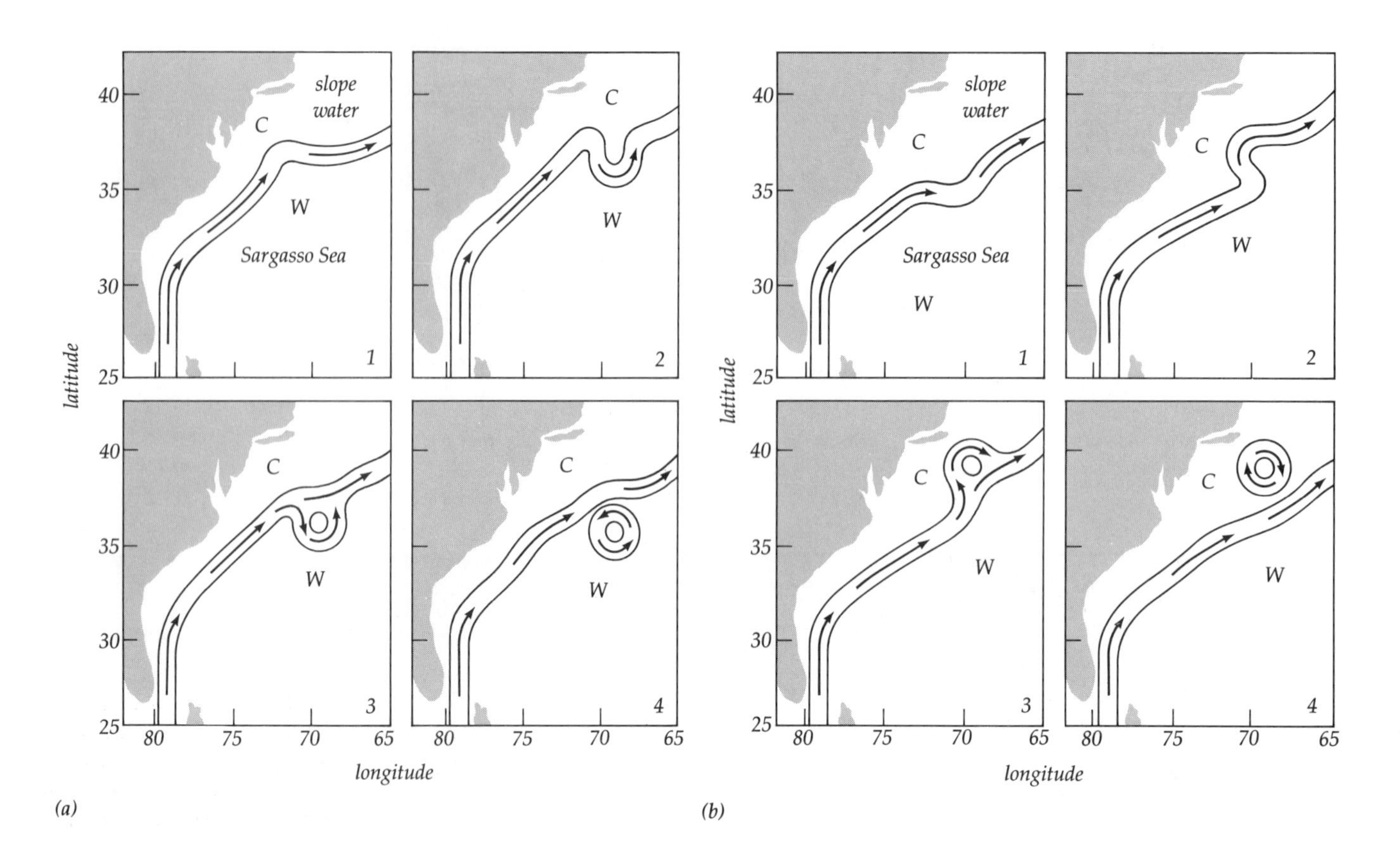

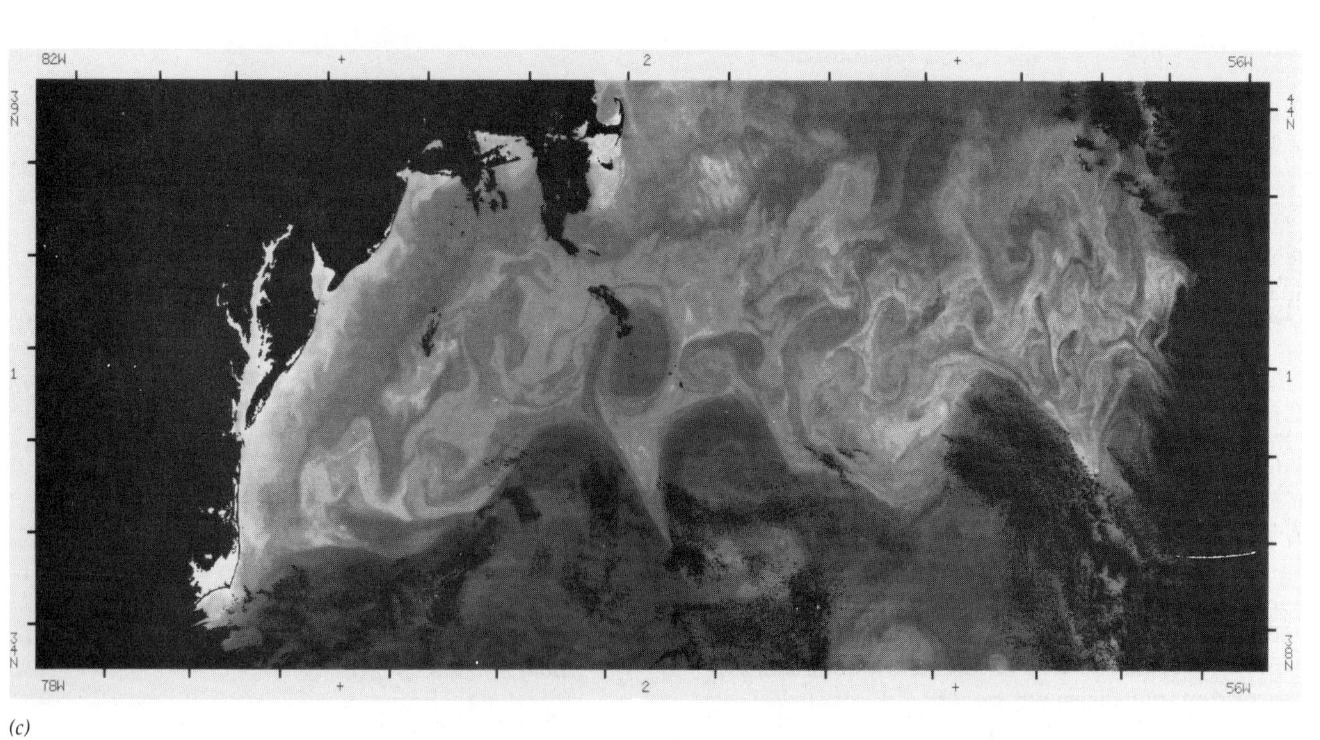

(c)

10.23 *(a) A meander in the Gulf Stream forms a counterclockwise cold-water ring. (b) The loop or ring formed here rotates clockwise and is a warm-water ring. (c) A* Nimbus 7 *CZCS image of the eastern North Atlantic in the 490-nm range taken on May 7, 1979. The image is actually measuring chlorophyll-related pigments. Away from the effects of the continents, the optical clarity of seawater is determined by phytoplankton concentration; the more phytoplankton present, the greater the amount of chlorophyll. The pigment patterns provide an excellent means of monitoring water movement. Here the complicated surface-current pattern with rings moving in both clockwise and counterclockwise directions is obvious.*

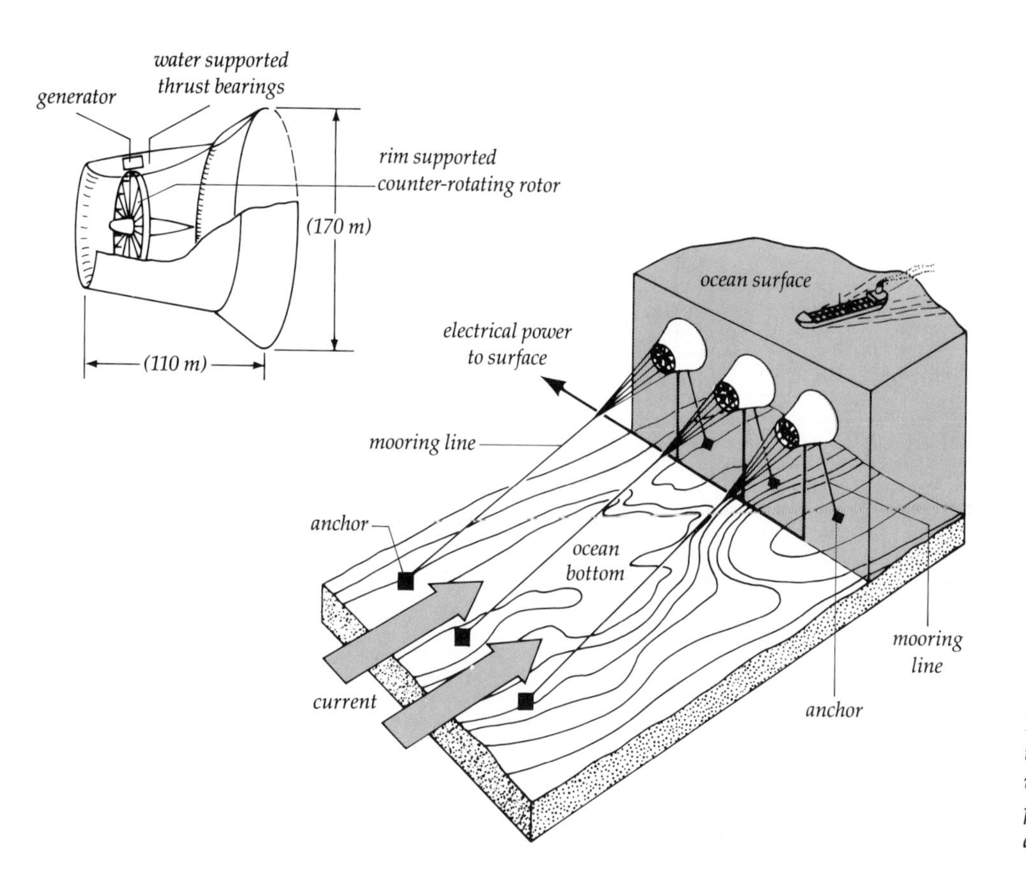

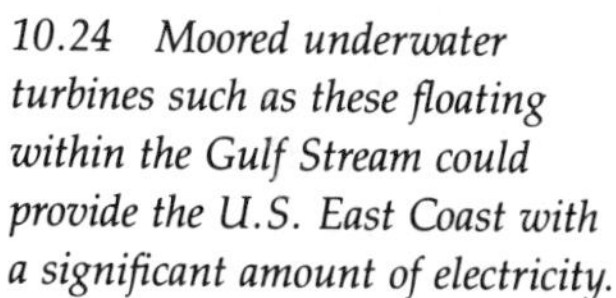
10.24 *Moored underwater turbines such as these floating within the Gulf Stream could provide the U.S. East Coast with a significant amount of electricity.*

turbines, each with a diameter of 170 m (558 ft) and each capable of generating 43 MW in an area 30-by-60 km (18.6-by-37.2 mi); the entire "field" would produce energy equivalent to that of 130 million barrels of oil.

Moulton predicted that each turbine would show a profit in about fourteen months, compared with the ten or more years it takes a nuclear power plant to show a profit. The turbines would be tethered deep enough to avoid creating a hazard to shipping. Although operating the turbines would pose serious technical problems, these problems could presumably be overcome in time. Unless some other economically feasible, nonfossil energy source is discovered within the next quarter-century, tapping the tremendous power of the ocean may prove feasible.

OCEANIC CIRCULATION: THE DEEP CURRENTS

The movement of the surface currents is related to the movement of water deep below the surface. Solar energy generates the winds, which are the main cause of surface currents, and it also affects the movement of surface waters directly. It warms waters at the equator, causing them to expand into a bulge a few centimeters high. Even this small slope accounts for some of the net transport of surface waters as they move downhill toward the poles. As they cool in the polar regions, these high-salinity waters increase in density, sink, and then slowly move along the bottom back to the equator. Figure 10.25 shows an idealized rendering of this pattern. The picture is complicated, however, by the earth's rotation, the Coriolis effect, and the fact that the equatorial region isn't quite the barrier to oceanic flow it was once thought to be.

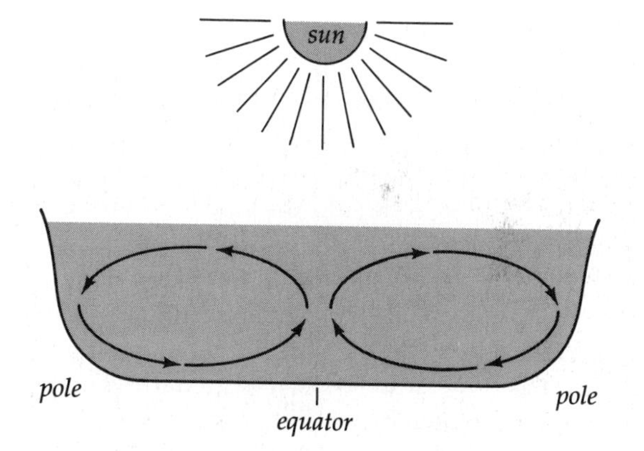

10.25 *Theoretical model for deep-ocean circulation.*

Actually, there are only a few regions in the oceans where surface waters sink to the bottom, spread, and flow toward the equator. In Figure 10.26, two well-defined deep currents are shown that flow north and south away from the two regions where they form

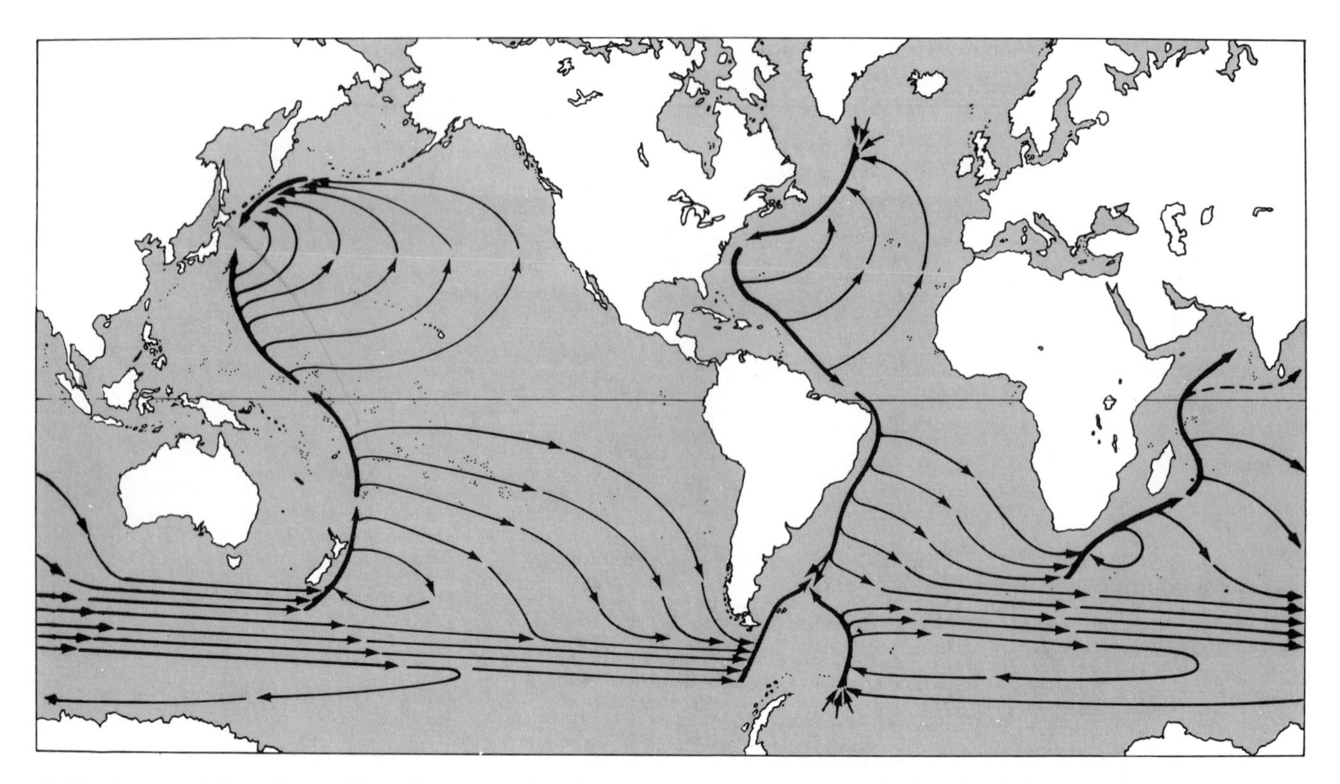

10.26 Deep circulation below 2,000 m. The regions where three short arrows converge (near Greenland and the Antarctic) are the input locales for sinking deep waters. In addition, there is a sinking of water at the Antarctic Divergence. (After Stommel, 1958, see Further Readings)

from sinking waters. These currents move along the eastern edges of North and South America, meeting in the South Atlantic and forming a broad flow eastward across the Atlantic into the Indian Ocean and finally into the Pacific.

The waters of the ocean depths are layered rather than well mixed. Figure 10.27 shows a north-south cross-sectional profile of the Atlantic, Pacific, and Indian oceans. Each of the layers shown has a narrow range of density determined by small ranges of temperature and salinity (see Table 10.1). Recall that the density of seawater is a function of temperature, salinity, and pressure. The motion of deep currents, both vertical and horizontal, is caused by variations in density. Because density depends on temperature and salinity, these currents are called **thermohaline** currents.

Table 10.1 gives the temperature, salinity, and density of several oceanic layers. So specific are these parameters that each layer has a specific name. Any body of water that has a characteristic temperature and salinity range is called a **water mass**, and oceanographers can identify and trace water masses by determining those two parameters.

Water masses are divided into five types: surface (to a depth of about 200 m, or 656 ft); central (to the bottom of the main or permanent thermocline); intermediate (from the bottom central waters to about 1,500 m, or 4,920 ft); and deep and bottom. Bottom water is a water mass that is in contact with the bottom and that has distinctly different properties from those of the deep water above it. Technically, surface water is not regarded as a water mass, because its range of temperature and salinity is too great and too variable to give it specific characteristics.

Temperature-Salinity Diagram

Photographs of the ocean bottom reveal ripple marks, which suggest that even at these depths the water is in motion. Water molecules move randomly in response to temperature. Although they move at fairly high speeds (a few hundred meters per second) they are constantly colliding with one another. Consequently, their net movement is very slow. This random molecular spreading is called **diffusion**. A cube of sugar carefully placed in a glass of still water will take weeks to become evenly mixed in the water, through diffusion. If we agitate the water, however, the sugar will dissolve in a minute or two. This process is called **advection**.

If the deep-water masses had not been constantly forming by ongoing advective processes, over time

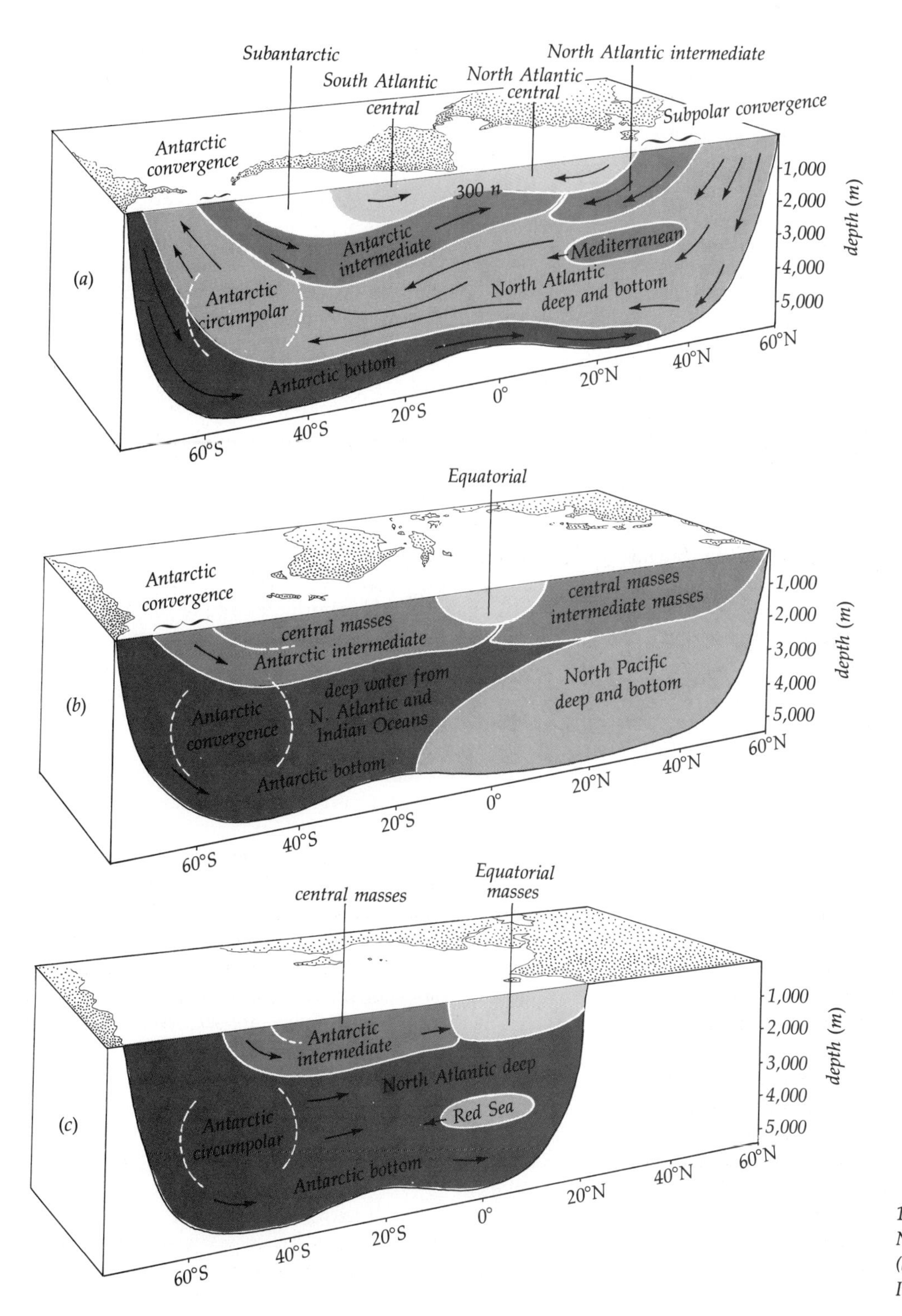

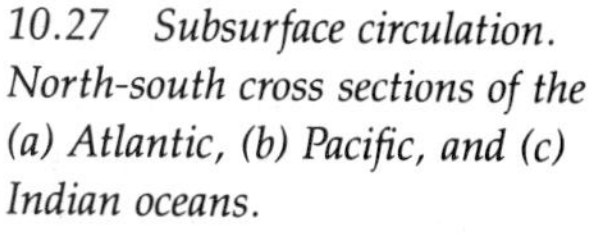

10.27 Subsurface circulation. North-south cross sections of the (a) Atlantic, (b) Pacific, and (c) Indian oceans.

they would have been totally mixed even by such a slow process as diffusion. The fact that these specific water masses exist at all shows that they are continuously being formed. It is difficult, however, to determine the rates at which the processes of advection and diffusion are acting in the deep ocean. The water molecules in a parcel of the Gulf Stream may move 100 nautical miles a day. In the deep ocean, however, the movement is usually so slow (about 1 to 2 cm/s) that conventional measuring devices are inadequate.

Deep currents are relatively weak, and diffusion may play a more important mixing role than has been believed. Mixing by diffusion occurs more readily across horizontal water masses of constant density than it

Table 10.1 *Characteristics of Water Masses*

Water Mass	Salinity (‰)	Temperature (°C)	Density (g/ml)	Location Depth (m)
Antarctic Bottom	34.66	−0.4	1.02786	4,000–Bottom
Antarctic Circumpolar	34.7	0–2	1.02775–1.02789	100–4,000
Antarctic Intermediate	34.2–34.4	3–7	1.02682–1.02743	500–1,000
Arctic Intermediate	34.8–34.9	3–4	1.02768–1.02783	200–1,000
Mediterranean	36.5	8–17	1.02592–1.02690	1,400–1,600
North Atlantic Central	35.1–36.7	8–19	1.02630–1.02737	100–500
North Atlantic Intermediate	34.73	4–8	1.02716–1.02765	300–1,000
North Atlantic Deep and Bottom	34.9	2.5–3.1	1.02781–1.02788	1,300–Bottom
South Atlantic Central	34.5–36.0	6–18	1.02606–1.02719	100–300
North Pacific Central	34.2–34.9	10–18	1.02521–1.02634	100–800
North Pacific Intermediate	34.0–34.5	4–10	1.02619–1.02741	300–800
Red Sea	40.0–41.00	18	1.02746–1.02790	2,900–3,100

does vertically across masses of varying density. Since the ocean is far broader than it is deep, it is essentially a horizontal formation.

Although conventional devices cannot be used to measure the movement of deep waters, chemical methods are effective in tracing deep currents. These methods are based on the fact that living organisms at various depths have predictable effects on the chemical composition of the water. For example, with increasing depth the amount of organic matter decreases and the amount of inorganic material increases. Moreover, concentrations of dissolved oxygen decrease and concentrations of carbon dioxide increase with depth, all as a function of the time that has elapsed since the water parcel left the surface. Radioactive dating with carbon 14 is especially useful in tracing these deep currents.

The **temperature-salinity** (**T-S**) **diagram** in Figure 10.28 gives lines of constant density. Note that a wide range of temperatures and salinities may yield the same density. At particular points at the sea-air interface, weather conditions, in combination with the water's own temperature and salinity, produce water of a specific temperature and salinity. Each such combination is represented by a unique point on the T-S diagram. Because the deep waters cannot change their salinity or temperature in situ except by mixing, they receive their basic characteristics only at the surface. Water that is truly homogeneous and that has maintained a specific salinity and temperature over an appreciable vertical distance is called a **water type**. Few water types persist for long in an unmixed state, however; the major exception is the Antarctic Bottom Water. A point on a T-S diagram represents a water type, whereas a water parcel with a temperature-salinity range is a portion of a curve on a T-S diagram.

Because most ocean water is the result of mixing, water masses are combinations of two or more water types. A straight line on a T-S diagram represents the mixing of two water types (Figure 10.29). The mixing may be lateral (along surfaces of equal density) or vertical (across regions of varying density, as when cold, dense water sinks to the bottom). The mixing of two or more water masses with the same density but different temperatures and salinities will produce a new mass of greater density. This mixing-and-sinking process is called **caballing**. If a parcel of water moves as a unit along a density surface, the movement, which is due to gravity, is advection.

The T-S diagram is a powerful diagnostic tool. A T-S diagram for hydrocast data from each oceanographic station indicates densities and identifies water masses. It also reveals the layering structure of the water and shows vertical regions of mixing and the mixing proportions. T-S diagrams also serve as a rapid check for bad data. Since denser water is usually beneath less dense water, the data ordinarily produce a smooth curve; any departure from such a curve is suspect. Finally, comparing T-S diagrams from neighboring stations can indicate the density structure and deep-current movement of the area.

Water Age

The oceans are continously renewing themselves at all depths, more slowly at great depths than at lesser depths. **Water age** is the length of time that has elapsed since a water parcel (now part of a deep mass) has descended from the surface. Maximum water ages vary even within the same ocean. The age of the North Atlantic Deep Water (NADW), for example, has generally been set at about 700 to 750 years, although recent information gleaned from ^{14}C studies indicates that it may be only 100 years. (Water age is different from **residence** [or **flushing**] **time**, which is the time it takes for all the water in an area to be replaced.) The

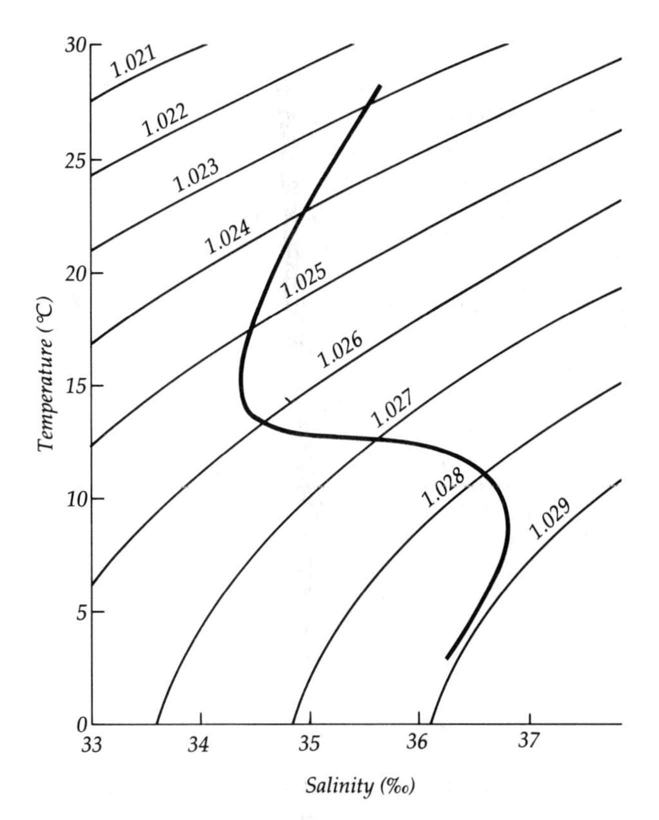

10.28 A typical T-S diagram with specific gravity curves added. The T° and S‰ data from a specific oceanographic station (from a hydrocast) are plotted and the points connected to form the wide line.

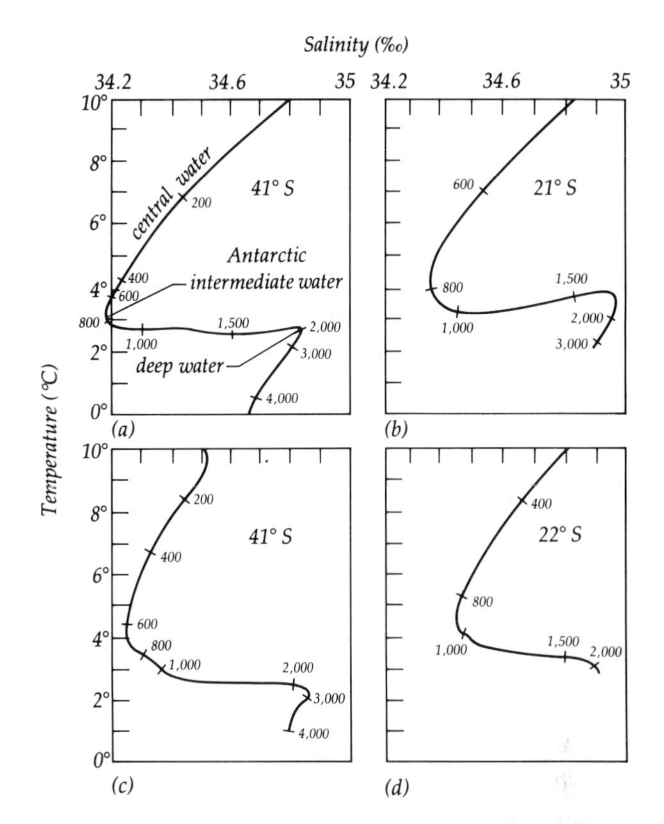

10.29 T-S diagrams for the western and eastern basins of the South Atlantic. (a) and (b) represent two locales in the western basin of the South Atlantic on the same line of longitude but different by 20° of latitude. (c) and (d) are in the eastern basin of the South Atlantic, again about the same 20° (21°) of latitude apart from one another but on the same longitude. The little numbers on each graph line represent depths in meters.

age of the deep waters in the Pacific seems to be several times greater than that of the deep waters in the North Atlantic, perhaps 1,000 years or more. Water age, however, gives only a slight indication of the water's average speed and says virtually nothing about the rate of formation of the water mass.

Atlantic Ocean

Figure 10.27a showed a cross section of the Atlantic Ocean. Keep in mind, however, that cross-sectional diagrams fail to give an adequate picture of the spread of deep waters across an entire ocean bottom. Even so, notice that one water mass forms the bottom water of the South Atlantic, most of the bottom water of the North Atlantic, and the bottom water of the Indian and the South Pacific to about 15°S latitude. This water mass, because of its low temperature and high salinity, has a greater density than any other water mass. Known as the Antarctic Bottom Water (AABW), it flows east around Antarctica in response to the surface current, the West Wind Drift (Figure 10.2). The AABW flows into the Atlantic along the bottom to a latitude about equal to that of New York.

Although the AABW forms the bottom waters of the South Atlantic, the waters of the eastern and western basins of the South Atlantic exhibit different properties. The densities are the same, but the minimum water temperature in the western basin is 0.4°C (32.7°F), whereas in the eastern basin it is 2.4°C (36.3°F). The salinity of the western basin is 34.7‰, and that of the eastern basin is 34.9‰. These variations are caused by the presence of the Walvis Ridge, which extends from Africa to the Mid-Atlantic Ridge at an average depth of 3,000 m (9,840 ft) and blocks the AABW flow. In Figure 10.29, note that (a), (b), and (c) all show the presence of deep and bottom waters, whereas (d) does not. A comparison of (c) and (d) shows that something has happened to the AABW. It is blocked by the barrier imposed by the Walvis Ridge.

For a long time, oceanographers thought that the North American Deep Water (NADW) formed off the southeastern coast of Greenland by the convergence and mixing of the cold Labrador Current from the

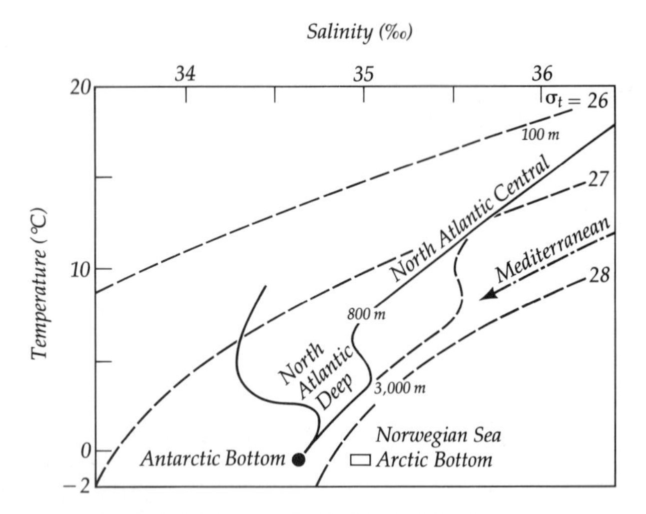

10.30 Typical T-S diagram for the North Atlantic.

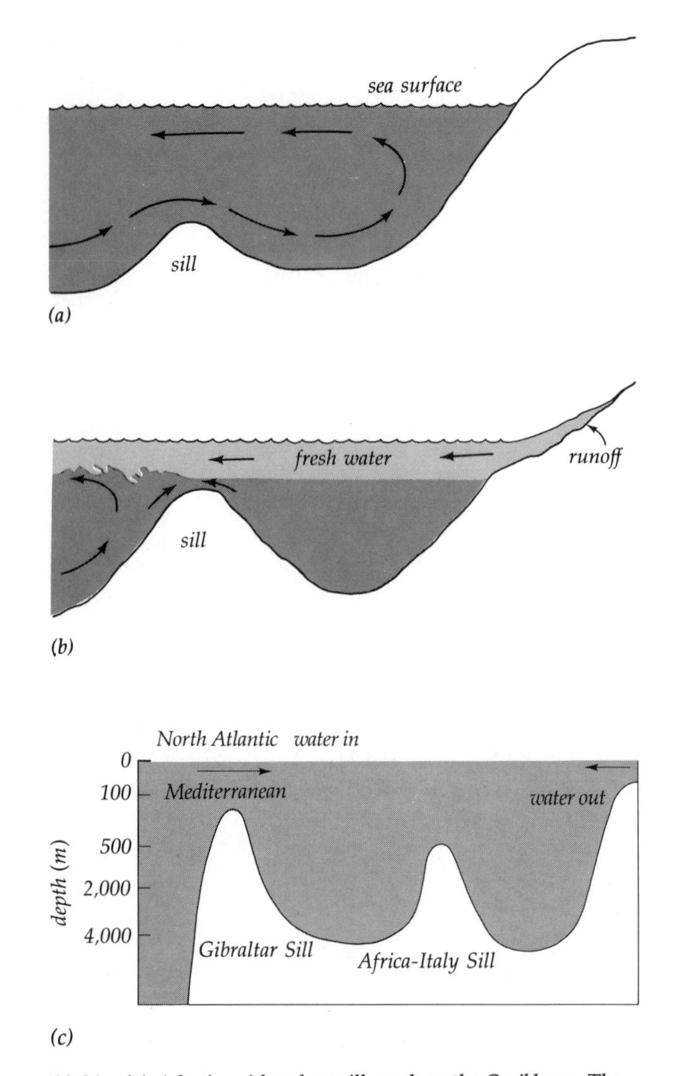

10.31 (a) A basin with a deep sill, such as the Caribbean. The waters are usually well mixed. (b) A basin with a shallow sill, such as the Baltic Sea and the Black Sea. A large amount of freshwater runoff, exceeding evaporation, results in deep waters with little circulation. (c) East-west cross section of the Mediterranean.

north with the warm, saline Gulf Stream. Here the winter temperature of the water varies only from 3°C to 3.25°C (37.4°F to 37.85°F) from the surface to a depth of 3,000 m (9,840 ft) over a region 150 km (93 mi) across. They thought that the high-density water formed at the surface by mixing sank to its equivalent density level and flowed south as NADW. Oceanographers now believe that about 80% of NADW flows into the North Atlantic from the Norwegian Sea (as Norwegian Sea Deep Water, or NSDW) by way of sills between Greenland and Iceland and between Iceland and Scotland. The Mid-Atlantic Ridge, of which Iceland is a part, divides the bottom flow of NADW into two parts: Northwest Atlantic Bottom Water (NWABW), which has a temperature of 1°C (33.8°F) and a salinity of 34.91‰, and Northeast Atlantic Bottom Water (NEABW), which has a temperature of 2.5°C (36.5°F) and a salinity of 35.03‰. The values differ because NSDW has mixed with some North Atlantic Central Water (NACW) to become NEABW. NWABW is formed by the mixing of NSDW with Labrador water and with some NEABW that has crossed the Mid-Atlantic Ridge. The bulk of NEABW flows southeast of the ridge and overrides the AABW. Figure 10.30 shows a generalized T-S diagram of the North Atlantic.

Figure 10.31c shows that Atlantic surface water flows into the Mediterranean, and that Mediterranean water intrudes into the North Atlantic at a depth of 2,000 to 3,000 m (6,560 to 9,840 ft). Warm, saline water flows out of the Mediterranean below the surface water, which is moving east into the Mediterranean (Figure 10.31). Water from the Mediterranean flows out over the colder water of the Atlantic, which is only a little denser. Its core enters the Atlantic at a depth of approximately 1,000 m (3,280 ft). The effects of this Mediterranean water on the North Atlantic water have been traced 2,500 km (1,550 mi) from Gibraltar into the central Atlantic. As it moves in a generally westward direction, it mixes gradually with waters of slightly different density above and below it. Ultimately it loses its identity. This influx of saline water from the Mediterranean is one of the reasons for the high salinities of the North Atlantic. The inflow of colder Atlantic water over the saltier water shows that salinity is the controlling variable.

During World War II, German and Italian submarines attempted to use the two-way currents in the Straits of Gibraltar interface to sneak past the British blockade. They coasted silently along with the current, their engines shut off. Even though the Straits of Gibraltar are only 29 km (18 mi) wide and 320 m (1,050 ft) deep, it was rumored that some submarines were successful in this passage.

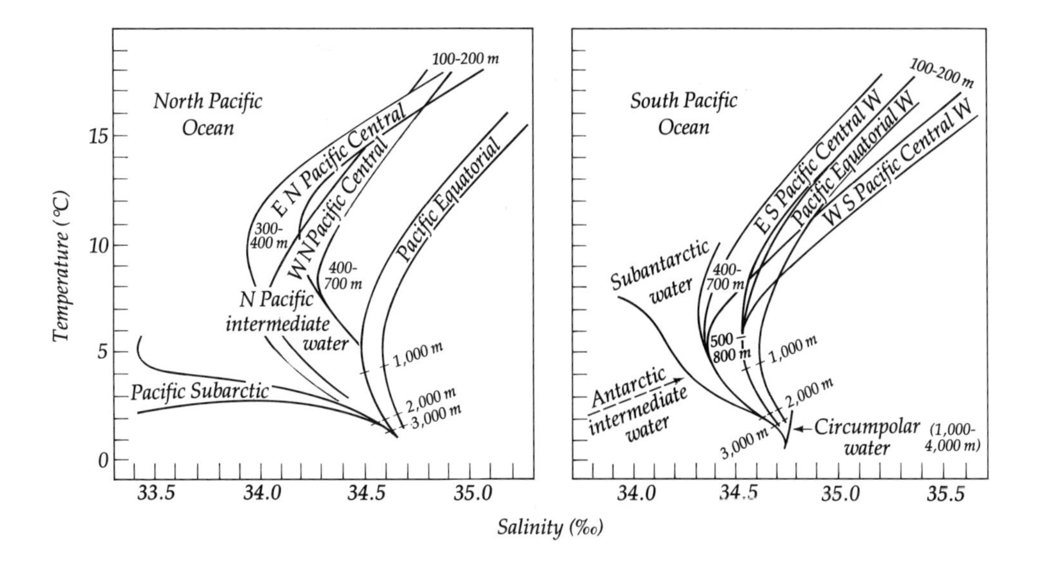

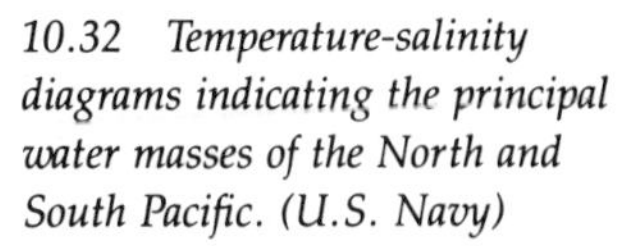
10.32 Temperature-salinity diagrams indicating the principal water masses of the North and South Pacific. (U.S. Navy)

Pacific Ocean

Unlike the Atlantic, the Pacific Ocean has only limited access to cold northern waters. Thus the extensive thermohaline convection typical of the North Atlantic is absent from the North Pacific, which affects the entire ocean. No really dense water masses appear in the North Pacific, the boundaries of its central and intermediate waters are not clearly defined, and the deep water is much more sluggish than that of the North Atlantic. Consequently, the Pacific is less dynamic than the Atlantic. No deep water is formed in it, and neither is there any intrusion of Mediterranean-like water. Since it is so vast, however, its water masses are more complicated.

The subsurface waters of the North Pacific constitute a single central water mass about 200 to 300 m (656 to 984 ft) thick and extending to about 40°N. This water mass has the lowest salinity of any central water mass. South of this North Pacific Central Water (NPCW) is equatorial water, which extends across the entire Pacific from 20°N to 10°S. On the eastern side, it is only 50 m (164 ft) deep; on the western side, it is 150 to 200 m (492 to 656 ft) deep. It is characterized by a very uniform T-S relationship and is separated from surface waters by a strong thermocline of 28°C to 15°C (82.4°F to 59°F). The southernmost limit of the South Pacific Central Water (SPCW), at 40°S, is regarded as the southern boundary of the Pacific. It extends to the Subtropical Convergence. SPCW varies more in temperature and salinity from west to east than does NPCW, and it is more difficult to distinguish from equatorial water.

The North Pacific and the South Pacific each contain an intermediate water mass below the central water mass. The intermediate water mass of the North Pacific is characterized by a salinity minimum; it is 500 m (1,640 ft) deeper in the west than in the east and displays a weak, clockwise circulation. The intermediate water mass of the South Pacific, called Antarctic Intermediate Water (AAIW), is formed at the Antarctic Convergence (see Chapter 17) by subsurface mixing. It has a low salinity of 33.8‰ and a temperature of 2.2°C (35.96°F). Its core is about 1,000 m deep (3,280 ft). AAIW can be traced to just north of the equator.

The waters of the Pacific below 2,000 m (6,560 ft) show very uniform properties. These deep waters are not formed in the Pacific; rather, they represent a kind of dead end of oceanic circulation. Montgomery showed in 1958 that 30% of the ocean's water has a temperature of between 1°C and 2°C (33.8°F and 35.6°F) and a salinity of between 34.6‰ and 34.8‰. Only 3% of the Atlantic falls within those ranges, but 44% of the Pacific and 25% of the Indian Ocean do. This, the world's largest water mass, is called Oceanic Common Water. It is formed primarily from NADW, AABW, and AAIW (see Figure 10.27), along with some minor mixing of Indian and Pacific waters. It enters the Pacific between New Zealand and Antarctica and then branches and spreads throughout the South and North Pacific. The flow is greater in the west, but even there it is very slow, no greater than 0.1 cm/s. This water is very uniform. The enormous volume of North Pacific Deep Water has a temperature of 1.5°C (34.7°F) and a salinity of 34.68‰. Very little deep water is exchanged between the North and South Pacific. Since the North Pacific is effectively closed at the Bering Strait, the water is trapped and circulates sluggishly. Figure 10.32 shows a T-S diagram for the North Pacific.

Indian Ocean

As Figure 10.27c indicated, the northern limit of the Indian Ocean extends only to about 25°N. The ocean

Table 10.2 *Characteristics of Semi-enclosed Bodies of Water*

Body of Water	Sill Depth (m)	Salinity of Outflowing Water (‰)	Residence Time of Water (yr)	Average Depth (m)	Maximum Depth
Baltic Sea	18	<10	1	100	460
Black Sea	40–100	11	3,000	1,271	2,245
Hudson Bay	60	25–32	1	90	200
Mediterranean Sea	320	37.8	70	1,500	3,400

has no northern basin, and all its bottom water is AABW. The source of the Indian Deep Water (IDW) is NADW, into which a tongue of Red Sea Water, with a salinity as high as 42.5‰, has intruded. Above the IDW, extending to a depth of 1,000 m (3,280 ft), is AAIW. Between the AAIW and the surface is Indian Central Water, with a salinity range of 34.5‰ to 36‰. The Indian Equatorial Water has a fairly consistent salinity of 34.9‰ to 35.5‰.

No deep or bottom waters are formed in the Indian Ocean. The deep circulation is slow, and the temperature and salinity are characteristic of Oceanic Common Water.

The boundaries of the water masses shown in the diagrams do not mean that the masses simply stop at those lines. They mix with other waters along these boundaries and gradually lose their identity. When vertical or horizontal subsurface mixing occurs, the characteristic temperature and salinity of each water mass changes.

The density of a water parcel is a function of temperature, salinity, and pressure, and it is the density that determines how a water mass moves. Denser masses will flow under less dense ones, and less dense masses will flow over denser ones. Dense water formed at the surface will sink until it either reaches its equivalent density or until it encounters the bottom and spreads out laterally. If two water masses have a large density difference, appreciable vertical mixing between them, even by diffusion, is unlikely. Conversely, as the density difference between the layers gets smaller, vertical mixing becomes more probable.

MARGINAL SEAS

In addition to the deep ocean basins, there are a number of semi-isolated seas, sometimes called **mediterranean** ("middle of the land") seas. These **marginal seas** commonly have sills near their entrances, which may hinder circulation (Figure 10.31). Among the world's marginal seas are the Mediterranean Sea, the Baltic Sea, Hudson Bay, the Black Sea, the American Mediterranean (the Gulf of Mexico and the Caribbean Sea), and the Red Sea.

Circulation in these seas is a complicated function of sill depth, basin depth, precipitation, runoff, and evaporation. Landlocked seas such as the Mediterranean are similar in structure to the ocean, but they differ appreciably in current profile. The Mediterranean has characteristic water masses and vertical circulation. The flow of Atlantic water into the Mediterranean is so great that the entire sea is renewed every 75 years. The Mediterranean is deep, has a deep sill, a surface inflow, and a subsurface outflow. Evaporation exceeds precipitation and freshwater runoff. The saline surface waters sink, constantly renewing the deep waters and keeping the oxygen content relatively high. In the Baltic, Hudson Bay, and the Black Sea, shallow sills impede the movement of waters in and out (Table 10.2). Precipitation and runoff exceed evaporation in these seas, keeping the surface waters fairly fresh. In the shallow Baltic and Hudson Bay, vertical mixing takes place.

In the deep Black Sea, however, the low-salinity surface waters are stable and do not mix vertically. Only small amounts of water enter the Black Sea through the shallow, narrow Bosporus Straits. The net effect is a lack of vertical circulation. Vertical movement of the deep water to the surface is impeded, and the water becomes stagnant and depleted in oxygen (anoxic). The turnover, or mixing time, for the Black Sea is 3,000 years. Its surface water is aerated and contains living organisms, but below 65 m (213 ft) the waters are stagnant and the conditions anaerobic, suitable only for certain species of bacteria that thrive in the absence of oxygen. The seafloor, covered with black sulfide mud, is anoxic and nearly devoid of life.

The American Mediterranean consists of the Gulf of Mexico and the Caribbean, which are separated by the Yucatán Peninsula and Cuba. The sill depth of the Gulf of Mexico between Florida and Cuba is 880 m (2,886 ft); that of the Caribbean is 1,500 m (4,920 ft). Though precipitation and runoff vary greatly from place to place and evaporation is high all year, the water is not appreciably altered between the time it enters and the time it leaves.

FURTHER READINGS

Armi, L. "Mixing in the Deep Ocean—The Importance of Boundaries." *Oceanus*, 1978, *21*(1), 14.

Baker, D. J., Jr. "Models of Oceanic Circulation." *Scientific American*, January 1970, pp. 114–21.

______. "Currents, Fronts, and Bottom Water." *Oceanus*, 1975, *18*(4), 8.

Craig, R. E. *Marine Physics*. New York: Academic Press, 1972.

Ebert, H. V. "El Niño: An Unwanted Visitor." *Sea Frontiers*, 1978, *24*(6), 347.

Gaskell, T. F. *The Gulf Stream*. London: Cassell, 1972.

Idyll, C. P. "The Anchovy Crisis." *Scientific American*, June 1973, pp. 22–29.

Knauss, J. A. "The Cromwell Current." *Scientific American*, April 1961, pp. 105–19.

______. *Introduction to Physical Oceanography*. Englewood Cliffs, N.J.: Prentice-Hall, 1978.

Mason, P. "The Changeable Ocean River." *Sea Frontiers*, 1975, *21*(3), 171.

Miller, J. "Barbados and the Island-Mass Effect." *Sea Frontiers*, 1975, *21*(5), 268–72.

Montgomery, R. "The Present Evidence of the Importance of Lateral Mixing Processes in the Ocean." *American Meteorological Society Bulletin*, 1940, *21*, 87–94.

"Ocean Eddies." *Oceanus*, Spring 1976, *19*(3).

"Ocean Energy." *Oceanus*, Winter 1979/80, *27*(4).

"Oceanography from Space." *Oceanus*, Fall 1981, *211*(3).

Picard, G. L., and W. J. Emery. *Descriptive Physical Oceanography*. Oxford: Pergamon Press, 1982.

Pirie, R. G., ed. *Oceanography: Contemporary Readings in Ocean Science*. 2d ed. London: Oxford University Press, 1977.

Pond, S., and G. L. Pickard. *Introductory Dynamic Oceanography*. Oxford: Pergamon Press, 1978.

Smith, F. W. G. "Measuring Ocean Movements." *Sea Frontiers*, 1972, *18*(2), 166–74.

______. "Power from the Oceans." *Sea Frontiers*, 1974, *20*(2), 87.

______. "Oceans of Energy." *Sea Frontiers*, 1980, *26*(4), 194.

______. "Turbines in the Ocean." *Sea Frontiers*, 1981, *27*(5), 300–305.

Smith, F. W. G., and R. H. Charlier. "Saltwater Fuel." *Sea Frontiers*, 1981, *27*(6), 313.

Stewart, R. W. "The Atmosphere and the Ocean." *Scientific American*, September 1969, pp. 76–105.

Stommel, H. "The Circulation of the Abyss." *Scientific American*, July 1958, pp. 85–93.

______. *The Gulf Stream*. 2d ed. Berkeley: University of California Press, 1965.

Sverdrup, H. U., M. W. Johnson, and R. H. Fleming. *The Oceans*. Englewood Cliffs, N.J.: Prentice-Hall, 1970.

Tchernia, P. *Descriptive Regional Oceanography*. Oxford: Pergamon Press, 1980.

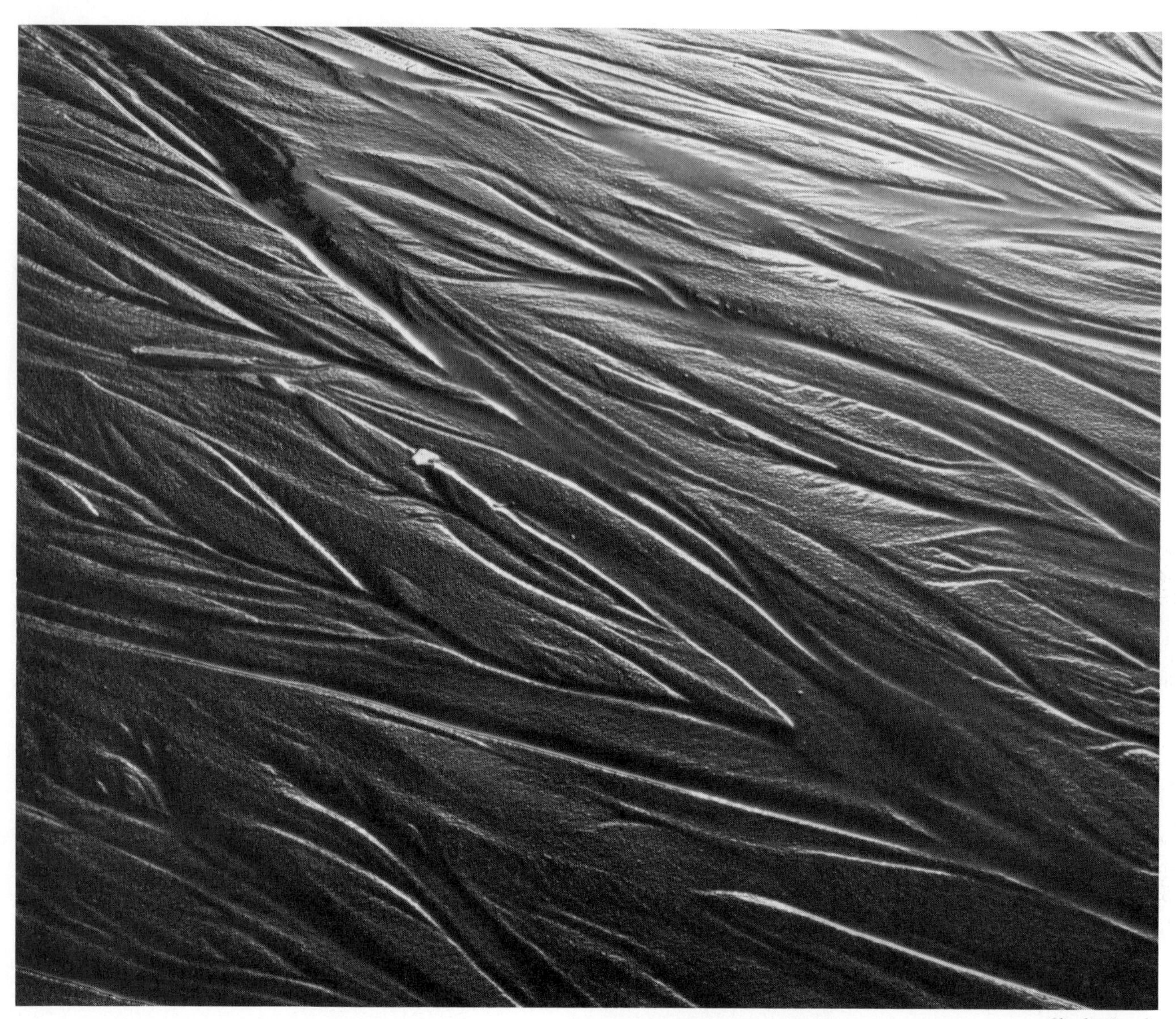

Charlie Gurche

The wind increased to a howl; the waves dashed their bucklers together; the whole squall roared, forked, and crackled around us like a white fire upon the prairie, in which unconsumed, we were burning; immortal in these jaws of death! . . .

Now, in calm weather, to swim in the open ocean is as easy to the practiced swimmer as to ride in a spring-carriage ashore. But the awful lonesomeness is intolerable. The intense concentration of self in he middle of such a heartless immensity, my God! who can tell it? Mark, how when sailors in a dead calm bathe in the open sea—mark how closely they hug their ship and only coast along her sides. . . .

An intense copper calm, like a universal yellow lotus, was more and more unfolding its noiseless, measureless leaves upon the sea.

Herman Melville

ELEVEN

Waves

How would you demonstrate what a wave is to someone who has never seen one? You might tie one end of a long rope to a tree and whip the other end up and down. Each up-and-down cycle moving along the rope is a **wave**. Neither the rope nor the particles that compose it move toward the tree; only the waveform moves in that direction. Or you might drop a pebble into a pool of still water. The water surface at the point of impact will vibrate vertically like a drumhead, and concentric circles will emanate from the center. These, too, are waves. They are about as close to the ideal wave as we can find in the physical world.

Various terms are used to define waves. Figure 11.1 illustrates the meaning of some of the terms defined here.

1. The **wave crest** is the highest point of a wave; the **wave trough** is the lowest point of a wave.
2. **Wavelength** is the distance measured from any point on one wave to the equivalent point on an adjacent wave. For example, wavelength may be measured from crest to crest or from trough to trough. Wavelength may be as small as 1.73 cm (0.67 in) or as great as half the circumference of the planet.
3. **Wave height** is the vertical distance from the crest to the trough of a wave.
4. Wave **amplitude** is the distance a wave moves the water above or below sea level. Wave amplitude is equal to half the wave height.
5. The **period** of a wave is the time it takes for one wave to pass a specified point. The velocity and length of a water wave are determined by its period. Table 11.1 shows the classification of waves on the basis of period.
6. Wave **frequency** is the number of waves passing a specified point in a given unit of time. For instance, waves that pass a point every 30 second(s) have a frequency of two waves per minute. However, it is more common to express frequency in waves per second; thus a wave with a period of 30 s has a frequency of 0.033 wave/s (or cycles/s).
7. The **propagation rate** of waves is the velocity at which they travel. The velocity of waves in relatively shallow water depends only on depth; hence, such waves travel at a constant rate. The velocity of waves in relatively deep water depends only on their wavelength. The formulas for the velocity of shallow waves and the velocity of deep waves are

$$V_{\text{shallow}} = \sqrt{gd} \quad \text{and} \quad V_{\text{deep}} = \sqrt{gL/2\pi}$$

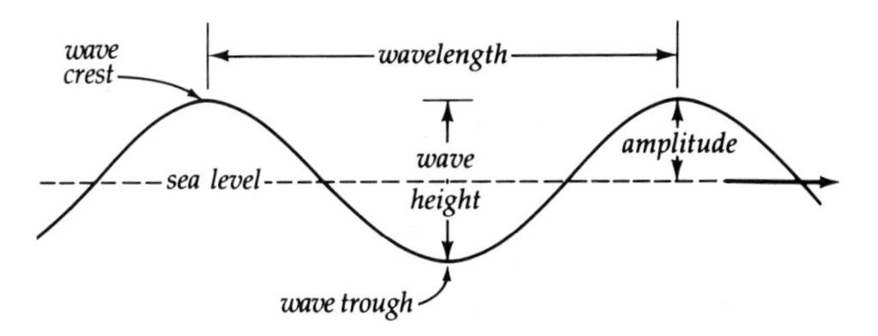

11.1 *Wave terminology as described in the text.*

Table 11.1 *Classification of Ocean Waves by Period*

Wave Classification	Period
Capillary	<0.1 s
Ultragravity	0.1–1 s
Wind	1–30 s
Infragravity wind	30 s–5 min
Long-period	5 min–12 h
Tidal	12–24 h

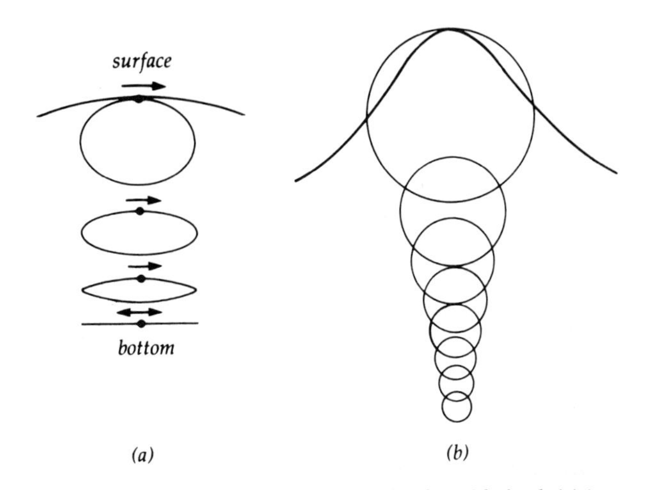

11.2 *Wave orbits of different water molecules with depth (a) in shallow water and (b) in deep water.*

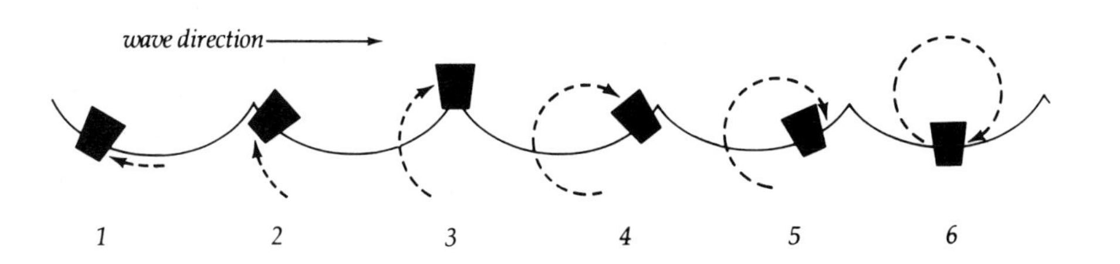

11.3 *Motion of a cork on a wave.*

where V is velocity, g is force due to gravity, d is water depth, and L is wave length. The terms *shallow* and *deep* as used here are relative to wavelength only.

Waves represent a form of kinetic energy, which is the energy of motion. A wave with a height of 1.3 m (4.3 ft) breaking over a kilometer of coastline releases about 22,000 hp. Waves once tossed a rock weighing 61 kg (134 lb) through the roof of Tillamook Rock Lighthouse in Oregon, which rises 46 m (151 ft) above the water! Waves tore loose a chunk of cement weighing 1,220 t (1,342 U.S. tons) along the coast of Scotland. The force (or, more precisely, pressure, since pressure equals force divided by area) of surf waves has been measured at more than 1,127 kg/cm^2 (16,335 lb/in^2). Some of the energy released when waves break against the shore can be harnessed to generate electrical power. In an experiment on the Algerian coast, waves are funneled through V-shaped cement troughs into a reservoir above sea level. The water is then released to flow over turbines.

Waves possess potential energy as well as kinetic energy. Kinetic energy is due to the orbital motion of the water molecules, as shown in Figure 11.2. Potential energy is due to the displacement of the wave above sea level. Potential energy is not concentrated at any one point; instead, it is spread throughout the entire wave. Thus waves of the same length but of different heights vary in energy content. The maximum height of a wave is one-seventh its length. For example, the maximum height of a wave 156 m (512 ft) long is 22 m (72 ft). This ratio of height to length defines the wave's steepness. When a wave exceeds this ratio (or when its internal angle is greater than 120°), the wave becomes unstable and begins to break. Large waves seldom achieve maximum height, however. In fact, smaller wind-generated waves are far more numerous in the ocean and represent the greatest total energy.

It is important to realize that it is energy, not water, that is transferred by wave motion. If you place a cork on the surface of a calm pool of water and then drop a stone into the water, the cork bobs up and down but does not move along with the waves (Figure 11.3). It may move slightly in the direction of wave travel because some energy is transferred from the wave to the cork, but such movement will be negligible. The water particles in a wave oscillate in this same up-and-

Table 11.2 *Wind Scales and Sea Descriptions*

Sea		Wind				
Description	Force	Mariners' Description	Velocity (m/sec)	Knots	Beaufort Scale	State of the Surface for Fully Developed Sea, Wave Heights
Smooth	0	Calm	0–0.2	<1	0	Mirrorlike, glassy
Rippled	1	Light air	0.3–1.5	1–3	1	Small ripples; no foam crests
Gentle	2	Light breeze	1.6–3.3	4–6	2	Small wavelets, crests show glassy appearance and don't break (0–.3m)
Gentle	2	Gentle breeze	3.4–5.4	7–10	3	Crests begin to break; foam has glassy appearance; occasional whitecaps (.3–.61 m)
Light	3	Moderate breeze	5.5–7.9	11–16	4	Waves small, becoming longer; fairly frequent whitecaps
Moderate	4	Fresh breeze	8.0–10.7	17–21	5	Moderate waves, taking more pronounced long form; mainly whitecaps; some spray (.61–1.21 m)
Heavy	5	Strong breeze	10.8–13.8	22–27	6	Large waves begin to form; crests break forming large areas of white foam; some spray (1.21–2.44 m)
Very heavy	6	Near gale	13.9–17.1	28–33	7	Sea heaps up, white foam from breaking waves begins to be blown in streaks along the direction of the wind (2.44–4 m)
—	–	Gale	17.2–20.7	34–40	8	Moderately high waves with crests of considerable length; edges of crests break into spindrift; foam is blown in well-marked streaks in wind direction (4–6 m)
High	7	Strong gale	20.8–24.4	41–47	9	High waves; dense streaks of foam in direction of wind; crests of waves begin to roll; spray may reduce visibility
Very high	8	Storm Whole gale	24.5–28.4	48–55	10	Very high waves with long overhanging crests; foam causes sea surface to appear white; tumbling of the sea becomes heavy and shocklike; visibility reduced (6–9.2 m)
Exceptionally high sea	9	Violent storm	28.5–32.6	56–63	11	Exceptionally high wave; edges of wave crests are blown into froth; spray reduces visibility (9.2–13.7 m)
Exceptionally. high sea	9	Hurricane	32.7–36.9	>64	12	Air filled with foam and spray; sea completely white; visibility seriously reduced (over 13.7 m)

Source: Courtesy U.S. Coast Guard.

down manner, but the paths traced are actually circular orbits.

If you weight several corks so that each is suspended at a slightly different depth, you will see that their pattern of motion changes with depth. In shallow water, the circular motion is flattened into an ellipse and ultimately becomes a back-and-forth motion, as shown in Figure 11.2a. In deep water, the motion is as shown in Figure 11.2b. At a depth of $L/2$, there is no movement at all. Thus the energy of a wave varies not only with length and height but also with depth.

Some slight amount of water does move along with the waves. This very slow transference of water in the direction of wave motion is called **mass transport**. Although mass transport is negligible in the open ocean, it is more significant inshore. In shallow areas, the water tends to pile up against the shore. As the water level rises, some of the water is transported seaward below the surface. Mass transport, especially during storms with high waves and onshore winds, contributes to the extensive erosion of beaches.

TYPES OF WAVES

Recall that Table 11.1 lists a wide range of wave types. The smallest waves, called **capillary waves**, have a period of less than 0.1 s and a maximum wavelength

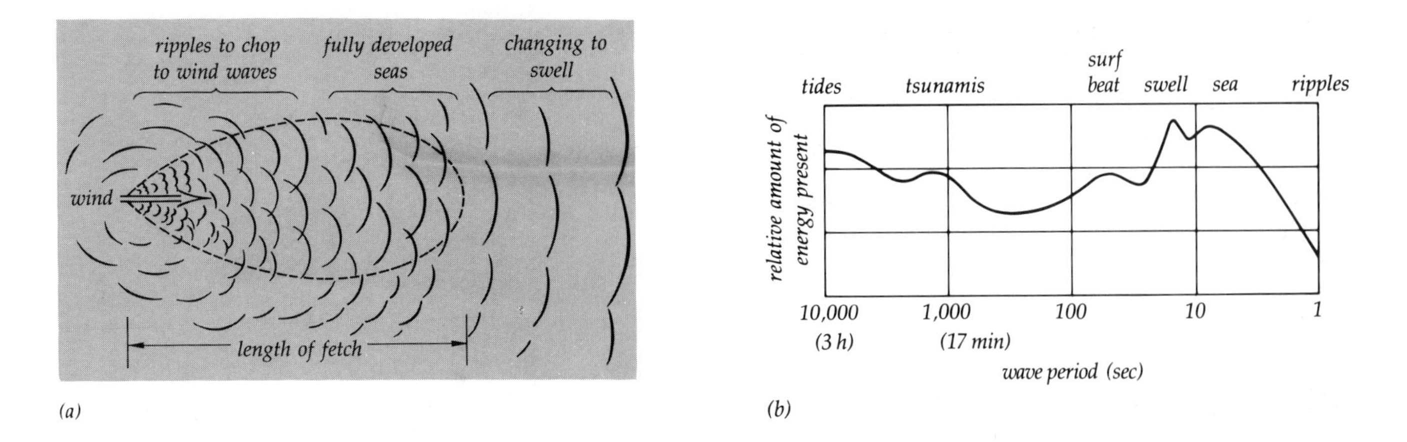

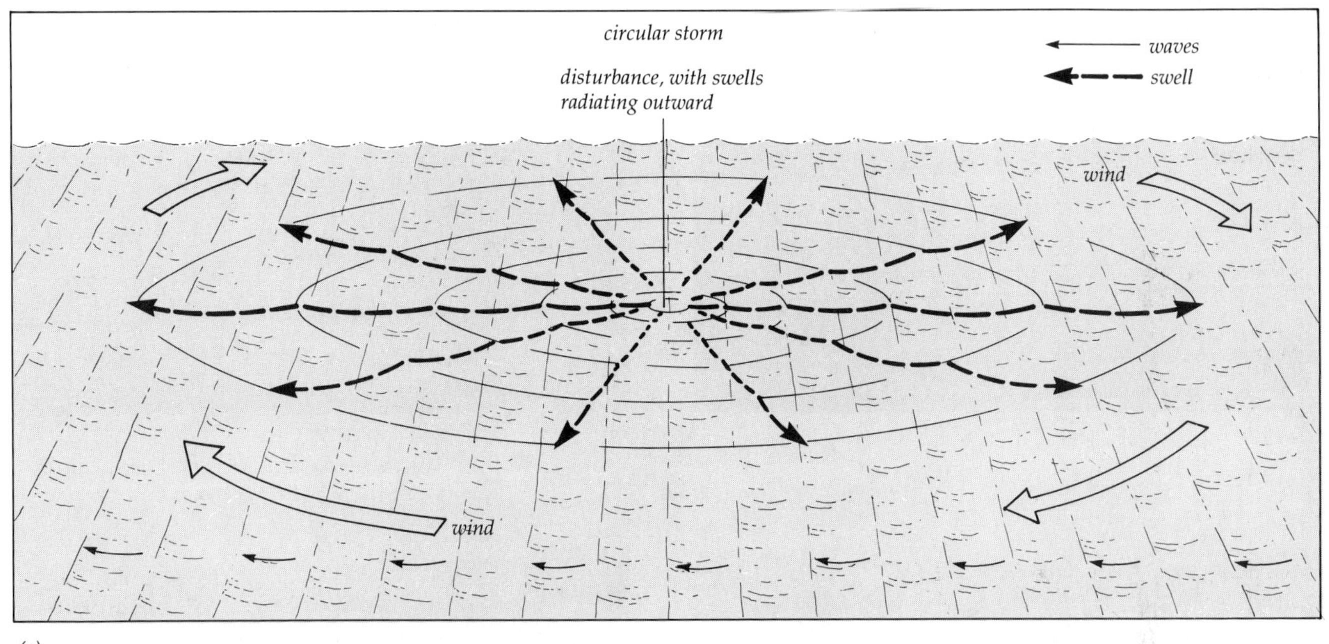

11.4 (a) Storm-generated ocean waves. (Because storms themselves are basically circular, the waves actually move away from the storm in all directions, not in the limited manner depicted here.) (b) Relative amount of energy present in ocean waves. (After Walter Munk) (c) Like the ripples from a handful of pebbles thrown onto a pond's placid surface, the waves generated from a storm at sea move out in all directions.

of less than 1.74 cm. Their size is limited by the surface tension of the water. Capillary waves appear on larger waves as tiny ripples.

Waves with periods between 1 and 30 s are called **wind waves**, or **gravity waves**. In the subsections that follow, we shall look at the characteristics of wind waves in their various phases of development, including sea, swell, shallow-water waves, and breaking waves.

Wind Waves

Wind is the primary energy source of ocean waves. Imagine a smooth ocean surface with no wind and no waves of any kind. Then imagine that the wind gradually picks up and ripples the surface (Table 11.2). A breeze of only 0.5 knot can cause surface ripples. The ripples form in response to the varying pressure of the turbulently moving air on the surface and to the frictional drag of the air against the water. As the ripples increase the surface area exposed to the wind, the resulting increase in friction and pressure gradually builds the ripples up into small waves. The surface of the ocean grows choppy, and the waves move roughly in the direction of the wind, as shown in Figure 11.4a.

In the region of a high wind or a storm (Figure 11.5d, e, f), the ocean surface is a jumble of waves of various sizes. Sailors often use the term *sea* to refer to

(*a*)

(*b*)

(*c*)

(*d*)

(*e*)

(*f*)

11.5 Wind and wave conditions. (a) Capillary waves, shown close to actual size, superimposed on larger waves. (W. J. Wallace) (b) Wind speed of less than 1 knot. (c) Wind speed of 4 to 6 knots. (d) Wind speed of 22 to 27 knots. (e) Wind speed of 37 to 44 knots. The waves are not fully developed. (f) Wind speed of 48 to 55 knots. The sea has a limited fetch, or open surface area, because the sea ice is only 30 nautical miles to windward. (Dr. F. Krügler, Hamburg)

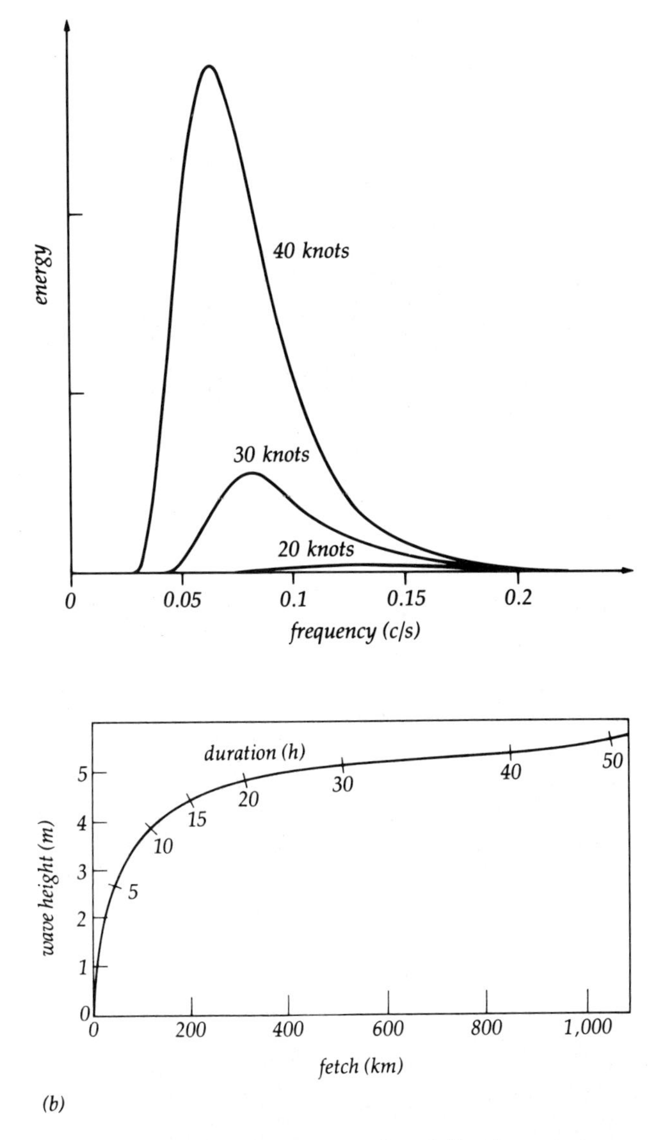

11.6 (a) Continuous wave spectrum for a fully arisen sea at wind speeds of 20, 30, and 40 knots. Notice the displacement of the optimum band (maximum energy) from higher to lower frequencies with increasing wind. (b) (top right) Increase in wave height as a function of increased wind duration and fetch. The wind speed is constant at 54 km/h (33.5 mph). (From W. J. Pierson, Jr., G. Neumann, and R. W. James, Practical Methods for Observing and Forecasting Ocean Waves by Means of Wave Spectra and Statistics, *U.S. Navy Hydrographic Office Publication 603, 1955)*

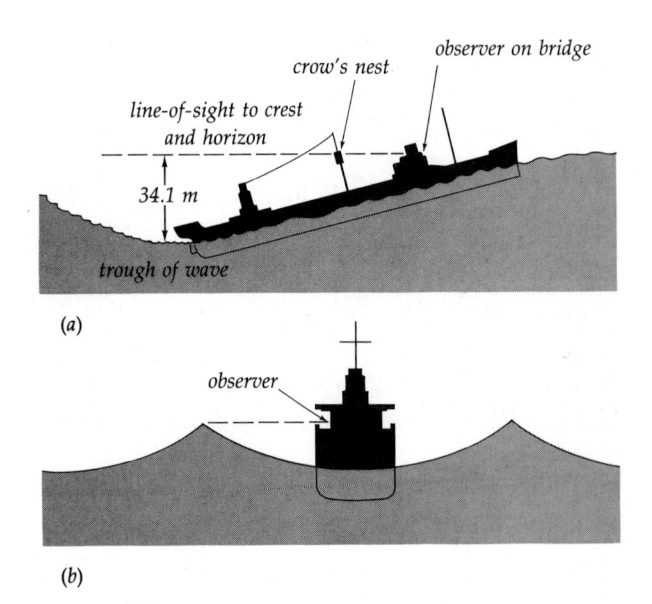

11.7 Measuring the height of a wave: (a) wave perpendicular to ship's line of travel and (b) wave parallel to the ship.

this confusing state. As the wind speed increases, the average wave height also increases. The length of time and the area over which the wind blows (fetch) also help determine wave size. Wind-generated waves increase in speed as they grow in period, eventually attaining the same velocity as the wind itself in some cases (see Figure 11.6).

To produce large waves, high-speed winds must move in the same direction over an extensive area for a considerable period of time. When a 20-knot wind blows for at least ten hours along a minimum distance (or **fetch**) of 125 km (78 mi), 10% of the waves generated will average 3.3 m (10.8 ft) in height. When a 50-knot wind blows for three days over a distance of 2,500 km (1,550 mi), 10% of the waves generated will be 33 m (108 ft) high. Such immense waves seldom occur in the ocean, however, where most waves are less than 3 m (10 ft) high.

Large waves at sea are generally not dangerous. When the crest of a storm wave breaks into whitecaps, however, the water may strike a ship with tremendous force. Standing on a ship at sea during a storm is not the easiest way to measure wave height, but the largest wave ever measured was observed in just that manner (see Figure 11.7a). On February 7, 1933, storm winds in the Pacific reached 68 knots. The USS *Ramapo*, a navy tanker almost 160 m (525 ft) in length, was en route from Manila to San Diego. Lt. Comdr. R. P. Whitemarsh calculated the height of one wave at 34.1 m (112 ft). From the *Proceedings of the U.S. Naval Institute*: "While standing watch on the bridge, he saw seas astern at a level above the main crow's nest and that at the moment of observation the horizon was hidden from view by the waves approaching from astern. Mr. Margraff is 5 feet 11¾ inches tall. The ship was not listed and the stern was in the trough of the sea."

Swell

What happens when the wind abates or when waves move out of a storm system? Then the waves sort themselves out and become known as a **swell** (Figures

(a)

(b)

(c)

(d)

11.8 *(a) Ocean swell. (W. J. Wallace) (b) Ocean swell "lifting" up and over a treacherous sandbar west of San Francisco known locally as the "Potato Patch." (U. S. Coast Guard) (c) A Coast Guard cutter in heavy ocean swell. (Jan Hahn, Woods Hole, Massachusetts) (d) Surf. (W. J. Wallace)*

Table 11.3 *Velocities of Individual Deep-Water Waves*

Wavelength (L)		Period (T)	Velocity ($V_{deep} = \sqrt{gL/2\pi}$)		
m	ft	s	m/s	km/h	mph
1.56	5.12	1	1.56	5.6	3.5
6.24	20.5	2	3.12	11.3	7.0
25	82	4	6.25	23	14
56	184	6	9.4	34	21
100	328	8	12.5	45	28
156	512	10	15.7	57	35
225	738	12	18.8	68	42
306	1,004	14	21.9	79	49
400	1,312	16	25.0	90	56
506	1,660	18	28.2	101	63
624	2,047	20	31.3	113	70
790	2,591	22.5	35.2	127	79
1,405	4,608	30	47	169	105

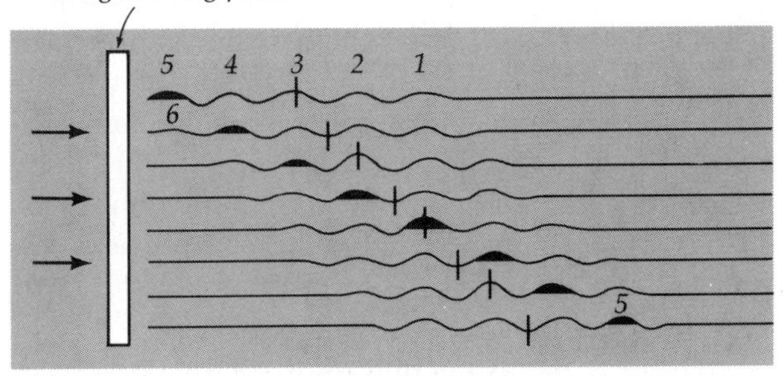

11.9 A group of waves traveling down a wave tank (as seen from the side). One wave crest (number 5) is shown in black in successive profiles.

11.4a and 11.8). At sea, a passing swell may be scarcely noticeable; often it does no more than rock a boat slightly. At other times, it may seem like a series of slowly rolling hills. Because storm winds move in circular paths, the waves spread out in all directions. In addition, since the waves generally move faster than the storm center, they may reach shore well ahead of the storm itself.

Waves of short wavelengths interfere with one another, dissipate rapidly, and break on the crests of larger waves. Consequently, they tend to die out soon after leaving the region of a storm. Other waves tend to lose height and to sort themselves out according to wavelength in a process called **dispersion**. As the newly formed swell moves from a storm region, the packet of waves spreads out.

Waves with longer wavelengths move more rapidly than waves with shorter wavelengths and are the first to reach land. Therefore, the farther the waves move from a storm region, the longer it takes for the packet to pass a given point. Waves generated by a storm in the South Pacific at 65°S latitude take two weeks to reach the shores of Alaska, while waves formed by a storm in the North Pacific 4,000 km (2,480 mi) from the U.S. coast will reach the coast in about three days.

Table 11.3 lists the velocities of individual **deep-water waves** of different periods. Long-period waves move faster and thus in advance of short-period waves. As the leading wave of a swell moves into undisturbed water, it transfers some of its energy to the water molecules and sets them into orbital motion. Consequently, the wave is continuously being worn away and is ultimately replaced by the next wave, which in turn is worn away (Figure 11.9). After a wave group or train has passed, the water molecules continue to oscillate for a short time, leading to the formation of a new wave.

Thus, individual waves travel twice as fast as the wave train itself. This constant addition of new waves creates an interference pattern throughout the wave train that has been measured experimentally and can be demonstrated scientifically. In a laboratory demonstration, a paddle is moved back and forth in a wave tank four times (eight movements of the paddle), producing a packet of four waves. If you stand at the paddle end of the tank, you see four waves moving away. However, if you stand alongside the tank, you see eight waves passing, and eight waves will break at the other end of the tank. So the waves striking a beach are the distant descendants of waves generated by a storm that may have been thousands of kilometers away.

Figure 11.10 shows a graph based on data taken at six research stations along a great circle route from the Antarctic (about 65°S) to Alaska. The study traced

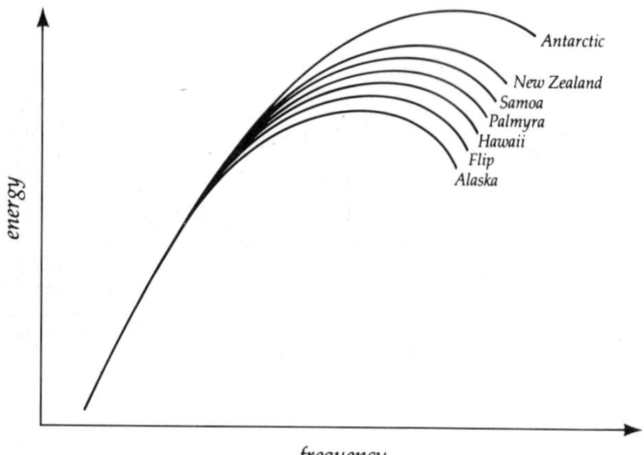

11.10 Energy loss for waves from the same storm as they move across the Pacific, as measured at various research stations.

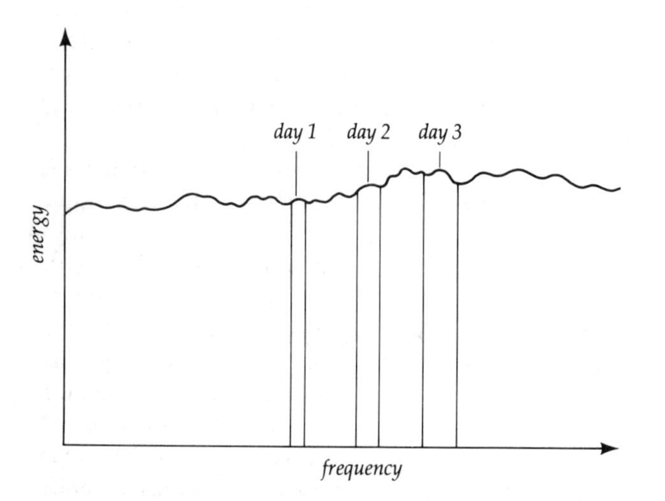

11.11 Energy profile for waves from a distant storm as they break on a beach at Palmyra over a three-day period.

the behavior of waves generated by a particular storm over a two-week period. Note the decline in wave energy from the storm area to New Zealand. This decline was caused by wave interference. But note that there was only a small decrease in wave energy from station to station and across the entire Pacific. The stations were chosen along a great circle route, which is the shortest distance from place to place on a globe. Since there is no mass transfer in the moving wave front, there is no Coriolis effect, and the wave advance is considered to be a straight-line traverse across the planet. The decrease in wave energy was mirrored by a decrease in wave height. The waves at Hawaii were noticeably higher than the waves striking the West Coast of the United States and Alaska. Similarly, storms in the eastern North Pacific produced higher waves on the shores of Washington, Oregon, and British Columbia than they did at Hawaii.

Table 11.4 *Velocities of Individual Shallow-Water Waves*

Water Depth (d)		Velocity ($V_{\text{shallow}} = \sqrt{gd}$)		
m	ft	m/s	km/h	mph
0.3	1	1.7	6	3.9
1.5	5	3.9	14	8.6
3.0	10	5.5	20	12
6.1	20	7.7	28	17
9.1	30	9.5	34	21
12.2	40	11.0	39	24
18.3	60	13.4	48	30

Figure 11.11 presents data for three days from one of the stations. Note that the greatest energy was recorded on the third day. In other words, the shorter the period—that is, the shorter the wavelength—the greater the total energy. This was true despite the fact that the waves arriving on the first day had greater heights than the waves arriving on the second and third days. Since deep-water waves are continuously changing and re-forming, we cannot compare individual waves. We must compare the total energy of the waves from one time to another. In this case, the great number of smaller-period waves on the third day contained greater total energy.

The total energy per unit width (of wavecrest) carried by waves is proportional to the product of the wavelength and the square of the height.

Shallow-Water Waves

When wind-generated waves reach a depth that is less than half their wavelength, they are called **shallow-water waves**. The velocity of these waves is a function of depth (Table 11.4); thus all shallow-water waves at a particular depth travel at the same velocity. Moreover, the velocity of individual waves and the group velocity are equal.

Breaking Waves

Waves breaking on the coast of California and Oregon may have formed near Antarctica and traveled for two weeks across the Pacific—about half the circumference of the globe—before finally spending themselves on the coast. Waves hitting the California coast may also have formed north of Hawaii or any place in the North or South Pacific along a great circle route. Waves striking the New Jersey coast may come from the South Atlantic or from waters off Iceland. Once a swell has moved out of a storm area (where its energy loss is

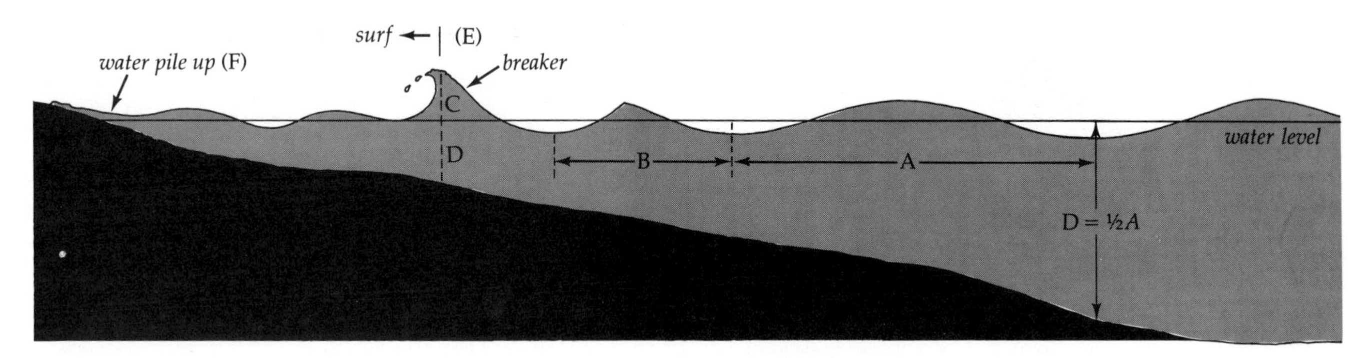

11.12 *Wavelength* A *in shallow water shortens to length* B. *The waves pile up in height (at* E*) when breaking occurs at a height-to-water-depth ratio of 3(*C*) to 4(*D*). From this point to the shore, the wave becomes surf, with foam and water pileup on the beach (*F*).*

greatest), it loses very little energy as it travels across the ocean.

Because depth decreases toward a beach, waves change from deep-water waves to shallow-water waves as they approach the shore. As the leading waves interact with the shallow bottom, incoming waves overtake them and crowd more closely together because their speed is proportional to depth. If the crests are 100 m (328 ft) apart, the waves "feel bottom" at 50 m (one-half their wavelength). The speed of the waves then decreases, causing their wavelength to decrease. Succeeding waves tend to pile up as the orbiting water particles are squeezed together. The orbital velocity of the molecules at the trough, but not at the crest, is reduced. The wave becomes more and more unstable, and the ratio of height to wavelength becomes greater than 1 to 7. At this point the crest of the wave moves faster than the trough, and it curls over at the top. Just before breaking, the orbital velocity at the crest may be twice the speed of the wave. At last the wave topples, and the water breaks into foam and surf (Figure 11.12).

Another way to explain the growth of shallow-water waves involves the transfer of energy from one form to another. As the wave speed decreases, the wave's kinetic energy decreases. In accordance with the law of conservation of energy, that energy appears as potential energy, which is proportional to height.

There is no clear-cut transition from a deep-water wave to a shallow-water wave. The transformation is gradual, growing more pronounced closer and closer to shore. A shallow-water wave begins to break at a depth equal to about one-twentieth of its wavelength, or when the depth is 1⅓ times its height. For instance, a wave 3 m high will break at a depth of 4 m; a 6 m wave will break at a depth of 8 m.

If you know the approximate slope and depth of a beach at various distances from shore, you can estimate the height of a wave at each distance. Walk toward the water's edge until the crest of the breakers is aligned with the horizon. The height of the next wave will equal the vertical distance from your eye level to the lowest level of the retreating water from the last wave.

Breaking waves are of two types: **spillers** and **plungers** (Figure 11.13). Spillers roll in evenly and slowly. Plungers (or *dumpers*, as surfers call them) break over a short distance, pounding the beach with a great roar and splash of flying water and foam. Spillers give surfers the longest ride. Although winds and currents have some effect on whether a swell becomes a spiller or a plunger, the main influence is the bottom topography. The steeper the bottom slope, the more quickly the wave will slow down and break. The composition of the bottom is also a factor, although its effect is not as pronounced as that of the slope. A smooth, steep slope tends to produce plungers, whereas a gentle slope, perhaps with slight irregularities, tends to produce spillers. The slope and the texture of a beach are affected by the endless bombardment of the waves and the release of large amounts of energy. Moreover, since weather conditions vary daily, the type of breakers found on any given beach is not constant.

Before the Allies undertook amphibious landings during World War II, the beach slopes were carefully studied. How close could landing craft come to the beach to dispatch thousands of men and their supplies? Aerial photography of the beaches and the offshore waters provided the answers. On the basis of the decreasing wavelength as waves approached the shore, specialists determined the depths and drew up slope profiles of the beaches.

Surfing and Rip Currents

Porpoises can ride the wake of a ship indefinitely. Boaters, too, can enjoy a free ride by positioning a small boat on the forward face of the first large wave in a ship's wake. In a similar way, surfboards are pro-

11.13 *Two surf wave types: (a) and (d) are plungers. (b) and (c) are spillers. (c and d, W. J. Wallace)*

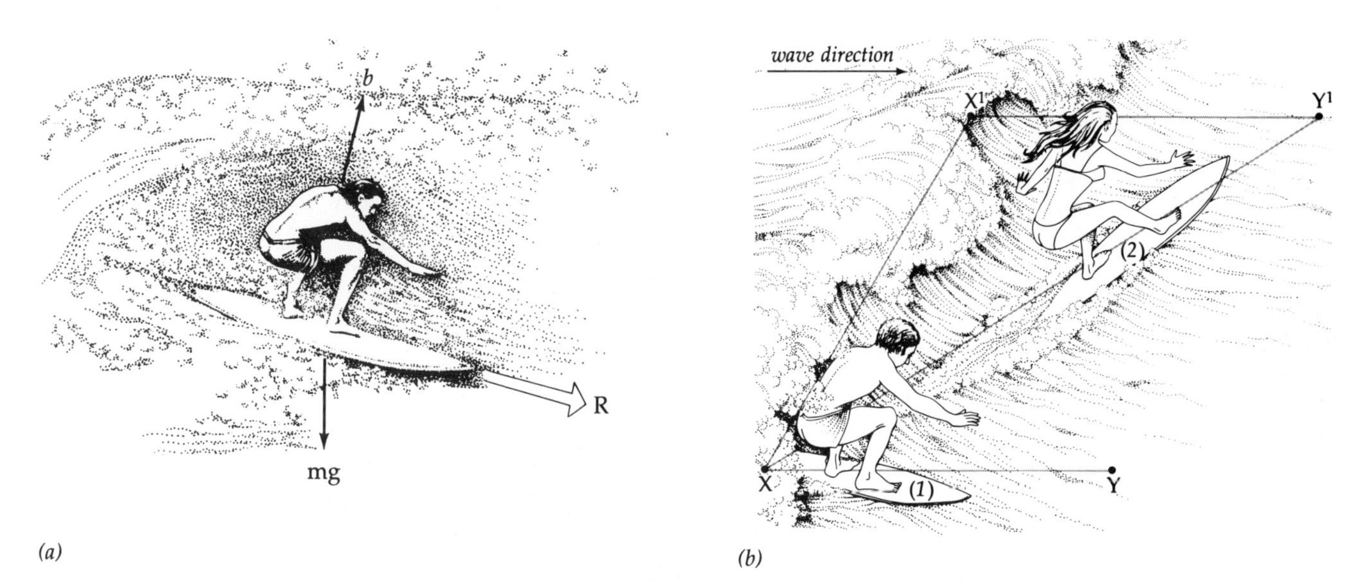

*11.14 (a) Forces involved in surfing. The upward buoyancy of the board (*b*) and the downward force of weight (*mg*) yield the resultant vector* R. *(b) The distances (and therefore the velocities) that surfers can travel on the same wave may be different.*

pelled down the forward surface of a wave by taking energy from the wave.

Figure 11.14 shows the forces involved in surfing. In effect, the surfer slips down the moving front "hill" of the wave under the influence of gravity. Correctly positioned and moving with the direction of the wave, the surfer travels at the speed of the wave crest. By traversing the face of the wave, the surfer may move even faster than the wave. The trick is to catch the wave as it passes so that the gravitational force on the board is greater than the resistance of the water. The wave's energy propels the board.

When a surfer passes through a wave, the circular motion of the water particles produces a brief downward pull. This is only of short duration and is not dangerous in itself, but surfers close to shore may be thrown forward and deposited (not always lightly) on the beach.

Surfers are sometimes puzzled by the variation of breakers on a beach, even over a short time. A dozen or so moderate or low waves will break and then a few high ones, and then there will be relatively few waves for a while. This irregular sequence is caused primarily by the simultaneous arrival of several swells from separate storms at sea. Larger waves form when the waves of separate trains merge and reinforce one another. Smaller waves form when the incoming waves are out of phase and interfere with one another. The resulting patterns produce a beat frequency, called **surf beat**, that is much longer than the period of the individual waves. A surf beat of two to three minutes may occur when wave periods are between ten and fourteen seconds (Figure 11.4). When surf beat occurs, longer waves strike the beach and the water level is raised.

As Table 11.4 indicated, the velocity of the shallow-water waves on which the surfer rides is not great, and it decreases rapidly with depth. Moving across the wave may produce noticeably higher speeds, however, as shown in Figure 11.14b. In the figure, surfer 1 must move from point X to Y at the speed of the wave, but surfer 2, moving from X to Y', covers a greater distance in the same time. Because velocity equals distance divided by time, surfer 2 achieves a greater velocity.

Waves at the edge of the sea are often associated with what is called "undertow." Actually, there is no real undertow that draws or sucks swimmers under. There are, however, **rip currents**—water movements away from the shore that occur whenever currents funnel through a narrow opening formed by the erosion of a channel or through a gap across the lowest part of an offshore sandbar. Rip currents may move at 3 km/h (2 mph). They are localized, and their position changes with changes in wave conditions. The higher the waves, the stronger the rip current. If you are ever caught in a rip current you should try to swim parallel to the beach rather than against the current. A rip current is narrow, and you will soon be out of it. If you tire, look for a sandbar where you can rest. Some sandbars are only waist deep in places, especially on the East Coast, although they may seem to be well over one's head. Foaming, light-colored water is generally a sign of a shallow sandbar.

11.15 Wave refraction. Ocean waves, approaching Point Loma in San Diego, as indicated in the photo, bend, or refract, around the peninsula. (W. J. Wallace)

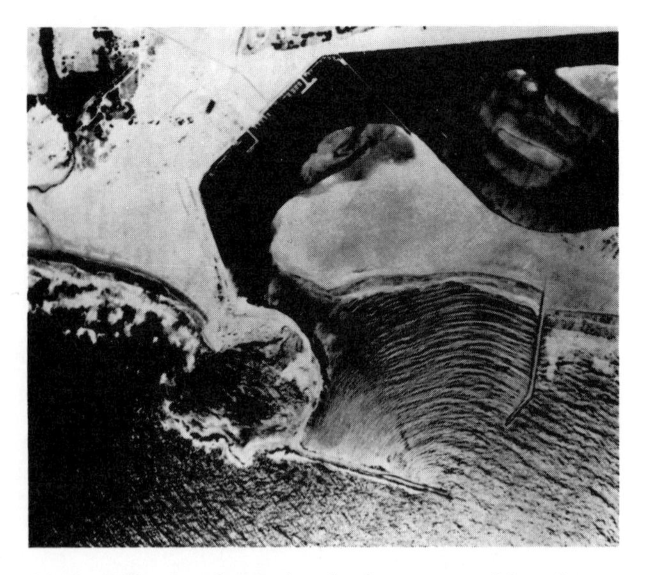

11.16 Diffraction of waves by a breakwater gap at Morro Bay, California. (From Oceanographical Engineering *by Robert L. Wiegel. © 1964. Reprinted by permission of Prentice-Hall, Inc., Englewood Cliffs, New Jersey)*

REFRACTION, DIFFRACTION, AND REFLECTION

When waves approach the shore, they undergo **refraction, diffraction,** and **reflection**.

Deep-water waves entering shallow regions at an angle are bent as they touch bottom. This refraction is caused by the reduction of wavelength as depth decreases. The part of the crest entering shallow water first moves more slowly than the part of the wave still in deep water, and the wavefront bends to fit the shore (Figure 11.15). As a result, the wave is almost parallel to the shore when it strikes. Numerous refraction patterns are possible, depending on the angle at which the wave approaches and on the bottom topography.

Diffraction accounts for the presence of waves in sheltered regions, such as harbors (Figure 11.16). Like light waves, water waves may be bent into shadow zones behind steep-sided obstacles. Diffracted waves in a harbor or inside a breakwater are smaller and choppy. Diffraction patterns observed from the air leeward of Pacific atolls often extend for several kilometers. Early Polynesian mariners interpreted these choppy waters as a sign that they were approaching an atoll. Since an atoll lies so low in the water, it cannot be seen from a canoe or outrigger more than 10 km (6.2 mi) away.

A vertical or nearly vertical seawall or cliff situated in deep water may reflect a wave train with little energy loss (Figure 11.17). The reflected waves exert very little force on the wall. This phenomenon, called **clapotis** (from the French for "standing wave"), occurs if the waves strike the wall perpendicularly. If they strike at an angle other than 90°, waves of equal and opposite angle are formed. The original waves and the reflected waves may interfere with each other, in which case their crests and troughs will cancel out, or they may reinforce each other. It is often difficult to tell which waves are coming and which are going.

PROBLEM WAVES

Seismic Sea Waves

Since so-called tidal waves have nothing to do with tides, oceanographers use the Japanese word **tsunami**. Actually, that term, which means "large waves in harbors," is less than accurate, but it helps to reduce confusion. Tsunamis are seismic sea waves caused by some sudden movement of the earth's crust (Figure 11.18), perhaps an earthquake caused by the vertical displacement of the ocean floor, or perhaps water is set in motion by an underwater landslide (avalanche), which may have been caused by an earthquake, by long-period earthquake waves, or by resonance in submarine trenches. A drop in the sea-surface level of only a few meters above a landslide may be enough to create a tsunami.

Tsunamis are very long waves that travel at considerable speed. Their wavelength may be as much as 240 km (149 mi). Because their wavelength is so great,

11.17 Wave reflection. The waves are approaching the coast parallel to the top edge of the photo. The diagonal wave in the foreground has reflected off a vertical cliff. (W. J. Wallace)

even relative to the 4,600-m (15,088-ft) average depth of the Pacific Ocean, they behave like shallow-water waves. Thus the velocity of a tsunami depends on depth. In the open ocean, a tsunami may attain speeds of 760 km/h (471 mph)—almost the cruising speed of a jet aircraft!

Some tsunamis consist of a packet of three or four waves spaced about fifteen minutes apart. The first is not necessarily the highest. In the open ocean, these waves may rise less than a meter above normal sea level, imperceptible to a person on a ship under which the tsunami is passing. As the waves approach shore, however, they slow down and the water piles up, possibly building up to 20 m (66 ft) on flat, low-lying shores. At the head of V-shaped inlets, the waves may form a powerful wall of water 30 m (98 ft) high (Figures 11.19 and 11.20).

In April 1946, a tsunami that struck Scotch Cap, Alaska, completely destroyed a two-story concrete lighthouse built 15 m (49 ft) above sea level (Figure 11.21). This tsunami consisted of a series of crests arriving fifteen minutes apart. At sea the crests had been only 0.6 m (2 ft) high and 203 km (126 mi) apart.

Before the arrival of a tsunami, the water along the coast sometimes withdraws out to sea. Between waves, a harbor may seem to dry up for a brief time. During a tsunami on April 1, 1946, the fishing fleet at Half Moon Bay, Alaska, was suddenly stranded on the bottom of the bay where it normally floated at anchor even at the lowest tide. The next wave drove the boats onto an asphalt road 4 m (13 ft) above sea level.

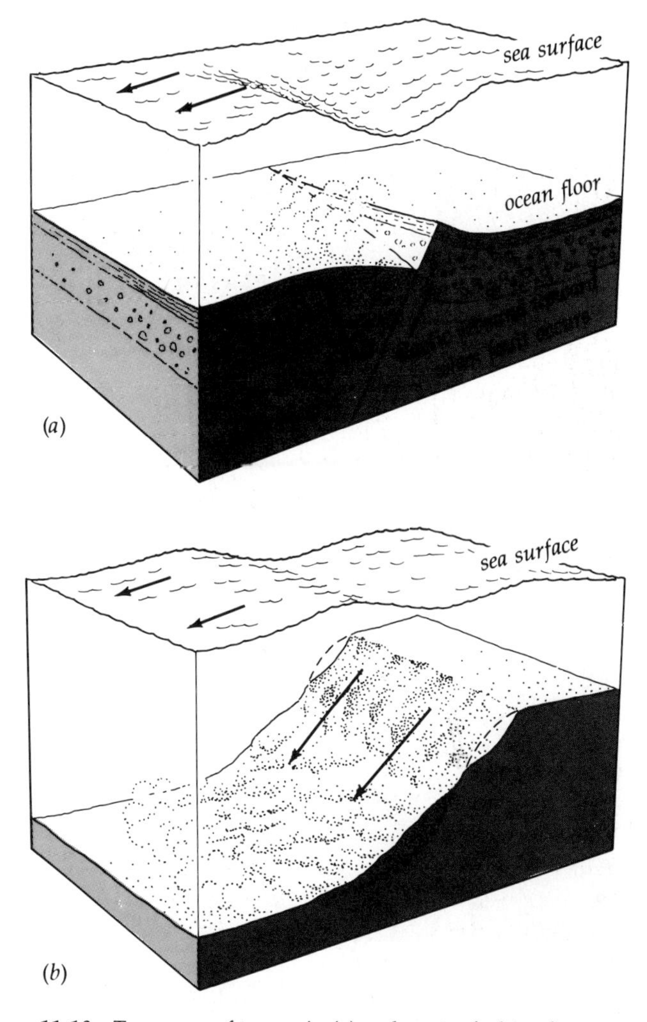

11.18 Two causes of tsunamis: (a) underwater faulting from an earthquake and (b) an underwater landslide.

On May 22, 1960, a major underwater fault generated a catastrophic earthquake that shook Chile (approximately 39°S, 74°W). In a region extending 800 km (496 mi) from the epicenter, hundreds of landslides occurred, killing 4,000 people. The earthquake produced one of the greatest tsunamis of modern times, causing loss of life and destruction of property across the Pacific. The piers and boats in Los Angeles and San Diego harbors sustained $1 million in damages. Waves reaching Honshu, Japan—14,500 km (8,990 mi) and 23 to 24 hours away—were 4.6 m (15 ft) high. The tsunami killed 180 people there and caused $50 million in damage. Fortunately, the Pacific, where most tsunamis occur, is so large that some warning is usually possible, and the warning system is being improved (Figure 11.22).

Tsunamis may also be caused by immense explosions. The explosion of the volcanic island of Krakatoa, west of Java, on August 27, 1883, blew away about

11.19 The Great Wave off Kanagawa, *by Katsushika Hokusai (1760–1849). A famous Japanese print depicting a large tsunami in shallow waters.* (The Great Wave off Kanagawa. *Courtesy of the Metropolitan Museum of Art, The Howard Mansfield Collection, Rogers Fund, 1936.) Used by permission.*

17 km^3 (4 mi^3) of the island. The sound was heard 4,800 km (2,976 mi) away. The dust thrown into the air was so thick that ships could not make their way through some areas where it had settled. The fine dust that remained in the atmosphere caused reddish sunsets around the world for several years. Some distance away, the resulting tsunami carried a warship almost 3.2 km (2 mi) inland to 9.1 m (30 ft) above sea level.

It seems likely that a violent explosion of Thíra off Greece around 1400 B.C. created an immense tsunami that overwhelmed the island of Crete and destroyed the Minoan civilization. That same tsunami briefly turned the Bardawil Peninsula east of Port Said into a land bridge. Such a tsunami might have enabled Moses to lead his people out of Egypt, and the ensuing tsunami shortly thereafter might have drowned the pharaoh's horsemen. See "The Volcano that Shaped the Western World," by John Lear (in *Saturday Review of Literature*, 1966, Nov. 5, pp. 57–66).

The detonation of a thermonuclear bomb might very well generate a tsunami. In fact, there was concern that the first thermonuclear explosion at the Eniwetok atoll in the Marshall Islands might create waves like those created by the eruption of Krakatoa. The energy of the bomb was similar in magnitude to that calculated for Krakatoa. The explosion did trigger a tsunami, but because the bomb crater did not breach the outer edge of the atoll reef, it was only a small one.

Rogue Waves

Great waves are generally the result of intense storms that blow steadily for several days and are accompanied by low barometric pressures. Gigantic waves, however, sometimes appear as isolated phenomena. Over the years, mariners have reported massive, solitary waves with an immense crest and a seemingly bottomless trough (Figure 11.23). Such reports were largely unheeded, however. Then in 1942 the *Queen Mary*, while serving as a troopship, was making her way through a winter gale off the coast of Scotland.

(a)

(b)

11.20 Tsunami damage caused by the Good Friday earthquake in Alaska, March 27, 1964. (a) Boats beached by the tsunami at Kodiak. (Official U.S. Navy photograph) (b) Spruce trees 0.7 m (2.3 ft) in diameter at elevations between 29 and 34 m (95 and 112 ft) above lower low water (the average of all the lower daily tide levels) that were broken by a landslide-generated wave. (c) Seward waterfront, showing damage from destructive waves and submarine sliding. (d) Whittier waterfront, showing damage and extent of wave run-up (note dark area washed clear of snow near shore). (George Plafker, Geological Survey, U.S. Department of the Interior)

(c)

(d)

(a) (b)

11.21 (a) Scotch Cap lighthouse, Unimak Island (of the Aleutians), taken before the tsunami of April 1, 1946. The foundation of the lighthouse is 15 m (49 ft) above sea level, the upper plateau 36 m (118 ft) above sea level. (b) The same place after the tsunami. The lighthouse and the radio masts are gone, the slopes are heavily washed almost to the plateau level, and debris has been deposited on the plateau. Notice the exposure of stratification on the slopes in the background, revealed by removal of the overlying sediment. (U.S. Coast Guard)

Already in heavy seas, she was struck broadside by a single, mountainous wave that rolled her over on her side until her decks were awash, further than she had ever listed before. After several agonizing moments, she righted herself. In 1966, the liner *Michelangelo* was struck by a huge solitary wave 1,300 km (806 mi) out of New York. Glass windows on the bridge 25 m (82 ft) above the waterline, along with steel railings and bulkheads, were smashed, and the bow was badly crumpled. Several passengers were injured, and three died.

These are only a few of the reports of great solitary waves suddenly rising up out of the ocean. Until twenty or so years ago, most marine traffic was relegated to narrow, well-defined shipping routes. Now, with the vast increase of worldwide shipping, rogue waves are encountered more often than they used to be and cause the loss of numerous ships. They are more likely to occur in particular regions, and they are more damaging to large ships than to small ones.

Sir Ernest Shackleton recorded these observations during his voyage to the Antarctic in 1907 (see also Figure 11.24):

> Suddenly one enormous wave rushed at us, and it appeared as though nothing could prevent our decks being swept, but the ship rose to it, and missed the greater part, though to us it seemed as if the full weight of water had come on board. We clung tightly to the poop rails, and as soon as the water had passed over us, we wiped the salt from our eyes and surveyed the scene. The sea had smashed in part of the starboard bulwarks and destroyed a small house on the upper deck, pieces of this house and the bulwarks floating out to the leeward; the port washport was torn from its hinges, so that water now surged on board and swept away at its own sweet will, and the stout wooden rails of the poop deck, to which we had been clinging, were cracked and displaced.
>
> And later, as I was standing on the bridge at 2 A.M., peering out to windward through a heavy snow-squall that enveloped us, I saw, in the faint light of breaking day, a huge sea, apparently independent of its companions, rear itself up alongside the ship. Fortunately, only the crest of the wave struck us, but away went the starboard bulwarks forward and abreast of the pony stalls, leaving a free run for the water through the stables. [The Heart of the Antarctic. Toronto, 1910 (The Musson Co.)]

The odds are about one in 300,000 that a storm will create a wave four times the average height of waves. Still, two or more giant waves may suddenly come together and reinforce one another. Also, in certain areas, such as the southeast coast of South Africa, well-defined ocean swells meet currents moving in the opposite direction and create great rollers.

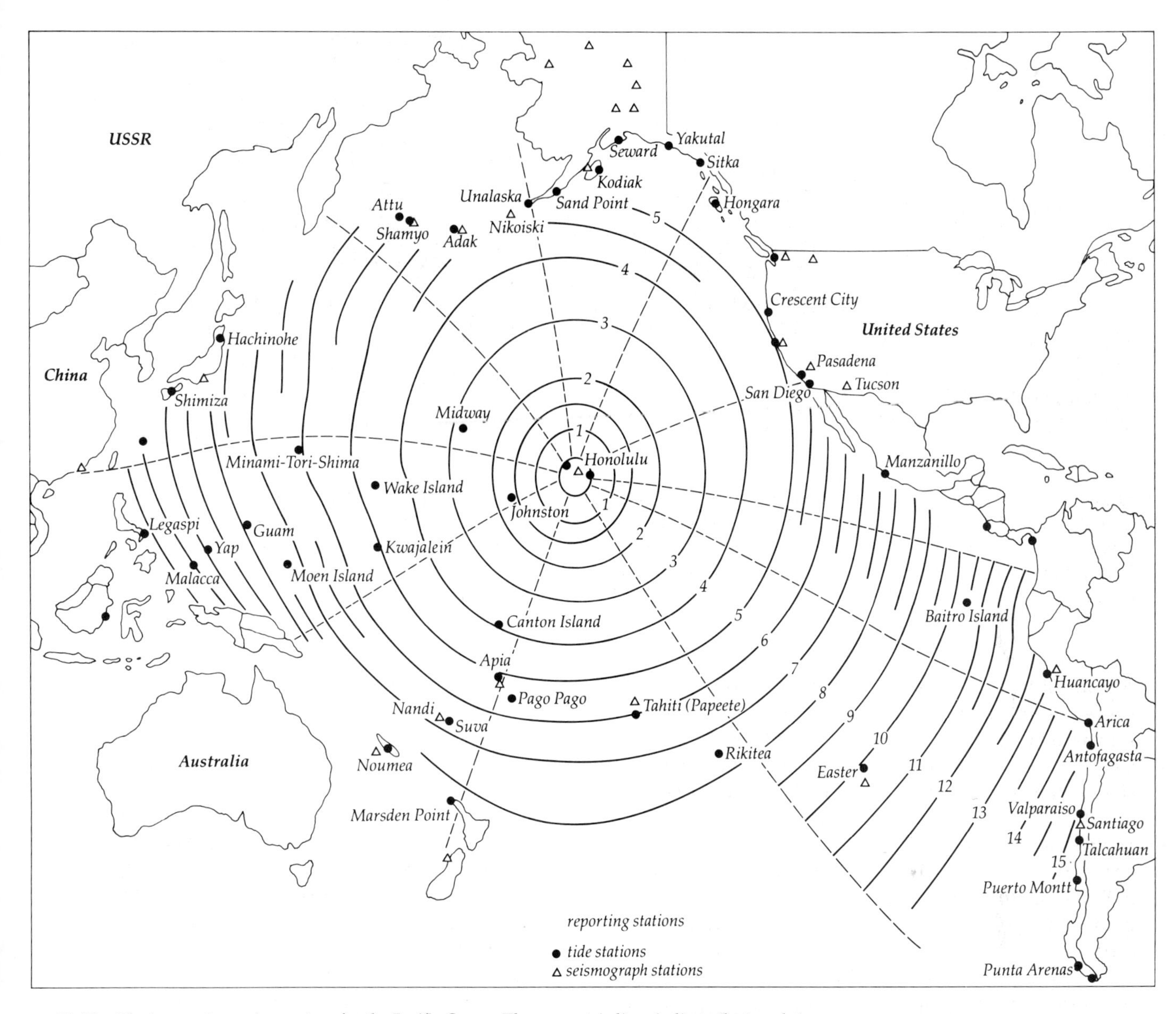

11.22 The tsunami warning system for the Pacific Ocean. The concentric lines indicate the travel time in hours for a tsunami to reach Hawaii. (National Oceanic and Atmospheric Administration)

INTERNAL WAVES

An old nautical term still used by Norwegian fishers is translated as "lying in dead water." The phrase refers to the plight of small vessels that become immobilized. Unable to make headway, they respond poorly to the helm. Sailing craft, which have a tendency to fall away to leeward (track downwind), are the most frequent victims of this predicament. Hours may elapse before the craft frees itself. Sometimes it breaks free for no apparent reason or because the wind suddenly picks up. This phenomenon is common along the Scandinavian coasts, where the tidal effect is not great and where copious fresh water runs off the land. The layering of fresh and salt water is common here, because the two are slow to mix.

So-called **internal waves** sometimes develop along subsurface density discontinuities. The waves move along the surface of the denser water, just as they do on the surface of the ocean. The internal waves once formed cause a very irregular discontinuity surface, and they travel at about one-eighth the speed of surface waves. They have an average height of 4.1 m (13.4 ft), and a rise and fall of 30 m (98 ft) is common

11.23 *Rogue, or gigantic, wave.*

11.24 *The giant wave mentioned by Shackleton in 1907.*

and may exceed 100 m (328 ft). They have a period of five to eight minutes and a wavelength of 0.6 to 0.9 km (0.4 to 0.6 mi). Their energy and velocity are small.

Internal waves may also be generated by tidal currents, by the pressure of short-lived surface winds, or by the propeller or bow of a slow-moving ship. This last situation results in the condition known as **dead water**, where all energy is directed into the lower layer of stratified water, producing internal waves, and none is left to propel the vessel itself (Figure 11.25). Powerful, motor-driven vessels find this less of a problem. Long, parallel bands of smooth, slowly moving water called **slicks** (which can sometimes be seen from the air) may be related to the presence of internal waves (Figure 11.26). Internal waves break on the coast much as surface waves do except that they break farther out.

Submarines in the region of a density discontinuity occasionally rise and fall in response to internal waves and may have difficulty maintaining a constant depth. The loss of the nuclear-powered USS *Thresher* in April 1963, including its crew of 129 men, may have been caused by internal waves. The subsequent search

11.25 *A boat "lying in dead water." The energy of the propeller (or given to a sailboat by the wind) generates internal waves on the boundary of a lower, denser layer.*

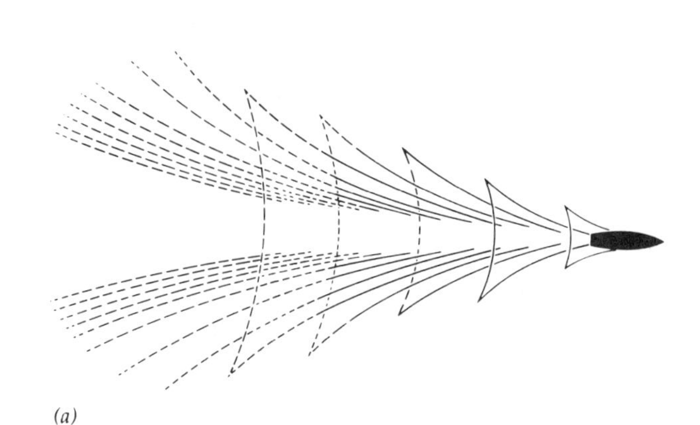

(a)

11.26 *Ocean slicks. (E. C. LaFond)*

(b)

11.27 *(a) A ship's wake in deep water. (b) The wake of a freighter in the normally calm waters of Rotterdam's inner harbor. (W. J. Wallace)*

for the vessel revealed how little is known about the deep ocean. Although the *Thresher* went down in only 2,600 m (8,528 ft) of water in an area where the ocean bed is relatively flat, months of searching with the best equipment available produced only pieces of the craft. The entire hull was never found.

The *Seasat* satellite has produced photographs of the activity of internal waves. Such satellite photosurveys may in time be able to detect the underwater wake of a submarine as it interacts with the surface, and it may also provide fresh information on underwater topography.

WAKES

As a ship's propulsive power overcomes the viscous resistance of deep water, it generates a wave system called a **wake**. The two V-shaped lines of waves consist of a number of short transverse waves (Figure 11.27). Another series of longer waves travels at the ship's speed. So long as the ship is in deep water with respect to the wavelength of the waves, the angle of the V-shaped lines is constant at 39°. In shallow water, the wakes travel with the same velocity regardless of wavelength. As the speed of the boat increases, the angle becomes progressively smaller until the V-shaped waves become parallel. Finally, at high speed, the wake can no longer maintain its position astern of the vessel. Small boats are often swamped by their own wake or, if they slow down too abruptly, by the high initial transverse wave.

The large bulbous bows of modern merchant vessels virtually eliminate wakes by decreasing the normal bow pressure gradients caused by the main hull. The bulbous bows reduce pressure by accelerating the motion of the water the vessel is passing through relative to the hull speed.

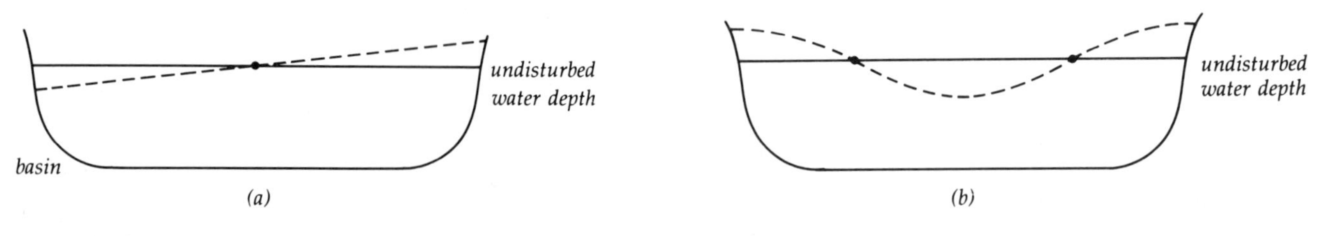

11.28 Seiches: (a) mononodal and (b) dinodal, or binodal.

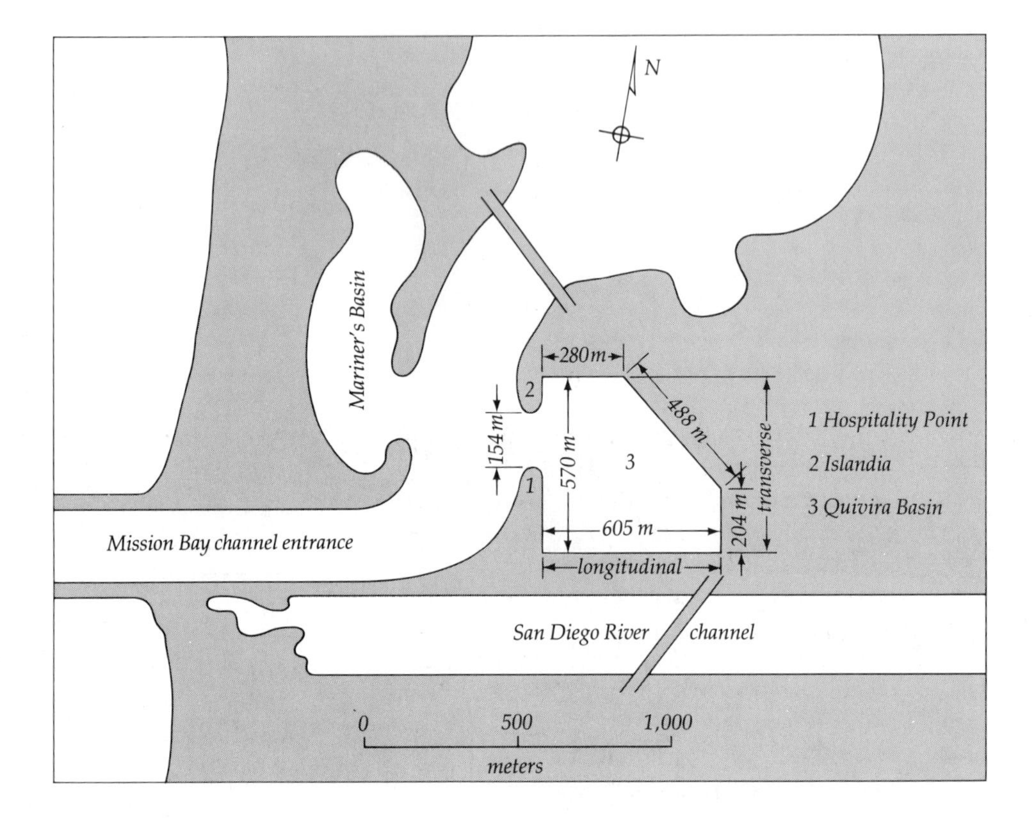

11.29 Mission Bay, San Diego, California: entrance channel and Quivira Basin. The distance from the channel entrance to Hospitality Point is 1,803 m (5,914 ft). During seiches, the entire basin and channel act as one unit with a period of 1,200 s. Within Quivira Basin, oscillations occur longitudinally with periods of 300, 100, 150, and 75 s, transversely with periods of 140 and 70 s.

SEICHES

Most of the waves discussed so far are **progressive waves**—that is, waves that move forward through the water. Other waves, like the waves set up in a vibrating guitar string, are called **standing waves**. One type of standing wave is the **seiche**. The water level in certain lakes rises and falls regularly with a period as short as an hour or so. As the water sloshes back and forth, long waves are set up. The action resembles what happens when you try to carry an ice-cube tray full of water to the refrigerator.

One way to visualize seiches is to half-fill a deep, rectangular pan, raise one end, and then set it down. The water moves back and forth, reflecting off each end. In this simple case, the center of the water, called the **node**, is stationary. The wave is called a *mononodal wave*. In a mononodal seiche, high and low water occur at the same time at opposite ends, as shown in Figure 11.28a. However, two or more nodes may exist in an oscillating water basin. In a dinodal (or binodal) seiche, high water occurs at each end at the same time with low water between the nodes (Figure 11.28b); when high water occurs between the nodes, the water is low at both ends.

Seiches are caused by a variety of factors. Strong winds may temporarily drive the water toward one end of a natural basin, or differences in barometric pressure at opposite ends of a basin may cause a surface imbalance that triggers the oscillation. In bays or marginal seas with open access to the ocean, long-period wave trains or tides often cause seiches (Figure 11.29). On rare occasions, a tsunami may cause seiches under such conditions. Seiches usually persist for a while even after the cause has subsided.

Many bays and harbors around the world, even those with highly irregular shapes, oscillate with remarkable regularity. Because the wave height of the

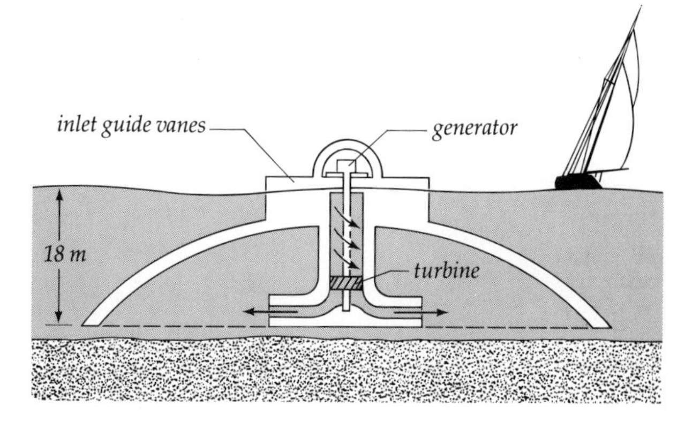

11.30 *Lockheed Corporation's Dam-Atoll designed to generate electricity from wave action. Water entering at the surface spirals down an 18-m (59-ft) central cylinder to turn a turbine located at the bottom of the cylinder. Each unit is designed to produce from 1 to 2 MW.*

oscillation is so low and the length so great, seiches generally go unnoticed. They occasionally cause problems in harbors, however. You may have noticed all the boats in a harbor straining against their mooring lines, even when no wind is noticeable. This is a common occurrence in Los Angeles harbor, where seiches are fairly pronounced. Often for no apparent reason, heavy ships strain against their mooring lines, sometimes snapping them and damaging pilings, piers, and other ships. The periods of these oscillations are three, six, and twelve minutes, corresponding to the natural periods of the basin.

ENERGY FROM WAVES

The energy released by waves breaking on a coast is in a sense wind energy that has temporarily been trapped in the waves. This great energy is potentially usable. Five-meter (16.4-ft) waves with a period of ten seconds release the equivalent of 74 kW per linear meter of coast, or 74,000 kW per kilometer. Even waves only 0.6 m (2 ft) high and 20 m (66 ft) long develop 0.9 kW per linear meter of wave front.

A wide variety of methods have been proposed to capture this energy from waves, including using floats that rise and fall with the passing waves and pump water or compressed air through turbines or lift weights. Figure 11.30 shows such a device currently in use.

FURTHER READINGS

Adams, W. A. *Tsunamis in the Pacific Ocean*. Honolulu: East-West Centre Press, 1970.

Barber, N. F. *Water Waves*. London: Wykeham, 1969.

Bascom, W. *Waves and Beaches*. Garden City, New York: Anchor/Doubleday, 1980.

Charlier, R. H. "Ocean-Fired Power Plants." *Sea Frontiers*, 1981, *27*(1), 36.

Iselin, C. 1963. "The Loss of the *Thresher*." *Oceanus*, September 1963, pp. 4–6.

Kinsman, B. *Wind Waves: Their Generation and Propagation on the Ocean Surface*. Englewood Cliffs, N.J.: Prentice-Hall, 1966.

Knauss, J. A. *Introduction to Physical Oceanography*. Englewood Cliffs, N.J.: Prentice-Hall, 1978.

Land, T. "Freak Killer Waves." *Sea Frontiers*, 1975, *21*(3), 139–41.

______. "Europe to Harness the Power of the Sea." *Sea Frontiers*, 1976, *22*(6), 346–49.

Mooney, M. J. "Tragedy at Scotch Cap." *Sea Frontiers*, 1975, *21*(2), 84–90.

______. "Tsunami." *Sea Frontiers*, 1980, *26*(3), 130.

Newmann, G., and W. J. Pierson, Jr. *Principles of Physical Oceanography*. Englewood Cliffs, N.J.: Prentice-Hall, 1966.

Pararas-Carayannis, G. "The International Tsunami Warning System." *Sea Frontiers*, 1977, *23*(1), 20.

Robinson, J. P., Jr. "Newfoundland's Disaster of '29." *Sea Frontiers*, 1976, *22*(1), 44.

______."Superwaves of Southeast Africa." *Sea Frontiers*, 1976, *22*(2), 106.

Smith, F. G. W. "The Simple Wave." *Sea Frontiers*, 1970, *16*(4), 234–45.

______. "The Real Sea." *Sea Frontiers*, 1971, *17*(5), 298–311.

Smith, F. G. W., and R. H. Charlier. "Waves of Energy." *Sea Frontiers*, 1981, *27*(3), 138.

Tricker, R. A. R. *Bores, Breakers, Waves and Wakes*. New York: American Elsevier, 1964.

Truby, J. D. "Krakatoa—The Killer Wave." *Sea Frontiers*, 1971, *17*(2), 130–39.

Dennis Brokaw

I edged back against the night,
The sea growled assault on the wave-bitten shore.
And the breakers,
Like young and impatient hounds,
Sprang with rough joy on the shrinking sand,
Sprang, but were pulled back slowly,
With a long, relentless pull,
Whimpering, into the dark.

Then I saw who held them captive;
And I saw how they were bound
With a broad and quivering leash of light,
Held by the moon,
As, calm and unsmiling,
She walked the deep fields of the sky.

Jean Starr Untermeyer

TWELVE

Tides

Unlike wind-generated waves, which are highly variable, tides come and go every day. No shoreline, not even the shoreline of a lake, is without tides, although they may be so weak or so obscured by local weather conditions that they go unnoticed. A **tide** is the vertical and horizontal movement of mass in response to the gravitational pull of the sun and the moon on the earth. On a steep or abrupt coast, the vertical movement is more pronounced. In a region of shallow water, the horizontal movement is more pronounced (Figure 12.3).

Although wind-generated waves affect only the ocean's surface, tides move the entire ocean (Figure 12.1). Ponds, lakes, and even the continents and the atmosphere experience tides. The tides of the Great Lakes measure only a few centimeters. The continents rise and fall about 0.2 m (0.66 ft) during a corresponding 3.3 m (10.8 ft) oceanic tide. The atmosphere bulges out a distance of many kilometers. Even the human body experiences tides. As the tides rise and fall, body weight is decreased and increased, respectively, by a few grams.

The terms **ebb** and **flood currents** describe the horizontal motions associated with the fall and rise of the tide in restricted regions along the coast. The terms **high tide** and **low tide** are self-explanatory, as are **high water** (HW), **low water** (LW), and **mean** (average) **sea level** (MSL), or mean tide. But more complicated terminology is also used to describe tides, especially the vertical and datum levels for surveys and charts. **Mean of lower low waters** (MLLW), a term that appears on tide tables for the U.S. Pacific coast, refers to the average height of the lower of two low tides per day. On Canadian charts MLLW is slightly lower than on U.S. charts because their hydrographers are more conservative in their datums. On European charts, LLWS stands for **low-low-water spring**, which is the lowest spring tide per month averaged over several years. The depth most charts use as a reference level is **mean low water** (MLW). A region with a **diurnal** tide has one high tide and one low tide in a 24-hour, 50-minute period; a region with **semidiurnal tides** has two high tides and two low tides in each such period.

PERIODICITIES

The earth is an oblate spheroid, which means that its equatorial diameter is greater than its polar diameter. It revolves on its axis once a day with respect to the sun. The mean solar day, then, is by definition equal to 24 hours, 0 minutes, and 0 seconds. With respect to the fixed stars, however, the length of the day, called the sidereal day, is 23 hours, 56 minutes, and 4.091 seconds. The reason that the solar day is longer by about four minutes is that the earth moves about 1° of its annual path around the sun each day, and the earth must rotate that additional degree to catch up.

Since the earth is not at the exact center of the lunar orbit, the moon's orbit around the earth is said to be elliptical. While the mean distance from the earth's center to that of the moon is 384,404 km (238,796 mi), at **perigee** (closest point) the moon is 356,400 km (221,400 mi) away and at **apogee** (farthest point), 406,700 km (252,647 mi) away. The length of time it takes for the moon to move from one perigee to the next, the **anomalistic month**, is 27.55 days. The interval between successive new moons, however, is 29.53 solar days, called the **synodic month**. For this time period the moon, making one complete revolution to the east, loses one revolution with respect to the earth. This is why we never see the dark side of the moon. The length of a **lunar day** must equal 24 hours, 50.47 minutes:

$$\frac{29.53}{28.53} \times 24 \text{ hours} = 24.84 \text{ hours, or 24 hours and 50.47 minutes}$$

Note that this is the observed average day-to-day lag in the semidiurnal tides.

The earth's orbit around the sun is also elliptical, with the distance from center to center varying from 153,000,000 km (94,800,000 mi) at aphelion in July to a perihelion of 147,000,000 km (91,400,000 mi) in January. The mean distance is about 150,000,000 km (93,000,000 mi). The length for one orbit—365 days, 5 hours, 48 minutes, and 46 seconds—is called the **tropical year**. Calendars are constructed to conform to this interval.

Added to these eccentricities is the fact that the plane orbit of the moon is inclined to that of the earth by an angle of 5°9′ (Figure 12.2) and that the earth's axis of rotation is not perpendicular to the plane of its solar orbit (it is canted 23°27′). So the moon's declination to the celestial equator ranges from 28°36′ to 18°18′, and the pattern of the tides varies with this declination (see Figure 12.4). The period for the intersection of the planes of the earth's and the moon's orbits is 18.6 years.

The tides are the result of complicated interacting forces, and over 150 factors must be considered to predict the tides precisely. Many of these factors, however, are too small to be of concern for most purposes. Tides are classified as diurnal, semidiurnal, and mixed. As Table 12.1 indicates, diurnal tides occur only once a day, and semidiurnal tides twice a day; **mixed tides** are combinations of diurnal and semidiurnal. Table

12.1 Free-falling instrument capsules, which record deep-sea tides and return to the surface on command. The capsules were developed to study the tide in the deep open ocean and are capable of detecting and recording a water-level change of a few millimeters in water 8,000 m (26,240 ft) deep. These capsules have proven that tidal action does cause movement of all the earth's waters. (Scripps Institution of Oceanography)

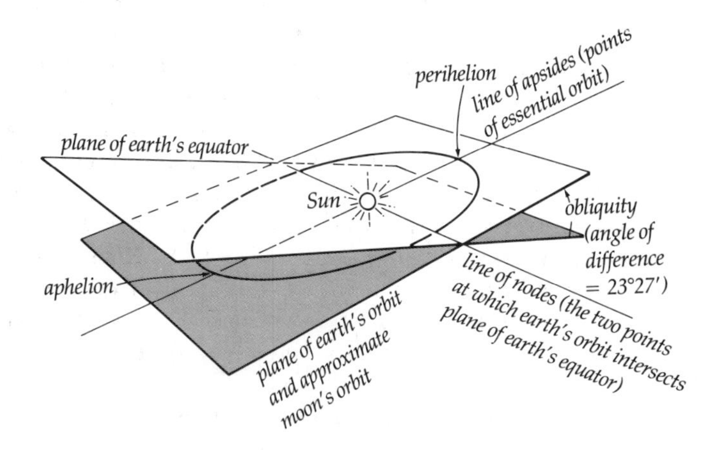

12.2 Declinations of the earth and moon.

(a)

12.3 *(a) The vast horizontal range of the tide at Mont-Saint-Michel in the Gulf of Saint-Maló, France. Here the tide has gone out over 15 km (9.3 mi). (W. J. Wallace) (b) Mont-Saint-Michel at high tide.*

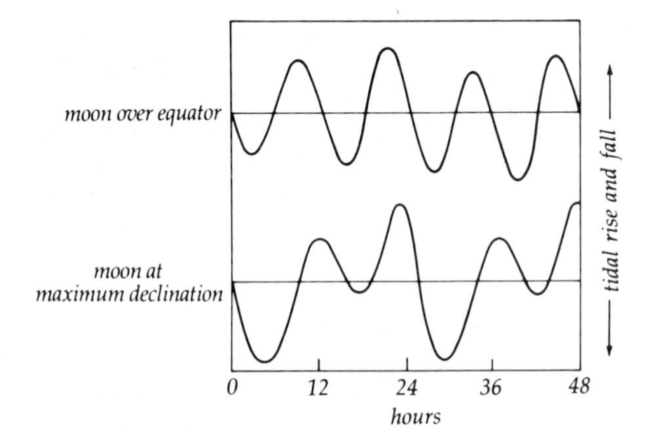

12.4 *Change in diurnal inequality due to different declinations of the moon.*

12.1 lists the major periodicities determining tidal frequencies. Very seldom do all the factors that contribute to the tides occur precisely at the same time; the last time they did was in the year 1400, and the next time will be approximately in the year 3300.

TIDE-GENERATING FORCES

Sir Isaac Newton explained that tides are caused by **gravitation**, the attraction that each mass has for each other mass. According to the theory of universal gravitation that he proposed in 1687, tides are bulges of land and water caused by the gravitational pull of the sun and the moon on the earth and its waters. Matters are simplified by assuming that the earth is uniformly covered with water of the same depth and that there

(b)

are no continents, protruding land masses, or submarine mountains to interfere. The gravitational attraction of the sun or the moon varies directly with its mass and inversely with the square of its distance from the earth. Water bulges on the side of the earth facing the sun, but another bulge equal in size occurs on the opposite side. Why? As the earth moves around the sun, it is held in orbit by two balancing forces: gravitational force (which tends to pull the earth and the sun together) and **centrifugal force** (which tends to pull them apart). The same can be said of the earth-moon system (Figure 12.5).

If you whirl a ball on a string above your head and the string breaks, the force that makes the ball fly off in a straight line tangential to the circle is centrifugal. The force that acts toward the center of the circle to keep the ball moving in a curved line around you rather than a straight line away from you is **centripetal** (Figure 12.6). The revolution of the earth about the center of the earth-sun system generates centrifugal force. (In this case, the center of mass of the system, for all practical purposes, is the center of the sun.) If this force did not equal the gravitational force, the earth would not stay in orbit.

SOLAR AND LUNAR TIDES

The sun has about 27 million times the mass of the moon, but the moon is 387 times closer to the earth. As a result, lunar gravitational effects are more pronounced and lunar tides are twice as great as solar tides. At the centers of the earth and the moon, the centrifugal force exactly balances the gravitational force,

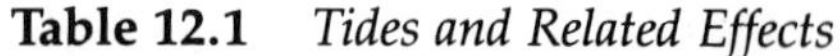

Table 12.1 *Tides and Related Effects*

Phenomenon	Period	Related Astronomical Cycle	Cause
Semidiurnal tide	12 h, 25 min, 23.5 s	Time between upper and lower transits of moon	Rotation of earth
Diurnal tide	24 h, 50 min, 47 s	Time between succeeding upper or lower transits of moon	Rotation of earth and declination of sun and moon
Interval between spring tides	14.76 d (average)	Time from conjunction to opposition of sun and moon or vice versa	Phase relation between sun and moon
Lunar fortnightly effect	13.66 d	Time for moon to change declination from zero to maximum and back to zero	Varying declination of moon
Anomalistic month effect	27.55 d	Time for moon to go from perigee to perigee	Ellipticity of moon's orbit
Solar semiannual effect	182.6 d	Time for sun to change declination from zero to maximum and back to zero	Varying declination of sun
Anomalistic year effect	365.26 d	Time for earth to go from perihelion to perihelion	Ellipticity of earth's orbit

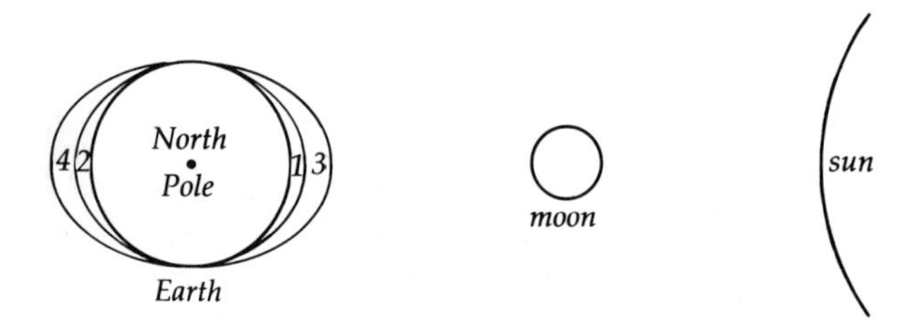

12.5 Tidal forces. The solar tides are 1 (gravitational) and 2 (centrifugal). The lunar tides are 3 (gravitational) and 4 (centrifugal).

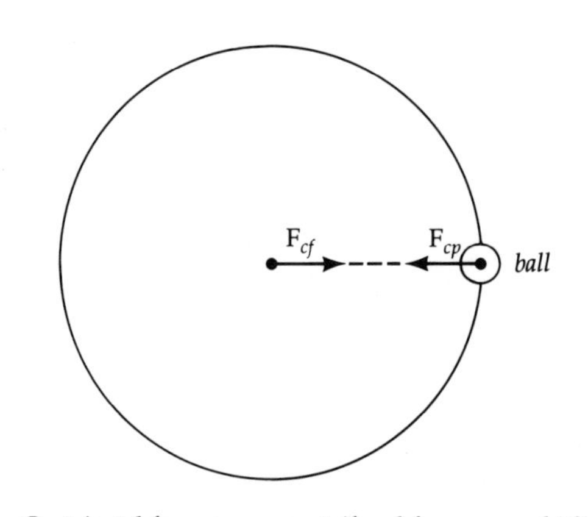

12.6 Centripetal force versus centrifugal force on a whirling-ball-and-string system.

but these two forces do not cancel each other everywhere on the surface of the two bodies. On the side of the earth toward the moon, the moon's gravitational attraction is greater than its centrifugal repulsion; the converse is true on the opposite side of the earth. The differences in forces are small, but they are enough to move the water. Therefore, water bulges on the side facing the moon, and another bulge of the same size appears on the opposite side. This second bulge, caused by centrifugal force, is generated by the rotation of the earth-moon system about the system's center of mass.

It is fairly easy to visualize a centrifugal force generated on the moon by its motion around the earth, but it is not as easy to imagine the earth generating a centrifugal force of its own. Usually we think of the moon circling the earth. The fact of the matter is, the earth and the moon both rotate as a system about a point that is their **center of mass**—which is not the same as the center of the earth. Consider what happens when a wrench is thrown through the air. It, too, rotates around its center—not its geometric center, but its center of mass: the point at which all its mass appears to be concentrated (Figure 12.7).

Unlike the sun-earth system, in which one body is much more massive than the other, the masses of the earth and the moon are not too dissimilar—the earth is only about 99 times more massive than the moon. Because the earth has the greater mass, the center of the system is actually 1,700 km (1,054 mi) below the earth's surface. The bulge opposite the moon is caused by the centrifugal force generated by rotation

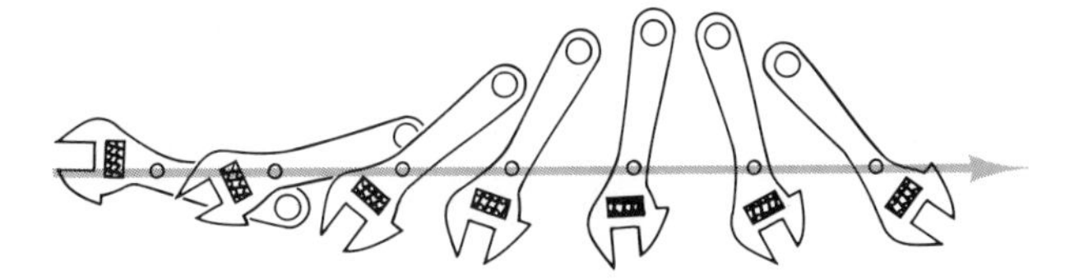

12.7 *A thrown wrench rotates around its center of mass.*

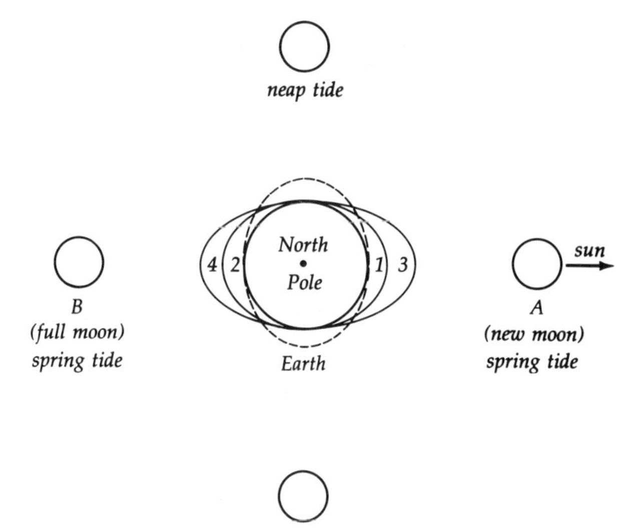

12.8 *Spring and neap tides, as described in the text.*

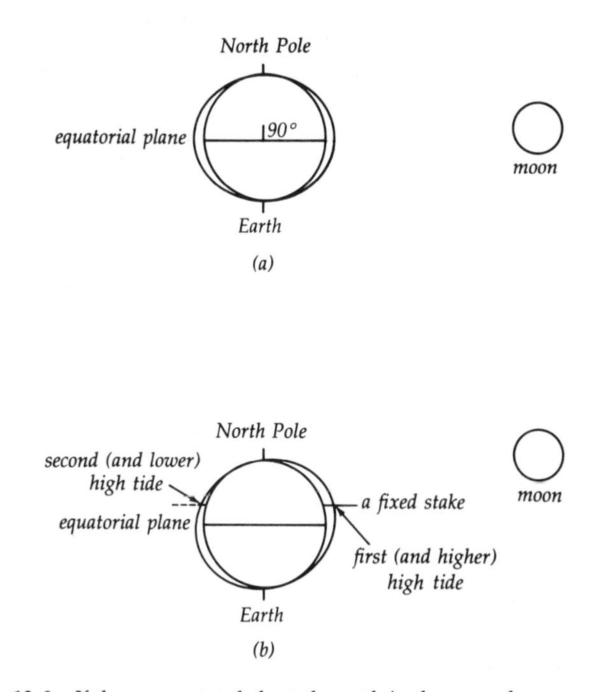

12.9 *If the moon rotated about the earth in the same plane as the earth, the position of the tides would be as shown in (a). Because of the moon's declination, the actual position of the tides is as shown in (b).*

about this point. The earth's geometric center, however, is 6,373 km (3,951 mi) from the surface.

Just as the relative positions of the moon and the sun vary with time, so do the sizes and positions of the bulges. In Figure 12.8, bulges 1 and 3 (and 2 and 4) occur together when the moon is at *A* or *B*, producing the highest and lowest tides, called **spring tides**. These tides have nothing to do with the spring season; they occur every fourteen days with the new or full moon. When the moon and the sun are at right angles to each other, tides of intermediate range (lower high and higher low), called **neap tides**, result.

If the earth rotated about its axis only once a year, the solar tide would always be in the same place. If the earth did not rotate at all about its axis, solar tides would occur once annually at any particular place, and lunar tides would progress around the earth. Thus there would be spring tides every fourteen days. This may not seem too different from what actually occurs until you remember that there would be no daily tides. Daily tides occur because the earth does rotate about its axis. As the moon is also moving in the same counterclockwise direction with respect to the earth, it takes a bit longer for a point on the earth's surface to return to exactly the same position relative to the moon. The time span for one complete rotation of the earth relative to the moon is 24 hours and 50 minutes, or one lunar day. Thus every point on the earth generally experiences two low tides and two high tides in a lunar day.

The moon does not revolve about the earth in exactly the same plane as the earth. It is said to have a declination (elevation) relative to the earth's equatorial plane (Figure 12.9). This effect is sometimes called declination-type tides. Imagine a stake fixed in the earth and extending well above the water's surface, no matter what the depths of water bulges are. As the stake moves with the rotating earth, it experiences water of different depths because it moves through different parts of the two bulges. Tides for a given point on earth are expected to have this diurnal inequality (Figure 12.10). Only when the moon's declination is zero will the bulges be equal.

THE TIDES ACCORDING TO NEWTON

Consider the earth-moon system. If you placed eight equal particles at locales on the earth's surface as shown in Figure 12.11, each particle would trace out a circle

(a)
(b)

12.10 *Low and high tides at Fisherman Bay Spit, Lopez Island, Washington. This region experiences a mixed tide. (a) −0.5 m low tide. Notice the darker regions where the very low water has exposed large amounts of marine algae. (b) High tide. Like most coasts, this region experiences two high and low tides daily, but here they are of unequal heights. (W. J. Wallace)*

12.11 *Eight equal masses or particles of the earth at eight locales on the earth's surface.*

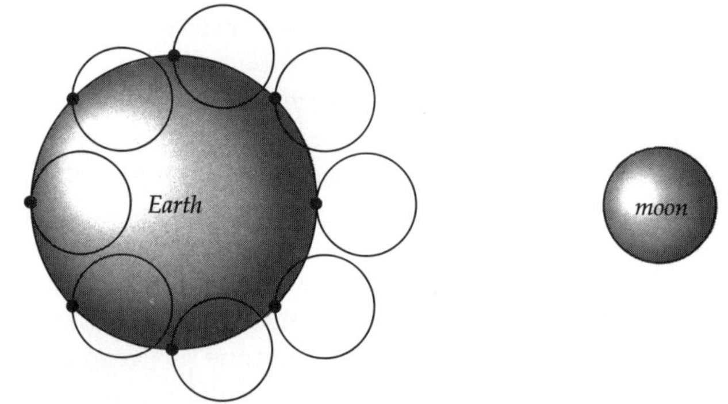

12.12 *Circular paths for each particle or mass as it moves monthly about the center of mass of the earth-moon system.*

or circular path as the earth-moon system moves about its common center of mass each month ($27\frac{1}{3}$ days—the time for a complete orbit or mutual revolution). This orbital motion is the same for each particle, regardless of the distance from the center of mass of the system (Figure 12.12). This means that the force necessary to make each of these particles move in its own circle is equal in magnitude and direction for all particles of the earth. This force, called centripetal, is equal all over the earth and is directed at right angles to the moon (Figure 12.13). The lunar gravitational attractive force on each of these particles, however, is a function of the particle's distance from and orientation to the moon (Figure 12.14). At the center of the earth, and only there, are the centripetal force and the moon's gravitational attraction equal in both strength and direction.

There is no place on the earth's surface where the force of the moon's gravity acting on an object is exactly the same in magnitude and direction as the centripetal

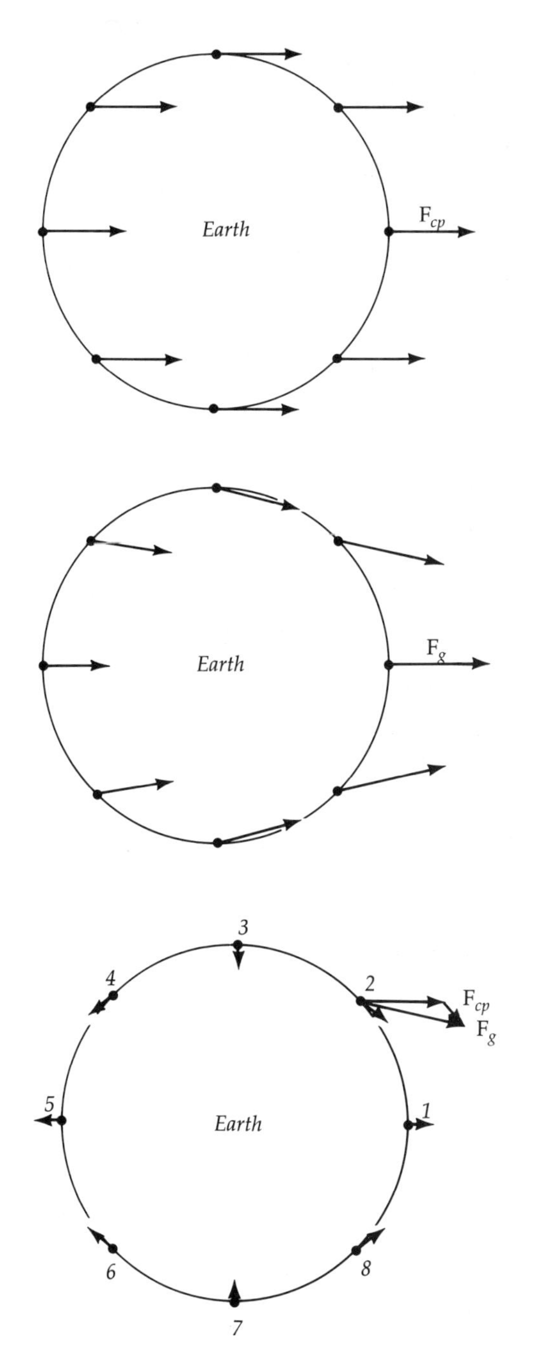

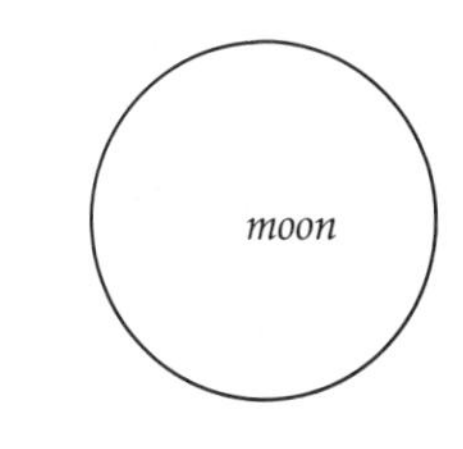

12.13 The centripetal forces on identical objects at various locales on the earth's surface as a result of the rotation of the earth-moon system about its common center of mass.

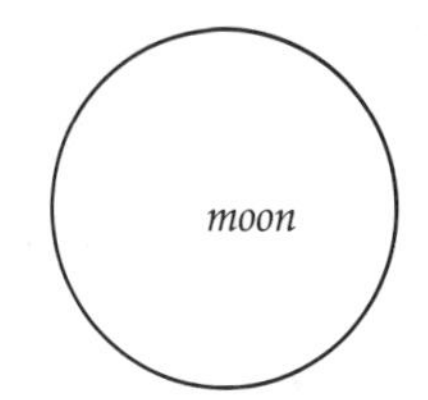

12.14 The moon's gravitational forces on identical objects at various locales on the earth's surface.

12.15 The resultant tide-generating forces produced by drawing a resultant line from the head of each F_{cp} and F_g arrow (as in point 2).

force. It is the difference between these forces that constitutes the tide-generating force at each locale. The tidal action over the entire earth is the result of the sum of all the individual differences.

In Figure 12.15, note that along a great circle between points 3 and 7, which are at right angles to the direction of the moon, the resultant tide-generating forces are turned toward the center of the earth. These forces are half as large as those at points 1 and 5, which are directed toward and away from the moon, respectively. Since the water at point 1 experiences the greatest lunar attractive force, it tends to be pulled slightly away from the earth, forming a bulge. At point 5 the water is less strongly attracted by the moon than by the earth as a whole, and the earth is pulled slightly away from the water at that point. This, then, is why there is another bulge on the earth at the side opposite the moon.

Since the vertical component of the **tide-generating force** is very small compared with the force of gravity at the earth's surface, it produces no noticeable motion of the sea. It is the horizontal tide-generating component that produces the tides on each side of the earth. This force is at a maximum at points 2, 4, 6, and 8 in Figure 12.15.

DYNAMIC TIDES

The foregoing description of tides is known as the **equilibrium theory** and was originally set forth by Sir Isaac Newton. It explains monthly and yearly changes in the tides, but it does not allow for the influence of the continental land masses on the tides.

According to the equilibrium theory, tidal bulges should remain stationary directly under the sun and the moon. As the earth rotates on its axis, the tidal bulge (which can be considered a traveling or pro-

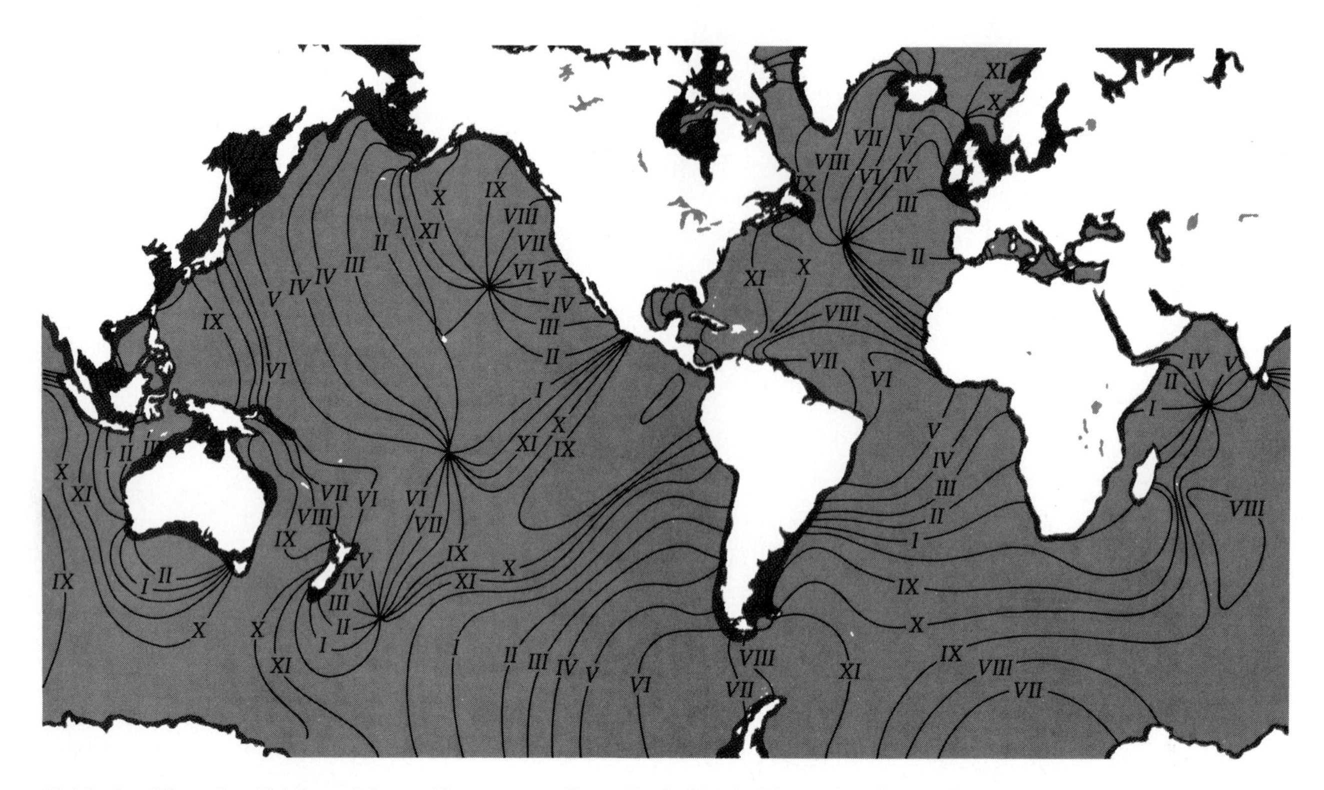

12.16 *Semidiurnal cotidal lines of the world ocean, according to R. A. Harris. The contour lines with roman numerals show the progression of the tide at successive lunar hours. Where these lines meet are the nodes, or amphidromic points.* (Manual of Tides. *Washington, D.C.: U.S. Coast and Geodetic Survey, 1907)*

gressive wave) at the equator would have to move at a velocity of 1,656 km/h (1,035 mph) to remain aligned with the moon. For it to move that fast as a deep-water wave, the water depth would have to be at least 22.5 km (14 mi). With an average depth of 4,000 m (13,120 ft) for the Atlantic and 4,600 m (15,000 ft) for the Pacific, such movement is clearly impossible. Consequently, the tides lag behind the earth's rotation (a condition called **tidal age**). Where angular velocities are smaller (above 66° of latitude) the bulges could in principle keep pace, but the geographic restrictions of the polar regions prevail. Also, according to the equilibrium theory, the tidal heights should be at a maximum at the equator (.8 m [2.6 ft]) and decrease toward both poles.

The equilibrium theory deals with the tides as though the water itself were standing still and only the tidal bulges were progressing through it. Although there is absolutely no tidal sensation in the open ocean, the motion of the water due to tides in coastal and semienclosed regions is obvious.

The continents present an effective barrier to the passage of the tidal bulges. Were it not for that barrier, the tidal bulges would tend to stay more or less closely aligned with the moon and the sun. When a continent directly faces the moon, there is no oceanic bulge (although there is a terrestrial tide), and the water level at the shores of the continent experiences a high tide. When the continent is not aligned with the moon but the ocean is, the bulge re-forms and the continental edges experience low tide. The net effect is a continual oscillation of the water, which may be considered as a very long wave with a wavelength one-half the planet's circumference and a period of 44,700 s (12 hours, 25 minutes). The crests and troughs of the wave appear as high tides and low tides, respectively. These tidal bulges may be considered as shallow-water waves. Their velocity, which averages about 200 m/s (656 ft/s, or 447 mph) in the deep ocean, depends on the water depth, and their period depends on the periodic motion of the sun and the moon. Since these bulges cannot travel at the apparent velocity of the sun and moon across the sky, at any given place there is a time lag between the positions of the sun and the moon overhead and the tide.

Only in the ocean areas around the Antarctic is the world's ocean continuous enough for the tidal bulges not to be significantly hindered. Here the tide moves in a manner that is said to be progressive, like the movement of a wave.

It has long been known that the currents generated by the tides in bays, estuaries, and certain river mouths, as well as in the English Channel and the North Sea, have a rotary nature. The gyroscopic motion

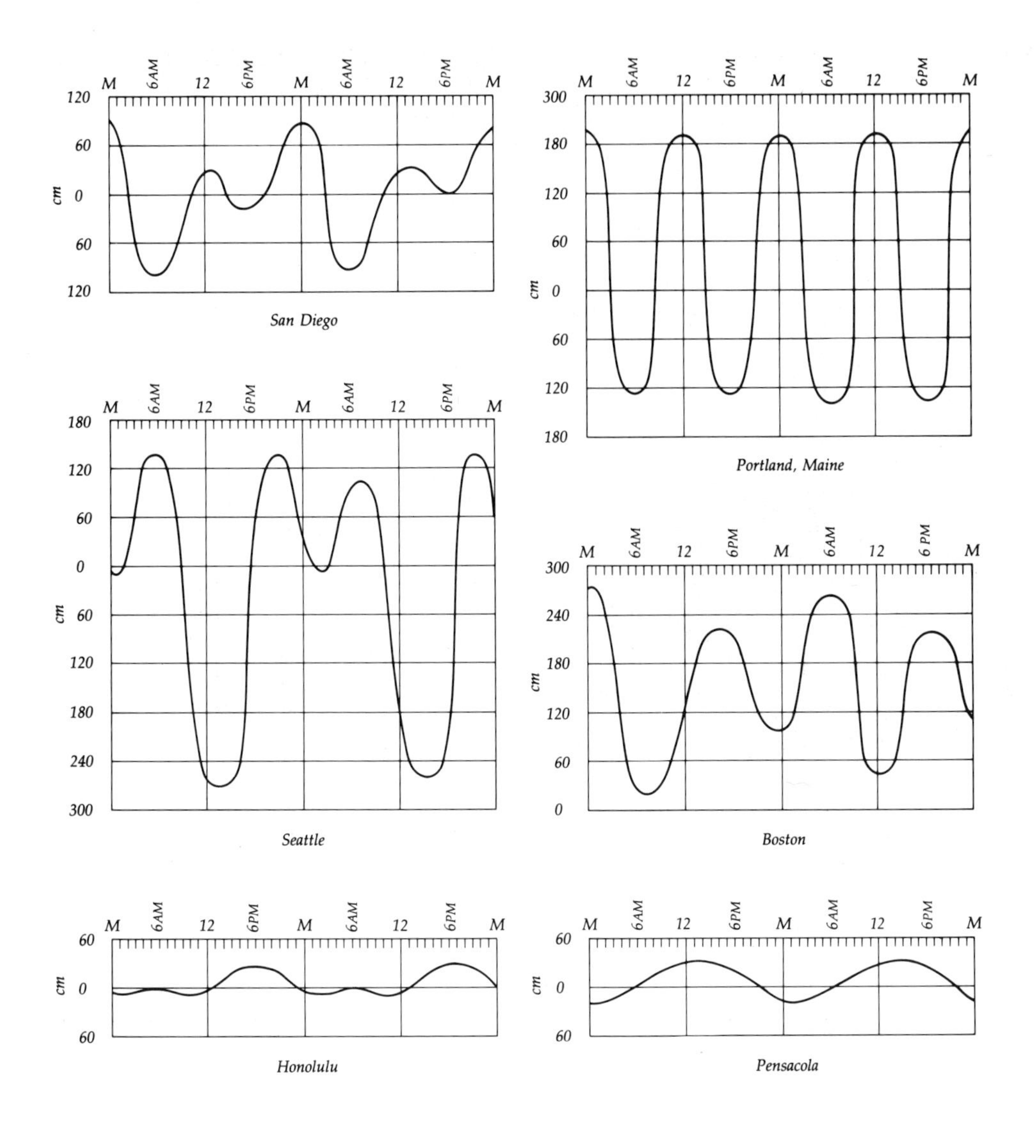

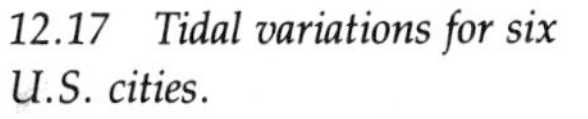
12.17 *Tidal variations for six U.S. cities.*

resulting from the earth's rotation on its axis causes the tides in these basins and in the ocean basins as well to oscillate in such a way that they appear to rotate about a fixed point, or node, called an **amphidromic point**, at which the tidal amplitude is zero. From these amphidromic points **cotidal lines** may be drawn that connect places where the high tides occur at the same time. The **tidal range** (or amplitude) becomes greater as distance from an amphidromic point increases along cotidal lines (see Figure 12.16). Thus an island close to an amphidromic point would experience only a small tidal range.

As Figure 12.16 pictures, the tidal wave in the South Atlantic is a progressive wave, much like a seiche, or a standing wave, in a channel, moving up the entire ocean. The continents of South America and Africa constrict the basin. Progressive tidal waves occur only at a few places in the open ocean, but they are fairly common in shelf and semienclosed regions like the Chesapeake Bay, where the progressive wave requires twelve hours to move from Hampton Roads to Baltimore at a speed of 10 knots.

TIDAL RANGES

From the preceding discussion alone, you might conclude that all shorelines experience two high tides and two low tides a day, with one high tide being somewhat higher (except at the equator, where both are the same). Thus, if high tide occurred at 10 A.M. at a particular point, you would expect the next high tide to occur at 10:25 P.M., and the high tide after that to be the next morning at 10:50 A.M. Unfortunately, such a generalization is too simple. Each ocean basin varies in shape, dimension, and depth. As a consequence, tides at different places deviate from what would be expected. The tides around the Atlantic occur twice a day (two high and two low) and have essentially the same range, or vertical distance from high to low water, but the regions around the edges of the Pacific have tides of unequal height, as would be expected. Figure 12.17, which charts the observed tides at six major cities, illustrates the tidal variances. Notice that San Diego, Seattle, Honolulu, and Boston have two high tides and two low tides per day, but Pensacola has an

Table 12.2 *Times and Heights of High and Low Waters for San Francisco (Golden Gate), California, December 1988*

Day	Time h m	Height ft	Height m
1	0616	5.0	1.5
TH	1205	2.7	0.8
	1704	4.0	1.2
	2327	1.2	0.4
2	0646	5.2	1.6
F	1301	2.1	0.6
	1832	3.7	1.1
3	0012	1.6	0.5
SA	0717	5.4	1.6
	1347	1.4	0.4
	1956	3.7	1.1
4	0058	2.0	0.6
SU	0743	5.6	1.7
	1429	0.8	0.2
	2108	3.9	1.2
5	0143	2.4	0.7
M	0815	5.9	1.8
	1507	0.3	0.1
	2207	4.1	1.2
6	0225	2.8	0.9
TU	0848	6.1	1.9
	1543	−0.2	−0.1
	2259	4.4	1.3
7	0304	3.1	0.9
W	0920	6.3	1.9
	1618	−0.5	−0.2
	2345	4.6	1.4
8	0346	3.4	1.0
TH	0955	6.5	2.0
	1653	−0.8	−0.2
9	0030	4.7	1.4
F	0425	3.5	1.1
	1033	6.6	2.0
	1732	−1.0	−0.3
10	0116	4.8	1.5
SA	0505	3.6	1.1
	1113	6.6	2.0
	1814	−1.1	−0.3
11	0158	4.8	1.5
SU	0551	3.6	1.1
	1154	6.5	2.0
	1857	−1.1	−0.3
12	0243	4.9	1.5
M	0644	3.6	1.1
	1243	6.2	1.9
	1943	−0.9	−0.3
13	0325	5.0	1.5
TU	0748	3.4	1.0
	1339	5.8	1.8
	2032	−0.6	−0.2
14	0411	5.1	1.6
W	0901	3.1	0.9
	1441	5.2	1.6
	2120	−0.1	0.0
15	0453	5.4	1.6
TH	1021	2.6	0.8
	1557	4.6	1.4
	2212	0.4	0.1
16	0535	5.7	1.7
F	1139	1.9	0.6
	1727	4.2	1.3
	2308	1.1	0.3
17	0617	6.1	1.9
SA	1246	1.0	0.3
	1903	4.0	1.2
18	0004	1.7	0.5
SU	0700	6.4	2.0
	1346	0.2	0.1
	2034	4.1	1.2
19	0101	2.3	0.7
M	0745	6.7	2.0
	1442	−0.4	−0.1
	2147	4.4	1.3
20	0158	2.8	0.9
TU	0828	6.9	2.1
	1529	−0.9	−0.3
	2249	4.7	1.4
21	0255	3.1	0.9
W	0914	7.0	2.1
	1615	−1.1	−0.3
	2341	5.0	1.5
22	0346	3.3	1.0
TH	0957	7.0	2.1
	1657	−1.2	−0.4
23	0030	5.1	1.6
F	0435	3.4	1.0
	1041	6.8	2.1
	1739	−1.1	−0.3
24	0112	5.1	1.6
SA	0524	3.4	1.0
	1122	6.6	2.0
	1820	−0.9	−0.3
25	0154	5.0	1.5
SU	0610	3.4	1.0
	1204	6.2	1.9
	1858	−0.7	−0.2
26	0235	5.0	1.5
M	0659	3.3	1.0
	1243	5.8	1.8
	1937	−0.3	−0.1
27	0312	4.9	1.5
TU	0752	3.3	1.0
	1326	5.3	1.6
	2014	0.1	0.0
28	0349	4.9	1.5
W	0851	3.1	0.9
	1410	4.7	1.4
	2053	0.5	0.2
29	0421	5.0	1.5
TH	1001	2.8	0.9
	1506	4.2	1.3
	2128	1.1	0.3
30	0453	5.1	1.6
F	1106	2.4	0.7
	1615	3.7	1.1
	2210	1.7	0.5
31	0529	5.3	1.6
SA	1209	1.9	0.6
	1757	3.4	1.0
	2259	2.2	0.7

Note: Time meridian 120°W. 0000 is midnight. 1200 is noon.
Heights are referred to mean lower low water, which is the chart datum of soundings.
Source: National Oceanic and Atmospheric Administration.

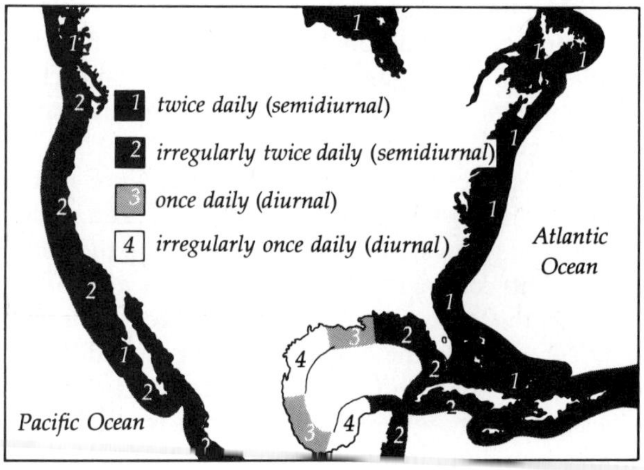

12.18 Tides along the U.S. coast. The numbers signify the following: 1, twice daily; 2, irregularly twice daily; 3, once daily; and 4, irregularly once daily.

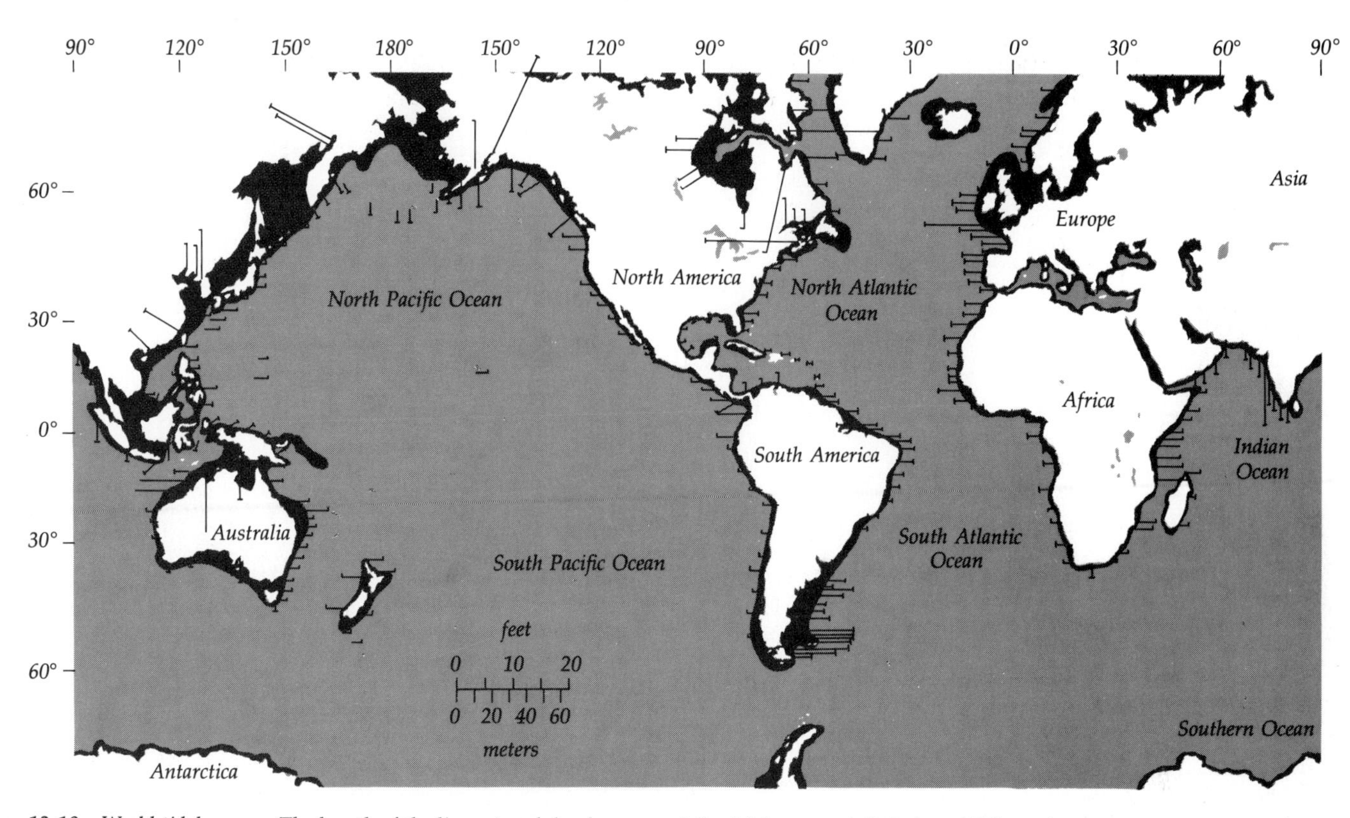

12.19 *World tidal ranges. The length of the lines at each locale represent the tidal ranges at that place. With a ruler, a set of dividers, or a compass, compare the length with the reference length given in feet and meters.*

entirely different profile. The harbor of Portland, Maine, is only 160 km (99 mi) from Boston, but notice the difference in their tide heights.

Tidal prediction tables (Table 12.2) are published annually by the U.S. government for most locales. Some areas, such as the Mediterranean, the Gulf of Mexico, and the Baltic Sea, are almost tideless. Islands away from the shore and nearer the center of tidal basins have small tides. Nantucket (off Massachusetts) and Tahiti have tides seldom ranging more than 0.3 m (1 ft). The tidal variances at the ends of the Panama Canal, separated by only 80 km (50 mi), illustrate the differences that may occur. The tides at Colón on the Caribbean end of the canal have a range of about 0.3 m (1 ft) and are diurnal. At Balboa, on the Pacific side of the canal (which is actually east of the Caribbean end), the tides are semidiurnal, with an average range of 4.3 m (14.1 ft). Figure 12.18 shows these tide types along the coast of the United States.

The world's tidal ranges are given in Figure 12.19. The tidal range varies from 0 m to over 15 m (49 ft) around the world, but the horizontal distance of land covered and uncovered by the tide can be as great as 16 km (10 mi), as at Mont-Saint-Michel (see Figure 12.1) or on the northern Dutch coast. In these places, the tide comes in as fast as a person can run.

TIDAL CURRENTS

Although tides move the entire ocean in a rotary or circular fashion, it is in the wide coastal indentations and semienclosed seas, where water movement is not restricted by topographic barriers, that the rotary nature of the **tidal currents** really becomes obvious, as in San Francisco Bay and the North Sea. In such locations, the **rotary current** is not confined to a definite channel but changes its direction continuously, never coming to a slack. In a tidal cycle of 12 hours, 25 minutes, these rotary tidal currents move through all the directions of the compass (see Figure 12.20). The speed of the current usually varies throughout the tidal cycle, passing through two maximum speeds in approximately opposite directions.

The tidally generated currents found in rivers, elongated bays, and narrow coastal indentations, where flow is essentially restricted, are of the reversing type. This means that they set in one direction for about a quarter of a lunar day, come to a **slack**, and then set in the opposite direction for another quarter-day. The terms *ebb* and *flood* apply specifically to these types of currents, which, if unaffected by nontidal flow like that of a river's discharge, will each last about six hours. The speed of the flow in each direction will vary from

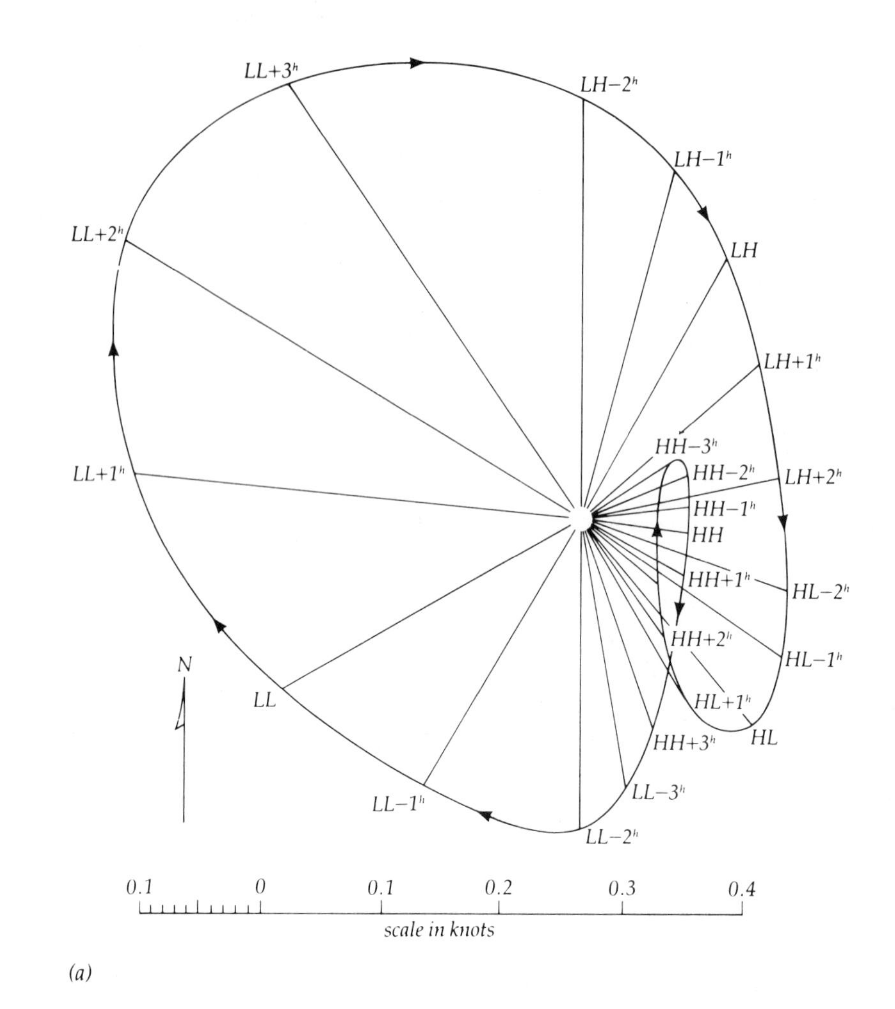

12.20 Rotary quality of the tidal currents at two places off the U.S. coast; (a) the Pacific (San Francisco) and (b) the Atlantic (Nantucket). In diagram (a) the time period is about 25 hours; in (b), only 13. The tides at the coast of Massachusetts for any given day have highs and lows of the same relative heights; therefore the curve formed by the rotary currents is the same for each tidal period (12 hours, 25 minutes). For the Pacific, there are two highs and two lows, each of unequal heights; therefore the complete rotary current takes a day to form a closed loop. The outside curve is the path a float would follow when tethered to an elastic line. Each arrow represents an hour interval; its length represents the relative current velocity.

zero at the time of slack water to a maximum about midway between the slacks.

Near land, a tide's horizontal flow and vertical rise of water are apparent, especially when viewed against a pier or seawall. The effects of local winds or weather, sometimes called meteorological tides, cause tidal anomalies that can be more pronounced than the tide itself. Storm surges (Chapter 9) illustrate this. Changing tides can set up currents, especially in restricted areas. The current under the Golden Gate Bridge in San Francisco during a change in tides is 6 knots—about the same as the flow of the East River in New York—and a value not uncommon in the world's straits and channels. The tidal current may reach a velocity of 10 knots in some places, such as the Strait of Georgia between mainland British Columbia and Vancouver Island. A ship entering the English Channel toward the North Sea as the tide comes in has the benefit of as much as 6 hours of favorable current. If the ship reaches the Strait of Dover just at high tide, it can continue with the outgoing tidal current to the North Sea.

Like the tides themselves, the tidal currents generated in restricted waters such as harbors, bays, estuaries, and embayments change direction continuously. However, the direction of the currents associated with an incoming tide is not simply opposite to the direction of the outgoing tide. Ebb and flood currents may actually be at right angles to one another or may even travel in the same direction. Tidal currents are weakest in shallow waters and strongest in deep waters. Many other factors, such as embayment shape, river flow, channel depth and shape, and friction, affect the nature of tidal currents. Like the tides, tidal currents can be predicted. Table 12.3 lists predictions for tidal currents in Deception Pass, shown in Figure 12.21.

Whirlpools

The **whirlpool** is another example of water movement. The rotary direction of water flow in a whirlpool is determined by coastlines, bottom topography, and local currents; it is independent of the earth's rotation

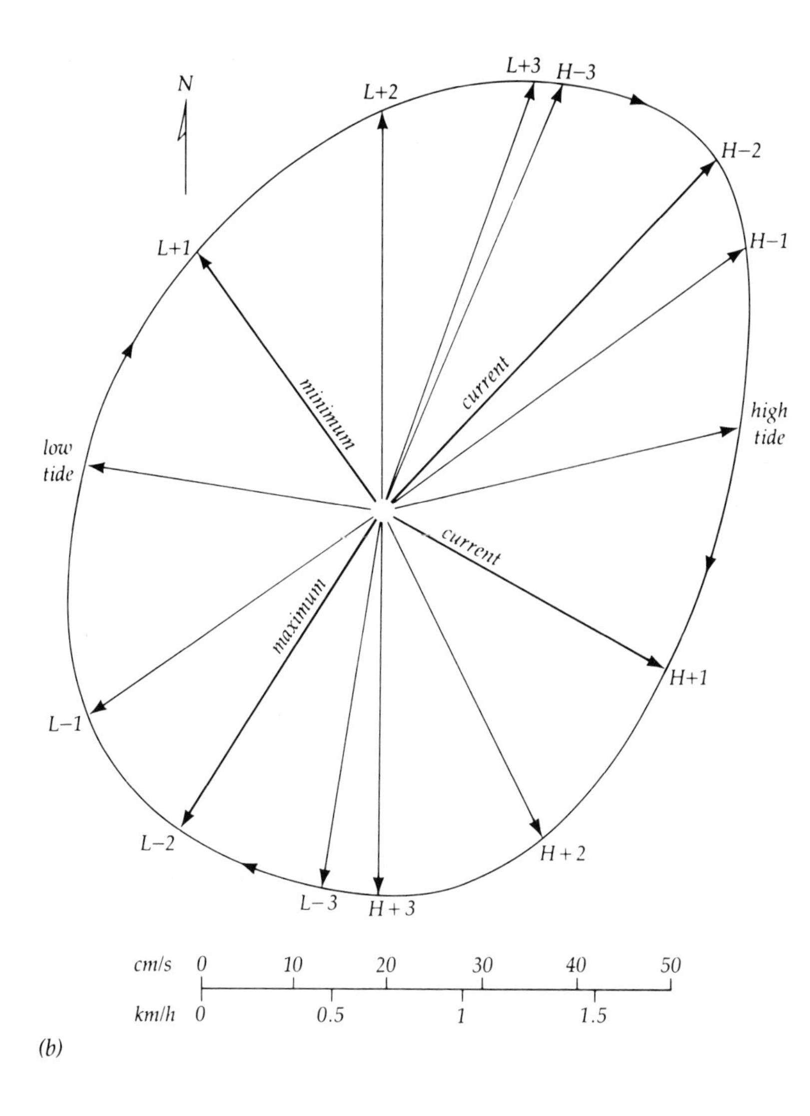

(b)

and is almost always associated with the tides and tide-generated currents. Such currents occur in semi-enclosed regions where waters moving rapidly between islands encounter countercurrents caused by the shift in tides or by currents passing one another with appreciable friction between them (Figure 12.22).

The fascination whirlpools hold for us is reflected in Edgar Allan Poe's "A Descent into the Maelström":

> The ordinary account of this vortex had by no means prepared me for what I saw. That of Jonas Ramus, which is perhaps the most circumstantial of any, cannot impart the faintest conception either of the magnificence or of the horror of the scene—or of the wild, bewildering sense of the novel which confounds the beholder. I am not sure from what point of view the writer in question surveyed it, nor at what time; but it could neither have been from the summit of Helseggen, nor during a storm. There are some passages of this description, nevertheless, which may be quoted for their details, although their effect is exceedingly feeble in conveying an impression of the spectacle.
>
> "Between Lofoden and Moskoe," he says, "the depth of the water is between thirty-six and forty fathoms; but on the other side, toward Ver [Furrgh], this depth decreases so as not to afford a convenient passage for a vessel, without the risk of splitting on the rocks, which happens even in the calmest weather. When it is flood, the stream runs up the country between Lofoden and Moskoe with a boisterous rapidity; but the roar of its impetuous ebb to the sea is scarce equalled by the loudest and most dreadful cataracts; the noise being heard several leagues off, and the vortices or pits are of such an extent and depth that if a ship comes within its attraction, it is inevitably absorbed and carried down to the bottom, and there beat to pieces against the rocks; and when the water relaxes, the fragments thereof are thrown up again. But

Table 12.3 *Tidal Currents in Deception Pass, Washington, December 1988*

Slack Water		Maximum Current	
Day	Time h.m.	Time h.m.	Velocity (Knots)
1	0235	0602	5.0F
TH	0937	1228	6.1E
	1636	1854	3.8F
	2210		
2		0035	5.1E
F	0349	0657	4.9F
	1024	1319	6.3E
	1719	1953	4.3F
	2317	0137	5.2E
3		0751	4.8F
SA	0456	1407	6.6E
	1107	2039	4.9F
	1758		
4	0012	0235	5.5E
SU	0555	0840	4.7F
	1146	1451	6.8E
	1833	2124	5.4F
5	0100	0324	5.8E
M	0648	0926	4.7F
	1222	1530	7.0E
	1906	2206	5.8F
6	0144	0412	6.1E
TU	0737	1009	4.6F
	1256	1609	7.2E
	1937	2246	6.2F
7	0226	0457	6.3E
W	0824	1053	4.5F
	1328	1650	7.2E
	2008	2326	6.4F
8	0307	0539	6.4E
TH	0911	1136	4.3F
	1400	1729	7.2E
	2040		

Slack Water		Maximum Current	
Day	Time h.m.	Time h.m.	Velocity (Knots)
9		0007	6.6F
F	0349	0624	6.5E
	0957	1217	4.1F
	1433	1809	7.2E
	2113		
10		0048	6.6F
SA	0431	0708	6.5E
	1045	1302	3.9F
	1510	1851	7.1E
	2150		
11		0132	6.6F
SU	0514	0755	6.5E
	1134	1350	3.7F
	1553	1936	6.9E
	2231		
12		0217	6.5F
M	0558	0843	6.5E
	1225	1441	3.6F
	1644	2027	6.7E
	2317		
13		0306	6.4F
TU	0643	0932	6.6E
	1317	1534	3.7F
	1745	2119	6.4E
14	0009	0357	6.2F
W	0729	1024	6.6E
	1410	1633	3.9F
	1858	2220	6.2E
15	0107	0448	5.9F
TH	0816	1117	6.8E
	1503	1732	4.3F
	2021	2321	6.0E
16	0213	0546	5.6F
F	0904	1209	6.9E
	1553	1832	4.8F
	2145		

Slack Water		Maximum Current	
Day	Time h.m.	Time h.m.	Velocity (Knots)
17		0026	5.9E
SA	0326	0644	5.4F
	0951	1302	7.1E
	1643	1933	5.4F
	2300		
18		0131	6.0E
SU	0442	0739	5.1F
	1039	1354	7.3E
	1731	2030	5.9F
19	0008	0236	6.1E
M	0554	0836	4.9F
	1126	1445	7.5E
	1818	2124	6.3F
20	0109	0337	6.2E
TU	0701	0930	4.7F
	1213	1536	7.6E
	1905	2218	6.6F
21	0205	0433	6.4E
W	0804	1022	4.5F
	1300	1625	7.6E
	1952	2307	6.8F
22	0257	0528	6.5E
TH	0902	1116	4.4F
	1348	1713	7.6E
	2037	2356	6.9F
23	0346	0617	6.6E
F	0956	1208	4.3F
	1437	1802	7.4E
	2122		
24		0043	6.9F
SA	0433	0706	6.6E
	1048	1256	4.2F
	1526	1848	7.3E
	2206		

Slack Water		Maximum Current	
Day	Time h.m.	Time h.m.	Velocity (Knots)
25		0130	6.8F
SU	0516	0755	6.6E
	1136	1347	4.1F
	1616	1936	7.0E
	2249		
26		0214	6.6F
M	0558	0837	6.6E
	1223	1433	4.1F
	1709	2022	6.7E
	2332		
27		0256	6.3F
TU	0637	0918	6.6E
	1308	1522	4.1F
	1803	2112	6.4E
28	0014	0341	6.0F
W	0715	1001	6.6E
	1352	1613	4.2F
	1900	2200	6.1E
29	0058	0426	5.7F
TH	0752	1045	6.6E
	1436	1702	4.3F
	2001	2250	5.7E
30	0144	0509	5.3F
F	0829	1130	6.6E
	1520	1751	4.4F
	2105	2343	5.4E
31	0236	0558	4.9F
SA	0906	1215	6.5E
	1603	1846	4.5F
	2212		

Note: Time meridian 120°W. 0000 is midnight. 1200 is noon.
F = flood; direction 090° true. E = ebb; direction 270° true.
Source: National Oceanic and Atmospheric Administration.

these intervals of tranquillity are only at the turn of the ebb and flood, and in calm weather, and last but a quarter of an hour, its violence gradually returning. When the stream is most boisterous, and its fury heightened by a storm, it is dangerous to come within a Norway mile of it. Boats, yachts, and ships have been carried away by not guarding against it before they were carried within its reach. It likewise happens frequently that whales come too near the stream, and are overpowered by its violence; and then it is impossible to describe their howlings and bellowings in their fruitless struggles to disengage themselves. A bear once, attempting to swim from Lofoden to Moskoe, was caught by the stream and borne down, while he roared terribly, so as to be heard on shore. Large stocks of firs and pine

(a)

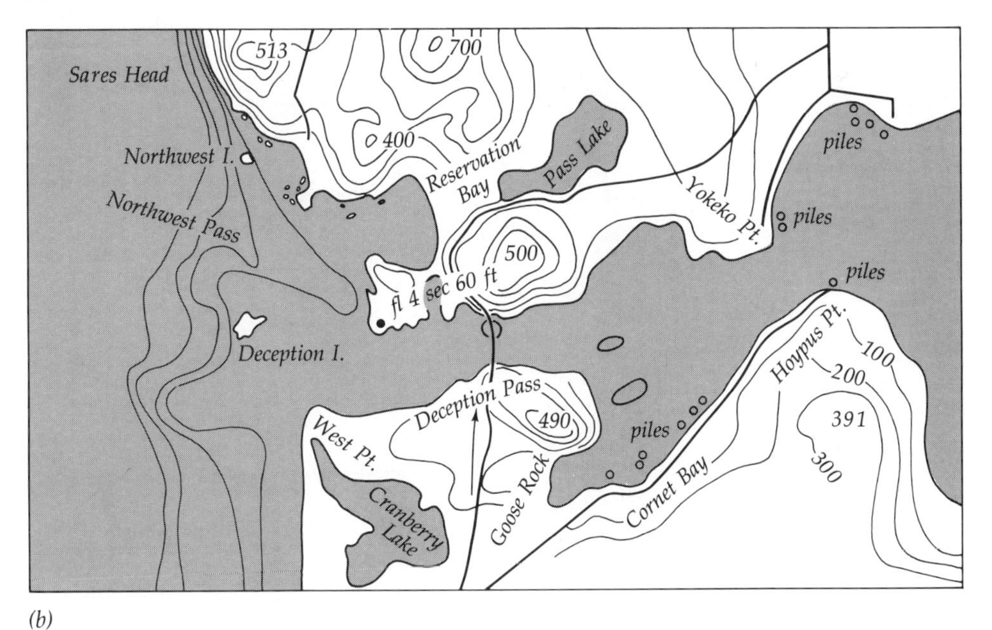

(b)

12.21 (a) Tidal currents in Deception Pass, Puget Sound, Washington. (W. J. Wallace) These currents can attain speeds of almost 8 knots. They occur only in regions where the waters are restricted by land masses, as indicated in (b).

trees, after being absorbed by the current, rise again broken and torn to such a degree as if bristles grew upon them. This plainly shows the bottom to consist of craggy rocks, among which they are whirled to and fro. This stream is regulated by the flux and reflux of the sea—it being constantly high and low water every six hours. In the year 1645, early in the morning of Sexagesima Sunday, it raged with such noise and impetuosity that the very stones of the houses on the coast fell to the ground."

Such is the famous Maelström. Although the name has been used for large whirlpools elsewhere, it originally applied to the Moskenstrom, a strong current (to 3 m/s [10 ft/s]) running south between the islands of Moskenesøya and Moskoe in the Lofoten Islands off the west coast of Norway. This whirlpool results from strong tidal currents that frequently change

12.22 *Whirlpool in an inland sea of Japan. Whirlpools are caused by the opposition of tidal currents in coastal (or semicoastal) waters. (Keystone Japan)*

12.23 *Tidal bore on the Araguari River, Brazil. Tidal bores are tidally generated waves that move up certain estuaries and rivers. They represent the incoming tide, which, being confined and amplified by the estuary, becomes a single, well-defined wave with several smaller waves behind. The bore may be breaking or smooth (undular), and after its passage the river flows upstream for up to an hour. Bores only occur where both high tides and gently sloping, slowly moving rivers occur together. (David K. Lynch)*

direction. The whirlpool is not always there, but it appears during certain conditions of wind and tide. The region of greatest violence in these hazardous waters is between the island of Moskoe and the coast.

The whirlpool Charybdis (now called **Galofalo**) was first described by Homer. This large, dangerous whirlpool, driven by current and winds, is situated along the Calabrian coast in the Strait of Messina between Italy and Sicily. A similar though smaller whirlpool appears near Black's Harbour, northeast of Passamaquoddy Bay in New Brunswick. This spectacular phenomenon occurs often enough for tourist brochures to list it as a point of interest.

Tidal Bore

Perhaps the most striking effect of local conditions on tides is the **tidal bore**. A tidal bore occurs when the incoming tide forms a single wave that moves up a river, estuary, or estuarine region as a foaming, churn-

(a)

(b)

12.24 (a) Tidal bore in Petitcodiac River, Bay of Fundy. (Courtesy of Canadian National Railways. National Copyright Reserved) (b) Bay of Fundy at low tide. Because of the extreme tidal range, special wooden platforms have been built to allow ships to rest on the bottom during low tide. (Courtesy of the Reid Studio and Minas Basin Pulp and Power Company Limited)

ing wall of water. In some locales bores may form a smooth wave followed by other waves (Figure 12.23).

Tidal bores are most common in Asia, but there are bores in French rivers, such as the Seine, the Gironde, and the Orne, and in the Severn and Trent rivers of England. The most famous in North America is the bore of the Petitcodiac River at the head of the Bay of Fundy between New Brunswick and Nova Scotia (Figure 12.24). Here a wall of water up to 1.2 m (4 ft) high during spring tides surges up the bay and fills all its rivulets twice daily. The tidal range in the Bay of Fundy is the highest in the world, at times exceeding 15 m (49 ft). Approximately 104 billion cubic meters of water come into the bay with each tide, an amount equal to all the water consumed in three months by all the people in the United States. The bore of the Amazon, the world's greatest river, moves upriver for more than 480 km (298 mi) at a speed of 12 knots. Its roar can be heard for 24 km (15 mi). The most famous of all tidal bores is that of the Fu-Ch'un

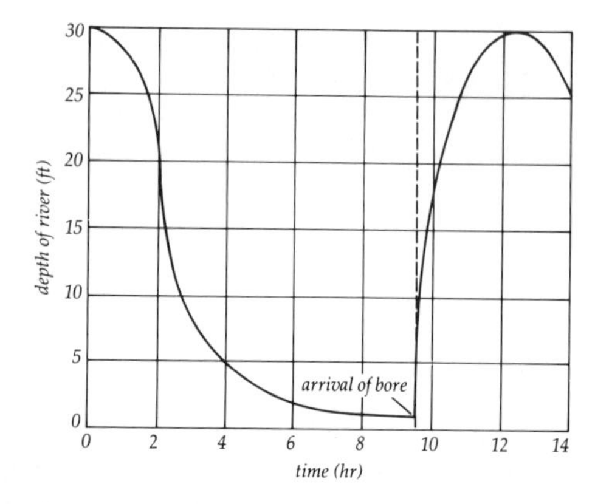

12.25 The rapid increase in depth of the Petitcodiac River with the arrival of the tidal bore. Instead of the usual six-hour ebb and six-hour flood of the tide, here the tide comes in during about two and half hours. It takes over nine hours to go out.

River estuary in northern China, which may attain heights of 7.6 m (25 ft) and actually moves at 12 to 13 knots! When this wall of water rumbles up the river, all boating stops. Boat operators either pull their boats out of the river or tie them up. Once the bore passes, the boat operators can ride rapidly upstream with the following current.

At least two conditions must exist for a bore to occur: The tidal range must be at least 6 m (20 ft) high and the river basin must be long and funnel-shaped, its wide end facing the sea, and quite shallow, its depth regularly decreasing upstream. Bores become noticeable about eight tides before the spring tides, and they occur for about four days before and after these tides. As a rule, the greater the range of the tide, the greater the bore.

Figure 12.25 shows a graph of a tidal bore in the Petitcodiac River. It takes over nine hours for the tide to go out but only about two and a half hours for the tide to come in as a bore. The bore wave is formed at the turn of the tide. Since the water seaward of the bore is higher than the water in front of it, the bore is actually flowing downhill as it moves upstream.

The tidal bore is analogous to the sonic boom produced when aircraft travel faster than the speed of sound. It is formed when the incoming tide forces the front of the tide into a narrow region at a speed greater than the speed at which it would normally move. The water in most natural basins oscillates with frequencies peculiar to the particular basin (see Chapter 11). If the period of resonance of the basin coincides with the tidal period, the incoming tide is reinforced and the resultant bore is higher and more pronounced than it would otherwise be. While any river with a high tidal range may have a bore, resonance is associated with only about half of the world's bores. In the case of the Bay of Fundy, the 11-hour period of resonance is similar enough to the tidal period of 12 hours and 25 minutes for the 15 m (49 ft) tide to be reinforced and to produce a significant bore.

As the bore forms, faster-moving, deep-water waves enter a shallow, sloping region and change gradually into shallow-water waves with shorter wavelengths, much as wind-generated waves change before breaking on a coast. As the leading wave is shortened, waves to the rear overtake it to form the bore. The elliptical orbit of the water particles flattens, adding more energy to the moving water.

TIDAL PREDICTION

The forces that affect the tides are much more complicated than we have suggested. For example, the moon is not equidistant from the earth at all times. The difference in distance between apogee and perigee is about 28,000 km (17,396 mi). About twice a year, perigee coincides with spring tide conditions to produce tides that are the highest and lowest of the year—about a 20% differential. Knowing the seven phenomena listed in Table 12.1, we can predict the tide with about a 10% accuracy for one year. To predict the tides in harbors for two to three years ahead, we would need twenty to thirty types of data; to predict tides in river estuaries, we would need as many as 60 or more. Since tides are the result of complex periods and amplitudes, the longer we make observations and obtain data, the more accurate our predictions will be.

Tidal predictions are based on empirical measurements, and the predicted tide for a particular day and place may not agree with the tide that actually occurs. Table 12.2, one of a series published by the National Oceanic and Atmospheric Administration late each year for the next year, shows that the lower tide for Thursday December 20, is −0.3 m at 3:29 P.M. PST. But on that day the actual tide at San Francisco may not be as predicted due to meteorological conditions, such as storms, winds, and changes in atmospheric pressure.

Might not predictions for tides at a particular place be based on tidal theory instead of on long-term observations for hundreds of stations? At present, tidal theory is not precise enough to generate accurate predictions. It has, however, helped in making predictions about the course of tides from place to place and in supplementing direct measurements with mathe-

12.26 The Maremotrice Tidal Power Plant on the Rance Estuary in France. (W. J. Wallace)

matical relationships based on physical laws and principles. Tides are now predicted by using digital computers.

ENERGY FROM TIDES

Before the tremendous energy of tides can be harnessed to generate electrical power, several conditions must be met. There must be a large basin with a narrow entrance open to sea and a minimum tidal range of 5 m (16 ft). A dam containing two-way turbines would have to be built across the mouth of the bay to contain the tide waters as they come and go. Such generators would have certain advantages. They would draw on a constant source of driving power and cause comparatively little environmental pollution, although they might upset regional sediment balances and local marine communities. In contrast, most conventional generators either burn large amounts of coal and oil and release noxious fumes into the environment or produce radioactive waste.

Tidal mills were listed in the *Domesday Book*, ordered by William the Conqueror in 1085 as a census of everything and everybody in his realm. As early as 1650, tides were used to run grist mills in New England. So far, however, only a few modern attempts have been made to use tides as a source of power. One tidal power station has been built at Kislaya Bay near Murmansk in Russia.

In Brittany, France, a tidal power station located in the Rance estuary has been in operation since 1966, with production rising from about 3,000 MW·h in 1966 to over 500,000 MW·h annually now (Figure 12.26). That amount is roughly equivalent to the energy produced by burning 500 million tons of coal each year. The plant houses twenty-four generating sets, each driven by a four-bladed turbine operating at 93.75 rpm with up to 275 m^3/s of intake discharge. The Rance estuary narrows at the power station site to 750 m (2,460 ft) and has an average tidal range of 11.5 m (37.7 ft) and a maximum at equinox of 13.5 m (44.3 ft).

There has been a smaller pilot tidal power plant in Annapolis Basin since 1986. Although Passamaquoddy Bay in northern Maine and Cook Inlet in Alaska

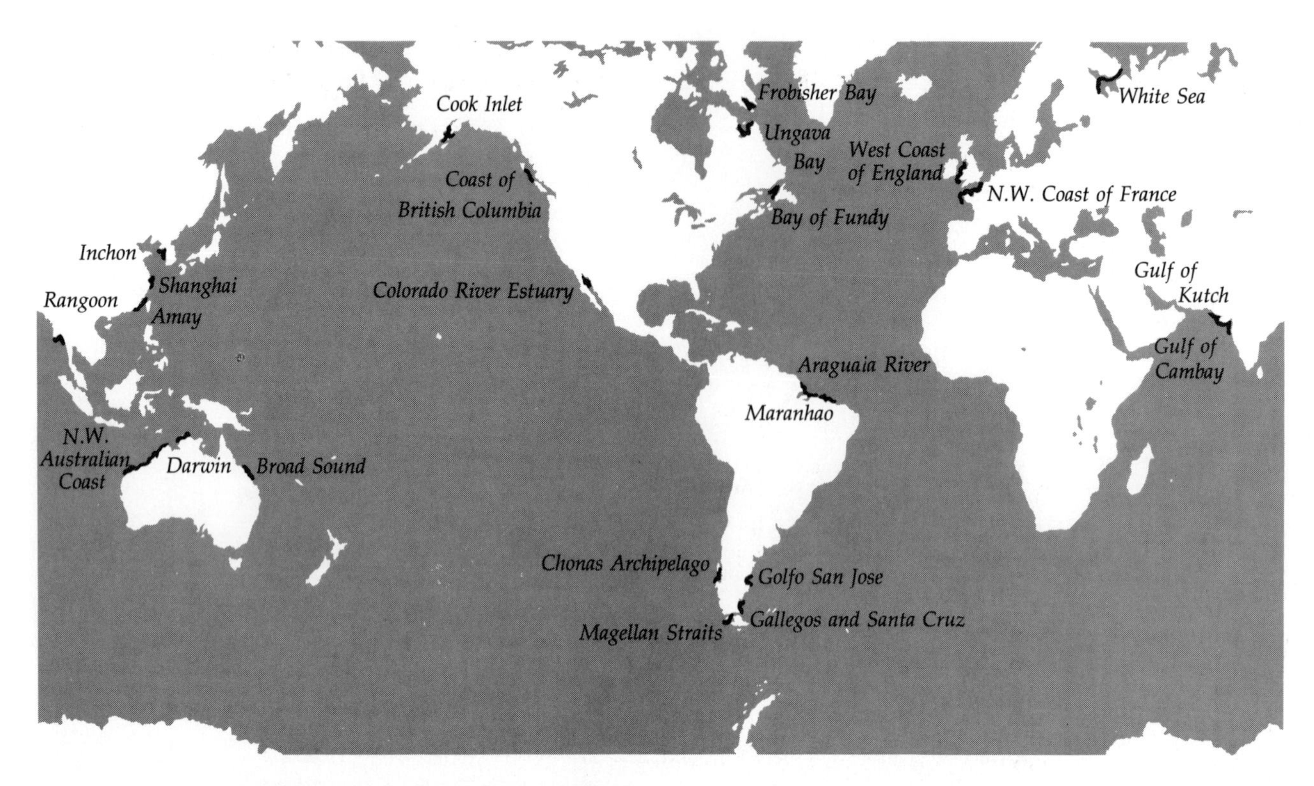

12.27 World sites with tidal ranges in excess of 5 m (in black), the only places where the development of tidal power is feasible. The only places where tidal power is now being exploited or tested are on the northwest coast of France at the mouth of the River Rance, the White Sea, and Digby, Nova Scotia.

would be ideal sites for such plants, there are few other areas where the tidal range is sufficient (see Figure 12.27). If all these regions were used, they could supply about 10% of the world's total electrical power requirements. Such dams are expensive to construct—more so than conventional or nuclear power stations—costing about $820,000 per linear meter of dam. At present the cost is not economically competitive with other methods. Still, the tides come and go faithfully and will offer a dependable source of energy as other energy sources dwindle.

Considerable friction is generated between the tides and the ocean bottom as water moves back and forth. In fact, it has been estimated that such friction is slowing the earth's rotation about its long axis by about one second every 120,000 years. The earth's day, in other words, is getting longer by a fraction of a second each century. The primordial molten earth (Chapter 2) may have rotated much faster than it does now, with one day lasting no more than a few hours. The kinetic rotational energy is dissipated into the ocean as heat. A recent calculation indicates that total use of the tide (2 billion kilowatts per year) would slow this rotation by about 24 hours in 2,000 years, but there is no reason why some tidal power stations cannot and should not be built.

FURTHER READINGS

Charlier, R. H. "Tides and Turbines." *Sea Frontiers*, 1980, *26*(6), 355.

Clancy, E. P. *The Tides*. Garden City, N.Y.: Doubleday, 1969.

Darwin, G. H. *The Tides and Kindred Phenomena in the Solar System*. San Francisco: W. H. Freeman, 1962.

Defant, A. *Ebb and Flow*. Ann Arbor: University of Michigan Press, 1968.

Goldreich, P. "Tides and the Earth-Moon System." *Scientific American*, April 1972, pp. 42–57.

Knauss, J. A. *Introduction to Physical Oceanography*. Englewood Cliffs, N.J.: Prentice-Hall, 1978.

Lawton, F. L. "Time and Tide." *Oceanus*, Summer 1974, pp. 30–37.

Nicholson, T. D. "The Tides." *Natural History*, 1959, *68*(6), 326–33.

Pond, S., and G. L. Pickard. *Introductory Dynamic Oceanography*. Oxford: Pergamon Press, 1978.

Redfield, A. C. *Introduction to Tides: The Tides of the Waters of New England and New York*. Woods Hole, Mass.: Marine Science International, 1980.

Tricker, R. A. R. *Bores, Breakers, Waves and Wakes*. New York: American Elsevier, 1964.

von Arx, W. S. *An Introduction to Physical Oceanography*. Reading, Mass.: Addison-Wesley, 1962.

Warburg, H. E. *Tides and Tidal Streams*. New York: Cambridge University Press, 1922.

Zerbe, W. B. "Alexander and the Bore." *Sea Frontiers*, 1973, *19*(4), 203–208.

Nature, to be commanded, must be obeyed.

Francis Bacon

T H I R T E E N

Coastal Processes and Estuaries

Throughout history, the abundant food and the agreeable climate of the coasts have been attractive to people. Major cities—Shanghai, Calcutta, Paris, Naples, London, Cairo—have grown up around natural harbors and along the rivers leading to the sea. Of the twenty largest cities of the world, only Moscow and Mexico City have no direct access to the ocean. Even in an age of air transportation, the port at land's edge remains the gateway for international trade.

The first European colonists in North America settled along the coast and used its estuaries and rivers to gain access to the interior. Those settlements are now major cities, and all the effects of urbanization are evident in these coastal areas. Consequently, much of our coastland has suffered environmental deterioration.

The United States has approximately 135,300 km (84,000 m) of shoreline. Of this land, 11% is federally owned, 12% is owned by states and municipalities, and 76% is privately owned. Coastal regions constitute only 6% of the country's land area, but 27% of the population lives in those regions and the percentage is increasing rapidly. Pressures on the coasts and their resources continue to mount.

COASTS AND COASTAL PROCESSES

The **shoreline** is the region bounded by high and low tides and by the farthest landward reach of waves (Figure 13.1). It exhibits the following features:

1. *Backshore*. The **backshore** is the area of the shoreline above the high-tide mark. It may contain large cliffs of exposed rock, sand dunes, or a **berm** (a flat upper beach).

2. *Foreshore*. The **foreshore** is the area that is exposed at low tide. The typical foreshore has a berm crest, marking the edge of the flat upper beach, and a beach face, which is a sloping section above the high-tide mark that is washed over by waves. The foreshore may have one or more beach **scarps** (a vertical slope produced by wave erosion) and a low-tide **terrace** (a broad, flat area exposed at low tide).

3. *Offshore Region*. The **offshore** region extends from the low-tide mark seaward beyond the wave-breaking zone. It has a **shoreface**, which is the slope below the low-tide mark. It has a **longshore trough**, which is a depression parallel to the beach between the low-tide mark and the wave-breaking zone, and a **longshore bar**, which is a sand ridge parallel to the beach (Figure 13.2). Waves often break on the longshore bar.

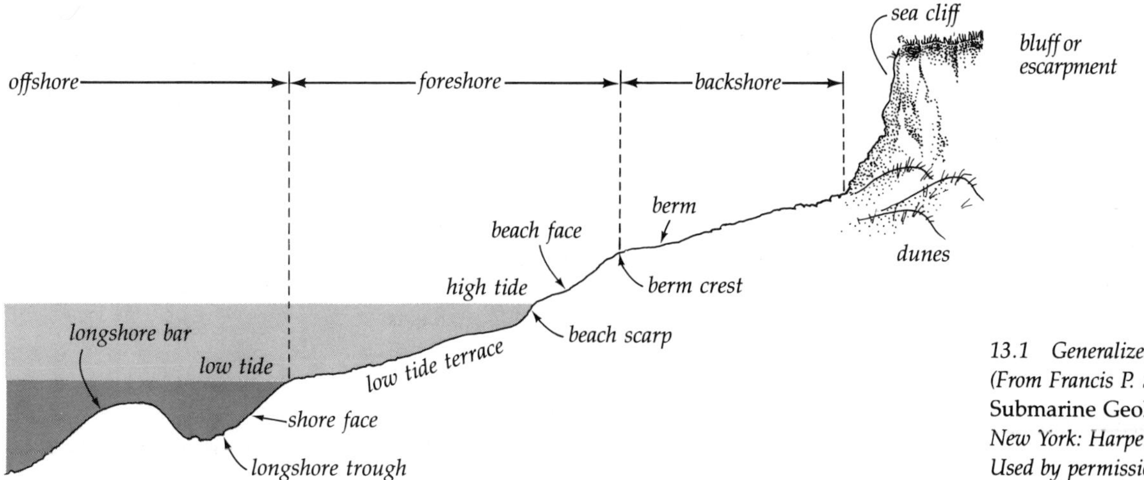

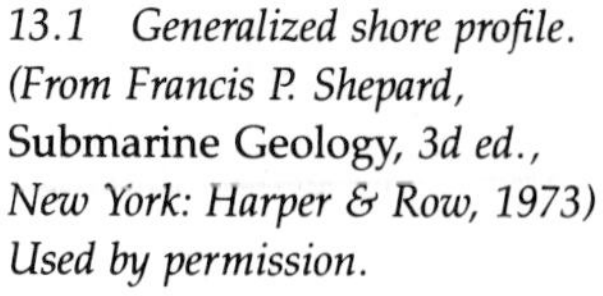

13.1 Generalized shore profile. (From Francis P. Shepard, Submarine Geology, *3d ed., New York: Harper & Row, 1973) Used by permission.*

13.2 The waves breaking in two bands parallel to this Oregon coast indicate the presence of two longshore bars. (W. J. Wallace)

The topography of the shoreline is intimately associated with the dynamic processes operating there. Several systems of classification are based on this interplay; Table 13.6 (p. 248) presents the system that we shall use in this discussion. Basically, five factors interact along the coastline: rock materials present along the shore, changes in sea level, the energy of certain natural processes, large-scale earth movements, and human activities.

Coastal Rock and Sand

Some materials along the coastline are more susceptible to weathering than others (Figure 13.3). **Granite,** a crystalline rock, is much more resistant to erosion than sandstone. Granite is composed mostly of the minerals quartz and feldspar, which are less likely to dissolve than olivine, which is common in basalt. Outcroppings of granite and other resistant rock produce rugged coastlines. The coast of Maine, for example, consists mostly of granite.

Regions made up of sand are quite susceptible to changes in configuration, which could be long-term erosion, or accretion, or just seasonal change. Deposits of unconsolidated sand produce smooth, shifting beaches. Quartz sand is particularly common in beaches along the continental shorelines. Some beaches, for example, those along the Gulf of Mexico from Tampa to Pensacola, consist of almost pure quartz sand. In southern California, several types of feldspar (a family of silicate minerals) are common, along with quartz and mica. At La Jolla, the oxidation of the iron present

(a) (b)

13.3 *(a) Sea cliff and boulder beach. Notice the alternating beds of shale and sandstone, which vary in resistance to erosion. Point Loma, San Diego, California. (b) Granite, weathering and eroding slowly. This outcrop is still sharp after thousands of years of pounding waves. Dyce's Head, Castine, Maine. (W. J. Wallace)*

in the mica gives the sand a gold cast. When the flakes of mica are picked up by waves and longshore currents, the water sparkles in the sun.

Sometimes minerals with high iron and magnesium content are present in beach sand. These minerals, which are generally black, brownish, or greenish, account for the dark patches often seen on beaches. The strong waves of the winter months leave heavy iron-containing minerals on the beach and remove quartz and feldspar. That is why many beaches appear darker in winter than in summer.

The sand of many tropical and subtropical beaches is predominantly calcium carbonate secreted by marine organisms. Beaches in the Florida Keys and the Bahama Islands, for example, consist almost entirely of crushed calcareous shells. Two types of algae, *Halimeda* and *Penicillus*, contribute small, flat platelets to the sand, forming a "lawn" of shallow "grass." The shells of foraminifera, mollusks, and echinoids form the bulk of calcareous sand. The black beaches of the Hawaiian Islands and Iceland consist of broken, abraded particles of basalt lava brought by rivers from the interior or by direct lava flow into the ocean.

Valuable minerals are sometimes found in deposits of beach sand. Gold has been mined from the beach sand at Nome, Alaska, and diamonds have been recovered from ancient, or relict, beaches in South Africa and Namibia.

Changes in Sea Level

Fluctuations in sea level directly influence coastlines. The tides produce short-term fluctuations. For example, along most of the U.S. coasts, tidal variations are less than 2 m (6.6 ft) daily, and along the western coast of Florida they are less than 1 m (3.3 ft). In some areas, however, they may be as great as 9 m (30 ft) and expose entire harbors or broad sand flats every day (Figure 13.4). Tidal fluctuations of that magnitude may cut channels in the bottom and carry off tons of sand and mud.

Powerful tsunamis and other major disturbances of the surface waters produce sudden changes in sea level, often leading to catastrophic flooding and changes in coastal configuration. Tsunamis have been altering the shorelines and bays of Japan throughout recorded history. Along the shorelines of the eastern United States and the Gulf of Mexico, the strong winds and high waves of hurricanes create great fluctuations. High water levels are also generated by extreme low-pressure systems, which may cause the local water level to rise from 1 to 7 m (3 to 23 ft) above normal. The flooding and changes in the shoreline resulting from such fluctuations in sea level are appreciable, especially in areas with nonresistant rocks and low topography.

Long-term changes in sea level are also important in determining shoreline features. In many places and at many times over the past million years, the sea level has been at least 100 m (328 ft) below its present level and at least 30 m (99 ft) above. These shifts are thought to have resulted from continental glaciation, interglacial episodes, and tectonic movements of the earth's crust. During widespread glaciation, when water was locked up in ice on the continents, the sea level was low; when the ice melted, the sea level rose. The rising water carved platforms in the sea cliffs and deposited shell fossils (skeletons) of marine organisms on the

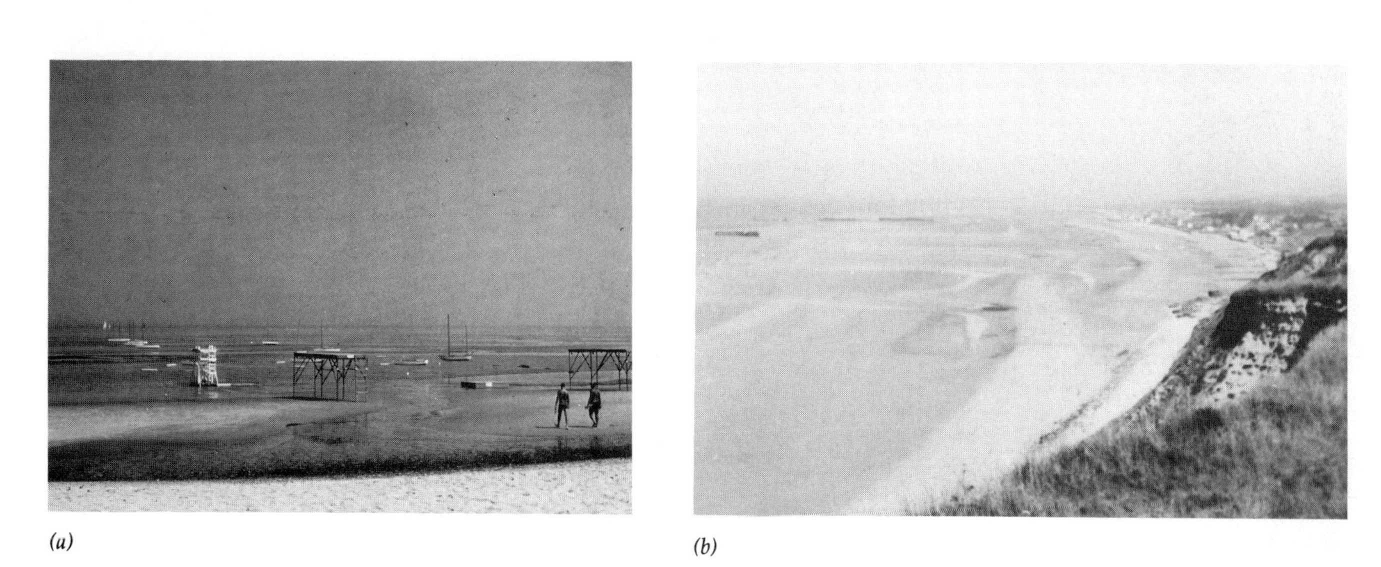

(a) (b)

13.4 *(a) Tidal flats at East Brewster, Cape Cod, Massachusetts. Notice the diving towers in the foreground, which are left high and dry at low tide. (D. Ingmanson) (b) The wide beaches of Normandy in France at low tide. The offshore structures are concrete remnants of the British artificial harbor Mulberry B, built during the D-Day invasion. (W. J. Wallace)*

terraces. When the sea level was low, plant remains, the teeth of mammoths, and Indian shell middens were left on land now covered by the ocean. In low-lying coastal areas, even a slight change in the water level will move the shoreline appreciably. For a coast with a 1° slope, a 20-cm (8-in) rise in sea level would move the shoreline 27.5 m (90 ft) inland.

During the past 15,000 years, after the most recent period of continental glaciation, the sea level has tended to rise steadily, with small-scale fluctuations (Figure 13.5 and Table 13.1). The ice sheet on Greenland continues to melt, and the average temperature of ocean water may be rising.

In the past 100 years, sea level has risen more than 30 cm (1 ft)—faster than at any rate in the past millennium. Recent studies predict a sea-level rise of between .5 and .9 m (1.6 and 3 ft) in the next century and between .6 and 3.3 m (2 and 11 ft) by the year 2100. The historically based prediction for rise in sea-level value for the year 2100 is only .1 to .8 m (.3 to 2.6 ft).

Energy Acting on Coasts

The approximate total length of the world's shorelines is 444,000 km (275,000 mi). As Table 13.2 shows, the length of coastal land increases greatly when the tidal shoreline and its irregularities are taken into consideration. The processes acting on the coastline are affected by topography, exposure to wave attack, accessibility of the coast, availability of sediment, tidal range, velocity of tidal currents, and weather. Most of the energy that drives these coastal processes comes from waves and tides. As waves formed by storms in the open ocean strike the coast, their energy is concentrated and dissipated as breakers.

In the continental United States, the percentage of shoreline that consists of beach varies from 2% in Maine to 79% in Washington (see Table 13.3). A **beach** is a region between the high and low waterlines that is covered by sand or some other unconsolidated material. Sand particles, classified as shown in Table 13.4, may be composed of anything from quartz or feldspar to shell fragments. In temperate latitudes, most beaches consist of very fine to very coarse sand eroded by rivers and streams from mountains. The erosion of local cliffs and rocks by waves and currents probably accounts for no more than 5% of beach sand, although the percentage varies greatly with the hardness of the rock and the availability of a sandy beach buffer.

Turbulent wave-driven currents along the shore usually remove mud from the beaches and hold it in suspension. As the fine particles of mud are transported to deeper or less turbulent water, they settle to the bottom. Beaches where wave action is very mild, however, may consist entirely of mud and may be termed **mudflats**.

A beach is described in terms of the average size of its sand particles, the range and distribution of those particles, the elevation and width of the berm, the slope of the foreshore, and the slope of the backshore. Generally, the larger the sand particles, the steeper the beach.

Waves are primarily responsible for moving sand away from the river mouths and along the coast. As a wave breaks, the sudden release of energy within a small area causes turbulence that dislodges sand particles (Figure 13.6). If all the waves rolled in perfectly

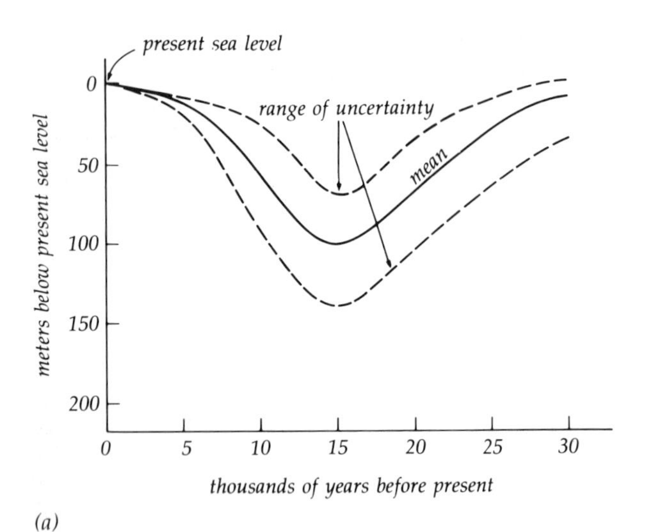

(a)

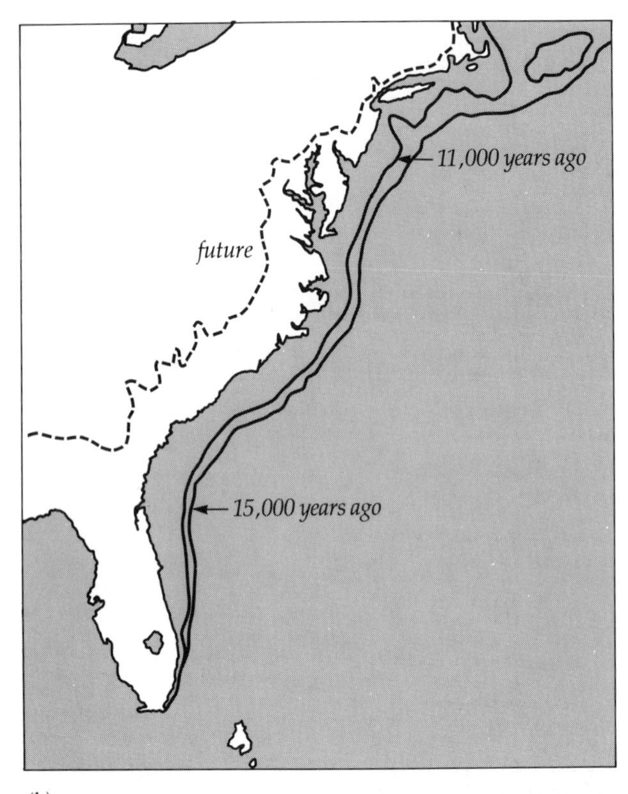

(b)

(c)

13.5 (a) Changes in sea level over geologic time. The dashed lines indicate the range of data from different geographic locations. Changes in sea level can be inferred from the radiocarbon ages of shallow-water marine organisms and the depth at which they were recovered. (b) The Atlantic coast shoreline has varied greatly in the past and will undoubtedly continue to do so in the future. This illustration compares the shorelines of 15,000 and 11,000 years ago with the probable shoreline if all the ice at the poles were to melt. Confirmation that the continental shelf was once laid bare is found in discoveries of elephant teeth, freshwater peat, and the shallow-water formations called oolites. (After K. O. Emery, in Late Cenozoic Glacial Ages *by Karl K. Turekian, ed. Copyright © 1971 by Yale University Press) (c) Wave-cut marine terraces on the western side of San Clemente Island off the southern California coast. (John S. Shelton) Used by permission.*

perpendicularly to the beach, the sand grains would only shift back and forth. However, waves approach the coast from almost any angle. When a wave strikes the coast from, say, the north, the water runs back to the sea in a southerly direction (Figure 13.7a), forming what is known as a **longshore current**. The movement of such currents provides the longshore transport (or littoral drift) that carries the sand along the shore. Beach sand is always in transit from one place to another.

Short, steep waves, usually formed by winter storm winds near the coast, tend to flatten, lower the slope of, or redistribute beaches (Figure 13.7b). However, the long swell that comes ashore from distant storms tends to rebuild them. Most beaches undergo alternating periods of erosion and rebuilding. On some beaches, erosion and accretion are seasonal; the winter storms erode them and the summer swells rebuild them. That is why beaches are often described as summer or winter beaches (Figure 13.8). Beaches may also follow long-term cyclic patterns, with several years of erosion followed by several years of accretion.

The direction and size of incoming waves determine the direction and magnitude of longshore trans

Table 13.1 *Changes in Sea Level Observed at Selected U.S. Coastal Points*

Location	cm/decade	in/century
Northeast		
Portland, ME	2.2	8.66
Eastport, ME	3.3	12.99
Boston, MA	2.8	11.02
Woods Hole, MA	3.3	12.99
Newport, RI	3.0	11.81
New London, CT	2.6	10.24
Montauk, NY	2.6	10.24
New York, NY	2.9	11.42
Sandy Hook, NJ	4.9	19.29
Atlantic City, NJ	4.1	16.14
Lewes, DE	3.7	14.57
Philadelphia, PA	2.8	11.02
Annapolis, MD	4.2	16.54
Solomons, MD	4.0	15.75
Norfolk, VA	4.7	18.50
Southeast		
Charleston, SC	3.8	14.96
Savannah, GA	3.1	12.20
Miami Beach, FL	2.6	10.24
Gulf		
Key West, FL	2.3	9.06
Pensacola, FL	2.7	10.63
Galveston, TX	6.3	24.80
Pacific		
La Jolla, CA	1.7	6.69
Los Angeles, CA	0.5	1.97
San Francisco, CA	1.3	5.12
Astoria, OR	−0.1	−0.39
Seattle, WA	1.9	7.48
Friday Harbor, WA	1.0	3.94
Juneau, AK	−13.4	−52.76
Ketchikan, AK	−0.2	−0.79

Source: Modified from Steacy D. Hicks, "An Average Geopotential Sea Level Series for the United States," *Journal of Geophysical Research*, 1978, *83*, 1377–79.

port. For instance, on a coast facing east, violent storm waves from the northeast produce a large littoral, or intertidal, transport toward the south. Conversely, mild wave action out of the southeast results in a much lower rate of littoral transport to the north. However, if the waves from the southeast persist over a long time, their effect may predominate. Although weather patterns affect the direction of longshore transport, longshore transport along most coasts runs consistently in one direction. Along the East Coast and the West Coast of the United States, where most waves come from northern storms, the sand is transported

Table 13.2 *Lengths of Coastline and Tidal Shoreline for the United States and its Possessions*

	Length (mi)	
Area	General Coastline	Tidal Shoreline
UNITED STATES	**12,383**	**88,633**
Coterminous only	4,993	53,677
Atlantic Ocean	**2,069**	**28,673**
Maine	228	3,478
New Hampshire	13	131
Massachusetts	192	1,519
Rhode Island	40	384
Connecticut	—	618
New York	127	1,850
New Jersey	130	1,792
Pennsylvania	—	89
Delaware	28	381
Maryland	31	3,190
Virginia	112	3,315
North Carolina	301	3,375
South Carolina	187	2,876
Georgia	100	2,344
Florida (Atlantic only)	580	3,331
Florida (Atlantic and Gulf)	1,350	8,426
Gulf of Mexico	**1,631**	**17,141**
Florida (Gulf only)	770	5,095
Alabama	53	607
Mississippi	44	359
Louisiana	397	7,721
Texas	367	3,359
Pacific Ocean	**7,623**	**40,298**
California	840	3,427
Oregon	296	1,410
Washington	157	3,026
Hawaii	750	1,052
Alaska (Pacific only)	5,580	31,383
Alaska (Pacific and Arctic)	6,640	33,904
Alaska (Arctic only)	1,060	2,521
EXTRATERRITORIAL		
Atlantic Ocean		
Nayassa	5	5
Puerto Rico	311	700
Swan Islands	8	10
Virgin Islands	117	175
Pacific Ocean		
Baker Island	3	3
Guam Islands	78	110
Howard Island	4	4
Jarvis Island	5	5
Johnston Island	5	5
Midway Island	20	33
Palmyra Island	9	16
Samoa Islands	76	126
Wake Island	12	20

Source: National Oceanic and Atmospheric Administration.

Table 13.3 *Length of Beach of the Continental United States*

State or Region	Length of Beach (mi)	% of Total Shoreline
Maine	60	2
New Hampshire	25	63
Massachusetts	940	78
Rhode Island	185	54
Connecticut	145	54
New York	331	52
New Jersey	215	46
Delaware	76	34
Maryland	46	2
Virginia	294	30
North Carolina	1,269	35
South Carolina	196	6
Georgia	102	50
Florida (Atlantic)	390	15
Florida (Gulf)	968	26
Alabama	227	65
Mississippi	97	39
Louisiana	835	43
Texas	377	15
California	412	23
Oregon	300	60
Washington	1,847	79
Total	**10,983**	**30**
Maine to Virginia	2,320	27
North Carolina to Florida (Atlantic)	3,600	25
Florida (Gulf) to Texas	2,504	29
Pacific	2,559	55

Source: U.S. Army Corps of Engineers, *The National Shoreline Study* (1971).

Table 13.4 *Beach Sediment and Average Beach Face Slopes*

Type of Beach Sediment	Size (mm)	Average Slope of Beach Face
Very fine sand	.0625–.125	1°
Fine sand	.125–.25	3°
Medium sand	.25–.50	5°
Coarse sand	.50–1.0	7°
Very coarse sand	1–2	9°
Granules	2–4	11°
Pebbles	4–64	17°
Cobbles	64–256	24°

Source: F. P. Shepard, *Submarine Geology,* 3d ed.

south at a rate of 150,000 to 1,600,000 m^3/yr. In such land-locked bodies as the Great Lakes, the rate of longshore transport is normally only about 114,700 m^3 (150,000 yd^3) per year. The rate of transport along any coast depends on local shore conditions and on the energy and direction of the wave action. At several places along the southern California coast (Figure 13.9), there are no beaches at all because submarine canyons divert the longshore drift offshore.

In semicontained coastal regions with moderate to high tidal ranges, such as Puget Sound, tidal currents may create longshore transport that produces both erosion and deposition, especially in shallow water. Strong local winds may produce the same results by picking sand up from tidal flats during low tide and moving it shoreward during high tide, depositing it as dunes or filling in lagoons.

13.6 *Suspended sand and sediments in the surf zone. (Scripps Institution of Oceanography)*

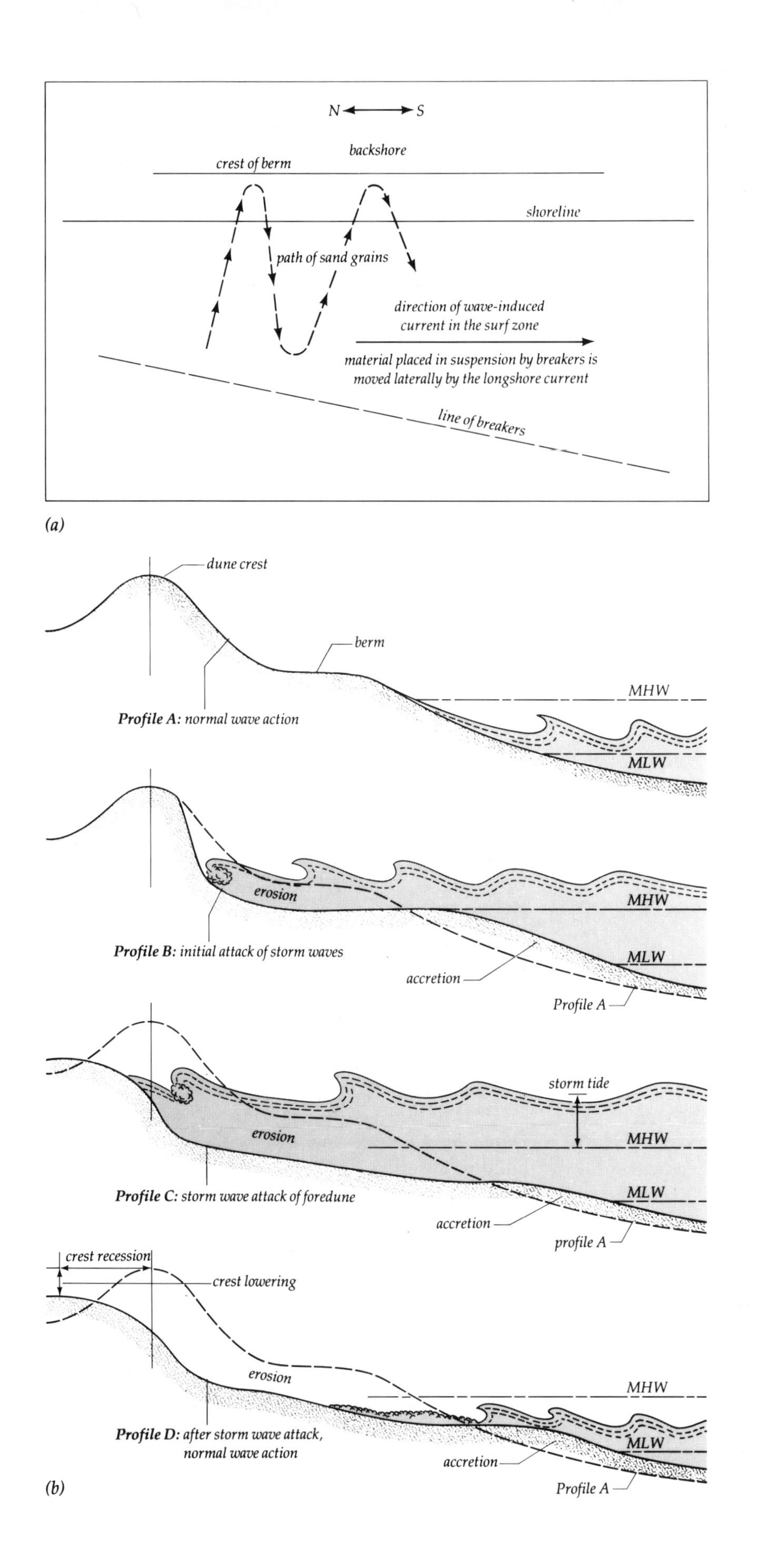

13.7 *(a) Sand-grain pathways along a beach as a function of the direction of wave approach. (b) Change in normal beach profile caused by storms.*

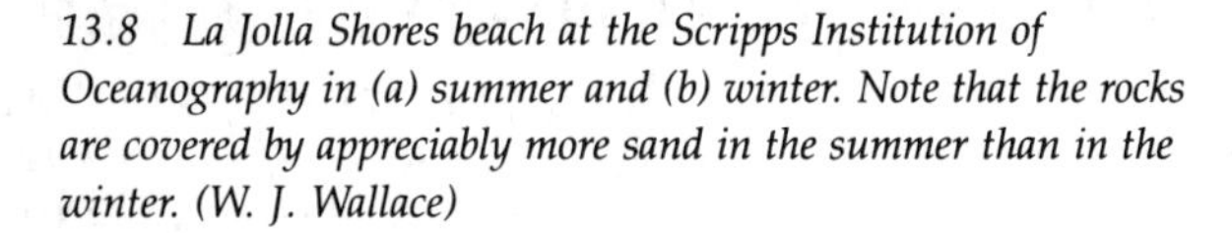

13.8 *La Jolla Shores beach at the Scripps Institution of Oceanography in (a) summer and (b) winter. Note that the rocks are covered by appreciably more sand in the summer than in the winter. (W. J. Wallace)*

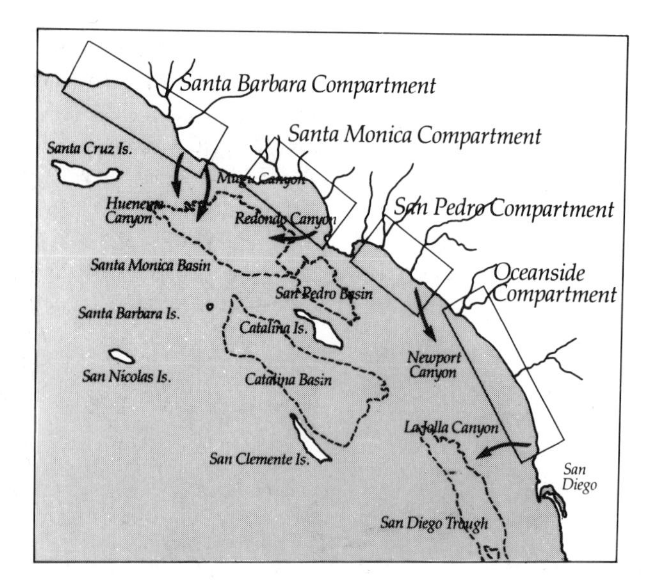

13.9 *The compartmentalization of beaches along the southern California coast, illustrating the loss of sand into the heads of submarine canyons at the southern end of each compartment. (From Inman and Frautschy, "Littoral Processes and the Development of Shorelines."* Coastal Engineering, *Santa Barbara Conference, 1966)*

Large-Scale Earth Movements

As we saw in Chapter 5, the earth, especially its surface, is in constant flux. The movement of crustal plates that affects entire continents affects their coasts as well. Inman and Nordstrom (1971) have classified coasts as collision coasts, trailing-edge coasts, and marginal coasts. **Collision coasts** (or subduction coasts) occur on the collision side of continents and island arcs. They are characterized by narrow continental shelves, volcanism, and active earthquakes. We find the world's youngest mountains along these coasts. **Trailing-edge coasts** occur where the coasts of continents and islands are moving away from mid-oceanic ridges. **Marginal coasts** occur on the protected landward side of island arcs.

Sometimes earlier classifications of coasts are used, such as, submergence coasts, emergence coasts, and neutral coasts. A collision coast may experience submergence when a sufficiently large change in sea level is superimposed on the downwarp or uplift. Emergence and submergence have a marked effect on coastlines.

Human Activities

Human activities have a significant effect on coastlines, sometimes beneficial, sometimes harmful. The activities that have the most profound effects are the damming of rivers, land reclamation programs, the dredging of inlets, the development of dune areas, and the construction of erosion control structures.

Dams Politicians and engineers who set about building reservoirs, hydroelectric power plants, and flood control structures often overlook the effect that damming a river will have on the shoreline (see Figure 13.10). Although a few rivers terminate in desert basins or in inland seas or lakes, most empty into the ocean. They deposit their sediment in the nearshore zone, where it is picked up by waves and transported by longshore currents. When a river is dammed, however, deposition occurs upstream from the dam. Thus the supply of sediments carried by coastal currents is less than normal, which results in **starved currents** (starved longshore transport). The starved longshore currents then tend to pick up sand from the foreshore and offshore zones, causing extreme erosion of the

13.10 Flaming Gorge Dam in Wyoming-Utah: a sediment trap. (U.S. Department of the Interior, Bureau of Reclamation photo by F. B. Slote)

beach and shoreline (Figure 13.11). Much of the Washington, Oregon, and California shorelines are undergoing extensive erosion largely as a result of river damming.

Land Reclamation Land reclamation is a costly endeavor, both ecologically and financially. In the Netherlands, hundreds of square kilometers of shallow coastal marshes and bays have been transformed into fertile farmland by means of dikes, dredging, sediment relocation, and pumping out salt water. Even so, storms occasionally break through the dikes and cause extensive flooding and loss of life. Although land reclamation may be an economic necessity in some places, it inevitably destroys marshes and bays—the regions that support the animal and plant life that provides much of the food in the sea. In addition, many commercial foods—shrimp, crabs, birds, and fish—depend to some degree on the stability of marshes and bays.

Economic needs must be carefully balanced against ecological needs when planning land reclamation programs (Table 13.5). In San Francisco, those who want to fill in certain areas of the bay have clashed with those who want to preserve it, resulting in a moratorium on landfill there. In New England, starting in the 1700s with the filling in of Boston's Back Bay, coastal marshes have been filled at a rapid rate. They are now protected in a number of communities, including Guilford, Connecticut.

Dredging In the past forty years, inlet dredging has become increasingly widespread. The popularity of pleasure boating has prompted many communities to seek the assistance of the U.S. Army Corps of Engineers in easing access to open water. Here again, economic or recreational needs must be balanced against ecological needs.

The dredging of a channel in San Diego Bay, which had no effect on known wildlife habitats, provided the

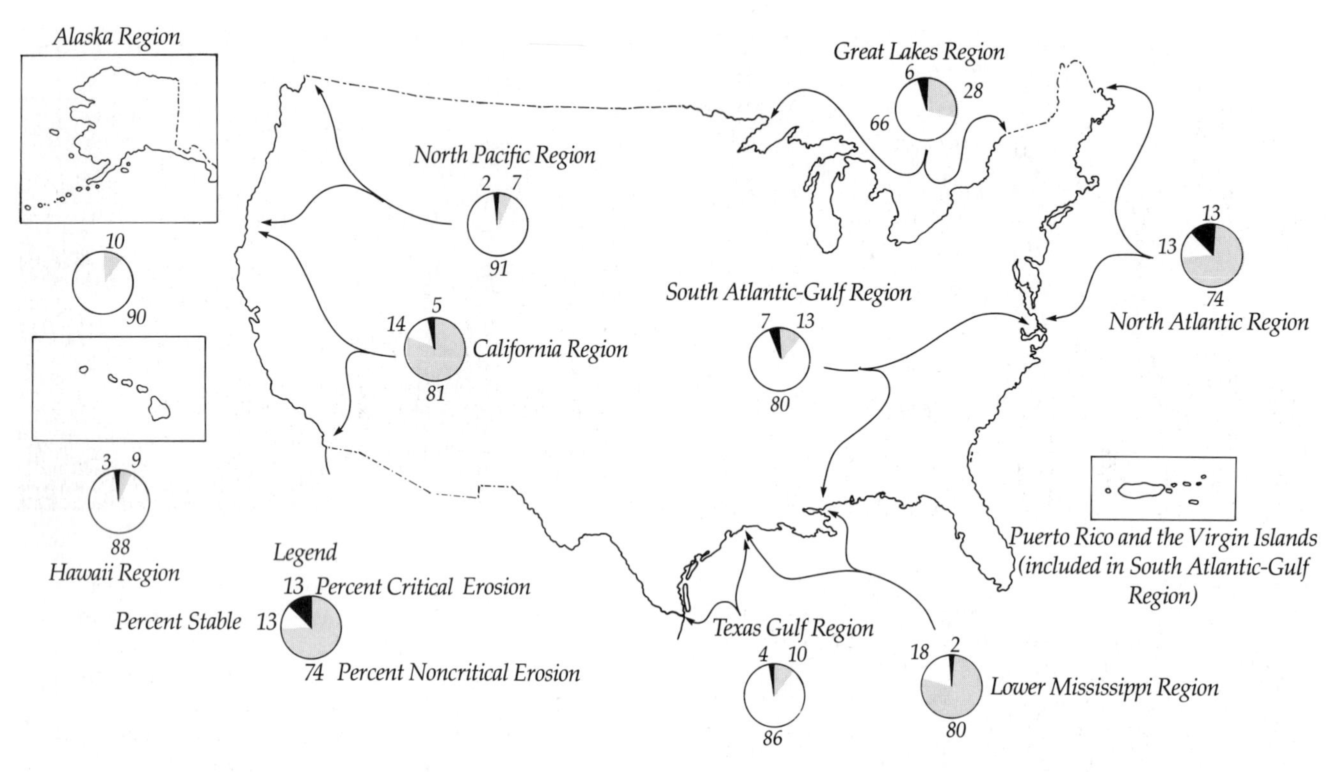

13.11 *Shore erosion by regions.*

Table 13.5 *Coastal Wetlands by Type: 1780, 1954, and 1978*

Wetland Type	Existing Wetlands (millions of acres)			Annual Loss of Wetlands (acres)	
	1780	1954	1978	1780–1954	1954–1978
12 to 14	6.0	4.0	2.0	11,500	83,000
15 to 18	4.5	3.7	3.2	4,800	20,800
20	0.5	0.5	0.5	0	0
Total	11.0	8.2	5.7	10,300	103,800

Source: U.S. Army Corps of Engineers.

U.S. Navy with its largest Pacific port facility and encouraged military and commercial shipping in the city (Figure 13.12). In contrast, the dredging of a channel in Matanzas Inlet, Florida (between Daytona Beach and St. Augustine) to enhance land values at Crescent Beach and Marineland had disastrous consequences. Tidal currents rushing from the new channel have diverted the strong longshore current, causing sediment to be deposited on the inlet's north shore. As the starved longshore current swept by the south shore, it rapidly eroded the shorefront property and a highway to boot! The beach has now been extended inland, but no further corrective action has been taken (Figure 13.13).

Development of Dune Areas The commercial or recreational development of dune areas has also deteriorated the ecology (Figure 13.14). Dunes are impermanent formations that rarely survive the building of houses and other structures. Moreover, the sparse vegetation that provides them with some slight stability is readily destroyed by dune buggies. As the dunes are destroyed, the shoreline migrates deeper and deeper into the land.

Erosion Control Structures **Jetties** built to prevent longshore currents from filling inlets or to bring about the deposition of sediments (Figure 13.15) benefit the

(a)

(b)

(c)

13.12 *Examples of the impact of human activities in coastal zones. (a) San Diego Bay, California, in 1969. The two barrier islands at lower left are landfill. The North Island Naval Air Station at center was at one time a saltwater marsh. (Courtesy City of San Diego) (b) Mission Bay, San Diego, in 1948. (Historical Collection, Title Insurance and Trust Co., San Diego) (c) The multimillion-dollar Mission Bay Aquatic Park in 1976—1,862 hectares of recreation land and facilities. (Courtesy City of San Diego) Compare the configuration of the bay entrance in (b) and (c). Much of the fill came from dredge spoils.*

(a)

(b)

13.13 *(a) Pumping station on the north side of Boynton Inlet, Florida. (b) Sand and water slurry draining into the south side of the inlet. (D. W. Lovejoy)*

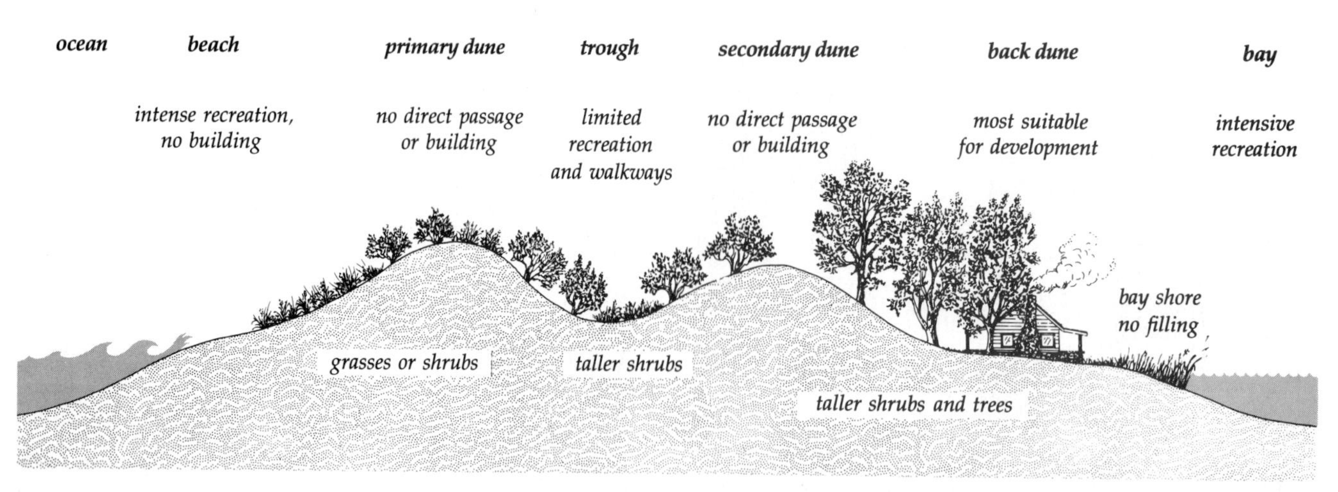

(a)

(b)

(c)

13.14 (a) Primary and secondary dunes offer natural protection from flooding. Ideally, construction and development should be allowed only behind the two dunes, and walkways should be built over the dunes to the beach. Dune grasses are intolerant of foot traffic, much less motor vehicles. (b) The Oregon Dunes area. Here the sand dunes are well back from the sea. (W. J. Wallace) (c) Detail photo of Oregon Dunes area. (Oregon Transportation Dept.)

property immediately upcurrent but often cause the downcurrent shoreline to recede. Property owners often make the mistake of building **groins**, small piles of resistant rocks, at right angles to the beach. Such groins transform the longshore current into a zigzag current that destroys the linear form of the coast. Wherever the current encounters nonresistant rock, it inevitably erodes the shore (Figure 13.16).

The precarious stability of the shoreline and of the transition zone between the terrestrial environment and the delicate marine environment suggest that the shoreline should be left undeveloped except where development poses no hazards. In most states, developers must now consider the long-term consequences of massive shoreline development programs before proceeding.

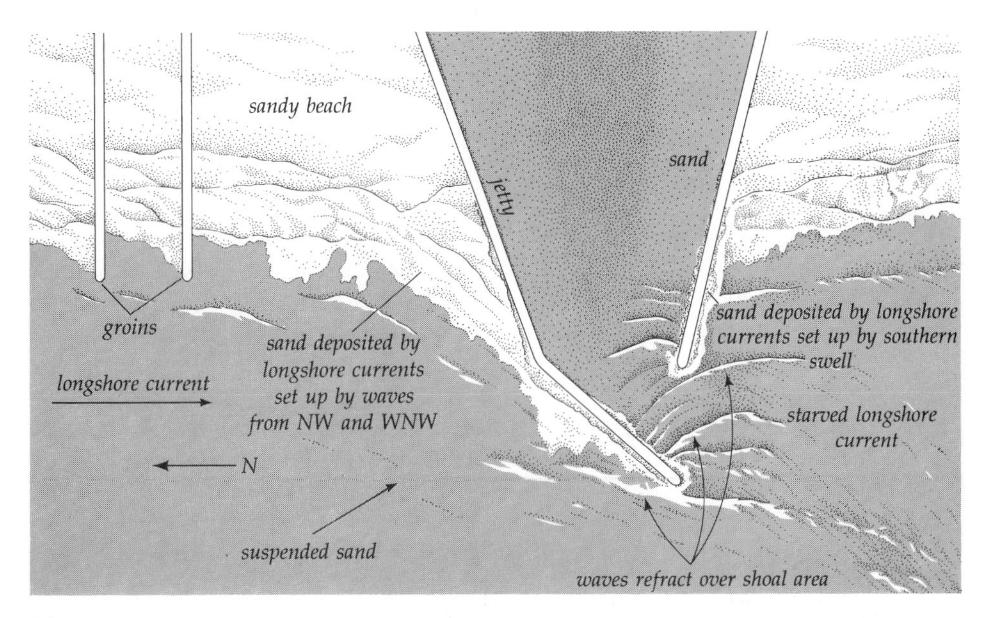

(a)

(b)

13.15 (a) Sedimentation and wave patterns around groins and a jetty system. Notice how the movement of sand by the longshore current has been altered—a pattern evident in (b), which shows the jetty at Provincetown, Massachusetts. (U.S. Department of the Interior, National Park Service)

CLASSIFICATION OF COASTS

Coastlines may also be classified according to dominant geological processes. Some coastlines are erosional, for example, and others are depositional. An emergent coast is a coast of rising topography, whereas a submergent coast is a **drowned coast**. Coasts can also be described as advancing or retreating. The scheme proposed by Francis Shepard (Table 13.6) classifies coasts according to two categories: A **primary coast**, or terrestrial coast, is a coast that was shaped by nonmarine processes and that is essentially the same as it was when the last rise in sea level came to an end. A **secondary coast**, or marine coast, is a coast that has been appreciably modified by such marine processes as wave action after the sea level stabilized.

(a)

(b)

13.16 *(a) Attempt to halt cliff erosion in San Diego, California. (W. J. Wallace) (b) Attempt to prevent wave damage to apartments built on dunes in Imperial Beach, California. (D. Ingmanson)*

Table 13.6 *Classification of Coasts*

I. Coasts Shaped by Nonmarine Processes (Primary Coasts)
    A. Land erosion coasts
        1. Drowned rivers
        2. Drowned glacial-erosion coasts
            a. Fjord (narrow)
            b. Trough (wide)
    B. Subaerial deposition coasts
        1. River deposition coasts
            a. Deltas
            b. Alluvial plains
        2. Glacial-deposition coasts
            a. Moraines
            b. Drumlins
        3. Wind deposition coasts
            a. Dunes
            b. Sand flats
        4. Landslide coasts
    C. Volcanic coasts
        1. Lava flow coasts
        2. Tephra (ash) coasts
        3. Coasts formed by volcanic collapse or explosion
    D. Coasts shaped by earth movements
        1. Faults
        2. Folds
        3. Sedimentary extrusions
            a. Mud lumps
            b. Salt domes
    E. Ice coasts

II. Coasts Shaped by Marine Processes or Marine Organisms (Secondary Coasts)
    A. Wave erosion coasts
        1. Straightened coasts
        2. Irregular coasts
    B. Marine deposition coasts (prograded by waves, currents)
        1. Barrier coasts
            a. Sand beaches (single ridge)
            b. Sand islands (multiple ridges, dunes)
            c. Sand spits (connected to mainland)
            d. Bay barriers
        2. Cuspate forelands (large projecting points)
        3. Beach plains
        4. Mud flats, salt marshes (no breaking waves)
    C. Coasts formed by biological activity
        1. Coral reef, algae (in the tropics)
        2. Oyster reefs
        3. Mangrove coasts
        4. Marsh grass
        5. Serpulid worm reefs

After Francis P. Shepard, *Submarine Geology*, 3d ed.

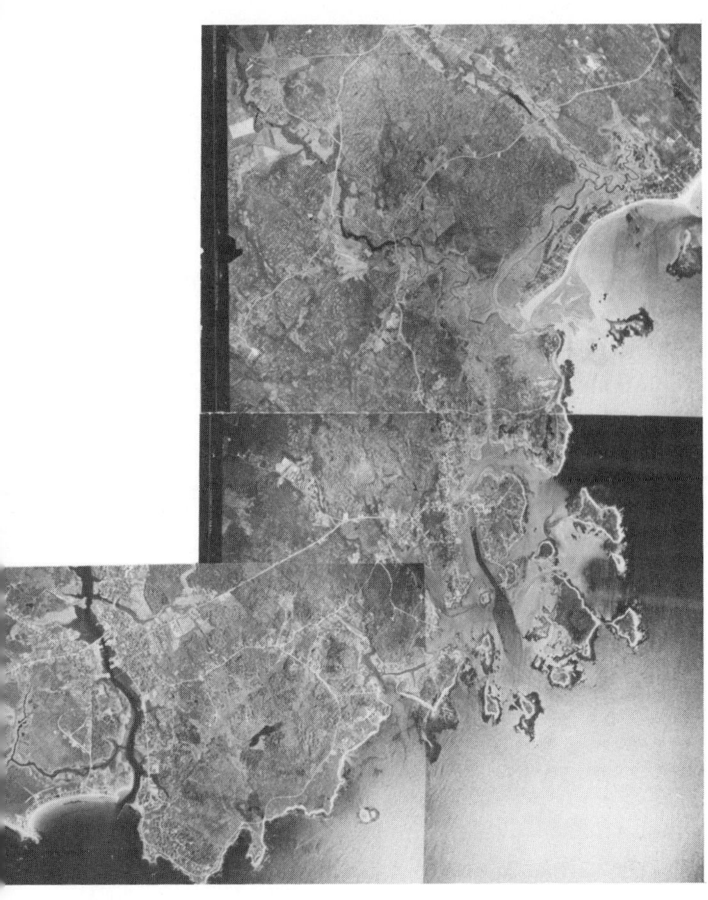

(a)

(b)

(c)

(d)

13.17 (a) Drowned river valleys, or estuaries, along the southwestern coast of Maine, Kennebunk–Cape Porpoise area. (Francis P. Shepard) (b) An estuary on Long Island behind a barrier island. (D. Ingmanson) (c) The Grays Harbor estuary in Washington at low tide with exposed mud flats. (W. J. Wallace) (d) The Nehalem River estuary in Oregon with a barrier, or baymouth, bar. (W. J. Wallace)

This system of classification is simple and entails relatively few technical terms. Moreover, it provides a convenient way of identifying coasts. We shall therefore use Shepard's designations.

Coasts Shaped by Nonmarine Processes

Fresh water flowing from the land to the sea can erode indentations into a coast, creating a river-cut coastal valley that can become flooded by the sea after postglacial melting. Such valleys, called **estuaries**, are usually fairly shallow, have a V-shaped cross section, and vary greatly in size (Figure 13.17).

Glacial ice carves steep U-shaped indentations usually perpendicular to the coast that may be as deep as 100 m (328 ft). Small, narrow indentations are called **fjords** (Figure 13.18); larger ones are called sounds (as in Puget Sound) or **troughs**. These indentations have

(a)

(b)

13.18 A cross-sectional, U-shaped valley cut by a glacier (a) and subsequently flooded by the sea (b) is called a fjord. (a) Steen Mountains of Oregon. (b) Sommes Sound in Maine, that state's only true fjord. (both photographs by W. J. Wallace)

13.19 The complicated coast of Maine, formed by glacial deposition and erosion. These islands are depositional. (W. J. Wallace)

a deep inner region and a shallow entrance, or sill. Large numbers of islands are frequently found seaward of deeply indented coasts (Figure 13.19).

Rivers and glaciers deposit sediment in coastal regions. River deposits extend the coastline by forming deltas, named after the Greek letter because of their triangular shape (Figure 13.20). Many of the world's great rivers deposit vast amounts of silt and clay every year and extend their deltas seaward in complex patterns.

Glaciers push large amounts of material in front of them. When they melt and recede, they leave extensive deposits of debris called **moraines**, ranging from sand to massive rocks. This process is analogous to a bulldozer leaving a pile of earth in front of its blade when it backs up. These deposits range in size from Long Island to the islands of Martha's Vineyard and Nantucket. **Drumlins** are small deposits such as the oval-shaped islands in Boston Harbor and a few of the coastal hills of eastern New England (Figure 13.21). Winds create extensive areas of sand dunes along the world's coasts (Figure 13.22), usually landward of the beach.

Many coasts are unstable, eventually succumbing to gravity and collapsing. A landslide (Figure 13.23) can often be recognized by the presence of an irregular terrace at or near the base of a cliff formed by a slumped mass.

Volcanic coasts have a characteristic profile. Most of the islands in the open ocean are of volcanic origin. An apron of solidified lava that surrounds a volcanic island forms a lava-flow coast. Coasts consisting of volcanic ash are called **tephra** coasts, which are particularly susceptible to erosion. A volcanic collapse or a volcanic explosion creates a coast with a circular indentation (Figure 13.24).

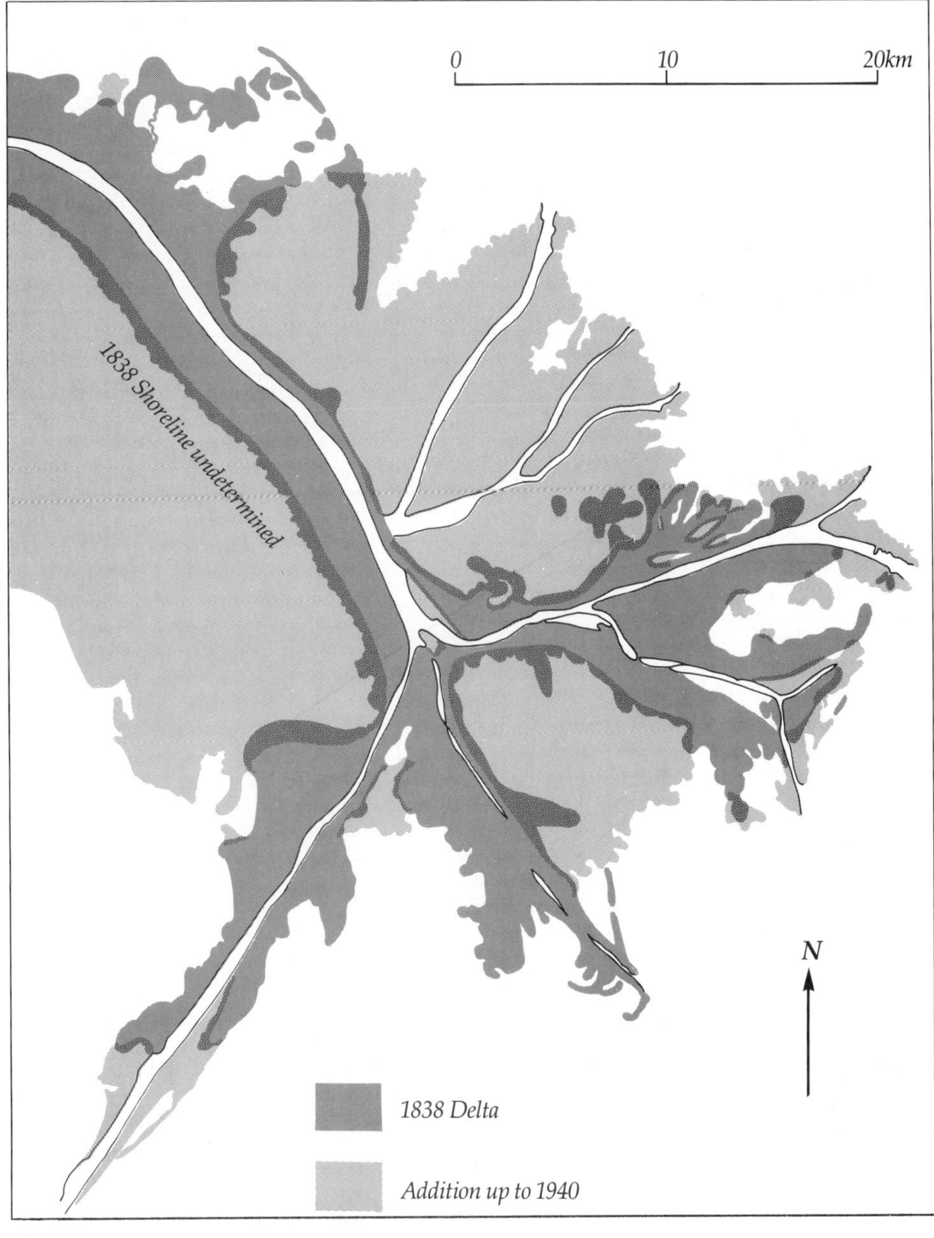

(a)

(b)

13.20 (a) Changes in outline of the Mississippi River delta between 1895 and 1944. (Francis P. Shepard, Geological Oceanography, *Crane, Russak & Company, New York, 1977) (b) Landsat satellite image of the Mississippi River delta (U.S. Geological Survey, courtesy of John S. Shelton)*

13.21 *A drumlin in Ipswich, Massachusetts. The characteristic shape is clearly seen through the leafless trees. In the summer when the trees have leaves, the shape will be retained but the drumlin will seem to have grown. (W. J. Wallace)*

13.22 *Plum Island, Massachusetts. A sand dune coast. (W. J. Wallace)*

13.23 *A landslide coast near San Simeon, California. Note the lean of the telephone poles. (W. J. Wallace)*

Coasts shaped by earth or major tectonic movements are difficult to recognize, especially where subsequent changes have masked the initial formations, as along the East Coast of the United States. We might expect California's shoreline, which is marked by relatively nonresistant rocks, strong currents, and high waves, to be broad and flat. Yet high cliffs formed by the uplifting of the land are common. For at least 20 million years, vertical land movements have been producing cliffs (Figure 13.25), and horizontal movements have been producing points or capes along the California coast. Figure 13.26 shows an example of coastal folding and faulting.

Mud lumps are small oval islands produced by the pressure of an advancing delta on underlying mud layers (Figure 13.27). They are soon washed away by waves and currents. A similar process may create salt domes from underlying salt layers.

Spreading glacial ice produces what is called an ice coast. In certain areas of Antarctica, for example, the continental ice mass has depressed the continent and spread out over the sea.

Coasts Shaped by Marine Processes or Marine Organisms

According to Shepard's scheme, a coast is classified according to the last identifiable process that acted on it. A wave erosion coast, for example, is a coast that was formed by earlier processes but that now shows the effects of wave erosion (Figure 13.28). Such coasts are marked by steep cliffs. Where waves have acted on material of relatively uniform composition, fairly regular cliffs are produced. Where waves have acted on material of varying hardness, the resultant cliffs are highly irregular. Much of the West Coast of the United States is a wave erosion coast.

Because of the seemingly negligible erosion along the southern California coasts, it may seem inappropriate to consider them wave erosion coasts. But the

(a)

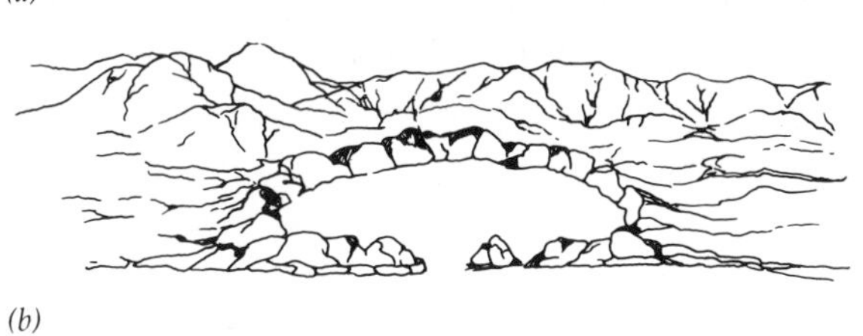

(b)

13.24 *(a) Prior to the May 18, 1980, explosive eruption, Mount St. Helens exhibited the beauty and symmetrical profile typical of many of the world's volcanoes. (State of Washington, Department of Natural Resources, Division of Geology) (b) The explosion or collapse of a volcano adjacent to the sea may cause this type of coast. (Francis P. Shepard)*

(a)

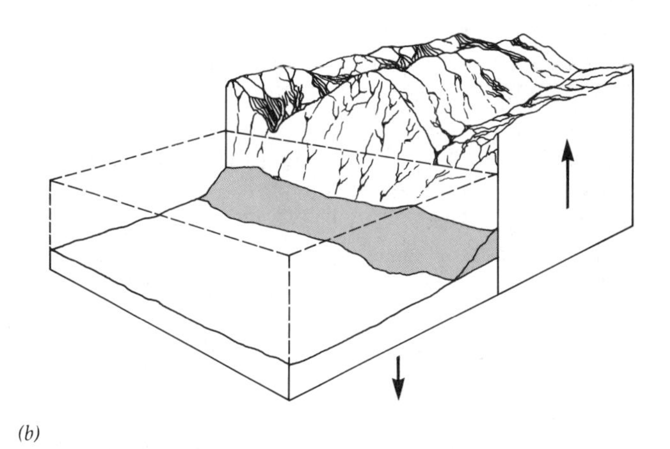

(b)

13.25 *(a) Torrey Pines in southern California. (Francis P. Shepard) A coast with vertical cliffs may have been produced by faulting, but later wave erosion usually masks the process. (b) The movement of land up or down along a fault will produce vertical cliffs. (Francis P. Shepard)*

13.26 *The Oregon coast. The more resistant rock layers indicate that the coast has undergone folding. Note the presence of a fault. (W. J. Wallace)*

13.27 *Louisana mud lump islands. (W. J. Wallace)*

(a)

(b)

(c)

13.28 *(a) Straightened wave erosion coast. The homogeneous volcanic rock of St. Paul Island, in the Pribilofs, erodes uniformly, creating cliffs that provide homes for a multitude of seabirds. (W. J. Wallace) (b) Irregular wave erosion coast. Composed of rock of varying hardness, this Oregon coast erodes unevenly. The characteristic outlying islands are called stacks. (W. J. Wallace) (c) The white cliffs of Dover. Made up of a very uniform chalk, these wave erosion cliffs are about as straightened as those found anywhere. (W. J. Wallace)*

(a) (b)

13.29 *While wave energies expended on southern California beaches are usually low, occasionally waves like this one occur. (a) Five-meter wave smashing the pier off Imperial Beach, California. (W. J. Wallace) (b) A low-energy beach such as Seahorse Key, Florida, has a typical compressed profile. (D. Ingmanson)*

(a) (b)

13.30 *(a) The beach at Carlsbad, California, April 4, 1978. (G. G. Kuhn) (b) The same beach on Feburary 5, 1983, during the severe winter of 1982–83. (G. G. Kuhn)*

winter of 1982–83 proved that short bursts of high energy can erode these coasts significantly (Figure 13.29). The winters in southern California are usually quite mild, with a few local storms. Occasionally, however, a series of severe local storms will strike, accompanied by waves and runoff that cause extensive erosion in only a few weeks. The rise in sea level along the California coast caused by the El Niño of 1982–83 increased the erosive effects of the waves (Figure 13.30).

Along very sandy coasts with a gentle slope, waves sometimes produce a long, low barrier island off the mainland coast, as shown in Figure 13.31. An island consisting of a single ridge may in time widen to seaward and form multiple ridges with sand dunes. Barrier islands make up approximately 47% of the coast of the forty-eight contiguous states (Table 13.7). From Maine to Texas, 295 barrier islands protect the majority of the coast. Galveston, Miami Beach, and Atlantic City are all located on barrier islands. By absorbing the energy of waves and storms, barrier islands protect the mainland from the direct ravages of the ocean. For this reason they are probably hazardous places for permanent human habitation. Since they are composed of sand, they shift continuously. Propelled by the overwash of storm waves, they may migrate landward and join the mainland, especially if wind-blown sand fills the lagoon between the barrier island and the mainland. At Monmouth Beach in New Jersey, Matagorda Peninsula in Texas, and Myrtle Beach in South Carolina, the barrier islands have become connected to the mainland, leaving the coast with no off-

13.31 Chappaquiddick Island, east of Martha's Vineyard. Note the presence of a large single-ridge barrier island as well as a double-ridge island. (W. J. Wallace)

13.32 A barrier sand spit on the Massachusetts coast. (W. J. Wallace)

Table 13.7 *Length of Barrier Islands by State*

State or Region	Number of Islands	Total Length (mi)	% General Shoreline
Massachusetts	2	18	9
New York	4	93	73
New Jersey	10	100	77
Delaware	1	6	21
Maryland	2	31	98
Virginia	9	67	60
North Carolina	20	285	95
South Carolina	18	96	51
Georgia	12	89	89
Florida (Atlantic and Gulf)	49	560	41
Florida (Atlantic)	17	283	49
Florida (Gulf)	32	247	32
Alabama	3	19	86
Louisiana	11	59	15
Texas	6	217	59
TOTAL	**147**	**1,610**	**47**
Massachusetts to Atlantic Florida	95	1,063	50
Massachusetts to Virginia	28	315	50
North Carolina to Atlantic Florida	67	753	65
Florida Gulf to Texas	52	542	35

Notes: Some islands are in two states.
Sources: Shoreline information from National Oceanic and Atmospheric Administration, 1975, *The Coastline of the United States;* U.S. Army Corps of Engineers, 1971, *The National Shoreline Study*. Adopted from John R. Clark and others, July 1977, *Review of Major Barrier Islands of the United States* (New York: The Barrier Islands Workshop).

shore barriers. Barrier **spits** are peninsula-like formations connected to the mainland, often extending into the mouth of an estuary or bay (Figure 13.32). Bay barriers are sand spits that completely block the entrances of bays.

Low-lying coasts such as barrier islands are especially susceptible to rises in sea level. Any rise in sea level yields 10 to 1,000 times as much horizontal shore retreat.

Cuspate forelands are a unique type of barrier island found almost exclusively in the United States. They form major barrier systems many miles across, with large, symmetrical sand points (Figure 13.33). Cape Hatteras, which extends along the coast of North and South Carolina for about 600 km (372 mi), is a cuspate foreland, as are Cape Canaveral and Cape San Blas in Florida.

Beach plains are flat tidal plains formed by the deposition of sediment borne by tidal waters. Mud flats and salt marshes are common types. Beach plains form along deltaic or other low-lying coasts where the offshore gradient is too small to allow the waves to break (Figure 13.34).

Finally, the activities of plants and animals may alter existing coasts or produce entirely new coastlines. Myriad coral polyps have built coral reef coasts in the Florida Keys and in the coral islands and atolls of the Pacific (the Carolines, Marshalls, and Gilberts). These coasts are only slightly higher than the high-tide mark, and they never develop in water with a temperature lower than 20°C (68°F). So the term *coral reef* actually means a formation of biogenic origin.

Mangrove trees (Figure 13.35) that grow in seawater may extend the land seaward and may, as in the Everglades, form new islands. In shallow tidal regions protected from breaking waves, the salt marsh grass *Spartina* develops and stabilizes new coasts by trapping and holding mud sediments (Figure 13.36). Mussels and oysters form beds in shallow waters by cementing new generations onto the remains of old generations. These structures can do serious damage to the propellers of outboard motors (Figure 13.37).

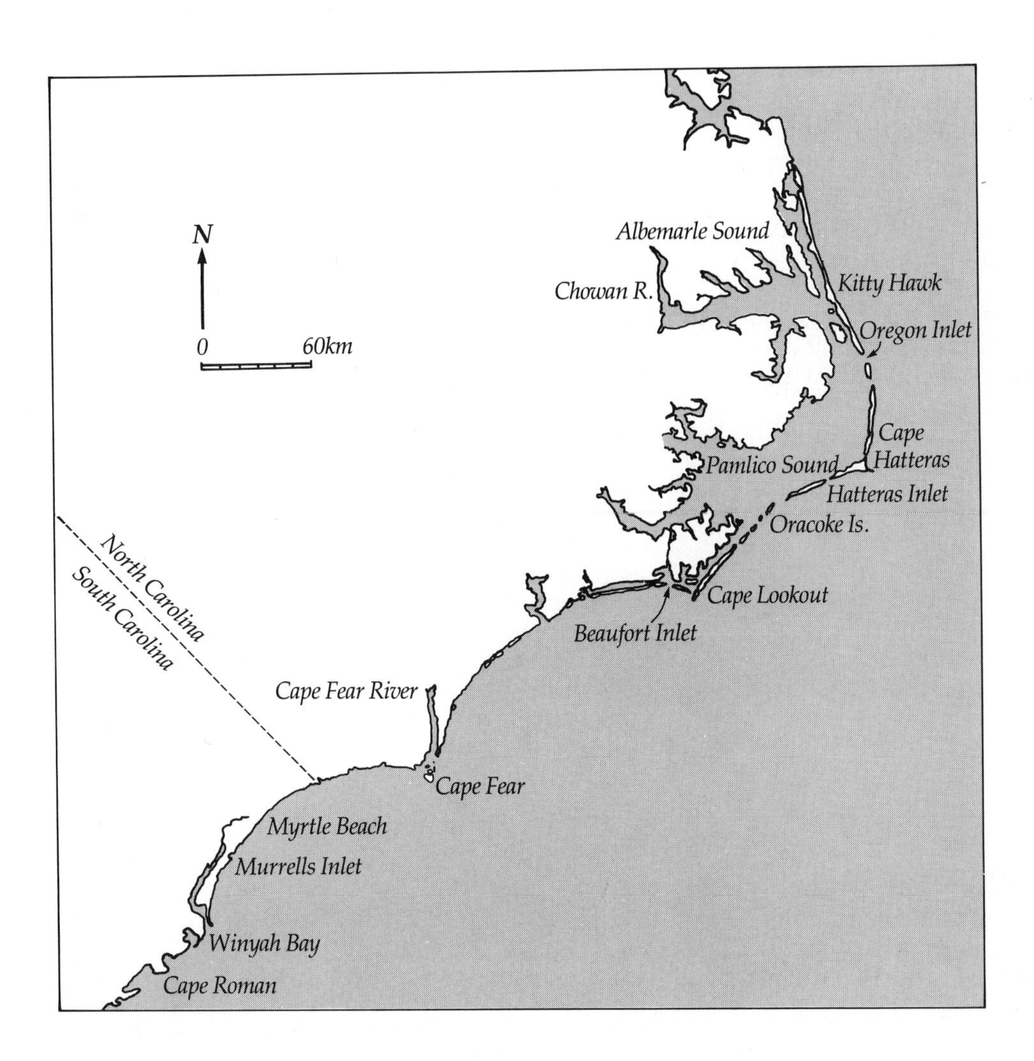

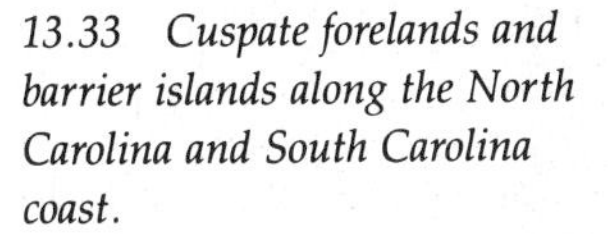

13.33 Cuspate forelands and barrier islands along the North Carolina and South Carolina coast.

13.34 A mud flat 16 km (10 mi) wide at Mont-Saint-Michel in France, formed by receding low spring tides. (W. J. Wallace)

13.35 Mangrove swamp along the coast of Baja California. (Reproduced by permission of L. James Grattan)

13.36 The marsh grass Spartina *protects the region by holding the soft mud together. Note some slumping at the channel's edge. (W. J. Wallace)*

COASTS OF THE UNITED STATES

Descriptions

The coasts of the United States are quite varied. The coasts of New England are marked by numerous rocky headlands. Most of them are stable, but some are made up of glacial till that is subject to erosion. Between the headlands are either extensive sandy beaches or short, stable beaches called pocket beaches. Such beach areas include the shore from Old Orchard Beach to Saco, Maine, the beaches from Hampton Beach, New Hampshire, to Annisquam, Massachusetts, the east shore of Cape Cod, the shores of Nantucket and Martha's Vineyard, and the south shore of Rhode Island from Point Judith to Westerly. The Maine coast is dotted with islands and inlets resulting from severe glacial activity.

The East Coast from New York to Florida and the Gulf Coast from Florida to Texas consist for the most part of long, straight reaches of beach. A few eroding headlands exist in New Jersey. Much of the shore con-

(a)

(b)

13.37 In both photographs, (a) the Ipswich River, Ipswich, Massachusetts, and (b) the North River, Salem, Massachusetts, mussel beds are visible at low tide. Once covered by the tide, they can present a hazard to boat propellers. (W. J. Wallace)

sists of barrier islands, and the straight reaches of beach are interrupted by many inlets. The migration of the inlets up and down the coast makes the intervening beaches unstable.

The coast of Washington, Oregon, and California consists of both rocky headlands, with pocket beaches and coves, and long, sandy beaches, such as those north and south of the Columbia River in Washington and Oregon, those of Monterey Bay, the Oxnard Plain near Santa Monica, the area from Long Beach to Newport Bay, and the area south of San Diego Bay. Rocky headlands and eroding cliffs are more common here than along the East Coast, and the coastline is more linear and abrupt, with fewer good natural harbors and bays.

Now let us examine the processes at work on these coastlines.

Active Processes

New England From time to time over the past 2 million years, continental glaciers have covered parts of the North American continent. The most recent ice sheet extended southeastward to what is now Cape Cod (Figure 13.38). As it melted, it dumped the unconsolidated sand, gravel, silt, and clay that now cover the surface of this region. Since these materials are unconsolidated, they are particularly susceptible to erosion, and the Cape Cod shoreline is impermanent. Wide beaches, extensive longshore movement of sand and silt, offshore bars, tidal channels, and other transient features typify the region. Wave action is particularly vigorous on the eastern side of Cape Cod, which is exposed to the Atlantic Ocean.

In the fall and winter, low-pressure systems form off Cape Hatteras and move on to New England as the formidable "nor'easters." As they advance, the sea level rises as much as 1 m (3.3 ft), the winds reach 40 knots, rain and snow swirl out of the storm clouds, and waves 3 to 7 m (10 to 23 ft) high pound the shorelines facing the ocean. The action of such high-energy conditions on the unconsolidated sediments not only causes but accentuates the impermanence of the shoreline.

Cape Cod is, however, a very stable land area. No major earthquakes or earth movements have disturbed it for over three centuries. The area may have arched by about 1 m (3.3 ft) in response to the retreat of the glaciers, although there is little evidence of this. The normal tidal range on the north side of Cape Cod Bay is about 5 m (16.4 ft), which means that the tidal currents are strong enough to cut channels in the offshore sand shoals and sandbars.

At Nauset Beach, facing the Atlantic Ocean, a cliff of unconsolidated sediments rises between 30 and 70 m (98 and 230 ft) above the shore. The cliff has been extensively undercut by waves and currents during fall and winter storms. In one area, it has eroded almost 70 m (230 ft) since 1911! Longshore currents have moved sediments south from the foreshore area to form Cape Monomoy. The hook at Provincetown is continuously being extended, and during the low-energy conditions of summer, wide, sandy beaches quickly form.

(a)

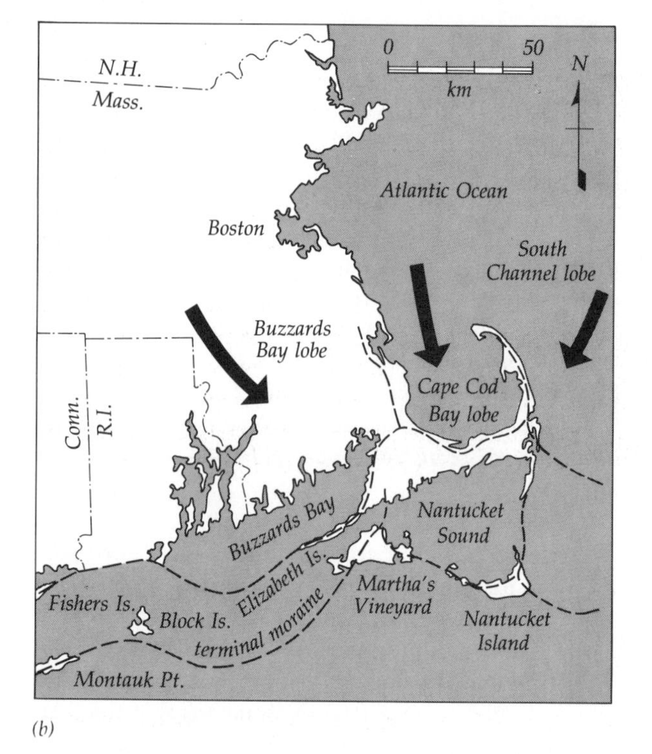

(b)

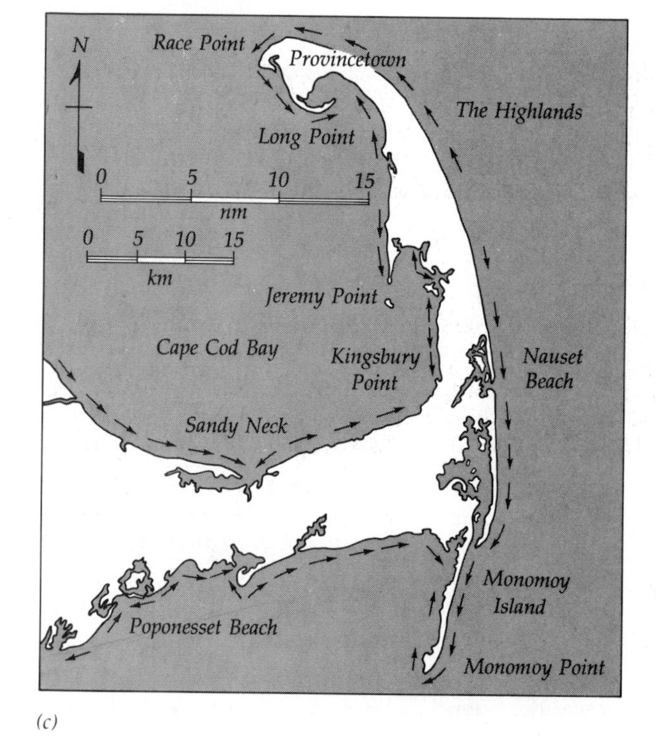

(c)

13.38 *A look at the continental margin in the Cape Cod region of New England. (a) Barrier coast spit at Long Point and Provincetown, Massachusetts. Compare this spit with the maps in (b) and (c). (U.S. Department of the Interior, National Park Service) (b) Surface sediments in the Cape Cod region were deposited by three ice sheet lobes during periods of glaciation. The dashed lines show two positions of the farthest extension of ice. (c) Direction of longshore currents in the region. (From A. N. Strahler,* A Geologist's View of Cape Cod, *New York: Doubleday and Co., 1966)*

13.39 Resistant rock outcrops at Mount Desert Island, Acadia National Park, Maine. (W. J. Wallace)

Mount Desert Island, most of which is now Acadia National Park, is located off the coast of Maine (Figure 13.39). Here the energy conditions, land stability, and sea-level changes are similar to those at Cape Cod. The major difference is that the outcropping granite bedrock is highly resistant. Along the steep slopes at the shoreline are bars that have stood unchanged for hundreds of years. These harbor valleys are fjords, gouged out by glaciers and then flooded by the rising sea level as the glaciers melted. Fjords are most common in North America from British Columbia to Alaska.

At some places along the shore, winter storm waves have eroded even these resistant cliffs. Wave-cut caves dot the foreshore, especially where there are fractures from ancient stresses. During the winter months, water seeps into the fractures, where it freezes and expands. This process, called frost wedging, slowly pries the granite apart along the fracture lines. At high tide, the waves loosen large pieces of the rock and eventually break them away. The boulders roll onto the foreshore and offshore zones, providing an ideal environment for Maine lobsters. Most of the Mount Desert Island shore remains largely unaffected by other shoreline processes because of the extreme resistance of the bedrock.

The New England shoreline consists of irregular crystalline rocks with occasional strips of loose sand deposited by glaciers. The continental shelf extends seaward 140 km (87 mi) east of the present shoreline. Its area is nearly equal to that of New England itself.

Long Island Sound, which lies between Connecticut and Long Island, is 5 to 35 km (3 to 22 mi) wide and connects at both ends with the Atlantic Ocean. At its deepest point (between Long Island and New London, Connecticut), it is only about 100 m (328 ft) deep; at its central part, it is less than 50 m (164 ft) deep. At no point from Bridgeport, Connecticut, to Port Jefferson, New York, is the sound deeper than 40 m (131 ft). The bottom is relatively smooth, consisting of mud with local patches of sand and gravel. These sediments form a thin veneer over ancient crystalline rocks, similar to those along the present Connecticut shoreline. In some parts, sedimentary rocks lie between the crystalline rocks and the glacial sediments. The sediments have been reworked and distributed to their present locations by currents and waves.

At one time, an ice sheet 100 to 300 m (328 to 984 ft) thick blanketed all of Connecticut and Long Island Sound. When the glacier melted, it dumped its load of unconsolidated sand, silt, and gravel in what is called a **terminal moraine**. Long Island is a remnant of that moraine. During the glacial epochs, what is now the continental shelf southeast of Long Island was a broad, gently sloping land area with a shoreline about 75 km (46 mi) southeast of Long Island's present Atlantic shore. Grass and trees covered the land, and streams meandered through it. Scattered remnants of preglacial sediments have been found on Long Island, Cape Cod, and Martha's Vineyard, vestiges of a rich geological past.

Cape Cod is similar in structure and origin to Long Island, but it is a complex of several moraines that were formed when three lobes of the ice sheet converged, as shown in Figure 13.38(b). The continental shelf directly east of Cape Cod is 140 km (87 mi) wide. In a few places east of the present shoreline, the shelf is only 10 m (32.8 ft) below sea level. On this shallow shelf, which is known as Georges Bank, cold water from the Labrador Current moving south along the New England coast mixes with warm water from the Gulf Stream. The cold current brings nutrient-rich bottom water to the surface, encouraging the growth of phytoplankton and zooplankton. Feeding on these marine organisms are crabs, lobsters, flounder, sole, and haddock, as well as tuna, bass, and herring. Georges Bank is one of the richest fishing grounds in the world.

North of Cape Cod is a deep basin called Wilkinson's Basin (Gulf of Maine). It was formed long before the glaciers, at a time when crustal stresses and plate movements were producing the now deeply buried tectonic dams farther south. Moving ice scoured the shelf around this basin, creating numerous valleys and strips of sand and gravel. A map of the submarine contours of this area is very similar to a map of the present land contours of Maine, New Hampshire, and northeastern Massachusetts.

Mid-Atlantic Shelf The gently sloping continental shelf from New Jersey to North Carolina did not experience glaciation. Although there are no shelf dams along this stretch, recent seismic profiles have revealed ancient tectonic dams buried under sediments eroded from the eastern slope of the Appalachian Mountains.

Longshore currents have created numerous barrier islands from Sandy Hook, New Jersey, to Miami Beach, Florida. In some places, the currents have built up shifting sandy shoals. In the vicinity of Cape Hatteras, the shoals are scattered across the continental shelf near the shoreline and several kilometers offshore. Many ships have run aground on these shoals during heavy storms and been pounded to bits by the strong waves. As mentioned earlier, mariners refer to this area as the "Graveyard of the Atlantic."

Florida Everglades National Park in Florida is a unique region populated by alligators, cougars, egrets, ibises, roseate spoonbills, and many more exotic forms of wildlife. Its remote shoreline is most unusual (Figure 13.40 and Figure 13.41).

The area is underlain by a hard, relatively nonporous limestone known as Tamiami limestone. At some time in the past, acidic groundwater and acidic rain water dissolved its surface. Then, as the water evaporated or flowed away, the surface recemented (Figure 13.41), producing a hard pavement over which the water now flows. The area slopes gently from north to south. Runoff trickles slowly down from the region of the Great Cypress Swamp and the region south of Lake Okeechobee. In the past, the runoff flowed across the countryside in a sheet more than 180 km (112 mi) wide and only a few centimeters to 1 m (3.3 ft) deep. Now the runoff is channeled into flood control canals built by the U.S. Army Corps of Engineers. The ecology of the region has been so severely altered that much of the plant and animal life is endangered.

The low topography and the resistant bedrock have produced a highly irregular shoreline with almost no beaches but with thousands of islands on which mangrove trees grow (as in Figure 13.35). Scattered on these islands are oyster bars and shell middens, some of which now form islands themselves.

Because Florida is one of the most stable land areas in the United States, it is an ideal place to study the sea-level changes that have occurred over the past million years and more. These fluctuations have been significant in the past and continue to be so. Long-term fluctuations have caused alternating inundation and exposure of the entire area and have produced a rise in sea level as great as 10 m (33 ft) several times in the remote past. Such a rise would completely flood Everglades National Park. In fact, if the sea level rose 100 m (328 ft) above its present level, all of peninsular Florida would be submerged.

Because of Florida's gently sloping topography, even short-term rises in sea level of about 1 m (3.3 ft), which often occur during hurricanes, flood most of the coastal islands and carry salt water several miles inland. During severe thunderstorms and hurricanes, high winds, torrential rains, and high tides batter the Everglades shoreline. Local thunderstorms often produce winds of over 50 knots and have dumped more than 15 cm (5.8 in) of rain in a three-hour period!

All of Florida, including its continental shelf, is built on top of a carbonate bank. Fragmented or broken sediments, called **clastic** sediments, occur at depths, however, and noncarbonate sediments cover the thick carbonate sediments. Reef dams ring the continental shelf, and reef deposits make up a large part of the peninsula. Broken shells and dissolved and subsequently reprecipitated calcium carbonate fill in the voids between the reefs.

East of Florida, the continental shelf narrows. East of Miami, the shelf gradient is fairly gentle, but from 500 m to 600 m offshore there is a series of steep scarps with indented terraces. Between 600 m and 24 km (15 mi) offshore, the water is over 600 m deep. Here the shelf ends steeply and abruptly. About 45 km (28 mi) off the coast, the depth decreases just as abruptly. This region, called the Bahama Bank, is made up of sediments similar to those found in Florida. Moreover, terraces appear on its shelf at about the same depth as those on the Florida shelf.

At the southern tip of Florida are the Florida Keys, which are the exposed peaks of a fossil reef. Small patches of coral and algal buttresses growing near the

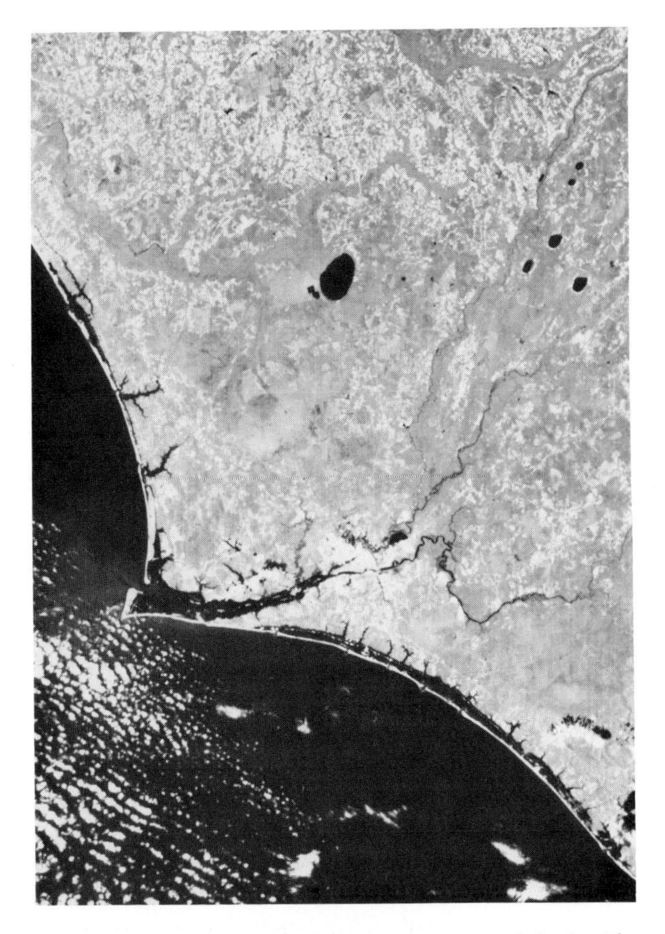

13.40 This aerial photograph shows calcareous mud shoals and cuspate foreland at the Florida Keys, Florida Bay, and the Everglades. (NASA)

13.41 A view of the Everglades, showing a labyrinth formed by mangrove plants. Part of the Everglades National Park is periodically flooded with fresh water during heavy rains and with salt water during dry seasons. Drought and reclamation projects to the north tend to upset this delicate natural habitat. (U.S. Department of the Interior, National Park Service)

southern edge may be a reef dam in the making. By directing the Florida Current into a deep channel between the keys and Cuba, the reef reduces the effect of the current on the shoreline. Because there are a few planktonic organisms in this region and no river sediments, the water is extremely clear, with a visibility of 60 m (197 ft) or more. Conditions are ideal for skin diving and scuba diving. At Pennecamp State Park at Key Largo, the first submarine state park in the United States, divers swim over magnificent reefs and banks, and nonswimmers view the undersea world through glass-bottomed boats.

In Florida Bay, a shallow body of water that separates the Florida Keys from the Everglades, currents are weak and temperatures reach high levels, especially in summer. Many microscopic organisms, such as foraminifera, thrive in the bottom waters. Because the water is saturated with calcium and carbon dioxide, the shells of dead organisms do not dissolve. Instead, they accumulate to form a fine calcium carbonate mud on the bottom. When winter winds from the northwest stir up the water, the tiny carbonate particles become suspended and give the bay a milky color. As the bay water flows south through inlets to the edge of the Florida Current, it creates a striking visual contrast of milky white and blue-green.

West of Florida, the continental shelf is 200 km (124 mi) wide, sloping at only about 1 m per 10,000 m offshore. During the Pleistocene, when great quantities of ocean water were locked up in the glaciers, this broad, flat bank was completely exposed. Mammoths, saber-toothed tigers, alligators, turtles, and other animals lived on the plain, and their bones and teeth are found today by fishermen and divers.

The bank is underlain by limestone and fringed by reef dams. Its topographical features resemble those found on the adjacent land. In past periods of low sea level, the same weathering processes that now shape

13.42 A warm, shallow lagoon in Texas with barrier bars and islands. (W. J. Wallace)

the land helped shape the shelf. Groundwater containing carbonic acid slowly dissolved the limestone, producing sinkholes and springs that pockmark both the shelf and the land. Freshwater springs still flow into the Gulf of Mexico far offshore. The extensive phosphate deposits found in central Florida can be traced onto the shelf off Fort Myers. About 200 km (124 mi) west of the present shoreline, the slope abruptly increases. Less than 10 km (6.2 mi) farther out, the water is more than 600 m (1,968 ft) deep. This is one of the deepest scarps known.

Gulf Coast The Louisiana shoreline, which is dominated by the Mississippi River delta complex, is exposed to the Gulf of Mexico. Under normal conditions, waves are less than 1 m (3.3 ft) high, and both longshore currents and tidal currents are weak. The primary sources of energy are the Mississippi River during flood stages and occasional hurricanes. The sediments that outcrop on the shoreline consist mainly of alternating layers of unconsolidated river sand, silt, and mud. They are nonresistant and highly susceptible to erosion, but the combination of low energy and low resistance prevents the large-scale shifting that characterizes coasts like Nauset Beach. Instead, it encourages the development of rich soil, and abundant plant growth extends almost to the berm in many places.

The land of southern Louisiana is very unstable. The Mississippi River is continually depositing sediments at the coast and building the shoreline out into the gulf. In this process of accretion, loose sediment gradually settles and is compacted by the weight of newly arriving sediment. But even as this accretion is taking place, New Orleans and the surrounding vicinity are sinking, partially due to the load of the material being deposited in the Mississippi River or to some tectonic subterranean process. The pumping of deep oil wells may be contributing to the subsidence. In any case, southern Louisiana is slowly sinking.

Sea-level fluctuations and flooding are also at work on the Louisiana coast. Long-term sea-level changes during glacial epochs have altered the coastline, and short-term changes caused by hurricanes continue to do so. Each hurricane cuts out terraces and redistributes large amounts of sediment.

The continental shelf off the Louisiana coast is presumably the result of the steady deposition of sediment that is being eroded from the entire North American interior, from the Rockies to Canada to the Appalachians. The sand, silt, mud, and gravel on the shelf have been brought to the delta over the course of millions of years.

Some of the most unusual features of the world's continental shelves are found off the Texas coast. The nearshore zone has barrier islands with shallow, warm lagoons (Figure 13.42). In summer, high air and water temperatures evaporate the water and increase the salt concentration above the saturation point. Salt then precipitates onto the bottom. Offshore sediments also contain extensive salt deposits, which may have had an origin similar to that of the lagoon deposits. The weight of the overlying sediment is so great that the salt becomes fluid, and because its density is less than that of its surroundings, it migrates upward to form salt domes. The domes fold the sediments and push them upward, forming dams. Sediment accumulates shoreward from these dams.

Oil and gas often seep through the permeable sediments against the impermeable salt domes, where they accumulate in commercially important reservoirs. Such domes have been found in the middle of the Gulf of Mexico as well as on the shelf and on the land itself.

(a)

(b)

(c)

13.43 A look at the southern California coast. (a) Looking northward from the National Marine Fisheries Service Building at La Jolla, California. Scripps Canyon originates in the surf zone in the foreground. (Thomas H. Foote) (b) Winter beach at Torrey Pines State Park near La Jolla, California. Notice the gravel left on the beach by waves that have separated the finer-grained material. (D. Ingmanson) (c) At the Scripps Institution beach, cobbles are being covered up as a summer beach forms. (W. J. Wallace)

Southern California The southern California coast between Oceanside and San Diego is dramatically different from the regions we have described, although it shares certain features with the Atlantic side of Cape Cod. Except in coastal bays and estuaries, the southern California coast is exposed to occasional high-energy conditions (Figure 13.43). Normally, waves are about 1 m (3.3 ft) high, but they frequently reach 2 m (6.6 ft). About two or three times each year, storms over the Pacific Ocean generate waves more than 6 m (20 ft) high that travel more than 1,800 km (1,116 mi) and eventually pound the coast. Even tsunamis occur here, although rarely. Waves also sweep from north to south along the coast. The size and the direction of the swells and surf movement are forever changing, affecting the amount of sediment suspended in longshore currents, the steepness of the beach, and the location of offshore bars. These high-energy conditions have produced wide beaches all along the coast and rapid transport in the offshore surf zone. In winter, the heightened activity of the waves removes sand from the foreshore to the offshore. Then in summer, when lower energy conditions prevail, the sand slowly piles up again in the foreshore (Figure 13.7).

The sediments outcropping along the coast between Oceanside and La Jolla range from unconsolidated barrier beaches across the mouths of estuaries to well-cemented sandstone cliffs. Where well-

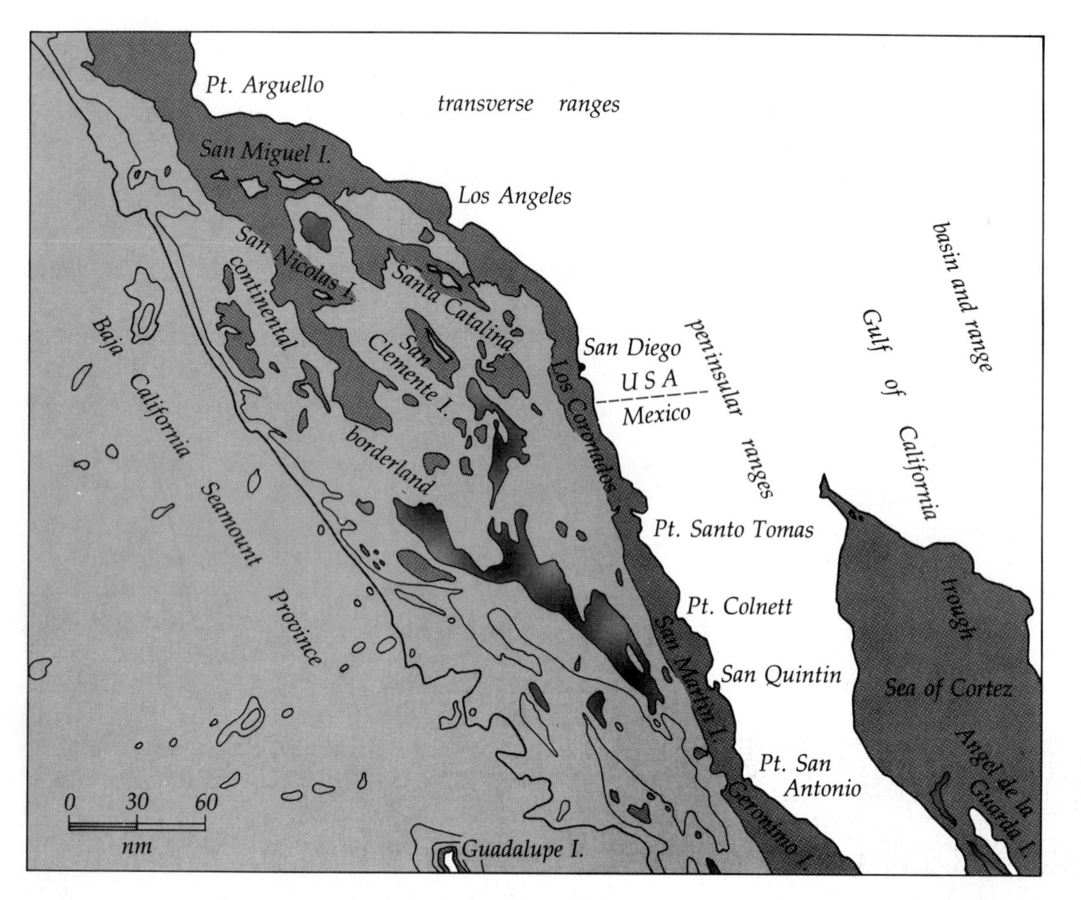

13.44 A portion of the continental borderland off southern California. Basins are shaded. (From D. J. Moore, 1978, "Reflection Profiling Studies of the California Continental Borderland," Geological Society of America paper 107)

cemented and poorly cemented sediments are mixed, precipitous cliffs with undercut arches have formed.

Southern California is extremely unstable geologically. About 50 to 180 km (31 to 112 mi) east of the coast, the San Andreas and the Elsinore faults produce weak earthquakes almost every day. Another fault, which has been dormant for two centuries, cuts through downtown San Diego and intersects the coast at San Diego Bay, Mission Bay, and La Jolla.

The continental borderland off California is a highly irregular region below sea level marked by depths much greater than those typical of a continental shelf. It is composed partly of sedimentary rocks that extend inland. These rocks outcrop in cliffs along the coast and beneath the surface. About 15 km (9.3 mi) west of the San Diego shoreline, the sediments form a steep, submerged cliff. Farther west another cliff rises to form a ridge. Most of the ridge is submerged, but parts of it break through the surface waters as islands. Between the cliffs is the San Diego Trough. The cliffs were produced by erosion and faulting, and faulting may also have led to the submergence of the wave-cut terraces now observed below sea level. Linear banks with islands and troughs are found throughout the region (Figure 13.44).

The topography of this region reflects plate tectonic processes. Some of the more recent sediments underlying the area were transported from mountains to the east and deposited in troughs and basins during the Pliocene and Pleistocene. In the offshore and foreshore zones, millions of tons of sand shift about. Submarine canyons near the shoreline act as funnels for the movement of sand from the shore to deep water. At some sites the sand is being transported 21 km (13 mi) out into the San Diego Trough to depths of more than 800 m (2,624 ft).

This one-way movement of sand is depleting the beaches, especially when the winter waves and currents increase their sediment load. In addition, dams across the few rivers in southern California decrease the replenishment of nearshore beach sand. The starved longshore currents have tended to accelerate shoreline erosion over the past twenty years. One suggestion for halting this massive erosion is to build submarine dams at the canyon heads to catch the sediment and to pump the sediment back to the beaches.

Oregon and Washington South of the Columbia River along the Oregon coast are numerous small rivers, the largest of which is the Rogue River. The Klamath Mountains and Coast Ranges, which closely parallel the coastline, seldom reach elevations greater than 1,500 m (4,920 ft). Their proximity to the coast precludes the formation of any major river system, but they do support many small rivers that carry modest amounts of runoff and sediment. Together, these rivers deposit abundant sediment over a wide area on the coast.

Extremely high energy conditions prevail along the Oregon coast. Winter storms in the North Pacific create high winds, low atmospheric pressure, and large waves. Indeed, the waves hitting the Oregon coast may be the largest to hit any point in the contiguous United States. Winter swells often exceed 5 m (16.4 ft). Although the zone of longshore drift may be as wide as 1 km (0.62 mi), the longshore current shows only a slight trend from south to north. High-energy winter waves flatten the beach slope, carrying sediment offshore, and the summer waves cannot bring it back to the shore. As a result, the continental shelf is widening. The width of the continental shelf and slope off Oregon and Washington ranges from 65 km to 115 km (40 to 71 mi). The base of the slope is at a depth of about 3,500 m (11,480 ft).

There are few submarine canyons in this region. Off the Columbia River, the Astoria submarine canyon cuts across the shelf and slope, terminating in the Astoria fan. To the north, other submarine canyons funnel sediments out to the Cascadia Basin. No trenches are evident, although they may have been filled in by sediments. The convex profile of the slope suggests the presence of trenches filled with sediments that are being compressed.

The coast of Washington and Oregon, together with the continental margin, consists mainly of sedimentary and volcanic rocks less than 45 million years old. Thus they date from the Eocene or more recent times. They have been faulted, and the coast has been uplifted. Apparently Oregon and Washington are being pushed westward over the deep-sea sediments in the Cascadia Basin. Many of the faults and volcanoes are still active. Although no trench is present, thick sediments along the coast suggest that the trench may have been buried by sediment eroded from the Coast Ranges.

MINOR BEACH FEATURES

The sand on a beach shows a wide range of minor features (Figure 13.45). The most noticeable are cusps, which are regular undulations 0.5 to 100 m (1.6 to 328 ft) in length that occur along the water's edge. The distance between adjacent cusps depends on the size of the waves and the size of the sand grains. Large waves and fine sand combine to form long cusps; small waves and coarse sand form short cusps. They are formed initially as water piles up on the berm. Some of the water sinks into the sand, and some recedes seaward, causing the crescent-shaped undulations. Once formed, the cusps are maintained by wave refraction. Figure 13.45 shows other minor beach features, including ripples, backwash ripples, and diagonal backwash.

ESTUARIES

An estuary is a semienclosed coastal body of water freely connected with the open sea and containing seawater measurably diluted with fresh water from land drainage. An estuary is part of a coast but does not form a coast; it is a coastal feature. The circulation patterns of an estuary are influenced by its lateral boundaries. Because the estuary and the sea are connected, there is a virtually continuous exchange of water between them, although barriers or bars may hinder this exchange.

Classification

Estuaries may be classified into four types: (1) drowned river mouths, (2) fjord type, (3) bar-built, and (4) estuaries produced by tectonic processes. Drowned river mouths are widespread throughout the world and are particularly common along the East Coast of the United States. The mouths of the York, James, and Susquehanna rivers are examples of this type of estuary.

Fjord-type estuaries are steep-sided and deep, often 300 to 400 m (980 to 1,312 ft). They have a shallower region, or sill, at their mouth formed of terminal glacier deposits. In fjords with shallow sills, little vertical mixing occurs below the sill depth, and the bottom waters become stagnant. In fjords with deeper sills, the bottom waters mix slowly with adjacent waters.

Bar-built estuaries are formed when a barrier island or a barrier spit builds up above sea level. Even when more than one river flows into a single bar-built estuary, the total drainage is not likely to be great. Since these estuaries are shallow and usually have only a small connecting inlet, tidal action is limited. Thus the waters are mixed mainly by the wind. Albemarle and Pamlico sounds in North Carolina are bar-built estuaries.

Estuaries produced by tectonic processes are coastal indentations formed by faulting or local subsidence that have an inflow of fresh water. San Francisco Bay is an estuary of this type.

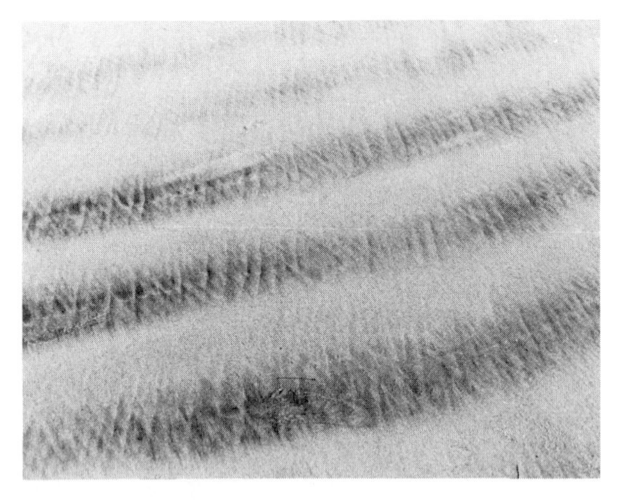

(a)

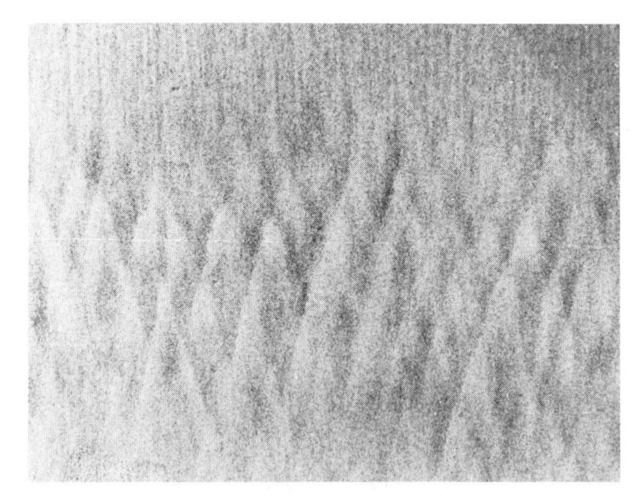

(b)

(c)

(d)

(e)

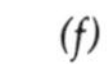

(f)

13.45 *A few of the many minor features common along sandy shorelines. (a) Backwash ripples, typically about 50 cm (20 in) apart, occur only on fine sand beaches and are caused by wave backwash setting up turbulent motion. (b) The diagonal pattern of backwash marks occurs as the wave backwash encounters small obstacles like pieces of shells or gravel, which divide its flow. (c) Ripples on an East Coast sandbar at low tide caused by the tidal and wave generated current. (d) A line of debris or mica deposited at the highest level the waves reach on a beach before the water sinks into the sand. (e) Giant ripples, often a meter or more in length, formed in larger channels in regions with strong currents. (f) Smaller ripples superimposed on giant ripples as a result of the ebbing of the tide in an estuary. (Photographs by W. J. Wallace)*

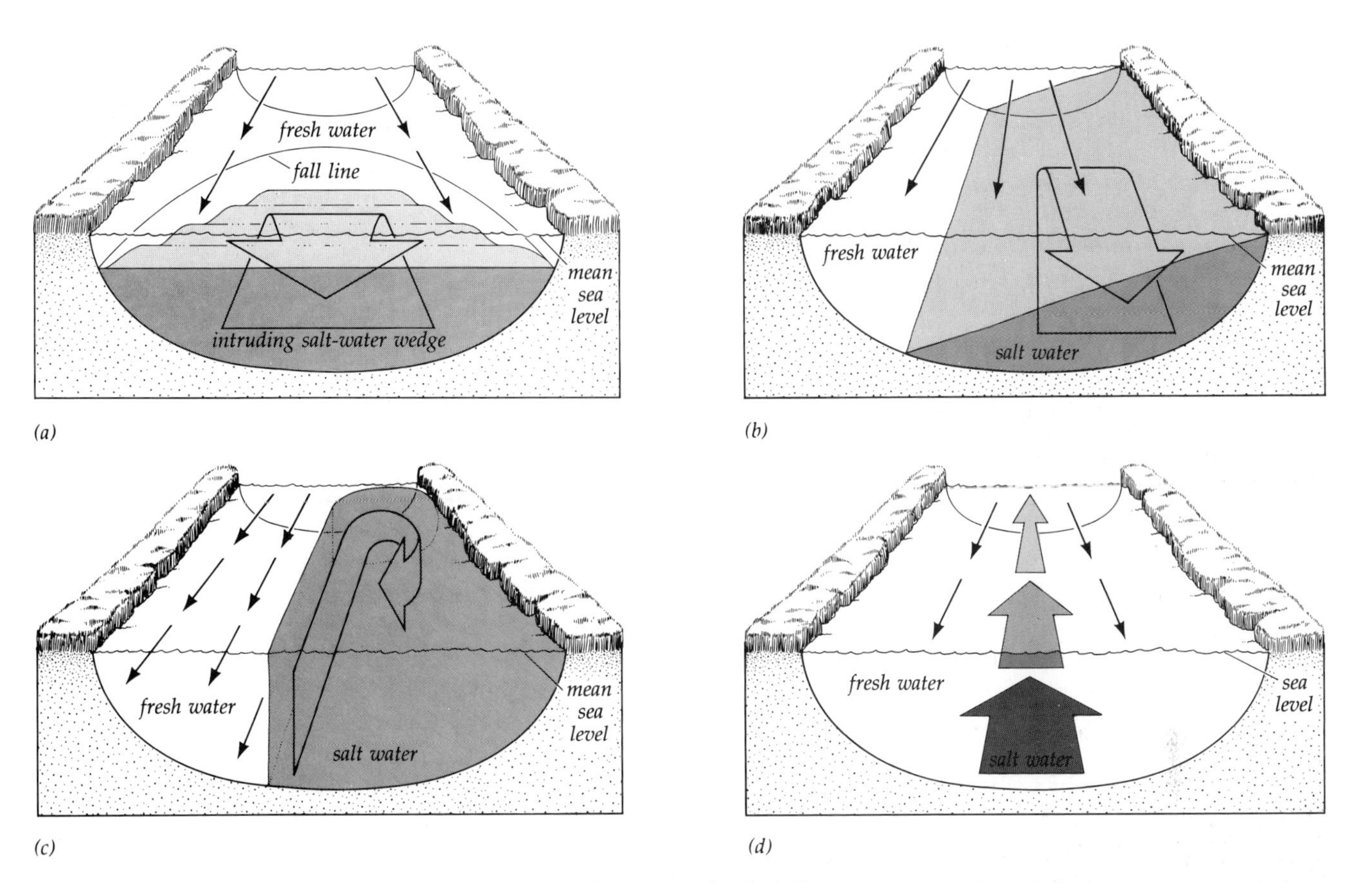

13.46 (a) A type 1 estuary is highly stratified. This type is the closest to the ideal. (b) A type 2 estuary is partially stratified. (c) A type 3 estuary is vertically homogeneous. (d) A type 4 estuary is sectionally homogeneous.

Estuarine Circulation

Estuaries may be divided according to circulation (Figure 13.46), into four types, which are really arbitrary points along a continuum. Estuarine circulation is caused by wind, tidal flow, and river flow. An idealized estuary would have a precise interface between fresh water and salt water. In actual estuaries, salt water brought by the tides forms a **saltwater wedge**, or **tidal prism**, beneath the river run-off (Figure 13.46), and there is always mixing at the interface as a result of wind, tides, and river flow. The circulatory pattern of an estuary depends on the rate of flow of the river water and the tides. In spring, when the freshwater discharge is greatest, the salt wedge does not extend very far up the river, the mixture of seawater and fresh water is at a minimum, and the fresh water may persist as a surface layer well out to sea. In late summer, when freshwater runoff is low, the salt wedge may extend into the river channel, and fresh and salt waters may be fairly well mixed for some distance upstream.

In a type 1 estuary (Figure 13.46a), the river flow pushes the salt wedge seaward. Internal waves are produced by the vigorous flow and the resulting friction. The waters are mixed and the boundary between them (called a **halocline**) becomes indistinct. The estuaries of the Mississippi, Danube, and Po rivers exhibit this circulation pattern.

In a type 2 estuary (Figure 13.46b), the tidal prism is larger than the river discharge. Tidal flow and the resulting turbulence mix the waters, so that no sharp halocline exists. The interface between the river discharge and the tidal prism is tilted due to the Coriolis effect. In the Northern Hemisphere, the outgoing river water inclines toward the right of the channel (looking downstream), and the incoming seawater also inclines to the right. Again, it is the difference between river flow and tidal flow that determines the pattern of circulation. During different seasons (at spring runoff, for instance, or during periods of lower tidal range), estuaries may vary between the type 1 and type 2 classifications. Examples of tide-dominated estuaries are the Penobscot, James, Merrimac, and Hudson rivers.

In an estuary that widens seaward but whose depth remains constant, the tidal flow becomes so vigorous that the halocline becomes nearly vertical, as in a type 3 estuary (Figure 13.46c). The salinity of the salt water is the same from top to bottom. As a result of the Coriolis effect, the fresh water and the salt water in a type 3 estuary incline to the same sides of the channel as in a type 2 estuary. The southern end of Delaware Bay is a type 3 estuary.

In a type 4 estuary (Figure 13.46d), the tidal influence is so great that there are neither horizontal nor vertical variations in salinity. However, salinity gradually decreases upstream.

Estuaries in which runoff plus precipitation exceed evaporation and in which the seawater is consequently diluted are called **positive estuaries**. Estuaries in which runoff plus precipitation approximately equal evaporation are called **neutral estuaries**. Estuaries that have almost no freshwater input, such as San Diego Bay and Laguna Madre in Texas, are called **inverse**, or **negative, estuaries**.

Estuarine Currents

The term *estuary* comes from the Latin *aestus*, meaning "tide." The currents in estuaries result from any combination of several factors: tides, river runoff, transport induced by wind-generated surface waves, direct stress exerted by the wind on the water surface, variations in density distribution, and internal waves.

Tidal forces, combined with limiting topographical features, cause three types of tidal currents in estuaries: (1) rotary; (2) rectilinear, or reversing, found in elongated estuaries (or bays); and (3) hydraulic, found in straits connecting two independent tidal bodies, such as the Cape Cod Canal or the Chesapeake and Delaware Canal.

Plants and animals possess different tolerances for such physical parameters as salinity. Consequently, an estuary may be divided on the basis of salinity, each part corresponding to certain biotic classifications.

Types of Estuaries

Several terms are used interchangeably with the term *estuary*: wetlands, lagoon, slough, salt marsh, marsh, and swamp. The term **wetlands** refers to land on which water dominates the soil development and consequently the types of plant and animal life that live in the soil and on its surface. Wetlands is a vague, all-encompassing term. The term **lagoon** is used for the region between a barrier island or spit and the mainland, although the term also refers to a shallow estuary essentially isolated from the ocean. A **slough** is a shallow estuary in which large areas of the bottom are exposed during low tides. The term is commonly used to refer to the Great Lakes and the Everglades. A **salt marsh** is a shallow, tidal estuary protected from ocean waves and inhabited by plants that can withstand limited submergence. The term **marsh** is often used interchangeably with *salt marsh*, but in precise usage it refers to a region with zero salinity upstream from a salt marsh. A **swamp**, like a marsh, is a lowland area saturated with fresh water, often quite large, and usually farther inland. Trees dominate the vegetation in swamps.

Origin and Fate

Estuaries were uncommon during most of the earth's geological history. About 6,000 years ago, when the sea level was approximately 6 m (20 ft) lower than it is now, estuaries were rare indeed. As the sea rose to its current level, it drowned the tributaries of entire valley systems and produced such large-scale estuaries as Chesapeake Bay and Delaware Bay. Smaller estuaries resulted from local conditions.

Over the last 3,000 years, sedimentation has been occurring in most estuaries, particularly those in flat, older areas. Fjords bordered by rock have changed the least. The most formidable natural enemy of estuaries is the laterally advancing delta. On the central Louisiana coast, Lake Atchafalaya, which was once an estuary 160 km (99 mi) long and 55 km (34 mi) wide and open to the gulf, has almost disappeared.

Estuaries tend to be destroyed by the encroachment of fresh or brackish water marshes at their heads, by the deposit of sediments from marine marshes, by the formation of barrier islands and bars near their mouths, and by the general coastal retreat caused by marine erosion. Erosion and accretion are continually at work, with accretion generally more pronounced. The estimated rate of sediment accumulation in estuaries in temperate climates is 2 m (6.6 ft) per thousand years, and half that value in arid and semiarid lagoons. The sediments that enter an estuary do not come solely from the rivers that flow into it. The salt wedge that moves up an estuary may return river sediments from the sea bottom and even introduce sediments of marine origin. Most estuaries are short-lived formations, and they may be quite rare a few tens of thousands of years from now.

FURTHER READINGS

Bascom, W. *Waves and Beaches*. Garden City, N.Y.: Anchor/Doubleday, 1964.

Bird, E. C. F. *An Introduction to Systematic Geomorphology*. Vol. 4, *Coasts*. Cambridge, Mass.: MIT Press, 1969.

Burk, K. "The Edges of the Ocean: An Introduction." *Oceanus*, 1979, *22*(3), 2–9.

Carr, A. P. "The Ever-Changing Sea Level." *Sea Frontiers*, 1974, *20*(2), 77.

"The Coast." *Oceanus*, 1980–81, *23*(4).

Coastal Engineering Research Center. *Shore Protection Manual*. 3 vols. Washington, D.C.: Government Printing Office, 1973.

Dolan, R., B. Hayden, and H. Lins. "Barrier Islands." *American Scientist*, 1980, *68*, 16.

Duxbury, A. *The Earth and Its Oceans*. Reading, Mass.: Addison-Wesley, 1971.

Dyer, K. R. *Estuaries—A Physical Introduction*. New York: Wiley, 1973.

Francis, P., and S. Self. "The Eruption of Krakatau." *Scientific American*, November 1983, pp. 172–87.

Fulton, K. "Coastal Retreat." *Sea Frontiers*, 1981, *27*(2), 82.

Inman, D. L., and R. A. Bagnold. "Littoral Processes." In *The Sea*, edited by M. N. Hill, vol. 3. New York: Interscience, 1963.

______, and C. E. Nordstrom. "On the Tectonic and Morphologic Classification of Coasts." *Journal of Geology*, 1971, *79*(1), 1–21.

King, C. A. M. *Beaches and Coasts*. London: Edward Arnold, 1972.

Komar, P. D. *Beach Processes and Sedimentation*. Englewood Cliffs, N.J.: Prentice-Hall, 1976.

Lauff, G. H., ed. *Estuaries*. Publication 83. Washington, D.C.: American Association for the Advancement of Science, 1967.

Lemonick, M. D. "Shrinking Shores." *Time*, Aug. 10, 1987, pp. 38–47.

Lowenstein, F. "Beaches or Bedrooms—The Choice as Sea Level Rises." *Oceanus*, 1985, *28*(3), 20–29.

Mahoney, H. R. "Imperiled Sea Frontier: Barrier Beaches of the East Coast." *Sea Frontiers*, 1979, *25*(6), 328.

______. "Dune Busting: How Much Can Our Beaches Bear?" *Sea Frontiers*, 1980, *26*(6), 322.

Picard, G. L., and W. J. Emery. *Descriptive Physical Oceanography*. Oxford: Pergamon Press, 1982.

Ringold, P. L., and J. Clark. *The Coastal Almanac*. San Francisco: W. H. Freeman, 1980.

Schwartz, M., ed. *The Encyclopedia of Beaches and Coastal Environments*. New York: Scientific and Academic Editions, 1982.

Shepard, F. P. *Submarine Geology*. 3d ed. New York: Harper & Row, 1973.

Shepard, F. P., and H. R. Wanless. *Our Changing Coastlines*. New York: McGraw-Hill, 1971.

Teal, J., and M. Teal. *Life and Death of a Salt Marsh*. Boston: Little, Brown, 1969.

Waters, J. F. *Exploring New England Shores*. Lexington, Mass.: Stone Wall Press, 1974.

Wiegel, R. L. *Oceanographical Engineering*. Englewood Cliffs, N.J.: Prentice-Hall, 1964.

Wiley, M. L. *Estuarine Interactions*. New York: Academic Press, 1978.

Dennis Brokaw

Cease searching for the perfect shell, the whole
Inviolate form no tooth of time has cracked;
The alabaster armor still intact
From sand's erosion and the breaker's roll.

What can we salvage from the ocean's strife
More lovely than these skeletons that lie
Like scattered flowers open to the sky,
Yet not despoiled by their consent to life?

Anne Morrow Lindbergh

FOURTEEN

Life in the Sea: Bacteria, Protists, Plants, and Invertebrates

On the rocky seashore of Washington state, there is a tidepool about 2 m long, 40 cm deep, and 1 m wide. Next to it there is a shelf of flat rock on which one can lie and observe the plants and animals that live in the pool. (The shelf is soaking wet and ragged with barnacles and limpets, but the sight in the pool is worth the discomfort.) Most of the time, the organisms in the pool live in the sea itself. The crashing waves and surging currents that wash over their hideaway in the rock during most hours of the day and night are their real home. It is only when the tide withdraws to one of its lower levels that they are contained in a much tinier sea.

The familiar world of air-breathing life stops abruptly at the waterline. In the tidepool, appearances are deceptive and the rules of life as we know them on land seem suspended. Beneath the surface, snail shells move busily across the rocks and patches of sand. Snails? A closer look reveals that the shells are actually occupied by hermit crabs. Big green sea anemones sit motionless with soft, deadly tentacles outstretched. How? When exposed (as they are in the wet crevices around the pool), anemones don't have enough strength to stand upright. On some days a sea star can be found resting in the pool. If you're lucky, you can see it glide effortlessly along the bottom, its purple rays flowing flexibly over the rock, following the contours exactly. It is a predator, a relentless hunter of mussels and limpets, biding its time here until the rising tide liberates it to prowl the shore upslope. Many of the creatures on these rocks could not live here if it weren't for this carnivore.

Plants in the pool? At first glance, they appear to be familiar: Brilliant green blades of a surfgrass lie submerged at one end of the pool, providing cover for small fish, crabs, and other animals. Red is also the color of plant life in the tidepool, however, and the tiny tufts of feathery red fronds that sprout here and there from the submerged rock are also plants. Strangest of all is the pink, rocky crust that is plastered tightly to the walls of the pool, the loose stones on the bottom, and even the shells of some of the limpets. This, too, is a plant—armored, stony, and utterly unlike anything that lives on land. In contrast, some things that look as though they *ought* to be plants are actually colonies of animals. Branching, brown, upright, and tiny, these delicate growths occur in crevices or even densely crowded on the back of a crab under the surfgrass. Small multicolored "flowers" flash suddenly out of sight if your shadow passes over them—worms.

The scene in the tidepool is a view in miniature of all life in the sea. The rules of life, as we know them on land, seem to be replaced in part by new rules that apply only underwater. Plants and animals take strange forms. The water lends its support to creatures that could not function or even stand up in air. Sea stars, sea anemones, snails, crabs, fish—the differences between organisms are much more striking here than the differences between animals on land. Color is here—brilliant reds, jet black, creams, iridescent purples—as well as forms of sinister beauty. What are they all doing here, and how do they relate to the sea and to each other?

The intent of the next three chapters is to examine the forms and relationships of marine organisms, the rules by which they live, and the interactions between them and the sea. Because the organisms that play prominent roles in the sea—kelps, sharks, lobsters, whales—are very diverse, two chapters (14 and 15) are dedicated simply to getting to know them. Chapter 16 examines important ecological processes in the marine world and relates these to the wealth of life on coral reefs, the bounty of fish on the U.S. Gulf Coast, and the surprising abundance of life in Antarctic waters. Together, the chapters present the "cast of characters" inhabiting the sea and the reasons for their abundance or scarcity in various parts of the world.

We begin with the necessary technicalities of names and definitions, and then we move on to descriptions of the organisms themselves.

CLASSIFICATION AND TERMINOLOGY

Classification

The names of marine organisms (and indeed all organisms) are established by complex rules and procedures. The naming system was invented by the Swedish botanist Carolus Linnaeus, who published the definitive version of his system in 1758. The names fall into nested categories. Each category is called a taxon (plural: taxa). At the practical, everyday level, where one needs to distinguish one "kind" of organism from another, the term *species* is used (plural: species). A **species** is a kind of animal or plant, in the ordinary sense. Most of us don't need to think too carefully about exactly what a "kind" or species of organism is, but biologists need to do so, and their definition requires (among other things) that two parent organisms be able to contribute genes to an offspring that is itself reproductively fertile, in order for the two parents to qualify as members of the same species. The biologists' definition of a species is complicated by the facts that some kinds of organisms reproduce by budding, some populations consist entirely of females that reproduce parthenogenetically (in which case the offspring has only one parent), and other conceptual difficulties. For present purposes, it is important only to recognize that a species is a kind of organism, usually in the popularly accepted sense.

If we trace the ancestry of a species far back in time, we usually encounter ancestors that we would not consider to be of the same species as their living descendants. For example, were we able to examine the parents, grandparents, great-grandparents, and so on, of the living organisms of some species *B* back through millions of years into the past, we would eventually find ourselves dealing with ancestors that are different enough from their descendants, species *B*, that we would think of them as some other species—say, species *A*. Gradually, over the ages (or perhaps abruptly at some point), enough changes had accumulated in the lineage that the descendants could no longer be considered of the same species as their ancestors. Not only that, but certain populations of ancestral species *A*, in times past, may have become isolated from the others. The descendants of those populations, living today, we might call species *C*. Thus ancestral individuals of species *A*, living a few million years ago, have given rise to two modern species, *B* and *C*. In the Linnaean naming system, two or more modern species that can trace their ancestry back to a single species living in the past are clumped together in a single taxon, called a **genus** (plural: genera). For example, the glaucous-winged gull and the California gull are two species that are placed, by biologists, in the same genus. This means that, in the opinion of authorities on these gulls, each of the two species is a somewhat modified descendant of some single species that lived in the near past. The genus name is *Larus*, and their taxonomical names are *Larus glaucescens* and *Larus californicus*. Thus the name gives a hint of ancestry, in addition to identifying an organism.

Just as the organisms of living species can be clumped into genera, the organisms of various genera can be clumped into larger, more inclusive categories, called **families.** Again, the presumption is that all of the organisms in the same family are related through a common ancestor. Kittiwakes, for example, are small gull-like seabirds that differ in significant ways from gulls in the genus *Larus*. The several species of kittiwakes are all given the genus name *Rissa*. But species of the genus *Rissa*, and those of the genus *Larus*, as well as terns, are similar enough in most ways that all are placed in the family Laridae, reflecting the zoologists' opinion that all of these modern birds are descendants of some single species that lived in the more remote past. In a similar manner, the families of the Linnaean system can be grouped into **orders,** the orders into **classes,** the classes into **phyla,** and the phyla into **kingdoms.** To follow our gull example, the family Laridae is included in the order Charadriiformes, a group of families related by ancestry that

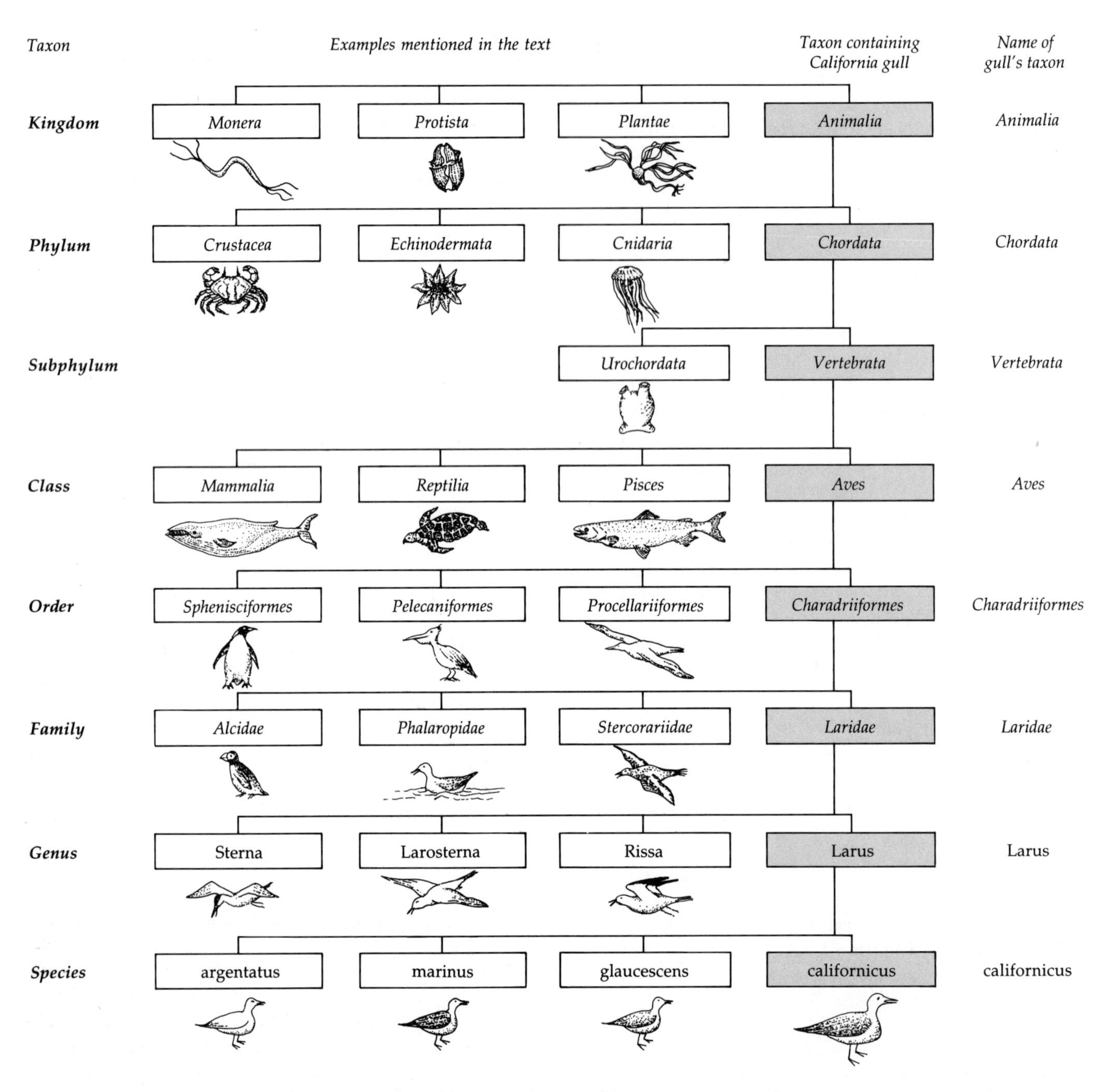

14.1 *The system of biological classification using the California gull* (Larus californicus) *as an example.*

includes (among seabirds) the family Alcidae—auks and their relatives. The order Charadriiformes is included in the class Aves, which includes all birds. The class Aves is included in the phylum Chordata, which includes all organisms with backbones, and the phylum Chordata is included in the kingdom Animalia, along with all other nonphotosynthetic animals on Earth. These relationships are illustrated in Figure 14.1.

For each of the successive taxa of related organisms, the ancestor of the taxon lived farther back in time. Ancestors of modern species that are placed by biologists in the same genus are usually considered to have lived only a few thousand to a few million years ago, whereas the common ancestors of organisms that we now place in the same phyla lived several hundreds of millions of years ago.

The Linnaean system works the same way for plants as it does for animals, with the exception that botanists use the term *division* where zoologists would say, *phylum*. There are five grand kingdoms of organisms: Animalia (animals), Plantae (land plants), Protista (single-celled creatures and related multicelled algae of simple architecture, whose cells are complex

Table 14.1 *Terms Used for Description of Marine Life and the Marine Environment*

Term	Definition	Example
Plankton	Organisms that drift passively with the currents. Nonswimmers or feeble swimmers.	jellyfish
Phytoplankton	Single-celled photosynthetic plankton organisms. Plant cells.	diatoms
Zooplankton	Animal members of the plankton.	copepod
Nannoplankton	Phyto- or zooplankton organisms that are between 2 and 20 micrometers in length.	microflagellates
Plankter	A term that refers to an individual organism of the plankton community. Phyto-plankter = a plant cell, zooplankter = a planktonic animal.	
Nekton	Those animals capable of swimming against a current.	fish
Benthos	The community of organisms that live on the sea bottom.	clams, worms
Infauna	Organisms that live buried in the ocean floor.	clam
Epifauna	Organisms that sit exposed on the ocean floor, on plants, or on pilings.	sea anemone
Benthic	Adjective that means "occurring on the bottom," as in "benthic organisms."	sea star
Pelagic	Adjective that means "occurs in water off the bottom," as in "pelagic fishes."	tuna
Neuston	The community of organisms that float at (or rest on) the sea surface.	man-of-war
Detritus	Dead organic matter.	
Eutrophic	Adjective that refers to nutrient-rich water in which plant growth is abundant; "eutrophic estuaries."	estuary
Oligotrophic	Adjective that refers to nutrient-poor waters in which growth of plants is sparse; "oligotrophic tropical seas."	Sargasso Sea
Deposit Feeder	An animal that eats benthic sediments. It lives by digesting the edible detritus in the sediments.	sea cucumber
Suspension Feeder	An animal that collects suspended edible particles of detritus from the water and eats them; uses tentacle-like structures to collect particles.	feather duster worm
Filter Feeder	An animal that uses filter-like structures to collect suspended edible particles from seawater.	mussel

or "eucaryotic"), Fungi (funguses), and Monera (bacteria, single-celled creatures with simple or "procaryotic" cells). Of these, the Animal, Protist and Moneran Kingdoms are the most prominent in the sea.

The names used in the following chapters (and in any biology text) are drawn from this system. The system provides an easy way of referring to whole groups of related organisms or to a particular species. (Larids, for example, refers to all of the closely related gull-like birds.) Because many of the organisms of the sea are unfamiliar and have no popular names, the Linnaean names are the only ones they have. The "correct" name of an organism, per the Linnaean system, is a two-word phrase. The first is the name of the genus (always capitalized); the second is the name of the species itself (never capitalized). Both words are italicized. The names are always "latinized," or written as though they were taken from Latin. Pronunciation is often a problem. These "official" names are usually used by biologists in preference to the common local names of organisms, to avoid confusion. The name "bullhead" can mean any of several different fishes, depending on whether you are on the Atlantic, Pacific, or Gulf coasts, or even whether you're fishing in a midwestern creek. But the name *Leptocottus armatus* is unmistakeable to biologists, be they from Japan, Pakistan, the United States, or elsewhere; it means the "bullhead" or staghorn sculpin of the U.S. West Coast, and there is no other organism in the world that bears this title. For all of these reasons, then—the absence of common names, the ease of referring to larger or smaller groups, the elimination of ambiguity, and as a handy way of keeping track of relationships—names from the Linnaean system are used in scientific works. Certain terms that are used in discussions of marine life, and that appear in Chapters 14, 15, and 16, are defined in Table 14.1.

BACTERIA: THE KINGDOM MONERA

Bacteria are single-celled organisms that differ from the rest of living things in profound ways. The internal architecture of their cells is much less complex than is that of "higher" cells. Bacteria are also much tinier

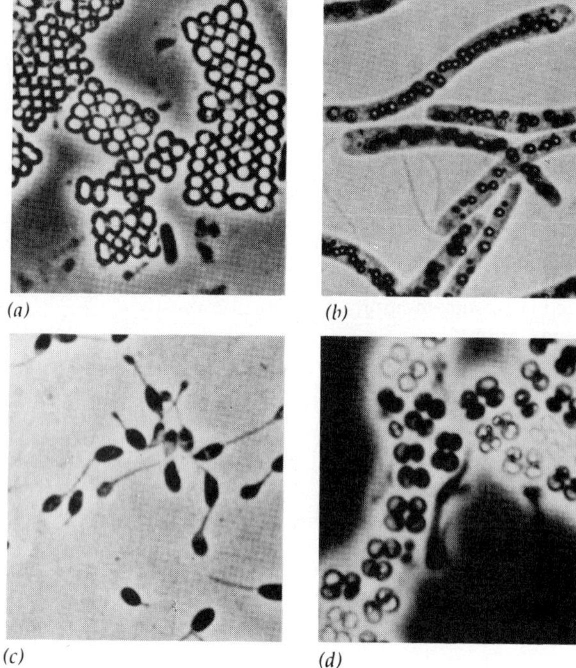

14.2 *Photosynthetic bacteria: (a)* Thiopedia, *(b)* Thiospirillum, *(c)* Rhodopseudomonas, *(d)* Thiocapsa. *(a), (b), and (d) are purple sulfur bacteria. Sulfur bacteria use various sulfur compounds in their metabolism and play an important role in the turnover of sulfur in the sea. (From W. Jensen and A. Salisbury,* Botany: An Ecological Approach, *Belmont, Calif.: Wadsworth, 1972)*

than protists or the cells of larger organisms. They are little more than complex living chemicals, but they are able to make dramatic changes in the chemistry of their environments when the right conditions prevail. They reproduce rapidly, dividing every 20 minutes or so when conditions are favorable. The simple rodlike or rounded forms of bacteria are illustrated by the photosynthetic marine bacteria shown in Figure 14.2.

Fossil evidence indicates that bacteria are by far the oldest life forms on Earth. They were present some 3.5 billion years ago, appearing 2 billion years earlier than the complex cells that eventually gave rise to protists and "higher" life forms. Such early photosynthetic forms as cyanobacteria (sometimes called blue-green algae, Figure 14.3) probably added the oxygen to the earth's atmosphere. During their long tenure on the earth, they have acquired a stupendous variety of chemical tricks—and therein lies a key to their importance. Higher life forms are limited in their abilities to digest organic materials. Bacteria can digest particles of cellulose and other **refractory** materials that defy the digestive tracts of most animals. These refractory materials are created by plants and animals themselves, in the course of their daily activities: growth, digestion, excretion, and the like. Bacteria provide the only easy means of transforming them back to material that is useful to the larger organisms of the ecosystem. Bacteria themselves are nutritious food for organisms that collect and eat them. By tapping the energy and materials in the abundant refractory material and serving as food for the rest of the marine community, bacteria return materials that would otherwise be lost to the higher organisms. Without them, marine systems would grind to a halt and suffocate in a litter of indigestible detritus.

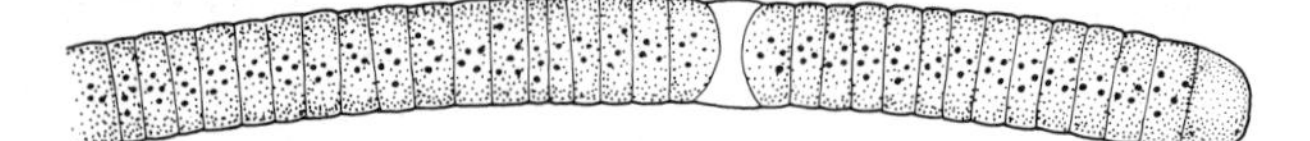

14.3 Oscillatoria, *a cyanobacterium, or blue-green alga. Many cyanobacteria are nitrogen fixers and play crucial roles in the global cycling of nitrogen compounds. Diameter: about 10 μm.*

Because of their small size and specific requirements, marine bacteria are difficult to study. It is not easy to estimate their contribution to the overall economy of the sea, but all indications are that the importance of bacteria is vast and far disproportionate to their small size. They are the prime retrievers of organic molecules that "leak out" of marine plants; this can amount to 35–50% of all the new material that the plants produce. They mop up the material spilled and wasted by feeding copepods. They convert refractory material back to edible material. The chemical wizardry of bacteria that inhabit anaerobic muds is crucial to the turnover of sulfur in the sea; other bacteria mediate the critically important cycling of nitrogen nutrients. Sulfur bacteria have recently been discovered manufacturing primary foodstuff at the sea bottom, in the vicinity of hydrothermal springs. These are important in sustaining the richly populated communities of larger organisms. Considering all factors, marine biologists find it entirely reasonable to suppose that fully 50% of all the new material produced by phytoplankton is consumed by bacteria before making its way to the rest of the living community. Bacteria probably process more than half of the organic material that settles to the bottom, before they, in turn, become food for the deposit feeders there. Invisible though they may be to the naked eye, they are the prime-moving forces behind all marine systems.

The roles of bacteria in nutrient cycling and hydrothermal vent ecology are discussed in detail in Chapter 16.

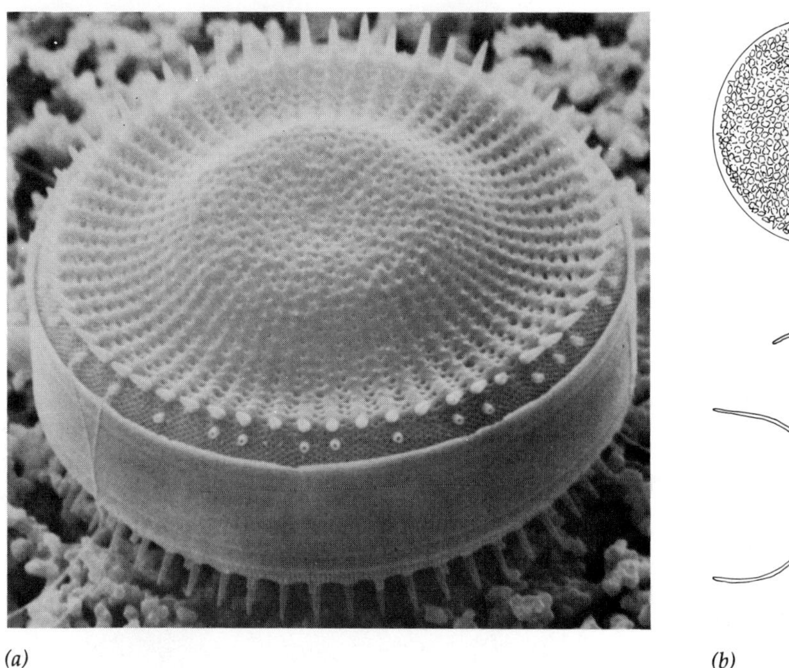

(a)

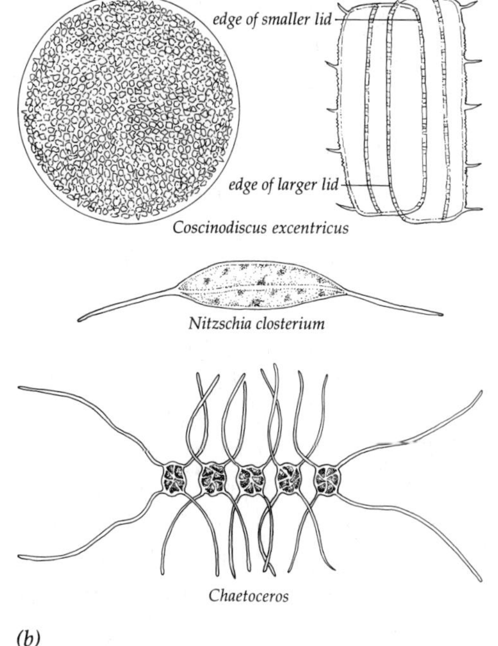

(b)

14.4 Diatoms. (a) Scanning electron micrograph of a diatom, showing the intricately sculptured silica test, or shell. 580×. (Starr, C., and R. Taggart, Biology: The Unity and Diversity of Life, *Belmont, Cal.: Wadsworth, 1981) (b) Views of other diatoms, showing a cylindrical shape* (Coscinodiscus), *an elongated form with spines* (Nitzschia), *and a chain of cells linked by their silica bristles* (Chaetoceros). *Sizes:* Coscinodiscus, *up to 1000 μm diameter;* Nitschia, *about 60μm total length;* Chaetoceros *cells, about 50μm wide. (After Lebour, M. V.,* The Planktonic Diatoms of Northern Seas, *London: Ray Society, 1930)*

SINGLE-CELLED ORGANISMS: THE KINGDOM PROTISTA

Although single-celled organisms are not conspicuous at first glance, they are among the most important players in the ecology of the sea. Of the plantlike unicellular protists, four groups, the diatoms, dinoflagellates, microflagellates, and coccolithophores, are particularly important. Considering the ocean as a whole, these phytoplankton organisms are responsible for about 98% of all marine plant growth. Many other marine protists, including forams and radiolarians, are less prominent in the ecology of the sea, but they are of great interest to students of the earth's recent biological and climatic history.

Diatoms

Diatoms (division: Chrysophyta) are single-celled photosynthesizers whose construction is strangely different from that of more typical plant cells (Figure 14.4). A diatom cell consists of a droplet of oil, surrounded by cytoplasm that is in turn encased in two tiny siliceous shells, or "tests." The tests overlap, one forming a lid over the other. These opaline tests are beautifully ornate and occur in a wide variety of forms. Some are solitary, others are linked in chains. When diatoms reproduce by simple fission, the two tests partially separate, the cell cytoplasm divides, and the cell nucleus creates a duplicate of itself. Two new tests are grown, abutting each other inside the original pair, before division is complete. When the two new diatoms finally separate, one carries the old lid (and is identical in size to the parent cell), the other's lid is the parent's former bottom shell, now acting as a lid that encloses a slightly smaller new bottom shell. This makes that cell smaller than its parent. This process is occasionally interrupted by a complex sexual process, one result of which is to reverse the progressive decrease in size of some of the offspring.

Benthic diatoms grow on the sea bottom and on the surfaces of shells, plants, and other objects. These tiny plants give an apparently "barren" sea bottom a biological productivity that can be astonishing. Hardly noticeable to the naked eye, they give a greenish-brown appearance to sand, rocks, eelgrass, and other objects

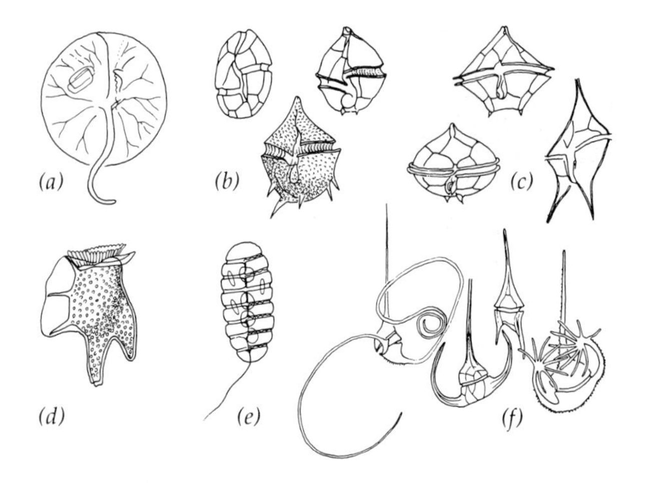

14.5 Dinoflagellates: (a) Noctiluca, *(b)* Gonyaulax, *(c)* Peridinium, *(d)* Dinophysis, *(e)* Polykrikos, *and (f)* Ceratium. *Sizes:* Noctiluca, *about 1000 μm; others, 40–300 μm. Most flagella are not shown.*

14.6 Photomicrograph of Gonyaulax polyedra, *magnified 500 times (arrows).* Gonyaulax *is a genus of luminescent marine dinoflagellates that cause red tides along ocean shorelines when they become abnormally abundant. Size: about 40μm.*

that they cover. They are capable of doubling their numbers every twenty-four hours. In the Grevelingen estuary of Holland, the "bare mud" produces as much new plant matter every day as the scattered patches of eelgrass do, thanks to its coating of benthic diatoms. Grazing snails, chitons, and limpets, moving about on apparently bare rock, are usually eating diatoms. Such is the phenomenal growth of diatoms that an animal can return to a place cleared of its diatoms on a previous day and find as much to eat as was there a few days before. In this way, diatoms contribute to benthic food webs in an amount that is far disproportionate to their size.

Many diatoms drift in the waters above the sea bottom. As drifters, they are members of the phytoplankton. Planktonic and benthic diatoms are usually of different species. The planktonic species face a number of problems unique to small, drifting plant cells, the greatest of which is the likelihood of sinking out of the sunlit surface zone into dark depths where they cannot photosynthesize. Like their benthic relatives, planktonic diatoms play an important role in marine ecology. Far from shore, beyond the continental shelves, they are often the most abundant (or only) plants available to herbivores. They sustain the marine food web in such regions of the offshore ocean. Like their benthic relatives, they can multiply rapidly when provided with light and nutrients. In an area teeming with fish, crustaceans, squids, and other large organisms, it is usually not evident that the whole community is being fueled by the photosynthesis of these tiny, sparsely distributed diatoms. Reproducing at a phenomenal rate, they are harvested by tiny crustaceans, who are in turn passed along the food web to the larger creatures. Diatoms tend to occur in greater numbers in cold and temperate seas, although some are found in all oceans.

Dinoflagellates

The role of diatoms in tropical warm waters is assumed by protists of a different sort, the dinoflagellates (division: Pyrrophyta; Figure 14.5). Most **dinoflagellates** are single-celled photosynthesizers. They take a variety of forms. Commonly, the dinoflagellate architecture involves minute plates of a cellulose-like material, joined at the edges to form a top-shaped cell with a groove around the middle. Unlike diatoms, dinoflagellates are equipped with two whiplike hairs, the flagella, whose lashing drives the cell through the water. Although these cells have a feeble ability to move themselves, they are still powerless to resist being swept away by currents and so are included among the phytoplankton. Some dinoflagellates depart from this basic body plan in bizarre ways. Those of the genus *Noctiluca*, for example, look like tiny spheres of jelly with

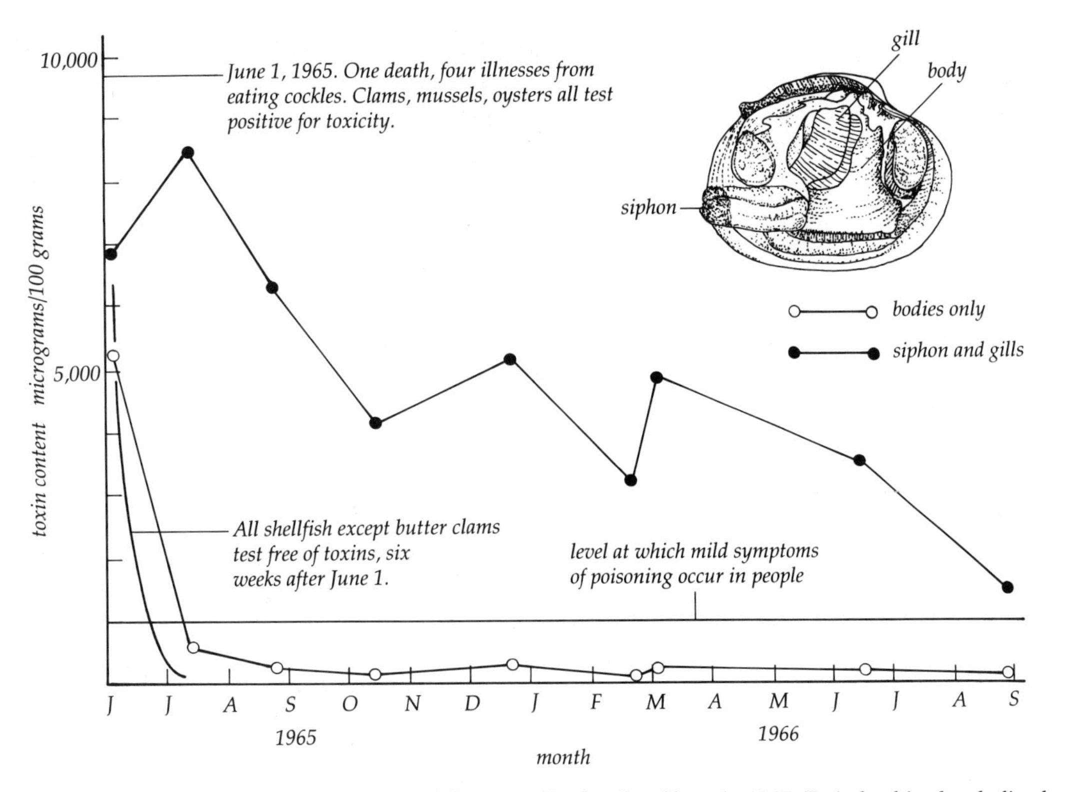

14.7 *Persistence of toxicity in butter clams following a dinoflagellate bloom in 1965. Toxin level in clam bodies dropped below 1000 μg/100 g after about 6 weeks. Toxicity in clam siphons and gills rose, then dropped slowly, still exceeded 1000 μg/100 g over a year later. Other shellfish reached safe level (less than 88 μg/100 g) within 6 weeks. No further dinoflagellate blooms occurred during this period. British Columbia. (After Quayle, D. B., and N. Bourne,* The Clam Fisheries of British Columbia, *Fish. Res. Bd. Canada, 1972, Bulletin 179)*

one of the two flagella swollen to resemble a tentacle (Figure 14.5a). As strange in its habits as it is in its anatomy, *Noctiluca* has no cholorophyll and cannot photosynthesize. It captures microscopic prey and engulfs it.

Most of the "ordinary" dinoflagellates are planktonic. Like planktonic diatoms, they provide food for small crustaceans and ultimately for the predators that eat the crustaceans. Dinoflagellates tend to be most numerous in warm water; they do not usually sustain rich fisheries. In some ways, however, dinoflagellates make their presence known more forcefully than diatoms do. Many species are luminescent. When disturbed, they emit a tiny flash of light. When water containing a lot of dinoflagellates is stirred, a flash of light from the dinoflagellates outlines the disturbing object, and a glow persists for a few moments. Thus oars of a boat rowed at night leave glowing phosphorescent circles in the water; the wake is outlined in greenish light, and one can see streaks of light marking the passage of fish startled by the boat. A scuba dive in phosphorescent water at night is a spine-tingling experience, with every movement of one's companions and those of the nearby fish lit up in what looks like eerie green fire. Dinoflagellates, including species of the genus *Noctiluca,* are largely responsible for this.

Dinoflagellates are responsible for "red tides" and a phenomenon called "paralytic shellfish poisoning." Certain species of the dinoflagellate genera *Gonyaulax* and *Gymnodinium* carry a poison that is toxic to vertebrates (Figure 14.6). When unusual conditions favor the reproduction of these normally scarce organisms, they become so numerous as to give the water a noticeable red color. Toxins released by the dinoflagellates under these conditions can kill fish by the thousands, causing them to wash up on shore with dead birds, sea mammals, and other organisms. Invertebrates are, for the most part, scarcely affected by the dinoflagellate toxin. Mussels, clams, and other shellfish consume the toxic species in great numbers, digesting the dinoflagellates and storing the toxins in their tissues. The toxin is not destroyed by cooking; persons collecting and eating shellfish that have been concentrating the toxin in this way can be poisoned and killed. Eventually, if undisturbed, the shellfish itself will neutralize the toxin and become edible again. Some species can do this relatively quickly after a red tide has passed, whereas others require years to detoxify themselves (Figure 14.7).

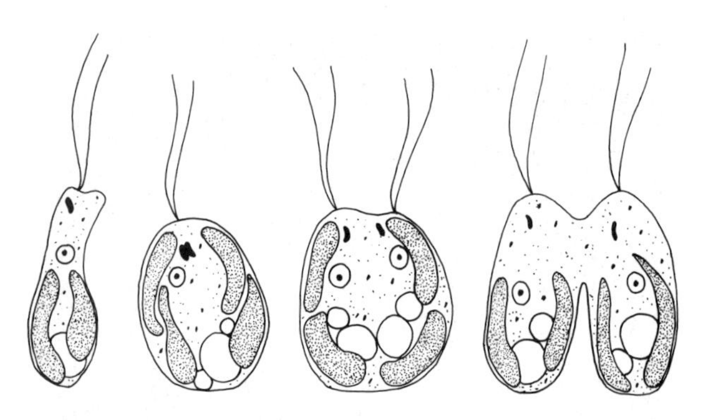

14.8 *Microflagellates. Minute cells similar to these can be more numerous in water than diatoms and dinoflagellates. Their small size makes them difficult to detect. They are important as food for invertebrate larvae and some other plankton animals. Species illustrated is* Isochrysis galbana, *shown dividing. (After Parke, M.,* Studies on Marine Flagellates, Jour. Mar. Biol. Ass. *28:255–86)*

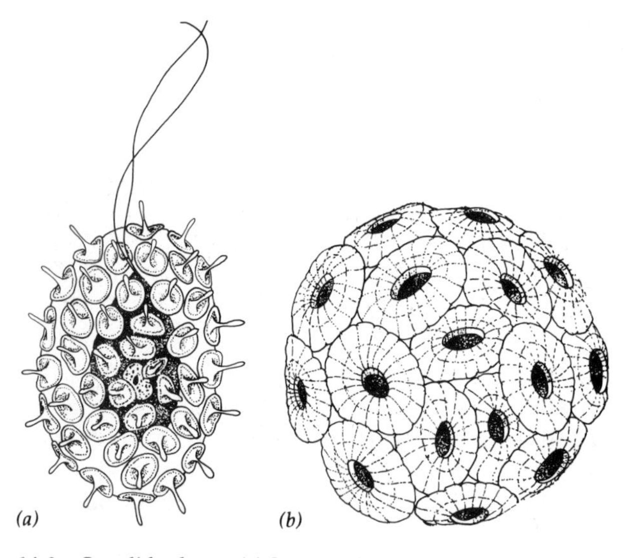

14.9 *Coccolithophores. (a)* Syracosphaera, *(b) illustration showing the calcareous plates carried by a single cell,* Emiliana huxleyi. *Sizes:* Syracosphaera *about 40 μm.* Emiliana *about 7 μm. Source: ([a] After Raymond, J.E.G.,* Plankton and Productivity in the Oceans, *New York: Pergamon Press, 1963. [b] After Itonjo, S., & K. O. Emery. In* Volcanoes & Techosphere, *H. Aoki & S. Iizuka, Tokai University)*

Perhaps the most peculiar dinoflagellates are those that have entered into permanent alliance with other organisms, to serve as symbiotic photosynthesizers. Such dinoflagellates are known by the general term *zooxanthellae.* These cells have taken up permanent residence inside the cells of other organisms, notably reef-building corals, giant clams, certain sea anemones, and even such single-celled organisms as forams. This association usually occurs in warm, nutrient-poor waters. The zooxanthellae, embedded in the cells of their hosts, use the waste nitrogen, phosphorous, and CO_2 of their host as nutrients, while supplying the host with oxygen and manufactured organic foodstuffs. Both host and zooxanthellae benefit. In some cases, the activity of the zooxanthellae is so prevalent that 100% of the host animal's requirements are met by the zooxanthellae. Such animals as the Australian "blue dragon," a beautiful sea slug, need never feed after their bodies become infested by zooxanthellae. These beneficial partnerships between zooxanthellae and larger animals are responsible for the high productivity of coral-reef communities (as discussed in Chapter 16).

The roles of dinoflagellates, then, are varied. Some provide ordinary planktonic fodder for small crustaceans, much in the same way as the diatoms. Others jar the marine system on occasion, producing toxic red tides. Still others operate as partners with much larger organisms, producing food in return for raw materials and protection. Together, these roles are significant in the economy of the sea.

Microflagellates

Microflagellates are tiny single-celled plants that differ from diatoms and dinoflagellates in significant ways. They comprise several taxonomic divisions. As their name implies, they are small, even when compared with other single-celled organisms. A large diatom or a *Noctiluca* dinoflagellate is about 1,000 μm (1 mm) in diameter; small but distinctly visible to the naked eye. By contrast, microflagellates range from about 2 to 60 micrometers in size and are difficult to see even under a microscope. They usually contain green cholorophyll and starches that differ from the comparable molecules in diatoms and dinoflagellates, and they are equipped with two flagella (Figure 14.8). It is now clear that microflagellates can be of extraordinary ecological significance. Careful studies show that their productivity, despite their small size, is comparable to (or greater than) that of the larger diatoms or dinoflagellates in the water around them. Indeed, waters that have low diatom productivity can nevertheless have high microflagellate productivity. This translates into availability of food for various animals, although in a somewhat biased way. Because the microflagellates are small, only specially equipped animals can collect and eat them. This includes filter-feeding animals that pass seawater through ultra-fine filters. Many benthic animals (for example, clams and tunicates) do this, as do certain swimming animals (such as larvaceans, discussed below). Other animals that eat microflagellates are tiny enough to make a reasonably full meal of individual cells. These include the larvae of many marine animals, which spend part of their early stages adrift in the plankton. Large diatoms, with their glass tests, are too formidable for such creatures. Thus the soft-bodied microflagellates (and small diatoms) are crucial to their feeding and development. By supplementing (or even surpassing) the productivity of diatoms and dinoflagellates and by providing food for the larvae of benthic animals, microflagellates play a leading role in the ecology of the sea.

Coccolithophores (division: Chrysophyta) are especially prominent among the microflagellates. Coccolithophores are tiny photosynthesizers that occur mainly in warm seas. Some have flagella and are motile, others have none. All secrete minute calcareous plates, which adhere to the sides of the tiny cells like a coating of shingles (Figure 14.9). (Their name means "little stone carriers.") These tiny cells can outnumber dinoflagellates and diatoms combined, especially in warm seas. In the warm Sargasso Sea, for example, they are the most numerous phytoplankton organisms throughout most of the year.

The plates secreted by coccolithophores are beautifully ornamented and easy to distinguish. These accumulate in seabed sediments, under the waters in which the coccolithophores live. Most species are restricted in their distribution by sea-surface temperatures. Thus a study of the distribution of the plates in seafloor sediments shows how sea-surface temperatures have changed in past times.

Forams and Radiolarians

Animal protists are less prominent in the ecology of the sea than plants are, although they are of interest for other reasons. Foremost among these are the order Foraminiferida and the subclass Radiolaria. Both are single-celled creatures that secrete hard skeletons. The skeletons of **forams** are mostly calcareous, those of **radiolarians** are usually siliceous. The skeletons of both types are often exquisitely ornamented (Figure 14.10). They sink to the sea bottom and can dominate certain sediments (Chapter 6). These are called "foraminiferan (or radiolarian) oozes." Like the coccolithophores, foram species living in a particular area are finely attuned to the prevailing conditions of temperature and salinity, and soon disappear, to be replaced by other species, if those conditions change. Changes in the species preserved in the seabed oozes, therefore, provide important clues to climate changes of the past. Important forams, often mentioned in studies of this sort, are those of the genus *Globigerina*.

The study of forams is of great importance to petroleum geologists, who rely heavily on fossil foraminiferans as indicators of strata in which oil may be found. Representative foram skeletons are shown in Figure 14.11.

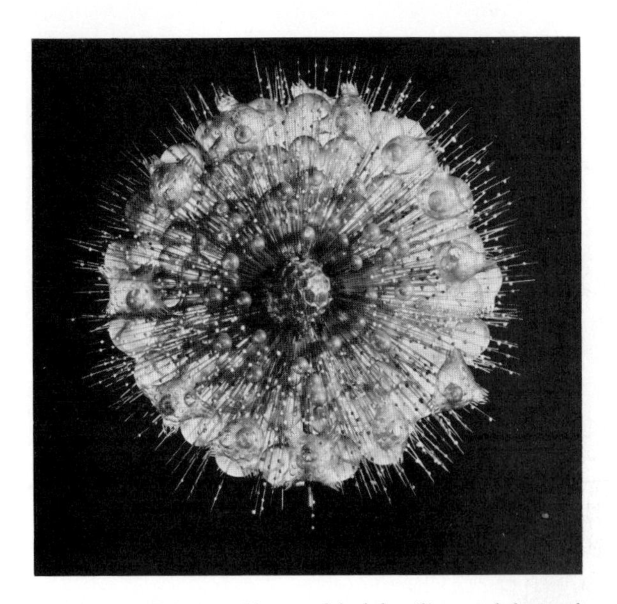

14.10 A radiolarian. Glass model of the siliceous skeleton of a colonial radiolarian, Trypanosphaera regina. *Single-celled radiolarians secrete skeletons similar to this in general appearance.*

14.11 Foraminiferan skeletons. The calcareous skeletons are expanded as the unicellular ameba-like organism grows older. Foram shells are abundant in some ocean sediments. Sizes: about 500μm.

MULTICELLULAR PLANTS: THE KINGDOMS PROTISTA AND PLANTAE

Multicellular marine plants include seaweeds, a few green invaders from the land, a multitude of stony encrusting algae that do not resemble plants at all, and a fringe of mangroves and salt-marsh grasses that line many of the marine shores of the world. These marine members of the botanists' kingdoms Plantae and Protista are overshadowed by the phytoplankton on a global scale, although they are often more important than the phytoplankton in providing food for animals in shallow water.

Seaweeds

The most familiar submerged marine plants are the **seaweeds.** These are members of three botanical divisions. They can often be distinguished by their colors

14.12 Nereocystis, *a large brown alga or "kelp," stranded on a beach. Fronds, bladder, and stipe are shown. Length: to 30 m. (Steve Renick)*

14.13 Macrocystis, *a kelp, forms dense forestlike canopies. Each "leaf" or frond is attached to a gas-filled float; the floats are attached to a long central stipe. The stipe can reach 30 m in length. (Photo by Kelco)*

alone: the green algae (division: Chlorophyta), the brown algae (division: Phaeophyta), and the red algae (division: Rhodophyta). They differ from each other in fundamental ways, including the makeup of their chlorophyll, the presence or absence of various plant pigments (these give them their colors but perform other functions as well), their manner of storage of the materials they produce while photosynthesizing, their life cycles, and other key details.

Compared with the straightforward life cycles of the larger land plants, those of seaweeds are complex. It is not unusual for the tiny, swarming reproductive units that leave an adult seaweed to grow into something that looks nothing at all like the parent plant. Thus the "offspring" of the bull kelp *Nereocystis* of the U.S. West Coast, a plant that can grow to 30 m in length (Figure 14.12), is a tiny, threadlike filament that only a specialist can find and recognize. These tiny plants are either male or female. They produce sperms or ova. The sperms swim to the female plants, where they fertilize the ova. These fertilized ova then develop into the large bull kelps. The large plants eventually release spores to complete the cycle.

Seaweed life cycles may be completed in one, two, or three steps. Brown and green algae typically have a two-step process (adult to offspring, and offspring to adult), whereas red algae sometimes use a three-step process (adult to offspring #1 to offspring #2 to adult). The two (or three) forms may have different ecologies. A seaweed at a tiny stage in its life cycle (as in *Nereocystis)* can be devoured by a grazing snail, whereas the large adult form may be immune to this sort of attack. The two forms may differ in their need for bright or dim light, and they may differ in other ways. Thus a range of favorable conditions, suitable for all stages of the life cycle, is needed if the species is to prosper in a particular location.

14.14 *Brown algae (division: Phaeophyta). From left to right:* Nereocystis, Pylaiella, Fucus, Laminaria, Sargassum, Ralfsia *(on rock). Lengths: to 30 m; 5 cm; 20 cm; 100 cm; 100 cm; 2 cm respectively.*

Members of the three seaweed phyla range in size from single-celled to very large. The brown algae become the largest. These plants, commonly called kelps, form the huge offshore kelp beds of the U.S. West Coast (Figure 14.13). Smaller brown algae include the familiar brown rockweed *Fucus* of Atlantic and Pacific shores, and the weed *Sargassum* from which the Sargasso Sea takes its name (Figure 14.14). Green algae include the familiar sea lettuce, *Ulva,* and a plant known to East Coast oystermen as the "oyster thief" (*Codium*). Some of these (Figure 14.15b) secrete hard calcareous chips as they grow. Red algae include the "Irish moss" (*Chondrus)* of the U.S. East Coast, and the plants from which the Japanese specialty dish, *nori,* is made (*Porphyra*). Red algae are typically small and often feathery plants (Figure 14.16). Brown algae tend to be characteristic of cold seas, whereas reds and greens are more numerous in warmer waters.

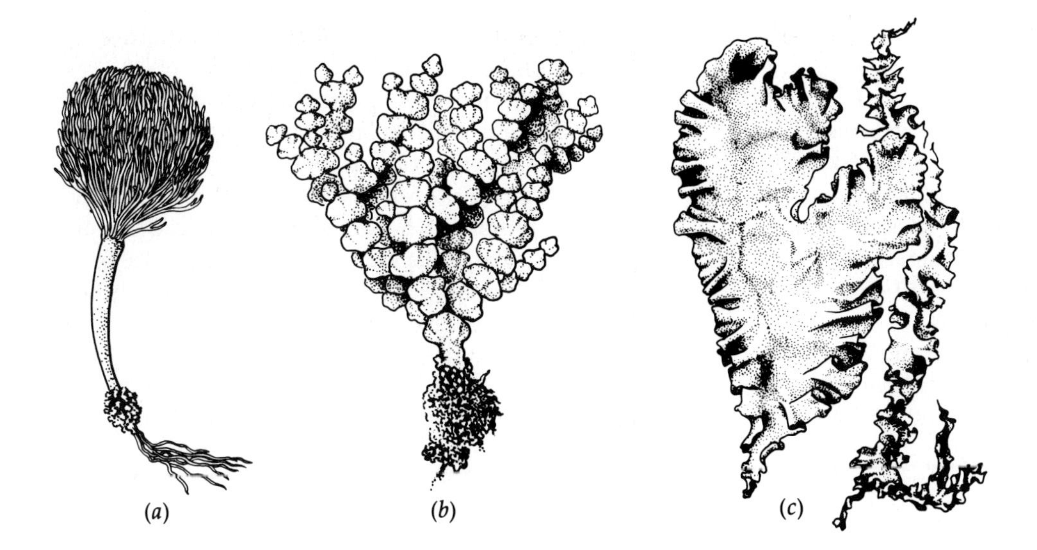

14.15 Green algae (division: Chlorophyta). (a) Penicillus *and (b)* Halimeda *contribute large amounts of calcium carbonate to shallow tropical sediments. The joints of* Halimeda *are stony and brittle. (c) Two species of* Ulva, *or sea lettuce, a noncalcareous green alga. Lengths:* Penicillus, *about 15 cm;* Halimeda, *about 30 cm;* Ulva, *to about 30 cm.*

14.16 Red algae (division: Rhodophyta). From left to right: Cumagloia, Polyneura, Callophyllis, Nemalion, Agardhiella, *and* Ptilota. *Lengths:* Cumagloia, *4 cm;* Polyneura, *up to 30 cm;* Callophyllis, *10 cm;* Nemalion, *10 cm;* Agardhiella, *6 cm;* Ptilota, *6 cm.*

Seaweeds are necessarily restricted to growing near the surface of the sea, where sunlight is available. For practical purposes, that means growing attached to the bottom in shallow water. While it is possible to envision floating seaweeds occurring everywhere in the sea (practicing lifestyles similar to the freshwater water hyacinth, for example), in practice, few seaweeds do so. Apparently, the problem is that ocean currents would inevitably carry floating weeds out of areas favorable to their growth, either to be eventually cast ashore or to be taken to a region with a hostile climate. One of the few regions in the sea hospitable to floating seaweeds is the Sargasso Sea, an area in the center of a gyre formed by the Gulf Stream and other Atlantic currents. Here, the water is dotted with patches of *Sargassum* weed. With this exception, it is accurate to say that the great majority of larger plants in the sea thrive only where they can attach themselves to the bottom in shallow water. Not all bottom conditions will do. Unlike land plants, marine seaweeds do not have rootlike structures that penetrate into soft sediment. They usually require firm surfaces—rock, logs, shells, another plant—for their attachment. In some cases, the plant attaches itself by means of a "holdfast," a tangled mop of thick filaments that is superficially similar to the roots of a land plant, yet very different in anatomy and function.

Where the water is shallow, the bottom is solid, and light and nutrients are in plentiful supply, beds of marine seaweeds form some of the most productive natural plant communities on Earth. Here, they are vastly more prolific than the single-celled phytoplankton. Seaweeds of the genera *Chondrus, Laminaria, Agarum,* and others were compared with the phytoplankton of St. Margaret's Bay, Nova Scotia, by Karl H. Mann and his colleagues. The biologists found that, for the bay as a whole, the large attached seaweeds produced about three times as much new plant matter in a year as the phytoplankton. And considering only the shallow shore zone where the seaweeds are able to grow, the seaweeds outproduced the phytoplankton of that zone by nearly ten to one.

The plant matter produced by seaweeds provides a food bonanza for marine animals in more ways than one. Most directly, live seaweed can be eaten whole. A number of marine grazers, including sea urchins, worms, certain snails, sea hares, and others consume some algae directly. In warmer waters, many fish (including many of the most colorful species on reefs) also consume algae. Much of the seaweed, at least in colder seas, feeds the animals in a less direct way. Studies have shown that large kelps are "leaky" plants. They photosynthesize at a rapid rate, but as they do so, much of the new material they create leaks out into the water, to be swept away in the form of isolated organic molecules. These dissolved organic molecules are difficult for the animal community to recover. Many marine organisms can absorb them directly from the water, but most studies suggest that these loose molecules provide only a tiny percentage of the nutritive needs of animals that can absorb them. Much of this escaped material is used, however, by bacteria, which in turn become food for larger organisms. This loss by leakage accounts for an astonishing 35–50% of all new organic matter produced by kelps.

Seaweeds enter the food chain by yet another route. Most of them are not eaten alive (at least, in colder waters). Their blades wear off, or they end up detached and adrift and eventually break up and sink to the bottom. There, as pulverized dead matter (detritus), they are consumed by such deposit feeders as sea cucumbers, burrowing worms, burrowing urchins, and other scavengers. In this respect, temperate seaweeds are like the leaves of trees. Most of them are not eaten by browsers directly, while alive, but are consumed after they end up detached and defunct. In tropical situations, however, the nibbling of fishes at seaweeds is so incessant that much of the weed is eaten while still alive and functioning.

Other Plants

Not all of the large plants in the sea are algae. In some cases, the marine plants are angiosperms, flowering species whose nearest relatives live on land. These include the familiar eelgrasses of shallow waters of all U.S. coasts, the surfgrass of tidepools, and a few others such as "turtle grass" and "manatee grass." Like marine algae, these plants form dense and productive beds. They serve as the **substrate,** or foundation, for many smaller plants and animals, and they provide cover for such adult organisms as crabs. These plants may be very abundant locally, but on a worldwide scale, they are far outnumbered by true marine algae.

Some of the most important plants in the sea are hardly recognizable as plants. These are the crustose or coralline algae. Most of them belong to the division Rhodophyta (red algae), and indeed the stony crusts that they form on surfaces where they thrive are often pink. A distinguishing characteristic of all crustose algae is that they deposit calcium carbonate as they grow. Their cells are imbedded in the stony matrix that they secrete, with the result that the "plant" looks more like a patch, or a lump, or a jointed string of beads of pink rock, than a seaweed. These encrusting algae are important rock-formers. They grow over the rubble that breaks loose from coral reefs, cementing it together as they grow. In some situations, they are as important as the coral animals themselves for building up massive rocky reefs. Where jointed crustose algae are common, the chips, joints, and beads that they form eventually break loose and form a major fraction of the sand on nearby beaches.

Only the most specialized and determined animals are able to feed on such plants. Sea urchins are among the few that can do it. And in turn, crustose algae are among the few marine plants that can stand up to an onslaught by urchins. Where urchins are abundant, most of the softer species of algae are completely consumed. The denuded rock on which the urchins rest, however, is often pink—encrusted by stony red algae that even the urchins cannot completely eradicate. The parrot fish of tropical reefs, and some other organisms, can also gnaw away the rock-like encrusting algae, but these armored marine plants are well adapted to resist being eaten by most other marine organisms.

Backwater seashores are often lined with land plants that make major local contributions to the economy of marine life. In temperate regions, the most important of these plants form tidewater salt-marsh communities. In the tropics, great forests of mangroves line the quiet marine backwaters. Both types of community have disproportionately large effects on the marine life in the nearby waters.

The grasses of salt marshes and the small mangrove trees that make up mangrove swamps are land plants that have become adapted to exposure to salt water. This is no trivial accomplishment. Land plants maintain internal fluids that are fresher than seawater. Immersion in seawater causes salt to leak into the plant and water to leak out. Very few plants are able to resist fatal wilting under such conditions, but salt-marsh plants have evolved mechanisms that enable them to survive. Generally, the survival mechanism is very costly in energy. For example, a common salt-marsh grass, *Spartina,* expends about 77% of the total energy it fixes via photosynthesis simply to maintain its cells in working condition—an outlandish expenditure for grasslike plants, whose respiratory requirements are normally only about 25% of the total energy captured. If a land plant is able to survive saltwater exposure, however, there are many benefits that compensate for this cost. One benefit is that most plants of other species cannot live in a salt marsh; thus competition by other species is drastically reduced. Another is that the daily tidal change delivers fresh nutrients to the salt-marsh plants on a regular schedule—a situation that occurs almost nowhere else except on farmers' fields. Thus the advantages of life on a tidal flat outweigh the disadvantages for those few land plants that are able to establish themselves there.

Salt marsh and mangrove communities are highly productive. A number of animals eat salt-marsh grasses, including marine snails and grasshoppers. Many others, mostly smaller organisms, devour the diatom films and algal mats that form on and among the grass stems, particularly at high tide. But these grazers are not able to consume all of the gigantic productivity of a marsh. About half of the grass and other organic plant matter produced each year dies, drifts away, and eventually enters the detritus food chain in nearby deeper water.

Mangroves provide a productive habitat for crabs, oysters, snails, fish (some of which scamper about the mudflats or catch insects by spitting at them), and other marine organisms, not to mention monkeys, insects, parrots, and other tropical land animals. Like the salt marshes, mangrove swamps contribute much of their productivity to sustaining the marine detritus food chain. In addition to providing large supplies of dead stems and leaves to their respective locales, salt marshes and mangrove swamps provide cover for organisms and nursery areas in which the early life stages of many species are passed. Some mangrove trees, for example, prop themselves up by means of external root systems that form dense, impassable palisades (Figure 14.17). Oysters attached to these roots are safe from predation by sting rays, which cannot enter the densely packed root refuges.

To summarize, many multicellular plants contribute to the productivity of the seas. The most conspicuous are seaweeds, crustose algae, descendants of a few land plants that have successfully invaded the sea, and land plants growing in quiet tidal areas around the sea margin. Where conditions are right for them, these plants form the basis of some of the most biologically productive natural systems on Earth. The phytoplankton may rival the productivity of the large plants in some areas, but on an acre-for-acre basis, phytoplankton productivity is generally much less than that of macroscopic plants. On a global scale, however, phytoplankton productivity is much greater than that of the large plants, simply because the oceanic areas that phytoplankton inhabit are so much larger than the shallow, sunlit areas suitable for the larger plants. Impressive as the larger plants are, they account for only about 2% of the ocean's productivity as a whole; microscopic photosynthesizers account for all the rest.

14.17 A mangrove (Rhizophora) *growing in salt water in Everglades National Park, Florida. The tangled roots descending from the main branch are called stilt roots. They provide anchorage and protection for oysters and other organisms.*

INVERTEBRATE ANIMALS: THE KINGDOM ANIMALIA

Invertebrates constitute most of the animal life on Earth. They are diverse and play many different significant ecological roles in the sea. The sections that follow give an overview of those that are most conspicuous, or most significant, while necessarily overlooking many (including whole phyla) that are interesting, beautiful, or otherwise worth examining. A complete listing of all of the phyla, including those omitted in this chapter, can be found in Appendix IX.

Sponges

Simplest of the multicelled marine animals that we will consider here are the **sponges** (Figure 14.18). These organisms (phylum: Porifera) are so inanimate and passive that zoologists couldn't decide for many years whether they were plants, animals, or minerals. The sponges that are most easy to visualize are shaped like vases (Figure 14.19). The walls of the "vase" are penetrated by thousands of tiny pinholes. Water is drawn into the vase through the pinholes and is discharged through the large opening at the top. The water current is set up by the beating of millions of tiny flagella on cells within the sponge's walls. These same cells, and others, remove bacteria and similar-sized edible particles from the inflowing water stream and share them with the other cells that make up the sponge body. Thus the sponge is a sophisticated pump-and-filter system for extracting microscopic edible material from the water. More complex sponges differ in the architecture of their body walls, in their sizes, and in the fact that there may be several large water-exit openings on the body, rather than the single opening characteristic of vase-shaped sponges.

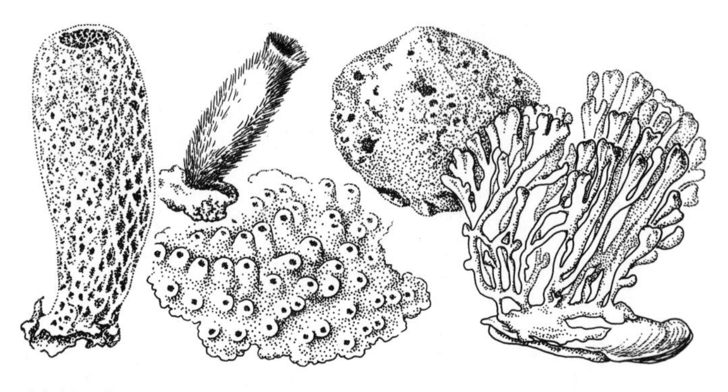

14.18 *Sponges (phylum: Porifera). Deep-sea glass sponge (left), shallow-water bath sponge (right rear), and branching sponge on a mussel shell reach 30 cm in length or width. Vase-shaped and encrusting sponges shown here project about 1 cm above the substrate.*

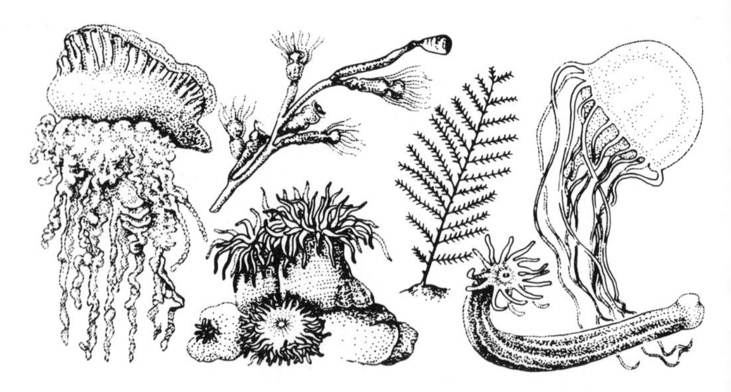

14.20 *Animals of the phylum Cnidaria. From left to right: man-of-war, sea anemones (lower), magnified view of hydrozoan polyps (over anemones), hydrozoan polyp colony, burrowing sea anemone (lower), jellyfish. Sizes: man-of-war tentacle lengths, to 3 m; anemones, to 20 cm tall; polyps, about 1 mm; polyp colony, 1 cm to 20 cm; burrowing anemone, to about 20 cm; jellyfish, 1 mm to more than 40 cm diameter.*

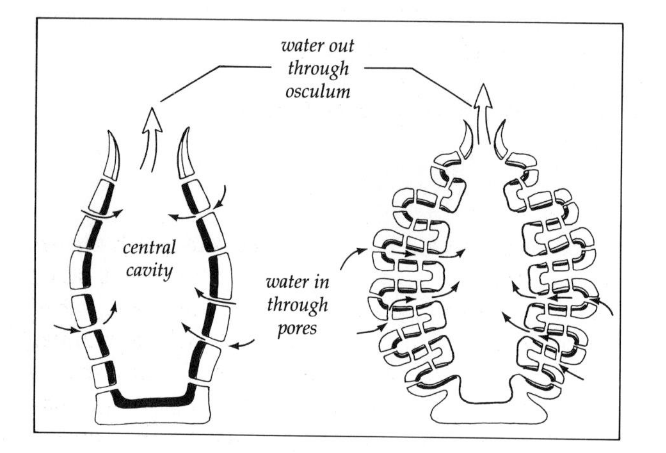

14.19 *Water currents generated by a simple sponge and a more complex sponge. Water is drawn in through microscopic pores. Edible particles are removed by specialized cells in the dark zones, and filtered water is ejected through the large opening (or osculum) on top.*

Sponges are minor elements of today's seas. They are relicts from much earlier times, when they were among the most sophisticated animals in the oceans. Today, little has changed in their lifestyle, but they have been largely overshadowed by other marine animals. In pursuit of their age-old task, however, they have adapted themselves to the new opportunities provided by more recently evolved organisms. Certain sponges are characteristically found, for example, on the upper shells of scallops. In this location, when an approaching predatory sea star touches the sponge, its tube feet withdraw for a critical instant—giving the scallop time to flutter its shells and swim away. The sponge provides the scallop with an early warning system, the scallop provides the filtering sponge with a competition-free surface and a free ride to other feeding grounds. Other sponges that begin by growing on shells occupied by hermit crabs end up dissolving the shell and providing a substitute coiled housing for the crab. The crab provides its filter-feeding host with mobility and stirred-up benthic sediment, and the sponge provides the crab with a "shell" that grows bigger with the crab. Sponges have few enemies (thanks to the fact that many of them are loaded with glassy spicules), although certain opisthobranch molluscs eat them. Some deep-sea shrimps make their way into the interiors of certain glassy sponges, where they become imprisoned (and sheltered) for life by the growing walls of the sponge. Other animals interact with sponges in other ways.

Like all filter-feeding animals, sponges have a cleansing effect on their surrounding water, if their population is large enough. One such instance was documented in the Caribbean, where zoologist Henry M. Reiswig showed that the sponges on a typical patch of reef were able to filter the whole volume of water, from surface to bottom (25 m deep in places) more than once every twenty-four hours.

There are about 5,000 species of sponges, almost all of them marine.

Cnidaria

The phylum Cnidaria consists of the jellyfish, sea anemones, corals, and related animals (Figure 14.20). The phylum consists of three classes: Scyphozoa (jellyfish), Anthozoa (anemones and corals), and Hydro-

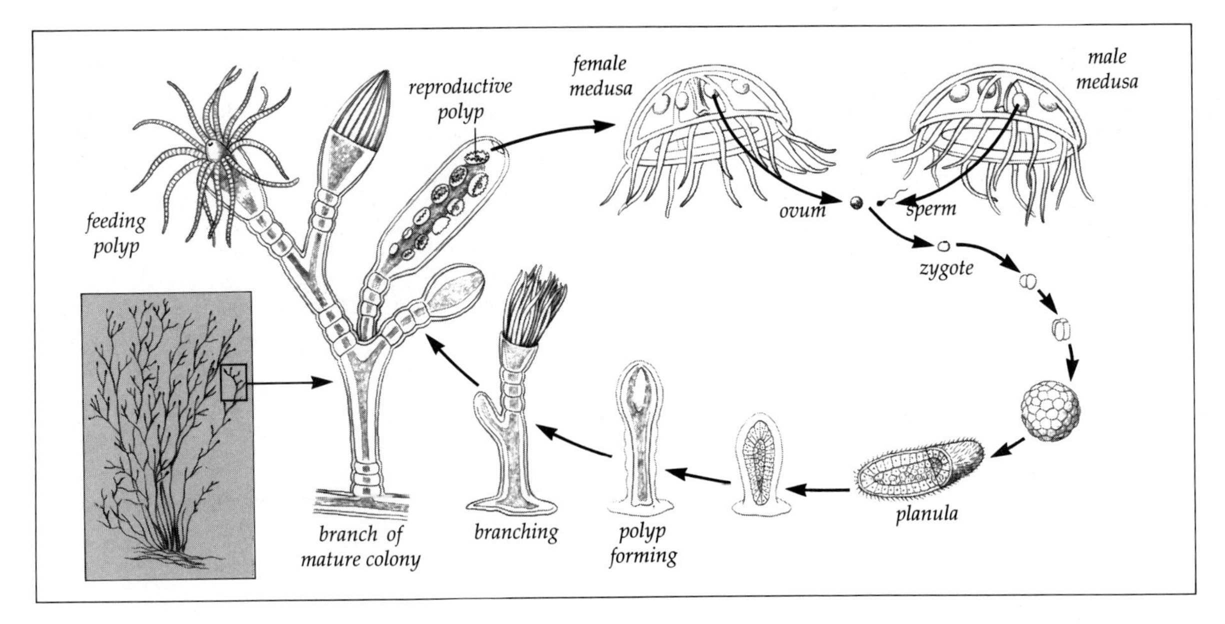

14.21 Life cycle of a hydrozoan (Obelia). *The reproductive polyp buds to form male or female jellyfish, which create fertilized eggs. The larva is planktonic but soon settles to the bottom. There it grows into a branching colony of polyps, completing the cycle. Inset shows a colony at about actual size. The polyp is about 1 mm long, the medusa is 2–10 mm in diameter.*

zoa (animals that may combine both jellyfish and anemone-like stages in the same life cycle). All **cnidarians** have in common the following features: They have no anus, their body makeup is structured around two main cell layers (in contrast to almost all other complex animals, which employ three layers), and they possess sophisticated stinging cells called cnidoblasts. Animals with an anemone body shape are referred to as polyps, those with a jellyfish form are called medusae (singular: medusa). There are about 9,000 species.

All cnidarians are carnivorous. An anemone sitting with tentacles spread, or a jellyfish trailing its tentacles, is a living trap, awaiting vulnerable animals that blunder into it. Some are equipped with particularly virulent stinging cells and can kill or paralyze fair-sized fish. Prey trapped by the tentacles is drawn to the mouth and stuffed inside to be digested. Because cnidarians lack anuses, they must excrete bones and other indigestible wastes back out of their mouths. Certain coral polyps hardly need to feed at all, since their metabolic needs are satisfied by the symbiotic zooxanthellae.

The cnidarians of the class Hydrozoa provide a clue to relationships among the other cnidarian classes, as well as a fascinating look at marine ecology on a smaller scale. A hydrozoan colony, appearing at first glance to be a brown, feathery clump on a mooring rope or piling, consists of several dozens of polyps attached to each other at their bases. The polyps are often encased in stiff, transparent tubes, with only their tentacles or crowns protruding. Each individual feeds on its own, but it also shares its catch via an internal digestive system with connections to all other polyps. A polyp colony is usually swarming with other small organisms, including copepods, ciliate protists, strange, mantis-like caprellid amphipods, and carnivorous opisthobranch sea slugs. Some browse on diatoms that grow on the polyps' tube system, others (the opisthobranchs) neatly nip off the crowns of the polyps.

In reproduction, a polyp colony produces buds, which develop into tiny jellyfish (Figure 14.21). These break free from the colony, swim away, and release eggs and sperms. The fertilized eggs develop into paramecium-like larvae ("planulae"; singular, planula) which eventually settle to the bottom and develop into the first polyps of new colonies. This polyp buds to form subsequent polyps. Hydrozoans of various species show variations on this theme, ranging from tiny polyps with large, long-lived medusas to large, long-lived polyps with tiny, short-lived medusas. In some hydrozoan species, the polyp stage is totally absent (as it often is in the jellyfish class Scyphozoa). In other hydrozoans, the medusa stage is totally absent (as it is in the anemone class Anthozoa). The reproductive methods of hydrozoans give evidence regarding the origins of the life cycles and body forms of the other two classes of cnidarians: Scyphozoa and Anthozoa.

Coral polyps are similar to sea anemones, except that they are smaller and occur as colonies (Figure 14.22). A common form of reproduction consists of

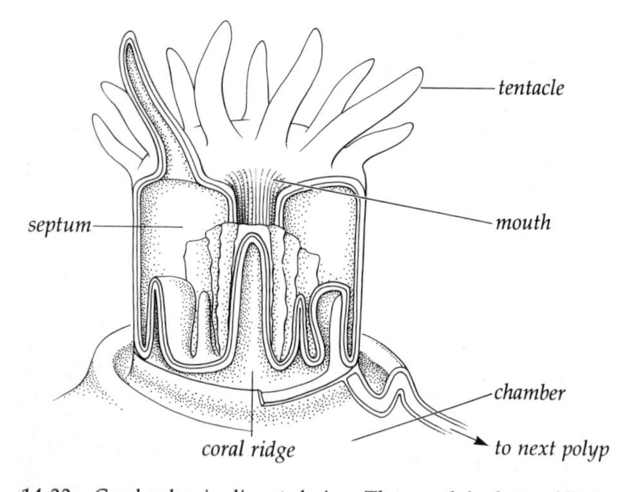

14.22 Coral polyp in dissected view. The mouth leads to a blind internal digestive cavity hung with vertical curtains, or "septa." The polyp occupies a stony chamber secreted by it, whose raised ridges fit the complex folds in the bottom of the organism. Connection with neighboring polyps is made via a "skin" that overlies the stony coral base.

14.23 Coral forms depend on water conditions and the growth pattern of each particular species. These fragile columns of Hawaiian Porites *coral grow in deeper quiet waters. (D. H. Milne)*

division down the middle. The two new polyps thus formed do not completely separate from each other but remain attached at their bases. The reef-forming ("hermatypic") coral polyps secrete calcium carbonate beneath their bases, thus building up a limestone platform on which the joined polyps sit. Pockets may be left in the limestone into which the polyps can withdraw. Different species reproduce in different ways and deposit different patterns of calcium carbonate. These result in the characteristic appearances of various coral structures: massive, boulderlike, fanlike, branching, or "brainlike" in appearance (Figure 14.23). These various coral forms are well adapted to the currents and surf conditions found at different depths and locations on the reef. Limestone deposition by corals is mainly limited to tropical waters that remain warmer than 23°C all year long. In such waters, an actively growing polyp community can lay down as much as 0.7 cm of massive new limestone every year. Branching corals can grow even faster, at least until they reach the sea surface. Polyps growing upward on a subsiding volcanic island can therefore remain near the sea surface, even though their substrate is slowly sinking beneath them, and they can form fringing reefs, barrier reefs, and atolls. This slow, patient limestone deposition by these tiny animals has created some of the most massive biological formations on Earth. The Great Barrier Reef of Australia, the world's largest contiguous reef system, stretches 1,950 km from its southern to its northern end, is up to 145 km wide in places, and consists of 120 m of limestone, the product of some 26 million years' deposition by coral polyps.

Some of the hydrozoans have particularly strange colonial forms. The infamous man-of-war of the U.S. East Coast and tropical regions is a colony of hydrozoan polyps (order: Siphonophora) in which life activities are shared by various specialized polyps (Figure 14.24). One individual is filled with gas, and it serves as a float. Others specialize in killing prey, digesting prey, or reproduction. This highly efficient collective is so well integrated that it appears to be a single organism. Its stings are dangerous; swimmers are advised to stay on the beaches when winds drive man-of-wars into shallow water.

Whether or not cnidarians (other than reef builders) are significant in the ecology of the sea is not clear. As watery creatures, jellyfish do not provide much food value to the animals that eat them. Thus, although jellyfish consume numbers of copepods, larval fish, barnacle larvae, and other prey, the food value of their prey appears to be "wasted" in the production of the "jelly." Investigation of marine food chains in which jellyfish and other gelatinous animals are prominent is now in progress. There is some reason to believe that food webs dominated by gelatinous animals ultimately produce large animals (such as the relatively unpalatable ocean sunfish) that are of less commercial value than those in which nongelatinous animals predominate, but it is too soon to be sure. From another viewpoint, jellyfish can be important as a marine hazard (Figure 14.25). Many have virulent stings. *Cyanea capillata*, a large yellow or brown jellyfish of both U.S. coasts, produces a distinct burning sensation when touched. Smaller jellyfish have more potent stings.

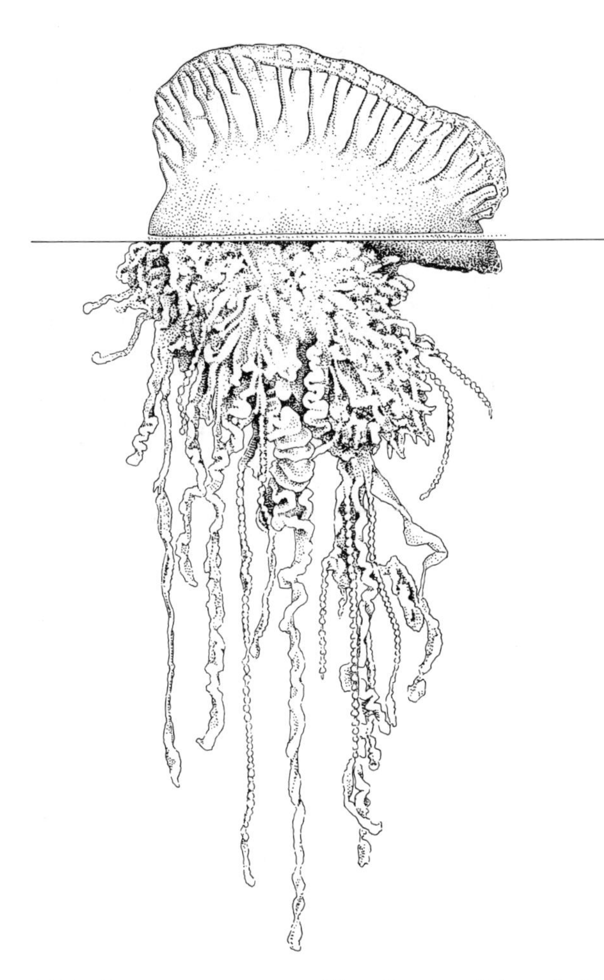

*14.24 Portuguese man-of-war (*Physalia physalia*), a colonial hydrozoan, composed of up to 1,000 specialized polyps and medusae integrated to act as an individual. Stinging cells on the tentacles contain a powerful nerve poison that immobilizes prey. Float length: about 20 cm.*

14.25 A large stinging jellyfish. Diameter: about 15 cm. (Steinhart Aquarium © 1976 Tom McHugh)

The small, nearly-invisible cubomedusae of Australian waters are known to have caused the deaths of swimmers within 10 minutes after contact.

Echinoderms

Another small but conspicuous phylum of marine animals is the phylum Echinodermata. The members include the sea star, brittle stars, sea urchins, sand dollars, sea cucumbers, and crinoids (Figure 14.26). Most of these animals are familiar, even to people who have not visited the sea. Crinoids are the exception; these animals (sometimes called sea lilies) resemble delicate ten-armed sea stars that are often attached to the sea bottom by stalks. They are mostly found in deep water, where they remain in flowerlike poses, practicing a suspension-feeding way of life.

The phylum name means "spiny skin." Almost all **echinoderms** have calcareous plates, spines, or ornaments embedded in their skins. These are usually small, but in the case of the sea urchins, the plates enclose the entire body in a rigid shell, and the spines are large and awesome (Figure 14.27). Another fundamental feature of echinoderms is their possession of an internal hydraulic system, unlike anything in animals of any other phylum (Figure 14.28). This hydraulic (water vascular) system shows itself externally, in the flexible tube feet with which certain echinoderms cling to rocks, aquarium glass, and other surfaces. Another common feature is their tendency to exhibit a fivefold radial symmetry, most evident in the five arms of a typical sea star. This unique group of animals is represented by about 6,000 species, virtually all of them marine.

Echinoderms practice diverse lifestyles. Most of the sea cucumbers, most brittle stars, many sea stars, and burrowing sea urchins are deposit feeders. Other echinoderms are suspension feeders. These sit poised

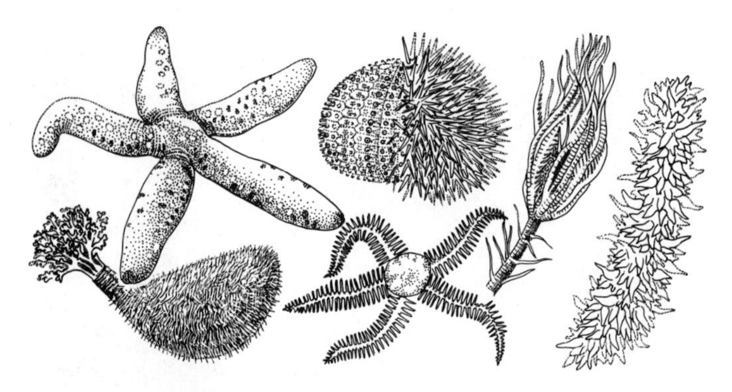

14.26 *Animals of the phylum Echinodermata. From left to right: sea star (top), sea cucumber (below), sea urchin with spines removed from left half, brittle star (below urchin), sea lily or crinoid, and sea cucumber. Sizes: sea star, to radius of 30 cm; cucumbers, 5–50 cm; urchins, to 20 cm diameter; brittle star, radius to 5 cm; sea lily stalk and crown, to 100 cm.*

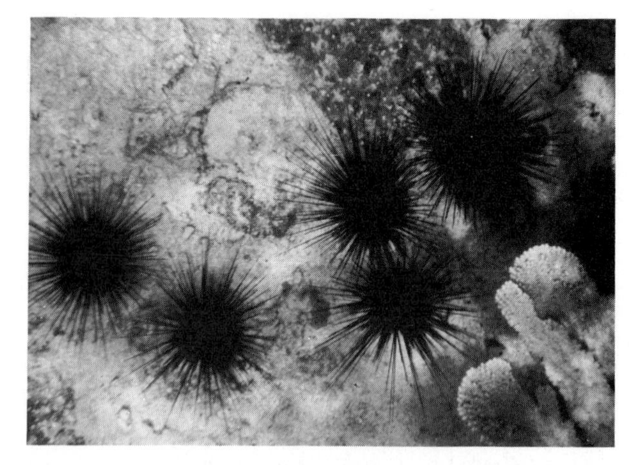

14.27 *Sea urchins (*Diadema*) from the Virgin Islands. Body diameter (excluding spines), about 6–10 cm. (Courtesy U.S. Department of the Interior, National Park Service)*

with tentacles or arms outspread, collecting edible particles that settle onto them and passing these to the mouth. Most crinoids feed in this way, as do some sea stars, a number of cucumbers, and many brittle stars. A few, such as some urchins and the bat star, are browsers on large seaweeds, and others, including many sea stars, are predators.

Echinoderms are capable of exerting profound influences on marine communities. The grazing of herbivorous sea urchins, for example, is often so intense that these animals have been called the underwater equivalent of forest fires. Urchins often live in dense herds, denuding the rocks of all large algae and leaving only a crust of resistant coralline algae in their place. In one study in Puget Sound by zoologists Robert Paine and R. L. Vadas it was discovered that urchins avoid eating a large seaweed (a species of *Agarum*), evidently because they find it distasteful. By browsing on other seaweed species that could compete with it, urchins make it possible for dense growths of *Agarum* to become established and to take over. Elsewhere, they prevent algae living in the intertidal zone from moving down onto subtidal rock, where the plants could prosper if it weren't for the sea urchins. As predators, sea stars are also known to alter or maintain the character of marine communities. Studies of a West Coast sea star (*Pisaster ochraceus*) also by Robert Paine showed that this species is responsible for the survival of some two dozen other invertebrate species on the bare rock of the mid-tidal zone. When the sea stars were experimentally removed, California mussels (the sea star's favorite prey) multiplied to such an extent that they occupied 100% of the available space and left no room for the anemones, goose barnacles, and other species that were able to coexist there prior to the experiment (Figure 14.29). A North Sea study showed that a clam (*Venus*) was assisted in its coexistence with

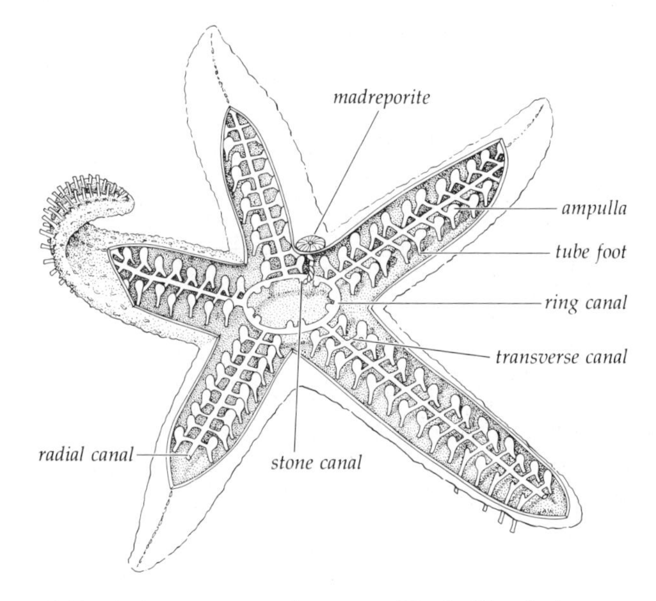

14.28 *Sea star water vascular system. The flexible tube feet (seen under the lower right ray) are connected to an internal system of tubes, squeeze bulbs ("ampullae"), and sieves. These enable the animal to climb vertical surfaces and to generate powerful enough suction to open the shells of a clam.*

a competitive species of clam (*Spisula*) by the feeding of sea stars. The star, *Astropecten*, swallows its prey whole. *Venus*, a slow-growing slow-metabolizing clam, is able to keep its shells shut for up to eighteen days, finally causing the sea star to give up and disgorge it unharmed. *Spisula*, a fast-growing, fast-metabolizing clam, opens up immediately and is digested. By selectively removing its competitors, the sea star helps to maintain the population of *Venus* clams on this particular stretch of sea bottom. Thus some of these slow-moving and (usually) scarce carnivores have been

14.29 The West Coast sea star Pisaster ochraceus *can keep intertidal rock clear of dense growths of mussels. This leaves space for other organisms that would be crowded out by the mussels if the sea star were absent.* P. ochraceus *stars are seen here at lower right. A clear space extends from the stars to a dense growth of mussels and goose barnacles, which are high enough up the rock to be protected from the stars by tidal exposure. (D. H. Milne)*

shown to play key roles in maintaining ecological balances within their communities.

Like many other invertebrates, many echinoderms begin life as larvae swimming in the plankton. The larvae, which have no resemblance to the adults, feed and drift. Eventually they settle, and metamorphose into the adult form. The sexes remain separate. Eggs and sperms are simply discharged into the water, where they unite to renew the cycle.

The echinoderm group is an ancient one and is now reduced to only a few of the classes that once occupied prehistoric seas. Whereas only six classes exist today, fully nineteen classes were present during the Paleozoic era.

14.30 Animals of the phylum Mollusca. From left to right: chiton (lower), clam, snail, octopus, sea slug (upper right), tusk shell. Lengths: chitons, up to 20 cm; snails, to 30 cm shell length; bivalves, to 60 cm; tusk shells, to 6 cm; sea slugs, to 30 cm; octopus, to 20 kg, arm spread, up to 7 m.

Mollusca

The animals of the zoologists' phylum Mollusca include the familiar snails, bivalves, chitons, octopuses, and squids (Figure 14.30). "Snails" (class Gastropoda) include limpets, sea slugs, abalones, and a few other noncoiled shelled or shell-less animals, as well as the familiar coiled ones. "Bivalves" (class Bivalvia) include oysters, mussels, clams, and scallops, but not the relatively rare bivalved "lamp shells," which belong to another phylum (Brachiopoda). The **molluscs** have several major anatomical features in common. These include a shell; a complex gill; a "radula," or scraping device for feeding; and characteristic stages through which they pass during their early development. Not all members of the phylum have all of these features.

It is usually a surprise to find out that zoologists consider octopuses and squids (class: Cephalopoda) to be related to the other molluscs. These animals have radulas and the same architecturally unique gill system that bivalves and most snails possess. Their ancestors in prehistoric eras were also equipped with shells. Despite the changes that have occurred in various members since the origin of the mollusc phylum, clues such as these attest to the close evolutionary relationships of the members of this ancient and successful group of organisms. There are about 100,000 species of molluscs; they are the most diverse of all marine animals.

Snails, Sea Slugs, and Chitons The snails of rocky seashores provide an easy introduction to marine molluscs. Many of them have coiled shells (Figure 14.31a).

(a)

(b)

14.31 Shells of (a) a moon snail and (b) limpets. Snail diameter is about 8 cm; limpet lengths, about 2 cm. (Steve Renick. Courtesy Steinhart Aquarium, California Academy of Sciences)

Others, the limpets, carry shells shaped like broad, shallow cones (Figure 14.31b). Interspersed among them are molluscs with a fairly similar appearance, the chitons, which bear a shell that consists of eight separate plates. On intertidal rocks, most of these organisms are engaged in similar business, rasping diatoms and other small plants off the rock surfaces with their radulas and eating them. Superficially, there appears to be more difference between the coiled snails and the limpets than there is between the limpets and the chitons. Both snails and limpets share fundamental features, however, that the chitons lack. They experience a severe, 180° twist of the part of the body that bears the shell, known as "torsion," during their early development. (This torsion is unrelated to the coiling or noncoiling of the shell design itself.) All of the molluscs that experience torsion in their development (or are descended from ancestors with torsion) are known as gastropods (class Gastropoda). The chitons are the most familiar of the creeping molluscs that do not undergo torsion. There are other differences between chitons and snails. For example, snails often have eyes and tentacles on their heads; chitons lack these structures. Gastropod larvae are also quite different from those of chitons. The chitons are placed in a separate class (Polyplacophora) for those reasons. Most of the shelled marine snails and limpets are members of the order Prosobranchia.

Grazing gastropods and chitons play important roles on rocky seashores. They are sometimes responsible for the fact that the rocks are kept clear of seaweeds. They rasp the rock with their radulas, clearing it of diatoms, settled barnacle larvae, and settled sporelings of larger algae. The importance of mollusc grazers in keeping rocks clean on the Cornish coast of England was demonstrated when oil washed ashore from the wreck of the tanker *Torrey Canyon* in 1967. Along with other animals, limpets were decimated. The next year, settled sporelings of a green alga, *Ulva*, were able to grow without interference, with the result that rocks that had been clean in previous years were suddenly lush with dense green algae. Intertidal molluscan grazers have other subtle effects on their communities, as well. The periwinkle snail of the U.S. East Coast, *Littorina littorea*, affects the seaweeds growing in tide pools by feeding preferentially on one of the green algae, *Enteromorpha*. Where the snails are abundant, *Enteromorpha* is scarce and *Chondrus*, another seaweed that is much less palatable to *Littorina*, is common. *Enteromorpha* is a tough competitor in tide pools, and where the snail is scarce, *Enteromorpha* is common—and *Chondrus* is scarce.

Moving into deeper water from the rocky shore, one finds gastropods practicing diverse feeding habits. Many of the snails there are predaceous. These include "drills," small (often ornate) whelks that bore holes in barnacles or bivalves and devour them. Scavengers with an uncanny ability to find and devour dead fish or other animal matter are there, as well. Some snails (for example, the slipper-shell *Crepidula*) are filter feeders. These gastropods sit motionless on a stone or shell (often the shell of another snail of their own species) passing water through a filter membrane that is located under the shell and consuming both filter and the edible particles caught on it. In some cases, filter-feeding snails have given up their mobility altogether. These "vermetid" molluscs begin life as an ordinary coiled snail, then cement themselves to a rock and continue shell growth in an irregular way. As adults, they resemble marine worms housed in stony tubes that stand upright from the rock.

The sea slugs are some of the most strikingly beautiful animals in the sea. These gastropods (order Opisthobranchia) usually lack shells. They appear to be experiencing a reversal, more or less, of the torsion event that affects all gastropods in their early development, and they also appear to be losing (or to have lost) their shells. They have varied styles of feeding, although a particularly common one is that of voracious carnivore. Sea lemons and other sea slugs devour sponges, colonies of hydroids, bryozoans, and other seemingly unpalatable prey with dependable regularity. One of the frequent and pleasant surprises of viewing a colony of hydroids through the microscope is finding a small opisthobranch among the animals, exquisitely ornamented with waving gill appendages and stripes and dashes of fluorescent color, attacking the hydroids. Other sea slugs are herbivorous, and a few are able to swim. One familiar swimmer, *Melibe leonina,* is a pale, translucent white creature that grows to a length of about 10 cm. It is such a bizarre and unrecognizable organism that observers frequently call biologists at local universities to ask what it is. (One perplexed person in British Columbia reported it as a sea monster.) Other opisthobranchs are far more competent swimmers. These organisms, known as pteropods, are equipped with winglike extensions on their bodies that can be flapped to propel them through the water. They spend their entire lives in the plankton, most of them in offshore tropical and subtropical oceans.

Many pteropods have lost their shells, although some still possess them. The shells accumulate on the sea bottom, under surface waters where the pteropods are abundant, and contribute significantly to the "pteropod ooze" there. Some opisthobranchs (usually tiny ones of the order Pyramidellacea) are parasitic, residing on the surfaces of clams or other organisms and continuously sucking their juices. Some of these have become so degenerate and modified that only a zoologist can detect clues to their opisthobranch ancestry. (The term *degenerate* refers to an animal that has changed so much, in adaptation to the parasitic way of life, that it has lost all obvious resemblance to the ancestors from which it descended.)

A few marine gastropods have most recently descended from land snails. The land snails and slugs (order: Pulmonata) are themselves descendants of marine snails. These have lost their gills and breathe by means of a vascularized mantle cavity that acts like a lung. This "lung" is present in a few intertidal gastropods (for example, the "limpet" *Siphonaria* of the Gulf Coast). This is strong evidence that such gastropods have returned to the margins of the sea from life on land.

The life cycles of the marine gastropods are diverse, yet there are common underlying themes. The cycle usually begins with the mating of two (or more) gastropods. The animals are often hermaphroditic—possessing the reproductive organs of both sexes; each transfers sperms to its mate, fertilizing it. More than two mates may be involved in a group mating, all at the same time. The fertilized eggs are spewed into the water, where they develop. The hatchlings do not resemble the parent gastropods at first, although later in their development they acquire a tiny coiled shell, one of the first clues to their gastropod parentage. The larvae swim about in the plankton, consuming nannoplankton. It is here that they experience torsion. They eventually settle to the bottom, where their growth continues. The shell either enlarges and continues to coil, enlarges without coiling (as in limpets), or disappears entirely (as in most opisthobranchs).

The planktonic phase of the life cycle is hazardous, since the helpless swimming larvae are vulnerable to young fish, filter-feeding clams, and a multitude of other predators. Therefore, gastropods with planktonic larvae compensate for the enormous risks by releasing vast numbers of fertilized eggs. In an extreme example, a sea hare observed by G. E. and Netti MacGinitie laid 478 million eggs over a seventeen-week period. Thus, even though the odds against a particular individual's survival are great, these numbers ensure that a few larvae will survive.

Clams and Their Relatives The bivalves (class: Bivalvia; Figure 14.32) are a less diverse group of molluscs than gastropods. Most of them are filter- or deposit-feeders, for which opportunities for diverse ecological role-playing are not as abundant. Nevertheless, as filter-feeders, they are important in the economy of the sea.

A typical bivalve, such as a mussel, is equipped with two shells and sophisticated apparatus for filter-feeding. Two gills occupy much of the space between the shells. By means of beating cilia, the mussel draws water into the posterior portion of the shell, forces it through the gills, and blasts it back out to the exterior, again at the posterior end of the shell. As edible particles encounter the gills, they are stopped by beating cilia and are moved down the surface of the gill toward the ventral edge. As they move, they become entangled in strands of mucus, which are secreted by the gill. Mucus and entrapped particles reach the lower margin of the gill, where the mucus is passed forward toward the mussel's mouth. There, a sophisticated sorting system directs the mucus toward the mouth, but only if the trapped particles are of the size pre-

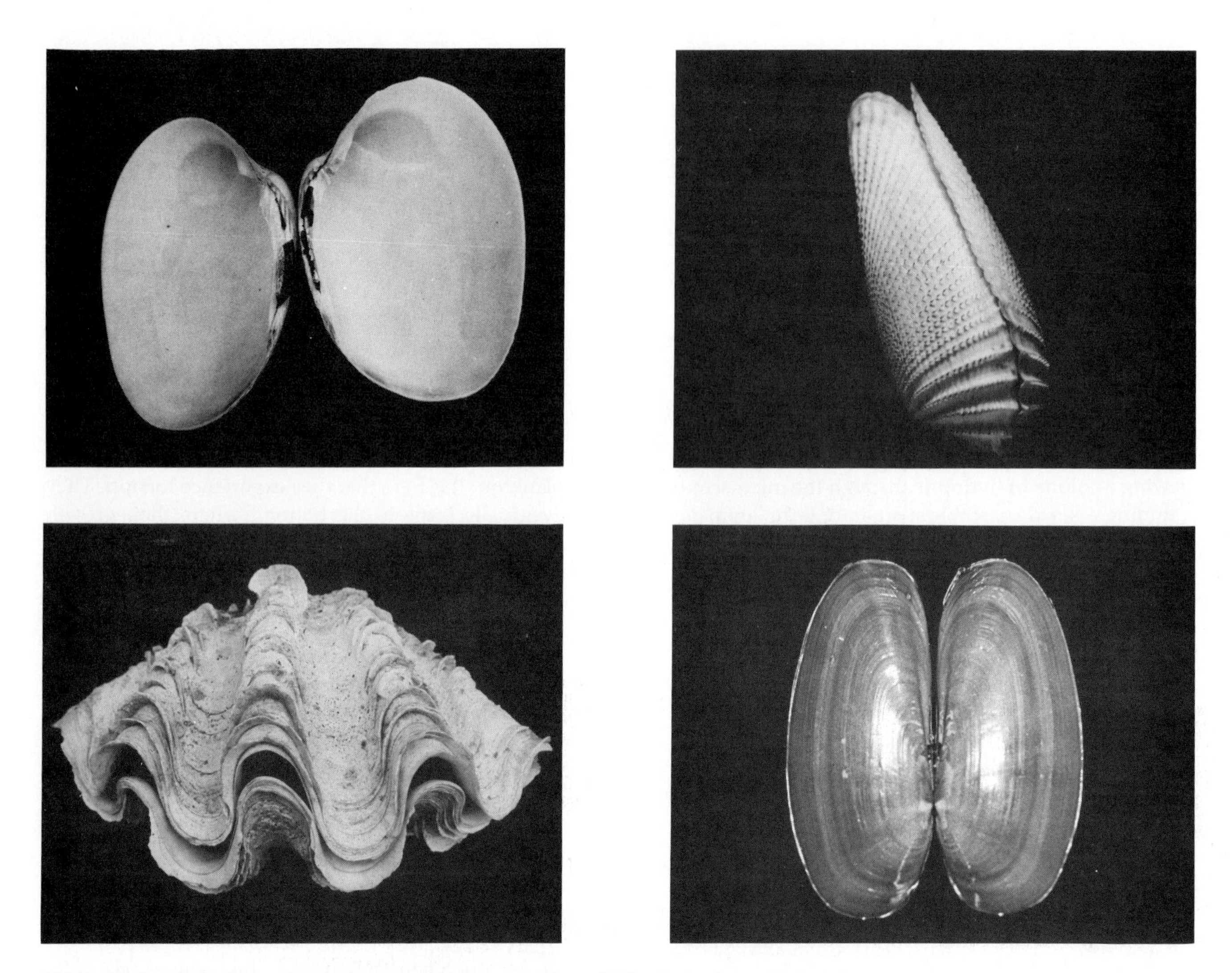

14.32 Bivalve shells. (Steve Renick. Courtesy Steinhart Aquarium, California Academy of Sciences)

ferred by the mussel. Mucus containing undesirable particles is discarded. This system can capture particles as small as one to five micrometers in diameter, making mussels capable of straining microflagellates and bacteria out of the water.

Not only does a mussel capture edible particles for its own consumption, it removes other particles (including living diatoms, snail larvae, and virtually everything else that is unable to escape its inflowing water stream) from the water as well. This is a significant accomplishment. Food in the sea is mostly dispersed in suspended or dissolved microscopic bits. Any animal that can collect and concentrate it, as the mussel can, makes it available to more conventional feeders that could not, by themselves, utilize the dispersed edible particles. The mussel itself becomes prey for other organisms, thereby transferring to the rest of the system some of the food energy that was once hopelessly scattered in the form of small, dispersed planktonic particles. The discarded mucus, itself rich with collected particles, serves as an important food source for organisms of the detritus food chain.

Most bivalves use variations on the mussel's style of collecting food (Figure 14.33). Many clams have elaborated on the posterior water-intake-and-discharge system, elongating it into a double-barreled tube known as the siphon, or neck. These clams are able to burrow deep into the bottom sediments, while continuing to draw water for filtration from above the sediment surface. Some clams of the genus *Macoma*, for example, use their siphons to vacuum up detritus on the sea bottom. These species collect edible debris that has settled around their burrows. Oysters and scallops also filter the water from incoming streams, although their filtering equipment is not as sophisticated as that of the clams and mussels. Some bivalves

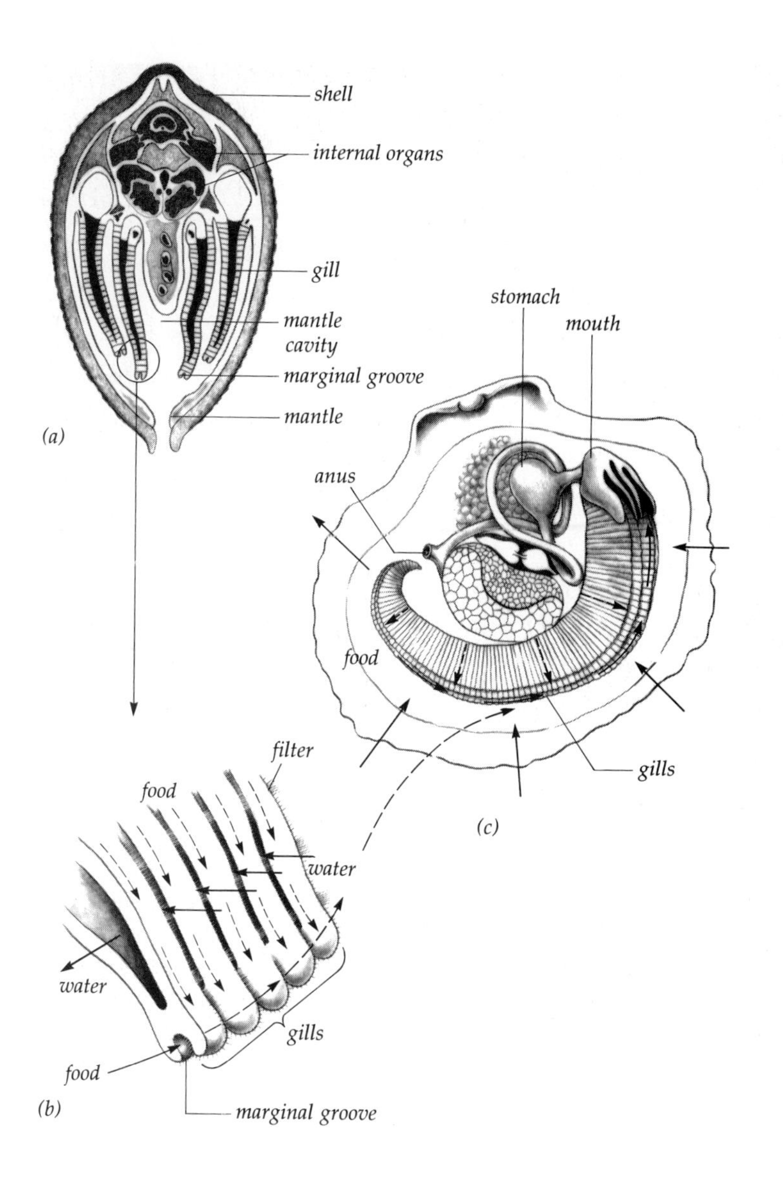

14.33 Bivalve water-filtering system. (a) Animal in cross section, showing gills. (b) Water is driven by cilia, through slots in the gill, into interior space (solid arrows). Food particles and a mucus filter are carried down the gill surface by cilia and collected at the marginal groove (dashed arrows). (c) Food and filter move along marginal groove of gill to mouth. Water exiting interiors of gills leaves bivalve at posterior edge. In this species, water drawn into the gills enters around the whole periphery of the shell.

(for example, clams of the genus *Nucula*) do not filter-feed. These animals sort edible particles from the sediment in which they lie buried.

In addition to the ecological significance of filter-feeding, there are a few points of direct economic benefit to people. First, filtering bivalves can process great quantities of water. It is frequently true that they are capable of passing the entire water volume of a bay or inlet through their filters every few days or so. For example, in the North Inlet estuary near Georgetown, South Carolina, the oysters alone are capable of filtering nearly 70% of all the water that covers them on every change of the tide. This is obviously helpful in areas where efforts to maintain high water quality are being made; the bivalves are keeping the water clean. The expense of an engineered water treatment system capable of duplicating their efforts would be astronomical. This is an example of what environmental scientists call an ecosystem service; a monumentally large beneficial effect, practiced obscurely and entirely without a monetary cost, that helps to maintain a liveable environment. Second, filter-feeding is potentially hazardous to people, since the bivalves collect and concentrate harmful substances. Hepatitis viruses from untreated sewage and red-tide dinoflagellates loaded with toxins are two important examples. For that reason, oysters, clams, and mussels are closely watched by public health officials for the first signs that hazardous deteriorations of water quality are underway.

Octopuses and Squids Cephalopods are the biggest and most intelligent of all invertebrates (Figure 14.34). All of them (there are about 650 living species) are predators. Cephalopods were more abundant in prehistoric seas than they are today. Those best represented in the fossil record were either housed in straight

14.34 A Pacific octopus. The more than 200 sucking discs are used for locomotion, food capture, and manipulation. Each disc reportedly exerts suction sufficient to lift 124 g. (Scripps Institution of Oceanography, University of California, San Diego)

*14.35 Chambered nautilus (*Nautilus macromphalus*) from New Caledonia. Shell diameter is about 15 cm.*

or coiled shells (ammonoids and nautiloids) or were equipped with massive, stony, bullet-shaped internal hard parts (belemnoids). The heyday of shelled cephalopods ended with the global extinction episode at the end of the Cretaceous period, about 65 million years ago. Only the chambered nautiluses, the cuttlefishes, and a few small deep-sea cephalopods still carry the shells or internal hard parts that were characteristic of their prehistoric forebears (Figure 14.35). All of the rest of the living cephalopods either lack skeletal hard parts entirely (for example, octopuses) or have had them reduced to a flimsy cartilaginous "pen" (for example, squids). Most living cephalopods still retain the major features of their ancestors, including grasping tentacles, large image-forming eyes, a parrotlike beak, and characteristic molluscan gills.

Altough cephalopods loom large in folk tales, their role in modern seas seems to be rather secondary. As predators, they are usually less numerous than the herbivorous animals, and they are less numerous than fish. Scarcity does not necessarily mean that an animal plays an insignificant role in marine communities (recall the earlier discussion of the sea star *Pisaster*), but thus far cephalopods have been identified as key characters in the dynamics of only a few food webs. They are, for example, a primary food for elephant seals. Individual octupuses certainly dominate the locality in which they live, capturing large crabs and piling their remains about the entrances to their hideaways. In a like manner, schools of small squids consume large numbers of shrimp and small fish while serving as food for such larger fish as bluefin tuna. Giant squids of the deep ocean waters, reaching 6 m in body length (with another 10 m for the tentacles), serve as food for sperm whales. In these ways, octopuses and squids are inextricably interwoven into the upper levels of the food webs of the sea.

The cephalopod life cycle is straightforward. The sexes are separate. Mating involves transfer of a packet of sperms to the female, often by means of a specialized tentacle that the male uses to put sperms into the female's mantle cavity. Female octopuses guard their eggs and keep them clean. The female usually ceases feeding during this period and dies after the eggs hatch.

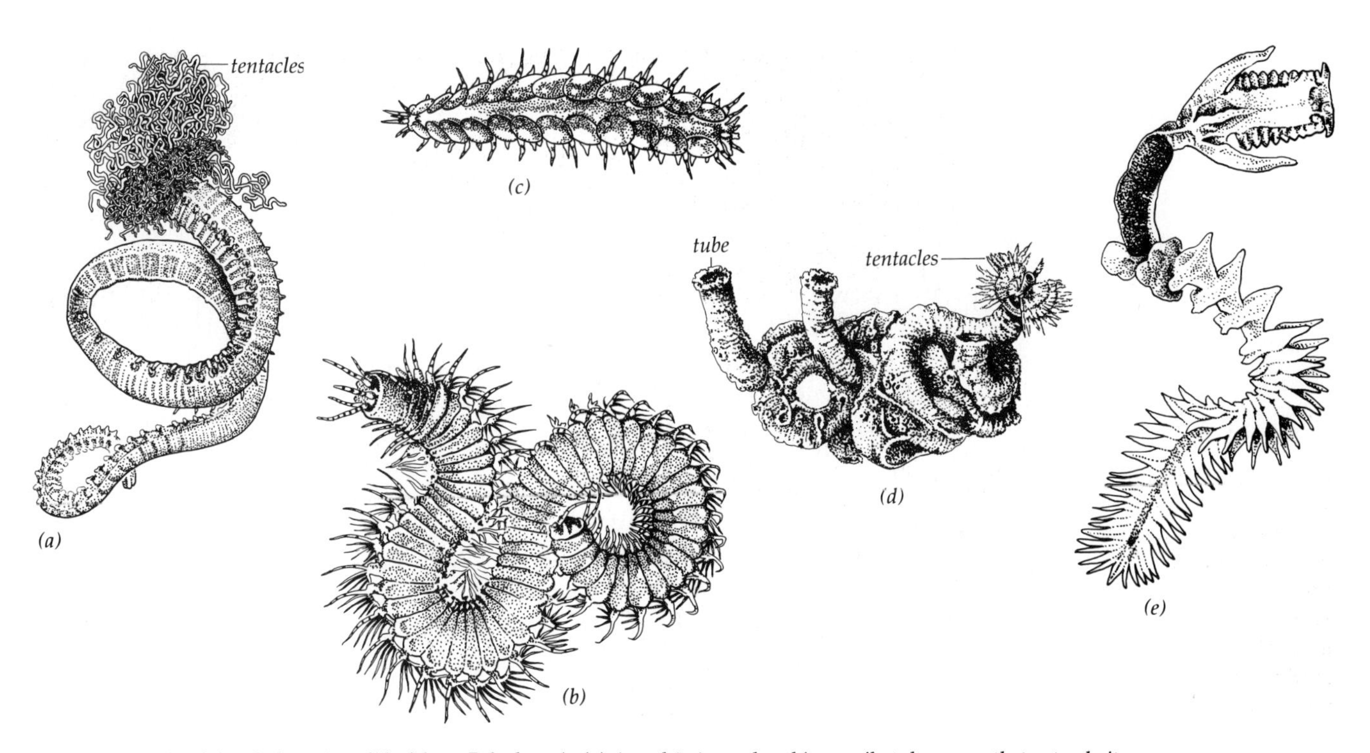

14.36 *Animals of the phylum Annelida (class: Polychaeta). (a)* Amphitrite, *a benthic, sessile tube worm that extends its tentacles across the sea bottom to obtain food. The worm remains buried in the bottom. (b)* Eunice, *an active crawling and swimming predator. (c)* Lepidonotus, *a scale worm, free-living, or commensal. (d)* Serpula, *which lives in stony tubes on the sides of rocks. (e)* Chaetopterus, *a benthic tube worm that draws water currents through its U-shaped burrow by waving its flaplike appendages. Lengths:* Amphitrite, *to 25 cm;* Eunice, *up to 3 m but typically a few centimeters;* Lepidonotus, *about 3 cm;* Serpula, *about 5 cm;* Chaetopterus, *about 10 cm.*

(Males do not sacrifice their lives for their offspring, and they live to reach larger sizes than females.) The young animals are often planktonic, drifting with the currents, preying on the other zooplankters, and falling victim to fish and other predators.

Octopuses are timid, retiring animals that do their best to hide from human intruders. As predators, they tackle formidable prey. Large marine crabs of the genus *Cancer* are among their favorite foods. These crabs are pugnacious, and are easily capable of nipping off tentacles. An octopus stalking a crab approaches cautiously, springs on it, quickly grabs and immobilizes its legs and claws with its suckered tentacles, and kills it by biting it. Venom is injected with the bite, and the crab quickly succumbs. The octopus then carries it off to a hiding place and consumes it. Other prey, including shellfish and dead fish found on the sea bottom, is also eaten.

The largest of all invertebrates are the giant squids of the deep ocean. Little is known about these animals. They apparently surface only when ill or dying, and provide little opportunity for study of their life cycles. They are thought to concentrate at depths of 300 to 600 m on continental slopes, although sperm whales capture them at much greater depths. Some of the large squids are formidable animals. After the troopship *Brittania* was sunk by German gunfire in the central Atlantic in 1941, men clinging to a small float were attacked by squids. In other action, wealthy sport fishermen off Peru, angered by the attack of 350-lb squids on their hooked marlins, began fishing for the squids themselves. The squids proved to be worthy antagonists, flailing their assailants with tentacles, deluging them with torrents of ink and water, and biting chunks out of a wooden-handled gaff. In response to the unexpected ferocity of the cephalopods, the fishermen took to wearing pillowcases with eye-holes over their heads for protection.

Annelid Worms

Marine worms have many different internal body designs and are placed in many different phyla, but a majority of the ones noticed on seashores are members of the phylum Annelida (Figure 14.36). **Annelids** are assembled in the manner of a string of identical beads. Each segment of the elongate, wormlike body contains a set of excretory organs, its own set of muscles, access to the digestive system, access to the nervous system, and other body elements, in much the same arrangement as in the segment ahead of it and the segment behind it. The digestive system, the main

(a)

(b)

14.37 (a) Tubes of essile polychaete worms project from intertidal sand, Baja California. Heights 2–3 cm. (© 1980 Tom McHugh) (b) A small sessile annelid worm, removed from tube, showing feathery tentacles on head, segmented body. Tentacle length 5 mm. (© R. F. Head)

nervous system, and the blood circulatory system all run through the chain of identical segments that make up the worm's body, supplying each segment with food and with nerve coordination. The segments that occur at the head end have coalesced to some extent, to concentrate essential apparatus (such as eyes and nerve ganglia) at the forward end of the worm. Various annelid worms depart from this basic body plan in several ways, but all of them show evidence of it in their internal architecture. Externally, they can usually be distinguished from worms of other phyla by the obvious segmentation of the body, visible as repeated transverse lines dividing the worm's length into many short segments. They usually overshadow the worms of other phyla in intertidal marine communities.

The marine annelids are mostly of two general types: errant (meaning mobile) worms and sessile worms (Figure 14.37). The errant worms are ordinary free-living individuals, the sessile ones usually inhabit fixed tubes or burrows. These are placed by zoologists in the class Polychaeta and are known familiarly as polychaetes (the name means "many bristles"). The most conspicuous polychaetes in marine communities are usually the sessile ones. They live in tubes (ranging from a few millimeters to about 20 cm in length) that are cemented to rocks, shells, pilings, or other substrates. The heads of these worms are often ornamented with tufts of colorful bristles, which, when spread, give the colony the appearance of a spectacular flower garden. The "petals" of their flowerlike apparatus serve as collecting devices. Cilia on these structures set up water currents that pass among the branches, and edible particles in the water are filtered out and passed down to the worm's mouth. The tubes are secreted by the worms. Some are made of stony calcium carbonate, others are parchmentlike or are made from inedible particles that the worm has collected. The worm is free-living inside the tube. Although it is possible to carefully remove it without injury, it will not leave the tube voluntarily.

Errant polychaetes may live buried in sediments or in well-defined burrows. As a rule, they are willing and able to leave these dwellings and move elsewhere. Some of these more mobile worms lead lives closely parallel to those of terrestrial earthworms, swallowing sediment and digesting the edible material. Others eat seaweed, and still others are carnivorous. Some errant worms have taken up commensal feeding methods. Such worms, usually scaly ones of the family Arctinoidae (Figure 14.36c), take up permanent residence in the shell of a hermit crab, on the underside of a sea star, or somewhere else close to the mouth of a larger animal. When the animal feeds, the worm shares the food.

The life cycles of annelid worms are straightforward. The sexes are separate, and males and females discharge their gametes directly into the water. Fertilization depends on the random rendezvous of eggs and sperms. The fertilized egg develops into a larva that is initially similar to that of the molluscs. The larva lives in the plankton, feeding, growing, adding new segments to the posterior part of the body, and becoming recognizably more wormlike. Eventually it settles to the bottom and matures into adulthood. Many worms, rather than leaving fertilization to blind chance, swim up into the plankton and discharge their gametes in the vicinity of a worm of the opposite sex. The worms usually burst in the process. In preparation for this one-way trip, various strategies are employed. In

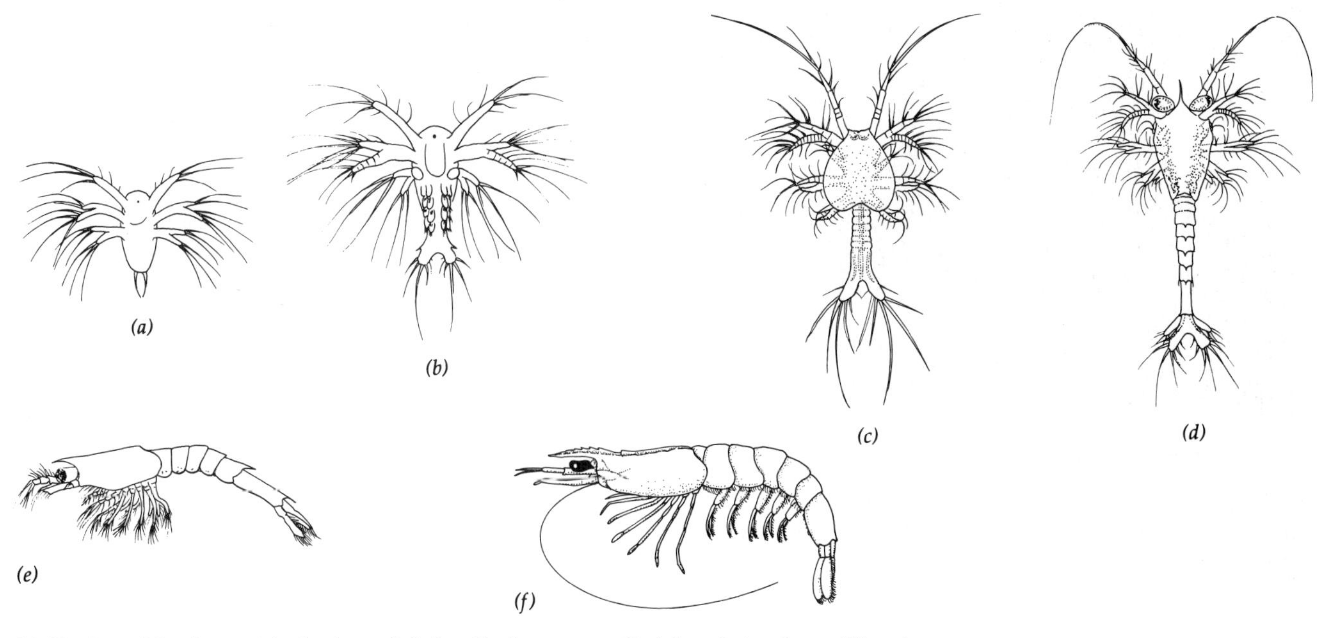

14.38 Larval development in the decapod shrimp Trachypeneus. *Each larval stage has a different name: (a) nauplius (0.26 mm long), (b) metanauplius (0.42 mm long), (c) early protozoea (0.7 mm long), (d) late protozoea (1.9 mm long), (e) mysis (2.8 mm long), and (f) adult (30 mm long).*

some cases, the whole worm develops large eyes and enlarged paddle-shaped bristles for swimming. In others, only a portion of the body is thus specialized. Carrying the gametes, equipped with paddles and newly grown eyes, and capable of going on the swimming mission by itself, this portion buds off and heads for the surface, while the parent worm remains under cover. The release of these swimming "epitokes," as they are called, is usually synchronized so that millions of them swim to the surface at the same time. During the brief period in which they mill about, discharging their gametes, they provide a bonanza for fish and other animals that eat them. The palolo worm of South Pacific waters (genus: *Eunice,* Figure 14.36b) is one of these. Its cycle is synchronized so that its epitokes invariably congregate at the surface during October and November, at daybreak on the day before the moon enters its last quarter. Islanders living near the reefs where this occurs have learned the timing of these releases, and they gather to collect and eat the fattened epitokes.

Polychaete worms are inconspicuous players in the economy of the sea. They are always present in marine communities and perform a major share of the work of converting plankton, detritus, and seaweed to edible meat for larger carnivores. There are about 5,300 marine species. Their relatives include the earthworms (class: Oligochaeta) and leeches: (class: Hirudinea), both of which are more prominent on land and in fresh water than in the sea.

Crustaceans

Arthropods are animals that have external hard parts and jointed legs. All such creatures were formerly considered to be members of one phylum (Arthropoda), but studies of fossils and lingering doubts about living joint-legged creatures suggest that these animals should be sorted into four phyla (Trilobita, Crustacea, Chelicerata, and Uniramia). The dominant marine group is the Crustacea, formerly a class of the old Arthropoda, but now considered by many to be a phylum in its own right. The term *arthropod* is useful for referring to animals from all four phyla.

Crustacean Anatomy and Life Cycle Most of the marine joint-legged animals are crustaceans. Despite their diversity, all **crustaceans** have several things in common. These include a nauplius larva in the early life cycle (or, if the nauplius stage is absent, reasonable certainty that it was once present in the animal's ancestors) and two pairs of antennae. The nauplius stage in development is a clue to crustacean ancestry that often persists even after all other clues are lost. It is useful to examine this part of the crustacean life cycle before discussing the adult animals.

In the shrimp *Trachypeneus,* which is a fairly typical crustacean, mating occurs and the fertilized eggs are released into the water by the female. The larva that hatches bears no resemblance to the adult shrimp (Figure 14.38). Instead, it is a tiny egg-shaped object with six legs and (often) one tiny eye. This nauplius

*14.39 American lobster (*Homarus*). These lobsters range along the eastern coast of North America. Life size: up to 60 cm. (Steve Renick. Courtesy Bodega Marine Laboratory, University of California, Berkeley)*

larva lives in the plankton community, where it feeds on microflagellates and sheds its skin ("molts") several times. The nauplius form is still recognizable after several molts. During one of the later molts, it assumes a new form, now more similar to the adult, and larval development continues. The rest of development consists of additional molts, each adding segments and appendages to the later larval forms (the protozoea and mysis forms). Whereas these later larval forms vary among different crustacean species, early development typically begins with the nauplius form. This early characteristic step in the crustacean life cycle makes it possible to deduce the relationships of animals that, as adults, hardly resemble crustaceans at all. Thus barnacles, once thought to be related to molluscs on account of their shells, were found to have nauplius larvae. Certain sac-like parasitic growths on crabs that are otherwise unrecognizable also prove to have nauplius larvae. The life cycles retain clues of crustacean ancestry, even though evolution has modified the adult animals almost beyond recognition.

In many cases (as with crabs, isopods, and amphipods), the nauplius stage is absent. The young animals hatch as a larva of some other form (or even as a miniature adult form). In many such cases, the embryo in the egg passes through a nauplius form before continuing to develop. Thus this evidence of crustacean ancestry is present even in these life cycles. The nauplius larva gives zoologists a powerful tool in deducing relationships of marine crustaceans. Unfortunately, the nauplius larva stage also exposes those crustaceans that still retain it to the extreme hazards of early life in the plankton.

All crustaceans must break out of the body armor encasing them from time to time during their life cycles, or they could not grow larger. These occasions, called molts, are crucial moments in the lives of crustaceans. The animal softens the inside of its external armor, a split develops, and the animal pulls itself out of its old skeleton, leaving the skeleton behind. (The amount of material thus abandoned is astonishing; the whole outer surface, internal muscle attachment points, segments of the digestive tract, gill linings, and other parts are so complete that a person finding an abandoned skeleton, or molt, often assumes that it is a dead crab.) The outer surface of the newly molted crustacean is soft for a few hours. During that time, it is vulnerable to attack. The molted crustacean pumps itself up with water, expands visibly, then hides to wait for its new "skin" to harden up. Often all of the crabs of a particular year-class will molt at the same time. Their skeletons then wash ashore, providing biologists with useful information about their growth rates and seasonal molting cycles—and the public sometimes shows undue concern that "something is killing the crabs!"

In species diversity, crustaceans are a distant second to the molluscs. There are about 31,000 species of crustaceans, most of them marine. The crustaceans in the sea are the ecological counterparts of the insects on land. They are diverse, numerous, found in virtually every habitat, practicing every conceivable feeding style, and are crucial to the ecology of the sea. They are usually not very large; fish and some other organisms outweigh them. Because their forms are diverse, their classification is complex. For our purposes, we may subdivide them as follows: decapods (including crabs, shrimps, and anomurans) and nondecapods, including several types. Crabs and shrimplike animals (including lobsters) are familiar forms; anomurans are a diverse set that include hermit crabs, burrowing "ghost shrimp," the Alaskan king crab, a few animals that are very crablike in appearance, and the mole crabs of surf-swept beaches. Nondecapods

(a)

(b)

(c)

14.40 *Crabs. (a)* Cancer. *(W. J. Wallace) (b) King crab* (Paralithodes) *of the northern Pacific. (Steve Renick. Courtesy Steinhart Aquarium, California Academy of Sciences) (c) A crab that lives among the gulf weed in the Sargasso Sea. (Penny Hermes) Carapace diameters: about 15 cm* (Cancer), *28 cm* (Paralithodes), *1 cm (gulf crab).*

include small but significant inhabitants of the plankton (copepods) or seashores (isopods, amphipods, and barnacles) and many obscure forms that are beyond the scope of this book.

Shrimps, Crabs, and Anomurans: The Decapod Crustaceans The shrimplike animals, which include lobsters, typically have a saddle-like carapace covering the head and front of the body, and a large, jointed abdomen (Figure 14.39). The part under the carapace is usually equipped with ten legs (counting the claws). The abdomen usually has swimming or egg-carrying appendages. In crabs (Figure 14.40), the body layout is essentially the same, but the carapace is expanded and flattened. The abdomen, much reduced in size, is folded forward and hidden beneath the carapace. The anomuran body form is similar, but with other variations on the basic plan. All of these animals are placed in the order Decapoda (the name means "ten legs"), and are related. The decapods are the most commercially valuable of the marine crustaceans. The annual global harvest of crabs, shrimp, lobsters, and related forms is about 3 million metric tons.

Commercially valuable shrimp species illustrate a commonly occurring crustacean life cycle and the vulnerability of some ocean resources to human destruction. One of the major species of the Gulf Coast fishery, the brown shrimp *Penaeus aztecus*, shows a common pattern for that area. After mating, the female shrimp releases her fertilized eggs, which produce nauplius larvae. The nauplii drift with the plankton, feeding on microflagellates and in turn being consumed by the predators of the plankton community. Like the larvae of *Trachypeneus*, they molt several times as they grow, eventually metamorphosing to a more shrimplike larval form. The drift of the Gulf Coast currents and a certain amount of active swimming carries the larvae landward. The larvae enter the back bays and estuaries. Here, in water of lower salinity, they continue to feed and grow, eventually working their way seaward again for the final molts to adulthood. This life cycle enables the young shrimps to tap the immense biological productivity of the sea-margin communities. It also exposes them to hazards created by people. Pesticides designed to kill insects are especially toxic to other arthropods; drift of pesticide sprays from agricultural fields near wetlands can kill the young shrimp, and eventually result in a much-decreased harvest of the adult animals. The draining and filling of coastal wetlands deprives the shrimp of this essential nursery habitat, also resulting in reduced shrimp harvests. Valuable offshore fisheries may therefore be damaged by spraying and drainage events that, at first glance, do not seem to be related to the collapse of a fishery.

Besides shrimp, four large crustaceans from the waters of the forty-eight contiguous United States are most familiar to us. These are the West Coast's Dungeness crab and spiny lobster and the East Coast's blue crab and Maine lobster. As adults, these crustaceans

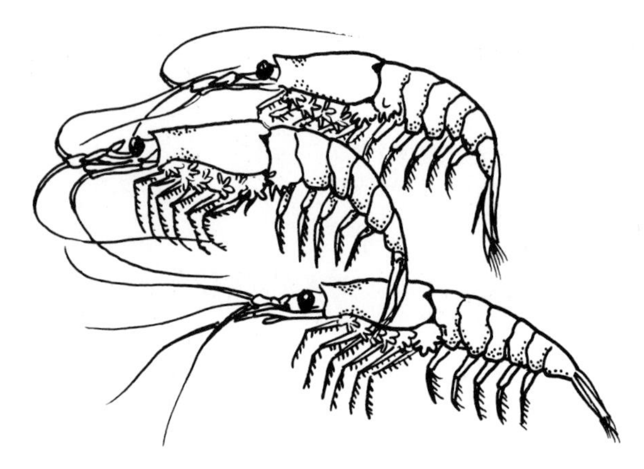

14.41 Euphausiids, or "krill." These shrimplike animals occur worldwide; they are particularly numerous in Antarctic seas.

tend to be predaceous. The Dungeness crab is a powerful predator of shellfish, breaking or crushing their shells with its claws. A large Maine lobster (Figure 14.39), reaching 25 kg in weight and 60 cm in length, is also a formidable predator, consuming sea urchins, rock crabs, and other prey. The fact that they are carnivorous and often cannibalistic in close quarters makes raising these crustaceans for consumption a more difficult undertaking than raising such omnivores as shrimp. Crabs and lobsters fight and eat each other and must be reared individually in separate containers—an expensive proposition.

Another crustacean familiar to seafood aficionados is the Alaska king crab (Figure 14.40b). The king crab differs in appearance from ordinary crabs, apparently having only six legs and two claws. This is a clue to its anomuran relationships. Like many anomurans, it has eight legs in addition to the claws, but two of them are tucked under the carapace and are not easily visible. Adult king crabs are large. The body reaches 28 cm in diameter, the legs span 100 cm, and the weight reaches 11 kg. They live in huge congregations in the cold waters off northern British Columbia and Alaska. They are fished by lowering baited crab traps. At times, the traps come up so crammed with king crabs that the fishermen, out of curiosity, find themselves unable to stuff all the crabs back into the same trap after the crabs have been cooked.

In addition to the larger decapods, there are myriad small crabs, hermit crabs, benthic prawns, and shrimp that inhabit the waters of cold seas. These animals are scavengers and predators. Hermit crabs, prawns, and many small crabs feed on the fine edible detritus accumulated on the sediments. Mud bottoms are often tunneled by ghost shrimp, which mine the buried detritus and eat it or flush water through their tunnels while filtering out edible plankton. In this way, these crustaceans concentrate finely dispersed food in their own bodies and are available themselves as bite-sized food for fish and larger organisms.

As predators, the larger decapods have had a subtle effect on marine communities over the ages. Heavy, ornamented shells on snails were rare before the Mesozoic era, when carnivorous crustaceans first became abundant. Since that time, molluscs have evolved thick shells, spiny shells, and effectively armored shells of other sorts, apparently as defenses against crustaceans that attack and break the shells. The crustaceans, in turn, have evolved progressively heavier apparatus for breaking shells. Unfortunately for themselves, even the largest crustaceans are not at the top of their food webs. Shrimp, prawns, crabs, hermit crabs, and their decapod relatives all serve as important food for fish and cephalopods. Even the biggest crabs are not safe. Full-grown Dungeness and Alaska king crabs have been found in the stomachs of large halibut, swallowed whole.

Krill, Copepods, Isopods, Amphipods, Barnacles: The Nondecapods These crustaceans are progressively more dissimilar to the decapods. One important subcategory includes the **krill** (Figure 14.41). These small, shrimplike animals are most well known as food for whales and other animals in Antarctic waters. Krill (order: Euphausiacea; "euphausiids") resemble shrimp, except that they have limbs in the positions where shrimp have auxiliary mouthparts. They therefore have more than ten legs and are classified separately from the decapods. They are smaller than commercial shrimp (reaching about 6 cm) but large and numerous enough to form the most important food web link in the Antarctic circumpolar seas. Krill filter the seawater with their basketlike forelimbs, straining out diatoms and eating them. Adult Antarctic euphausiids occur at the surface in enormous shoals, hundreds of meters in length and width, 5 or so meters deep, with up to 63,000 individuals in every cubic meter of water.

Baleen whales scoop krill up in gigantic mouthfuls, thus feeding only one step removed from the primary diatom productivity—an unusual trophic position for large marine animals. The demise of the great whales that once utilized this vast krill resource has been accompanied, first, by a huge increase in the numbers of krill and, second, by large increases in the numbers of penguins and seals that also depend on krill. In recent years, Russian and Japanese krill-fishing vessels have attempted to harvest Antarctic krill for sale as a shrimplike seafood. It has been estimated that the yield of krill alone could match that of all the world's other marine fisheries combined. Fears have been expressed that large-scale harvesting of these

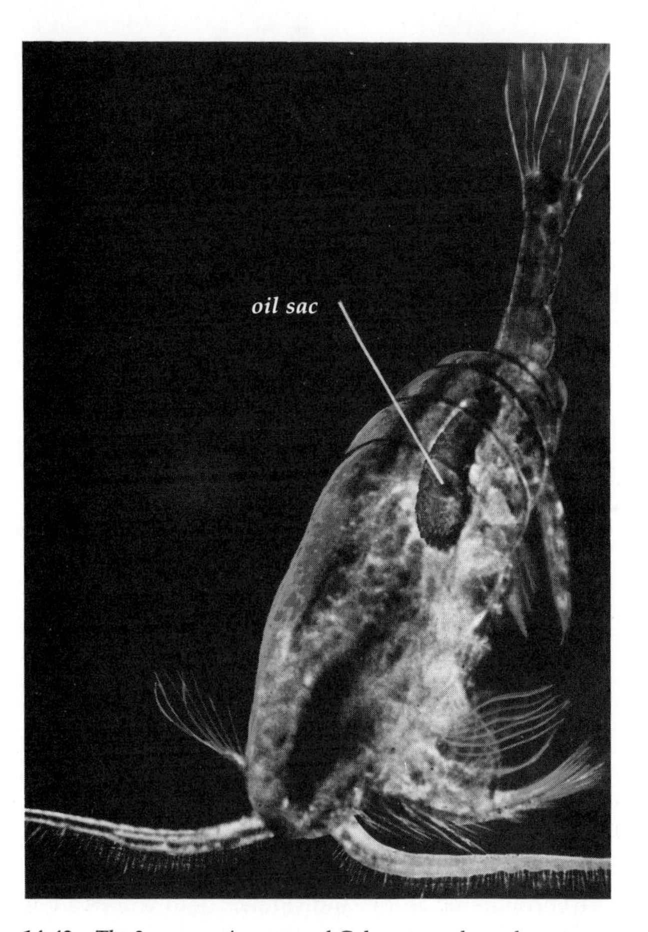

14.42 *The 3-mm marine copepod* Calanus, *perhaps the most common animal in the sea. This animal is an important herbivore in the plankton. Copepods are often filled with oils from the diatoms on which they feed. (Scripps Institution of Oceanography, University of California, San Diego.)*

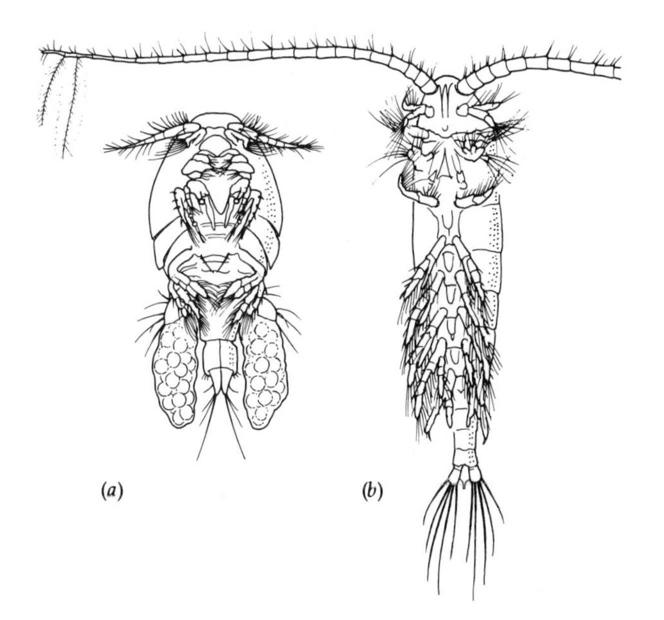

14.43 *Copepods. (a)* Clausidium, *a copepod that is commensal on the surfaces of some kinds of burrowing shrimps. Size: about 2 mm long. (b)* Calanus, *a planktonic copepod, about 3–5 mm long. Ventral views of both.* Clausidium *is carrying egg sacs.*

animals would endanger the Antarctic food webs that now depend on them and make recovery of the endangered whales difficult or impossible.

Unlike euphausiids, which are generally familiar to people in connection with whales, the copepods (Figures 14.42 and 14.43) are largely unknown and unappreciated. They deserve wider recognition, for these small crustaceans are the animals that sustain most of the animal economy of the global seas. They are the prime converters of planktonic diatoms, dinoflagellates, and other photosynthesizers to living matter that can be harvested by such larger animals as herrings. In terms of keeping the marine system going, they are the most important of all of the animals.

Copepods are usually small (less than 1 cm long), and usually go unnoticed in seawater. They may attract one's attention as a sudden flash of motion in a jar of seawater held up to sunlight, but they must be observed under a microscope to be seen in detail. Under the microscope, a copepod's body appears to be divided into two main parts; an oblong forward unit shaped like a grain of rice (the cephalosome) and a trailing tail piece with a terminal fork (the urosome). Most of the appendages are on the cephalosome. Most conspicuous are two large antennae, located on the front end. On the underside, a complex of finely whiskered limbs serves as a filtering basket, used by the copepod to strain its planktonic food from the water through which it moves. The mouth is on the underside near the filter basket. In action, a copepod moves steadily forward, generating circulating currents that carry plankton into the whiskers of its filter apparatus. Its powerful jaws, working continuously, crunch up the glassy diatoms that it captures. Most of the edible stuff goes into its mouth, but a significant slip stream of spilled organic material trails behind the animal as it moves. With its limbs flickering at fully 10 to 44 beats per second and its body moving steadily forward at a speed of a few body lengths per second, a feeding copepod is a marvelously efficient collector of finely dispersed edible particles.

There are about 1,000 to 1,500 species of marine planktonic copepods (and about 3,500 additional species that live on the bottom, in fresh water, or as parasites). By comparison with insects on land (which are represented by somewhere between one and 14 *million* species), these are not impressive numbers. In numbers of individuals, however, and in global sig-

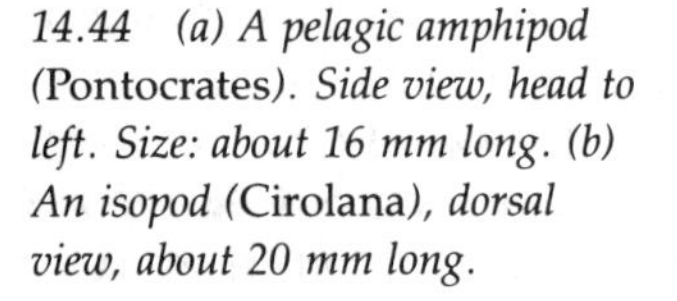

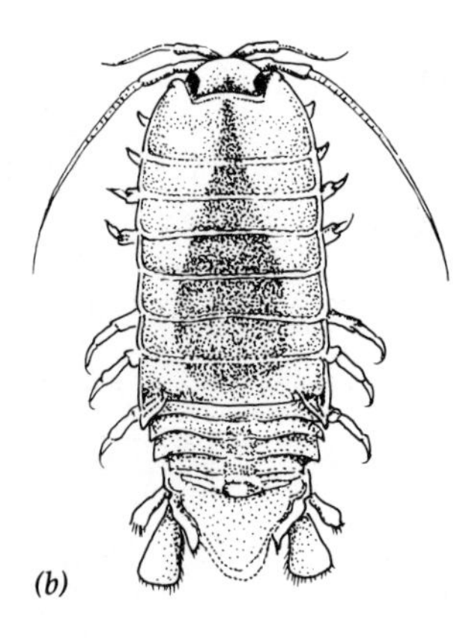

14.44 *(a) A pelagic amphipod* (Pontocrates). *Side view, head to left. Size: about 16 mm long. (b) An isopod* (Cirolana), *dorsal view, about 20 mm long.*

nificance, the copepods are contenders for first place. They are usually the first animal links between plants and carnivores in ocean waters far from shore. They are the prey of jellyfish, ctenophores, arrow worms, and many other small and unfamiliar oceanic carnivores. They are also a main food source for herring, anchovies, sardines, pilchards, and other small fish that serve as food, in turn, for the commercially important larger species of fish. Although the significance of copepods may be outweighed in shallow marine water by benthic food chains that originate with seaweeds, or in areas dominated by euphausiids, in most of the ocean copepods are the crucial link between plants and larger animals.

Planktonic copepods have complex life cycles. The sexes are separate, and mating produces fertilized eggs. In one representative copepod of the North Atlantic, the common and ecologically important *Calanus finmarchicus,* the eggs hatch as nauplius larvae. These feed on nannoplankton, molting as they grow older. The sixth molt converts these larvae into forms (called copepodites) that resemble the mature copepods in outline, though not in detail. After another five molts, the copepodites become mature adults. During the summer, in temperate seas, this life cycle is completed in about two months. As winter approaches, the copepodites go into arrested development. They survive the winter in a state of low activity in the last copepodite stage, molting to become full adults in late winter, just before phytoplankton growth resumes.

Not all copepods are planktonic (Figure 14.43a). Many of them live on the sea bottom. Some live on the surfaces of other organisms, and others are parasitic. Some of them are so grotesquely transformed that it is hardly evident that they are copepods. One example is the copepod parasite *Mytilicola intestinalis.* This elongated, wormlike creature lives in the intestines of mussels. One would hardly know that it is related to copepods, except that its eggs produce nauplius larvae that go through a second nauplius stage and then a copepodite stage before making their way into a mussel's intestine for the transformation into adulthood.

Further removed from the shrimplike body form, but still noticeably similar to it, are the beach fleas, slaters, and other members of the orders Amphipoda and Isopoda (Figure 14.44). Both groups consist of animals with obvious segmentation of the bodies and a pair of legs on every segment. These animals are mostly herbivorous, although there are conspicuous exceptions. They can be phenomenally abundant in piles of seaweed on the beach. More commonly they are encountered as lone individuals. Both groups show no trace of the nauplius larval stage in their life cycles; their eggs tend to hatch as forms that are already similar to the adults. Some have adopted bizarre life cycles. The "whale lice" found on the skins of whales are flattened, modified amphipods. Some isopods have taken up parasitic feeding habits, occurring as more-or-less degenerate, blood-sucking lumps on the roofs of fishes' mouths, under the carapaces of shrimps, and in other places on the host's body. Usually the females are more degenerate than the males.

An important clue that the female parasite is in fact an isopod is provided by the males, which are unquestionably isopod and which can be found mating with the unrecognizable females. Isopods are among the few crustaceans that have succeeded in establishing a foothold on land. A few species (locally known as "pill bugs," "sow bugs," or "potato bugs") can usually be found under logs or in other damp places. They are far outnumbered, however, by insects and other true terrestrial arthropods in almost all situations.

Amphipods and isopods provide basic "meat" for many small and medium-sized marine organisms. Amphipods are an important part of the diet of such shorebirds as sandpipers.

Of all the familiar crustaceans, barnacles are farthest removed from the shrimp body form. Barnacles

14.45 Stalked barnacles. The animal is fastened "head first" inside its white, calcareous shell, with its "hind limbs" showing through the gap. The feathery limbs trap edible particles and pass them down to the animal's mouth. A rubbery stalk connects the shell to driftwood or some other substrate.

were indeed thought to be molluscs, on account of their shells, until the discovery of their nauplius larvae made their crustacean affinities clear. One can think of a barnacle as a much-modified shrimp that stands on its head inside its shell, using its hind legs to snare passing food and draw it into the shell toward the mouth. The animal is firmly attached to the inside of the shell and cannot get out. The feathery appendages, or cirri, visible from the outside are the food-gathering limbs. Acorn barnacles are fixed to rocks or other objects by their shells; stalked, or goose, barnacles attach themselves by means of a flexible stalk (Figure 14.45).

Barnacles are hermaphroditic. Each individual produces both eggs and sperms. Because they usually grow tightly packed on rock (and other surfaces), they are able to transfer sperms to their neighbors via long penises. The fertilized eggs are brooded within the shells until they hatch, whereupon the larvae swim away for a brief period of feeding in the plankton. The larvae have the nauplius form, with an additional feature that makes them easy to recognize: a characteristic triangular shield on the back. After six molts, they assume the "cyprid" form. The cyprid, a tiny swimming animal encased in a bivalved shell, soon settles to the bottom, actively seeking correct light levels and evidence that adults of its own species have lived there in the past. Finding proper conditions, it attaches itself to the rock, begins secretion of the familiar shell, and assumes the adult barnacle form. Tiny at first, and vulnerable to dislodgement by grazing limpets and other disturbers, its growth eventually protects it from some of the smaller accidents that can kill it. As adults, however, barnacles are subject to furious competition for living space, predation by carnivorous snails and sea stars, crushing by drifting logs, desiccation by hot sun, and many other hazards.

Certain barnacles are narrowly specialized. Some of them are found only on the surfaces of whales. Others occur only on barnacles that occur on whales. A fair number are parasites on other crustaceans. One, in particular, is found in the body cavity of crabs, occupying most of the interior space as an amorphous mass of tissue, often with the interesting result that the crab, albeit sterilized by the parasite, looks and acts as though it has changed its sex. Again, one would hardly know that the parasite is related to barnacles, except that its larvae go through typical barnacle nauplius and cyprid stages before settling on the surface of a crab.

Barnacles can be an expensive nuisance to operators of ships. In the days of sailing ships, a thick coating of barnacles on the hull slowed the ship and reduced its ability to sail to windward. Crews on long voyages routinely beached their vessels to scrape the barnacles off the hulls. Barnacles can slow the passage of modern ships so significantly that great effort is invested in developing techniques (including toxic anti-fouling paints) to keep them from surviving on hulls. Some of the modern anti-fouling paints (including those containing tributyl tin) are so toxic—killing everything on and around the boat hull—that they have been banned in some areas.

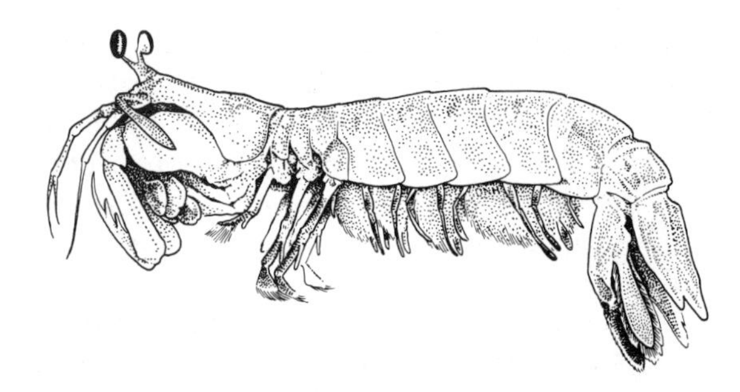

14.46 A California mantis shrimp, Pseudosquilla. *Length: about 8 cm. These pugnacious predators have been described thus: "They are the thugs of crustaceandom, hiding in their runs and warrens by day and consummating their murderous deeds under cover of darkness." (See Schmitt, in the Further Readings.)*

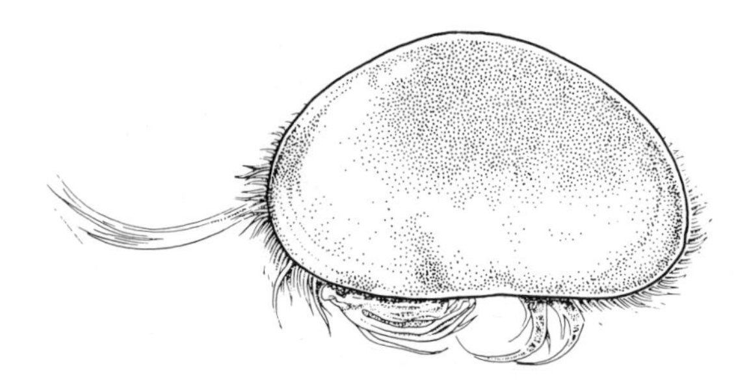

14.47 An ostracod (side view). Body and limbs are enclosed by a bivalved shell. Head is to the left. Size: about 1–3 mm.

14.48 Horseshoe crab (Limulus). *Length: to 60 cm. (W. J. Wallace)*

Other Marine Arthropods Many marine arthropods—also crustaceans—are members of groups that are usually overlooked by casual observers of the sea. There are about a dozen such groups. They include some fairly large animals (the "mantis shrimp," or stomatopods of the tropics, Figure 14.46), other animals that can be quite numerous and locally significant as food for fish (the "possum shrimp," or mysids), animals that contribute to the fossil record (ostracods, Figure 14.47), and many others. Insects are not crustaceans; a few of these (including marine water striders) also inhabit the sea. Discussion of these less conspicuous arthropods is beyond the scope of this book.

The only conspicuous and fairly familiar marine arthropod that is not a member of the phylum Crustacea is the horseshoe crab of the U.S. East Coast and the Pacific coast of Asia (Figure 14.48). These strange creatures are members of the phylum Chelicerata. Their nearest living relatives are spiders and scorpions. Under the large, shield-shaped carapace is a spiderlike body, equipped with two pairs of grasping claws and six pairs of walking or shoving legs. Totally lacking are antennae and mouthparts, the hallmarks of the crustacean phylum. Trailing the forward part of the body are a midsection equipped on the underside with flaplike gills and a spinelike tail, or telson. Horseshoe crabs have not changed much from fossil ancestors that date far back in the earth's history. They show in their planktonic life cycle a non-naupliar development that resembles the development of trilobites. It may be that these animals are the nearest living relatives of the long-extinct trilobites.

Adult horseshoe crabs are predators that feed on clams and other shellfish. They are fairly common where they occur and can be of significance in local

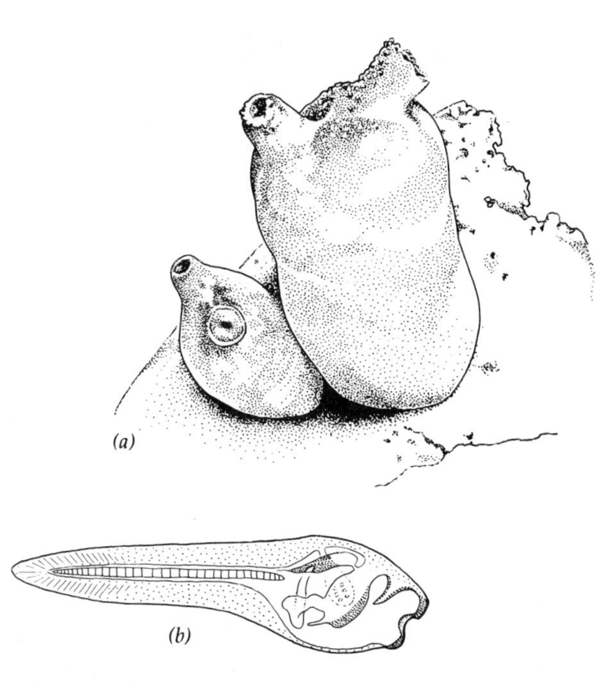

14.49 An adult tunicate and a tunicate larva. (a) the North Atlantic sea peach, Tethyum pyriforme. *(b) Larva of* Clavelina. *Sizes: sea peach, about 4–6 cm tall; larva, about 2 mm long.*

food chains. At Delaware Bay on the U.S. East Coast, for example, the eggs provided by massed mating horseshoe crabs supply migrating shorebirds with tons of food every spring, at an essential stage in the birds' northward seasonal migration.

To summarize, most marine arthropods are members of the phylum Crustacea. Some of these (copepods and euphausiids) are the most significant animals in the sea, in that they provide the essential link between tiny planktonic plants and the carnivores. Many others (crabs, hermit crabs, prawns) are also significant in that they collect small, dispersed particles of food and make it available and digestible to larger carnivores. Many crustaceans are of great commercial significance, either as food (crabs, shrimp, lobsters) or as marine fouling organisms (barnacles).

The Invertebrate Chordates

The best-known animals of the phylum Chordata are the vertebrates (discussed in Chapter 15). Zoologists place vertebrates in the subphylum Vertebrata; the remainder of the animals of this phylum are assigned to other subphyla. At first glance, one would hardly imagine that the bloblike invertebrate chordates have anything in common with fishes, sharks, and other vertebrate rulers of the sea, but close examination reveals some fundamental similarities. **Chordates** have gill slits, a dorsal nervous system, a post-anal tail, and a few other anatomical features that distinguish them from other animals. Vertebrates also possess these features, either as adults or during embryological development.

The most commonly encountered invertebrate chordates in northern waters are the tunicates (Figure 14.49). Only the larvae of these animals give hints of their relationships to the vertebrates. The larva, a tiny tadpole-shaped creature, has a dorsal nerve system, a pharynx with gill slits to the exterior, and a tail. After a brief life in the plankton, it settles to the bottom, attaches itself, and undergoes metamorphosis. The pharynx grows to occupy most of the interior of the animal. Water is drawn into a large opening on top, driven through a mucus sheet spread over the interior of the pharynx, and propelled out a second large opening near the first. The mucus sheet, moving over the interior of the pharynx, carries captured edible particles to the intestine.

Certain chordates (class: Larvacea) appear to spend their entire lives in arrested larval development. They remain in the plankton, where they secrete complex membranous structures equipped with filters, passageways, and escape hatches. The animals live inside these structures, filtering water through mucus funnels. For many years, these funnels were the finest plankton-gathering devices known to researchers. Capable of capturing the nannoplankton on which larvaceans feed, the funnels provided early plankton specialists with one of the few reliable means of obtaining microflagellates for study.

Salps (class: Thaliacea) are yet another group of invertebrate chordates. They are more commonly encountered in tropical waters than in cold seas. They are planktonic, gelatinous creatures, living singly or as swimming colonies with varied forms. Like many other invertebrate chordates, they filter nannoplankton for food. Certain salps are little more than self-propelled filters. One form, *Doliolum,* is shaped like a hollow barrel. It moves slowly forward by pumping water in through one end and out through the other. As the water passes through the "barrel," it is forced through a moving sheet of mucus that is continuously secreted along the dorsal interior and collected and passed on to a "mouth" on the ventral interior. Complete with a tiny vertical tail fin at the rear, this salp looks and acts like a small animated jet engine.

Many salp species are brilliantly luminescent and give the warm waters a breathtaking phosphorescent brilliance at night. Conspicuous as they are, and very common at times, their significance in marine food chains is not well understood.

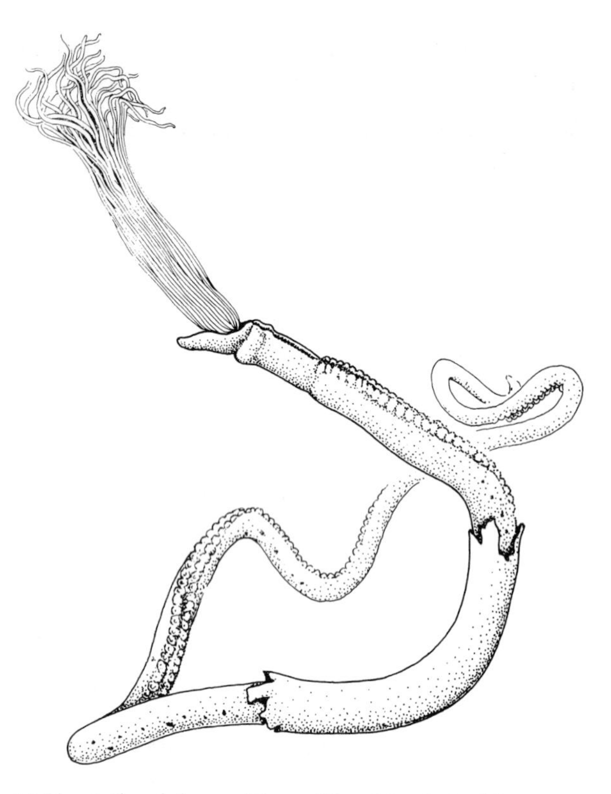

14.50 A "bearded worm," Lamellibrachia, *dredged from abyssal mud. Part of its cellulose tube is shown. It is an unexpectedly large new member of the phylum Pogonophora. Length: about 65 cm.*

The Rest of the Invertebrates

The invertebrate phyla discussed in the preceding sections are those whose organisms are large or conspicuous or of known ecological significance in the sea. Many marine organisms, members of other phyla not discussed here, do not fall into these categories. (For a complete listing of the marine phyla, see Appendix IX.) These animals are small or inconspicuous or scarce. They include both organisms that are sometimes noticed by observant visitors to the sea (for example, sea gooseberries, phylum: Ctenophora) and other organisms that are seldom seen except by the most determined specialists. Some of them (brachiopods, priapulids) have venerable family histories. Once abundant and dominant, they have in modern times become rather scarce. Others have never been abundant in the sea. All of them are zoologically interesting; many are exceptionally beautiful. As study of the seas proceeds, some of them suddenly become much more interesting to zoologists for various reasons. One recent example is provided by the "bearded worms" (phylum: Pogonophora, Figure 14.50). Prior to 1977, these obscure animals, interesting partly because they have no mouths or digestive systems, were known only from deep-sea dredge hauls and were mostly very small. The discovery of hydrothermal vents in the deep sea brought with it the parallel discovery of communities of strange animals living around the vents—communities that included pogonophorans 1.5 m long with the diameter of a garden hose. It appears that these worms have established symbiotic relationships with bacteria, which are responsible for keeping them fed. (The vent environments and this symbiotic relationship are of great interest to zoologists and are described in greater detail in Chapter 16.) Thus animals that were, only a few years ago, regarded as interesting but minor inhabitants of the marine scene have now moved to zoologists' center stage.

In this chapter, we have emphasized the roles of key invertebrates, plants, protists, and bacteria in the marine realm, with attention to important features of their anatomy, elements of their life cycles and evolutionary histories, and outlines of the ecology of some of the conspicuous organisms. Much remains unsaid about the "minor" phyla. Readings listed at the end of this chapter provide more information in all of these areas.

FURTHER READINGS

Barnes, R. D. *Invertebrate Zoology.* New York: Saunders College/Holt, Rinehart & Winston, 1980.

Buchsbaum, R. *Animals Without Backbones.* Chicago: University of Chicago Press, 1948.

Dawson, E. Y. *Marine Botany.* New York: Holt, Rinehart & Winston, 1966.

Dring, M. J. *The Biology of Marine Plants.* London: Edward Arnold, 1982.

Gates, David A. *Seasons of the Salt Marsh.* Old Greenwich, Conn.: Chatham Press, 1975.

Hardy, A. C. *The Open Sea: The World of the Plankton.* Boston: Houghton Mifflin, 1957.

Hardy, A. C. *Great Waters.* New York: Harper & Row, 1967.

Isaacs, J. D. "The Nature of Oceanic Life." *Scientific American,* September 1969, pp. 65–79.

MacGinitie, G. E., and N. MacGinitie. *Natural History of Marine Animals.* New York: McGraw-Hill, 1968.

Margulis, L., and K. Schwartz. *Five Kingdoms: An Illustrated Guide to the Phyla of Life on Earth.* New York: W. H. Freeman, 1982.

Nybakken, J. *Marine Biology: An Ecological Approach.* New York: Harper & Row, 1988.

Pearse, V., J. Pearse, M. Buchsbaum, and R. Buchsbaum. *Living Invertebrates.* Pacific Grove, Calif.: Boxwood Press, 1987.

Schmitt, W. L. *Crustaceans.* Ann Arbor: University of Michigan Press, 1965.

Sumich, J. L. *Biology of Marine Life.* 4th ed. Dubuque, Ia.: William C. Brown, 1988.

Thorson, Gunnar. *Life in the Sea.* New York: McGraw-Hill, 1971.

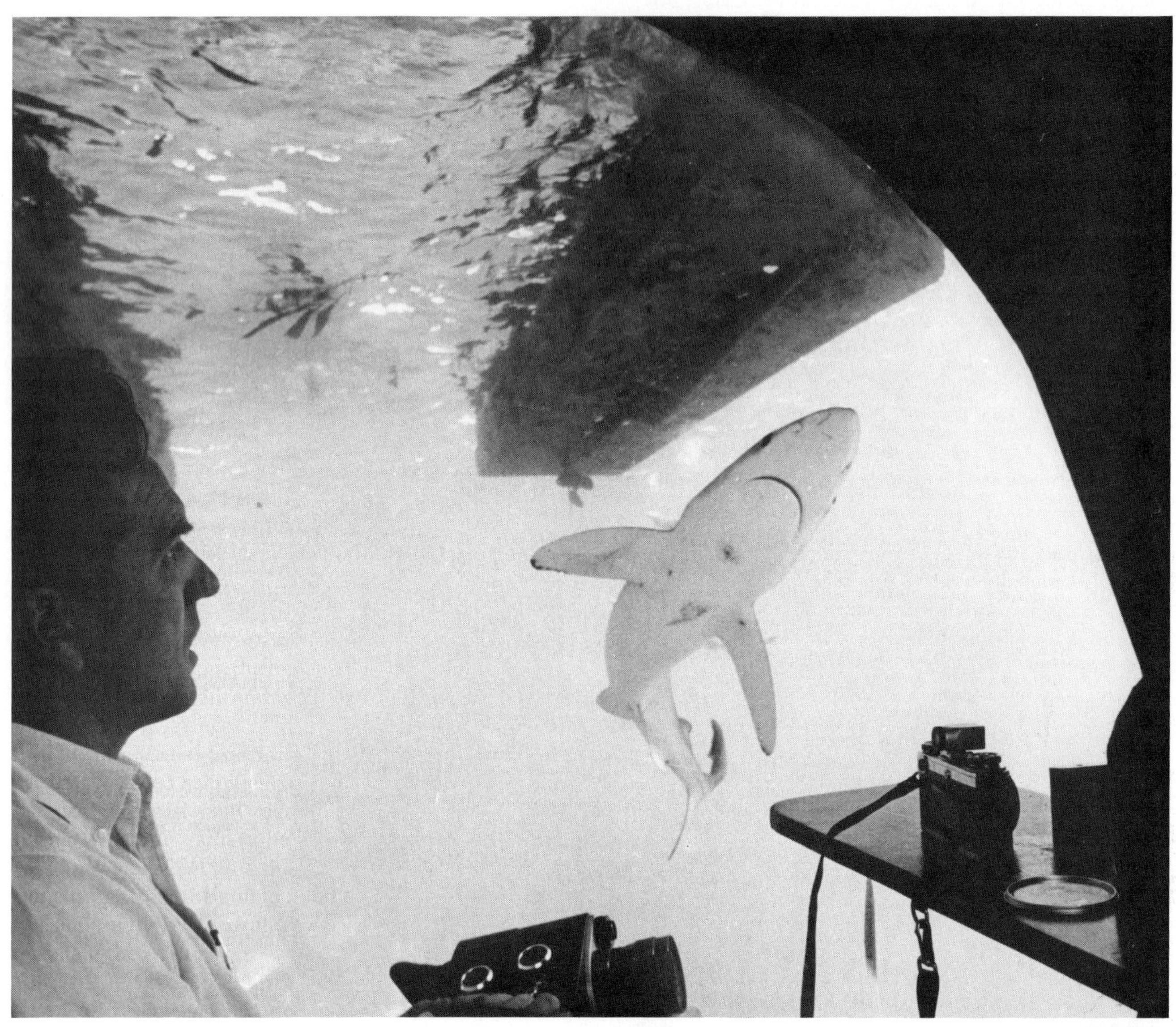

U. S. Navy

"Oh! Ahab! Not too late is it, even now, the third day, to desist. See! Moby Dick seeks thee not. It is thou, thou, that madly seekest him!"

Herman Melville

FIFTEEN

Life in the Sea: Vertebrates

Most marine vertebrates are fish, or fishlike forms. The latter include a few jawless, eel-like animals of ancient ancestry (lampreys and hagfishes), and sharks and their relatives. Most of the rest are "true" fishes, equipped with bony skeletons, scales, and other anatomical features that distinguish them from the sharks. The air-breathing marine vertebrates—mammals, birds, and a few reptiles—are all relatively recent invaders from land. Some, such as sea otters, show little modification from forms best-suited for life on land. Others, such as sea lions, walruses, seals, sea cows, and whales are progressively more modified for life in the water.

The fossil record suggests that even the fish have invaded the sea from fresh water. The oldest fish fossils appear to be marine, but the record hints that early fish may have moved into fresh water, diversified there, then returned to the sea in force, to dominate that environment. If so, then the sharks are the only conspicuous marine vertebrates of genuine uninterrupted marine ancestry.

Fish appeared in the seas in great numbers, perhaps as invaders from fresh water, about 400 million years ago in Devonian times. Ancestors of the other vertebrate marine organisms of modern times invaded the sea much later. The earliest marine birds are late Cretaceous (about 70 million years ago), whereas the ancestors of marine mammals entered the sea later, during Eocene times (about 50 million years ago). Invertebrate life in the sea is much older, dating back at least 240 million years earlier than the first widespread occurrence of sharks or fishes.

The rise of the fishes permanently upset many balances that had been established earlier in the sea and probably resulted in declines in the abundance of certain invertebrates of earlier ancestry, such as trilobites. Today the fishes dominate most marine food webs and provide important food for the top predators of marine systems, the birds and mammals.

Estimates of the number of living fish species (freshwater plus marine) vary wildly (15,000–40,000 species). A concerted 7-year effort to arrive at an accurate figure, by Daniel Cohen of the former U.S. Bureau of Commercial Fisheries, gives about 12,000 marine species. According to Cohen, sharks, rays, skates, and their miscellaneous relatives are represented by about 550 species. About 500 species of birds spend all (or part) of their lives at sea or as members of shoreline communities. Marine mammals comprise about 115 species, and reptiles (about 60 species) make up the rest of the marine vertebrates. These numbers are not great, compared to the diversity of life on land, where land mammals comprise 4,400 species; land reptiles, 4,900; land birds, 8,500; freshwater fish, 8,000; and amphibians, 2,000 species. Recall that invertebrate

15.1 Sharks and their relatives. Left side, top to bottom: manta ray (wing span: to 6 m), chimaera (1 m), sting ray (1 m). Right side, top to bottom: horn shark (1 m), great white shark (to 11 m), dogfish shark (1 m).

15.2 An angel shark cruising the bottom at a depth of 20 m near Los Coronados Islands, Mexico. Length: about 1.5 m. (Cal Messner)

animal life in the sea is represented by about 150,000 species.

Vertebrates are the largest and most commercially significant animals in the sea. By virtue of their size, abundance, intelligence, and mobility, they usually dominate the marine communities in which they live. The sections that follow examine some of the more conspicuous and significant representatives of the marine vertebrates.

SHARKS AND THEIR RELATIVES

Few marine organisms inspire the fear and fascination that sharks do. They are among the few animals left on the earth that routinely attack human beings. The dread reputation that sharks have earned is due to the activities of only a few species; fully 80% of the species are small (less than 2 m in length) and relatively inoffensive, and some of the largest ones (whale sharks, for example) have never been known to attack people.

Sharks and rays (collectively known as elasmobranchs), together with similar fishes, called chimaerids, are placed by zoologists in the class Chondrichthyes (Figure 15.1). The name means "cartilaginous fishes," and it refers to the fact that these animals have cartilaginous skeletons. This distinguishes them from the "true" fishes, whose skeletons are made of bone. Elasmobranchs differ from true fishes in several other significant ways. Whereas fishes have gills that are covered by a single external plate, or operculum, in the cheek area, elasmobranchs usually have a series of holes or slits along the pharynx, visible from the exterior and not covered by any external structure. Whereas fishes must expend energy to separate salt from water (see Chapter 16), elasmobranchs have a simpler and more efficient system. In the elasmobranchs, a high concentration of urea in the blood keeps the blood salinity at the same level as the sea salinity, thereby eliminating any tendency for seawater to dehydrate the animals. The scales of sharks are formed by the same processes that form their teeth. In fact, the teeth are simply enlarged versions of scales that happen to occur in the mouth. In fishes, the scales and the teeth have different embryological origins. There are other differences that confirm that living elasmobranchs and fishes are different sorts of animals whose ancestries on Earth have long been separate. Chimaerids (Figure 15.1) have an operculum covering the gill area but are more similar to the elasmobranchs than to true fishes in most other ways.

The elasmobranchs consist of sharks, skates, rays, and sawfishes. Although the flattened skates and rays appear at first glance to be quite different from the sharks, many intermediate living forms exist to show their relationship and to show how sharklike animals could become transformed into raylike animals. Many sharks, including the nurse sharks, spend a great deal of time lying on the sea bottom. Some benthic sharks—the angel sharks (Figure 15.2)—have markedly flattened bodies and expanded fins. The rays and skates, which are also mostly bottom dwellers, show extreme body flattening and fin expansions that are simple continuations of the trends seen in the others. Chimaeras are more remotely related to the elasmobranchs. They are grotesque-looking animals,

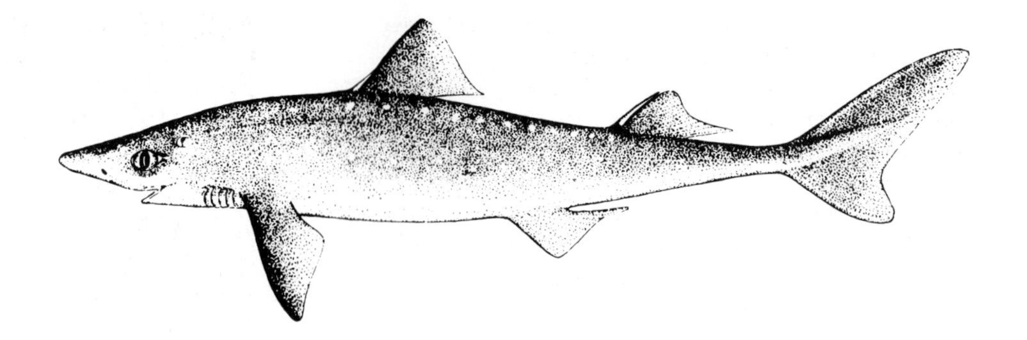

15.3 *Dogfish shark,* Squalus acanthias. *Length: to 1.5 m. (Hart, J. L.* Pacific Fishes of Canada. *Used by permission.)*

about a meter long, that live on the bottom and prey on shellfish.

A typical pelagic shark is a sleek, streamlined animal with one or two dorsal fins, a pair of large pectoral fins, a pair of pelvic fins to the rear, and often a single ventral anal fin. A pair of wide, staring eyes, five or more gill slits, a large powerful tail fin, and mouth full of teeth complete the design. Variations from this basic pattern may include a much elongated dorsal lobe on the caudal fin (thresher sharks, for example), or lateral expansions of the head with the result that the eyes are located on massive stalks (hammerhead sharks).

Almost all elasmobranchs are predators. Most pelagic forms take fish and/or marine mammals as their prey. The largest of all pelagic sharks are the whale sharks (*Rhincodon typus*) and the basking sharks (*Cetorhinus maximus*), which are plankton feeders. These species reach 15 and 12 m in length, respectively. The gills of both sharks are equipped with strainers. The strainers of the basking shark are spaced at intervals of 0.75 mm, which allows them to easily capture copepods and larger zooplankton. The shark swims at or near the surface with its cavernous mouth wide open, flushing water through its gills and straining out the plankton. A feeding basking shark can process about 2,000 tons of water per hour. Whale sharks inhabit warmer seas and are rather rare; basking sharks inhabit temperate and cool waters and are more common. Although they have not been implicated in attacks on human beings, sightings of these plankton-feeding sharks are usually thrilling occasions. The crew of Thor Heyerdahl's *Kon Tiki* reported a whale shark following their raft in the tropical Pacific in 1947 whose length exceeded the width of the raft (about 40 ft).

The dogfish shark, *Squalus acanthias,* of northern Atlantic and Pacific waters is a typical small pelagic shark (Figure 15.3). Adults reach a length of 130 cm. Dogfish inhabit shallow waters, ranging from the intertidal zone at high tide down to about 730 m. They are voracious feeders, eating herring, smelt, other small fishes, and invertebrates. They take bait intended for salmon and other sport fish and are considered a nuisance by fishermen. Dogfishes have spines just ahead of the dorsal fins, whose cross sections show annual growth rings. Using these rings, we can deduce parts of their life cycles. They are very slow to mature, taking about 20 years to become sexually mature. Their reproductive rate is also very low. Females are fertilized internally and carry about eight live young for nearly 2 years before giving birth to them. There is no placental connection to the mother throughout this long gestation. Instead, the huge yolk of the egg provides nourishment to the young shark throughout its development. (In some sharks a placental connection exists; in others the developing juveniles eat unfertilized eggs that enter the uterus from time to time.) Dogfishes have been harvested for their livers, which are high in oils and vitamin A. Dogfish meat is also used for fish and chips in England.

Many elasmobranchs reside on the sea bottom. These include such predaceous sharks as the nurse shark and such shellfish eaters as horn sharks, skates, and rays. As noted, rays are much-flattened elasmobranchs. Some catch and eat fairly active fish; others shuffle through the sediments in search of clams, worms, and other edibles. The biggest benthic rays reach a width of about 150 cm, although their largest pelagic relatives, the plankton-feeding mantas, can measure 6 m from wingtip to wingtip and tip the scales at 1,300 kg. Benthic rays usually lay large eggs, one to five at a time, in elaborate horny capsules. The young ray remains within the capsule for 18 weeks to 15 months, neatly tucked in with its pectoral fins folded over its back, attached by its belly to the egg yolk and nourished by it. Immediately on leaving the egg capsule, it starts with a diet of amphipods and other small crustaceans as early food, then it moves to larger prey. Where rays occur in large numbers, they can have a major impact on shellfish populations. Clam beds must sometimes be protected from them by building palisades of stakes to keep the rays out.

Some rays (stingrays) make a nuisance of themselves at tourist beaches by injuring swimmers. These rays (family: Dasyatidae) have a large, sharp, defensive spine on the tail. They are partially covered with sand as they shuffle about on the bottom searching for shellfish, and they are hard to see. When stepped on, stingrays lash out with their tails, driving the spine into the leg of the hapless tourist. The wound is painful, but seldom fatal. These rays occur mostly along warm-water beaches throughout the world. Shuffling one's feet, rather than taking large steps, is the best way to scare them away without incident.

Stingray assaults are rather insignificant in comparison with the attacks launched by large predaceous sharks (Figure 15.4). These attacks are among the most horrible occurrences in human experience. The sharks responsible for them are usually big ones and include such well-known species as the great white shark (*Carcharodon carcharias,* length: to 11 m), the tiger shark (*Galeocerdo cuvieri,* 5+ m), and the gray nurse shark (*Odontaspis arenarius,* 4+ m). One of many such incidents during World War II illustrates why the U.S. Navy has had a longstanding interest in studying shark attacks and finding ways to prevent them. The *Cape San Juan* was sunk by Japanese torpedoes in the South Pacific, in November 1943. Sharks attacked the men in the water with particular savagery, hurling themselves up on the liferafts to get at them, even as the men were being rescued. Hundreds of men died in this frenzied assault, and even some of the rescuers were bitten as they tried to assist the men in the water.

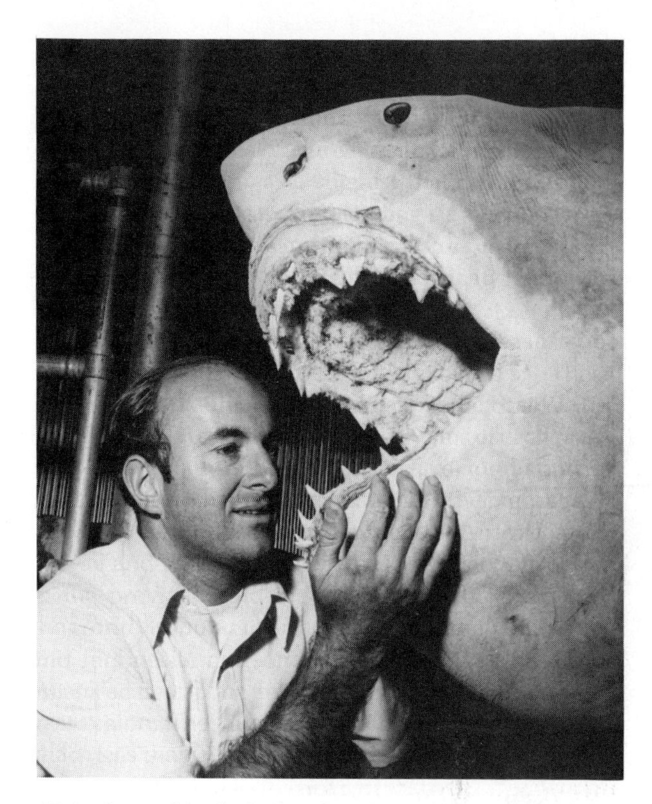

15.4 Great white shark. Length: to 11 m. (Charlie Van Valkenburgh, Sea World, Inc.)

Unfortunately, studies by the Navy and almost everyone else reveal only that sharks are unpredictable and that once they begin an attack, they are largely unresponsive to any attempts to ward them off. Shark repellants are mostly ignored (or eaten); shouting or slapping the water will not deter sharks; porpoises will not provide protection; and other defensive techniques are usually ineffectual. One antishark weapon with a modest success record is the diver's "bangstick," a long pole with an explosive charge at one end. This weapon of last resort is thrust at the head of a charging shark and the bomb is detonated via a trigger. Several divers owe their lives to this device. An extraordinarily effective repellant is known to be secreted by a flatfish, the "Moses sole" of the Red Sea. Sharks avoid eating the sole and have made strenuous efforts to avoid its secretions in experimental tests. Unfortunately, the potency of the secretion quickly diminishes when it is stored, a problem that shark researchers are attempting to overcome. Other promising chemicals are known, and the search for an effective shark repellant continues.

Certainly one of the best ways of preventing attack is to avoid antagonizing or attracting sharks. Nearly 20% of all shark attacks occur on victims who are in possession of hooked or speared fish. Many shark attacks begin when the "victim" molests an otherwise "harmless" shark in some way, often by pulling its tail. In unprovoked attacks, some intriguing patterns have been noted. For example, it is usually, although not always, true that a courageous person swimming to the rescue of a swimmer under attack by a shark will not, him- or herself, be attacked. Another interesting finding is that, despite the fact that 244 shark attacks on divers were recorded as of 1974, only one of these was an attack on a woman. (In that case, she was spear fishing.) Women do not dive as often as men do, but this discrepancy between numbers of attacks on males and females is much more than can be expected from the disparity in numbers of divers. In many cases, the bite of an attacking shark indicates that it struck with a glancing slash—not a full mouth-closing crunch. This, together with the rarity of attacks on female divers, has led to speculation that sharks may be simply attempting to drive off intruders, rather than to eat them, and that male divers move in some manner that makes them appear more intrusive to the turf-defending shark. However, it can be argued that a glancing bite is a standard method of attack for some feeding sharks. The great white shark, a predator on large sea mammals such as sea lions, deliberately strikes, then withdraws to wait for its prey to bleed to death from its glancing, slashing blow. In any event, few methods are known for safeguarding swimmers from shark attack, other than building and maintaining protective fences in the water, as is done in Australia.

A shark attack is among the more horrible fates that can befall a person. In one instance, Australian

Henri Bource was diving with about two dozen companions in water 9 m deep, examining and photographing seals. He was wearing a wetsuit, as were the others. Bource and two companions were maneuvering near a bull seal, about 90 m offshore, with a group of seals about 14 m away on the shoreward side. An observer on the beach saw a big dorsal fin tearing through the water, heading directly for the divers, bypassing many seals, and cutting right through the group of seals nearest the divers. About 6 m from the divers, the fin submerged, and the shore observer saw Bource burst upward from the sea surface. Bource later reported watching the bull seal dive to the bottom; he maneuvered himself to watch its return to the surface. Then something grabbed his left leg. Without seeing it, he knew that it was a shark. The force of the attack hurled him to the surface, where he shouted, "Shark," and struggled to escape. His mask and snorkel were knocked off, and he was dragged about 6 m underwater and shaken furiously by the shark. He grabbed at the shark's eyes and kicked at it, then became aware that his leg was being torn off. (He later surmised that, had the shark not taken it, he would have drowned.) He rose to the surface in a torrent of blood. The shark, some 4 m long, was seen dashing away with his leg, but it quickly returned to the attack. The other two swimmers rallied around Bource and used their spears to fend it off. The shark made repeated passes at them through the bloodstained water. A boat approached the scene of the attack, and Bource's companions pulled him toward it. The shark followed the three men until they were hauled into the boat and rescued. (Bource's left leg had been bitten off cleanly below the knee; he was saved by quick application of a tourniquet. Undaunted, he resumed diving after his recovery. Four years later, he was again attacked by a shark. Again, the shark made off with his leg—this time, an artificial one.)

15.5 Fishes. Clockwise from upper left: Scorpionfish (to 30 cm), triggerfish (15 cm), Siamese fighting fish (4 cm), lanternfish (5 cm), sea horse (6 cm), oarfish (to 300 cm), butterflyfish (15 cm).

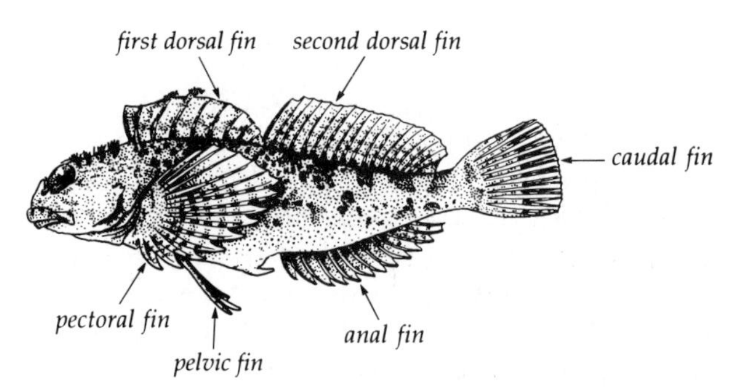

15.6 Tidepool sculpin, illustrating typical set of fish fins. Size: 5 cm.

FISHES

Fishes are the third most abundant group of animals in the sea. Numbering some 12,000 species, they are often large, numerous, and dominant in marine communities (Figure 15.5). Many of them are of great value to people. Fishes, particularly those of tropical reefs, are among the most beautiful animals on Earth.

The elongate body of the tidepool sculpin (Figure 15.6) illustrates the basic exterior structure of all fish. The tail, or caudal fin, is symmetrical and stiffened by a number of bony rays. There are two dorsal fins on the back, a pair of pectoral fins just behind the gill covers, and a pair of narrow pelvic fins on the underside of the fish. A single anal fin is also situated on the underside, just posterior to the anus. In a fast-swimming fish, the tail drives the fish, the dorsal and anal fins serve as keels to keep it on a straight course, and the pelvics and pectorals are flipped out or folded to brake or turn it. Evolution has modified this basic design in many ways, however. Some general trends are for fusion of the two dorsal fins into a single elongate fin (often with two distinct components), loss of one or more of the dorsal fins (Figure 15.7a), transfer of the pelvic fins from a position at the rear of the body to a more forward position nearly beneath the pectoral fins (as in the sculpin), addition of a third dorsal and/or a second anal fin (Figure 15.7b), and fusion of both dorsal and anal fins with the tail to provide a continuous fin encircling the entire rear portion of the body, as in many eel-like fishes. Examination of the ways in which the basic fin plan has been modified provides a useful way of looking at the diversity of marine fishes (Figure 15.8).

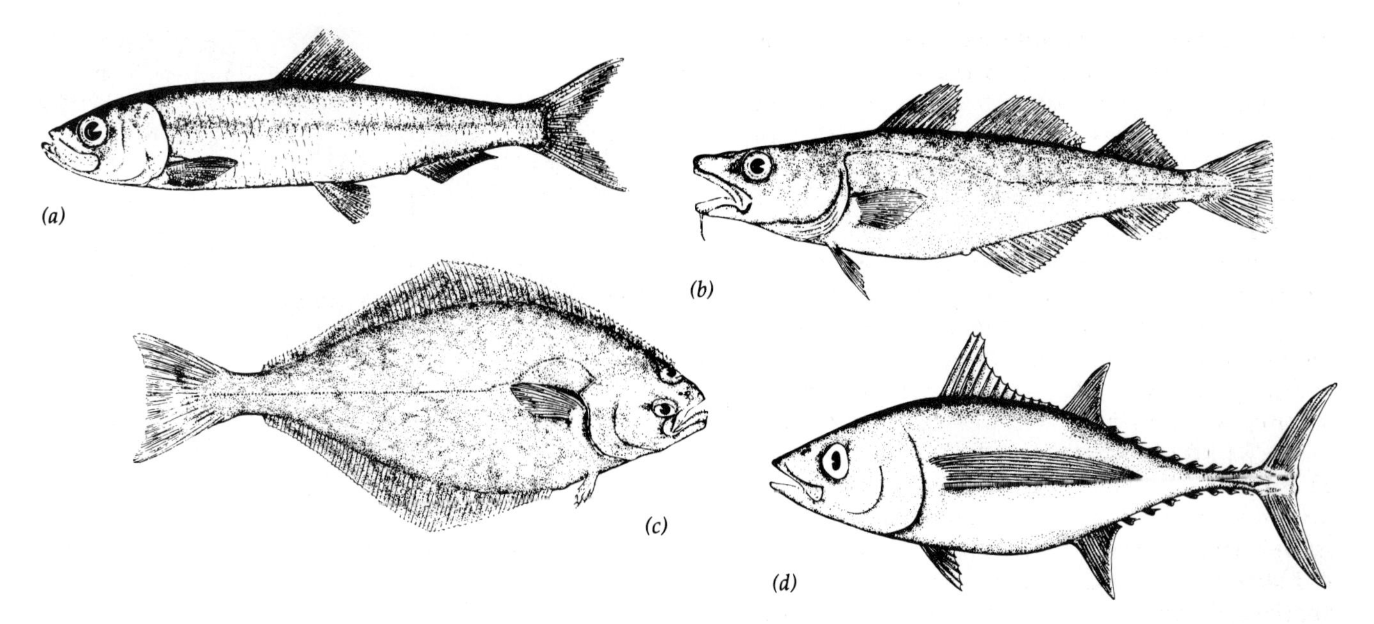

15.7 Fish of commercial significance: (a) herring, (b) cod, (c) halibut, (d) albacore. Maximum sizes: about 25 cm, 180 cm, 260 cm, 125 cm, respectively. (Hart, J. L. Pacific Fishes of Canada. *Used by permission.)*

Adaptations

Although the fastest fishes propel themselves by means of their tails, not all fish propel themselves in this way. Wrasses, which are common reef fish, use their pectorals as oars, rowing themselves forward or backward with stroking movements of these fins. (They are also able to swim by using their tails.) This makes for slow but precisely controlled motion, appropriate to their habits of nipping at the crustaceans, algae, and coral polyps situated in crevices in reefs. Certain other fish propel themselves via undulating movements of the dorsal (and sometimes the anal) fins. These fish are sometimes rather bizarre-looking creatures that, at first glance, do not appear to be fishes at all: for example, sea horses and pipefishes. Pipefishes are long, slender creatures whose colors, body shapes, and slow, swaying motions cause them to resemble eelgrass. The limited motility that they possess is all they need to maintain their camouflage and to approach the crustaceans and other small organisms that they eat. The triggerfish of tropical reefs (Figure 15.5) also moves by undulating its dorsal and anal fins. This gives it close control of its movements and an ability to sink vertically (or backward) into reef crevices.

Many fish hardly swim at all. These include such eel-like forms as morays, gunnels, pricklebacks, and others that can be found snaking their way about amid crevices and hideaways on the sea bottom. In such forms, the dorsal, anal, and tail fins tend to be fused all the way around the dorsal-to-ventral margin of the body, and often the pectoral or pelvic fins are missing. These fish behave more or less like snakes.

The most powerful swimmers include tunas and their relatives (albacores, mackerels, bonitos), swordfish, sailfishes, and marlins. In these fishes, the tail fin is huge and crescent-shaped, the body shows the ultimate in fish streamlining, and the dorsal and anal fins are sometimes accompanied by a number of extra finlets strung along the back or caudal peduncle (Figure 15.7d). The caudal peduncle (stump attaching the tail to the body) is narrow and often equipped with small, stiff lateral stabilizing keels.

Tests of model fishes have shown that the pelvic fins, when flipped out for braking or turns, tend to cause a speeding fish to rock upwards, unless the fins are positioned approximately underneath the pectorals. Fast swimmers such as the tunas usually have the pelvics in this forward position. These fish pursue herrings, mackerel, and other fast-moving prey. Generally, larger fish are faster than smaller ones, gaining more in muscle power than they lose in added surface area friction with the surrounding water.

Some of the most interesting adaptations of fish fins have little or nothing to do with swimming. The flying fishes of tropical seas use their fins to become airborne. These small fish have elongated pectorals that can be flipped to stand out from the body like the wings on a glider. Some species have expanded pelvic fins, as well. When alarmed or pursued, they dash to the surface, burst out into the air and glide away like tiny sailplanes, a foot or so above the waves, for glides

(a)

(b)

(c)

(d)

(e)

15.8 *Various fishes. (a) A school of small tropical fish. (b) Big-eyed tuna from the Pacific. (c) Anchovy* (Engraulis mordax) *are a major ocean fishery resource. This species is the most common anchovy from Cabo San Lucas, Mexico, to British Columbia. (d) The Pacific mackerel* (Scomber japonicus), *found from Chile to Alaska to Japan, is an important commercial fishery resource. (e) The Pacific hake* (Merluccius productus), *like the mackerel, has a wide distribution and is a valuable fishery resource. (Courtesy of [a] Penny Hermes, [b] U. S. Department of the Interior, Fish, and Wildlife Service, [c]–[e] National Marine Fishery Service)*

15.9 A remora, or sharksucker. These fish are found clinging to the undersides of sharks and other large organisms. Studies of remora embryology reveal that the suction-creating pad on the fish's head is a much-modified first-dorsal fin.

lasting up to 30 seconds. On descending, many flying fish lash the sea surface with the lower lobe of their caudal fin, which is much longer than the upper lobe, and prolong the glide. Thus propelled, they can evade pursuers in the water by vanishing into another dimension, as it were, and dropping back into the sea far from where their predator last saw them. (Unfortunately for the flying fish, this exposes them to airborne attack by seabirds, which are much better flyers than they.)

A number of fish have adapted their fins to serve as suction devices. The clingfish and some other fish are equipped with pelvic fins that form a shallow suction cup on the underside of the body. Using the suction developed by this device, these small inhabitants of tropical waters are adept at clinging tightly to rocks in areas washed by heavy seas. Many species live on kelps or surfaces in less rigorous environments. The sharksuckers or remoras also possess a fin equipped for suction, in this case the first dorsal fin (Figure 15.9). The fin in these fish forms a pad near the back of the head, with which the various species cling to the undersides of sharks, turtles, whales, billfish, boats, and other large objects. A sharksucker clinging to a shark is transported effortlessly about. It detaches itself to feed on the scraps provided by the shark's meal. Some have been shown to feed on other organisms (including parasites) that inhabit the surface of the shark. The suction developed by the fish is unexpectedly strong. Columbus observed the inhabitants of the West Indies fishing for sea turtles by tying lines to the tails of remoras, allowing them to swim to nearby turtles and attach themselves, and then carefully reeling them in. This "fishing" technique is still practiced in some parts of the world.

Yet other fish use elongated spines from their pectoral and caudal fins to support themselves on the sea bottom, tottering about on these spines as though on stilts. Perhaps the most bizarre use of a fin for purposes other than swimming is shown by the anglerfish of deep waters. These small, rounded fish live in perpetual darkness, lighted only by the flashes of luminescent creatures about them. In these dark depths, they apparently hunt by luring victims within range of their large, tooth-studded mouths. The luring device is a fishing pole, modified from the first spine of the first dorsal fin. This apparatus consists, in some cases, of a stiff "rod" with a flexible "line" attached to the tip and a "lure" (usually bioluminescent and sometimes equipped with barbless hooks) at the end of the line. The fish is able to move the rod, and it presumably whisks the luminous lure back and forth in front of its cavernous mouth, attracting small would-be predators to their doom. Whether they actually "hook" and "play" their victims is not known (and perhaps unlikely); it is amusing to note that certain anglerfish are found lacking their lures, as though one of their intended victims got away with the bait.

Certain adaptations of marine fish are not focused on modifications of the fins. The flatfish provide an example. These commercially important fish, including flounders, halibut, sole, plaice, and sanddabs, live on the bottom, essentially lying on one side. They begin life as ordinary-appearing fish larvae and usually feed on and dwell in the plankton. Early in their development, one of the eyes (usually the left) begins to move, and it migrates over the top of the head to the other side of the body. The fish takes up life on the bottom, lying on its blind side and watching the water above it with both eyes on the "up" side (Figure 15.7c). The "down" side usually becomes white and colorless, while the "up" side is camouflaged with sandy colors and gravelly patterns. These predaceous fish cruise the sea bottom, swallowing shellfish whole or biting off their siphons, consuming crabs, and devouring other benthic animals. From their positions of ambush, they sometimes make sudden dashes into the water overhead, snatching fish and other swimming prey.

As these examples illustrate, fish have acquired many adaptations to survive in the sea. The following sections examine life cycles of some of the commercially significant fish and other fish of unusual interest.

Fish of Commercial Significance

Herrings and Their Relatives One of the most important groups of marine fishes is the order Clupeiformes, comprising herrings, anchovies, sardines, pilchards, and their relatives. These are usually small fishes, reaching lengths of 30 cm or less. They hold a significant albeit lowermost position among fishes in marine food webs. This makes them primary transformers of copepod and diatom **biomass** into food suitable for larger organisms. Most of the solar energy fixed by phytoplankton is eventually dissipated as it is passed from one consumer to another up the food chain, but because the herrings and their relatives are among the first to utilize the solar energy trapped by the phytoplankton, a relatively large fraction of that energy finds its way into their biomass before it becomes dispersed. This fact makes it possible for them to become very abundant. The clupeiform fishes therefore serve as primary food for a vast array of other organisms, some of them commercially important. They are also harvested directly for human use.

The Atlantic herring (*Clupea harengus*) reaches a length of about 30 cm. It is a sleek, large-scaled fish, lacking a second dorsal fin. It is deep blue-green on the back and silvery along the sides and belly (Figure 15.7a). This species occurs widely in the North Atlantic, ranging from Newfoundland and the British Isles north to Iceland and Norway. A subspecies is similarly widespread in the North Pacific. The gills of the adults are equipped with stiff, forward-projecting filaments (the "rakers," not visible unless the gill cover is removed) that sieve copepods out of the water. These structures make it possible for the adult fish to feed on these tiny zooplankters. Herring will eat such medium-sized prey as sand lances (small fish), but their main mode of feeding involves swimming through swarms of copepods, barnacle nauplii, and other small organisms and catching them in a partially random way. Rather than simply swimming with their mouths open, as the basking sharks do, herrings dart from copepod to copepod, engulfing them in their mouths, flushing the water through their gills, and catching the small crustaceans on the gill rakers. Herrings feed in large schools. They tend to avoid patches of phytoplankton, concentrating instead on areas where zooplankton abound.

Because the North Sea has been so intensively studied by British fisheries workers, the life cycles of herring are better known there than anywhere else. The fish appear to circulate between four different types of sea bottom during their lifetimes. Adult herrings of the "Downs" stock feed in dispersed schools in the west-central North Sea. They migrate south to the Straits of Dover for spawning between November and January. Here they congregate in shallow water (40 m deep or less), where each female lays some 10,000–60,000 eggs. These are attached to stones, weeds, and other benthic objects. The adults then drift and swim back to the feeding grounds, and then disperse into smaller feeding schools. The eggs hatch within 10 days, and the larvae take up life in the plankton. Feeding on copepod larvae, diatoms, and other small prey, they drift into specific estuaries on the coasts of Denmark, Germany, Holland, Britain, and France, where they remain in water shallower than 20 m for their first year of life. At the age of 1 year, they migrate to a nursery area ("Bloden ground") in the central North Sea, where they feed until they are sexually mature. Reaching maturity at 4 to 5 years, they move off to join the stock in the adult feeding area. Thus, a spawning ground, a first-year inshore ground, an adolescent feeding (or "nursery") area, and an adult feeding ground are used by the herrings during their life cycle.

Though the adult feeding ground is shared by three separate stocks—the Downs, Buchan, and Dogger Bank stocks—the spawning and first-year regions used by each stock do not overlap very much, and the three stocks spawn at different times. The herrings of these stocks live to a maximum age of 11 years. Thus adult fish that escape the intensive North Sea fishery make several spawning migrations during their lifetimes.

Anchovies (family: Engraulidae) are the tropical and subtropical counterparts of the herrings. Like herrings, anchovies are small, silvery, large-scaled fishes (usually less than 10 cm long) that lack a second dorsal fin (Figure 15.8c). They are equipped with gill rakers capable of sieving the plankton, which constitute most of their diet. Like herrings, they occur in huge schools. There are about 135 species, worldwide.

Peruvian anchovies rose to commercial importance during the 1960s, when fishermen first began to appreciate the numbers of these small fishes inhabiting the waters off western South America. Research and preliminary fishing revealed that a stock constituting about 20 million metric tons of one species, *Engraulis ringens,* inhabited the nearshore waters there. (1.0 metric ton = 1,000 kg = 2,200 lb = 1.1 ton.) Three decades of relentless exploitation made this species the most heavily harvested fish on Earth. Boats operating out of Peru at one time harvested more than 10 million metric tons per year of Peruvian anchovies—a catch that exceeded the yearly U.S. harvest of *all* fish species, and that of every other nation, as well. For a few heady years, harvests of the Peruvian anchovy constituted nearly 20% of the *global* fish catch. Predictably, the fishery crashed. An occurrence of an El Niño in the early 1970s changed sea conditions and reduced the anchovy populations. Thanks to the El Niño and overfishing, the harvest dropped from 11.2 million metric tons in 1971 to 4.8 million in 1972. The fishery went from bad to worse in subsequent years, dropping to less than 1 million metric tons in 1980. The El Niño of 1982–83 finished a business that greed and overfishing had begun. The recent catch has been a consistent 100,000 metric tons per year, from 1982 through 1985. Thus a resource capable of providing daily protein for hundreds of millions of people was destroyed. People are not the only losers. The former anchovy populations supported some of the densest seabird populations on the planet. Adult birds abandoned their nests and their young, and the juveniles starved by the millions.

Herrings, anchovies, and their relatives are eaten by larger fish, cephalopods, marine mammals, and birds. They are a staple in the diet of cod, salmon, tuna, dolphins, seals, pelicans, cormorants, and other familiar creatures. In 1985, the global catch of all fish was about 75 million metric tons. Herring, anchovies, sardines, and other clupeiform fish constituted 21 million metric tons, or about 28% of this catch. No other group of fish approaches this level of harvest. Although some anchovies and herrings appear on pizzas and elsewhere in edible form, the majority of them are made into fish meal and are fed to poultry and livestock. As expected from food-chain dynamics (see Chapter 16), the livestock consuming this fish meal metabolize most of the fish protein, producing only about half to a tenth as much animal protein as they eat. These fishes are prime sources of protein in a protein-starved world; feeding them to livestock is essentially a waste of this bounty of the sea.

The Cod Fishes Much larger than the herrings, and next in significance as fisheries resources, are the fishes of the cod family (family: Gadidae). These include codfish, pollocks, hakes, haddocks, and tomcods. Most of these fish are distinguished by the fact that they have three dorsal fins, two anal fins, and pelvic fins that occur slightly forward of the pectorals (Figure 15.7b). Cods and their relatives usually have a characteristic barbel on the lower lip, squared-off tails, and other similarities that set them apart from other fishes. (Hakes are slightly different; they lack the barbel, and their paired anal fins and second and third dorsal fins are fused [Figure 15.8e].) Some of the species reach very large sizes. The Atlantic cod, *Gadus morhua,* is reported to have reached 1.8 m in length and 96 kg in weight, earlier this century. Those caught today are much smaller. They are fishes of cold seas, occurring mostly in the Northern Hemisphere, but with some species in cold southern waters.

The life cycle of the Atlantic cod begins when the adults move to spawning grounds. In the North Sea, the adult cods begin to congregate at central locations from January through April. The fish appear to be seeking depths of about 60–100 m at this time. Males and females pair off, the males release sperm into the water, and each female releases about 3–7 million eggs. Fertilization occurs by chance encounter between eggs and sperms, a random process that nevertheless results in relatively few unfertilized eggs. The eggs float upward and accumulate near the sea surface. The young cods hatch after about 10 days (up to 20 days in colder water) and hang upside down for the next few days while they absorb the yolk sacs to which they are still attached. After the yolk sac is absorbed, the larval fish right themselves and begin feeding in the plankton. Copepod nauplii are the staples of their diet, and these are usually abundant, since the copepods themselves have recently reproduced in response to the rapid new growth of phytoplankton that occurs each spring. At the time of hatching, cod larvae are only 4 mm long. They remain adrift as members of the plankton community for about 10 weeks after they begin to feed, gradually shifting their diet to adult copepods as they increase in size. At the end of this time, having reached about 2 cm in length, they descend to the sea bottom and take up residence there. During their early residence on the bottom, they feed primarily on small crabs, amphipods, isopods, and other crustaceans. As they mature, they switch to larger and more active prey, such as herring and squids. By the time they reach sexual maturity (4–5 years, at a length of about 70 cm), they have become active predators, hunting in the water above the sea bottom as well as on the bottom itself. Throughout their adult life they continue to remain close to, or on, the bottom—a lifestyle that is termed *demersal*. They complete the reproductive cycle by congregating in spawning shoals, heading for shallower water at those times.

Two relatives of the cod, the hake and the haddock (*Merluccius merluccius* and *Melanogrammus aeglefinus*, respectively), have similar life cycles in and about the North Sea. The cods, hakes, and haddocks appear to use the resources of their habitats in such a way as to reduce direct competition between them. For example, the spawning areas of the adult fish, while showing some overlap, are mostly separate. Most hake spawn on the west side of the British Isles; while the haddock seem to prefer the deeper water of the North Sea, to the north of the cods' spawning grounds. Like the cods, the haddocks begin their lives as tiny larvae drifting in the plankton, but the young haddocks remain in the plankton longer than the cods, moving to the sea bottom only after they reach 5 cm in length. Once there, the growing haddocks continue to feed on the benthic resources to a much greater extent than the cods do, consuming crabs, worms, and molluscs right on into adulthood and seldom utilizing fish as prey to the extent that cods do. In a like manner, hakes begin life as larvae adrift in the plankton, but this species remains in the plankton for the first 2 years of life, moving to the bottom only after reaching a length of about 20 cm. As adults, hakes are primarily predators on cephalopods and other fish, seldom utilizing benthic prey. The hakes of British waters reside on the bottom during the day and migrate upward into the water for feeding at night. Thus the fish of the three related species share similar life cycles and ecologies, but the timing of phases in their growth and the way in which they partition their hunting and spawning grounds is distinct for each species. The result is that each makes use of a slightly different subset of the environmental opportunities. The species do not steer completely clear of each other; cods eat haddock and hakes eat cod.

Of the 75 million metric tons of fish caught in 1985, cods and their relatives constituted about 12 million metric tons, or about 16% of the total. They are second to the herrings in worldwide harvests.

Halibuts One of the most important commercial and sport fish of the United States and Canadian Pacific coasts is the Pacific halibut, *Hippoglossus stenolepis* (Figure 15.7c). These giant flatfishes are among the largest of their kind, reaching 2.5 m in length and 227 kg in weight. Like others in their family, they have both eyes on the right side of the head, and they lie on the sea bottom on their left sides. Females are the larger of the two sexes. Pacific halibuts are confined to the waters of the North Pacific, ranging from the Sea of Japan and the Bering Sea down to northern California. (A few other large flatfish, also known as halibuts, are found over parts of the same area). These big fish live in waters ranging in depth from surface level to 1,100 m below the surface. They are formidable predators, eating fish, clams, squids, and even whole king and Dungeness crabs.

Halibuts of the British Columbia coast spawn between November and January. The female releases some 2–3 million eggs, which are fertilized in the water by free-swimming sperm. The eggs drift at depths between about 40 and 935 m, away from the bottom, in the perpetual darkness or frigid twilight of those depths. Like many fish, each larva is still attached to the yolk sac of the egg when it hatches. The yolks nourish the larvae until they reach a length of about 18 mm, at which time the yolk sac is fully absorbed. By this time, the left eye has begun to migrate up the side of the head and over the snout. After absorption of the yolks, the young fish begin feeding on small crustaceans and other planktonic prey. By the time the young fish are about 30 mm long, they are recognizable as halibuts and have the familiar flatfish form. Then, some 3 to 5 months after hatching, they congregate at a depth of about 100 m and move toward shore. Encountering the bottom for the first time at ages of about 6 to 7 months, they become demersal and remain benthic in their habits for the rest of their lives.

Maturation requires between 8 and 16 years for females; males mature at a younger age. During their immature years, they spread out over the bottom, eventually concentrating, as adults, at depths between 55 and 422 m. Coastal currents help to scatter the halibuts widely during the course of maturation. As adults, they make regular migrations to and from the spawning areas, covering up to 1,600 km each way. Halibut caught by commercial fishermen are much smaller than the biggest fish on record. Most are between 60 and 90 cm long. Growth studies by biologist G. M. Southward in British Columbia's Hecate Strait indicate that halibut can reach 125 cm after 15 years, and that the 60- to 90-cm fish are between 5 and 7 years old. Evidently, the occasional giant halibut is more than 15 years old.

In global fisheries' harvests, halibuts and related flatfish are rather low on the list, representing only about 2% of the world catch in 1985.

Albacores and Tunas A commercial fish quite unlike the demersal cods and halibuts of northern waters is the albacore (*Thunnus alalunga*) of warmer seas (Figure 15.7d). Albacores are powerful, fast-swimming fish that spend their entire adult lives in the high-speed pursuit of prey. The travels of individual fish take them over the entire reach of the Pacific Ocean. These sleek, rapacious predators reach a maximum length of 135 cm. They can reach speeds of 11 km per hour and have

15.10 Spinner dolphins, (Stenella longirostris). *Length: about 2 m. (© Robert Hernandez)*

been known to move 24 km per day. Albacore spawn in the western part of the North Pacific, between Hawaii and Japan, and evidently patrol a regular beat that ranges all the way from the Japanese coast to North America in their ceaseless search for prey. They follow the North Pacific gyre, swimming westward along the equator, northward in the Kuroshio current, then eastward and south, taking about a year to circumnavigate the Pacific. Adults eat herring, pilchards, anchovies, squids, and other swimming prey; they themselves are caught and eaten by even more impressive predators, the marlins and wahoos.

Albacores feed by night and by day. They are among the very few fish whose bodies approach the "warm blooded" condition. Hard-driving as they are, their muscles generate a great deal of heat as they swim, and their blood systems are arranged to retain that heat within the body. Such is the furious tempo of their metabolism that, when hauled from the water, they warm up before they die.

Albacores are mainly found in warmer waters, ranging over the mid-Atlantic, Indian Ocean, and mid-Pacific regions. They are among the southern fishes that appear off the northern U.S. West Coast when an El Niño weather anomaly sends warm waters farther north than usual.

Cousin to the albacore, and like it in habits, is the much larger bluefin tuna (*Thunnus thynnus*). Bluefins reach a length of 4.3 m. More so than albacores and other tunas, bluefins range into temperate seas as well as tropical waters. Individual fish undertake immense migrations, moving from California to Japan and back. Like the albacores, tunas are high-speed predators, taking sardines, jack mackerels, mackerels, anchovies, and squids as prey.

Commercial fishing for the smaller yellowfin tuna (*T. albacares*) once consisted of trolling baitless hooks at high speed through a pack of furiously feeding fish. For large tunas, each hook was trailed on a line running back to two or three stout fishing poles. The huge hooked fish was hauled out of the water by three fishermen, and it was sent flying through the air over their heads into a bin where it fell off the barbless hook. The hook was then cast back into the water without a moment's delay.

A widespread modern technique of fishing for yellowfins involves encircling a school with a purse seine net, then hauling the net on board. Two boats, the main vessel and an auxiliary skiff, drag the ends of the seine around the tuna school to trap it. Yellowfins often travel accompanied by dolphins (frequently *Stenella longirostris*, the spinner dolphin [Figure 15.10]). The most successful seining technique is to find a school of dolphins, chase them into the vicinity of the seiner and skiff with speedboats, and then wrap the seine around the dolphins. This practice is called "fishing on porpoise."

Prior to 1958, no effort was made to save the dolphins. Almost all were killed when the net was closed and brought on board. The dead dolphins were separated from the tuna and thrown overboard. After 1959, tuna skippers began using a technique called "backing down" to allow the dolphins to escape. It involves maneuvering the tunas and dolphins into separate sections of the net, backing the seiner for a few moments, dragging the floating rim of the net under the surface, and allowing the dolphins to escape over the top. This resulted in a much lower dolphin mortality rate, although the United States tuna fleet was still killing 173,000–194,000 dolphins per year

15.11 *A green turtle* (Chelonia). *(Steve Renick. Courtesy Steinhart Aquarium, California Academy of Sciences)*

during the mid-1970s. Further changes in gear and techniques, pressure from environmentalists and the National Marine Fisheries Service, court interventions, and legal limits on the number of kills have continued to improve the situation, further reducing the mortality rate to 18,400 dolphins in 1979.

These brief descriptions of important commercial fish only hint at the significance of fish in the overall ecology of the sea. In almost every marine community, fish play dominant roles. A few large-scale trends in their activities are evident. In cold seas, fish act mainly as predators. (Even plankton feeders such as the herrings concentrate heavily on animal prey.) Predaceous fish abound in warm seas as well, but there a great number of fish have also become herbivores. Although fish are known to inhabit the mid-oceanic water column, from the surface down to the greatest depths, most fish live near the surface of the sea, since that is where most plants grow and where food is most abundant.

Fish are ancient denizens of the sea, first appearing in marine waters in Devonian times some 400 million years ago. They have changed markedly since that time, becoming less heavily armored and more mobile and versatile. Most modern marine fish are members of a group called the teleosts, whose earliest appearance dates back to Cretaceous times, 90 million years ago. Since that time, the number of teleost species has steadily increased, along with their dominance of the sea. This record of survival and increasing dominance testifies to the great success and superb adaptation of these ubiquitous and beautiful vertebrates.

REPTILES

Least numerous of the air-breathing marine vertebrates are the reptiles. The most familiar marine reptiles are the turtles. Marine snakes, one species of marine iguana lizard, and a few seagoing crocodiles are the other reptiles that inhabit the sea.

There are seven species of marine turtles. They are moderate to large animals, ranging in shell size from 80 to 180 cm. Unlike the turtles of fresh waters, the marine species have limbs modified as flippers (Figure 15.11). Their shells are usually smooth, plated, and lightly built, although one marine turtle (the leatherback) has a soft corrugated "shell" of leathery skin set with bony ossicles. Like all of the marine reptiles, the turtles are inhabitants of warm seas. Adults of one species, the green turtle (*Chelonia mydas*) are herbivores. Those of the Caribbean graze the turtle grass that carpets the shallow sea bottoms in places, and they also eat a few species of algae. Other sea turtles are predaceous, eating fish, crustaceans, and jellyfish. Loggerhead turtles have been observed feeding on man-of-wars, their eyes swollen nearly shut from furious stings. Their inclination to eat floating membranous organisms has created an unexpected and fatal hazard for marine turtles when they sometimes mistake plastic trash, floating balloons, and other jetsam for food.

Sea turtles must come ashore to lay their eggs. The life cycle of the green turtle illustrates this practice. This species ranges throughout the tropical and warm, temperate waters of the world. The sand beaches on which the adults of the Caribbean populations once nested are located in Florida, the Bahamas, and elsewhere throughout the Caribbean. A typical fertilized female turtle, about 1.3 m long and weighing about 130 kg, waits offshore until nightfall, then crawls ashore. Digging at first with all four flippers, then turning to face the sea and digging with its hind flippers alone, the turtle excavates a hole, deposits about 100 fertile eggs in it, covers the eggs with sand, and returns to the sea. She waits for about 2 weeks and then comes ashore for another nesting. This procedure is repeated for a total of 3 to 7 nests. After finishing her egg laying, she encounters waiting males, who mate with her within a few hundred meters of the nesting beach. Since she will not return to lay eggs for another 2 or 3 years, the sperm acquired in this mating must be stored for that long before fertilizing the next batch of eggs. The eggs, meanwhile, incubate for about 2 months before they hatch. The young turtles, each about 6 cm in length, dig their way to the surface, scramble frantically to the water, and swim away. Turtles that survive to the hatchling stage are often attacked by gulls as they make their way down the beach. They feed on fish and other animal prey initially, then gradually switch over to a diet of plants.

As armored adults, turtles are relatively free from attack by most marine organisms, but the vulnerable locations of their nests result in a large number of mortalities. Females leaving the water are beset by poachers, who slaughter them for food and for sale of the "calipee." Calipee is cartilage from the turtle's lower shell, and it is used in preparing turtle soup. A 140 kg turtle is killed, stripped of 3 kg of calipee, and left dead on the beach. Nests are also dug up and eggs taken, in some instances to feed hogs.

Where turtles have been abundant, they have provided important food resources for coastal people over the centuries. This resource has been squandered. One beach in the Gulf of Mexico was once visited by 40,000 Kemp's ridley turtles per day during the nesting season. The turtles were so densely packed that it was possible to travel much of the beach by walking on their backs. Now, the species has been reduced to nesting on a single 9-mi stretch of beach, where less than 30 turtles per mile can be seen at the height of the season. The food resource represented by these animals is gone. With respect to another species, herpetologist Archie Carr notes, "There are very hungry people in places where green turtles were once abundant."

The outcome of years of unrestrained exploitation is that, by 1955, all of the green-turtle nesting localities in the Caribbean were destroyed except for one, Tortuguero Beach, in Costa Rica. The Costa Rican government declared the beach first a refuge and later a National Park and voluntarily gave up the modest income it made by leasing the beach to turtle hunters. Thus conservationists were able to mount an effort to re-establish green-turtle colonies throughout their former range.

15.12 Pelamis platurus, *the yellow-bellied sea snake, which ranges from the Sea of Cortez to South America. (William A. Dunson)*

Other species of turtles are not so exposed to human predation. Their nesting sites are far from civilization, and as adults they can sometimes avoid contact with people. Accidental encounters with fishing boats occur, nevertheless. On the U.S. Atlantic and Gulf coasts, shrimp trawlers accidentally catch and drown Kemp's ridley turtles in their nets, in numbers that are large enough to threaten the species with extinction. Equipping a shrimp trawl with a "turtle exclusion device," an arrangement of slotted bars with a trap door that allows netted animals to escape from the net, costs about $450. Citing economic disadvantage, questioning the device's effectiveness, and blaming the turtle's decline on nest poachers, shrimpers resist using this device.

Sea snakes are confined to the tropical Pacific and Indian oceans. They are distant relatives of cobras, and are highly venomous. Some species must leave the water to deposit eggs in nests on the beach; others give birth to live young in the water. In both groups, the tail is flattened in the vertical plane and serves as an oar for propelling the snake. Sea snakes are typically about 1 or 2 m long, although 3-m specimens have been recorded. They are sluggish and slow-moving if removed from water. They represent an occasional hazard to fishermen, who are sometimes bitten and killed by snakes captured in their nets.

There are about 50 species of sea snakes in the western Pacific and Indian Ocean waters. Only one species, the yellow-bellied sea snake (*Pelamis platurus*, Figure 15.12) is found on the west coasts of South

15.13 Caspian terns (Sterna caspia). *Males and females cooperate in the care and feeding of the young. (© 1977 Bill Dyer)*

America and Central America. This attractive snake (purple back, yellow belly, striped yellow-and-purple tail) has been cited as a reason for not building a sea-level canal across Central America. Should it spread through a new canal into the Caribbean, it could pose a minor hazard to tourists.

A marine iguana confined entirely to the Galápagos Islands and saltwater crocodiles mainly centered about Indonesia and northern Australia make up the remainder of the marine reptiles. The iguana is a herbivore; the crocodile is definitely carnivorous. The crocodiles of earlier times were more numerous and larger than those living today. Once crocodiles reached lengths of 9 m, but modern specimens seldom live long enough to grow beyond 6 m. These formidable animals inhabit mangrove swamps and sea coasts, and will attack people if the opportunity arises.

BIRDS

Although most birds are land animals, great numbers of them have taken to life on the sea. Their involvement with the ocean varies, as do their methods of using the sea's resources. Wings, feathers, and the fact that they must return to land or ice to raise their young (Figure 15.13) all result in behaviors and adaptations that differ in some significant ways from those of the other terrestrial invaders of the sea, the mammals and reptiles.

Many "seabirds" spend parts of the year at locations that are far from the sea. Among these are small and moderate-sized shorebirds, including some of the sandpipers, tattlers, surfbirds, willets, plovers, dowitchers, and their relatives. Many of these birds congregate along North American coasts during their fall and spring migrations, and they use the coasts as routes to and from their destinations at these times. At other times of year (particularly during the nesting season), they live far inland, often around the margins of pools and streams. Loons, grebes, many sea ducks, many gulls, and a few terns also winter on bays, estuaries, and open coasts; then they move inland for nesting in the spring. Many marine birds, including cormorants, puffins, auks, and some gulls never go far from the sea, raising their young on marine shores, or sea cliffs and stacks, and ranging to sea to catch fish for the nestlings. Such birds as albatrosses and petrels fly completely out of sight of land, sometimes spending years at sea while they mature, before returning to shore.

Seabird Feeding

All of the birds associated with the sea are predators. The various seabirds use a multitude of techniques to catch fish. Cormorants, puffins, penguins, guillemots, grebes, auks, murres, and mergansers pursue their prey underwater, propelling themselves with powerful strokes of their feet and (in many cases) by flapping their wings. Others such as pelicans, ospreys, kingfishers, boobies, and terns spot their prey from the air, then drop into the water to catch it. Skimmers fly along the surface of quiet water with their beaks gaping, trailing the lower mandible in the water. When the beak touches a fish, the skimmer performs a spectacular flip and grabs it, while remaining airborne. Many diving birds, including sea ducks, prey on such invertebrates as crustaceans and shellfish. Certain of these birds (for example, the white-winged scoter) grope along the bottom, uprooting clams and swallowing them whole. Shorebirds concentrate on invertebrates inhabiting the rocks and sand beaches. One of the larger birds, the oystercatcher, prowls the rocks, watching for mussels with shells agape. The bird jams its beak into the gaping shell, disables the shellfish with a quick bite, then eats it. The sandpipers and other small birds of the beach often travel in busy, fast-moving flocks, poking at the sand for amphipods, eggs, and other edible bits. These species, ranging from tiny to moderately large birds, scour everything from exposed mud to belly-deep water in their quest for prey.

Other hunting methods are practiced by various seabirds. In addition to a certain amount of active hunting for stranded crabs and other live prey, many gulls operate as scavengers, eating dead material found on the shores or inland. Tropical frigate birds, and the skuas and jaegers of cold seas, steal fish from other birds. Cruising jaegers watch small gulls or terns. When one catches a fish, the jaeger speeds up and goes after it. The tern, an excellent flier, takes evasive action. The jaeger pursues the tern through a series of breathtaking loops, dives, turns, and reverses, always just a tail's length behind it, until the tern drops the fish. The jaeger catches the fish in midair, then returns to watching the terns for another opportunity. In addition to this method of fishing, jaegers make themselves unwelcome at seabird nesting colonies by swooping down on unoccupied nests and eating the eggs or nestlings.

Where seabird populations are large, their impacts on marine communities can be significant. The pelicans, boobies, and cormorants nesting off the Peruvian coast provide a good example. During the 1950s, it is estimated that some 30 million of these birds nested on the offshore islands—the largest concentration of seabirds in the world. One island alone had 5 million birds on it, whose estimated catch of anchovies was 1,000 tons per day. By one estimate, the total catch by all of the birds in that area was about 20–25% of that of the human harvest of anchovies.

In another example, the spring migration of shorebirds (principally red knots, sanderlings, turnstones, dunlins, and sandpipers) along the U.S. East Coast concentrates nearly 1 million newly arrived birds in Delaware Bay for 1 or 2 weeks in May. During this time the sanderlings alone consume about 27 tons of eggs laid by horseshoe crabs. On a global scale, birds provide one of the few mechanisms by which phosphorus, a crucial nutrient that is carried to the sea via rivers, can return to land. Bird guano accumulated over the ages on nesting islands off Peru piled up to heights of 150 feet before people began to use it as fertilizer. Elsewhere birds defecate or die on the mainland, where the return of nutrients is more likely to affect inland terrestrial ecosystems. This return of nutrients by birds to the land is not significant compared to the drainage of nutrients to the sea, but it is one of the few rapid-return mechanisms in the generally slow-moving phosphorous cycle.

Penguins

Penguins are among the most superbly adapted of all seabirds. There are about 16 species, all of them restricted to cold waters of the Southern Hemisphere. (One species inhabits the equatorial Galápagos Islands, which lie in the mainstream of the cold Peru Current.) They have lost their powers of flight and instead "fly" by flapping their wings underwater. The feathers of penguins are much modified, consisting of the usual central stalk of bird feathers combined with softer, closely packed side filaments resembling densely packed hairs or fur. This arrangement, together with a thick coat of fat under the skin, gives the penguin the insulation needed for survival in frigid water.

The Adélie and Emperor penguins are the only two species that nest on the Antarctic continent. All others use sub-Antarctic islands, the Galápagos Islands, or continental coasts. Adélies are small penguins (60 cm) with classic tuxedo markings. They spend the Antarctic winter at sea, catching krill. In October (late southern winter) they return to Antarctica, sometimes walking as far as 96 km over ice to reach their rookeries. There they scrape together stones on which the female lays two eggs. The male broods the eggs, while the female returns to sea for about 2 weeks. Thereafter male and female take 2-week turns brooding the eggs, then guarding and feeding the young, until the young go to sea in February. Somehow, each adult is able to find the rookery amid the trackless ice and its own mate amid acres of densely packed look-alike brooding penguins. Later the young are herded together by a few dozen adult penguins, who guard the whole batch from raids by skuas while the other parents feed. The returning parents find and feed their own young amid these hordes of near-identical youngsters.

All penguins subsist on fish, squid, or shrimp. Through evolution, their wings have become atrophied, incapable of supporting the birds in flight. The wings survive as short, flat, strong flippers. By moving them up and down, the birds manage to "fly" in the water. While most seabirds use their webbed feet for propulsion, penguins use their wings as propellers and their webbed feet as rudders. They spend most of their time on land in an upright position, but they can "toboggan" on their bellies, using their feet and wings for propulsion. The largest species of penguin is the Emperor, which is nearly 1.3 m (4.3 ft) tall and may weigh as much as 41 kg (90 lb). Its average weight is 30 kg (66 lb).

15.14 *Courtship behavior of albatrosses. Posturing, spreading their wings, pointing their bills skyward, and touching bills at the nest side precede copulation. (Irene Vandermolen, c/o Leonard Rue Enterprises; © 1984 Frans Lanting; © W. E. A. Schreirer)*

Penguins generally feed near the surface not far from land. Beneath the short, dense feathers that cover their entire body is a thick layer of fat. In addition to providing insulation, this layer provides a mechanism for storing food, enabling the Emperor and Adélie penguins to survive the long periods of starvation they undergo during the incubation and rearing of their young. Penguins protect their eggs from freezing by keeping them on their feet and further warming them by snuggling down over them.

Penguins are extraordinary divers, swimmers, and hunters. Experiments by Dr. Gerald Kooyman with telemetered Emperor penguins have shown that they can dive to depths of 270 m while fishing. Unlike the Weddell seal, which exhales before it dives, the Emperor penguin inhales. Some of its bones contain the hollow air spaces typical of birds, and it can ascend at a rate of 120 m per minute, at the end of its dive. By conventional wisdom, any animal ascending from such depths at such speeds with air in its body ought to get a fatal case of the bends; why these penguins can defy the laws of safe diving is not well understood.

Albatrosses

A superbly adapted bird of another sort is the wandering albatross (*Diomedea exulans*) of the Southern Hemisphere. This bird, tipping the scales at 9 kg and having a wingspan of about 3.3 m, is essentially a living glider. These albatrosses spend much of their lives at sea, far from land, soaring effortlessly with only an occasional movement of the wings or head. They watch for fish, squid, or garbage floating on the surface and drop to the water to feed. The wandering albatross inhabits latitudes 30°–60°S, where surface winds blow from west to east, all the way around the world. Soaring with these winds, some of the birds are known to circumnavigate the globe. Those from South Georgia Island, in the South Atlantic, are known to congregate off Australia's New South Wales coast for feeding. Albatrosses tagged in Australia have likewise appeared at South Georgia, and a few have made a complete round trip. It is likely that all wandering albatrosses circle the earth many times in the course of their lives.

The albatross nesting season starts in December. At that time, adult birds return to the rookery islands. There, the birds go through elaborate courtship rituals, pair up, and mate. The female lays a single egg and incubates it for about 2 months. The ungainly youngster is fed by both parents for about 10 months. Both parents forage far to sea during the latter part of this time, and the chick is typically fed a large meal only once every 3 days. When it is ready to take to the air, the young bird launches itself from the cliff where it was reared and glides away—to remain at sea for the next 3 years. Immature birds, 4 years old and older, find their way back to the isolated islands on which they were born. They observe the courtship activities for several years before beginning to partic-

15.15 Brown pelicans at Pebble Beach, California. (Steve Renick)

ipate in them, finally maturing and beginning to breed when they are 10 years old (Figure 15.14).

There are 13 species of albatrosses, inhabiting both hemispheres but absent from the North Atlantic. In addition to their mastery of effortless gliding, they (and related birds of their order, the Procellariformes) have other adaptations for near-perpetual life at sea. An important one is their ability to drink seawater and separate out the salt, thus freeing them of any need to seek fresh drinking water.

Pelicans

Pelicans are familiar birds that illustrate important negative interactions between marine organisms and people. Two species (*Pelecanus erythrorhynchos* and *P. occidentalis*, the white and brown pelicans, respectively) inhabit the United States. There are 4 other species, all of which are Old World residents. Both American species occur on the California and Gulf coasts. The brown pelican (Figure 15.15) nests in coastal areas; the white pelican migrates to the interior of the continent for nesting during the summer. Both are large, ungainly birds whose lower beaks are modified as softened expansible pouches. Brown pelicans watch for fish while flying. When one is spotted, they fold their wings and plummet headfirst into the water in pursuit of their prey. These swift dives start at heights of 10–15 m. White pelicans, by contrast, swim on the surface and dip into the water for fish. Brown pelicans establish crude nests on rocky offshore islands and in other safe locations. The female lays two or three eggs. Both parents take turns incubating the eggs, one sheltering them from the sun and predators (such as gulls), while the other feeds. The young hatch after about 30 days' incubation. They remain in the nest for about 5 weeks, and walk about in the vicinity of the nest for another few weeks, while the parents take turns feeding them. Each parent catches fish during this period, carrying the captured fish back to the nest in its stomach. The young birds feed by plunging their bills down the throats of the parents to retrieve the partially digested fish. If all goes well, the youngsters leave the nest, following the parents and begging for food for a few weeks, then taking up independent lives of their own.

All has not gone well for brown pelicans in recent years, however, since their populations have been seriously depleted by DDT. This long-lived pesticide, applied in various situations on land, inevitably found its way into the sea. There it was concentrated by successive organisms in marine food webs. Pelicans, eating fish with relatively high concentrations of DDT, were affected. The eggs of affected birds had abnormally thin shells (often resulting in their collapse under the weight of the brooding parent), and the embryos within the eggs appeared to be harmed as well. These effects were seen in many fish-eating species, including terns, cormorants, and ospreys. Partly as a result, the use of DDT was banned in the United States in the 1970s. Since then, DDT concentrations in pelicans have decreased, nesting success has improved, and brown-pelican populations have made a limited recovery.

Perhaps because of their beauty, their small size, and the fact that they eat organisms that are of little interest to fishermen, seabirds have not attracted the same negative attention some sea mammals have

15.16 *Northern fur seals* (Callorhinus ursinus) *at St. Paul Island, Pribilofs, Bering Sea. The herd numbers about 1.2 million animals. (W. J. Wallace)*

earned. Indeed, most American coastal communities value the scavenging activities of gulls, and these birds are protected. But the birds are nevertheless harmed by their contact with civilization. DDT and its derivatives were once detectable in the tissues of almost every living vertebrate, including tunas, gray and sperm whales, and people. DDT has not been banned in other countries, where it is heavily used. High concentrations of DDT were found in Bermuda petrels in the 1960s. These birds (among the world's rarest, with a population of only 100 at the time) feed entirely at sea, at least 650 mi from the mainland United States where the DDT contamination is believed to have originated. Antarctic penguins also concentrate DDT, even farther from the pollutant sources. Certain seabirds such as gulls seem extraordinarily resistant to it; others experience the same devastating effects that are seen in pelicans.

A scarcity of fish, caused either by people or by nature, inevitably reduces the seabird population. In 1965, an El Niño devastated the Peruvian anchovy population and affected the millions of birds that feed on these fish. Deprived of their food, the birds abandoned their nests. The juveniles starved, as did millions of adults. The seabird population failed to make a significant recovery after this devastating weather anomaly—evidence, perhaps, that too many anchovies were being removed by the fishing industry.

To conclude, birds participate in the marine economy in many ways: as consumers of shoreline invertebrates; as catchers of fish, crustaceans, and squids; and as consumers of floating debris. Where they are numerous, their impacts on marine populations are very large. They play a small role in slowing the one-way global drainage of phosphorus to the sea, and they add beautiful and interesting activity to the life of the seashore.

MAMMALS

The mammals are relative latecomers to the marine realm. Fossils of the earliest known aquatic ancestors of whales and sea cows occur in Eocene sediments and are about 50 million years old. Other marine mammals (seals, walruses, sea otters) are apparently descended from ancestors that entered the sea later, in Miocene times (22 million years ago). Consequently, marine mammals vary in their degree of adaptation for life in the water. Certain animals that venture into the sea, such as polar bears and river otters, spend a great deal of time out of the water and have retained the familiar four-legged stance of terrestrial mammals. Others, such as sea otters, sea lions, walruses, and seals, show progressive modification of the limbs to finlike structures and spend relatively more time in the sea. The most highly modified marine mammals are sea cows, whales, and dolphins. These retain only the front limbs as enlarged flattened flippers, swim via undulations of their bodies and expanded tail flukes, and never leave the water. With these modifications of the body have come increases in size and changes in behavior that make marine mammals the dominant animals in many marine communities.

(a)

(b)

15.17 Pinnipeds. (a) Elephant seals (Mirounga) *at Guadalupe Island off Baja California. (b) Harbor seals* (Phoca) *swim in the holding pool at the Physiological Research Laboratory Pool Facility at the Scripps Institution. (Scripps Institution of Oceanography, University of California, San Diego)*

Pinnipeds

Seals and sea lions (order: Pinnipedia; **pinnipeds**) are large carnivorous animals that eat fish and invertebrates. They are distinguished as follows: Sea lions and fur seals (Figure 15.16) have small external ears and are able to rotate their hind flippers forward, an ability that enables the animal to move about on land with an awkward, loping gait. Seals (Figure 15.17) lack external ears (only the auditory holes in the sides of the head are visible) and are unable to rotate their hind flippers forward. On land, seals are constrained to wriggle along on their bellies, rather like enormous slugs. Walruses are different still. These huge Arctic marine animals are equipped with tusks that they use for digging up the shellfish on which they feed. Like seals, walruses lack external ears; like sea lions, they can rotate their hind flippers forward for movement on land. They have fewer teeth than either seals or sea lions. Worldwide, there are 13 species of sea lions and fur seals (family: Otariidae), 18 species of seals (family: Phocidae), and 1 species of walrus (family: Odobenidae). Most of these animals are found in cold seas, although a few species of monk seals (now rare and endangered) live in Hawaii, the Caribbean, and the Mediterranean. Walruses are found only in the Arctic regions; the other families inhabit both hemispheres.

Pinnipeds must return to land or ice for reproduction. They use rocky islands and shores for mating and bearing their young undisturbed. On breeding shores, many species exhibit a dominance system, whereby large males drive off smaller males and accumulate "harems" of females. A dominant male occupies a segment of shoreline with a herd of females, while the younger males are forced to congregate elsewhere, in nonbreeding "bachelor" groups. The struggle to defend the shoreline from intruding young males, or to displace an older male and take possession of his shoreline, is often intense and ongoing. An example of this is provided by the elephant seals (*Mirounga angustirostris*, Figure 15.17a) of California and Mexico. The males are equipped with a short, flexible proboscis that resembles an elephant's trunk—hence, their name. Elephant seals are among the largest of all pinnipeds. Females reach 3.6 m in length. As is usually the case with pinnipeds with harem dominance behavior, males grow to larger sizes than females, in this species reaching 6.1 m. Between December and March, the bulls attempt to occupy short stretches of shoreline on the few islands and beaches that serve as their rookeries. Cows landing on the beach mate with whichever bull dominates that stretch of shore. At any one time, a bull may be seen guarding about thirteen cows, although the individual cows are free to move from bull to bull. Intruding bulls are met by the defender, and a ponderous pushing, threatening, and biting contest erupts. Usually the defender wins. The contests are often decided without major bloodshed, although they can escalate to result in serious injuries.

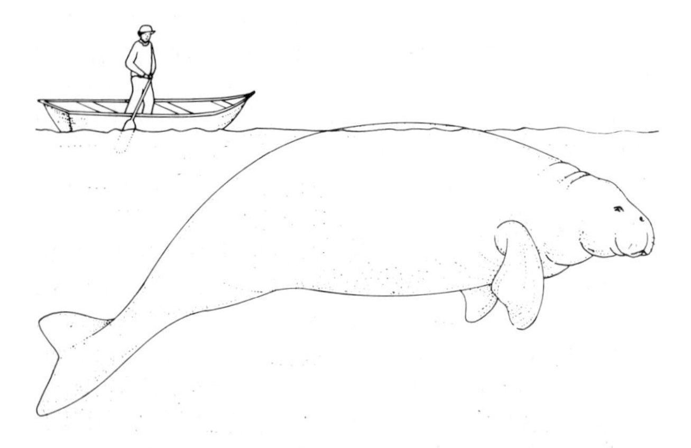

15.18 Seller's sea cow, Hydrodamalis gigas. *Grazers of kelp beds in the far North Pacific waters, the last of these large animals were exterminated by hunters in 1768. Length: about 8 m. (Adapted from Savage and Long,* Mammal Evolution, *New York:* Facts on File, *1986)*

In the Antarctic, seals are among the most important and impressive predators. The Weddell, Crabeater, and Ross seals prey on the fish and krill of the region. The leopard seal is more voracious than that, eating other seals and penguins and threatening human explorers on the ice. These animals shake penguins out of their skins before swallowing them. Weddell seals have been used for studies of seal diving ability. A telemetered animal released at a hole chopped in the ice miles from the nearest open water will often return to that hole to breathe, providing investigators with an opportunity to retrieve their apparatus and to measure the physical effects of the seal's dive. Their diving abilities are astounding. Weddell seals making shallow, exploratory dives under the ice typically hold their breath for 20–40 minutes and can make a round trip of nearly 10 km underwater. The longest dive recorded lasted for 70 minutes. On the deepest dives, seals have reached the sea bottom—600 m below the ice.

Despite their size, pinnipeds are not immune to predation. Their enemies are great white sharks, killer whales, and people. Smaller predators are lice—air-breathing insects much modified for life on an aquatic host. These small ectoparasites may be descendants of lice that lived on the remote ancestors of the pinnipeds, when those animals first took to the water.

Sea Cows

Sea cows (order: Sirenia; "sirenians") rank somewhere between pinnipeds and cetaceans in their adaptation to life in the sea. These mammals are placed in two taxa, the families Trichechidae (manatees) and Dugongidae (dugongs). The dugongs appear to be slightly more specialized for life in the ocean than the manatees.

Manatees inhabit the Caribbean Sea, the coast of South America from Venezuela to Brazil, and the west coast of Africa from Senegal to Angola. Although adults are usually about 2–3 m long, the largest animals (inhabiting coastal Florida and the Gulf of Mexico) reach lengths of 4 m and weights of about 900 kg. Manatees resemble gigantic sausages. The immense tail is flattened in the horizontal plane and rounded. Hind limbs are lacking; the front limbs consist of flippers, with claws marking the positions of the toe bones. Females have mammary glands, located near the "armpits" of the flippers. Manatees have ear openings but no external ears. The mouth contains large grinding molars, but canine and incisor teeth, typical of land mammals, are conspicuously lacking or are vestigial. Tiny eyes, nostrils that can be closed to exclude water, and a huge cleft lip adorned with bristles complete the manatee's rather cowlike muzzle.

Dugongs are similar to manatees in overall appearance. They are smaller than the largest manatees (about 3.5 m long, weighing up to 675 kg). Dugongs have a cleft tail more nearly similar to that of a whale, lack claws on the flippers, have less-developed molars, and are equipped with tusklike incisor teeth. Dugongs live about the margins of the Indian Ocean, inhabiting the Red Sea, the east coast of Africa, and the coasts and archipelagoes of southeast Asia to northern Australia. An extinct species of dugong (Steller's sea cow, Figure 15.18) is known to have lived in the Bering Sea. There are only 4 living species of sea cow; 3 manatees and 1 dugong.

Sea cows are vegetarians. Manatees eat a whole range of aquatic vegetation. They inhabit lushly vegetated jungle coasts and often move into rivers in search of various plants. The diet of dugongs is largely confined to marine algae and sea grasses, and they are less frequently found in fresh water. The animals have other subtle differences in their ecology and anatomy. For example, manatees are known to rest on submerged portions of shallow river banks, whereas dugongs never haul themselves up onto submerged ground. Manatees use their front flippers (and even their claws) to gather vegetation toward their mouths, whereas dugongs use their flippers only for swimming. Hair, found sparsely sprinkled over manatees, does not grow on dugongs. Dugongs use their teeth less when feeding than manatees do, using instead a set of tough plates, located in the mouth, to grind most of their vegetable food. These differences between the two sea-cow taxa suggest that dugongs seem to be a few evolutionary steps closer to complete adaptation to life in the sea than the manatees.

Fossil ancestors of both manatees and dugongs are found in strata dating from Eocene times. They were large tropical animals. Many of the adaptations seen in modern sirenians were already present in these earliest known ancestors. Several clues suggest that these creatures are descended from the same stock that gave rise to elephants. For example, the molars of manatees have a peculiar detailed (bilophodont) construction that is most similar to the molars of the ancestors of modern elephants. As in elephants, the molars of manatees emerge from the gums during the lifetime of the animal, are used for chewing for a while, then drop out, making room for new molars. As in elephants, the small, peculiar "tusks" of dugongs develop from the incisor teeth. These and other clues suggest that the sirenians have an ancestry quite different from the other large sea mammals. Details of the fossil record suggest that sirenian ancestors were herbivores that fed on sea grasses at a time when those grasses were just beginning to flourish in the sea.

Sea cows are the original "mermaids." The term *sirenian* refers to the supposed resemblance, in the feverish minds of homesick sailors, between sea cows and Sirens, the beautiful young women of Greek mythology who lured sailors to a presumably delightful doom on rocks throughout the Mediterranean. This historical misperception has combined with modern indifference, ignorance, and superstition regarding sea cows and has brought them to the brink of extinction. Sirenians are being rapidly killed by hunters for all of these reasons. The copious salt "tears" of captured dugongs are said to be an aphrodisiac. The meat is delicious, and each dugong contains about 10 gallons of a clear oil that is hailed as a remedy for everything from headaches to constipation. Mercilessly hunted, injured by accidental collisions with boats, and beset in a dozen other ways, these slow, inoffensive creatures have been utterly unable to defend themselves. Their numbers have dwindled drastically everywhere, and in some large sectors of their former range (for example, the whole island of Madagascar), they have vanished. They will soon be extinct without public attention, concern, and conservation.

Cetaceans

The largest of all marine mammals are the whales and dolphins (Figure 15.19). These mammals are sorted by zoologists into two categories, the suborders Odontoceti (toothed whales) and Mysticeti (mustached, or baleen, whales). The odontocetes include porpoises and dolphins, larger beaked whales, pilot whales, blackfish, narwhals, belugas, killer whales, and sperm whales. The mysticetes include the right, or bowhead, whales, sei whales, blues, grays, rorquals, humpbacks, and finback whales. Toothed whales feed mainly on fish and squids. Teeth are absent in adult mysticetes. Instead, large, whiskery fringes of baleen hang from their upper jaws. These are used as strainers for capturing plankton. The baleen bears a slight resemblance to a mustache, thus the name of the suborder. There are in all about 77 species of whales and dolphins.

Partly because humpback whales inhabit both coasts of the United States and are easier to observe than most other large whales, a great deal is known about their habits. Humpbacks (*Megaptera novaeangliae*) are medium-sized whales, reaching 12–15 m in length. They live all around the world and can be distinguished by their long pectoral flippers with irregular lumpy surfaces (Figure 15.20). Like many mysticetes, the "throat" is furrowed on the exterior, allowing it to expand greatly when the whale fills its mouth with water. The throat and pectoral flippers are white; the skin elsewhere is black. These whales serve as settlement sites for large barnacles, which form encrusted lumps on various parts of the whale's body. The arrangements of these clumps of barnacles, various scars on the flukes and body, and other distinguishing marks make it possible for researchers to photograph the whales and then find the same individuals later.

Study of the humpbacks of both coasts of the United States reveals that these whales undertake large seasonal migrations. They feed in northern waters during the summer, then move to equatorial waters for breeding and birthing during the winter. Those on the southeast coast of Alaska migrate to the Hawaiian islands, arriving in December and January. There they mate and give birth. During their stay in the islands they do not feed. Beginning about April, the whales depart for Alaskan waters, where they set about feeding and rebuilding energy reserves that were depleted during their long migration and winter fasting.

The feeding whales sometimes capture their prey (krill and small fish known as capelins) in a remarkable way: When a school of small fish is encountered, the whale dives, spiralling downward around the school and releasing a cloud of bubbles as it goes. At the bottom of its dive, it turns and swims back up through the center of the cylinder of bubbles. The fish, reluctant to swim through the bubbles, remain surrounded and concentrated by them as the whale comes up, filling its mouth. The whale breaks the surface headfirst, its mouth and throat distended by the water and fish it has taken in. The water is then forced out through the baleen while the whale is still upright, and the fish are swallowed. Humpbacks may feed cooperatively in this way, with eight to ten whales rising through a bubble net some 23 m in diameter.

Humpbacks apparently live to an age of about 25 years. They are about 5 m long at birth. The mothers

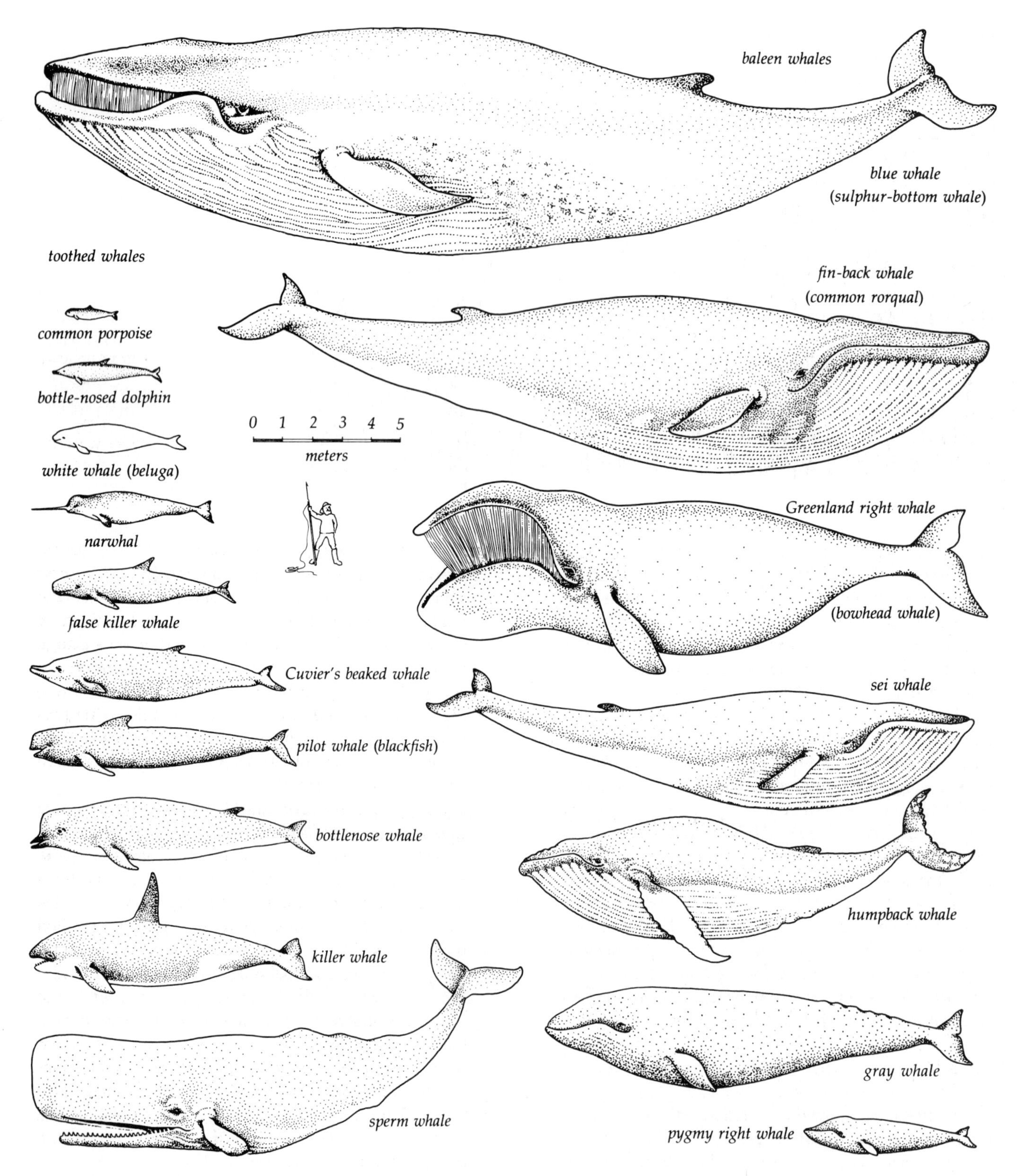

15.19 *Whales. Toothed whales (odontocetes) on the left, baleen whales (mysticetes), top and right.*

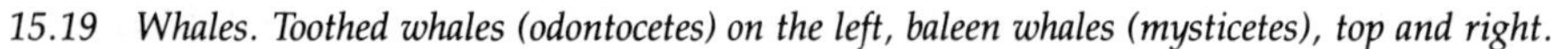

15.20 Humpback whale jumping out of the water (or "breaching"). This species is identified, in part, by the long white pectoral fins. (John Dominis, Life *Magazine © Time, Inc.)*

nurse the calves for up to 10 months. Both sexes reach sexual maturity in about 4½ years, and full adult stature at 10 years of age. The gestation period of the females is about 11 months. Thus those that mate in Hawaii in a given year give birth to their calves there a year later. Most humpback females give birth every 2 or 3 years.

Prior to 1920, humpbacks were subject to heavy attack by whalers. In 1910, fully 95% of the whales killed near South Georgia, in sub-Antarctic Atlantic waters, were of this species. Because they were abundant and easy to catch, the whalers sometimes found it more convenient to kill additional whales than to waste time looking for ones killed and left to drift earlier in the day's slaughter. Predictably, they decimated the whales. About 10,000 humpbacks were killed near South Georgia from 1910 to 1911. By 1926, the kill was down to less than 100 per year. As the whales became scarce, the whalers shifted to other species. Whaling for humpbacks ceased during the mid-1960s. Their world population prior to the onset of whaling numbered 120,000; today 10,000 remain. Hunting of humpbacks and other species is discussed in Chapter 18.

Sperm whales (*Physeter catodon*), the largest of the odontocetes, range widely throughout the world, reaching their southern and northern limits at the Antarctic and Arctic circles. They are fairly large (12–18 m long) and are distinguished by their squared-off snouts, nostrils at the front of their heads (evident when they blow), the absence of a fin on their backs, and their narrow tooth-studded jaws. These whales perform the deepest dives of any air-breathing animal, reaching depths of 2,250 m (and perhaps even 3,000 m). They feed mostly on squids, although tunas, sharks, and even seals have also been found in their stomachs. There is some speculation that they are able to stun large squids by emitting thunderous blasts of focused sound. The head of the whale contains a huge reservoir of a liquid substance called spermaceti, an oil that may help to provide buoyancy to the huge beast. This oil, one of the best lamp fuels of the pre-electric era and one of the finest lubricants known today, is an important reason why sperm whales have been hunted for centuries.

The whale of Herman Melville's *Moby Dick* was a sperm whale. Melville borrowed his whale's name from an actual whale of the nineteenth century, a white sperm whale named Mocha Dick, which ranged from the South Atlantic to the coast of Japan. Mocha Dick went out of his way to attack whalers. He was embroiled in more than 100 fights with men in whaleboats, sinking 14 of the boats and killing 30 men during the course of his career. Whether he was ever captured is not known. The ship of Melville's story, the *Pequod*, was patterned after a real whaling vessel, the *Essex*, that was rammed and sunk by a sperm whale in the equatorial Pacific in 1820. Other rogue sperm whales—Timor Tim, New Zealand Jack, Don Miguel—attacked whaleboats; the ship *Ann Franklin* was also sent to the bottom by an enraged whale.

The smaller toothed whales—the dolphins, porpoises, and killer whales, or orcas—have risen to public attention in recent years because, while maintaining them in oceanariums, we have begun to discover how intelligent these animals are. Although the terms *dolphins* and *porpoises* are used loosely and interchange-

15.21 *Rough-tooth dolphin,* Steno bredanensis. *Length: about 1.8 m. (James D. Watt/ Earthviews)*

ably, there is a difference. Dolphins have conical teeth and a pronounced "beak" (family: Delphinidae; Figure 15.21), porpoises have spatulate teeth and no beak (family: Phocaenidae). Dolphins are the animals usually seen in oceanarium shows; porpoises are smaller, more timid, and less receptive to captivity. Dolphins and porpoises eat fish and squids. Some (but not all) killer whales (*Orca orcina*), reaching about 9 m in length, also prey on seals and porpoises. There are about 39 different species in the 2 families. Most of these animals are fast-swimming predators. Far from shore in subtropical latitudes, for example, the spinner dolphins range alongside tuna, hunting in packs or family groups in pursuit of herring. In bays and estuaries in higher latitudes, harbor porpoises conduct rather solitary pursuits of fish. The orcas of the U.S. West Coast travel in family groups, or "pods," ranging over 120 km in a day's hunting and traveling. A pod, consisting of perhaps 12 females, 3 or 4 males, and 6 calves, generally remains in its own stretch of water, mingling with other pods on occasion but returning to familiar haunts for the most part.

Recently, scientists have been intrigued by the possibility that dolphins might possess intelligence comparable with that of people. The dolphin's convoluted brain looks remarkably like that of a person and is one of the largest brains per unit of body weight to be found in any animal. The nerves from the eyes and ears appear to be capable of carrying as much information to the brain as our own do. Dolphins "see" by emitting blasts of high-frequency sound, then processing the echos returning from the environment, and they are apparently able to form sonic images of the environment around them that are comparable to the "pictures" seen by human eyes. Tests of dolphins in silty pools (where vision is impossible), and of dolphins with their eyes covered, show that they are easily able to distinguish between different species of fish, surfaces of different textures, and objects of slightly different sizes. They also communicate by means of squeaks and clicks, and it is possible that these sounds are the basis of a spoken language. Efforts to decipher that "language" have revealed some interesting details of dolphin signalling ability, but, thus far, there is no firm indication that dolphins use language as we understand the term.

Some of the most intriguing examples of dolphin intelligence arise from their behavior, rather than from their use of sonar. Observations by Karen Pryor, the chief trainer at Sea Life Park in Hawaii, provide some startling examples (see also the Further Readings). Malia, a rough-tooth dolphin (*Steno bredanensis*; Figure 15.21), appeared to understand the concept, "You will be rewarded with a fish if you invent a new stunt, one that you have never done before." Malia then put on demonstrations of dolphin swimming and jumping that no one would have thought possible. In another example, two cetaceans, an 8-foot false killer whale and a Pacific bottlenose dolphin, enjoyed each other's company. They would cooperate to remove a heavy cover from the dolphin's tank so that the dolphin could leap over into the whale's tank—but only when they thought no one was looking.

In yet another example, Pryor noticed that two experienced rough-tooth dolphins, Malia and Hou, were performing clumsily during a show. They were exhibiting uncharacteristic "stage fright" and were only too eager to rush back into their pens after their stunts were over. After the show Pryor realized that each dolphin had been forced to do the other's stunts.

15.22 *The California sea otter,* Enhydra lutris. *A typical adult otter consumes 12 abalones, 20 sea urchins, 11 rock crabs, 60 kelp crabs, and 112 turban snails every day. (James A. Mattison, Jr.)*

Someone had inadvertently mixed up the two dolphins, so that when the curtain went up, each was presented with props for stunts that that animal had never performed. Each dolphin had, however, "seen" the other perform on many occasions. Thus each knew what was expected of it, attempted to do the stunts, and stumbled through the act. During her embarrassing moments under the floodlights, Hou managed to leap clumsily into the air and pass through a hoop suspended 6 feet above the water on her first attempt—a stunt that would normally have required many weeks' training to accomplish. Observations like these have left many with the conviction that dolphins are comparable in intelligence with human beings. However, some experienced observers (including Pryor) conclude that dolphins are not much more intelligent than dogs.

Marine Mammals and People

Marine mammals are at the center of many conflicts over uses of the resources of the sea. Because many of them eat fish, they run afoul of fishing gear and catches, causing hard feelings and (in some cases) outright economic hardship. Few seals can resist the temptation to tear into a net full of fish, causing damage to both the gear and the catch. Few fishermen can resist the urge to blame declining catches on marine mammals, which do, after all, have astonishing appetites. Sea otters, for example, may consume 20% of their body weight in invertebrates every day (Figure 15.22). Since an adult sea otter can reach 40 kg, and since the invertebrates they eat include abalone, sea urchins, and others prized by divers and the commercial fishing industry, the daily haul of a few sea otters can excite real controversy. Sea otters, under government protection, are increasing in numbers on the California coast, where they had been thought to be extinct as a result of hunting prior to 1911. Where their numbers are growing, sea urchins are declining in numbers and so are abalones, much to the outrage of some local people. At present, otters and all other United States sea mammals are under the protection of the Marine Mammal Protection Act, which makes it illegal to kill, disturb, or even approach marine mammals of any sort unless they are problem individuals with a proven history of raiding fishing gear.

Real or supposed damage to the fishing industry is only one dimension of the problem in United States waters. Alaskan native whalers, claiming historical and traditional privilege, assert a right to kill about a dozen bowhead whales each year. Bowheads (Figure 15.19) are scarce, thanks to the depredations of Yankee whalers during the 1850s. The Bering Sea population of this species (*Balaena mysticetus*) is now known to consist of at least 4,400 members, perhaps more. (A few hundred others live in the eastern Arctic ocean and the Sea of Okhotsk.) When the controversy began, the whales were mostly adults; the small number of calves prompted concern that the population was not reproducing. If so, it seemed possible that even the limited hunt proposed by the Alaskans might hasten the whales toward extinction. Yet the capture of bowheads is seen by native Alaskans as one of their most treasured traditions, even though their hunting techniques have changed. Concern for the Alaskan whaling tradition caused United States negotiators on the International Whaling Commission to refrain from calling for total cessation of commercial whaling, with the result that Japan, a whaling nation, has been encouraged to keep seeking loopholes in whaling moratorium proposals.

Whales have fared badly over the years. Populations of the larger whales have been decimated by reckless overexploitation. In certain cases, cessation of hunting has allowed whale populations to recover some of their numbers. (This is true, for example, of the Pacific gray whale, a species that spends most of its life within the protective boundaries of Mexico, the United States, and Canada.) In others (as in the case of the world's largest animal, the blue whale), protection has not resulted in immediate increases in whale populations. Those species may already be past the point of no return. The International Whaling Commission (IWC), a body of scientists and negotiators from most whaling nations, recommends catch limits based on studies of previous catches and other information. The commission lacks enforcement powers, but in recent years, because of pressure of public opinion, most whaling nations have increasingly abided by its decisions. Japan, to the contrary, continues to insist that whales are not endangered and to bypass recommended restrictions on commercial whaling by killing whales "for scientific research."

People have already been responsible for the extinction of some marine mammals. The party of explorer Vitus Bering, wrecked in 1741 on previously unexplored islands in what is now the Bering Sea, encountered a great marine mammal that is now vanished. Steller's sea cow, as the animal is now known, was named after George Wilhelm Steller, Bering's naturalist. It was widespread throughout the North Pacific during Pleistocene times, but by Bering's day these large, inoffensive sea cows had retreated to the only region where native hunters could not get them—the Bering and Copper Islands. Hunters descended on these last few thousand animals after 1741, and by 1768 they were exterminated. These animals, which were the only sea cows known to inhabit cold seas, were about 6 m long and weighed about 3,200 kg. They were grazers on the luxuriant kelp forests of the shores that they inhabited.

OTHER MARINE VERTEBRATES

Seas of earlier eras were dominated by air-breathing marine vertebrates that were unlike today's birds and mammals. In the Mesozoic Era (235 to 65 million years ago), seagoing mammals were non-existent, and the largest marine vertebrates were reptiles. Some of these (the ichthyosaurs) were remarkably similar in size and appearance to modern dolphins and small whales. Others, such as mosasaurs and plesiosaurs, were unlike anything in the seas today. The fossil remains of these animals, as well as such evidence as mosasaur bite marks on the shells of ammonoid mollusks, indicate that these large vertebrates were predators. Plesiosaurs may have practiced swimming and feeding techniques unlike that of any living marine vertebrate. The largest ones were up to 12 m long. Some of them had long, slender necks, small heads, mouths equipped with numerous teeth, heavy bodies with four flippers, and long tails. In certain plesiosaurs, the neck was twice as long as the body. Some zoologists doubt that the animal could have maneuvered rapidly underwater without breaking its neck and speculate that it may have floated at the surface, craning its neck far above the water as it watched for fish, then dipping down rapidly to seize them.

The Mesozoic world of these sea monsters was populated by a blend of modern-appearing and archaic-appearing animals of other kinds. Fish were numerous and were assuming familiar forms, yet the seas were also inhabited by large swimming shelled cephalopods—the ammonoids and nautiloids. Flying reptiles of those times were equipped with teeth that almost certainly identify them as fish catchers or plankton strainers. Fossils of Mesozoic birds are few, yet those that have been discovered give evidence that birds may already have begun their association with the sea, even in those early times. One of these, *Hesperornis,* was a big, flightless bird with a toothed bill and large feet, about 1 m long, evidently a swimmer similar to today's cormorants. Sea turtles were present; one of these was larger than any living turtle today.

As far as we know, all of the Mesozoic marine reptiles except turtles became extinct at or before the time of disappearance of the last dinosaurs. Is it possible that some might have survived to give rise to stories of modern "sea serpents"? Living coelacanths, large fishes of a kind believed to have been extinct for 70 million years, were discovered in the Indian Ocean during the 1930s (Figure 15.23). It is natural to wonder whether other large creatures, remnants of the prehistoric past and unknown to us at present, await discovery in the oceans.

Some of the most intriguing stories of the sea involve purported sightings of large unknown animals. In August of 1817, for example, an unknown marine animal appeared in Gloucester harbor, Massachusetts. The creature was snakelike, black, smooth-skinned, and had a head about the size of that of a horse. It was seen by hundreds of people, from the shore and from boats, during the 22-day period in which it remained in the area. Boatloads of stalwarts set out to capture or kill it on many occasions, and some parties evidently managed to wound it with musketballs. Estimates of its size varied widely. Most observers thought it was between 14 and 30 m (45 and

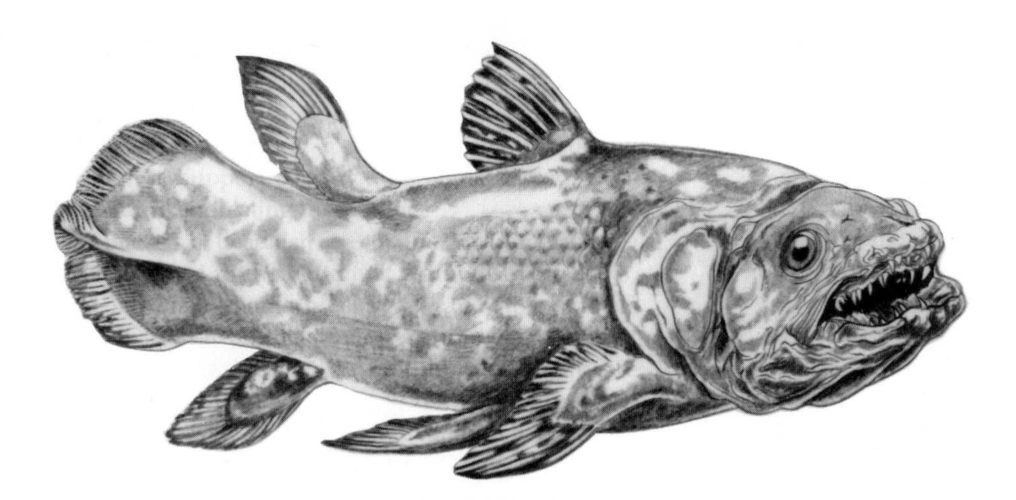

15.23 The coelacanth. Fish of this group were thought to have been extinct since Cretaceous time, until specimens similar to this one were found living in the Indian Ocean, near Madagascar, during the 1930s. Length: about 1.5 m.

100 ft) long. However, a party of gentlemen who saw it only 6 m from shore reported that it was much larger than commonly supposed, nearer to 45 m (150 ft) long. Many witnesses said that its body was similar in girth and appearance to a line of floating kegs and emphasized that it made vertical undulations as it swam, unlike the lateral undulations of water snakes. The witnesses included many seasoned mariners, who insisted that the animal was unlike anything they had ever seen.

Other reports of encounters with sea serpents, as recent as the 1970s, are surprisingly common. Some involve credible witnesses, close range, plausible detail, and little opportunity for a hoax. Others, like sightings of UFOs and the Sasquatch, are highly suspect. Given present knowledge, we can only say that there may well be large animals living in the sea whose existence is only dimly suspected.

FURTHER READINGS

Ashmole, N. P. "Sea Bird Ecology and the Marine Environment." In D. S. Farner and J. R. King (eds.), *Avian Biology.* Vol. 1. New York: Academic Press, 1971.

Baldridge, D. *Shark Attack.* New York: Berkeley, 1974.

Bjorndal, K. (ed.). *Biology and Conservation of Sea Turtles.* Washington D.C.: Smithsonian Institution Press, 1981.

Bonner, W. N. *Whales.* Poole, Dorset, England: Blandford Press, 1980.

Bonner, W. N. *Seals and Man.* Seattle: Univ. of Washington Press, 1982.

Cousteau, J. Y., and P. Diole. *Dolphins.* Garden City, N.Y.: Doubleday, 1975.

Croxall, J. P., P.G.H. Evans, and R. W. Schreiber. *Status and Conservation of the World's Seabirds.* ICBP Tech. Publ. 2: Int. Council Bird Preservation. Norwich, England: Paston Press, 1982.

Dorst, J. *The Life of Birds.* New York: Columbia Univ. Press, 1974.

Haley, D. *Marine Mammals,* 2d ed. Seattle, Wash.: Pacific Search Press, 1978.

Hardy, A. C. *The Open Sea: Fish and Fisheries.* Boston: Houghton Mifflin, 1959.

Harrison, C.J.O. *Bird Families of the World.* New York: Harry N. Abrams, 1978.

Kirkevold, B. C., and J. S. Lockard (eds.). *Behavioral Biology of Killer Whales.* Zoobiology Monographs. I. New York: Alan R. Liss, 1986.

Mathews, L. H. *The Natural History of the Whale.* New York: Columbia Univ. Press, 1978.

Nelson, J. S. *Fishes of the World.* New York: John Wiley, 1976.

Newberry, A. T. (ed.). *Life in the Sea.* San Francisco: W. H. Freeman, 1982.

Pryor, K. *Lads Before the Wind.* New York: Harper & Row, 1975.

Scammon, C. M. *The Marine Mammals of the Northwest Coast of North America.* New York: Dover, 1968. (Originally published 1874.)

Steel, R. *Sharks of the World.* New York: Facts on File, 1985.

The world below the brine,
Forests at the bottom of the sea—the branches and leaves,
Sea-lettuce, vast lichens, strange flowers and seeds—the thick tangle, the openings, and the pink turf,
Different colors, pale gray and green, purple, white, and gold—the play of light through the water,
Dumb swimmers there among the rocks—coral, gluten, grass, rushes—and the aliment of the swimmers,
Sluggish existences grazing there, suspended, or slowly crawling close to the bottom . . .

Walt Whitman

S I X T E E N

The Distribution and Abundance of Life in the Sea

Chapters 14 and 15 dealt with the cast of characters that make up modern marine life. As those chapters point out, the kinds of plants and animals that inhabit the sea are very diverse; as different in their anatomies and behaviors as jellyfish and sharks, and as varied in their sizes and specialized needs as diatoms and kelp. Interesting and diverse as they are, the individual plants and animals themselves are not the only subjects of study by marine biologists. Their occurrence in the sea in distinct communities—kelp beds, coral reefs, animals clustered about deep-sea hydrothermal vents, schools of tuna—poses intriguing intellectual puzzles for marine ecologists and brings up important questions that, if answered, could improve human food supplies and management of marine resources. Many modern research efforts have therefore focused on the relationships between marine plants and animals and the relationships between plants, animals, and the sea itself. These relationships provide important clues to the patterns of abundance and scarcity of marine life that we observe in different parts of the earth. They are the subject of this chapter.

At first glance, marine life seems to be distributed throughout the oceans in a rather puzzling way. For example, the frigid and apparently inhospitable waters of Antarctic seas support vast shoals of krill and huge populations of penguins, seals, whales, and other organisms that eat them. By contrast, the warm, mellow waters of open tropical oceans are almost devoid of marine life. (Indeed, the tropical open seas have been termed oceanic deserts.) Why are pelagic organisms abundant in waters that seem to be so physically challenging, while warmer, sunny seas lack such exuberant outbursts of life? Coral reefs provide another puzzle. The reefs in the shallow waters of the western Pacific are built and occupied by many more species of corals, algae, sponges, clams, sea cucumbers, and other characteristic tropical organisms than are the reefs in the Caribbean. Yet the Caribbean seems to be equally suitable as a habitat for reef organisms. What is the reason for this difference?

In general, we may ask why the organisms of one region are different in numbers and species makeup from those of other regions, and why the organisms in any particular region occur in the abundance and arrangement that we see. The answers, as marine ecologists currently understand them, have to do with the following natural laws:

1. Limits are imposed on living organisms by properties of the sea itself. Some of these are direct physical or chemical challenges (for example, the stressful daily changes of salinity occurring in an estuary). Others, such as shortages of light and

nutrients, inhibit plant growth. This, in turn, reduces animal abundance and diversity.

2. Food-chain dynamics place serious restrictions on the abundance of living organisms. Fundamental thermodynamic laws dictate that much of the food energy provided by plants for the rest of the living communities must eventually be lost or "wasted" as heat. Activities of the animals also disperse energy. This seriously reduces the numbers of animals in all living communities and affects the ratios of predators to prey.
3. In many cases, the activities of certain organisms (such as predators or strong competitors) impose limits on the abundance of other organisms.
4. The distribution of marine organisms in modern seas is related to the patterns of distribution in the prehistoric past, and it is also related to prehistoric environmental upheavals (such as the Pleistocene ice ages).

The objective of this chapter is to describe ways in which the factors just listed regulate the numbers and distributions of marine organisms. This chapter also describes marine communities in which these factors are particularly evident.

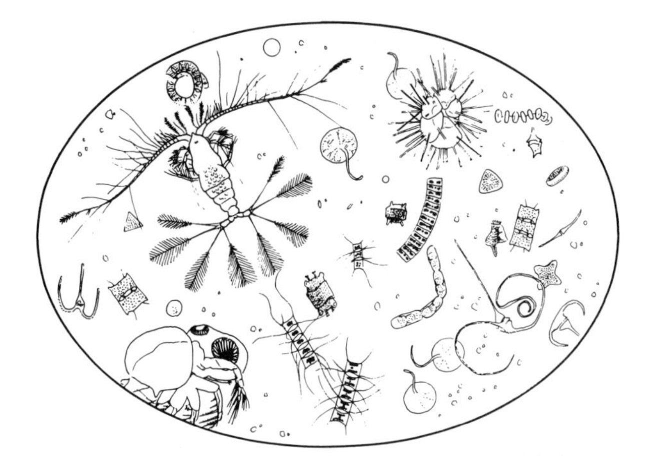

16.1 Planktonic plants and animals are tiny and often have flattened or spiny forms. These features make use of the viscosity of the water to slow the sinking of the organisms. Width of oval: 5 mm.

SEAWATER PROPERTIES: THEIR EFFECTS ON MARINE LIFE

Many properties of seawater, and of the sea as a whole, influence the distribution and abundance of marine life and the forms of plants and animals. Seawater properties that affect organisms are salinity, temperature, density, viscosity, heat capacity, the ability of water to transmit light, and the water's capacity for dissolving such vital substances as oxygen and carbon dioxide. Considering the sea as a whole, its great depth results in enormous pressure on the organisms of deep water, total absence of sunlight at the deep-sea bottom, long cycling times for nutrients and particles of organic matter, and increased solubility in deep water of the calcium carbonate materials from which many marine organisms build their shells. In the following sections we briefly examine the ways in which these properties of seawater affect marine life.

Density and Viscosity

Life in the sea is much more three-dimensional than it is on land; the density and viscosity of seawater (described in Chapter 7) make this possible. It is easy for organisms ranging from the most powerful shark to the most fragile jellyfish to move themselves forward through water by simply pushing the water backward. In addition to ease of movement, most planktonic and nektonic animals are essentially weightless in water. This makes it as easy for them to go up as it is to go down or sideways. This near- or perfect weightlessness results from the fact that the density of organisms is approximately (or exactly) the same as that of the seawater. The animals encounter frictional drag as they move (a consequence of the viscosity of water), but this drag can also be utilized to their advantage, since it retards sinking. Consequences of life in a dense medium that retards motion are as follows: fast-moving animals are streamlined—even the smallest swimming animals are able to move through huge vertical distances much more easily than animals that live on land—and many midwater organisms are tiny and/or flattened and/or spiny, as needed to slow their sinking rates (Figure 16.1).

Because of the viscosity of seawater, tiny edible particles of organic matter remain suspended for very long time periods. This, in turn, makes possible the important suspension-feeding and filter-feeding mechanisms of some marine animals who catch, concentrate, and eat these suspended particles. Sea anemones, tube-dwelling worms, and others that use this food-gathering strategy are often flower-shaped. This morphology is non-existent among land animals, partly because air has a low viscosity that cannot support much suspended edible "dust." Thus the three-dimensionality of life in the sea, the forms of the organisms, and the widespread adoption of suspension-feeding techniques are all consequences of the density and viscosity of water.

Salinity

A subtidal crab dies almost immediately when immersed in fresh water. A sea cucumber experiences a more ghastly death, slowly swelling up until it resembles a turgid balloon full of water. A crucial requirement of these and many other marine organisms is that they live in a place where the salinity of the seawater around them is roughly (or exactly) the same as the "salinity" of their internal fluids. If a permeable membrane (such as the "skin" of a sea cucumber) separates saltier water from less salty water, the water molecules seep through the membrane, moving from the fresher side to the saltier side. (This process is called osmosis.) Most marine invertebrates are quickly killed by this process, either by loss of their body water or by flooding of their tissues, if they are artificially transferred to water whose salinity is greatly different from their familiar environment. The internal "salinities" of open-ocean invertebrates match that of the water in which they live, and they have little or no ability to adjust if external salinity changes. In vast reaches of the open sea, salinity does not change, either daily or seasonally, during the lifetimes of the organisms, and they are spared this hazard.

Because of this potential for osmotic assault on their tissues, conditions become tougher for marine organisms as the salt content of seawater becomes lower. But even in conditions where the sea is abnormally dilute (as, for example, in the Baltic, where bottom salinities normally hover between 12 and 16‰), many marine organisms can survive as long as the level of low salinity remains relatively constant. The harshest conditions attributable to salinity are those in which salinity changes on a short-term basis. This is typically the case in estuaries, where river runoff, evaporation, rainfall, and the changes of the tides subject the organisms to daily or even hourly salinity changes. Some organisms have strategies for coping with this problem (for example, the marsh grass *Spartina*, discussed in Chapter 14); these plants and animals prosper in estuaries. Many do not, however, and simply cannot inhabit estuaries.

Fresh water represents the harshest extreme; it is a stressful environment in which all organisms are subject to continuous osmotic assault. Plants and animals that live in fresh water must be able to cope with this threat, either by preventing their tissues from being flooded or by continuously bailing out the water that invades them. The latter activity requires sophisticated physiological abilities, and comparatively few of the phyla and classes of animals are thus equipped. Many more fundamentally different kinds of animals live in the salty sea than in fresh water, mainly for this reason.

From an organism's point of view, the fatal problem with having its internal salinity change is that its cells stop functioning. Those that have become adapted to living in water of low or varying salinity have hit on two strategies for coping with this problem. Some organisms have cells that, in effect, ignore the internal salinity change and continue to operate. Others actively expend energy to pump water or salts out of their bodies, thereby maintaining the proper internal salinity. It is possible to distinguish organisms using the two strategies by measuring their internal blood salinities. In the group using the first strategy (called osmoconformers), internal salinity goes up or down, staying about the same as the changing external salinity, without killing the organism. In the group using the second strategy (called osmoregulators), internal salinity stays about the same, despite changes in the water salinity. Some organisms are able to use elements of both strategies (Figure 16.2). Bony fish are the most familiar marine osmoregulators. They maintain a blood salinity that is a little less than that of the sea. As a consequence, they continuously lose water (mainly through osmosis through their gills) and continuously drink seawater, separating the salt from the water and excreting the salt while retaining the water (Figure 16.3). As noted in Chapter 15, sharks employ a strategy that appears to be less costly in energy; they simply maintain an internal salinity that matches that of the seawater around them. Freshwater fish, of course, are assaulted by osmotic flooding and must extract water from their tissues and get rid of it in the form of dilute urine.

Temperature

Changing sea temperatures sometimes result in the deaths of organisms. For example, the lagoons behind the Texas barrier islands sometimes approach freezing temperatures during the winters, and many organisms die as a result. More often, however, water temperature acts in a more subtle way, as a moderator or governor of the tempo of activity of marine plants and animals. Cold-blooded (poikilothermous) organisms experience significant changes in their metabolic rates when the water temperature changes. Warm-blooded (homeothermous) animals are much less affected by sea temperature changes. As a rough rule of thumb for the cold-blooded creatures, an increase in temperature of 10°C causes a twofold increase in various rates of activities. Barnacles speed up their rates of sweeping for food, crabs are able to move faster, organisms use more oxygen and require more food, and many other accelerations occur when water temperature rises.

Perhaps for some of these reasons, organisms of the same species tend to live longer, mature later, and

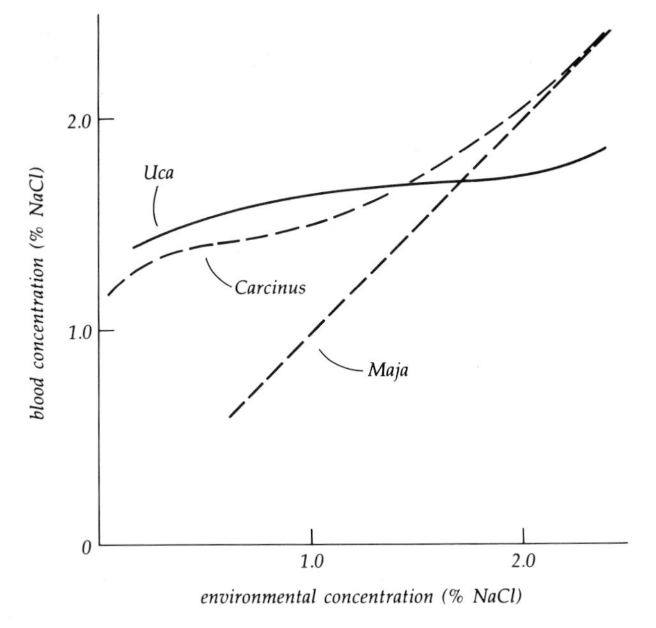

16.2 The internal concentration of salts in the blood of crabs versus the external concentration in seawater of different salinities. The subtidal spider crab, Maja, *is unable to adjust its internal salinity; the salinity of its blood is the same as the external salinity at all concentrations. The fiddler crab,* Uca, *can keep its blood salinity at approximately the same level, regardless of the external salinity. The crab* Carcinus *is intermediate between* Uca *and* Maja *in its abilities; able to keep its blood salinity approximately constant when external salinities drop but allowing its blood salinity to remain at the same level as the external salinity when the external salinity goes up. Only part of the total salinity picture for both crabs and seawater (the NaCl fraction) is shown here.*

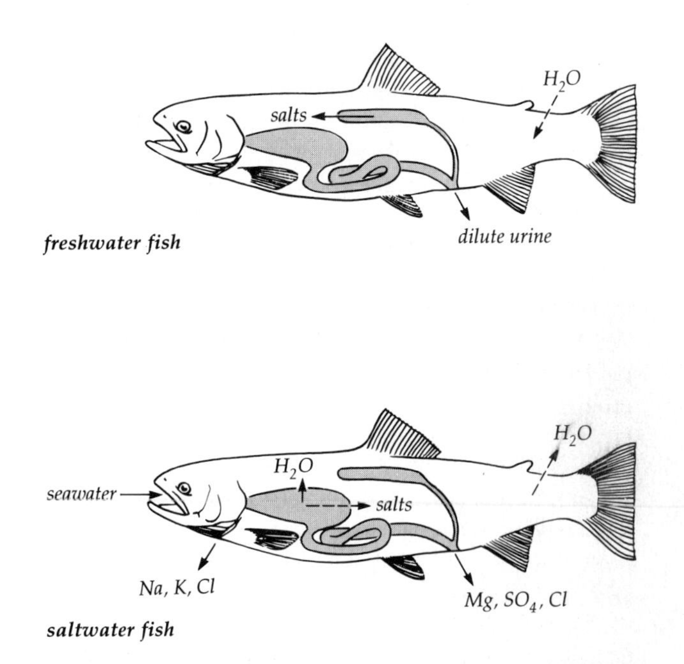

16.3 Fish in the sea maintain a body fluid salinity that is lower than sea salinity. They lose water via osmosis and must recover it by drinking seawater, removing the salts from the water, and excreting the salts (through the gills and concentrated in urine). Fish in fresh water are subject to osmotic flooding and must remove excess water from their tissues. Their kidneys isolate excess water, which is excreted as dilute urine.

grow to larger sizes in the colder parts of their ranges than in the warmer parts. On the basis of temperature alone, then, one would expect invertebrates in the 3°C water at the bottom of the sea to have lower overall metabolic rates and needs than those in 13°C water at the surface. Similarly, winter cooling of the sea (as in the North Atlantic) reduces the need for food of copepods and other animals just at a time when lack of light and other factors make much less food available. Not all temperature changes in the sea work to the advantage of organisms in this way. The warming of water increases organisms' need for oxygen, for example—but it also lowers the ability of water to hold dissolved oxygen. In this instance, the temperature change increases the demand for oxygen while reducing the supply.

Seasonal changes in water temperature provide marine organisms with important clues to the onset of spring or winter. In many animals, spawning is triggered by the warming or cooling of the water to certain critical temperatures. If something prevents the water from reaching that critical temperature (say, a warm discharge from a power plant or an unseasonably cool summer), such organisms may survive, but they are unable to spawn.

Because of the great heat-holding capacity of water (see Chapter 7), changes in sea temperature are seldom comparable with the wild swings in air temperature that occur hourly or daily in temperate latitudes on land. The most severe temperature changes are experienced by organisms of intertidal shores, as they are alternately subjected to the temperatures of both the air and the water. The subtidal organisms of estuaries are subject to temperature changes that are also relatively severe. The semienclosed marine waters of estuaries gain or lose heat in concert with the weather, cooling or warming on intertidal mudflats, and mingling with river water that may be warm or cold in different seasons. The temperature variability in estuaries contributes to the stress exerted on organisms by the changing salinities there.

Dissolved Gases

One feature of the sea that does not compare favorably with the land environment is the availability of oxygen. Oxygen makes up about 21% of our air, and terrestrial organisms never run short of this vital substance. Water holds much less oxygen than air (only about 10 parts per million, at best), and marine and freshwater creatures are always close to the threshold of shortage. Warm water holds less oxygen than cold water; salt water holds less oxygen than fresh water. As a result, the inhabitants of warm, saline tropical waters have only about half as much oxygen available to them as the organisms that inhabit cold polar waters. The decomposition of even a minute particle of organic debris (18 mg or less) by bacteria, can consume all of the oxygen in an entire liter of water, and respiration by marine animals similarly depletes oxygen rapidly. Pollutants (including sewage) often contain organic substances that are decomposed by bacteria. One result is to seriously deplete the dissolved oxygen content of the polluted water, with consequent deaths of fish and other organisms.

The supply of oxygen in seawater is maintained by the photosynthesis of phytoplankton and larger marine plants and by the movement of oxygen from the air into the sea surface. The deep-sea floor has neither photosynthesis nor contact with the atmosphere to replace the oxygen consumed by the animals living there. At present, oxygen is carried to the deepest regions by the sinking of well-oxygenated surface water near Antarctica. This well-oxygenated water flows slowly northward over the sea bottom (see Chapter 17). As a result, most of the modern seafloor has adequate oxygen for animal life. The waters with the least oxygen are those that lie at mid-depths, about 1,000 m below the sea surface. These mid-depth waters are depleted by the decomposition of particles sinking from the surface regions and by the respiration of animals. Even this "oxygen minimum" depth appears to have sufficient oxygen for animal life in most places.

Carbon dioxide (CO_2) is produced by the respiration of plants, animals, and bacteria on land and in the sea. It is also produced by the burning of wood and fossil fuels. Carbon dioxide is easily absorbed from the atmosphere by seawater because it reacts chemically with water in the following manner:

$$CO_2 + H_2O \leftrightarrow \underset{\text{(carbonic acid)}}{H_2CO_3} \leftrightarrow H^+ + \underset{\text{(bicarbonate)}}{HCO_3^-}$$

$$\leftrightarrow 2H^+ + \underset{\text{(carbonate)}}{CO_3^=}$$

This chemical reaction allows the sea to take up much more of this gas than it would if CO_2 were a relatively unreactive gas such as nitrogen (N_2). As a result, the oceans contain about sixty times as much CO_2 as the atmosphere, where CO_2 is very scarce. Once in water, CO_2 is transformed to bicarbonate and carbonate (HCO_3^- and $CO_3^=$). These are essential to shell- and coral-forming organisms, whose hard parts are largely built of calcium carbonate ($CaCO_3$). Carbon dioxide and bicarbonate are also used by plants for photosynthesis. Unlike the situation on land, where CO_2 is so scarce that land plants are actually starved for it, the sea contains much more than enough of this essential material for plant growth.

Carbon dioxide affects marine life (and indeed all life on the earth) in an important way. Visible sunlight passes easily through the atmosphere and warms the earth's surface. The earth radiates as much energy back toward space as it receives, but this radiation is in invisible (infrared) wavelengths. The CO_2 in the atmosphere retards the escape of this radiation. The consequences are complex, but the result is that, when more CO_2 is added to the atmosphere, the earth warms up. This is termed the **greenhouse effect.** (It takes its name from the fact that the glass in a greenhouse acts in the same way as the CO_2 in the atmosphere, allowing visible light to enter but retarding the escape of the re-radiated infrared wavelengths.) More CO_2 in the atmosphere strengthens the effect. Atmospheric CO_2 levels have risen by about 26% since measurement began in 1860, as a result of the burning of fossil fuels. This has prompted concerns that global warming and large-scale climatic changes are forthcoming. The oceans of the prehistoric past have usually been warmer than they are today, perhaps as a result of greenhouse effects in times past. If so, the atmospheric abundance of CO_2 has had profound effects on the distribution and abundance of organisms of all sorts, marine and terrestrial.

About half of all of the carbon dioxide added to the atmosphere since the widespread burning of fuels began has been absorbed by the ocean. Thus the oceans have helped to soften the impact of CO_2 on climatic changes. This dissolved CO_2 increases the acidity of the sea, however, in a complex way. One effect is to speed up the dissolution of coral, shells, and other calcareous hard parts of organisms. This and other possible negative effects of increasing CO_2 in the biosphere are discussed in Chapter 19.

Light

Light levels decrease rapidly with depth in the sea, restricting plant photosynthesis to the top 40 or 50 m of water (Figure 16.4). Seasonal changes in light also regulate plant productivity. Winter at high latitudes brings reduced light levels. This, in concert with deep mixing of the surface water, is responsible for the slowdown of photosynthesis by phytoplankton, which in turn results in a drastic slowing down of ecological activity and eventually reduces the number of phyto- and zooplankters. Seasonal changes in light levels, then, exert a powerful influence on the abundance of planktonic organisms at these latitudes. Light affects the abundance of organisms in other ways, as well. In the open ocean, many planktonic and small nektonic animals concentrate in swarms that inhabit the day-time waters between about 200 and 800 m below the surface. These organisms, which include small fish, copepods, ephausiids, and many others, often adjust to the depths at which they reside by seeking certain light levels. As each day progresses, the change in elevation of the sun changes the light levels in deep water. The organisms follow the waning light to the surface, where they feed at night, then they return to their daytime depths when the sun rises in the morning. This phenomenon is known as vertical migration. It is widespread in the sea and important in the sea's overall economy. (Vertical migration is discussed in more detail in the section "Life in Deep Water.")

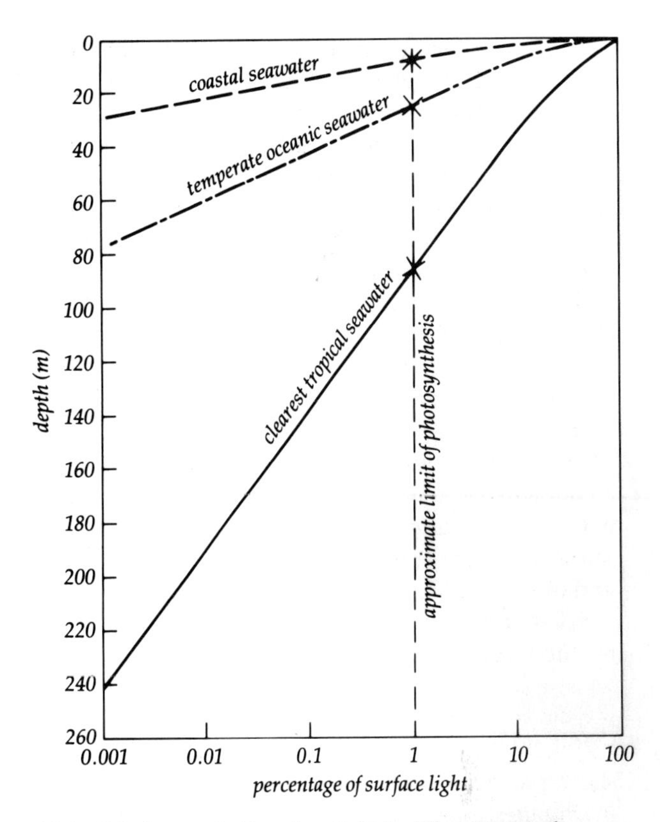

16.4 Relative penetration of sunlight in different types of seawater. Note logarithmic scale for percentage of surface light. An X marks the approximate depth limit of net photosynthetic productivity (the compensation depth) for each of the water types.

Pressure

The average depth of the sea bottom is 3.8 km, and the deepest trenches go down fully 10.8 km. Water pressure increases at the rate of about 1 atmosphere per 10 m of depth, and the pressure on an object at the deepest sea bottom is about 1,000 kg per square cm (or 7 tons/in^2)! It is likely that this enormous force is not even noticed, however, by the creatures that live there, just as we do not notice that 1 kg of air bears down on every square centimeter of our skins. Living biomass is relatively incompressible, and it exerts an equal pressure on the water (or air) surrounding it that neutralizes the pressure the water exerts on it. In certain ways, however, pressure places important restrictions on things that marine organisms can and cannot do. Any gas bubble confined within an organism in the sea will try to expand if the organism rises, as a result of the decreased pressure at lesser depths. The expanding bubble will damage the organism, unless the animal has a means of preventing it. A deep-sea fish with a gas-filled swim bladder must ascend slowly, and it must absorb some of the gas in the bladder, transferring it into dissolved form in its blood, to prevent the bladder from exploding as the fish swims upward. Likewise, it must be able to take the gas out of solution from its blood and put it back into the bladder when descending, to keep the bladder from being crushed. Fish that change their depth frequently are able to do this. Many have dispensed with the gas bladder entirely and have, instead, a reservoir of buoyant oil or fat (which does not expand or contract when the pressure changes) in its place. The expansion of depressurized gas is of foremost concern to scuba divers, who must exhale continuously while ascending to prevent fatal damage to their lungs.

Great hydrostatic pressure evidently inhibits deep-sea bacteria in ways that are not fully understood. An apple, a sandwich, and a thermos full of bouillon, all of which went to the New England sea bottom in the submersible *Alvin* when it was accidentally dropped overboard in 1,540 m of water, were astonishingly fresh when the submersible was recovered about a year later. Subsequent experiments showed that both deep-sea and shallow-water bacteria, under great pressure, decompose organic matter at only about 2% of the rate at which surface bacteria do their work. Consequently, bacterial decomposition of organic matter on the deep-sea floor must be very slow—a factor that exacerbates the general food shortage at the deep-sea bottom. Other effects of pressure are suggested by experiments that show that certain deep-sea bacteria don't grow well at low sea surface pressures. Pressure effects on the enzymes of other organisms are also suspected. It is possible, but not entirely certain, that some deep-water organisms are restricted to certain depths by their particular water-pressure needs.

One effect of pressure on deep-water organisms is to increase the rate at which their shells dissolve. This effect becomes important at depths below 4,500 m. This "calcium compensation depth," as it is called, has also been referred to as the "snow line." The white calcareous shells of forams and other pelagic creatures, settling to the bottom after their owners die, are less readily dissolved if they come to rest on the tops of seamounts and other bottom features that poke up above the snow line. Below that depth, sinking calcareous fragments dissolve more rapidly, with the result that calcareous sediments are less widespread below depths of 4,500 m than they are in shallower waters. For deep-water creatures, the increased ability of water to dissolve calcium carbonate makes it difficult for them to form and maintain shells and hard parts. Compared with their counterparts in shallow water, they tend to have small, fragile shells and bones—also due, perhaps, to the general scarcity of food in deep water as much as to the effects of pressure.

Depth

The depth of the oceans influences their suitability for life (and indeed it influences the suitability of the entire earth for life) in more subtle ways than pressure effects alone. First, the great depth of the sea excludes light from the majority of the sea bottom. This prevents plants from growing there, and it forces the entire economy of the sea to depend on the plant productivity that takes place in the top few hundred meters of water. The vast, continuous deep habitat in which permanent darkness prevails is something that is paralleled only marginally (in caves) on land. Second, the deep sea is a reservoir for much of the earth's available carbon. Particles sink slowly in the sea. On the average, 200–600 years are required for a particle starting at the surface to eventually reach bottom. Thus the deep, dark waters are sprinkled with bits of microscopic debris that have been sinking for centuries.

The ocean waters also carry vast quantities of dissolved organic molecules that do not sink at all. This particulate and dissolved carbon outweighs the carbon present in living organisms in the sea by a ratio of at least 50 to 1. Much of it is apparently in refractory form that larger organisms cannot use, even if they could find it. This dissolved and suspended carbon is essentially removed from the realm of living things, at least until the right bacteria encounter it. Throughout the water column, but especially at the bottom, bacteria slowly metabolize this material back to carbon dioxide or forms that larger organisms can digest, thereby making it accessible, once more, to the rest of the ecosystem. The time delay experienced by carbon, between its use at the sea surface by organisms, its long-term storage as particles or molecules in the water, and its ultimate return, regulates the cycling of food on a global scale. If the seas were shallower, the carbon could return to living systems more rapidly and life on Earth might well be more abundant than it is at present.

This depth effect is similar for other essential elements. Nitrate and phosphate, two of the most crucial nutrients for phytoplankton and seaweeds, are much more abundant in the deep, lightless sea than they are at the surface. That is because organisms, having incorporated these nutrients near the sea surface, ultimately tend to swim downward or sink, decomposing, defecating, or otherwise transferring the nutrients from the surface to deep water. The return of the nutrients from deep water is a slow process. The result, again, is that large stores of the earth's essential plant nutrients reside in deep, lightless water, where they are inaccessible to plants. Thus the depths of the sea regulate the abundance and character of its life, first, by providing dark, quiet environments unlike anything occurring on land and, second, by hoarding vast quantities of essential biochemicals.

The properties of seawater and the sea therefore affect marine life in many ways. They provide a very supportive environment for the most part; exactly the salinities that most organisms need, protection from rapid temperature change, an overabundance of carbon dioxide for plants, physical support, ease of motion, and a wealth of suspended edible particles. The physical and chemical properties of the sea become challenging to organisms in its marginal waters, and essential oxygen, while adequate in the sea, is not overly abundant. The depth of the sea creates problems: shortages of food and light at the bottom, shortages of nutrients at the surface, pressures to be accommodated, and a vast storehouse of most of the ocean's carbon.

PLANT PRODUCTIVITY AND FOOD-CHAIN DYNAMICS

Photosynthesis

The food-chain transactions that ultimately determine the abundance of marine plants and animals in an area begin at the molecular level in the cells of plants. There, sunlight is trapped and forced to give up its energy, and the machinery of the cell stores the solar energy in the bonds of a glucose (sugar) molecule. This conversion of the diffuse energy of sunlight into the tangible chemical form of digestible, energy-rich molecules is called **photosynthesis.**

The details of photosynthesis are quite complex. Chlorophyll, the main "green" molecule of plants that starts the process, traps the sunlight falling on the plant surface and manages to wrest some of the energy out of it, storing that energy in the bonds of an "acceptor" molecule. The energy, now stored in chemical form, is then passed along in bucket-brigade fashion from molecule to molecule (a number of different kinds are involved). Meanwhile, elsewhere in the plant, a furious manufacture of new large molecules takes place, using smaller molecules to build the bigger ones. The two assembly lines converge; the solar energy is passed to one of the new molecules, which, in this its final conversion, becomes glucose. If we trace all of the complex interactions to their sources, we find that the only raw materials used by the plant in starting and running the whole operation are simple water and carbon dioxide. Early in the process, a molecule of oxygen is formed and escapes as a waste byproduct. The plant cell uses some energy in running the operation, but the amount of energy that is finally stored in the glucose molecule is much greater than the amount used by the cell. Thus there is a net energy gain to the plant. Disregarding the internal complexity of photosynthesis, the process can be summarized in terms of its starting materials and end products as follows:

$$6H_2O + 6CO_2 \rightarrow 6O_2 + \underset{\text{(glucose)}}{C_6H_{12}O_6}$$

After glucose is made, different plants change it to such different forms as oils or starches for storage; these and glucose itself become the foods eaten by animals that eat the plants.

Glucose itself is as important to living things as gasoline is to automobiles. Almost every living organism, from bacteria to plants to animals, can metabolize it and use its stored energy. Other organic molecules are also needed by living organisms, but glucose, the main product of photosynthesis, is the energy mainspring that makes it possible for other metabolic processes to function. The more of it that plants can manufacture, the more plant, animal, and bacterial life a region can support. Thus anything that limits photosynthesis and the manufacture of glucose ultimately limits the abundance of marine life.

Several factors limit the rate at which marine plant photosynthesis can proceed. One is cosmic and fundamental: *It is impossible for any energy-capturing process to trap all of the energy available and convert it to useful form.* Some is always wasted and lost as heat; this loss is inevitable. The italicized statement is the physicists' second law of thermodynamics; it appears to apply to all systems, living and nonliving. In the case of plants, the energy "waste" in photosynthesis is shocking. Usually only about 0.5% of the solar energy falling on plants ends up being made into glucose. Thanks in part to this thermodynamic bottleneck, plants "harvest" only a very small fraction of the torrent of solar energy that falls on the oceans or on the earth as a whole.

A second factor that restricts photosynthesis by marine plants is the fact that water absorbs light very quickly. Below a depth of about 50–200 m, ocean water is usually too dim to sustain photosynthesis. Plants living just below these depths are still bathed in dim light, but the small amount of available and usable light yields less energy than the plants need to run their own internal machinery. They continue to photosynthesize, but their life processes require more energy than the dim light at those depths can supply. Consequently, the plants lose weight or die or go into suspended animation. The depth at which plants can acquire enough light to barely meet their own energy needs is called the **compensation depth;** only those plants growing in water shallower than this depth can gain enough weight to prosper and to feed animals. The compensation depth is only 40 or 50 m deep, whereas the average depth of the oceans is about 3.8 km. Because of the shortage of light at greater depths, then, we see a situation wholly unlike that in land ecosystems. Photosynthesis by plants in the *upper 1%* of the sea must feed all of the rest of the vast sea system.

In some parts of the world, the oceans are subject to shortages of light that inhibit plant growth even at the surface. The polar oceans experience long, dark nights, lasting for fully 24 hours a day poleward of the Arctic and Antarctic circles. Even in temperate latitudes, winter sunlight is barely sufficient, or not quite sufficient, to sustain plant growth. Light becomes plentiful during the long summer days, with the result that the dark-winter/light-summer alternation of seasons imposes a seasonality on plant and animal growth of temperate latitudes that is much less marked in the tropics.

Not all sunlight is equally suitable for photosynthesis. The "best" wavelengths are those of blue-violet or red light; green light is not especially effective (Figure 16.5). (Plants appear green because they absorb light of all the wavelengths in the color spectrum except green.) Because red solar wavelengths are absorbed within the first few meters of seawater, marine plants are deprived of some of the most useful wavelengths of light. They appear to compensate for this by producing auxiliary colored pigments, which give them their characteristic golden-brown or red colors when they are exposed to air. These pigments assist their chlorophyll by trapping light of suboptimal wavelengths that penetrate water more easily than red light. Actually, plants need other substances in addition to sunlight, carbon dioxide, and water, in order to photosynthesize. Their metabolisms (like all metabolisms) use more than just glucose. In particular, they need nutrient molecules, including nitrogen and phospho-

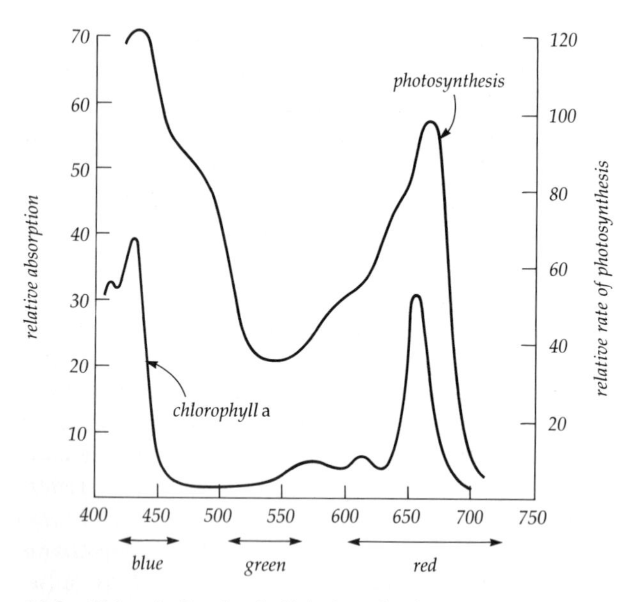

16.5 Chlorophyll a absorbs light best that has wavelengths in the blue-violet and red ranges of the spectrum (peaks in lower curve). Maximum photosynthesis is accomplished by plants exposed to those wavelengths (peaks in upper curve), although some photosynthesis is accomplished at all visible-light wavelengths.

rus compounds. These are typically scarce in the surface waters of the sea. Without them, photosynthesis and plant growth grinds to a halt.

To summarize, photosynthesis is the process that ultimately makes existence of marine plant and animal communities possible. Photosynthesis in the sea is restricted by thermodynamic limitations, by the fact that suitable sunlight penetrates only 40 to 50 m into the sea surface, and by seasonal shortages of light. Shortages of nutrients can also inhibit plant growth; the effects of these limitations ultimately govern the abundance of marine plant and animal life in different localities.

Primary Productivity

Primary productivity refers to the amount of new organic material created by plants. It varies by season and by geographic region and type of plants involved. Most of the productivity of the oceans is contributed by phytoplankton, although seaweeds and the plants of salt marshes and mangrove swamps can be more productive than the phytoplankton, in coastal areas. The continents are mainly occupied by large rooted plants, whose productivity is similar to that of the large seaweeds, mangroves, and salt-marsh grasses. The area occupied by the most productive land-plant communities (forests, grasslands, and agricultural land) constitutes about 100 million km^2. The productive sea-plant communities of reefs, estuaries, and marine seaweed beds, in contrast, are restricted by their need for shallow, nutrient-rich water, and occupy only about 2 million km^2. Although these marine communities are about as productive as the most productive continental plant associations in new plant growth per square kilometer per year, the fact that the continental communities occupy about fifty times more of the earth's surface partially accounts for the greater total productivity of the continents.

Over a 24-hour period, plants consume some of their own glucose and other carbohydrates, while simultaneously producing more. Consumption during the process of their own respiration occurs both in the dark and in sunlight, whereas production by means of photosynthesis is restricted to daylight hours. In favorable situations, the production of new material during a typical 24-hour day can outweigh the consumption of existing carbohydrate by a ratio of 4 to 1. Macroscopic plants gain in size and weight during such intervals and/or produce such new reproductive structures as seeds or gametes. Phytoplankton cells manifest the new gains in biomass mainly by dividing to form additional cells. This plant matter, added by seaweeds, marsh grasses, phytoplankton cells, and others, is the food or biological fuel that powers the herbivores, carnivores, bacteria, and other consumers of the sea. Where productivity is high, marine animals are generally abundant; where it is low, they are usually scarce. The abundance and distribution of animal life in the sea is ultimately controlled by the factors that govern plant productivity.

Biologists distinguish between gross primary productivity (*GPP*) and net primary productivity (*NPP*). The former is the total quantity of new material synthesized by the plants. The latter is what is left after the plants have utilized the amount they need for their own respiration (r). The quantities are related as follows: $NPP = GPP - r$. Since *NPP* is the amount of new plant material actually available to herbivores as food, discussions of "productivity" usually refer to *NPP* rather than *GPP*.

In quantitative terms, the net productivity of phytoplankton is the gain in weight of the plants, divided by the time required for the weight gain to occur, all divided by the amount of water containing the cells. Thus 1 L of seawater initially containing 4.0 mg of phytoplankton that increases to a biomass of 6.4 mg in 1 day has a net productivity of 0.1 mg/L-h. Since the weight of phytoplankton includes inedible materials (such as the silica tests of diatoms), only the carbon fraction of the plant biomass is usually reported. (The carbon fraction is usually about half of the total dry weight.) The calculation is also modified to report the productivity per unit area of sea surface, much as crop yields on land are given per acre or other area unit. Thus a productivity figure of 237 gC/m^2-yr for Pacific neritic waters states that, averaged over the whole year, the seaweeds under a typical square meter of sea surface (regardless of its depth) produced new biomass in excess of their own needs totalling 237 g of carbon in edible carbohydrate. This system enables scientists to make comparisons with other sea areas (for example, the Sargasso Sea at about 50gC/m^2-yr) and with land areas (for example, a Midwestern corn field in 1926, at 669 gC/m^2-yr). Very productive waters are said to be **eutrophic;** sparsely productive waters are termed **oligotrophic.** The terms are not rigid definitions, but, generally speaking, phytoplankton and/or larger plants of oligotrophic waters produce about 50 gC/m^2-yr or less, whereas those of eutrophic waters produce about 100 gC/m^2-yr or more. Oligotrophic waters occupy about 51% of the total ocean area, eutrophic areas constitute about 12%, and the rest of the ocean is of intermediate productivity. For comparison, about 49% of the land surface is covered by ice or is occupied by low-productivity plant communities such as deserts and tundra. Thus the oceans and the continents are similar in that about half of each is poorly suited to luxuriant plant growth.

Measuring Productivity The measurement of productivity in the sea is a sophisticated and demanding task. Because phytoplankters are small and sparsely distributed in the water, it is not possible to weigh them before and after an exposure to sunlight to estimate their change in weight. Instead, their productivity must be estimated by other means. There are several measuring techniques, each with its drawbacks and advantages. One of the earliest is the "light bottle/dark bottle" (LD/DB) technique. This makes use of the fact that phytoplankters produce oxygen as they photosynthesize and consume it as they respire. From chemistry, the exact relationship between carbohydrate and oxygen produced or consumed is known (and is calculated for photosynthesis in the earlier "Photosynthesis" section). Since changes in oxygen can be easily measured by chemical means, the associated changes in carbohydrate can then be calculated. To use the LB/DB technique, the investigator retrieves samples of water from various depths before sunrise. Each sample is divided among three bottles; an "initial bottle," a "light bottle" (ordinary clear glass), and a "dark bottle" (painted black). The water in the initial bottle is immediately tested for its oxygen content. Each pair of light and dark bottles is then lowered to the depth from which the sample was taken and left there until noon. At that time, the bottles are retrieved, and the oxygen content of each bottle is measured. The initial bottle shows how much oxygen was in the water sample when the experiment started (*IB*). In the light bottle, the phytoplankton used some oxygen for respiration (*r*) but also produced new oxygen via photosynthesis (*p*). The light bottle's oxygen content at noon (*LB*) is the result of these processes: $LB = IB + p - r$. In the dark bottle, from which light was excluded, the phytoplankton consumed oxygen via respiration (*r*) but produced no new oxygen. The dark bottle's oxygen content at noon (*DB*) is given by $DB = IB - r$. From the three measurements *IB*, *DB*, and *LB*, therefore, the amounts of oxygen produced and respired by the phytoplankton in the sample can be calculated: $p = LB - DB$, and $r = IB - DB$. Further calculations extend the 6-hour observation to include the whole 24 hours, and the net new carbohydrate produced by the phytoplankters of the sample is then calculated from the oxygen-carbohydrate relationship mentioned earlier.

An alternative method of measuring productivity is the "carbon-14 technique." This procedure measures the carbon dioxide consumed by phytoplankters as they photosynthesize (rather than the oxygen produced). A light and a dark bottle containing water samples are each injected with a certain amount of sodium bicarbonate ($NaHCO_3$) containing the radioisotope carbon-14. Phytoplankton cells are able to take up the bicarbonate in the same way that they take up CO_2. As in the LB/DB technique, the bottles are left suspended at the depth from which the sample was taken, from sunrise until noon. On retrieval, the water in each bottle is passed through a fine filter that retains the phytoplankton. The filters are placed in a counter, which measures the radioactivity of the phytoplankton. In theory, only the plankton in the light bottle should become radioactive, since they are the only ones able to photosynthesize. A small amount of radioactivity is almost always acquired by those in the dark bottle, however, for various reasons that have nothing to do with photosynthesis. Hence, the measurement of the plankton in the dark bottle is subtracted from that of the light bottle to give the amount of productivity actually due to photosynthesis. This measurement is then converted to the 24-hour productivity of the phytoplankton, using the known relationship between carbon taken in and carbohydrate produced.

The LB/DB technique works best when there is a lot of phytoplankton in the water and productivity is high. The technique is not sensitive enough to detect the small changes in oxygen that occur in oligotrophic water. Also, any zooplankton or bacteria inadvertently included in the bottles complicate the outcome, since they consume oxygen. (The bottles are left suspended only until noon, rather than all day, to avoid a buildup of bacteria on their inner surfaces.) The carbon-14 technique is more sensitive, and it is now the most popular method of estimating productivity. It is not without its problems. There is evidence that microflagellates and other minute photosynthesizers, sometimes responsible for as much as 60% of the productivity in oligotrophic waters, have passed through the filters undetected. Both techniques are impractical in that a research ship cannot sit idle, waiting half a day for one productivity measurement. In practice, the light and dark bottles are placed in a lighted incubator on board the ship, where light filters and temperature controls simulate the conditions that the bottles would encounter at the depths from which their samples were taken.

As is evident, techniques for measuring the productivity of marine phytoplankton are difficult to perform and are based on several assumptions. Despite some difficulties, they appear to give reliable information on the relative productivities of organisms at different depths and in different regions.

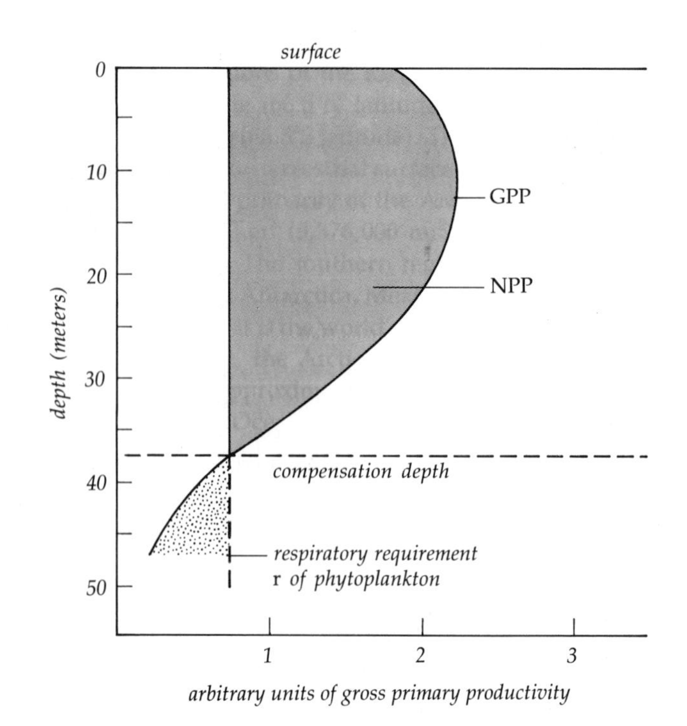

16.6 Gross primary productivity (GPP) *at various depths, mid-latitude, open ocean, on a cloudless summer day.* GPP *(curved line) is inhibited at the surface by bright sunshine and is higher a few meters below the surface, then it decreases with increased depth and decreased illumination. The respiratory rate* r *of phytoplankton (vertical straight line) is the same from surface to below compensation depth. At compensation depth,* GPP *is just enough to meet phytoplankton respiratory needs. Here, the* GPP *and* r *lines intersect. Above compensation depth, net primary productivity is the difference between* GPP *and* r *and is positive (shaded area). Below the compensation depth, where* GPP *is less than* r, NPP *is negative (dotted area).*

The Distribution of Productivity A typical vertical profile of 24-hour productivity, as measured by either of the techniques just mentioned, resembles that shown in Figure 16.6. If the plankton are evenly distributed, the organisms' respiratory rates are roughly the same from the surface to the deepest depth measured. Photosynthesis depends on sunlight, however, and sunlight becomes progressively less suitable for photosynthesis at greater depths. Not only does the amount of light decrease, but some of the most suitable wavelengths are absorbed and eliminated within the first few meters of water. Thus photosynthetic production of new carbohydrates varies with depth. Near the surface, phytoplankters may actually be inhibited by too-bright sun, and their productivity is not as high as it is a few meters deeper. Below this depth of maximum productivity, their production falls off as sunlight diminishes. At some depth, the total amount of new glucose produced (*GPP*) is barely enough to match the plants' own respiratory needs (*r*). At this depth, net production is zero, and $GPP = r$. Below this depth, gross production is less than the plants' respiratory needs, net production is negative, and any plants there, in effect, are starving. As noted earlier, this important depth is known as the compensation depth. As a rule of thumb, the compensation depth is reached when the sunlight is diminished to about 1% of its 24-hour average value at the surface (Figure 16.4). It typically lies between about 10 and 100 m, depending on the season and the clarity of the water, but it may occur at the surface during the temperate and polar winters.

Studies of marine primary productivity have shown that plant activity is unevenly distributed on a global scale (Figure 16.7). High productivity is seen mainly along coasts, including that of the Antarctic continent, and in a belt in the equatorial Pacific. The central areas of the oceans are mainly regions of low productivity. Eutrophic polar regions are inhibited, in part, by the fact that winter days are short and dark. The most important factor contributing to this unevenness of productivity, however, is the availability of plant nutrients. In addition to light, marine plants require phosphorus and nitrogen compounds for successful growth. The former occur mainly as phosphate (PO_4), the latter as nitrate, nitrite, and ammonium (NO_3^-, NO_2^-, and NH_4^+, respectively). Oligotrophic surface waters are usually characterized by abundant sunlight but suffer from crippling shortages of these nutrients. Other materials needed by marine plants are usually present in unlimited supply.

A problem basic to all open-ocean systems is that the necessary nutrients tend to sink to depths where there is not sufficient sunlight to sustain plant productivity. The dissolved ions themselves do not sink, but the plants and ultimately the animals into which they are incorporated tend to gravitate downward.

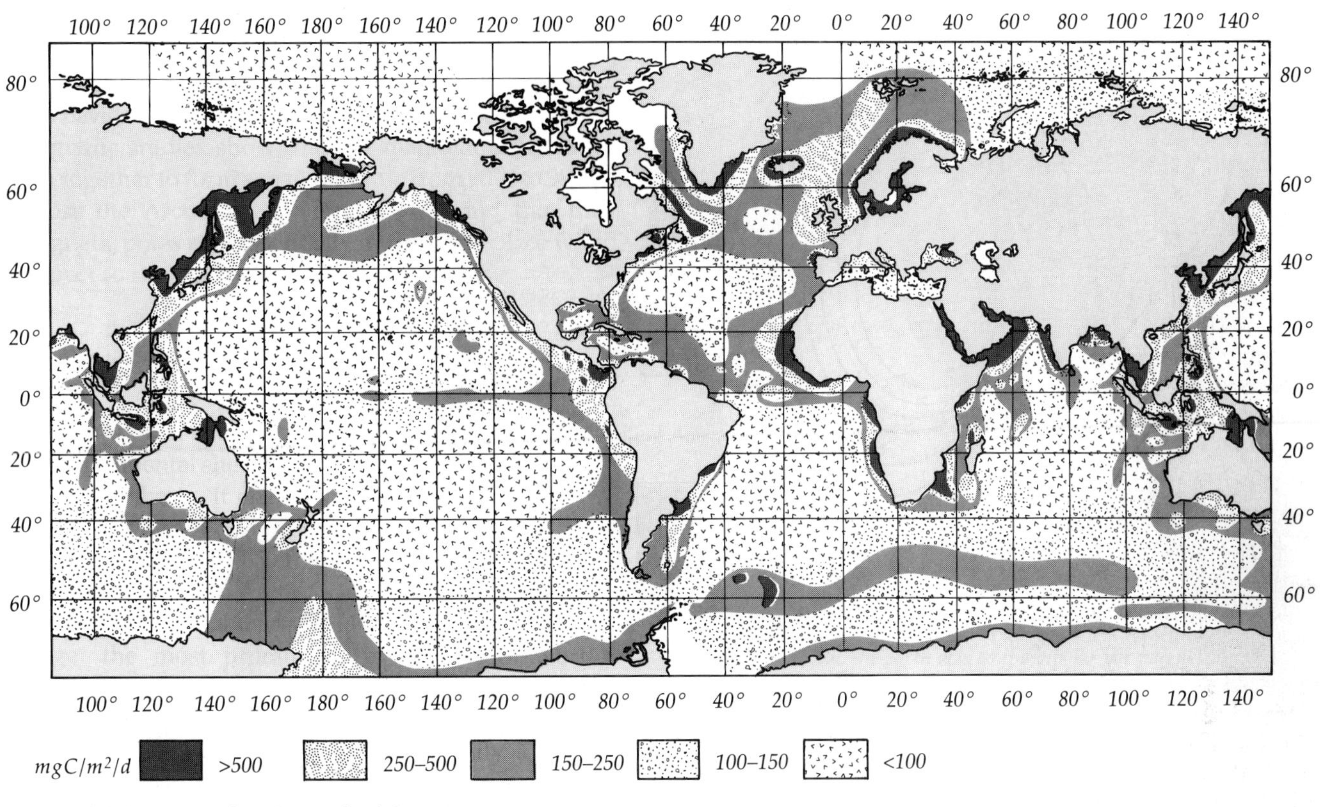

16.7 *Global patterns of marine productivity.*

Phytoplankton cells may sink slowly, carrying their captured nutrients with them. Many zooplankters (certain copepods, for example) live deep in the water during the day and swim to the surface at night to feed. They carry the nutrients they have eaten downward when they return to their daytime depths in the morning. Fecal pellets and the carcasses or parts of herbivores that die in deeper water sink, entraining nutrients downward. The vital nutrients are therefore much more abundant in deep water than they are above the compensation depth.

Only where the nutrients are unable to sink far below the surface for very long, as in shallow water, or where oceanic processes systematically return them to the surface, as in upwelling areas, does marine plant productivity remain high. The eutrophic waters of the sea are usually areas in which nutrients are either kept at, or return to, the surface. On the continental shelves, the relatively shallow bottom prevents the downward loss of nutrients. Neritic waters are also enriched by nutrients entering the sea from rivers, and winds along many coasts force the upwelling of deeper, nutrient-rich water to the surface. These factors are responsible for high coastal productivities throughout the world.

Representative productivities for open-ocean areas, coastal waters, estuaries, upwellings, and reefs are shown in Figure 16.8, with agricultural systems also shown for comparison. The most productive agricultural areas are more productive than the richest marine areas, partly because of the immense investment in fertilization and herbivorous pest control.

In the open sea, the waters around Antarctica and along the equatorial Pacific are zones of perpetual upwelling. In these two belts of ocean, rising nutrient-rich water sustains the high productivity of the phytoplankton communities. This is the reason for the contrast between the lush, thriving communities of Antarctic seas and the barren waters of open tropical oceans. In Antarctic seas, nutrient-laden water is continually rising to the surface, fueling the most exuberant large-scale growth of phytoplankton on our planet. In tropical seas, by contrast, nutrients have long since sunk from the surface waters and are unable to make any sizeable return. The harsh climate of Antarctica and the mellow climates of the tropics are of little importance to marine organisms. To them, the presence or absence of nutrients makes the crucial difference.

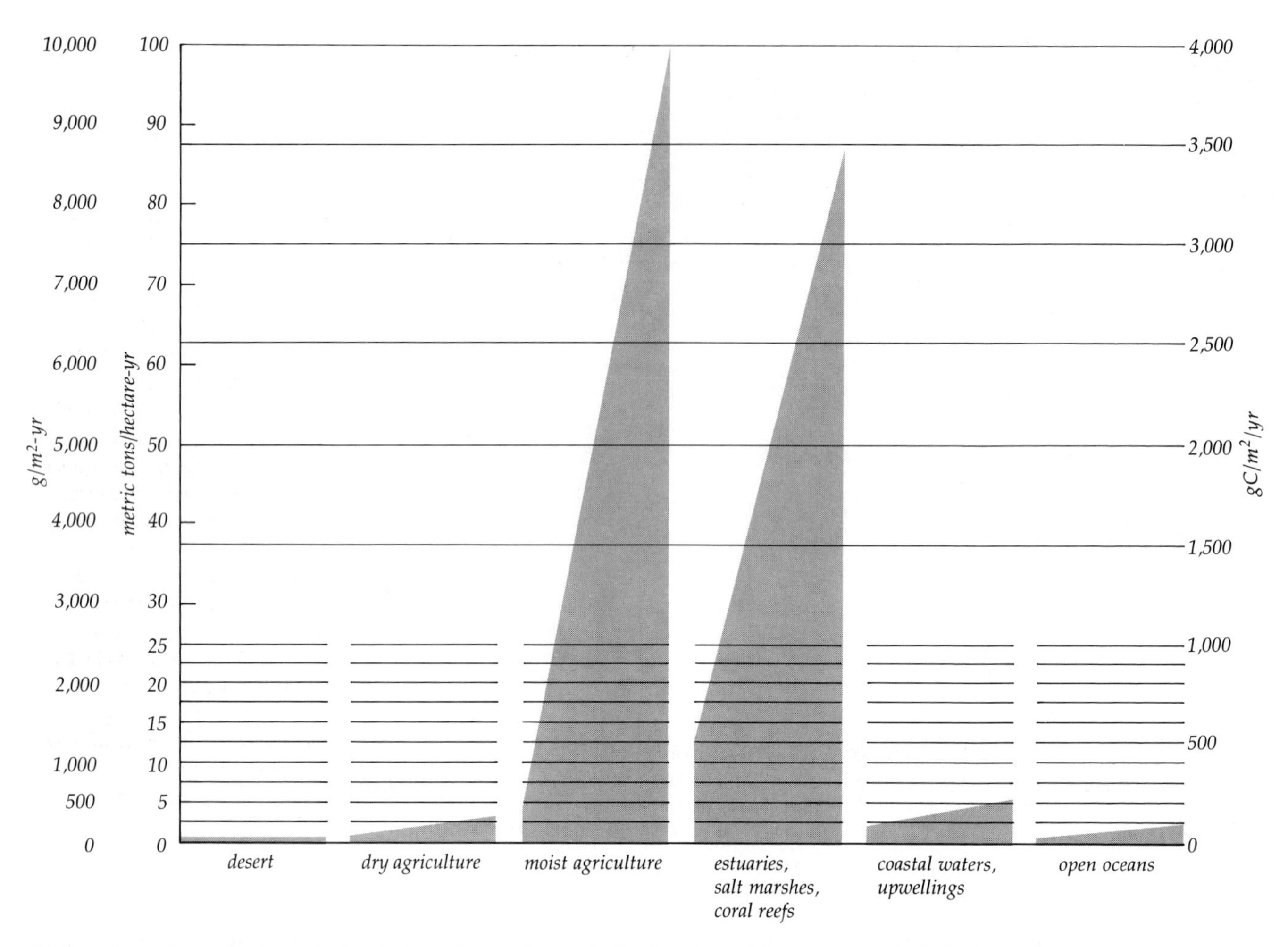

16.8 Comparative production rates for six types of natural or agricultural systems. (Net primary productivity.)

Nutrient Cycles

The shortage of nitrogen and phosphorus in the sea is a result of causes that range from cosmic to local. During the early history of the universe, when hydrogen and helium were the predominant elements, heavier atoms were generated in the interiors of stars. The primary atomic construction process built atoms whose atomic weights were multiples of 4: carbon (12), oxygen (16), neon (20), magnesium (24), silicon (28), and sulfur (32). These are now the most common elements in the universe, after hydrogen and helium. Nitrogen (14) and phosphorus (31) resulted from secondary atomic construction processes, which produced very limited numbers of these atoms. They have been scarce since the origin of the earth, partly because of their general scarcity in the cosmos. In terrestrial environments, the scarcity of nitrogen is further aggravated by the fact that most nitrogen exists in a form that organisms cannot use. The bulk of the atmosphere consists of nitrogen gas (N_2), but this molecule is resistant to the biochemical abilities of all but a handful of living things. To use the relatively inert nitrogen gas molecule, an organism must first convert it to nitrate (NO_3^-), nitrite (NO_2^-), or ammonium (NH_4^+) forms. Few organisms can do that. Because of this bottleneck in the nitrogen cycle, those organisms that can convert atmospheric nitrogen to one of these forms are crucially important to the economy of the sea and indeed to that of the entire earth.

This section examines the pathways by which both nitrogen and phosphorus make their way through the living and nonliving parts of the marine biosphere, and it highlights the bottlenecks and dead ends that restrict the tempo of marine productivity.

The Nitrogen Cycle Nitrogen in the air is converted directly to useful nitrate by such abiotic agents as discharges of lightning and photochemical processes. Nitrogen-fixing bacteria, however, make a more significant contribution. These organisms, including

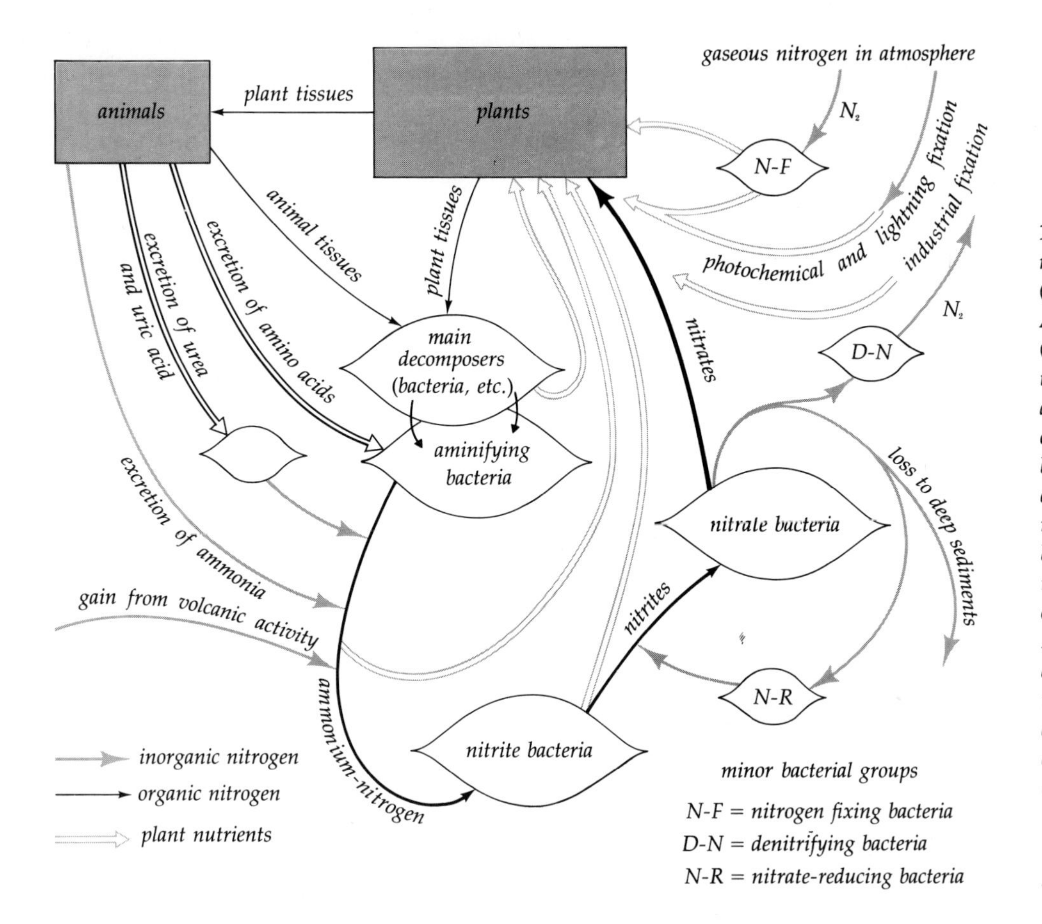

16.9 The biogeochemical cycle of nitrogen compounds in aquatic (and other) environments. Ammonium (NH_4^+), nitrate (NO_3^-), and nitrite (NO_2^-) are taken up as nutrients by plants and converted to organic compounds. These are consumed by animals and eventually by decomposing bacteria. The organic nitrogen compounds are converted back to ammonium, nitrite, and nitrate by other bacteria, completing the biological cycle. Nitrogen-fixing bacteria convert atmospheric nitrogen (N_2) to inorganic forms useful to plants; abiotic and industrial processes also fix nitrogen. Denitrifying bacteria convert some useful fixed nitrogen back to the atmospheric unusable form, completing the global cycle.

cyanobacteria, are able to perform the chemically difficult task of converting nitrogen gas molecules to nitrate. In the process, they produce more than they can consume themselves. As a consequence, they are often found in symbiotic association with plants that make use of the spare nitrate. On a global scale, bacterial conversion of nitrogen outweighs abiotic "fixation" (as the N_2 to NO_3^- conversion is called) by a ratio of 6 to 1. Most nitrogen-fixing bacteria are found on land, although a few occur in the sea.

After its conversion to nitrate by bacterial fixers, a nitrogen atom undergoes a bewildering flurry of transformations (Figure 16.9). Nitrate is taken up directly by algae or phytoplankton. The nitrate (NO_3) is then stripped of the three oxygen atoms and the nitrogen atom is incorporated into a complex molecule. This may be an amino acid molecule (which in turn becomes part of a larger protein) or a DNA molecule or some other vitally important biochemical. There it serves the plant until the plant dies or is eaten. If eaten, the nitrogen is simply incorporated into some comparable molecule in the herbivorous animal. Eventually, dead plants or animals containing the nitrogen are exposed to the activities of decomposer bacteria. Some of these (the "aminifying bacteria"; Figure 16.9) release the atom in the form of ammonium (NH_4^+). The ammonium may be used by "nitrite bacteria," which convert the nitrogen atom to the nitrite form. Nitrite is, in turn, used by "nitrate bacteria," which convert the nitrogen atom back to the nitrate form, completing one revolution of the biological cycle. Marine plants are able to use all three forms of nitrogen: ammonium, nitrite, and nitrate. Therefore, they can use some of these materials before bacteria get to them. Certain bacteria (nitrate reducers) complicate matters by converting nitrate back to nitrite; others (denitrifying bacteria) shortchange the whole community (but complete the cycle on a global scale) by converting nitrate back to molecular nitrogen, N_2. Nitrogen moves within the biological community in a few other ways. For example many marine animals excrete ammonia or nitrogen-containing uric acid, or urea. These are quickly used by bacteria. Thus the natural nitrogen cycle involves a huge reservoir of inaccessible nitrogen gas in the atmosphere, a relatively slow conversion of nitrogen to a useful form by nitrogen-fixing bacteria, rapid use and re-use complicated by many bacterial conversions once the nitrogen enters the biological community in useful forms, and a relatively slow return of nitrogen to the atmosphere

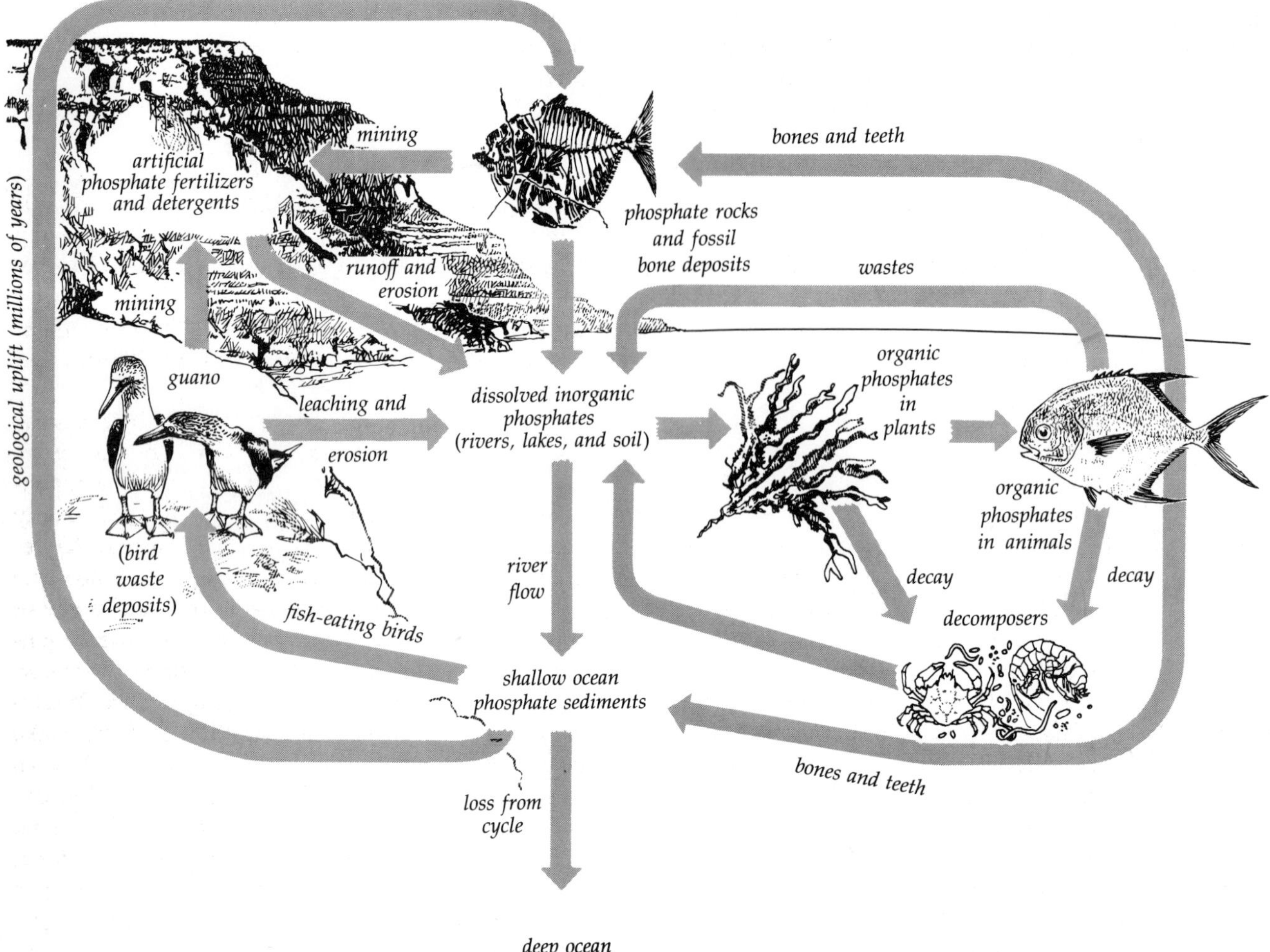

16.10 Phosphorus cycle. Phosphorus in the form of inorganic phosphate is eroded from land soils and rocks and enters the sea. The phosphate is used by plants and converted to organic form (in DNA, RNA, ATP). Plant organic phosphate is transferred to animals and eventually to decomposer bacteria, which liberate the phosphate in inorganic form. Some phosphate becomes locked in sea-bed deposits, leaving the biosphere until it is liberated millennia later by geologic uplift and erosion. (After Miller, G. T., Living in the Environment, *Belmont, Calif.: Wadsworth, 1975)*

by denitrifying bacteria. Small volcanic additions of nitrogen compounds to the atmosphere, and an equally small annual loss of nitrogen to sediments in the sea bed, complete the picture in the natural world.

Currently, a major modification of the nitrogen cycle has begun. Nitrate is important for use as agricultural fertilizers and in explosives, and the industrial fixation of nitrogen now occurs on a gigantic scale. The Haber process, invented by a German chemist in 1914, uses atmospheric nitrogen, a nickel catalyst, pressures of several hundred atmospheres, a temperature of 500°C, and hydrogen gas to accomplish what a few species of nitrogen-fixing bacteria are able to do easily under ordinary conditions. The result is that people now add nearly as much fixed nitrogen to the land and water as the nitrogen-fixing organisms do.

The Phosphorus Cycle The phosphorus cycle differs from the nitrogen cycle in several important ways. Phosphorus is not found in the atmosphere, its movements through the biosphere do not require extensive mediation by bacteria, and it experiences significant losses to the oceanic sediments. Indeed in the short run, phosphorus does not cycle—it makes a one-way trip from land to the sea, spins through the biosphere for a number of turns, then ends up locked in rock or shell on the sea bottom (Figure 16.10). Only in a very roundabout way does it eventually return to the land, either by means of uplift or by subduction of the seafloor, where erosion or volcanic activity makes it available to living organisms once again.

Phosphorus is liberated from rock in the form of inorganic phosphate (PO_4). That is the form in which it is found dissolved in water, and that is also the form in which organisms can use it. Plants take it up and convert it (as do the animals that eat the plants) to organic phosphate for use in DNA molecules and other important structures. When the organism containing it decomposes, the phosphorus returns to the water as organic phosphate. Decomposition by bacteria is almost immediate; the phosphorus is converted back to the inorganic form and is immediately available for another round of uptake by plants.

The tempo of phosphate use by marine organisms can be awesome. In oligotrophic seas, a phosphate ion escaping from a live or decomposing organism remains in the water for only a few minutes before it is used again by a phytoplankter. Grazing copepods are profligate wasters of phosphate. By one estimate, the copepod *Calanus* excretes 60% of the phosphorus it acquires in its diet (and loses another 23% via its fecal pellets). This excreted phosphorus is immediately useful to the phytoplankton, and the grazing of copepods, while fatal to the diatoms that get caught, is beneficial to those that escape being eaten.

Phosphate is an essential ingredient in the shells of certain brachiopods and clams, in sharks' teeth, and in the teeth, bones, and shells of other organisms. The formation of these hard parts, as well as chemical entrapment of phosphate in manganese nodules and other sediments, eventually results in the removal of phosphorus from the living part of the marine system and its quasi-permanent deposition on the seafloor. Only two mechanisms operate to return phosphorus quickly from the sea to the land. The fish, shellfish, crustaceans, and other edible products taken from the sea each year by people contain about 60,000 metric tons of phosphorus. All of this is returned to the land or fresh waters via the body wastes, clamshells, and other discards of the human consumers. Likewise, sea birds defecating on islands and coastlines also act to move an estimated 350,000 metric tons of phosphorus from the sea back to the land, each year. On a global scale, these reverse movements are not enough to balance the loss from the land. Each year, about 2 million metric tons of phosphorus enter the sea from the continents. At present the "cycle" is probably out of balance, with the removal of phosphorus to the oceans exceeding the return movement to the continents.

The Loss of Energy in Food Chains

The same thermodynamic bottleneck that prevents plants from harvesting more solar energy also restricts the communities of animals that eat the plants. To illustrate, imagine a simple mix of animals consisting of copepods, herrings, mackerel, and tuna. The feeding of these animals can be illustrated as shown:

organism	diatoms →	copepods →	herring →	mackerel →	tuna
trophic level	1	2	3	4	5

A simple linear feeding relationship such as this is termed a **food chain.** The organisms are said to occupy successively higher **trophic levels.** In almost all cases, the first trophic level is occupied by plants (diatoms in this example), and the copepods, herring, mackerel, and tuna occupy the second, third, fourth, and fifth trophic levels, respectively.

All of these organisms are restricted by the same thermodynamic law that affects plants and photosynthesis: It is impossible to transform all the energy consumed into useful forms. Much of the food energy taken in by these and other animals is "wasted." Rather than transforming itself into energy in a fat molecule or a protein or some other useful molecule, some of the energy in food eaten by copepods or other organisms simply escapes as heat. Heat is almost always an unusable form of energy for living things. Warm-blooded animals such as whales and seabirds make better use of it than invertebrates do, but in all cases it represents a loss of valuable food energy. This loss is unavoidable. As a consequence, copepods that eat 1 g of diatoms will not gain an entire 1 g of additional weight. Since some of the energy in this meal escapes as heat, the copepods will gain less than 1 g of weight. Of the weight actually gained (which shows up as stored fats, glucose, new eggs, and other forms), some will then be metabolized by the animals as they go about their daily business, consuming energy for motion, digestion, the repair and replacement of cells, and other functions. Herrings catching those copepods at some later time will get very much less than 1 g of weight from the original gram of diatoms that their victims ate earlier. Only about 10% (sometimes perhaps as much as 20%) of the food eaten by an animal in its lifetime ends up as available energy, in the form of its own biomass, for consumption by predators.

The consequences of this inevitable loss of energy from living communities can be astonishing. Consider, for example, the amount of plant productivity needed to fill a standard 6-oz can of tunafish. In order to gain 6 oz of weight, the growing tuna would need to eat about 60 oz of mackerel. (The energy in about 54 of the 60 oz would be lost, either as heat immediately after feeding or as heat later in the tuna's life as it swims, metabolizes, and reproduces.) In order for the mackerels to produce the 60 oz needed by the tuna, they would have to eat some 600 oz of herring. (The energy in about 540 of these ounces would also be lost for the same reasons.) The herring, in turn, would have to eat some 6,000 oz of copepods, and the copepods would need to eat some 60,000 oz of diatoms. Thus the amount of phytoplankton needed to produce a single can of tunafish is well over a ton. Figure 16.11 illustrates this principle, using a human consumer as the uppermost predator in the food chain.

In marine systems, the relationships between plants, predators, and prey is seldom as simple as that shown in the model food chain of our example. In reality, each animal is usually able to feed on many other species of plants and/or animals, and each is subject to attack by many different kinds of predators. A more realistic illustration of the typical maze of feeding relationships is shown in Figure 16.12. This complex pattern is termed a **food web.** The trophic levels are not as easy to identify as levels in a food chain, since some animals may feed in ways that place them in more than one trophic level. (For example, certain krill that eat mostly diatoms will also eat copepods, if given the chance. This makes them second- and third-level consumers.) The principles that apply to food chains also apply to food webs, however, and useful food energy is lost as heat whenever animals feed.

Thus plants use up much of the energy they capture (which, after all, is why they capture it), and animals both waste and use up much of the food energy that they consume. The result in all ecosystems, on land and in the sea, is that only a small fraction of the energy trapped by plants is available to support animals at higher trophic levels. In some cases, this translates into an obvious scarcity of predators and an abundance of herbivores. This is particularly true on land, where large populations of plants; modest populations of mice, rabbits, and other herbivores; and small populations of foxes and other carnivores are the rule. We would thus expect animals to be scarcer than their prey in the sea, as well, but here several complicating factors arise. It is often true, for example, that copepods and other zooplankton animals appear much more abundant than the phytoplankton that they live on—a situation that is exactly the reverse of the usual relationship between plants and animals on land. In numerical terms, the plankton community seems grossly out of balance, but in energy terms, nothing is amiss. The few phytoplankton diatoms seen at any moment are photosynthesizing and reproducing at such a furious pace that they manage to supply all of the energy needed by the swarms of animals feeding on them. Elsewhere in the sea, most of the conspicuous animals are long-lived, are equipped with slow metabolisms, and are much bigger than the highly-productive plants that ultimately sustain them, and these attributes contribute to the impression that predators are not much scarcer than their prey. Their numbers are influenced by many features of their life cycles as well as by their feeding relationships—but an ironclad upper limit to the numbers of all organisms is set by the availability of energy to the plants and its profligate waste and loss as it is passed from trophic level to trophic level in marine food webs.

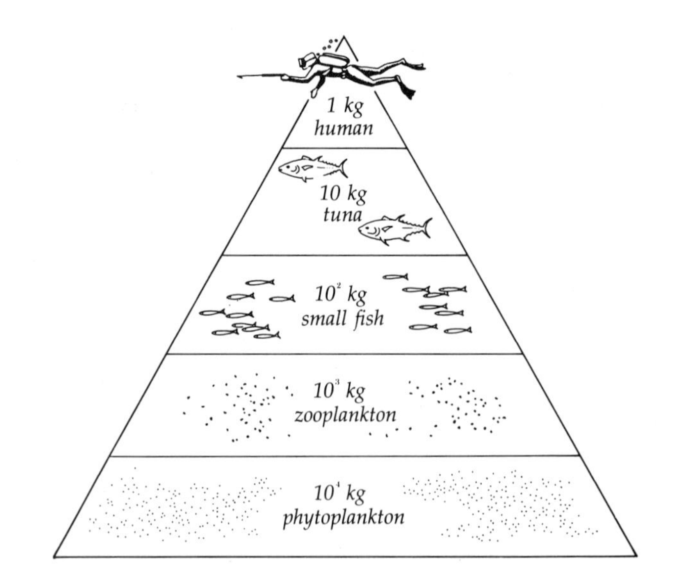

16.11 Model of biomass transfer in the sea. The amount at each level is the approximate amount of new biomass that can be produced by eating all of the amount shown at the next lower level.

LIGHT, NUTRIENTS, UPWELLING, SYMBIOSIS: LIFE IN FOUR MARINE SYSTEMS

The most favorable situations for marine plant growth and animal abundance are ones in which (1) the plants are able to remain in the sunlit upper reaches of the sea, (2) nutrients are delivered from below or are prevented from leaving the surface waters, and (3) sufficient light is always available. Only in a few oceanic areas do all three conditions occur simultaneously. This section describes four marine ecosystems, three of which depart from optimum conditions in one or more ways: In the open North Atlantic, light becomes scarce during the winter, nutrients dwindle during the summer, and the phytoplankton are seasonally carried to unfavorable depths. In the subtropical Sargasso Sea, light is always available, but lost nutrients are slow to make their way back to the surface. In Antarctic seas, nutrients are continually delivered to the surface by upwelling—but light is absent in the winter. On coral reefs, the fourth system we shall consider, all favorable factors converge to produce the most luxuriant communities of marine organisms on our planet.

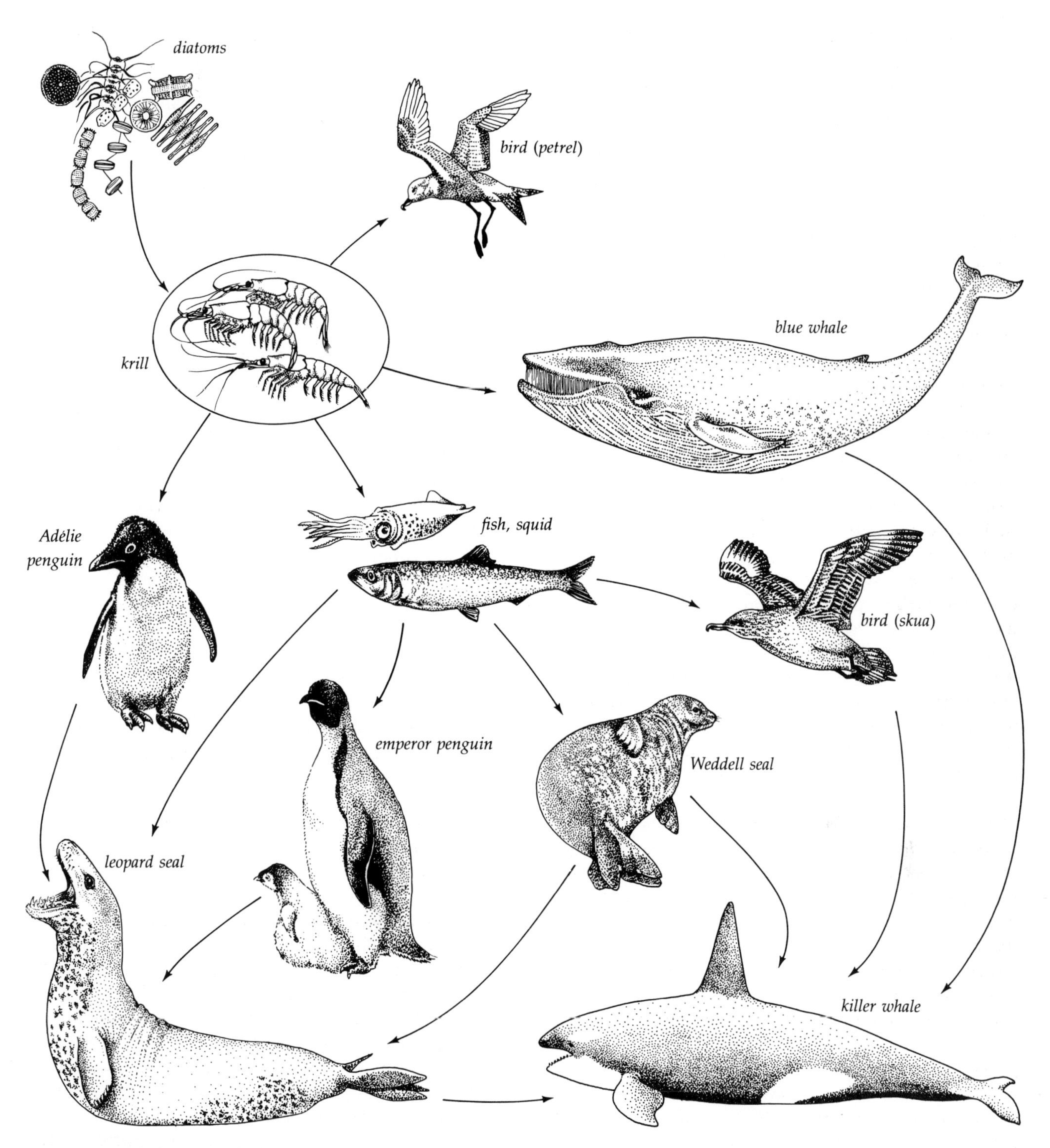

16.12 A simplified Antarctic food web. Note that even in this simplified perspective, several animals feed simultaneously at more than one trophic level. The leopard seal is a fifth-level consumer when it eats Emperor penguins and a fourth-level consumer when it eats fish and squid.

The North Atlantic Ocean

Climatic conditions in the open North Atlantic ocean are strongly seasonal and are best appreciated by examining a typical passage of the seasons. In an open-ocean location far offshore in winter, the days are short and dark, the sun is low in the southern sky, winds are strong, and the sea surface is cold, rough, and turbulent. The effect of winds and the winter chilling of the surface is to drive surface waters, with their resident phyto- and zooplankters, down to depths of 200 m or more. Once there, they circulate back upward

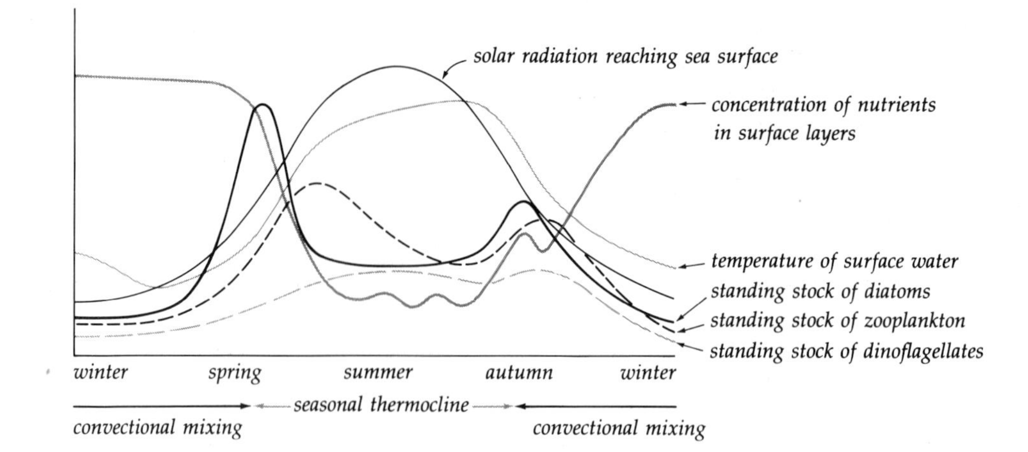

16.13 Diagram illustrating seasonal changes of temperature, nutrients, phytoplankton, and zooplankton in the surface layers of temperate seas. (After Tait, R. V., and R. S. DeSanto, Elements of Marine Ecology, *New York, Springer-Verlag, 1972)*

in a slow but steady turnover. At this dark time of year, the compensation depth (CD) is within a few meters of the surface. As we have seen, phytoplankton can only multiply if they spend more time above the CD than below it. The deepest depth to which they can circulate, while still spending enough time above the CD for reproduction, is called the **critical depth**. Vertical circulation during the winter carries them far below this critical depth; too little time is spent above the CD, and phytoplankton reproduction is halted. Copepods and other herbivorous zooplankters are scarce and live in a near-dormant state, awaiting better times. Because deeper water is stirred to the surface in this general vertical circulation, however, nutrients that sank during the preceding seasons now return to the surface. Conditions are favorable for phytoplankton growth, except that the vertical circulation, the short days, and the low sun prevent the phytoplankters from getting enough sunlight. This combination of low light levels, high surface nutrient levels, and low stocks of zooplankters and phytoplankters is shown in Figure 16.13 ("winter").

As spring advances, light levels improve. The sun climbs higher in the sky, days become longer, and the compensation depth becomes deeper. As the winter storms die down, the stirring of the sea surface diminishes. Planktonic organisms continue to rise and sink with the vertical water displacements driven by the winds. As wind strength diminishes, some of these organisms remain closer to the surface for longer periods of time. Sometime in March or April, the surface water becomes a few degrees warmer than the deeper waters below. A seasonal thermocline forms at a depth of about 50 m. All future stirring by the wind is thenceforth confined to the surface water over this thermocline. After the thermocline forms, those phytoplankton cells that happen to reside in the shallow surface layer find all of the sunlight they need for luxuriant growth, as well as abundant nutrients brought to the surface during the winter's stirring. All factors favorable to phytoplankton growth converge, and the phytoplankters increase rapidly in number. This characteristic outburst of luxuriant phytoplankton growth is called the "spring bloom," or spring diatom increase (SDI). The zooplankters respond almost immediately. Many copepods molt to their adult stages, begin to feed, and immediately lay eggs. Within a few days, the waters become populated by large numbers of juvenile zooplankters. Thus early spring finds the open sea surface warmer and better lighted than during the winter, and it is crowded with new zoo- and phytoplankters (Figure 16.13, "spring").

With summer begins a steady loss of nutrients from the surface, due to sinking of live and dead organisms and fecal pellets. The net effect, as summer progresses, is to deplete the surface waters of nutrients. This and the feeding of the herbivorous zooplankton reduces the phytoplankton population to a lower level than that seen during the spring bloom episode. Ample light, warmer water, diminished nutrients, substantial populations of zooplankton and diminished populations of phytoplankton therefore characterize the surface waters during the summer (Figure 16.13, "summer").

As autumn advances, the days grow shorter and colder, and the sea surface cools. Eventually, the cooling is sufficient to lower the surface temperature to that of the deeper water below, and the seasonal thermocline disappears. If a stormy period that stirs new deep water to the surface is followed by a few days of calm at this time, the renewed surge of nutrients, coupled with still-adequate sunlight, is enough to trigger a "fall bloom" of phytoplankton, followed by a brief increase in zooplankton ("autumn," Figure 16.13). As the season progresses, however, light levels fall, increased stirring of the sea circulates the plankton organisms below the critical depth, and net production ceases. The system has completed its annual cycle.

Thus during a typical temperate Atlantic season, productivity is limited by a shortage of light during the winter, by the fact that the phytoplankton organisms are carried away from the surface during the winter, and by a shortage of nutrients during the late summer. Productivity is exuberant during the spring and (sometimes) the fall. The effect of these factors is to give open Atlantic water a productivity of the order of 75 gC/m^2-yr—a respectable figure for open waters.

Marine ecosystems elsewhere in the world depart from this pattern in several ways. First, in tropical seas, brilliant sunlight is available all year long, whereas in polar seas the winters are absolutely dark. Second, in shallow water, the sea bottom prevents the escape of nutrients and keeps them near the surface where plants can continue to use them. Third, where seasonal surface cooling does not occur, the thermocline persists throughout the entire year, and the return of nutrients lost to deep water is feeble, at best. Fourth, in certain upwelling areas, where winds and/or topographic features ensure that deep water is always rising to the surface, a continued return of essential nutrients persists year-round. The net effect of these factors is to ensure that upwelling regions and shallow waters are the most productive of any on Earth; that temperate seasonal waters, such as those of the North Atlantic, are reasonably productive of marine life; and that warm, blue, tropical seas are low in biological productivity.

The Sargasso Sea

The Sargasso Sea is a mid-oceanic region of warm water, centered in the warm-temperate Atlantic Ocean. The sea is circled by the main Atlantic currents (see Chapter 10, Figure 10.12). Because it is reasonably close to oceanographic institutions of the U.S. East Coast and Europe, it has been studied more frequently than comparable tropical and subtropical seas of other areas. The Sargasso Sea's surface is populated by large floating brown algae (species of the genus *Sargassum*), which serves as cover for about 50 different species of fish, crabs, opisthobranchs, snails, and other creatures (Figure 16.14). Despite the plants floating in patches on its surface, the sea is oligotrophic. The surface varies in temperature between 20°C in winter and 28°C in summer. The water below the thermocline (which lies 600 m below the surface) is at 13°C. Thus "winter" is never cold enough to cool surface water to the temperature of the deeper water. The thermocline is therefore permanent. It acts as a year-round barrier to the recovery of nutrients that have sunk from the surface layer to the sea bottom.

The water of the Sargasso Sea is remarkably clear. The compensation depth can be as deep as 150 m. Surface nutrient levels are low; the concentration of nitrate is 14 mg of nitrogen per cubic meter of water, or 14 mgN/m^3 (as opposed to 150–200 mgN/m^3 in the wintertime in the North Atlantic). For this reason there are few phytoplankton. The *Sargassum* weeds floating at the surface grow very slowly. They evidently live for decades or centuries; some plants seen by Columbus during his first voyage might still be alive. Slight cooling of the surface during the area's mild winters, coupled with an occasional windy spell, has been observed to stir nutrients near the top of the thermocline back toward the surface. Because the water is so clear, this stirring, though it carries the phytoplankton below the compensation depth, does not take them to the much deeper critical depth. As a result, low productivity continues even through the winter, and a modest increase in productivity occurs in the spring, following the stirring.

Certain diatoms (genus: *Rhizosolenia*) have found a way to adapt to one of the nutrient shortages in the Sargasso Sea. These diatoms contain smaller cells, symbiotic cyanobacteria, that are able to convert atmospheric nitrogen to useful nitrate. (The diatoms are still starved for phosphate.) They were discovered only recently. Their existence shows that productivity in the Sargasso Sea may be higher than was previously supposed.

A representative productivity figure for the Sargasso Sea is about 50 gC/m^2-yr. This is surprisingly high (about one-sixth that of the most productive neritic waters), considering the chronic nutrient scarcity of the sea. Here, the fact that light is available even during the "winter" compensates, in part, for the shortage of nutrients. Productivity continues in the sea at a low level throughout the entire year, whereas many more nutrient-rich waters become too dark for phytoplankton growth during the winter.

The Antarctic Ocean

One of the richest pelagic communities on Earth is that inhabiting the waters around Antarctica (Figure 16.12). Diatoms, copepods, and teeming shoals of krill all support penguins, seabirds, fish, seals, whales, and other creatures in abundances that amazed many early explorers of this hostile region. Sunlight is limited or lacking during the long Antarctic winter, but the sustained upwelling of nutrients is more than enough to compensate for the seasonal absence of light. Antarctica is the beneficiary of the nutrients that have sunk out of the surface waters everywhere else on Earth. In the Atlantic, those nutrients that sink from the surface and do not return (as, in the Sargasso Sea or deep under the North Atlantic) find their way into a gigantic slow subsurface flow, called the North Atlantic Deep Water, that moves slowly south at a depth of about 1,000–4,000 m (see Figure 17.5). Eventually, this water is forced upward to the surface at a latitude of about 64°S, where it divides into one surface flow

1 *dolphins*, Delphinus
2 *tropic birds*, Phaëthon
3 *paper nautilus*, Argonauta
4 *anchovies*, Engraulis
5 *mackerel*, Pneumatophorus, *and sardines*, Sardinops
6 *squid*, Onykia
7 Sargassum
8 *sargassum fish* (Histro)
9 *pilot fish* (Naucrates)
10 *white-tipped shark* Carcharhinus
11 *pompano*, Palometa
12 *ocean sunfish*, Mola mola
13 *squids*, Loligo
14 *rabbitfish*, Chimaera
15 *eel larva, leptocephalus*
16 *deep sea fish*
17 *deep sea angler*, Melanocetus
18 *lantern fish*, Diaphus
19 *hatchetfish*, Polyipnus
20 *"widemouth,"* Malacosteus
21 *euphausid shrimp*, , Nematoscelis
22 *arrowworm*, Sagitta
23 *amphipod*, Hyperoche
24 *sole larva*, Solea
25 *sunfish larva*, Mola mola
26 *mullet larva*, Mullus
27 *sea butterfly*, Clione
28 *copepods*, Calanus
29 *assorted fish eggs*
30 *stomatopod larva*
31 *hydromedusa*, Hybocodon
32 *hydromedusa*, Bougainvilli
33 *salp (pelagic tunicate)*, Doliolum
34 *brittle star larva*
35 *copepod*, Calocalanus
36 *cladoceran*, Podon
37 *foraminifer*, Hastigerina
38 *luminescent dinoflagellates*, Noctiluca
39 *dinoflagellates*, Ceratium
40 *diatom*, Coscinodiscus
41 *diatoms*, Chaetoceras
42 *diatoms*, Cerautulus
43 *diatom*, Fragilaria
44 *diatom*, Melosira
45 *dinoflagellate*, Dinophysis
46 *diatoms*, Biddulphia regia
47 *diatoms*, B. arctica
48 *dinoflagellate*, Gonyaulax
49 *diatom*, Thalassiosira
50 *diatom*, Eucampia
51 *diatom*, B. vesiculosa

(*a*)

16.14 The subtropical Atlantic ocean—plankton and nekton. (a) Key. (b) Microcosmic view. Deep-sea fishes and the sargassum fish (8) are shown at actual size; epipelagic fishes are at least twenty times larger than shown in the figure. Note that this stylized representation shows the organisms to be much more crowded than is the case in real life. In the oligotrophic Sargasso Sea, organisms are usually few and far between.

going north and another going south. The belt of ocean where the North Atlantic Deep Water rises is called the Antarctic Divergence. Here, nutrients lost from the surface of the North Atlantic over hundreds of years finally make their way back to the surface. The northward surface flow, rich in these nutrients, fuels a luxuriant growth of phytoplankton, which supports the rest of the oceanic community in offshore Antarctic waters. Eventually this water reaches the edge of a warmer surface layer to the north and slides beneath it, along a broad front known as the Antarctic Convergence. A representative productivity figure for this region is 150 gC/m^2-yr. Were it not for the seasonal shortage of light, it is likely that this productivity would be even greater—perhaps twice its present level.

Coral Reefs

From a productivity standpoint, a reef is a puzzling community. The animals are fantastically diverse and numerous, but plants appear at first glance to be few and far between. In Figure 16.15, for example, which provides a glimpse of a reef community, almost every plantlike structure is actually an animal (or a colony of animals). Many reefs occur in coastal water where the bottom prevents the escape of nutrients. Marine life of all sorts, whether associated with a reef or not, is generally abundant in such areas. It is also true, however, that many reefs stand on isolated islands in deep tropical water (Chapter 4). The warm surface waters bathing these reefs, separated as they are by a permanent thermocline from the nutrients in deeper water, are typically nutrient-poor. How do reefs with an apparently small amount of plant life and nutrient-poor water maintain their exuberant riot of corals, fish, sponges, sea cucumbers, crabs, urchins, and other animals?

Part of the answer is that reef seaweeds, unlike the large, conspicuous plants of colder seas, tend to be small and inconspicuous. Reef rock is carpeted by an algal "turf"; low, fuzzy, and usually unobtrusive, which provides food for myriad herbivorous fish. Many of the reef algae are stony encrusting forms, more closely resembling pink rock than a plant. Finally, as noted (Chapter 14), many reef animals carry symbiotic algae (zooxanthellae) in their cells. Most coral polyps (no. 4, 9, 11, 15, and 29 in Figure 16.15) and many larger animals harbor these symbionts. The bril-

(b)

liant green mantles lining the rims of giant clams (no. 20, Figure 16.15), for example, are colored by their zooxanthellae. Thus plants are abundant on the reef—but they are either inconspicuous or disguised as rock or hidden inside animals.

Although coral polyps may obtain 50–95% of their food from their zooxanthellae, the polyps also continue to catch and eat zooplankton. The nitrogen and phosphorus in the zooplankton then enter the polyps. The metabolic wastes of the polyps, containing these

1 black-capped petrel
2 sea nettle
3 angelfish
4 lobed corals
5 sea whips and soft corals
6 triggerfish
7 sea fans
8 tube anemone
9 orange stone coral
10 bryozoans
11 brain coral
12 butterfly fish
13 moray eel
14 cleaner fish
15 tube corals
16 muricid snail
17 nudibranch
18 sponges
19 colonial tunicate
20 giant clam
21 purple pseudochromid fish
22 cobalt sea star
23 soft corals
24 barber pole shrimp
25 sea anemones
26 clown fish
27 worm tubes
28 cowry
29 sea fan

(a)

16.15 A coral reef habitat. (a) Key. (b) Microcosmic view. The organisms are as crowded and diverse in real life as they appear in this illustration.

elements, are used as nutrients by the zooxanthellae. The escape of these elements back into the water is therefore retarded or prevented because they are quickly consumed by zooxanthellae. Thus a coral polyp with its zooxanthellae is an effective trap for nutrients. Patiently extracting nutrients from the water in this way, the reef system builds up a store of nutrients, even in nutrient-poor water. Parrot fish, wrasses, butterflyfish, the crown-of-thorns sea star, surgeonfish, and many others that eat coral polyps harvest the productivity of plants by eating the animals that contain them.

The productivity of coral zooxanthellae, at a reef in Australia's Great Barrier Reef system, amounts to 300 gC/m^2-yr—three times that of the phytoplankton over and around the reef. The whole plant community on a reef at Eniwetok atoll was shown to produce about 3,500 gmC/m^2-yr—more than twenty times that of the eutrophic Antarctic waters and indeed one of the largest marine productivities ever measured.

Part of the secret of the success of coral reefs, then, relates in three ways to the productivity factors already discussed. First, the organisms themselves, by using a powerful cooperative arrangement that is not common in other marine communities, trap nutrients and prevent their departure from the system. Second, they live in tropical waters, where daily and seasonal sunlight are always adequate. Third, the fact that coastal reefs occur in shallow water, where the bottom prevents the escape of nutrients, also contributes to their success.

It would be misleading to suggest that these factors are the only reasons for the diversity and abundance of reef life. Other factors, not related to plant productivity, also affect reef organisms and those of other communities. The next sections examine communities in which the depth of the sea, interactions among organisms, and prehistoric events have promoted or inhibited the abundance and diversity of modern marine life.

DEPTH, DARKNESS, SHORTAGE OF FOOD: LIFE IN DEEP WATER

In this section we will imagine a slow descent to the bottom of the ocean, giving special attention to midwater organisms and the conditions of light, pressure, and food shortages that influence their feeding habits. Figure 16.16 illustrates terminology that is useful in discussions of deep water zones.

(b)

The Pelagic Realm

At the surface of the sea, the fish and other organisms are familiar ones (Figure 16.14). Anchovies, mackerel, copepods, diatoms, and a host of other creatures, large and small, make up an active community, photosynthesizing and metabolizing at a furious pace. As we go deeper in the sea, this world of relative abundance recedes overhead, and we enter progressively stranger domains, where inky darkness, deep chill, still waters, increasing pressure, and above all the scarcity of food

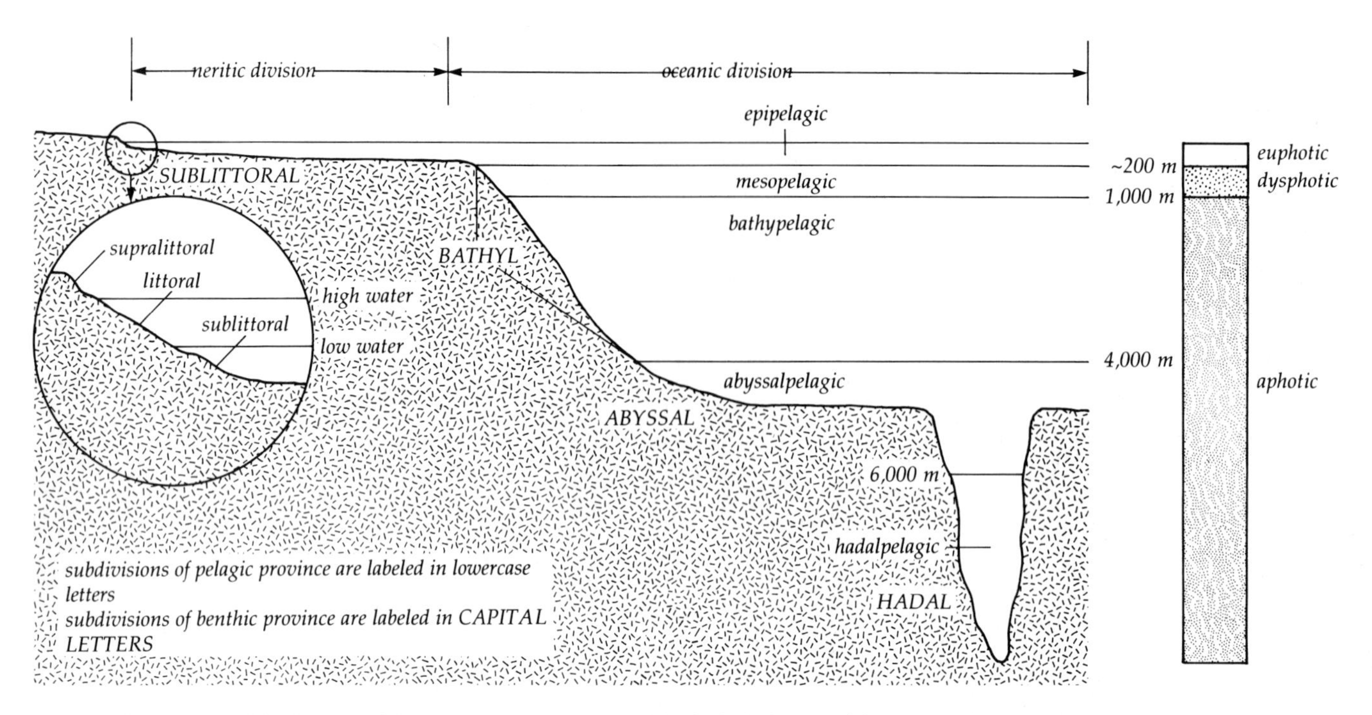

16.16 *Classification of depth zones of the oceans. A classification of depths by light availability is also shown (right).*

cause negative effects on the exuberance of life. Many (perhaps most) features of marine life at these greater depths are still unknown, partly because of the great effort and expense required to retrieve samples from the deep sea, and partly because the sampling techniques (deep dredging, photography, and occasional dives in submersibles) provide very limited views of this vast sector of the earth's surface.

The changes in organisms that live at progressively greater depths are manifold and complex. Generally, a descent into the open sea reveals the following features: The abundance of living creatures decreases with depth (with certain exceptions at shallow depths where fish and crustaceans congregate in layers during the day). Some organisms (particularly copepods) increase in species diversity to a depth of 1,000–2,000 m; thereafter the diversity of all species decreases (Figure 16.17). Certain organisms characteristic of the surface (for example, pteropods and larvaceans) become scarce in deep water, whereas certain organisms that seldom occur at the surface (for example, pelagic nemertean worms) become more abundant at increased depths. The proportion of luminescent organisms increases with depth to about 2,500 m, then decreases with depth thereafter. Dramatic changes in the colors of organisms begin at a few hundred meters, and many organisms begin to show reduced eyesight at depths below about 2,500 m. Finally, there is a tendency for certain organisms of the bathypelagic and abyssopelagic zones (defined in a later section) to become weak, flabby, and gelatinous. A few forms (particularly crustaceans) are unusually large, for their taxa.

The colors of surface organisms of the open sea are (with exceptions) fairly standard. Fish are mostly blue-green on the dorsal surfaces and silvery-reflective on the sides and bellies, whereas many of the invertebrates—jellyfish, arrowworms, ctenophores, copepods, and others—are transparent. These colors protect them from visual predators. Seen from above, a fish blends into the dark water below it; from the side or below, its scales reflect as much ambient light as the sunlit water behind or above it. Seen from any perspective, transparent invertebrates are well camouflaged at worst and are practically invisible at best.

At a depth of only a few hundred meters, life begins to take on deep-sea characteristics. With the exception of such peculiar photosynthetic forms as "olive-green cells" (a type of cyanobacteria), viable plants disappear. The animals decrease in abundance with depth and become more colorful. Down to a depth of about 500 to 800 m, most animals maintain contact with the surface by migrating upward each night to feed. This nightly vertical migration is exhibited by animals of virtually all phyla, including such small and feeble swimmers as jellyfish and copepods. Why don't they simply remain at the surface? That question is not easily answered. Apparently there are many different reasons for vertical migration that do not apply at all times to all animals. These may include hiding from predators in deeper, darker waters by day and conservation of energy by spending part of their time

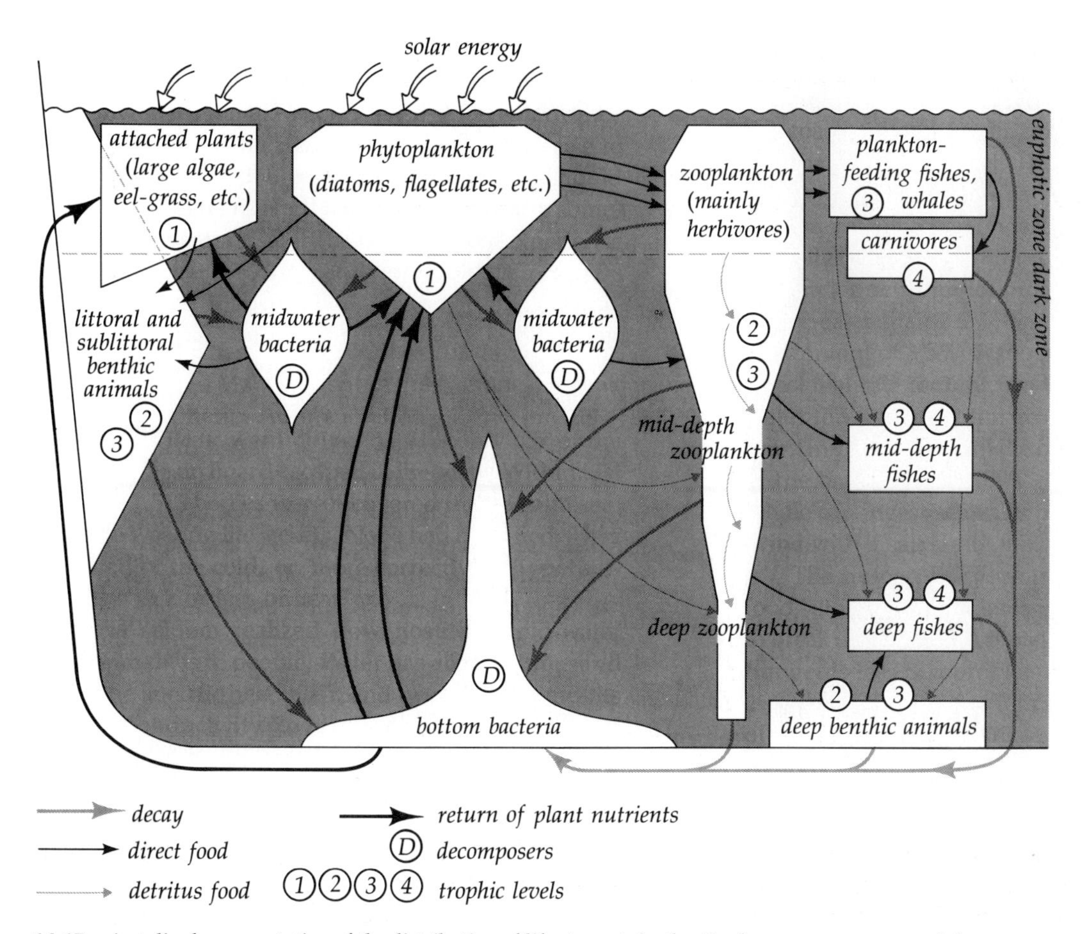

16.17 A stylized representation of the distribution of life at great depths. Producers, consumers, and decomposers are organized according to their trophic relationships. The widths of the bars show the relative abundances of the organisms at the various depths. In the oceans, the net primary productivity contributed by attached plants is much less than the net primary productivity of the phytoplankton. The dominant animals at the second trophic level are zooplankton organisms (mostly copepods) feeding as microherbivores. Note the importance of detrital food chains and the distribution of bacteria below the levels of greatest plant production. (Depth of surface lighted zone is greatly exaggerated.)

in the cold, deeper water where metabolism is slow. Or daily descent to deeper water may make it possible for animals to feed in fresh new surface waters every night, in areas where deep and surface currents travel in different directions. The actual patterns are complex. Not all individuals migrate every night. Some species migrate only to the base of the thermocline; others go through the thermocline into the warm surface water. Migratory behavior changes with the ages of some organisms, and sometimes the males and females differ in their migratory movements. Starting at depths of about 500 m, a change in the vertical migration habit becomes noticeable. Certain organisms at the greater depths make diurnal migrations all the way to the surface and back, but, by and large, the deeper creatures either stay put or migrate only partway to the surface and back. In any case, this massive nightly movement to the surface and back constitutes an important transfer of surface nutrients and food biomass to deeper water.

Sonic exploration of the community in the upper 1,000 m of water reveals that certain organisms often concentrate in layers during the day. One, two, or three dense layers of animals can almost always be detected hovering at different depths during daylight hours. For example, these layers were present at 290 m, 427 m, and 518 m during a study of offshore California waters reported by Robert Dietz. The organisms can be seen, on echo sounders, to swim away when a net or camera is lowered to sample them; consequently, their identification has not been easy. Studies suggest that these layers are usually composed of schools of lanternfish, sergestid shrimps, and euphausiids. These **deep scattering layers** or "DSLs," as they are called, were mistaken for the sea bottom in early work with sonar and may have resulted in the mapping of "shoals" in the Pacific that do not really exist. The organisms that compose the DSLs migrate to the surface at night. The layers can be seen rising gradually as nightfall approaches, then fading away as the creatures enter

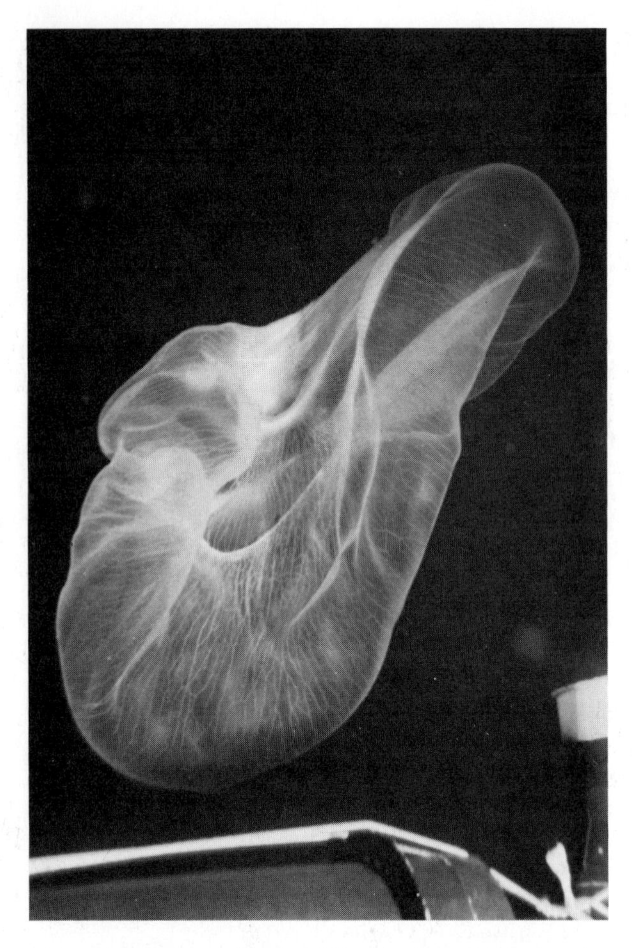

16.18 An organism sighted at 723 m in the San Diego Trough, moments before its capture by a "fish slurper" device. Later identified as a previously unknown species of jellyfish, it was named Deepstaria enigmatica.

the surface water around the ship. The animals eventually reassemble and descend to their characteristic daytime depths when sunrise approaches.

Lanternfish (family: Myctophidae) are common in the daytime sea at a depth of about 500 m. They are seldom seen by casual observers of the sea, but they are among the most abundant fishes on earth. These small (5–20 cm) silvery fish are named for the glowing photophores sprinkled over their undersides (see no. 18, Figure 16.14). There are about 250 species, worldwide, many of which are widely distributed. Like many other oceanic organisms, they remain in the deep sea during the day and rise to the surface to feed at night. Many lanternfish have gas bladders, which require constant adjustment of the amount of gas in the bladder, as the fish swims upward, to prevent the decreasing water pressure from causing a rupture of the bladder. One lanternfish of the U.S. West Coast, *Stenobrachius leucopsarus*, ranges in daytime depths from 150 to 600 m off British Columbia, and down to nearly 2,000 m off California. This small (9 cm) fish makes excursions to the surface and back, traveling distances of at least a kilometer and perhaps up to 4 km (2.5 mi) on each round trip. They feed at the surface on copepods, euphausiids, and smaller fish and are in turn eaten by salmon and other large fish.

Lanternfish are named for their ability to generate light. The spots on their undersides create a soft glow, whereas photophores near the tail are capable of flashing more brilliantly. Their possession of light-generating organs is one of the hallmarks of life at a few hundred meters of depth; approximately two-thirds of all organisms in the mesopelagic zone (between 200 and 1,000 m) are capable of producing light. Myctophids use their lights to make themselves less conspicuous to predators. Many predators of the mesopelagic depths hunt for prey by watching the dimly lighted sea surface above them for telltale shadows and silhouettes of prey swimming overhead. (Some fish of those depths have eyes that are permanently oriented upward, as in the hatchetfish (see no. 19, Figure 16.14). Lanternfish use their ventral photophores to create just enough dim light along their undersides to match that streaming down from the sea surface. The result: their shadow is obliterated, and predators watching from below can't see them.

Other organisms use light for many other purposes at meso- and bathypelagic depths. These include luring prey, spotlighting prey, signaling to other individuals of the same species, setting off brilliant flashes to startle or dazzle predators, engaging in courtship, and leaving luminous decoys to distract predators.

At about 500 m, the colors of organisms begin to change. Fish shift over to gray and darker colors (often black). Shrimps, euphausiids, mysids, and other crustaceans are often deep red, and jellyfish and other invertebrates take on brown, deep-purple, orange, and other hues, although some remain colorless (Figure 16.18). A species of worm that is as transparent as glass at the surface is replaced by a similar species (both of genus *Tomopteris*) of deep-crimson color at 750 m. Individuals of the jellyfish species *Atolla bairdi*, tinted in pale reds, creams, and purples at 500 m, are replaced by darker individuals of the same species at 750 to 1,000 m. These flamboyant colors cannot be seen at these depths. Red wavelengths are quickly filtered out of sunlight by the surface waters, and observers in submersible vehicles have noted that colored pictures of red deep-sea organisms, viewed in the last dim light in deep water, appear to be jet black.

Many of the mesopelagic fish are of quite ordinary appearance (see no. 16, Figure 16.14). These include deep-sea smelts and similar small, elongated fishes. Others are strangely modified, as their names suggest: "dragonfish," "viperfish," "daggertooths," "snaggletooths," and "telescopefish." The monstrous appearances and habits of these species, first encountered at about 500 m and becoming more conspicuous in deeper water, have brought them to public attention. Dragonfish (of several families) range in color from silvery-sided through gray to black, depending on their daytime depths. They usually possess one or two lines of photophores along their bellies, other lights elsewhere on the body (typically near the eyes), dorsal and anal fins far to the rear of the body, and mouths full of needle-like fangs. Many are able to expand their mouths to astonishing gapes (Figure 16.19), and swallow victims that are as large as themselves. (The prey in such cases is curled up in the predator's distensible, thin-walled stomach.) The fish *Malacosteus* (see no. 20, Figure 16.14) is possessed of one of the most incredible snap-trap jaws of any living thing. The bones of the lower jaw are attached to a hinge apparatus that makes it possible for its owner to shoot the jaw forward to nearly a third of the fish's total length. The jaw, armed with awesome fangs, is mostly skeletal bone; there is no flesh enclosing the cheeks or the floor of the mouth. The ability of this and many other deep-water fish to swallow large prey is thought to be related to the scarcity of prey. The strategy that they have adopted is to be able to eat everything they encounter, since opportunities to catch any prey at all in these sparsely populated waters are few and far between. These are not large fish; all known specimens are only a few centimeters long.

Some of the most highly adapted of all deep-water fish occur at a depth of 2,000 m. These are the anglers. Females are typically black, blob-shaped, and equipped with gaping mouths and "fishing" apparatus (Chapter 15). Victims are presumably lured to the waiting fish by means of the luminescent lure on the "fishing line," and swallowed. Certain anglers can eat large prey. *Melanocetus* females can swallow victims three times their own length (no. 17, Figure 16.14, depicts one that has recently done so). Male anglers are usually small and unequipped with fishing apparatus. They attach themselves to the females by biting them and then fusing their jaws and tongues with the tissues of the female. Thus they become parasitic. Nourished by means of a "placental" connection with the female's bloodstream, their only role thenceforth is to supply sperm for her millions of eggs. This permanent attachment is seen as another indication of the rare encounters between organisms in deep, dark waters. Having found a mate, the female takes permanent possession of him.

From 2,500 m and deeper, the production of light by organisms begins to dwindle. Certain creatures can be found generating light all the way to the bottom, but this is not as common a practice as it is between about 500 and 2,000 m. At greater depths, except for the faint luminescence of a few scattered organisms, the water is utterly lightless. Also beginning at about

16.19 A deep-sea dragonfish in action. Size of victim: about 6 cm.

2,500 meters, some groups of fish show a reduction in eye size. The pattern is not a simple one. Although some fish at these and deeper depths have rudimentary eyes or no eyes, others actually have enlarged and very effective eyes, even at the deep-sea bottom. Generally, more fish of the deepest sea have "ordinary" eyes than either reduced or exaggerated eyes. This surprising finding suggests that there is enough light, even in the deepest water, to make possession of eyes advantageous. (The light seen by these animals is almost certainly not natural sunlight; rather, it must be produced by luminous organisms.) By contrast, organisms that live in caves on land are often sightless.

Sporadic reduction in eye size is seen in other organisms than fish. Crustaceans tend to become sightless to a greater degree than the fish do, as depth increases. The small cephalopods tend to retain their eyesight, although one tiny pelagic deep-sea octopus (*Cirrothauma murrayi*) is sightless.

A trip to the bottom, then, takes us through realms where both light and food grow increasingly scarce.

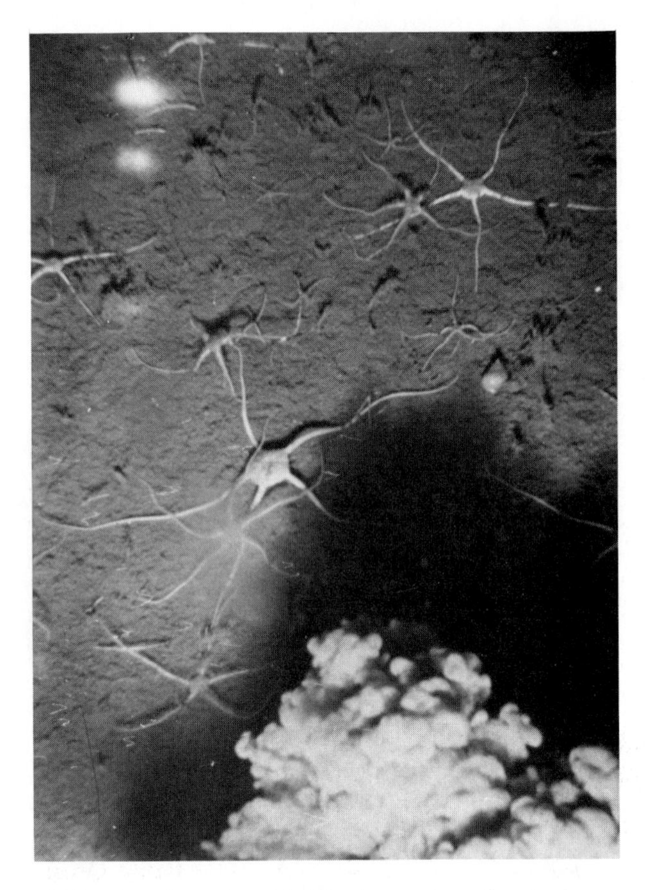

16.20 Fine silt, mud, and brittle stars on the seafloor in the San Diego Trough, at a depth of more than 1,000 m. At greater depths, most views of the abyssal sediments show mud with few or no organisms. Mud cloud at lower right was stirred by weight hitting the seafloor, triggering the camera.

The scarcity of light is apparently responsible for a host of adaptations that involve color, countershading, eye reduction or enlargement, and animal luminescence. The scarcity of food is reflected in the increased scarcity and progressively smaller sizes of animals at greater depths in the sea and by many bizarre adaptations for seizing and swallowing prey. On the deep-sea floor, food scarcity dominates the ecological scene and is paramount in determining the abundance and distribution of benthic animals. On the deep-sea bottom animal life is only about 1% as abundant as it is on the continental shelves.

The Deep-Sea Floor

The abyssal seafloor is one of the most widespread, uniform, and difficult-to-study habitats on Earth. The benthic substrate there consists of fine sediment, fields of manganese nodules, and the rocky flanks of mid-oceanic ridges, all spread over vast distances. The water is cold (0.5–2.5°C), temperature remains nearly constant over the entire year, pressure is immense, and sunlight never penetrates these dark depths. Photographs of the abyssal plains, taken in mid-latitudes, typically show a monotonous expanse of soft ooze marked with a few holes or low mounds and crisscrossed by a few tracks—and little or no life at all. Only about 10% of such photos show a vertical relief on the bottom of more than 3 cm, and only one photo in fifty shows an animal. When an animal is seen, it is usually a brittle star (Figure 16.20) or a sea cucumber. The rocky deep-sea bottom is often just as devoid of sessile life, although here the possibilities are better for attachment by such life forms as sponges, crinoids, barnacles, and bryozoans. The bottom strewn by manganese nodules is also very sparsely populated. Slow-moving and sessile organisms tend to be slightly more numerous on the deep bottom beneath areas of high surface productivity, and in a few areas (for example, the sea bed at about 3,000 m beneath Antarctica's Bellingshausen Sea) the organisms are as numerous as those on the deeper portions of continental shelves.

The poverty of the deep-sea environment is caused by a chronic shortage of food. That which arrives in the form of small sinking particles tends to settle uniformly over the vast, featureless abyssal plains, forming a dusting of thinly-distributed organic material that is best harvested by deposit feeders. But there is a problem. Much of this settled material is refractory—it cannot be digested by larger organisms. This is material that has escaped the attention of organisms in the water overhead for decades or centuries. Much of it has succeeded in sinking to the bottom only by virtue of the fact that it has no food value. Only bacteria can digest it. The bacteria, in so doing, become themselves food for the deposit feeders, which are among the most numerous large organisms on the deep-sea bottom. Sea cucumbers, brittle stars, and other deposit feeders eat the film of surface sediment and bacteria, digesting the sparse edible material and leaving the rest. Suspension feeders, while present, are much less numerous on the deep-sea bottom than deposit feeders, probably because the concentrations of digestible particles in the water near the bottom are so low. Organisms that live within the sediment are scarce, and by comparison with those of shallow sediments, extremely small. Shells are thin and fragile, perhaps because the pressure of the water makes it difficult to form and maintain them.

Mechanisms other than the slow, gentle "snowfall" of organic particles also operate to bring food to the seafloor. Turbidity currents, avalanching down the continental slopes and canyons, then gliding for hundreds of miles across the abyssal plains before finally dropping their loads of fine suspended particles on the bottom, must transport significant pulses of edible material from shallow water to the abyssal depths. Life on the abyssal plains is slightly more abundant close to the continents than it is toward the oceanic centers, perhaps as a result of these flows. Also, the occasional sinking of dead fish, whales, and other large food items apparently occurs frequently enough to support a sea-bottom fauna that survives by finding and eating these items.

Buckets of dead fish accompanied by cameras have been lowered to the bottom at many deep locations, and many fish, sharks, crustaceans, and other animals quickly appear on the scene to eat the bait. Some of these (such as the sablefish and the sleeper shark) are species known from shallow water; many are exclusively deep-water species. These animals begin arriving within half an hour of the appearance of the bait on the bottom. They devour it within a few hours and then disperse. This suggests that large edible items drop into their neighborhoods with dependable regularity. Some of them (for example, certain sharks) are so rigidly constrained to searching for their food on the bottom itself that, if the bait bucket were suspended a mere meter above the sediments, they could not find it. Not all deep-sea bottoms are populated by organisms with these feeding habits. Surprisingly, bait buckets lowered to the bottom under productive surface areas may sit for very long periods before scavengers show up. Motile scavengers appear sooner and in greater numbers and diversity under surface areas of lesser productivity.

The bait bucket experiments suggest that there may be more large animals in the deep sea than we now realize. The largest representatives of some groups live in deep water. Ostracod crustaceans, for example, are usually smaller than a grain of rice, but one deep-sea species is the size of a cherry. The largest isopods (20 cm long) inhabit the deep sea, as do the largest mysids, sea pens, and hydroids. Fish captured at great depths, however, are mostly very small. In recent years, a large shark of a species previously unknown to biologists was caught at a depth of 150 m near Hawaii. (It had grabbed the parachute sea anchor of a research vessel and become entangled in it.) This shark, informally known as "Megamouth," was about 3.5 m long. Although it was taken near the surface, it had many features of deep-sea animals, including a flabby body and a luminous mouth. Other large deep-sea fish have been sighted but never captured. Pioneer William Beebe, who had himself lowered to nearly 900 m in a steel sphere with windows in 1934, reported one such sighting in his book, *Half Mile Down*. Beebe saw two long fish, each with a row of blue belly lights, enormous eyes, luminous fangs, barbels trailing red and blue lights, and dorsal and anal fins at the rear of the body, glide through the searchlight beam 2 m from his bathysphere. He recognized them as dragonfish—2 m long. These and other clues suggest that more large organisms await discovery in the deep sea.

The fish of the deep-sea floor show a strong tendency to assume eel-like forms. Although some of them are true eels, deep-sea fishes of other taxonomic groups also have long, slender eel-like bodies. Many are members of the families Macrouridae and Brotulidae. Both macrourids and brotulids have a "rat-tailed" look. The head is large and often armored, and the body tapers back to a narrow pointed tail. The dorsal and anal fins are often long and continuous. Macrourids usually have very large eyes; in brotulids, the eyes are smaller and more reduced (or even absent). The fish of both groups have sensitive, well-developed lateral line systems and are apparently able to detect slight water movements created by prey. Both types of fishes also have excellent hearing. Thus equipped, macrourids are able to detect the occasional flashes of luminescence that occur in the dark water, and both macrourids and brotulids are quite able to find prey by means of other senses than sight. They feed on sponges, crustaceans, worms, and other invertebrates that they find in the benthic ooze.

Hydrothermal Vents

Among the most remarkable findings of recent decades is the discovery of hot-water vents and their associated organisms at great depths in the Pacific Ocean. Seafloor hot springs were first sought by investigators from Woods Hole Oceanographic Institute (WHOI) in an area near the Galápagos Islands in the mid-1970s. There, photos of the bottom taken by a camera on a sea-bed sled (the ANGUS system) showed clusters of large clams and yellow and orange stains on the basaltic rocks. Temperature measurements revealed that the bottom water was a few degrees warmer than would be expected at a depth of 2,700 m, and water samples opened on shipboard released quantities of hydrogen sulfide ("rotten egg") gas. Exploration of the Galápagos site with the submersible *Alvin* proved that warm water was issuing from the sea bottom. Explorations by WHOI and French investigators at a site to the north (the "21° North Site") in 1979 revealed more spectacular thermal vents, ones in which superheated water was gushing from "chimneys" of mineral deposits that were up to 20 m high. The hot water issuing from the chimneys (and elsewhere) had temperatures near 350°C—and only the gigantic pressure of seawater overlying these vents prevented the superheated water from exploding into steam. Subsequent to these early discoveries, warm- or hot-water vents have been located in the Sea of Cortez and at other sites on the Pacific coast of North America. It seems likely that many more will be found.

Hydrothermal vents may consist of simple holes in the sea bottom or of "chimneys" of mineral deposits from which dark, mineral-laden waters gush in con-

16.21 Crabs, clusters of tube worms, and other organisms near a hydrothermal vent in the Guaymas Basin. The white cylinders are tubes. Each is inhabited by a beard worm (phylum: Pogonophora) that displays a bright-red plume at the entrance. A temperature probe is seen at the lower right; chemical dye streaming upward at the left shows a warm current from the vent. (James J. Childress, University of California, Santa Barbara)

tinuous plumes, or they may consist of mounds of mineral material with hot water escaping through the pores. The water issuing from the vents is rich in sulfide compounds, which are toxic to most animals. The vents are typically surrounded by thickets of tube worms, clusters and patches of large mussels and clams, crabs, mats of bacteria, tangles of "spaghetti worms," and other animals. Needless to say, the striking abundance and large sizes of these organisms suggest that these are not typical food-starved deep-sea communities. Away from the vents, life dwindles to its typical deep-sea scarcity so abruptly that the pilots of submersible vehicles, descending in the vicinity, can easily find the vents by going "up the crab gradient"—in the direction in which crabs become more abundant.

The source of food for the vent communities and the reasons for the animals' ability to live in the warm toxic vent environment were not immediately obvious at the time of their discovery. Compounding the mystery was the discovery that the large tube worms had no mouths or digestive systems at all! Intrigued by such puzzles, biologists have studied the vent communities as carefully as the limited availability of specimens allow, and they have come to the realization that these are like no other animal communities on Earth.

The vent animals rely on **chemosynthesis** (rather than photosynthesis) to provide the foodstuff that serves as the basis for their food webs. This in itself is astounding; all other known communities of animals depend on plant photosynthesis to sustain them. The chemical raw material used by the chemosynthesizers is provided by the sulfide-rich seawater that gushes or seeps from the vents. Sulfide-oxidizing bacteria use the sulfides for their own metabolism, and they multiply and prosper. In turn, they become food for the larger vent animals that are able to collect and eat them. The most prominent vent animals (the large worms and clams) live in mutual partnership with these bacteria. The "beard worms" that dominate some vent neighborhoods (Figure 16.21) are filled with a core of special tissue in which sulfide-oxidizing bacteria reside. The worms have astonishing enzymes that enable their circulatory systems to safely deliver highly toxic sulfides to these bacteria. The bacteria metabolize it, and the worms use the excess metabolites created by the bacteria. This, at least, explains how a worm about 1.5 m long and as thick as a garden hose can survive without a digestive tract. The large vent clams also contain patches of sulfide bacteria. Evidently, they have established similar partnerships with the bacteria.

The beard worms, as mentioned in Chapter 14, are members of an obscure worm phylum (Pogonophora) that was previously known mainly from tiny, string-like specimens dredged from typical cold deep waters. A few larger specimens were known before the discovery of the vent communities.

There are many unanswered questions about the hydrothermal-vent communities. For example, it is evident that the vents eventually stop producing hot water. In this event, the chemosynthetic bacteria run out of "food," and the community must go elsewhere or starve. New vents start up as existing ones fade away, and the community must have a means of colonizing new sites, perhaps hundreds of meters or many miles away. Exactly how they do it is not known.

These surprising communities are atypical of the deep sea. Lush, crowded, dependent on chemosynthesis, independent of the rest of the earth's ecosystems except for the oxygen they respire, and scattered as oases of plenty across the sparsely inhabited abyssal depths, they represent a surprising alternative to the cold, austere landscape of the deep sea.

Although much is now known about the deep sea, it is clear that much remains to be learned. The organisms themselves have revealed that the deep-sea realm is not the eternally unchanging still-water environment that was once envisioned. The presence of currents at the sea bottom has often been indicated by the bent postures, in photos, of such suspension feeders as glass sponges, crinoids, and sea pens. Certain deep-sea isopods are known to breed seasonally—a fact that suggests that seasonal changes in the abyssal water are detectable by them and are used as a cue.

SALINITY FLUCTUATIONS: LIFE IN ESTUARIES

The shallow depths and circulation patterns of typical estuaries guarantee that these will be productive waters. In an ordinary estuary with significant river flow, a deep saline current enters from the ocean and moves landward along the bottom, compensating for a reverse flow of diluted seawater moving seaward at the surface. This system acts as a nutrient trap. Particles containing nutrients sink from the surface, enter the bottom current, and move landward, thus keeping the nutrients in the estuary. The deep flows from the ocean, as well as the freshwater river flows, bring new nutrients to the estuary. Nutrients therefore accumulate in estuaries and tend to remain there. This, combined with the shallowness of the water, makes estuaries some of the most productive waters on Earth.

However, estuaries are harsh environments because the salinities are always changing. The salinity at any particular point on the bottom changes daily and seasonally with tidal flows and changes in rainfall and river runoff. These temporal changes are superimposed on a permanent salinity gradient, from the ocean through the estuary and up the river. Many marine species cannot cope with this drastically fluctuating regime, and marine species diversity decreases rapidly from the mouth toward the head of an estuary. From the full complement of sea anemones, seaweeds, limpets, sea stars, nudibranchs, sea urchins and other familiar organisms seen at the coast, the shore fauna becomes progressively diminished to a handful of barnacles, mussels, crabs, and other resistant creatures at mid-estuary. At a point in the estuary where average annual salinity is about 5‰ or less, the diversity of species reaches a minimum (about 30% of the diversity found in coastal waters). Continuing upriver from this point, freshwater species begin to take over, and the diversity of species begins to increase. Thus estuaries are zones where nutrient levels are high, and harsh changes in salinity restrict the number of kinds of organisms that can live there. For those animal species that can survive an estuary's fluctuations in salinity, there are food resources in abundance and a general scarcity of predators or competitors of other species. For these reasons, estuarine organisms often tend to be few in species but many in number.

Interactions between the organisms themselves can make the estuarine environment a tough one to live in. One important species in this regard is the edible blue crab, *Callinectes sapidus*, of the Chesapeake Bay estuary. Males of this species reach a width of about 20 cm; females are somewhat smaller. They are fast, active swimmers and are significant predators of clams, worms, and other benthic organisms. One assessment of their impact was conducted by Robert Virnstein of the Virginia Institute of Marine Science. Virnstein placed crab exclusion cages over sections of the shallow, muddy bay bottom. The cages quickly became choked with burrowing benthic animals, including, in one case, so many clams of the species *Mulinia lateralis* that they were piled two layers deep. Studies such as this indicate that predation by blue crabs keeps the prey organisms scarce, that the diversity of benthic species is reduced as a result of the crabs' activities, and that most prey organisms do not survive long enough to grow to their full size. Thus the crabs (in company with predaceous fish) dominate the bottom communities and help determine their makeup.

Like many other organisms, blue crabs use the estuary for only part of their life cycle. Their eggs and early stages of their larvae cannot survive the low-salinity regime of an estuary. Thus female crabs carrying masses of fertilized eggs leave the bays and swim out to sea, where their eggs hatch. The larvae then work their way back to the estuaries, where they grow and mature. This life cycle—offshore spawning, movement of juveniles back into estuaries, then (often) departure of the mature organisms for life out in the open sea—is one of the most common patterns in the U.S. south Atlantic and Gulf coasts. The commercial pink, brown, and white gulf shrimp all spawn offshore. Their larvae, like those of the crab, must then find their way into enclosed inshore waters of low salinity for feeding and growth, before eventually returning to the sea (Figure 16.22). Menhaden, the fish of greatest commercial importance along the Gulf Coast, similarly spend their early lives in estuaries. Fully 97% of the catch of Gulf Coast marine species (which constitutes, in turn, 40% of the entire United States commercial fish catch, including Alaska) consists of organisms that make use of estuaries at some time in their lives.

The reduced salinity of estuaries makes it possible for organisms to avoid predators by living there. This is one benefit obtained by the larvae of marine species that migrate into estuaries to complete their life cycles. Many invertebrate predators (arrow worms, certain jellyfish, and ctenophores, for example) do not penetrate waters of low and fluctuating salinity and cannot follow their prey there. Oysters evade considerable predation by living in estuaries as adults. These shellfish can survive reduced and fluctuating salinities, whereas the sea stars and drills that eat them cannot. Adults do best in salinities between 10 and 30‰. Although oyster larvae attach themselves to shelly or

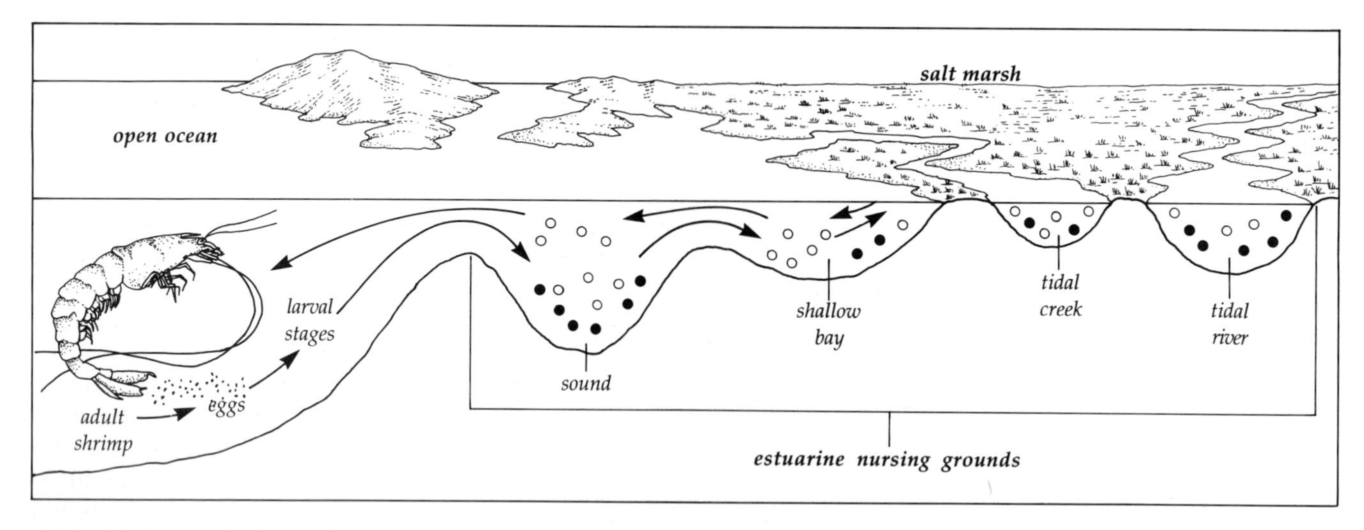

16.22 Life history of a shrimp that uses estuaries as a nursery grounds. Adult shrimp spawn offshore, and the larval stages move shoreward. They find protection and the food needed for rapid growth during the earlier stages (filled circles) and later stages (open circles). As they mature, the shrimp move back into deeper waters near the shore, then they move to the open ocean.

stony bottoms virtually everywhere, those that land on appropriate bottoms in water of reduced salinity are more likely to survive to adulthood. Larvae of the Virginia oyster, *Crassostrea virginica*, have been observed to settle on the bottom during ebbing tides and to rise and swim during incoming tides—a strategy that improves their chances of remaining in the estuary until the time comes for attachment to the bottom. The fact that lowered salinities keep many predators at a distance is occasionally demonstrated by episodes during which an estuary becomes more saline than usual. In Mobile Bay, droughts occasionally allow high-saline water from the gulf to flood the bay. This brings invasions of the oyster drill, *Thais haemostomum*, which devastate the oysters.

Like the oysters, certain barnacles also evade their enemies by settling far up the reaches of estuaries, where salinity changes exclude the enemies. It is often true that barnacles, oysters, and other organisms that employ this strategy can survive and reproduce in undiluted seawater (indeed, often better than in estuarine water), but the abundance of their enemies there reduces their numbers in the open sea. As a result they are more numerous in estuaries than elsewhere.

The abundance of nutrients in estuaries and the fact that nutrients are continuously restored to the surface by rising water usually result in high plant productivity. The organisms that spend parts of their lives in estuaries exploit this productivity and owe their own abundance to it. A representative figure for productivity in a temperate estuary (in Georgia) is 300 gmC/m^2-yr, higher than that of most oceanic sites.

INTERACTIONS AMONG ORGANISMS: LIFE ALONG THE SHORE

Communities of the shorelines are affected by plant productivity, which in turn is affected by salinity and temperature conditions, as are all marine communities. Shoreline communities also reveal particularly striking examples of ways in which interactions between the organisms themselves affect the makeup of the community. Unlike the situation in Chesapeake Bay, where the activities of the blue crabs give the whole bottom a rather featureless and uniform appearance, interactions among organisms along the shore result in zones, stripes, and patches of organisms that are obvious even to a casual observer.

Intertidal shorelines commonly exhibit a striking zonation of the plants and animals that occupy the shore. Here, organisms are often arranged in distinct belts or zones that run parallel to the shore. On certain rocky shores, four fairly well-defined zones of organisms can be found (Figure 16.23). Isopods, littorine snails, and certain limpets inhabit the uppermost zone; below them we find shore crabs, the brown alga *Fucus*, densely packed barnacles, and more limpets of other species. Lower still, mussels, goose barnacles, chitons, and predatory snails are prominent. The lowest zone, exposed only rarely by the lowest tides, may consist mainly of seaweeds, sponges, sea slugs, anemones, and other creatures. Generally speaking, the organisms of the lowermost belts are least resistant to exposure to air, sunlight, and rain, whereas those of the uppermost zones are most resistant to exposure. It is often the case that the uppermost limit of each

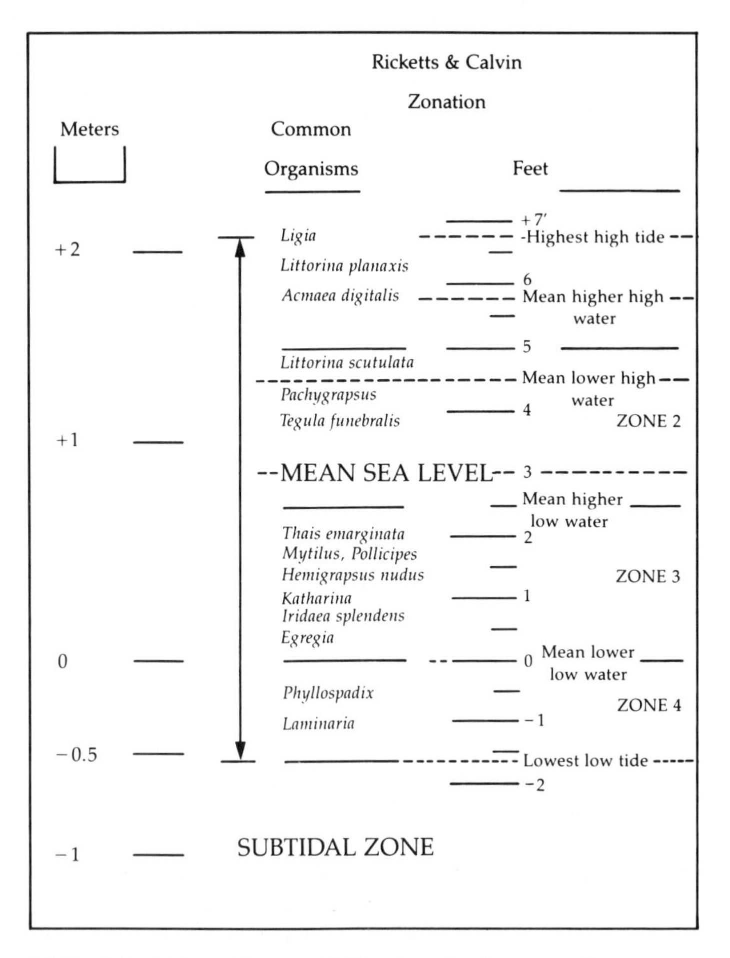

16.23 Intertidal zonation on a California rocky shore, according to the Ricketts and Calvin system.

species' range in the intertidal zone is set by some physical factor (usually desiccation), whereas the lower limit of the zone is set by some biological interaction such as predation or competition.

Because intertidal shores have diverse topography, the action of physical and biological factors is not uniform. The idealized arrangement of organisms in parallel zones is often broken up into a mosaic of patches (as in Figure 16.24). Wetting by waves is an important protection against desiccation. Where exposed rocks are washed by surf, intertidal organisms range higher onto the shore than in sheltered nearby areas. Algae can survive at higher elevations on the shaded north sides of boulders than on the sunny south sides. In certain exposed locations, surf is so heavy that organisms that would dominate their respective zones under less strenuous conditions are absent, leaving others to take their place. These factors tend to blur the boundaries of the intertidal zones, to shift the zones higher or lower on the shore, and to broaden or narrow the zones. Nevertheless, the belts and patches of organisms living together on intertidal rocks are often very sharply defined. It is not uncommon to find a densely packed belt of seaweeds, mussels, or other intertidal organisms ending abruptly and giving way to another belt of organisms living just below them.

The plants and animals themselves are partially responsible for this sharp zonation, as is illustrated by the following examples. Adult barnacles of the species *Chthamalus stellatus*, found in Millport, Scotland, live only in the highest intertidal zone. Their larvae settle everywhere on the intertidal and subtidal shore, however, colonizing many zones in which adult *Chthamalus* barnacles are never found. In a now-classic study, researcher Joseph Connell demonstrated that the *Chthamalus* barnacles can actually survive to adulthood at lower locations—but only if they are protected from their competitor, a second barnacle, *Balanus balanoides*. When living together, the larger *Balanus* barnacles invariably overtop and crush the smaller *Chthamalus* as they grow. *Balanus*, however, cannot withstand exposure as well as *Chthamalus*. Thus *Chthamalus* survives in the only place where its competitor cannot establish itself, the highest intertidal zone. The lower limit of *Chthamalus*'s range is therefore set by a biological interaction—competition with *Balanus*—and its upper limit is set by a physical factor—desiccation (Figure 16.25).

In a comparable study in Australia, researchers P. Jernakov and A. J. Underwood of the University of Sydney showed that 2 species of limpets were prevented from moving lower on intertidal rocks by the growth of dense seaweed there. The seaweed prevents the growth of the diatoms that constitute the usual food of the limpets. The limpets are unable to eat the seaweed. Any limpets that stray into the seaweed zone, therefore, starve. Thus the lower limit of the limpets' range on this exposed shore is set by a biological agent—the seaweed—whereas the sharp upper limit of the seaweeds' range is set by a physical factor—desiccation. The seaweeds are confined to a zone of their own, reaching their lower limit just below the level of lowest tide. They are prevented from growing at a lower zone by the heavy grazing of sea urchins. The urchins are prevented from moving higher up the shore by their vulnerability to exposure. Again, a physical factor is responsible for the shoreward edge of the urchin zone, and a biological interaction sets the seaward limit of the seaweed zone.

The arrangement of organisms in belts parallel to the shore sometimes occurs below the low-tide mark, as well. For example, sea pansies (related to hydroids and sea anemones) are mostly found in a narrow zone running parallel to the surf-swept beaches of the California coast, at depths of about 3 to 5 m. Immediately seaward of the sea pansies is a dense belt of sand dollars, also running parallel to the shore. The pansies

(a)

16.24 A Pacific coast tide pool and intertidal shore. (a) Key. (b) Microcosmic view. (c) An intertidal shore. (L. James Grattan)

1 *busy red algae,* Endocladia
2 *sea lettuce, green algae,* Ulva
3 *rockweed, brown algae,* Fucus
4 *iridescent red algae,* Iridea
5 *encrusting green algae,* Codium
6 *bladderlike red algae,* Halosaccion
7 *kelp, brown algae,* Laminaria
8 *Western gull,* Larus
9 *intrepid marine biologist,* Homo
10 *California mussels,* Mytilus
11 *acorn barnacles,* Balanus
12 *red barnacles,* Tetraclita
13 *goose barnacles,* Pollicipes
14 *fixed snails,* Aletes
15 *periwinkles,* Littorina
16 *black turban snails,* Tegula
17 *lined chiton,* Tonicella
18 *shield limpets,* Acmaea pelta
19 *ribbed limpet,* Acmae scabra
20 *volcano shell limpet,* Fissurella
21 *black abalone,* Haliotis
22 *nudibranch,* Diaulula
23 *solitary coral,* Balanophyllia
24 *giant green anemones,* Anthopleura
25 *coralline algae,* Corallina
26 *red encrusting sponges,* Plocamia
27 *brittle star,* Amphiodia
28 *common starfish,* Pisaster
29 *purple sea urchins,* Strongylocentrotus
30 *purple shore crab,* Hemigrapsus
31 *isopod or pil bug,* Ligia
32 *transparent shrimp,* Spirontocaris
33 *hermit crab,* Pagurus, *in turban snail shell*
34 *tide pool sculpin,* Clinocottus

(b)

(c)

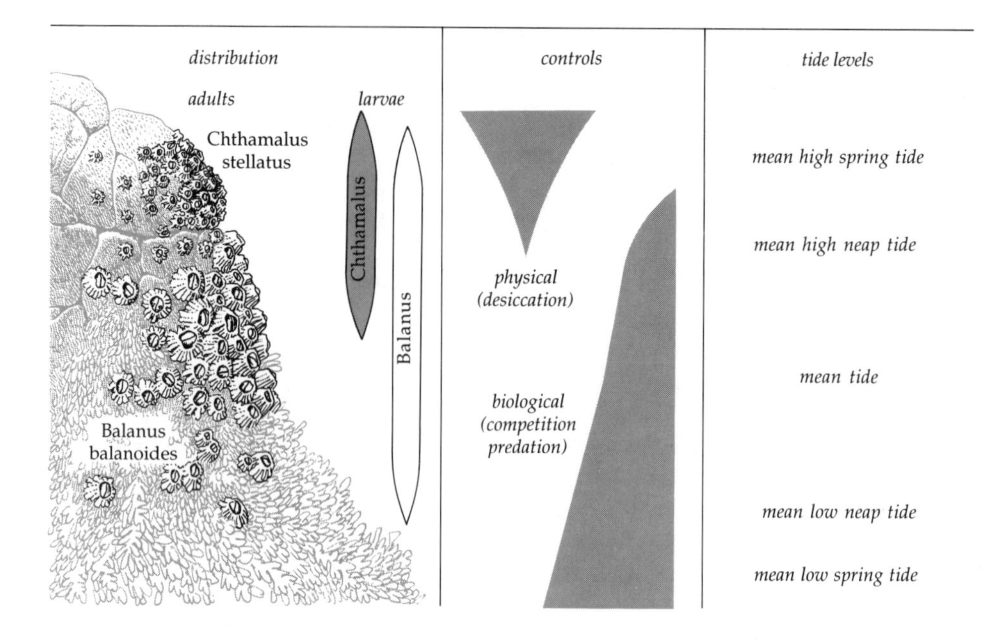

16.25 Adult barnacles, Chthamalus stellatus, *occur high in the intertidal zone; adults of* Balanus balanoides *occur in lower zones. Larvae of both barnacles settle over a wide range of depths.* Balanus *eliminates* Chthamalus *in competition by crowding in the lower zone, but it cannot survive desiccation in the upper zone as well as* Chthamalus. *The result, after mortality takes its toll of juveniles each year, is a segregation of the populations of adult barnacles into two zones with little overlap.*

are prevented from living closer to shore by the turbulence of the water and are severely restricted in their ability to inhabit deeper water by competition with the sand dollars. The shallow water in which the sea pansies can live is too turbulent for sand dollars. Like the *Chthamalus* barnacles, the sea pansies are hemmed in—confined to a marginal shoreward zone where wave action prevents their main competitors, the sand dollars, from establishing a foothold. In this situation, as in the intertidal zone, the shoreward limit of each species is set by a physical factor, while the seaward limit of the sea pansies is heavily influenced by a biological interaction.

DEEP-OCEAN ZONES

In deeper water, inferences about zones must be based mainly on the catches of grab and dredge samples. For oceanic depths, the zones illustrated in Figure 16.16 reflect researchers' early impressions that the organisms of different depths were characteristically and recognizably different from each other. Additional research usually tends to blur the boundaries between the zones, however. For example, research in the 1950s by Soviet investigators and others prompted them to propose that there is a distinct difference between organisms of abyssal and hadal depths. The hadal organisms were thought to be less likely to have eyes, more likely to be colorless, and less heavily calcified than their abyssal counterparts. By the 1970s, it was evident that few dependable differences existed between the organisms of the two depth zones. Among isopod crustaceans, for example, virtually all of those at both abyssal and hadal depths in two locations were shown to be eyeless by researchers R. J. Menzies, R. Y. George, and G. T. Rowe. These investigators showed that most (if not all) isopods from both depths are white, and those at abyssal depths have skeletons that are as fragile as those at hadal depths. Thus they concluded that the differences between organisms of the two zones are not as significant as earlier investigators believed. This has proven to be true of all of the deep-sea zonal boundaries. The mesopelagic zone corresponds to the depth over which the permanent ocean thermocline is found, and the hadal depths occur in ocean trenches, which are as isolated from each other as isolated mountains are on a prairie. It is possible that the differences in organisms from some of the zones may be traceable to unique hydrographic properties of those depths.

BROAD GEOGRAPHIC PATTERNS: THE LEGACY OF PREHISTORIC EVENTS

On a broad scale, oceanic processes of the past and present serve to separate whole communities of organisms. At Cape Hatteras, for example, the Gulf Stream leaves the continental shelf and moves out over deep water. The warm current supports an association of warm-temperate organisms to the south of the cape, whereas cold water meandering down the coast from the north supports cold-adapted organisms to the north. Oceanic conditions at the cape therefore make it a boundary between communities of different organisms. *Mytilus edulis*, for example, a mussel of the mid-Atlantic coast, is unable to establish permanent colonies south of the cape. About 28% of the decapod crustaceans that live on the south Atlantic coast reach the northward limits of their distributions at the cape. (The rest reach their northward limits at various other places to the north or south, but nowhere do so many of them reach the limit of their distribution as at Cape Hatteras.)

The boundary is not a sharp one. Many southern organisms move north of it during the summer, and many organisms are permanently established both to the north and to the south of the cape. The change in the makeup of the marine communities is more abrupt than at any point hundreds of miles to the north or south, however, thanks to the configuration of the coast and the trajectory of the Gulf Stream. A similar situation occurs at Point Conception, in California, which marks the northernmost limit of many warm-adapted species and the southernmost limit of many cold-adapted species. Here the faunal boundary is maintained by the convergence of a cool northern current and a warm southern one.

It is usually true that more species of marine organisms live in warmer waters than in colder seas. There are exceptions: For example, more species of seastars live in the cold coastal waters of British Columbia than anywhere else in the world, warmer or colder. In general, however, diversity of species is greater in the tropics. On the Washington state coast, for example, there are 6 families of crabs containing 31 species. One of these families, Xanthidae, is represented by only 1 species. This family alone is represented by about 111 species in Hawaii, and the Hawaiian waters are populated by 9 additional crab families that collectively contain an additional 99 species of crabs. About 80 taxonomic families of fish inhabit coastal Washington and British Columbia; the comparable number for Hawaii is about 150.

Exactly why warm seas should support so many more species is much debated by marine ecologists. One hypothesis is that tropical organisms were relatively unaffected by the global climate changes of the Pleistocene, whereas polar regions (which were much warmer in early Eocene times) were decimated by the ice age. In this view, there are fewer cold-water species because not enough time has elapsed to allow marine organisms to evolve and fully repopulate the decimated polar waters. It is also possible that high temperatures permit behaviors that low temperatures prohibit. For example, warm-temperate sand beaches are inhabited by myriad small clams (genus: *Donax*). These clams lie buried in surf-swept sand. When a wave breaks, they actively wiggle to the surface, feed while the water sloshes upward, then quickly rebury themselves before the receding wave can carry them down the beach. As poikilothermic animals, they can only succeed at this because the warm water allows them to be frantically active. At lower temperatures, they would be too sluggish to make the rapid movements needed to exploit the surf in this way. Sand beaches of cold seas have no such animals. The awesome running and swimming speeds of tropical crabs, compared with the slow movements of their counterparts in cold climes, also suggest that certain opportunities for active escape and feeding simply don't exist in cold regions. These and other factors may explain why warm seas are so much more richly endowed with species than the colder seas.

On a grand scale, life in the tropical seas is subdivided into four "zoogeographic provinces." These are the East Atlantic, the West Atlantic, the East Pacific, and the Indo-West Pacific. The organisms of each province are generally (although not always) found only in that province. The Indo-West Pacific is the biggest and richest by far; it stretches from the East Africa coast to Hawaii and Polynesia. The richest diversity of species in the province (and indeed in all of the oceans of the world) is found in an area between the Philippines and Indonesia, but some of the species of the Indo-West Pacific can be found ranging across the entire vast region. The "lucifer shark," for example, a small shark less than 1 m long, inhabits depths between 300 and 600 m, all the way from Hawaii to the Red Sea.

Surprisingly, the marine life along the eastern Pacific shores of Central America has more in common with that of the Caribbean than it does with that of the western Pacific. The similarity of eastern Pacific and Caribbean faunas results from several causes. The tropical Atlantic and Pacific were connected, as recently as Pliocene times, by a seaway through the Isthmus of Panama. This sea connection allowed marine life to move freely between the two oceans. Since the closure of this seaway, the West Atlantic and East Pacific provinces have been separated by the Isthmus of Panama. The East Pacific has also been isolated from the Indo-West Pacific area by the vast open water of the eastern Pacific Ocean, which has served as a barrier to the movement of larvae and adult shallow-water organisms between the eastern and western Pacific. This "East Pacific Barrier," as it is called, together with the isthmus, has isolated the eastern Pacific to such an extent that species unique to that province have evolved. But since they are descendants of organisms that once ranged across the area occupied by the present-day isthmus, they are today much more closely related to their cousins in the Caribbean than to the tropical marine life of the distant Indo-West Pacific province.

The organisms of temperate and boreal seas are also loosely segregated into zoogeographic provinces. These, too, result from evolution and climatic and geographic changes of the past. The borders of these provinces, and of subprovinces within them, are usually defined by the kinds of hydrographic conditions seen at Point Conception and Cape Hatteras.

Why are the tropical marine organisms of the Caribbean so much less diverse than their counterparts in the tropics of the Indo-West Pacific? We began this chapter with this question and must look to events of the past to provide a glimpse of the answer. Shortly after the Isthmus of Panama arose to separate the Caribbean from the Pacific, the earth experienced the great Pleistocene ice age. The Pacific, isolated from the Arctic Ocean by the Siberian and Alaskan land masses, was sheltered from its effect. The surface waters of the Atlantic are broadly connected with those of the Arctic Ocean, and they experienced devastating cooling. The Caribbean tropical organisms, trapped in the cul-de-sac formed by the Gulf of Mexico and adjacent waters with no southerly route for retreat, were mostly wiped out by the chill. No such disaster overtook their contemporaries in the Indo-West Pacific tropics. The last major glaciation ended only a few thousand years ago; evolution has not had sufficient time to rebuild Caribbean species diversity back to levels comparable with those of the Pacific.

Evolution, the movements of organisms, the opening and closure of seaways, events such as the Pleistocene glaciation, and other phenomena of the distant past have left their distinctive marks on the distributions of modern marine organisms in these and in other ways.

SUMMARY

All marine communities are powerfully influenced by constraints on productivity, certain limitations imposed by properties of seawater and the sea, and interactions among organisms. In addition, their species makeup and diversity, their geographic distributions, and the physiological abilities of the organisms reflect events and evolution in prehistoric time. No two communities are exactly alike; their histories differ, the climatic and oceanic constraints of different regions are not identical, and their species compositions are not identical. Marine communities are complex, and their responses to disturbance are hard to predict. They are composed of species with strikingly different forms and functions, woven together in a bewildering tangle of cooperative, competitive, and predatory relationships, and those communities at the sea surface, at least, are influenced by weather events that are themselves only broadly predictable.

One moral of this story is that it is not easy to predict exactly how any marine community will respond to such natural or human disturbances as an El Niño or overfishing. Another is that no two marine communities are likely to respond in exactly the same way to similar disturbances. The changes in species populations and diversity following an instance of pollution in an East Coast estuary, for example, are likely to be different in some (perhaps major) ways from the changes that would result from a similar event in an apparently similar estuary elsewhere. In practice, every marine community needs to be studied; lessons learned in one region are likely to apply only in part to the marine shores in another.

Certain practical generalizations can be made, of course. A major disturbance of a marine community, such as an oil spill, will certainly have predictable and widespread lethal effects. Perhaps of more immediate importance is the evidence that marine communities have a very limited ability to provide food for an overpopulated Earth. Return to the example of the plant productivity needed to provide one can of tunafish and you will find that 99.99% of the food energy first available in plants has been lost by the time a tuna reaches a human consumer. This staggering loss results from the fact that these fish are fifth-level consumers. Cattle are second-level consumers, and land plants are big enough for people to consume them directly. Thus much more of the productivity of the land is directly available to us. For this reason, and because phytoplankton are too small and thinly dispersed to harvest economically, land agriculture and ecosystems have much more potential for supplying food than the sea does. The sea's contribution in quality is important—it provides about 10% of the crucially important animal protein consumed by people—but in quantitative terms, its bounty is very small. Fully 99% of human food comes from the land; only 1% is taken from the sea.

LIFE IN THE SEA: A NONSCIENTIFIC PERSPECTIVE

Aristotle (348–322 BC) surmised, some twenty-two centuries earlier than scientists of the Western World, that sponges were simple forms of animal life. His early observations of the marine life of the Mediterranean are among the first on record in Western civilization. Many people have studied the life in the sea since Aristotle's time and have contributed to the great body of knowledge that is presently available. As microscopes, radioactive analysis, improved chemical techniques, and other tools became available throughout the development of Western civilization, the studies became increasingly sophisticated and often increasingly large in scope. The most comprehensive modern research now involves teams of researchers and technicians and equipment, ranging from ships to submersibles to satellites and computers. The preceding chapters have summarized some of the extensive scientific knowledge of sea organisms and marine biotic systems that has been accumulated by these studies. Chapters to follow discuss the vulnerability of marine life to increasing human encroachment on the sea.

Not discussed in this text, and properly the subject of a different approach to the study of marine life, are the deep cultural and aesthetic ties between people and the living things of the sea. Marine creatures add an aesthetic dimension to life that is not comparable to life on land. Whales, dolphins, sea lions, sharks, giant squids, crabs, and fish all have important roles, aside from any commercial or scientific significance, in the present-day lore of human beings and in the heritages of peoples who have lived close to the sea. Their beauty, their mystery, their very existence contributes to the quality of human life, and to the extent that they are endangered or have been exterminated, human experience is impoverished. Awareness of their plight and significance, respect for their contributions to our lives and for their own rights to existence, and a determination to restrain human population growth, frivolous uses of chemicals, and other excesses that degrade the marine world are all necessary elements of their preservation—and our own.

FURTHER READINGS

Barnes, J. N. *Let's Save Antarctica.* Victoria, Australia: Greenhouse Publications, 1982.

Cushing, D. H., and J. J. Walsh. *The Ecology of the Seas.* Philadelphia: W. B. Saunders, 1976.

Delwiche, C. C. "The Nitrogen Cycle." *Scientific American,* September 1970, pp. 136–46.

Ekman, S. *Zoogeography of the Sea.* London: Sidgwick and Jackson, 1953.

Endean, R. *Australia's Great Barrier Reef.* St. Lucia, Australia: Univ. of Queensland Press, 1982.

Green, J. *The Biology of Estuarine Animals.* Seattle: University of Washington Press, 1968.

Gunther, K., and K. Deckert. *Creatures of the Deep Sea.* New York: Scribner's, 1956.

Heatwole, H. *A Coral Island: The Story of One Tree Island and Its Reef.* Sydney, Australia: Collins Publ. Co., 1982.

Herring, P. J., and M. R. Clarke. *Deep Oceans.* New York: Praeger, 1971.

Idyll, C. P. *Abyss: The Deep Sea and the Creatures That Live in It,* rev. ed. New York: Thomas Y. Crowell, 1971.

Jessop, N. M. *Biosphere: A Study of Life.* Englewood Cliffs, N.J.: Prentice-Hall, 1970.

Ketchum, B. H. *Ecosystems of the World. 26: Estuaries and Enclosed Seas.* New York: Elsevier Scientific Publishing Co., 1983.

Lauff, C. H. (ed.). *Estuaries.* Publication 83. Washington, D.C.: American Association for the Advancement of Science, 1967.

Levinton, J. S. *Marine Ecology.* Englewood Cliffs, N.J.: Prentice-Hall, 1982.

Lewis, J. R. *The Ecology of Rocky Shores.* London: The English Universities Press, 1964.

Marshall, N. B. *Deep Sea Biology: Developments and Perspectives.* New York: Garland STPM Press, 1980.

Raymont, J.E.G. *Plankton and Productivity in the Oceans.* New York: Pergamon Press, 1963.

Russell-Hunter, W. D. *Aquatic Productivity.* New York: Macmillan, 1970.

Stanley, S. M. *Earth and Life Through Time.* New York: W. H. Freeman, 1986.

Tait, R. V., and R. S. DeSanto. *Elements of Marine Ecology.* New York: Springer-Verlag, 1972.

Teal, J., and M. Teal. *Life and Death of the Salt Marsh.* New York: Audubon/Ballantine, 1969.

Thorson, G. *Life in the Sea.* New York: McGraw-Hill, 1971.

Vernberg, W. B., and F. J. Vernberg. *Environmental Physiology of Marine Animals.* New York: Springer-Verlag, 1972.

Wiens, H. J. *Atoll Environment and Ecology.* New Haven, Conn.: Yale University Press, 1962.

Nina Lisowski

No pencil has ever yet given anything like the true effect of an iceberg. In a picture, they are huge, uncouth masses, stuck in the sea, while their chief beauty and grandeur— their slow, stately motion, the whirling of the snow about their summits, and the fearful groaning and cracking of their parts—the picture cannot give. This is the large iceberg, while the small and distant islands, floating on the smooth sea, in the light of a clear day, look like little floating fairy isles of sapphire.

Richard Henry Dana

SEVENTEEN

Polar Oceanography

The polar regions of the earth are located north of the Arctic Circle (66.5°N latitude) and south of the Antarctic Circle (66.5°S latitude). They are ice-covered regions with little terrestrial surface life. The northern region consists primarily of the Arctic Ocean. Covering 13,986,000 km^2 (5,376,000 mi^2), it is the world's smallest ocean. The southern region consists mainly of the continent Antarctica. Measuring 14,245,000 km^2 (5,476,000 mi^2), it is the world's fifth-largest continent. As you can see, the Arctic Ocean and the Antarctic continent are approximately the same size.

The Arctic Ocean, which is nearly landlocked, is surrounded by Greenland, Canada, Norway, Alaska, and the Soviet Union. It is ice covered throughout the year to a depth of 0.6 to 4 m (2 to 13 ft) except during the summer, when the ice of the fringe areas melts. Antarctica is entirely surrounded by the so-called **Antarctic** (or Southern) **Ocean**, the world's stormiest ocean. Less than 5% of Antarctica is ever free of ice, and then only in a few places at the end of summer. Along the seaward margins, large sections of the continental ice cap break off as icebergs.

The Arctic summer occurs during the Antarctic winter. Around December 22, no sunlight penetrates above the Arctic Circle, but on that same day below the Antarctic Circle, the sun shines for 24 hours. On about June 21, the seasons are reversed. In winter, temperatures on the Arctic sea ice fall as low as −51°C (−60°F). Average surface temperatures on the Arctic ice are −35°C (−31°F) in the winter and 0°C (32°F) in summer. The winters are long and cold, the summers are short and cool, and precipitation is only about 51 cm (20 in). Temperatures in the Antarctic coastal regions are seldom higher than −18°C (0°F) in the summer and are as low as −40°C (−40°F) in the winter. High wind velocities and frequent storms create the rapidly changing weather typical of a coastal region, bringing about 25 cm (10 in) of precipitation per year. The interior of Antarctica is the coldest place on earth, with typical winter temperatures as low as −57°C (−71°F). The coldest temperature ever recorded on the earth was −88.3°C (−126.9°F), at Vostok. The dry interior receives annual precipitation of only 5 cm (2 in).

MAGNETISM

A magnetic compass points north in the Northern Hemisphere and south in the Southern Hemisphere. Although the compass needle appears to be pointing to the magnetic North and South poles, it is not the poles that are attracting it. The needle is actually aligning itself with lines of magnetic flux (sometimes called meridians of magnetism), which are parallel to the earth's surface at the equator and vertical to the earth's surface at the poles (Figure 17.1). As early polar explorers discovered, the magnetic compass is useless as a navigational instrument in the vicinity of the poles. Magnetic studies show that the magnetic meridians run together to form a magnetic line from 90°E to 90°W across the Arctic Ocean (Figure 17.2) and that the magnetic poles stay essentially in the same place with respect to the polar axis of rotation.

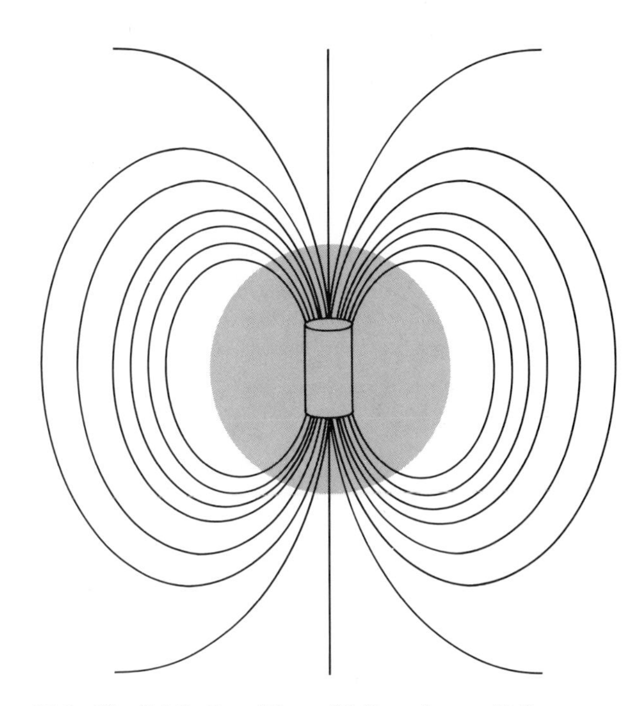

17.1 The distribution of the earth's lines of magnetic flux. Magnetically, the earth acts as if a bipolar bar magnet were suspended within a hollow sphere.

THE ARCTIC OCEAN

The continental shelf of the Arctic Ocean is the widest of all continental shelves, extending out 1,210 km (750 mi) from Siberia. It encloses a deep, oval basin with an average depth of 3,700 m (12,000 ft). The greatest depth is 5,441 m (17,850 ft), just north of the Chukchi Sea. The floor of the Arctic Ocean (Figure 17.3) is traversed by three submarine ridges. The Lomonosov Ridge, the most prominent, extends from North Greenland to the Laptev Sea and divides the basin in two.

Since the Arctic Ocean is permanently ice covered, conventional methods for determining currents cannot be used. The ice itself, however, gives an indication of surface drift. Beginning in 1893, Fridtjof Nansen studied the currents by allowing the vessel *Fram* to drift frozen in the ice for 3 years. It is believed that the surface circulation of the Arctic Ocean forms three systems (Figure 17.3).

The ocean is connected to the Pacific by the Bering Strait, a narrow passage less than 100 m (328 ft) deep. Water entering through this passage flows into the Arctic Ocean and across the pole. Large amounts of fresh water flow into the ocean from the Siberian rivers Ob, Lena, and Yenisei. This water mixes with Arctic surface water and the inflow through the Bering Strait, moves across the Arctic Ocean, and exits through the Greenland Sea to create the cold East Greenland Current.

Some of this outflow is deflected south around the northern end of Greenland into Baffin Bay, creating a smaller current that emerges into the North Atlantic through the Davis Strait as the Labrador Current. It is the Labrador Current that carries icebergs formed on Greenland out into the Atlantic. The cold Labrador and East Greenland currents give the shore of northeast North America a considerably colder climate than that of Britain and Western Europe.

A third surface current is formed by waters that are deflected to the north of Greenland into the Beaufort Sea, where they form a circular flow associated with a thick ice cover. The surface water throughout the Arctic Ocean is readily identified by its low temperatures—approximately −1.8°C (28.8°F) in winter and −1.5°C (29.3°F) in summer—and by its low salinity, which varies from 28‰ to 33.5‰.

A strong, warmer, more saline subsurface flow from the Atlantic into the Arctic basin replaces the outgoing Arctic Ocean surface waters, but submarine ridges bar any exchange of the deeper water. Consequently, the trapped Arctic Bottom Water (ABW) is the coldest water in the ocean, with temperatures as low as −1.4°C (29.5°F).

THE ANTARCTIC

Although the continent of Antarctica influences current flow and the regions of water formation, it is the waters of the area that are of greatest interest to oceanographers. The northern limit of the Antarctic Ocean is not precisely delineated. The most widely accepted boundary is the **Antarctic Convergence** (or Antarctic Polar Front [APF]) zone. Here, where the regional surface waters come together, temperature and salinity change abruptly over a region about 40 km (25 mi) wide that encircles the entire Antarctic continent between 50°S and 60°S (Figure 17.4). To the north (at about 40°S) is another transition zone, the **Subtropical Convergence,** which marks the northern limit of drift

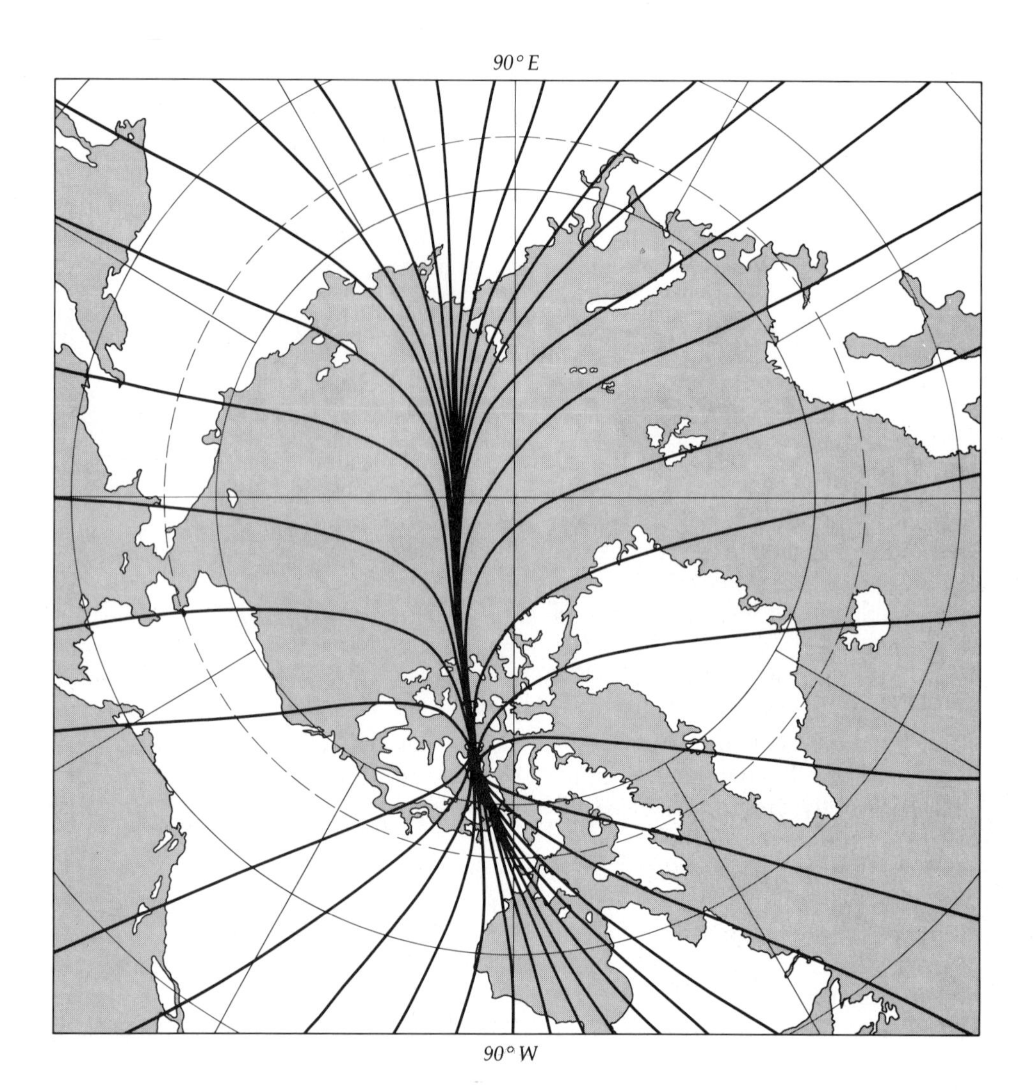

17.2 *The magnetic North Pole.*

ice from the Antarctic. To the south (from 65°S to 70°S) is the **Antarctic Divergence.** Between these two divergences is the Antarctic Circumpolar Current (AACP). This is the largest current on earth, with an estimated flow as great as 200 million cubic meters (7 billion cubic feet) per second (200 sv). Its maximum flow is in the region of the Antarctic peninsula (Figure 17.4). The strong, almost constant westerly winds drive the current to the east. (Recall that an east wind blows from the east but that an easterly ocean current moves toward the east.) The surface movement of AACP is referred to as the **West Wind Drift**, although the current actually extends to the bottom, where it is influenced by the topography. Current meters positioned 300 m (984 ft) from the bottom at Drake Passage record a speed of from 3 to 9 cm/s (1.2 to 3.5 in/s). The water temperature drops from around 9°C (48°F) to about 2°C (36°F) from the Subtropical Convergence to the Antarctic Convergence, where the colder Antarctic waters meet the waters from the north. Near the continent, the surface temperature drops from 2°C to 0°C. Between the Antarctic Divergence and the continent, easterly winds propel a surface flow called the **East Wind Drift**.

Figure 17.5 shows the surface and vertical movement of the Antarctic waters. Antarctic Bottom Water (AABW) is produced in the Weddell Sea. There shelf water from beneath the ice sheet associated with ice formation, which has a temperature of −1.9°C (28.6°F) and a salinity of 34.66‰, sinks and mixes with deeper, warmer water with a temperature of 0.5°C (32.9°F) and a salinity of 34.68‰. The resulting water mass at 0.3°C (32.5°F) and 34.66‰ flows northward into the world's ocean basins, especially into the western basin of the South Atlantic. Antarctic surface water sinking at the Antarctic Convergence forms a water mass of about 2°C (36°F) and a salinity of 33.8‰. Called Antarctic Intermediate Water (AAIW), it descends to a depth of 1 km (0.62 mi) and moves northward.

The northward flow of the AABW from the Antarctic region (800 sv) is about four to five times that of

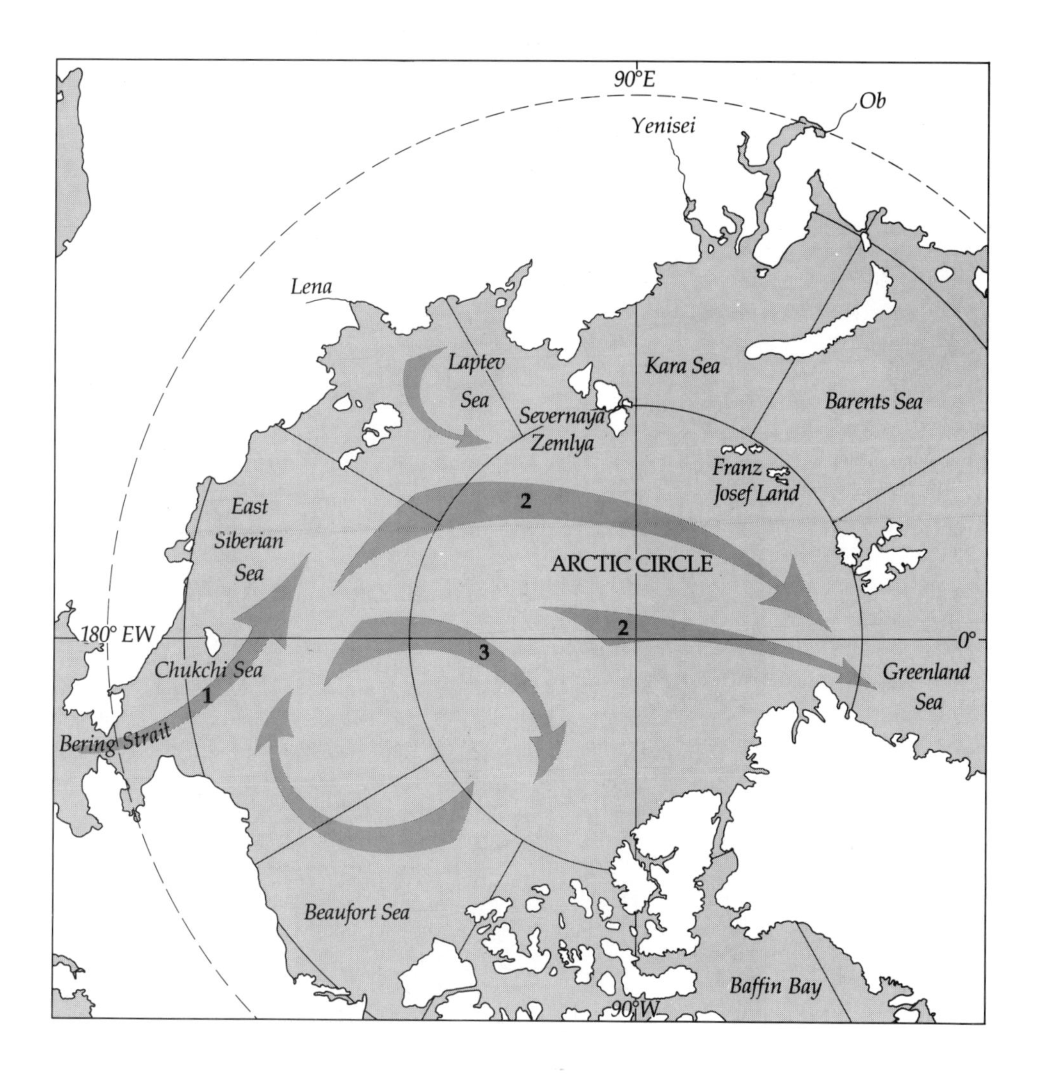

17.3 Arctic Ocean surface currents comprise three major systems (1, 2, and 3).

the AACP flow. These flows, coupled with the AAIW, constitute a vast flow of water out of the Antarctic Ocean. The outflowing water is replaced by North Atlantic Deep Water (NADW), which has a temperature between 2°C and 3°C (36°F and 37°F) and a salinity of 34.7‰.

THE HEAT BUDGET

The Antarctic Ocean annually transfers about twenty times more heat to the atmosphere than it receives from the atmosphere (3×10^{22} g•cal outflow, 1.5×10^{21} g•cal inflow). Heat is also transported from the ocean by the northward-flowing AAIW (5×10^{21} g•cal) and AABW. The annual melting of winter ice and the warming of summer waters require 4×10^{21} g•cal of heat. The Antarctic Ocean receives some heat from the ocean floor and from the atmosphere, but most of the heat it receives comes from North Atlantic Deep Water (3×10^{22} g•cal). The total net outflow of heat from the Antarctic Ocean is approximately 3×10^{21} g•cal per year.

Although the Antarctic Ocean represents 22% of the world's total oceanic surface area, it contains only 10% of all oceanic heat. It acts as a major heat sink by absorbing massive amounts of heat and then transferring it to the atmosphere, thereby influencing the atmospheric circulation of the Southern Hemisphere. The Arctic Ocean, in contrast, exercises far less influence on weather and climate, although it also acts as a heat sink. The ice that covers the polar oceans in winter serves as an insulator for the ocean; most of the heat lost is through open-water areas and areas of thin (less than 50 cm) ice.

Recent studies indicate that the Antarctic has been frozen for at least the last 20 million years but that a tropical environment existed there 250 million years ago.

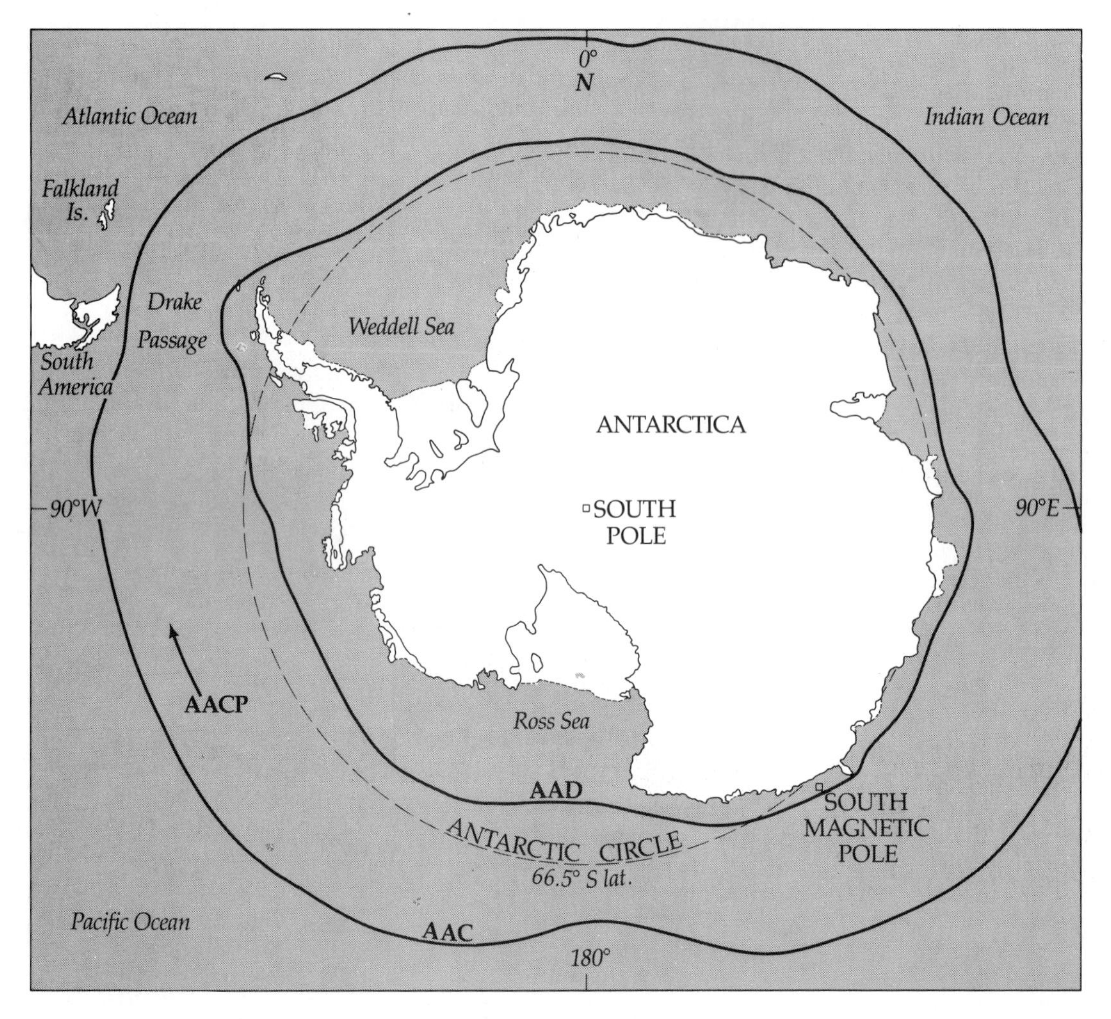

17.4 The Antarctic Continent, Antarctic Convergence (AAC) (or Antarctic Polar Front, APF), Antarctic Divergence (AAD), and Antarctic Circumpolar Water (AACP).

LIFE IN THE POLAR REGIONS

For a long time it was thought that the Arctic ice cap and the Antarctic continent supported no life at all. Since 1926, however, the Amundsen-Ellsworth flyover of the Arctic and subsequent expeditions have revealed that the Arctic supports hares, gulls, guillemots, polar bears, and seals as high as 88°N, as well as an indigenous human population. Twenty species of land animals inhabit the Arctic, and some survive on the pack ice the year round. Polar marine life in the Arctic abounds. The water beneath the ice teems with microorganisms and has pronounced blooms in the spring and summer. The fauna on the Pacific side of the Arctic Ocean differ from the fauna on the Atlantic side, and the fauna also vary from layer to layer in response to variations in temperature and circulation.

Conditions in Antarctica are too inhospitable for large animals to survive, although several species of birds nest and hatch their young there, including the skua gull and the penguin (Figure 17.6). All eighteen species of the penguin live in the Southern Hemisphere; none are found in the Arctic. Two species, the Emperor and the Adélie, reside seasonally in Antarctica, and several live on islands within the Antarctic Circle. One species is found on the warm, equatorial Galápagos Islands.

Beyond the birds, land life on Antarctica is limited to microfauna in summer melt pools, patches of moss and lichens, and small wingless insects. But marine life in the Antarctic is even more abundant than it is in the Arctic. All the marine animals, from penguin to seal to whale, directly depend on a small, reddish, shrimplike crustacean, *Euphausia superba*, or krill. Krill and some species of zooplankton use the Antarctic currents during their breeding cycle. Adult euphausiid shrimp release their eggs at the Antarctic Convergence at a depth of about 200 m (656 ft); the eggs sink with the Antarctic Bottom Water. Nine months later, as adolescents, the young krill rise farther south

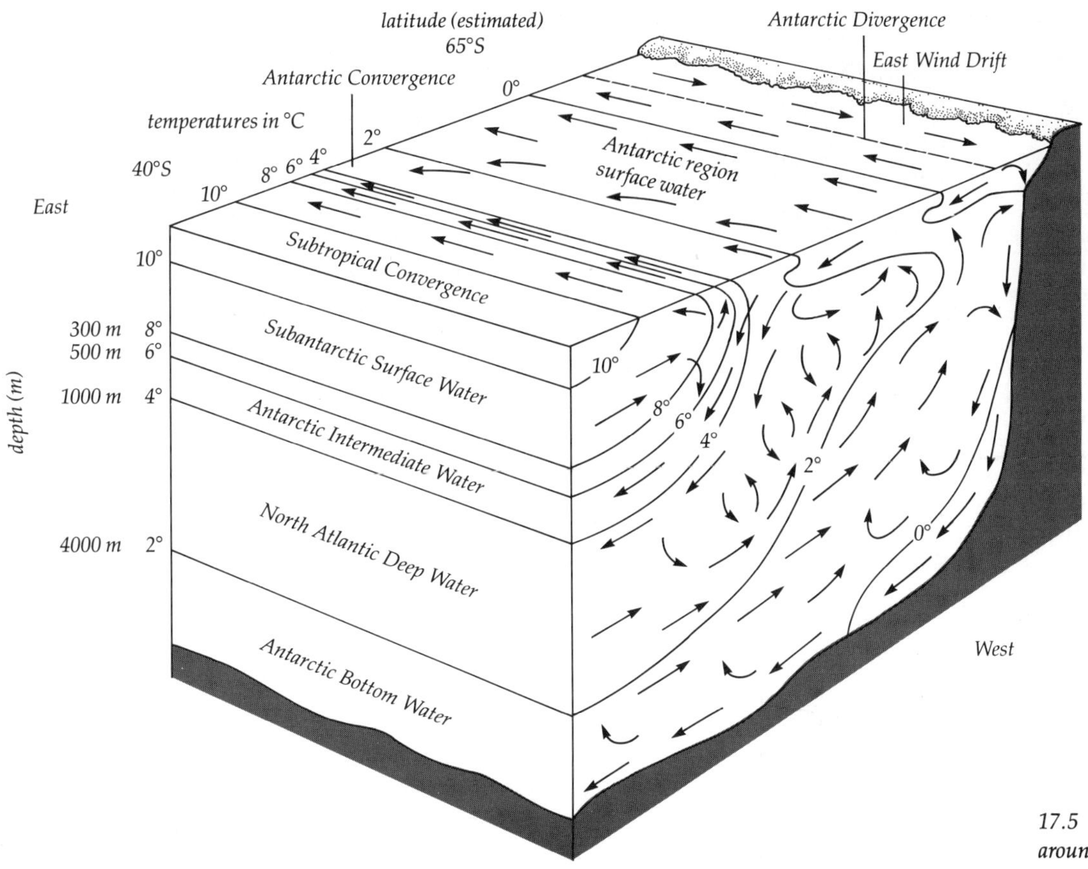

17.5 Cross section of waters around Antarctica.

17.6 An Adélie penguin silhouetted against the sun in the frozen pack ice of the Weddell Sea. This photograph was taken from the Coast Guard icebreaker Glacier *during its oceanographic expedition as part of Operation Deep Freeze 1970. Notice the unevenness of the ice caused by expansion pressure when the water froze, as well as some trapped small icebergs. The "halo" appearance of the sun, called "sundogs," is caused by light hitting ice crystals in the air. (Official U.S. Coast Guard photograph)*

17.7 Icebergs in Arthur Harbor, Antarctica, seen from the Coast Guard ice breaker Glacier *during Operation Deep Freeze 1970. A tabular berg appears in the middle background and a pinnacle berg on the right. The pack ice has broken up, "releasing" the icebergs. By grinding into one another, pieces of pack ice become almost circular. (Official U.S. Coast Guard photograph)*

at the Antarctic Divergence and then drift north with the surface currents. They reach the Antarctic Convergence as adults and repeat the cycle anew.

The zooplankton themselves feed on phytoplankton, mainly diatoms. The amount of plankton in the water is largely a function of the water's nutrient content, and there is no shortage of nutrients in Antarctic waters. The constant overturn of deep, nutrient-rich waters around Antarctica provides a rich supply of essential nutrients, such as nitrates, phosphates, and silicates. These are present in such quantity that even in late summer the supply is never exhausted. The lack of sunlight, especially in the winter, is the only limiting factor on phytoplankton productivity.

ICE IN THE SEA

Sea Ice

Sea ice is a collective term for ice in a number of different forms. As sea ice begins to form, the surface looks oily or greasy, much like oil and water that have been beaten together, an indication that ice crystals are forming. These crystals are flat, thin, pointed, and only a few centimeters long. They join to form oblong plates. As more and more crystals form, the sea surface becomes slushy. It is quiet, with few ripples, and takes on a gray tinge. A ship passing through the slush makes almost no sound.

Unlike ice in freshwater lakes and ponds, sea ice does not coalesce into a clear sheet. Instead, the crystals form a hard ice rind, which is usually less than 5 cm (2 in) thick. If the surface of the sea is agitated, the rind forms disclike cakes of ice. This **pancake ice** is usually 0.5 to 1 m (20 to 39 in) in diameter. On further freezing, the cakes form into a continuous ice sheet, or **ice floe**. Ice floes range in size from 10 m (32.8 ft) to about 10 km (6 mi) (Figure 17.6). They are found in northern harbors during winter and in the Greenland straits. It is not uncommon to find various types clustered together (Figure 17.7).

Ice in the Arctic Ocean is known by several names. Extensive polar **pack** or old **ice** covers about 70% of the Arctic Ocean. This ice is typically 3–7 years old, and the annual cycle of summer surface melting and winter bottom growth produces hills and ridges, and the tremendous pressure forms ridges or hillocks of ice that extend perhaps 10 m (32.8 ft) above sea level and 40 to 50 m (150 ft) below (Figure 17.8a). Pack ice that is attached to the shore and that extends seaward is known as **fast ice** (Figure 17.8b). Rafting may occur as the ice is forced together, some pieces sliding over and against others (Figure 17.8c). The thickness of the ice varies with location and with time. The average winter thickness is 3 to 3.5 m (9.8 to 11.5 ft) (with an extreme record of 9 m). The thickness decreases in summer to about 2.5 m (8.2 ft). Local patches of open water occasionally appear, called **polynyas** (a Russian-Eskimo word that means "regions of very thin ice"), which are much more common in summer than in winter. A nuclear submarine in polar waters would choose to surface in a polynya.

Around the outside 25% of the pack ice perimeter lies seasonal or first-year ice, since it has not survived a summer (also called *drift ice*). Its limits vary greatly with the seasons. It is this ice, not the old ice, that can be readily penetrated by icebreakers (Figure 17.9). First-year ice seldom exceeds a thickness of 2 m, and it usually melts entirely during the summer months.

The hull of an ordinary ship traveling through extremely hard ice would be slashed open. Thus ships designed to penetrate ice have an entirely different design (Figure 17.9c). The rugged hull of an icebreaker is rounded and dish-shaped. Instead of cutting its way through the ice, its bow moves constantly up and onto the ice, its weight crushing the ice below it. If it fails to break through, the icebreaker backs up for another attack. Its propellers are carefully protected. Most of the icebreakers owned by the U.S. government are now under control of the U.S. Coast Guard. Most of them were built during World War II, and only the few new ones are in the best condition.

The salinity of sea ice varies considerably. As ice forms from needlelike crystals of water, a fraction of the salt tends to separate out in minute pockets of trapped, concentrated salt solution, and the remainder is rejected into the sea. As a general rule, the faster the ice is formed, the greater its initial salinity is (Table 17.1). Sea ice is not quite solid, because liquid brine is trapped in the hexagonal spaces between the crystals. Age is a primary factor in determining the salinity and density of sea ice. Over time, the tiny brine deposits in the ice structure tend to migrate downward and outward. Old ice (sea ice is seldom older than 5 or 6 years) that has lost most of its salt appears mottled and honeycombed; in summer as it melts it is called "rotten ice." Because air takes the place of the departed brine, the resulting density may be as low as 0.85 g/mL. New sea ice, in contrast, has a density of 0.925 g/mL; pure water ice has a density of 0.917 g/mL.

Ice more than 1 year old is harder than younger ice because it contains less salt. New ice may have a salinity of 5.5‰ to 10.2‰, whereas the salinity of old ice is usually less than 2‰. Old ice and puddles of melted water on its surface have a low enough salinity to be consumed by people, and melted sea ice has often been drunk by explorers either by plan or by necessity. Although the Arctic pack ice is always present, some of it melts each summer, and some is continuously being re-formed.

Icebergs

During spring in the Northern Hemisphere, appreciable amounts of sea ice float out of the Davis Strait and Arctic Ocean. Since this ice is fairly thin and melts rapidly, however, it presents little danger to shipping in the North Atlantic. The real menace to oceangoing vessels is the **iceberg**. Although icebergs represent very little of the ice floating in the sea, they are the most dangerous form of all. Icebergs produced in the two polar regions differ in physical appearance, structure, and manner of formation.

For a long time, icebergs had been thought to be broken or fragmented pieces of the pack ice that had simply floated into the world's oceans. Actually, icebergs form above land from fresh water, which is why

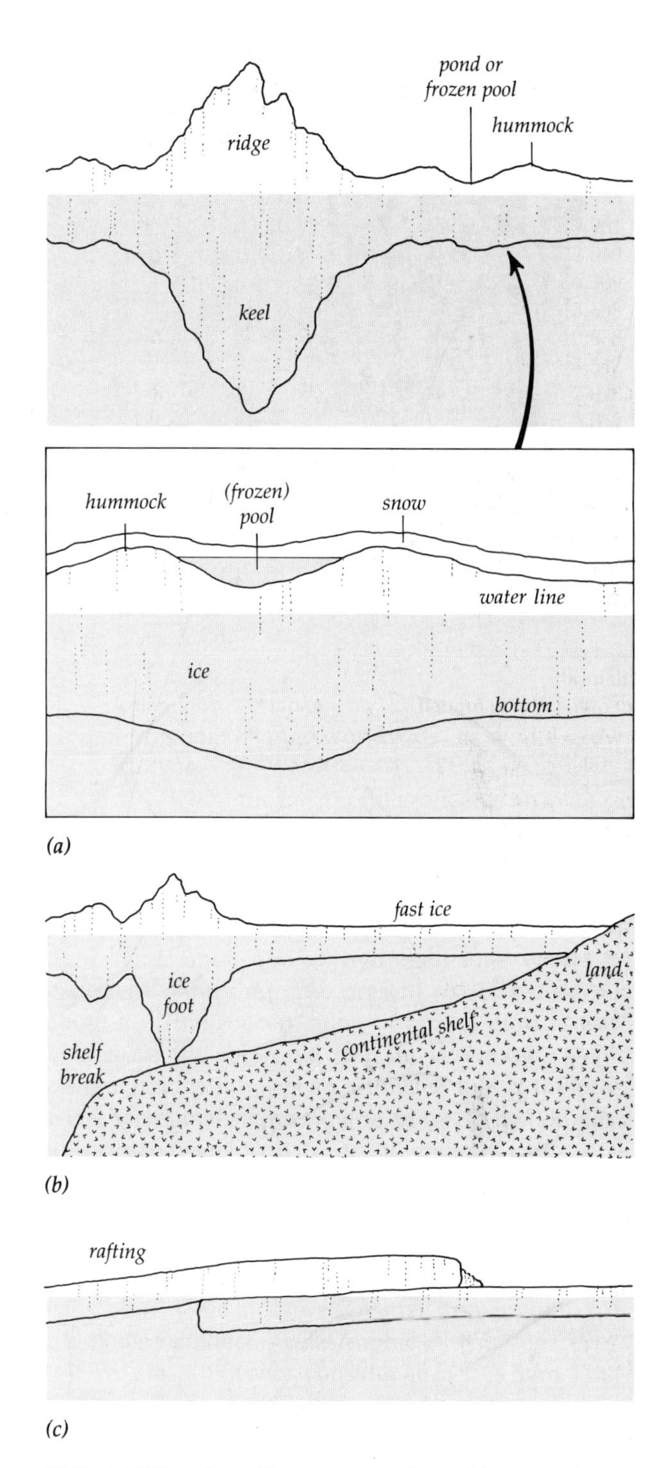

17.8 *(a) Ridges formed by summer surface melting. As winter progresses, the collection of keels and ice-feet becomes more extensive (inset). (b) In fall, the fast ice forms before the adjoining pack ice forms. (c) Rafting of sea ice.*

Table 17.1 *Salt Content of Sea Ice as a Function of Air Temperature During Ice Formation*

Air Temperature (°C)	−16	−26	−30	−40
Salinity (‰)	5.6	8.0	8.8	10.2

(a)

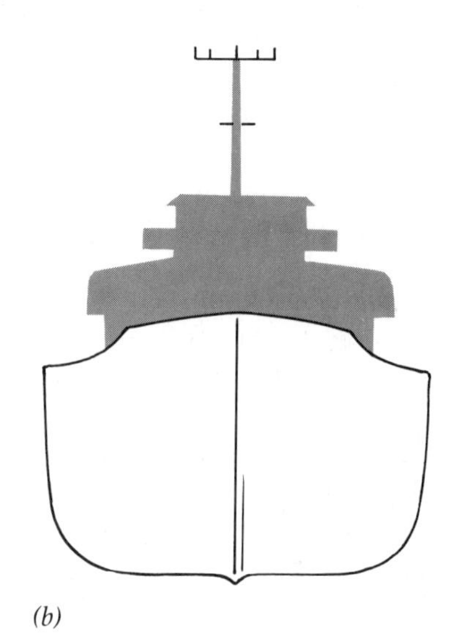

(b)

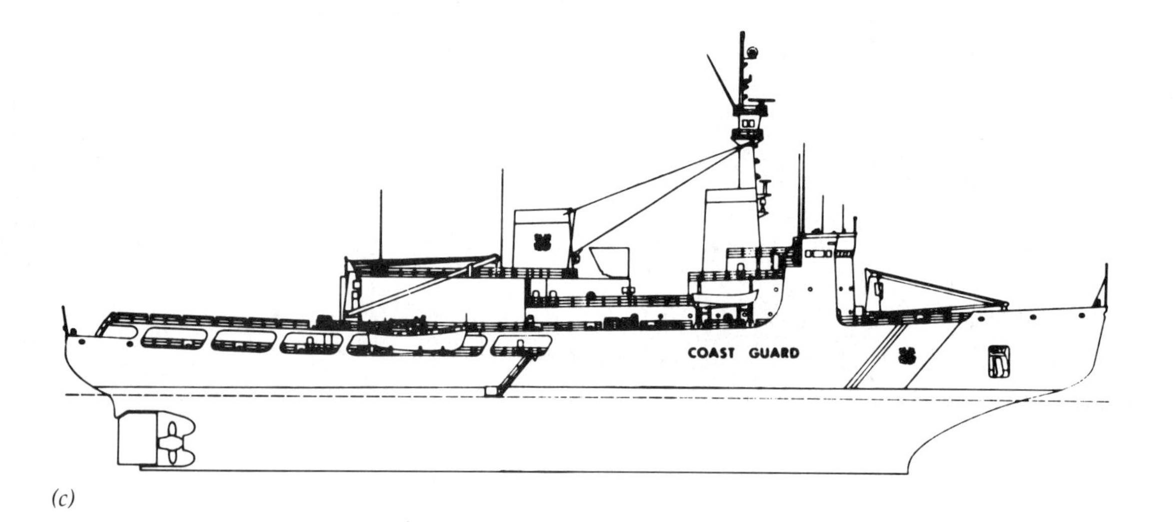

(c)

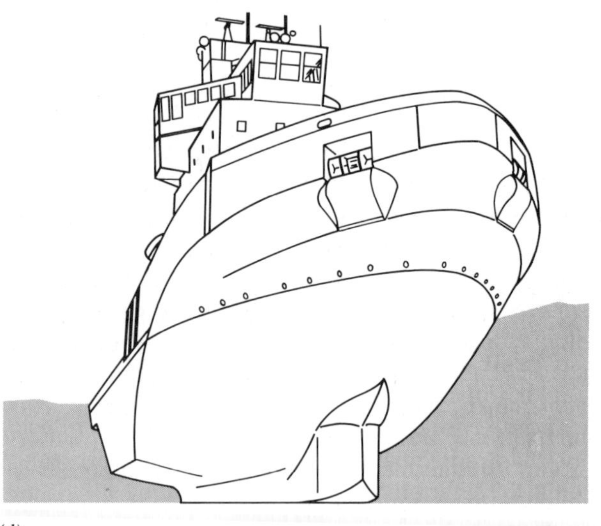

(d)

17.9 *(a) The U.S. Coast Guard icebreaker* Polar Star *moving through pack ice in 1983 to break into McMurdo Station on its 12,537-km (7,773-mi) circumnavigation of the Antarctic Continent, the first U.S. vessel to do so south of 60° latitude. The volcano Mount Erebus looms in the background. (Ed Moreth, U.S. Coast Guard) (b) Hull design for a conventional ship. (c) Hull configuration of the U.S. Coast Guard icebreaker* Polar Sea, *which is 122 m (399 ft) in length. (U.S. Coast Guard) (d) The rounded, sloping, heavily reinforced bow of a modern icebreaker. This allows the vessel to ride up onto the ice, breaking the ice with its weight.*

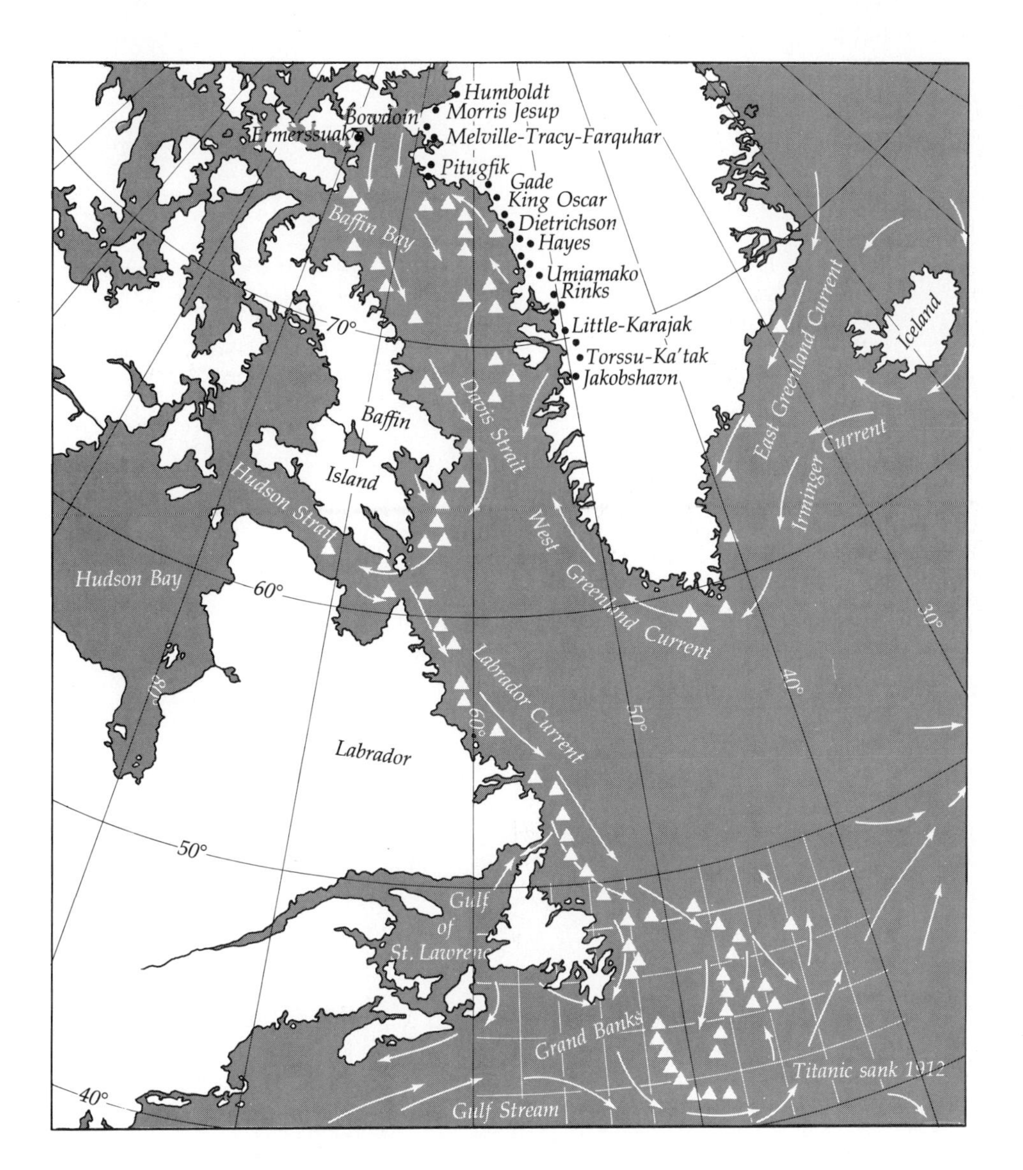

17.10 The drift of icebergs from their source into the North Atlantic. The map shows the main ice-producing glaciers of western Greenland (•). Of the many thousands of bergs (▲) that break off these glaciers each year, about 390 ultimately drift south of Newfoundland (48°N). (From the U.S. Navy Hydrographic Office and the U.S. Coast Guard)

they contain no salt. The only other freshwater ice in the ocean is a small amount of river ice from some of the northern rivers.

The massive glaciers covering most of Greenland are the source of most North Atlantic icebergs. Some icebergs break off Greenland's eastern coast, but by far the greatest number break off the western coast in the Davis Strait (Figure 17.10). **Glaciers**, formed from compacted snow, are rivers of ice that flow steadily downhill. The average flow rate for Greenland glaciers is about 10 m (33 ft) per day. When the front of a glacier reaches water, it simply slides in and breaks off, an action called **calving**. Depending on local water conditions and on the breakup of pack ice, currents then carry the icebergs out to sea.

As a result of long-term navigation over North Atlantic shipping lanes, an extensive terminology has developed for icebergs in this region. Here, the peaks of irregularly shaped icebergs are called pinnacle bergs (Figure 17.11a and 17.12a). A small iceberg about the size of a house is known as a "bergy bit." Smaller icebergs are known as "growlers," since they make a fair amount of noise as they bob in the water and melt.

Because most of an iceberg lies below the water's surface, the exposed portion is no indication of its total size (Figure 17.13). Many icebergs have underwater projections called **rams** that can split a ship's hull. On an average, between four-fifths and six-sevenths of an iceberg's total volume is submerged; seven-eighths is an upper limit. North Atlantic icebergs normally are no more than a few hundred meters long and 50 m (164 ft) high, but a height of 80 m (262 ft) is not uncommon, and every so often an iceberg 100 m (328 ft) tall is sighted. In 1882, the largest North Atlantic iceberg was sighted, which was 11.3 km (7 mi) long and more than 4.8 km (3 mi) wide.

The number of icebergs reaching the North Atlantic varies from year to year. Most arrive in April, May,

(a)

(b)

17.11 *(a) Icebergs assume fantastic shapes after they calve from the mother glacier and start the long, slow journey through Baffin Bay toward the North Atlantic. (Official U.S. Coast Guard photograph) (b) An iceberg propelled by a subsurface current. (Official U.S. Navy photograph)*

June, and July. Slightly more than 1,000 icebergs were sighted in 1912, but in 1924 only 11 of those sighted were big enough to cause concern. In 1929, there were 1,305 icebergs sighted and tracked. When the *Titanic* and an iceberg made their ill-fated rendezvous off Newfoundland on April 14, 1912, there had been 395 sightings that spring, a particularly large number. The *Titanic* was considered unsinkable, because even if 4 of the semi-watertight compartments making up its hull were ripped apart, the ship would still remain afloat. The iceberg ripped through 5 of them.

The loss of the *Titanic* and more than 1,500 lives tragically underscored the need for more effective iceberg tracking and prompted the formation of the International Ice Patrol by international treaty in 1913. The U.S. Coast Guard continues surveillance of icebergs, using reconnaissance ships and aircraft as well as sightings from vessels in iceberg areas. Seventeen nations contribute funds to maintain the International Ice Patrol.

An iceberg calved from the western coast of Greenland may take months to reach the North Atlantic shipping lanes. Unlike free-floating sea ice, which is wind-propelled, icebergs move with ocean currents (Figure 17.11b). On occasion, ships have been moored to an iceberg and towed through pack ice. Most icebergs float directly east across the Atlantic, some passing along the coast of the British Isles and penetrating as far south as 30°N. They may have a total life span of 2 years. An average iceberg with a mass of about 1 million tons formed, say, around Baffin Bay, will be reduced to one-eighth of its original size by the time

(a)

(b)

17.12 (a) An iceberg from the North Atlantic. Often these icebergs are irregular and "pinnacled." (National Oceanic and Atmospheric Administration) (b) Dwarfing the U.S. Coast Guard icebreaker Westwind, *this tabular iceberg measures 1.2 km (0.75 mi) long, 0.8 km (0.5 mi) wide, and 15 m (49 ft) above sea level. It extends as deep as 170 m (558 ft) below the surface. (Official U.S. Coast Guard photograph)*

it reaches southern Newfoundland. Icebergs travel about 9 to 15 nautical miles a day.

An iceberg may remain frozen in pack ice well over a year before it calves from the mother glacier. Once in the open ocean, however, it melts fairly rapidly, especially if it drifts near the Gulf Stream (which it will do sooner or later if it comes from Greenland). In the warmer waters of the Gulf Stream, melting is so rapid that 3 to 3.5 m (9.8 to 11.5 ft) in height may be lost each day. Seldom do icebergs proceed very far down the western side of the Atlantic. Even so, an iceberg was sighted near Bermuda in June 1926!

Once an iceberg reaches the open ocean, its life span is less than 3 months. It may break into parts, which wind, rain, and waves erode and scour down. Melting speeds their demise. The heat comes from the water, the air, and solar radiation, and it is not unusual to see small waterfalls cascading down the side of an iceberg.

There is little that can be done about an iceberg except to track it and perhaps mark it with some type of bright coloring so that it can be readily identified. Certainly, one cannot shoot and sink it. That has been tried, believe it or not, on the premise that the smaller fragments would melt faster. The results were not particularly successful. Black material (such as carbon black) has been sprayed over icebergs to make them absorb more solar energy and therefore melt faster. Although this does increase the melting rate enormously, the results apparently are not worth the effort involved. As with snow in many cities in the United States, nature ultimately solves the problem.

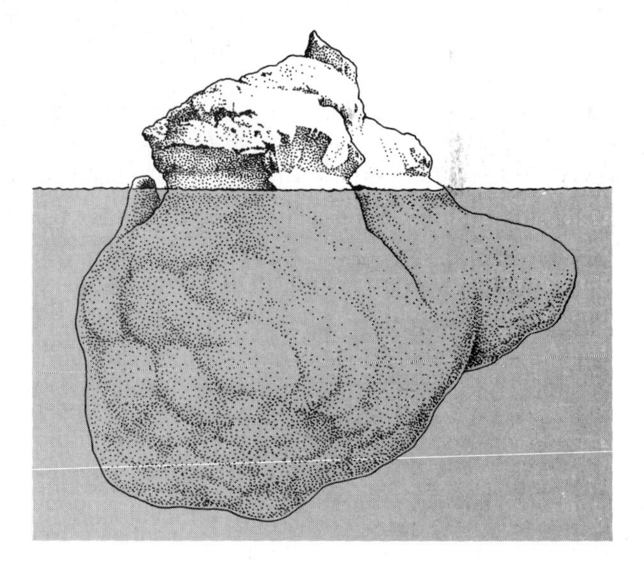

17.13 An iceberg above and below the waterline.

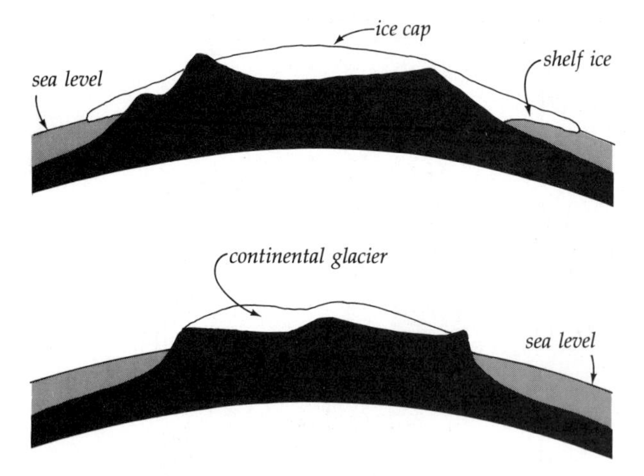

17.14 *(a) Cross section of the Antarctic continent. (b) East-west cross section of Greenland.*

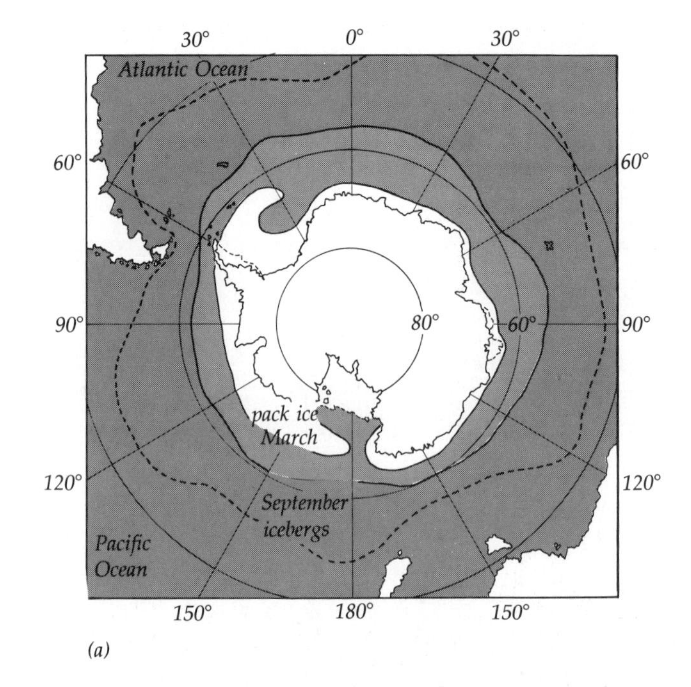

(a)

17.15 *(a) Average boundaries of pack ice for winter (September) and summer (March) in the ocean around Antarctica. (b) A computed fraction of the sea ice cover around the Antarctic for 12 months. Each figure is produced from the monthly averages of thermal emissions from the passive microwave radiances recorded by* Nimbus 5. *The legend in the December square indicates the percentage of ice coverage. In the images for July through November, a polynya is visible just above the Antarctic continent. It is by far the largest such feature ever observed.*

There are very few icebergs in the North Pacific. The narrow Bering Strait gives the waters of the Arctic Ocean little access to the North Pacific, and the spring current moves south to north in the strait. Furthermore, the land adjacent to the North Pacific lacks major glaciers and thus has no massive iceberg-producing regions. Only a few valley glaciers occur in Alaska, and they calve only a small number of icebergs each year.

Icebergs in the southern oceans are more regular in shape than those in the northern, even though they, too, are land ice. Most commonly they are long, flat, and tablelike; they are called **tabular icebergs** (see Figure 17.12b). A margin of ice fringes the continent of Antarctica, which contains more than a trillion (10^{12}) tons of ice with an average thickness of 2,200 m (7,200 ft). The ice forms on the continent but extends out into the sea (Figure 17.14). As the shelf moves northward about 100 m (328 ft) a year, icebergs break off along the way. This Antarctic pack ice at its maximal extent covers 10% of the world ocean. This shelf calves to form the large, flat icebergs of the southern Atlantic, Indian, and Pacific oceans. The largest tabular iceberg ever recorded was 350 km (217 mi) long, 100 km (62 mi) wide, and more than 33 m (108 ft) high. The dimensions of the shelf ice vary with the seasons (Figure 17.15). In winter, tabular icebergs may be trapped in the pack ice until spring, when melting finally frees them.

For many years little was reported about icebergs in the southern oceans because sea traffic was fairly uncommon there. Recently, however, supertankers too large to travel through the Suez Canal regularly travel around the Cape of Good Hope on their way between the Middle East and Europe or the United States. Antarctic tabular icebergs are a common sight off Cape Horn and the Falkland Islands and have been sighted in the South Atlantic as far north as 30°S. With rising oil transport and increasing ship size, tanker and supertanker traffic will increase in the world's southern oceans. Fortunately, ice is less dangerous in the Southern Hemisphere than in the Northern Hemisphere because weather conditions are more consistent. Although radar is now commonly used to detect and track icebergs, visual sighting is still the most reliable means.

Ice Islands

In the Northern Hemisphere, the counterpart to tabular icebergs is the **ice island**—a homogeneous, flat sheet or patch of ice. Once located, ice islands are easy to distinguish from pack ice, although weather conditions make it difficult to spot them. The existence of ice islands has been known since 1946, when such an island showed up on the radar of a U.S. Air Force reconnaissance aircraft. Ice islands are often much thicker than polar pack ice. They originate on land, apparently when pieces of shelf ice break off from

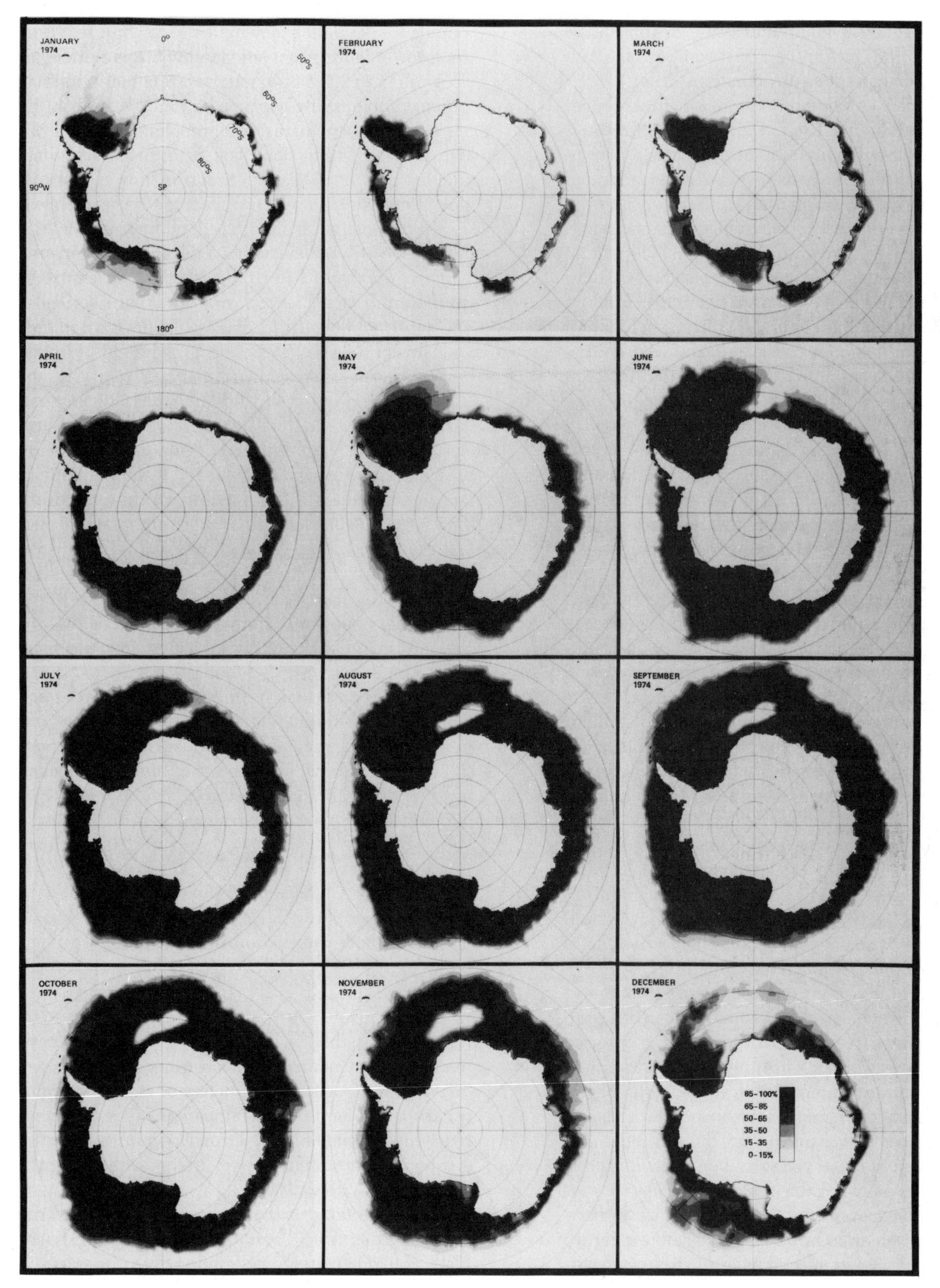
JANUARY
1974
0°
50°S
60°S
70°S
80°S
90°W
SP
180°
FEBRUARY
1974
MARCH
1974
APRIL
1974
MAY
1974
JUNE
1974
JULY
1974
AUGUST
1974
SEPTEMBER
1974
OCTOBER
1974
NOVEMBER
1974
DECEMBER
1974
85-100%
65-85
50-65
35-50
15-35
0-15%

(b)

northern Greenland, Ellesmere Island, or some other northern island. They drift clockwise around the Arctic Ocean with the prevailing currents.

Unlike ice floes, ice islands persist for much longer periods. The first one discovered, called T-1 (for "Target 1, radar determined"), has been followed sporadically since 1946. Such islands have been used for staffed meteorological and geophysical stations by both the United States and the Soviet Union. Fletcher Island (T-3), discovered in July 1950, was 16 km (10 mi) long and 8 km (5 mi) wide. It drifted at a rate of about 4 km (2.5 mi) a day, with a maximum rate of 14 km (8.7 mi) a day. It had been staffed by the United States many times since 1952, usually for several months at a time, but it drifted out of the Arctic Ocean in 1983 to break up and melt in the Davis Strait.

Putting Ice to Use

Ice is an unusual substance. It is extremely hard, and as long as the temperature remains below freezing, its durability makes it an excellent building material. Air Force and distant early-warning stations in Greenland and the northern parts of Alaska regularly use ice as a construction material. However, anyone who has ever struck ice with an ice pick knows that it is brittle.

Perhaps the most unusual use of ice occurred during World War II, when Geoffrey Pyke suggested making ships from ice. To overcome its brittleness, about 2% fiber content from newsprint was mixed in as the ice set. A pilot program was actually set up by the British navy in northern Canada to test the feasibility of ice ships. Several small vessels were constructed from the ice, which was set into molds around cooling equipment. The engine room and crew quarters were insulated from the ice. Buoyancy was hardly a problem. If a ship was hit by a torpedo, the cooling system could be turned on at the hit side, water sprayed onto the fractured parts, and the hull rebuilt in place.

Tests indicated that ice ships were feasible and that even in waters near the equator their life span would be appreciable. A ship could pass from a northern to a southern ocean with no apparent loss in the hull. However, by the time the technology had been largely perfected, American liberty ships were being produced extremely quickly. Had this not been the case, the British quite possibly would have built a number of ice ships for war use.

With shortages of fresh water becoming an ominous reality, attention is turning again to an earlier suggestion by J. Isaacs of Scripps Institution of Oceanography: Why not tow tabular icebergs from the southern ocean to large cities, such as Los Angeles, and use the melting waters as a source of pure fresh water? With the likelihood of major shifts in world climate, coupled with an exponentially growing human population, this idea may not be as fanciful as it once seemed. A Saudi-sponsored United States conference was held in 1977 to consider the feasibility of using this method to bring fresh water to Saudi Arabia. But there were too many technical problems, and the 125-day trip would be too long since a ½-mi^2 iceberg would melt by day 104. (Western Australia, however, might be a destination.) It is more practical today, to recycle readily available water. Rather than dumping 380 million liters (100 million gallons) of primary treated sewage (most of which is water) into the ocean every day—which is the average for San Diego alone—most of the water could be reclaimed at a lower cost than desalination. The thought of drinking sewer water may not be appealing, yet water that has gone through secondary and tertiary treatment would be no worse-tasting than the water that Los Angeles and San Diego use now. But this solution is not without cost. Sewage treatment facilities are costly to build and operate, and existing ones are operating at capacity. However, water from icebergs and water recycled from sewage may not seem expensive compared with having no water at all.

SURVIVAL IN THE SEA

Let's take a quick look at the question of safety and survival in the marine environment. With more and more people taking part in water sports, the number of boating accidents is on the rise, even with greater effort and vigilance by such organizations as the U.S. Coast Guard and its auxiliary.

Drowning

People who end up alone in the sea sometimes assume that they will be able to float until they are picked up. However, the human body has a high water content. In fact, its average density is very close to that of seawater. As a result, 90% of us sink unless we have some type of flotation device. When the remaining positively buoyant 10% grow unconscious, they will float; but, with only a few centimeters of the head above water, breathing becomes impossible. To survive, we must either swim or tread water. However, swimming is exhausting to all but the well trained and physically fit.

Once we fall asleep or grow unconscious in the water, we die swiftly, usually within 3 to 4 minutes. Water is first swallowed in large amounts, after which it may be inhaled, depending on the efficiency of the protective muscular reflex of the larynx. Once water is inhaled, which occurs in 60% to 80% of all cases of immersion, the chance of recovery is lowered. Water in the lungs sets up a reaction in lung tissue, which may cause death from progressive lung damage even after a victim has apparently recovered.

Obviously, it makes good sense to have buoyancy devices handy when at sea, and even better sense to wear them. They should be easy to put on and should support the body in a face-up position, which is something a water skier's safety belt will not do. Experience has shown that the newer inflatable and foam jackets are better than solid cork or kapok-filled ones.

Hypothermia

When the *Titanic* went down in 1912, only 687 of the 2,200 people on board escaped in lifeboats. Within just 2 hours, 1,513 bodies were recovered from a calm sea. All were wearing life jackets. None had drowned. They had died of the cold, or, more correctly, of **hypothermia**, the lack or loss of heat.

It is seldom realized how hostile a cold-water environment is to people. Water has about a thousand times the specific heat of air, and twenty-five times its thermal conductivity. So heat loss is much more rapid in water than in air of the same temperature. Wind-chill-factor tables show how dangerous cold air alone can be. A person clad only in a bathing suit can be comfortable in air of 23°C (73°F), but in water a temperature of 34°C to 35°C (93.2°F to 95°F) is necessary for comfort. A lightly clad person cannot survive in 0°C (32°F) water for more than 20 minutes. At 15°C (59°F) a person can last for 6 hours; at temperatures exceeding 20°C (68°F), a person can last indefinitely.

A fully clothed person who happens to fall overboard in most American coastal waters in winter cannot expect to live more than an hour unless clothed in a wet suit. Contrary to popular opinion, it is temperature rather than pressure that is the primary limiting factor in deep scuba diving. Even with very thick wet suits, loss of heat is the major problem.

Tolerance to cold differs widely from person to person. In general, the physiological response to cold at different deep-body temperatures follows this sequence:

36°–34°C Shivering, increased oxygen consumption; increased metabolic heat output; constriction of peripheral blood vessels.

33°–32°C Shivering stops; metabolic heat output falls; dizziness and nausea set in.

31°–30°C Voluntary motion lost; body collapses and person becomes unconscious; eye and tendon reflexes stop; cardiac muscle action becomes irregular.

26°–24°C Ventricular fibrillation begins; death follows.

Thirst, Hunger, and Exposure

Depending on how far a boating accident occurs from the coast, thirst, hunger, and exposure will influence the chances of survival. The average person needs at least a half-liter of water each day to remain alive. With no water at all, a person rarely lives more than 5 days. Seawater contains too much salt to be drinkable: The human kidneys cannot produce urine with a salt content greater than 2.2%, and seawater has an average salt content of 3.5%. If too much seawater is taken in, a residual salt content results. The kidneys use body water to wash away the excess salt, causing dehydration, and a human will die once the salt concentration in the body exceeds 0.9% of body weight. (A baleen whale can ingest seawater because its kidneys secrete urine with a salt content of 4%.)

The human body can exist for 6 weeks without food, although tissue damage occurs as the body begins to break down its own proteins to stay alive. Obviously, hunger is less of a problem than thirst.

The term *exposure* is often used to explain the death of mountaineers and hikers and sometimes to imply hypothermia. Actually, exposure is the cumulative weakening and damaging effects, both physical and mental, of the elements—cold, blistering heat, salt spray abrasion, sleeplessness, and fatigue. Good physical condition and good morale help ensure survival—the ordeals of Captain Bligh and Thor Heyerdahl confirm this. Many people die in small boats close to shore just before help arrives because they were inadequately prepared. Warm clothing should be available in a boat. Also, because much solar energy is reflected off the water, some form of head covering must be provided, even on a warm day, to guard against severe sunburn and sunstroke.

FURTHER READINGS

Boling, G. R. "Ice and the Breakers." *Sea Frontiers*, 1971, *17*(6), 363–71.

Burton, R. "Icebreaking by Hovercraft." *Sea Frontiers*, 1980, *26*(2), 78–83.

Harbron, J. D. "Modern Icebreakers." *Scientific American*, December 1983, pp. 49–55.

______. "Arctic Issues Coming to Fore." *Oceanus*, 1985, *28*.

______. "The Arctic Ocean." *Oceanus*, 1986, *29*.

Radok, Y., N. Streton, and G. E. Weller. "Atmosphere and Ice." *Oceanus*, 1975, *18*(4), 16.

Sobey, E. "Ocean Ice." *Sea Frontiers*, 1979, *25*(2), 66–73.

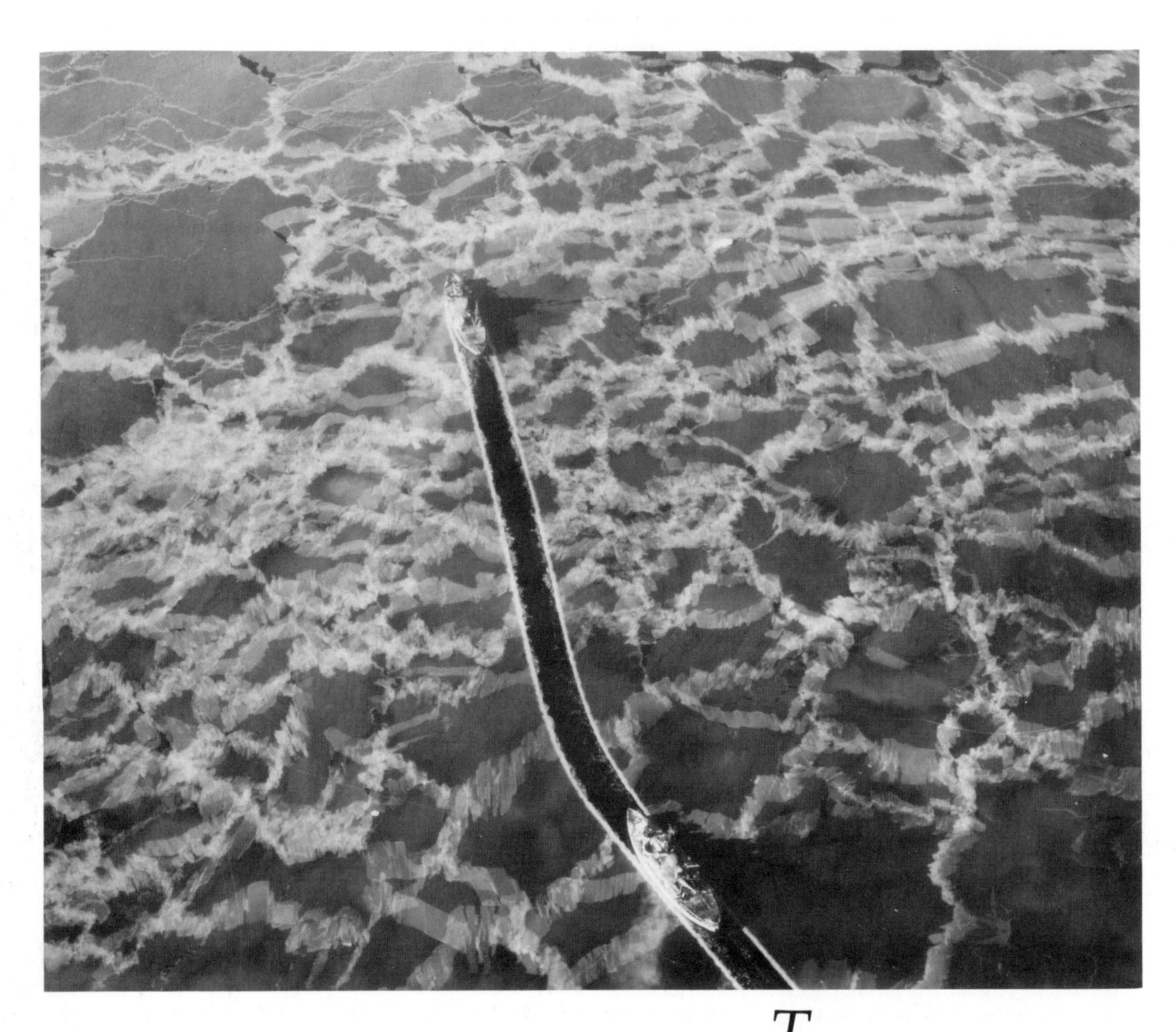

There is a tide in the affairs of men
Which, taken at the flood, leads on to fortune;
Omitted, all the voyage of their life
Is bound in shallows and in miseries.
On such a full sea are we now afloat,
And we must take the current when it serves
Or lose our ventures.

William Shakespeare

EIGHTEEN

Marine Resources and Ocean Technology

Seventy percent of the surface of the earth lies beneath the ocean. We shall inevitably turn again and again to the ocean for transportation, research and exploration, resource development, and, regrettably, military ventures. The ocean today is what the vast American continent was to the early pioneers—a huge domain ready for exploitation. However, we are living with the consequences of the mindless exploitation of many generations—polluted rivers, polluted skies, scarred hills stripped of their mineral resources, plains denuded of sod and soil and exposed to the winds. After two centuries, we are beginning to realize that uncontrolled exploitation impoverishes the future.

How shall we chart our course over the ocean? Will we wrest everything we can from a vulnerable environment? Or will we be sensitive to that vulnerability? Will we accept our responsibility to safeguard the charted and uncharted seas—the wellspring of weather and climate, of the oxygen we breathe, of life itself?

SHIPPING

The technology of marine transportation has advanced dramatically in recent years, mainly in commercial and military transportation but in short-haul passenger service as well.

This chapter begins with shipping because the world economy is so dependent on it. Most international trade is transported on ships. This has been the case for some 400 years, since the New World and the Orient began trading with Europe. Shipping is also cheap. The costs per ton of transporting goods via ship are about 2.5% of the costs per ton via air. Similar savings occur even within countries. Shipping via rivers, lakes, and canals has always been and still is cheaper than shipping via trucks or railroads. Career possibilities are excellent in the shipping business. The key word in the shipping business is *international*. The trade is between countries; ships are built in one country and registered in another. Seafarers come from many different countries.

The world shipping fleet has an interesting profile. The present fleet has a capacity of almost 405 million gross tons, or about 800 million deadweight tons. The principal merchant fleets are registered in Liberia (52.65 million gross tons), Panama, Japan, and Greece (see Table 18.1). The United States ranks only sixth in gross-ton shipping capacity. This figure is, however, misleading because many corporations from the United States, England, and Japan register their ships in Liberia, Panama, Greece, and Cyprus. By doing this

Table 18.1 *Principal Merchant Fleets, 1986*

Country	Gross Tonnage (× 1,000)
Liberia	$52,649
Panama	41,305
Japan	38,488
Greece	28,391
USSR	24,961
USA	19,901
China (PR)	15,840
UK	11,567
Cyprus	10,617
Norway	9,295

Source: Lloyd's Register of Shipping by country and gross tonnage.

Table 18.2 *Ship Types, 1986*

Type	Gross Tonnage (× 1,000)
Oil Tankers	124,140
Ore/Bulk Carriers	111,641
General Cargo	73,245
Non Trading	32,302
Container Ships	19,609
Liquified Gas Carriers	9,832
Other Trading	5,028
Oil/Chemical Tankers	4,286
Chemical Tanker	3,560
	404,910[a]

[a]Down from 1985 by 11,385,000.
Source: Lloyd's Register of Shipping.

(flying "flags of convenience"), the owners avoid many costly safety and union regulations.

The predominant ship types are listed in Table 18.2. Note that the most tonnage is found in oil tankers and ore or bulk carriers (coal, rice, and so on).

Many of the oil tankers now plying the ocean can carry more than 300,000 metric tons of oil (Figure 18.1); 20 years ago, the largest tanker could carry only 30,000 deadweight tons. These supertankers greatly increase the potential danger of pollution, and oil spills were more frequent during the decade of the 1970s. Marine life has been disastrously affected both by the spills themselves and by the dispersants used in the cleanup. Over time, bacteria, mussels, oysters, snails, and many other organisms can break oil down, and barnacles, bryozoans, and other organisms can attach themselves to tar clumps without harm. However, many diving birds, including bald eagles, ospreys, cormorants, and pelicans, die when their feathers are soaked with oil. Clearly, tankers must be equipped with more precautionary devices to reduce the incidence of oil spills.

18.1 The supertanker Universe Ireland *has a capacity of 300,000 tons. (Gulf Oil Corporation)*

In other matters, technological advances have been impressive. Navigational techniques have been greatly improved, and many harbors have adopted stringent rules for ship travel. Pressure from insurance companies has led to more rigorous requirements for training personnel and for maintaining safety and navigational aids. Finally, new techniques for tracing oil spills make it possible to hold ship operators accountable for their errors, and improved cleanup procedures have reduced the impact of spills on the environment.

Overall, 81% of the world shipping fleet is powered by diesel fuel. Some ships are powered by coal. Because of anticipated cost increases and diminishing world reserves, long-range plans call for more coal-powered ships and, perhaps, an increase in nuclear-powered and wind-assisted ships.

In the 1980s, there has been a decline in the number of very large ships, those larger than 100,000 gross tons. There were 564 in 1984, only 512 in 1985, and 447 in 1986. The main reason for this trend is changes in oil transportation arising from a lack of adequate port facilities to handle ships with deep draft and limited maneuverability.

A ship is an excellent long-term investment, but competition limits profitability. About 58% of the world fleet is more than 10 years old, and 32% of the United States' fleet is more than 20 years old. In fact, about 5% of the world fleet is over 25 years old. The Federal Republic of Germany (West Germany) has the newest fleet, with 68% less than 10 years old.

The United States' fleet peaked in size during World War II. In 1948, shortly after the war, the United States had 5,225 ships, with a gross tonnage of 29,165,000.

18.2 *Diver among fish examining a sand channel off the northern coast of Jamaica. (Official U.S. Navy photograph)*

18.3 SeaLab II *habitat, which was used by the U.S. Navy off La Jolla, California. (Official U.S. Navy photograph)*

18.4 *A cutaway model of the* Tektite I *habitat. (U.S. Navy)*

Table 18.3 *Deep-Diving Submersible Vehicles Used in Research*

Vessel (Type)	Operator	Depth Limit (m)
DSRV (Rescue vehicle)	U.S. Navy	1,600
Deepstar 4,000 (Submarine)	Westinghouse	1,200
Alvin (Submarine)	Woods Hole Oceanographic Institution	2,000
Star I, II, III (Submarines)	General Dynamics	2,000
Deep Quest (Submarine)	Lockheed Aircraft Corporation	2,200
Beaver (Submarine)	North American Rockwell	1,800
Aluminaut (Submarine)	Reynolds Metals	5,000
Trieste (Bathyscaphe)	U.S. Navy	12,000+
Ben Franklin (Mid-Water Drifter)	Grumman Industries	≈750

Our shipping industry went into gradual decline in the ensuing years, with ship owners switching registry and fewer ships being built in the United States. A postwar minimum was reached in 1974. Since then, there has been a gradual increase to 6,496 ships, with a gross tonnage of 19,900,000 in 1986.

Ship-building is a major part of the shipping industry and reflects the expectations for growth in international trade. There has been a gradual increase in the number of ships built recently. There has also been a shift in the dominant ship-building countries. This part of the industry, dominated by the United States 40 years ago, is now mainly a Japanese industry. Japan, in 1985, built 42% of all the new ships larger than 1,000 gross tons; the Federal Republic of Germany was second with 7%; South Korea had 6%; the United States had only slightly more than 3%.

SUBMERGENCE

Scuba Gear and Underwater Habitats

Divers can now descend into the marine environment for many purposes, under a variety of conditions, and for varying periods of time. Self-contained underwater breathing apparatus (scuba) consists of a tank of

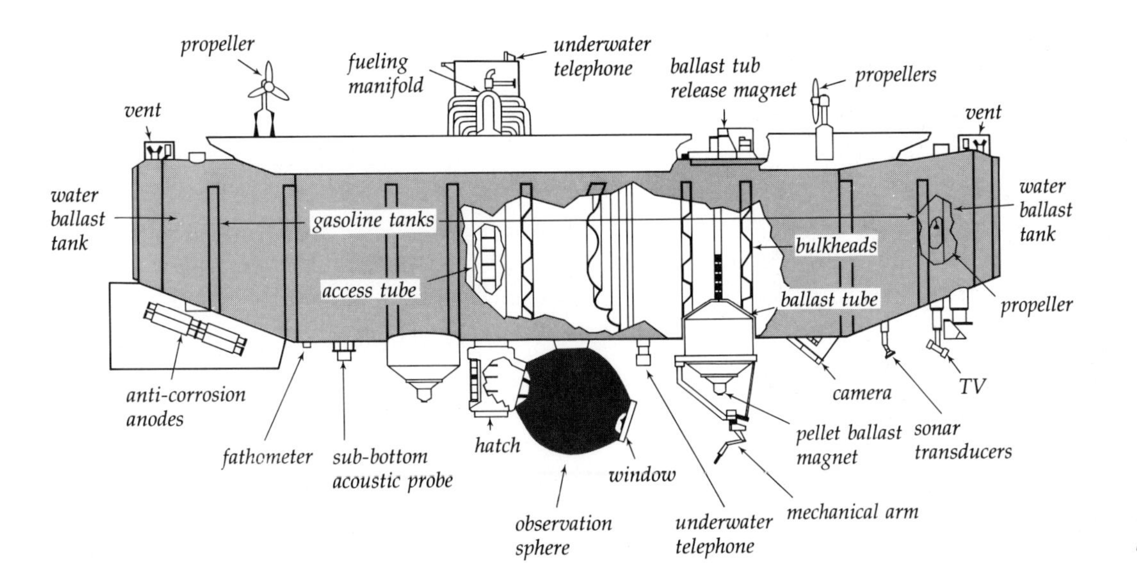

18.5 *A diagram of the bathyscaphe* Trieste.

compressed air, an air regulator attached to the tank, a mouthpiece, a face plate, rubber flippers, a rubber-insulated wet suit, a weight belt, and an inflatable life preserver. With this equipment, an experienced diver can explore the nearshore ocean to a depth of 40 m (131 ft; Figure 18.2).

Such recent developments as recycled-air units and special gas mixtures have enabled professional divers to reach depths of more than 300 m (984 ft) on test dives and to work for as long as an hour at about 200 m (656 ft). These activities are still in an experimental stage, and several experienced divers have died testing the apparatus. Nevertheless, amateur divers using scuba gear are now a common sight in coastal waters around the world.

Underwater habitats for use in research represent another recent advance. The first two phases of the SeaLab Project (Figure 18.3), sponsored by the U.S. Navy, were conducted in the 1960s off the coast of La Jolla, California. Crews spent 2 weeks inside an underwater habitat at a depth of approximately 65 m (213 ft) working on various scientific and technological problems.

The Musée Oceanographique (founded in 1910) in Monaco sponsored the Conshelf Project in the 1960s under the direction of Jacques Cousteau, the co-inventor of scuba. The main purpose of the project was to test human manipulative skill and physical endurance at a depth of about 65 m. Divers trained onshore to assemble an oil-well valve system successfully assembled the device on the bottom of the sea.

The Tektite Project (Figure 18.4) was a joint effort of private business firms, governmental agencies, and academic institutions. Its main purpose was to make observations and to conduct scientific experiments on marine life at a depth of 20 m (66 ft) in Lameshur Bay at St. John's, one of the U.S. Virgin Islands. *Tektite II*, conducted in part by an all-female diving team, carried on research activities with an underwater habitat through the 1970s.

Finally, the National Oceanic and Atmospheric Administration has recently launched an underwater laboratory program in association with the University of Hawaii and the University of Southern California.

Deep-Sea Submersible Vehicles

Submersible vehicles now enable scientists to enter the marine environment and make observations and measurements with almost no disturbance of the environment. These deep-diving submersibles make every part of the ocean accessible to observation.

There are four types of submersible vehicles: bathyscaphes, submarines, rescue vehicles, and midwater drifters (Table 18.3). Bathyscaphes are deep-diving chambers with little or no capability to move horizontally. They move vertically by adding or dropping ballast. The research bathyscaphe *Trieste*, operated by the U.S. Navy, set the world record for manually operated deep-sea dives in 1960 when it carried two men to a depth of about 12,000 m (7.4 mi) in the Challenger Deep, southwest of Guam (Figures 18.5 and 18.6).

Submarines capable of vertical and horizontal propulsion generally have observation portals, externally mounted mechanical arms, and attachments for collecting plankton, water, and sediment and for mounting television and photographic equipment (Figure 18.7).

Since 1975, attention has centered on unstaffed vehicles, called remotely operated vehicles (ROVs), which cost less to operate and pose no danger to crews (Tables 18.4 and 18.5 and Figure 18.8). These vehicles

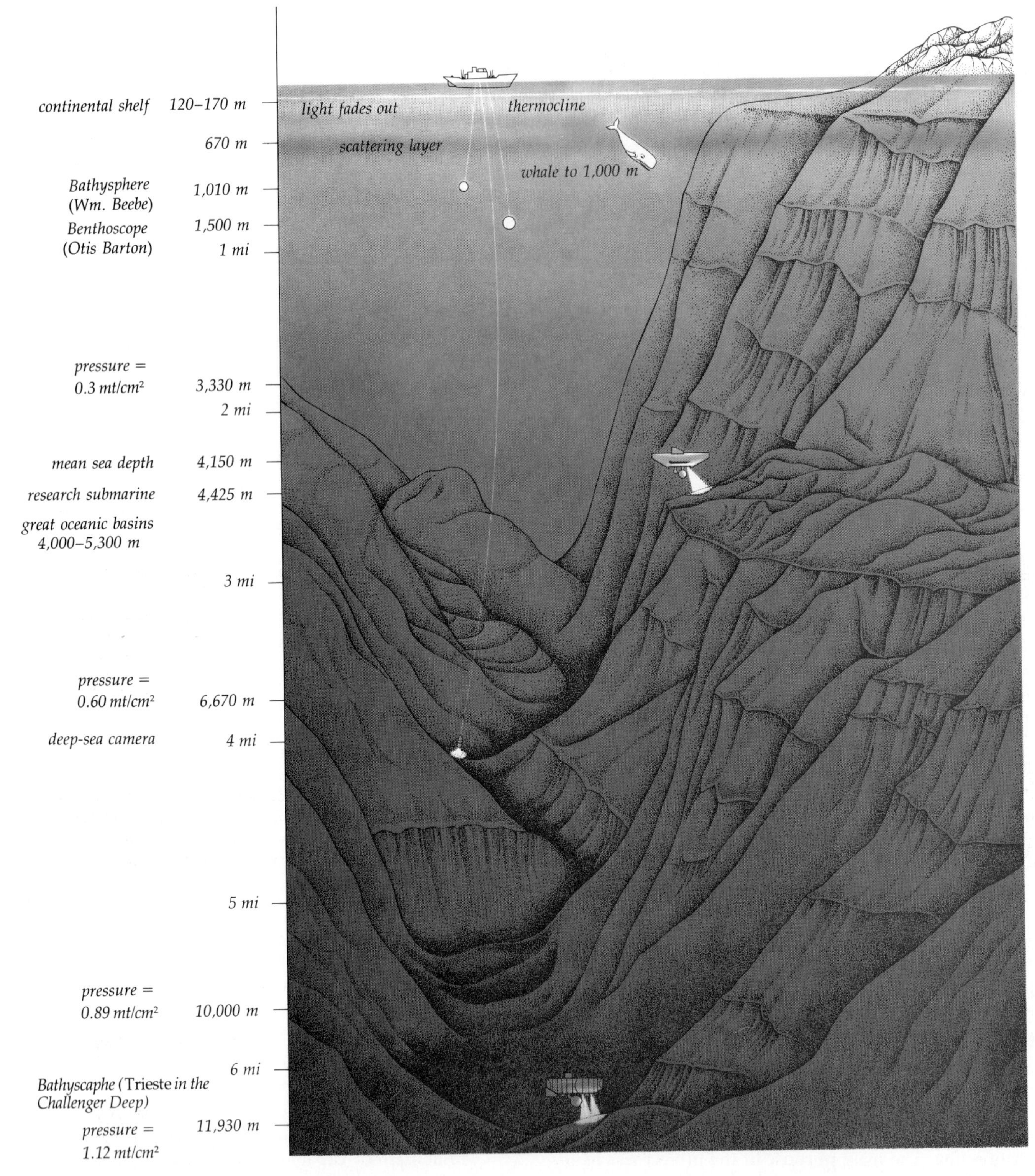

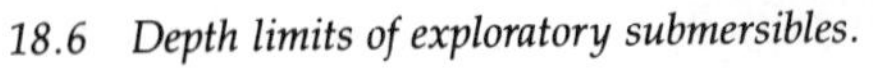

18.6 *Depth limits of exploratory submersibles.*

18.7 The Aluminaut, *a research submarine. (Reynolds Metals Company)*

Table 18.4 *Towed, Unmanned Submersible Vehicles*

Name	Owner	Operating Depth (m)	Dry Weight (kg)	Maximum Speed (km/h)	Work Equipment
ARGO-JASON	Woods Hole Oceanographic Institution	6,000	1,818	2.8	TV, still cameras, side-scan sonar, computer control, ROV Jason sub-system
BATFISH	Bedford Institute of Oceanology, Halifax, N.S.	198	71	9.3	TV, still camera, side-scan sonar
CRAB	Institute of Oceanology, Moscow	4,000	702	N.A.	TV
DEEP TOW	Marine Physics Lab., San Diego, Calif.	6,096	907	2.8	TV, still camera, magnetometer, side-scan sonar, subbottom profiler, echo sounder, differential pressure gauge
DSS-125	Hydro Products, San Diego, Calif.	6,096	2,313	2.8	TV, still camera
NRL System	Naval Research Lab., Washington, D.C.	6,096	998	5.6	TV, still cameras, magnetometer, side-scan sonar, water sampler, sub-bottom profiler
RUFAS II	National Marine Fisheries Services, Washington, D.C.	731	453	11.1	TV, still camera, scanning sonar
S^2	University of Georgia, Athens, Ga.	1,829	317	11.1	TV, side-scan sonar, magnetometer, subbottom profiler, dredge
SEA PROBE[a]	Alcoa Marine Corp., Washington, D.C.	3,048	—	0.9	TV (2), still camera (35 mm), side-scan sonar, scanning sonar
STOVE	Submarine Development Group One, San Diego, Calif.	6,096	N.A.	N.A.	still cameras, side-scan sonar
TELEPROBE	NAVOCEANO, Suitland, Md.	6,096	1,588	5.6	TV, side-scan sonar, magnetometer, stereophotography

[a] *Sea Probe* consists of a surface ship, drill string, and equipment pods; the drill string and pods can only operate from its specially designed ship.

Table 18.5 *Self-Propelled, Tethered Submersible Vehicles*

Name	Owner	Operating Depth (m)	Dry Weight (kg)	Maximum Speed (km/h)	Work Equipment
ANGUS	Heriot-Watt Univ., Edinburgh, Scotland	300	386	3.7	TV, 16-mm cinecamera, directional hydrophone, wide-band hydrophone, transponder interrogator
CONSUB 1	British Aircraft Corp., Bristol, England	610	1,360	4.6	TV (2) color and black and white, rock drill, stereo cameras
CONSUB 2	British Aircraft Corp., Bristol, England	610	1,996	4.6	TV (2) color and black and white, stereo cameras
CORD	Harbor Branch Foundation, Ft. Pierce, Fla.	457	327	9	TV, current meter, scanning sonar, temperature sensor, echo sounder
CURV II[a]	Naval Undersea Ctr., San Diego, Calif.	762	1,565	7.4	TV (2), still camera, scanning sonar
CURV III	Naval Undersea Ctr., San Diego, Calif.	3,048	1,814	7.4	TV (2), still camera, scanning sonar
CUTLET	A.U.W.E., Portland, U.K.	305	N.A.	N.A.	TV, directional hydrophone
DEEP DRONE	U.S. Navy SUPSAL	610	544	6.5	TV with 320° pan and 190° tilt, sonar, altimeter
ERIC	French Navy, Toulon, France	1,000	2,000	3.7	TV, still camera, echo sounder
MANTA 1.5	Institute of Oceanology, Moscow	1,500	998	N.A.	TV, various manipulator claws
RCV-225	Various[b]	2,012	82	3.1	TV, still camera, automatic constant depth control
RCV-150	Various[c]	1,829	204	3.7	TV on pan/tilt device, automatic constant depth control

[a] A second *CURV II* is operated by the Naval Torpedo Station, Keyport, Wash.
[b] Seven vehicles total: Seaway Diving, Bergen, Norway (2 vehicles); Martech International, Houston, Tex. (2 vehicles); SESAM, Paris, France (1 vehicle); Taylor Diving and Salvage, Belle Chase, La. (1 vehicle); Esso Australia Ltd., Sale, Australia (1 vehicle).
[c] Two vehicles total: Martech International, Houston, Tex.; and Scandive, Stavanger, Norway.

can remain submerged indefinitely, or at least as long as their equipment continues to function. They are being used for a wide variety of industrial, scientific, and military purposes, including the inspection of oil pipelines and submarine cables, the measurement of physical properties of the ocean floor, and the recovery of equipment. Deep-sea rescue vehicles (DSRVs) are carried by U.S. Navy nuclear submarines for rescuing crew members.

One of the most remarkable submersibles is the mid-water drifting vessel *Ben Franklin*, designed by Jacques Piccard (who also designed the *Trieste*) to drift

Table 18.5 *Continued*

Name	Owner	Operating Depth (m)	Dry Weight (kg)	Maximum Speed (km/h)	Work Equipment
RECON II	Perry Ocean Group, Riviera Beach, Fla.	457	281	4.6	TV on pan/tilt device, current meter
RUWS	Naval Undersea Ctr., Hawaii	6,096	2,268	1.8	TV, scanning sonar
SCARAB I and II	AT&T, Inc., New York, N.Y.	1,829	2,268	0.9	TV, camera, magnetometer, sonar, altimeter
SEA SURVEYOR	Rebikoff Underwater Products, Ft. Lauderdale, Fla.	200	175	9.3	TV, still camera (35 mm), side-scan sonar, magnetometer, depth indicator
SNOOPY	Naval Undersea Ctr., San Diego, Calif.	457	68	1.8	TV, 8-mm cinecamera
SNOOPY	NAVFAC, Washington, D.C.	457	136	1.8	TV, 70-mm still camera
SNURRE	Royal Norwegian Council for Scientific Research	600	1,200	2.8	TV, echo sounder, compass, depth gauge, still cameras, 8-mm cinecamera, directional hydrophones
TELENAUTE 1000	I.F.P., Paris, France	1,006	1,097	5.6	TV, constant depth control, magnetometer
TROV	Canadian Ctr. for Inland Waters, Burlington, Ont.	366	513	1.8	TV, magnetic compass, echo sounder, transponder, transponder interrogator
TROV-01	Underground Location Services Ltd., Stonehouse, Glasgow, Scotland	366	907	2.8	TV, magnetic compass, echo sounder, transponder, transponder interrogator

at relatively shallow depths. On its maiden voyage in 1969, Piccard and his crew drifted with the Gulf Stream from Florida to a region off Long Island, New York. In addition to numerous military submarines, almost one hundred submersibles are either in use or in the design stage.

Perhaps the most spectacular adventures to date involving a submersible have been with the deep-submersible vehicle *Alvin*. The first came in March 1977, when scientists discovered the Galápagos hot springs. This led to major changes in our understanding of ecology, evolution, the origin of life, and chemical cycles

18.8 RUM III *(remote underwater manipulator) is an ROV designed to operate at the end of 10,000 m (30,000 ft) of double-armored coaxial cable above and on the seafloor. The cable supplies power to the ROV and also serves as a communications link for control and observation. Two full-width tracks are capable of positioning the ROV on the very soft sediment of the deep-sea floor, and rotatable counter weights maintain the ROV's balance. Vehicle control and data transmission are fully computerized. (Scripps Institution of Oceanography, University of California, San Diego, 1985)*

(a)

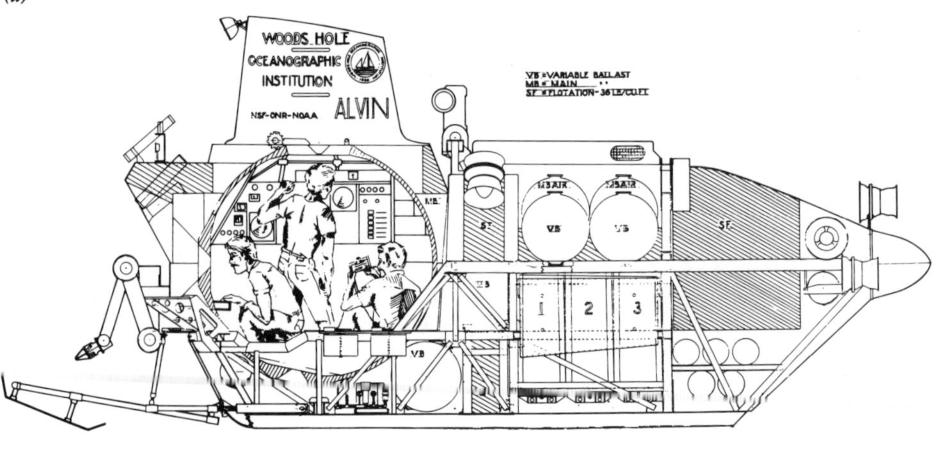

(b)

18.9 The DSV Alvin is the most active manned submersible, having made almost 2,000 dives. It can dive to a depth of about 4,000 m. © 1986 Woods Hole Oceanographic Institute. Used by permission.

July 13, 1986—Dive #1705
Ballard reported today that the *Atlantis II* arrived at the *Titanic* site at about 9:30 p.m. Saturday, and set the three-transponder net around the wreck. Weather is "excellent." They found the ship with no problem.

Ballard said there was about a ½ knot current at the bottom and reported seeing "a huge black wall" (the starboard side of the ship). The dive was very brief and the sub surfaced early in the afternoon because of a saltwater leak into *Alvin's* battery pack. About 3 p.m., ANGUS (Acoustically Navigated Geophysical Underwater Survey) was deployed to do a 33 mm picture run through the night. (Ballard would later liken the side of the *Titanic* to a giant sequoia. "Being in *Alvin* was like having your nose pressed up against the bark. You couldn't see the forest for the tree.")

Table 18.6 *Data on Key Nonliving Ocean Resources*

	Crude Oil	Natural Gas	Phosphate Rock	Manganese
Seabed Production (10^3 metric tons [MT])	788,834	246,670		
World Mine Production (10^3 MT)	2,788,913	1,296,405	159,000	23,406
Estimated Average Price ($ per MT)	70	95	24	141
Seabed Revenues[a] ($ in millions)	55,218	23,434		
World Revenues[b] ($ in millions)	195,224	123,158	3,816	3,300
Seabed Share of World Revenues[c] (%)	28	19		
Seabed Reported Potential Resources (10^3 MT)	>61,429,000	>60,000,000	7,939,000	706,000–2,600,000
World Onshore Resources (10^3 MT)	181,857,000	228,214,000	129,500,000	10,886,400
Seabed Comparison to World Resources [d](%)	34	26	6	6–24
"Resource Life Index"[e] (years)	65	176	814	465
Projected Onshore Depletion by Year 2030[f]	100	45	12	17

[a]Seabed production times estimated average price.
[b]World mine production times estimated average price.
[c]Seabed revenues times 100, divided by world revenues.
[d]Seabed reported potential resources times 100, divided by world onshore resources.
[e]World onshore resources divided by world mine production.
[f]Based on low growth case for developing economies.

in the sea. The second was the fantastic discovery and photographs of the Royal Mail Service (RMS) *Titanic* by Dr. Robert Ballard and others on the DSV *Alvin* using the tethered vehicle *Jason Jr.* (see Figure 18.9).

MINERAL EXPLOITATION

Hydrocarbons

Of the many minerals and other natural resources known to exist in ocean-bottom sediments, only a few are being mined commercially. Among those few are oil and natural gas, which are being produced from wells located on the continental shelves in many areas of the world (Table 18.6). Large fields are located off northern Australia, in the Persian Gulf, and in the North Sea. Bordering the United States are three major offshore producing areas: the coast of the Gulf of Mexico, the southern California coast, and the southern Alaska coast (Figure 18.10). The most productive of these is the Gulf Coast.

Another major oil and gas field has been found at Prudhoe Bay on the Arctic coast of Alaska. Conservationists are concerned about potential ecological effects of the trans-Alaska pipeline, which went into operation in the summer of 1977, and about the possibility of oil spills off the northern coast of Alaska. They are also concerned about damage to the pipeline system from earthquakes, glaciers, and permafrost heave. To date, however, the safety record has been excellent.

In 1969, a drill core taken in the middle of the Gulf of Mexico, at a depth of more than 2,000 m (6,560 ft) was found to contain oil-soaked sediment. This was the first time an oil reservoir of potential commercial significance had been located in deep-ocean sediment. The amount of oil contained in such sediments throughout the ocean is unknown.

In an effort to reduce reliance on imported oil, oil reservoirs on the continental margin of the United States will probably continue to be developed. Commercial developers will be guided by the presence of

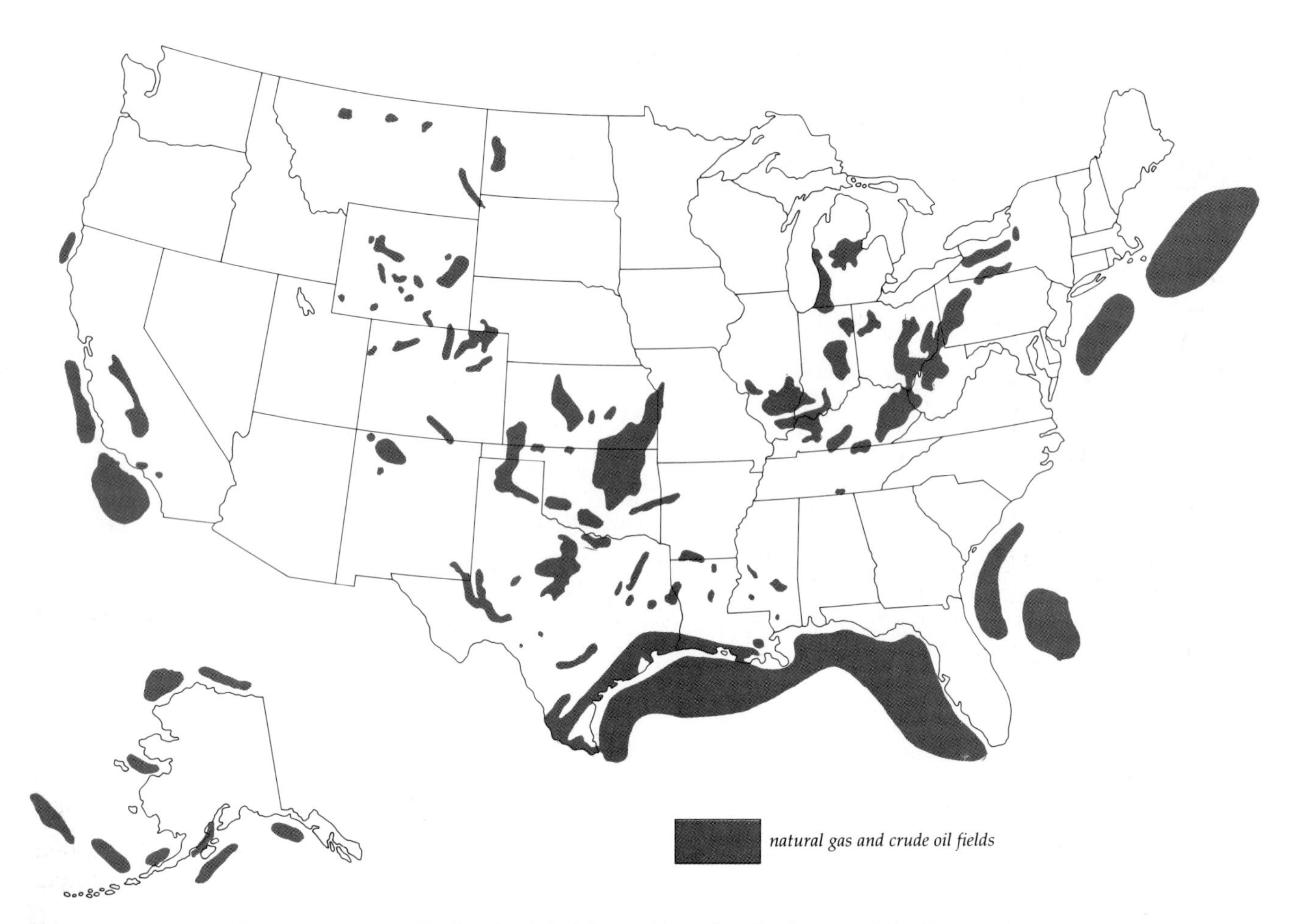

18.10 Major deposits of natural gas and crude oil in the United States. (Council on Environmental Quality, 1983)

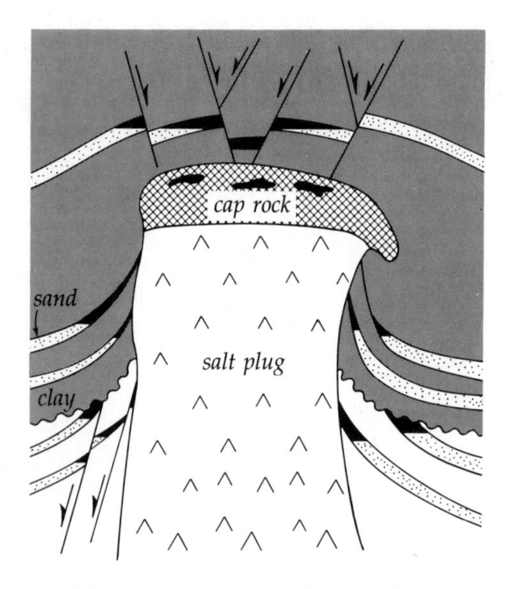

18.11 Idealized cross section of a Gulf Coast salt dome field showing some of the common types of pools (black) found in traps associated with salt intrusion. Many of the flanking traps are wedge shaped.

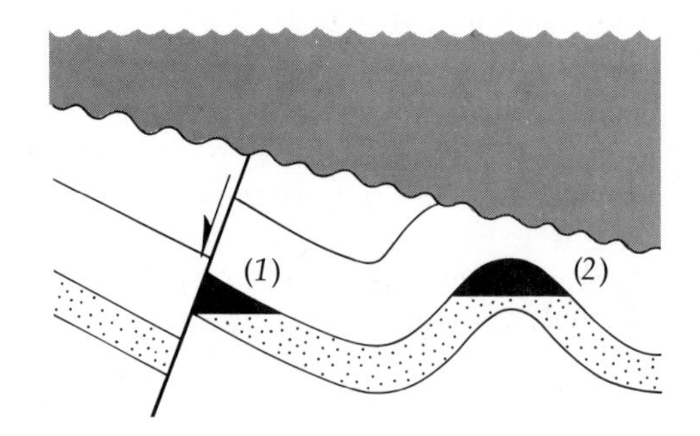

18.12 Idealized oil and gas traps: (1) a trap against a fault and (2) a trap in an anticline. The dark area is the oil and gas; the dotted layer is a permeable sand; the light area includes impermeable layers of clay.

18.13 Oil platform in the North Sea. (Mobil Oil Corporation)

oil in sedimentary layers, the depth of the overlying seawater, and the direction of prevailing winds and currents. Since the federal government receives about $10 billion annually from leases and royalties, we can expect it to continue to encourage offshore exploration.

To be commercially valuable, oil must accumulate in the pore spaces of sedimentary rock. First, water must enter the pore spaces of the rock and force the oil to migrate upward. There must be an impermeable cap that halts the movement of the oil and a structure in which it can accumulate. Cap rocks are typically shale or salt domes. The cap is formed where shale overlies sandstone in an anticline (a fold that is convex upward), where sandstone grades into shale within a layer or where shale adjacent to the sandstone has moved along a fault. In reservoirs associated with active faults, the oil may migrate along the fault zone to the seafloor (Figures 18.11 and 18.12), producing natural seeps.

Offshore developers must answer these questions before undertaking drilling activities:

1. Is the oil reservoir large enough to warrant commercial development?
2. Are there active earthquake faults in the area?
3. Will the prevailing winds and currents carry oil spills away from the shore?
4. Is the water shallow enough to accommodate a drilling platform?
5. What would the ecological, social, and economic impacts of drilling activities be on the community?
6. Would drilling conflict with some of the current uses of oceanic resources (such as fishing and recreation)?

The offshore platforms built on many of the continental margins of the world are impressive examples of specialized equipment (Figure 18.13). Designed to operate in over 3,000 m (9,840 ft) of water, some of them rise 100 m (328 ft) above the water line, weigh about 200,000 metric tons, and control forty producing wells. Such massive facilities are needed to withstand rigorous weather and sea conditions and to maximize efficiency. In 1981, there were 592 offshore rigs in operation: 358 jack-ups (rigs mounted on the seafloor), 120 semisubmersibles, 58 ships, 28 submersibles, and 28 barges. Of these, 217 were located in North America, most of them along the coast of the Gulf of Mexico.

Offshore platforms have had an excellent safety record. Although a few have caught fire and a few have toppled over, causing a serious loss of life, almost all of these accidents were attributed to human error rather than to errors of design.

Manganese Nodules

In response to reports that numerous manganese nodules lie on the ocean floor at depths greater than 2,000 m

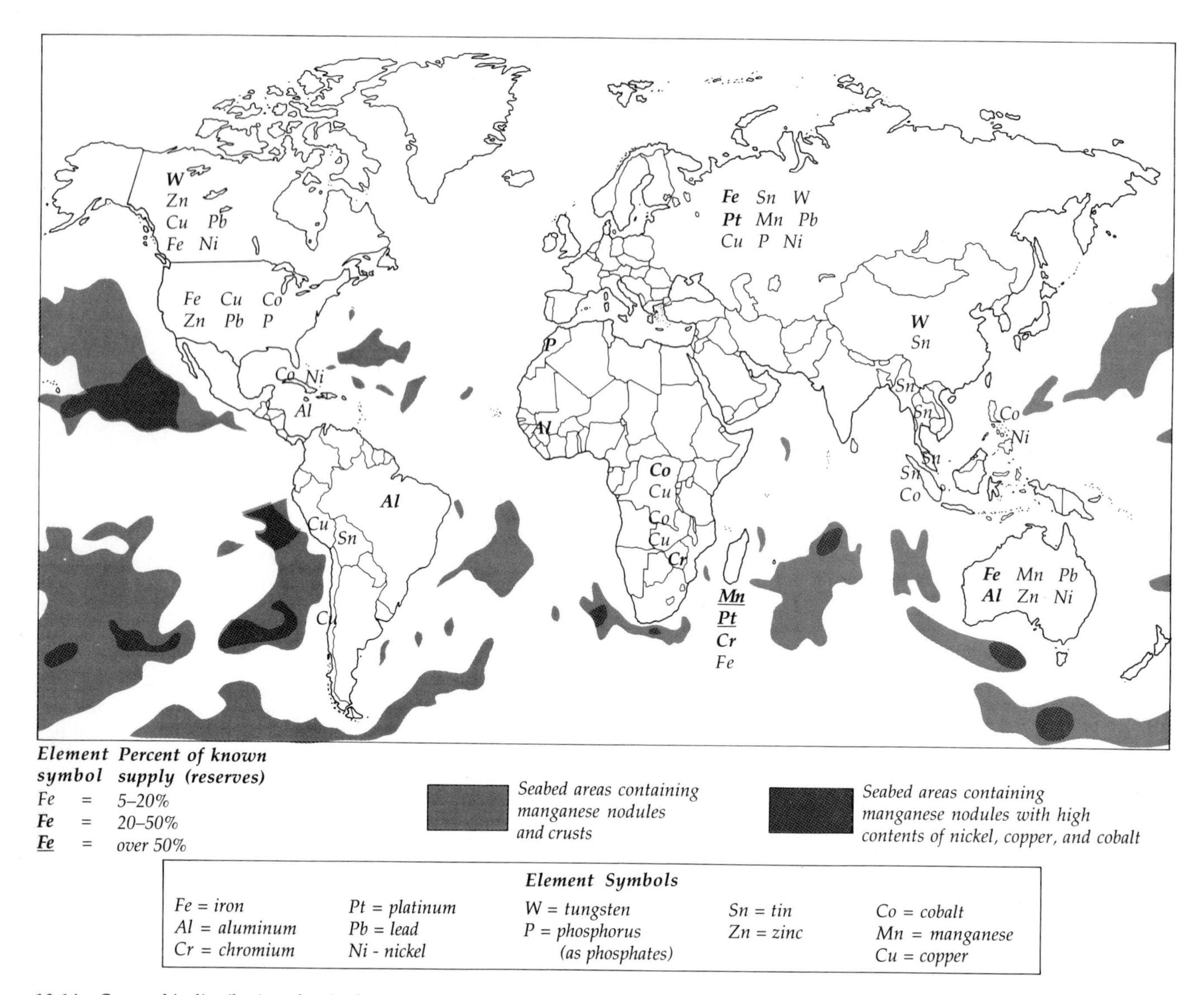

18.14 Geographic distribution of major known supplies (reserves) of thirteen key nonfuel minerals. Positions of symbols do not indicate location of deposits within countries. (Data from U.S. Bureau of Mines, 1983)

(6,560 ft), several companies have tested processes to recover nodules at a cost low enough to return a profit. None of these processes is now economically feasible and will not become so in the foreseeable future. Nodules seem to be most common in the equatorial Pacific (Figure 18.14).

Manganese nodules contain large amounts of iron and manganese and lesser amounts of nickel, copper, cobalt, and other elements. The United States imports 98% of its manganese (which is essential to the steel industry), 90% of its nickel and cobalt, and 20% of its copper. With successful commercial mining of these nodules, the United States might become able to export these elements instead of importing them.

Several independent corporations, along with federally supported research groups at oceanographic institutions, have explored regions of the ocean bottom known to contain nodules. One ship, the *Prospector,* has been especially designed for mining the seafloor, and at least one corporation has filed a mining claim on a section of the Pacific Ocean.

The chemical composition and growth rates of manganese nodules are related to the biological productivity of surface waters. Many nodules contain fossil remains, which may provide the "seed" for precipitating organic remains produced by bacterial decomposition. The activity of mud-eating organisms may keep the nodules, once formed, on the surface of the seafloor.

Phosphate

Phosphate compounds have accumulated on submarine terraces on many continental shelves (Figure 18.15).

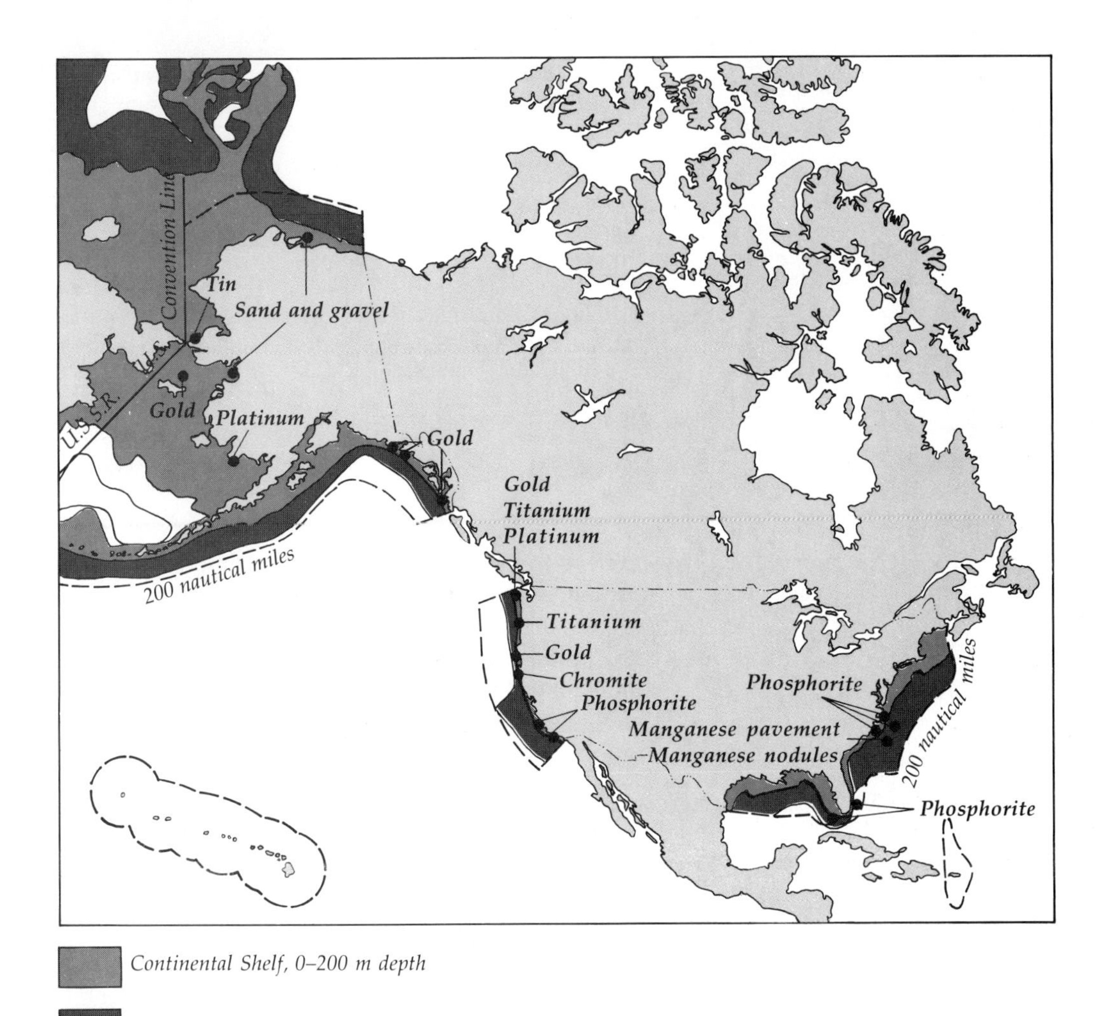

18.15 Areas of the United States coastline having resources or showing resource potential for the commodities indicated. (Office of Technology Assessment, 1987)

These terraces were formed by erosion during periods of continental glaciation, when the sea level was lower than it is now. The three largest known offshore phosphate deposits are off the coast of Florida, off the coast of southern California, and off the Moroccan coast.

Gold

The U.S. Geological Survey regularly conducts programs to explore for gold in all parts of the country (Figure 18.15). One of its main prospecting regions is in river delta sediments on continental shelves that are downstream from gold-producing mountains and rivers. Gold has been found off the Columbia River near Washington State and in many delta regions along the western coast of the continental United States and Alaska. Many deposits show commercial potential.

Diamonds

Diamonds are formed in crystalline rocks. After the host rocks erode, diamonds are transported to deltas by streams. Diamonds are mined in delta sediments on the continental shelf off South Africa, and rich deposits may exist off Brazil and Venezuela.

Salt

Salt deposits are mined around the world from Salzburg, Austria, to Michigan. These deposits were formed naturally by the evaporation of an enclosed bay or sea. Since these sources do not satisfy the total salt needs of the world, salt is also produced by flooding marshes along coasts and letting the water evaporate (Figure 18.16). The salt thus produced is then refined and sold.

18.16 *Salt evaporation ponds at the southern end of San Francisco Bay in California. The ponds are flooded, the water evaporated, and the remaining salts removed for separation and purification. (George Knight)*

Table 18.7 *Gross Value of Metals in Sediments Collected from the Atlantis II Deep*

Metal	Average Assay (%)	Metric Tons	Value[a] (millions)
Zinc	3.4	2,900,000	8,600
Copper	1.3	1,060,000	12,700
Lead	0.1	80,000	200
Silver	0.0054	4,500	2,800
Gold	0.0000005	45	500
Total			$24,800

[a] Based on 1983 metal prices.
Source: Bischoff, J. L., and Manheim, F. T. "Economic Potential of the Red Sea Heavy Metal Deposits": In Degens, E. T., and Ross, D. A. (eds.), *Hot Brines and Recent Heavy Metal Deposits in the Red Sea.* New York: Springer-Verlag, pp. 535–41, 1969.

Importance of Undersea Mining

Among the most recent discoveries on the seafloor are muds rich in metallic minerals near hot springs on the ocean ridges. The value of such muds in the Atlantis II Deep in the Red Sea, for example, has been estimated at about $25 billion (Table 18.7). Research suggests that valuable muds of this sort are common along oceanic ridge systems. Discoveries have also been made along the East Pacific Rise, the Gordo Rise, and the Juan de Fuca Ridge. These muds are rich in copper, nickel, iron, cobalt, zinc, and other metals.

In a 1975 report titled "Mining in the Outer Continental Shelf and in the Deep Ocean," the National Academy of Sciences predicted that by the year 2000 the United States will be deficient in the following

Table 18.8 *United States Reliance on Minerals That Could Be Recovered from the Oceans, 1983*

Manganese	99%	Silver	61%
Cobalt	95%	Iron Ore	37%
Platinum group metals[a]	90%	Vanadium	26%
Chromium	76%	Copper	19%
Nickel	75%	Gold	19%
Zinc	65%	Sulfur	15%

[a] Includes platinum, palladium, rhodium, iridium, osmium, and ruthenium. Only platinum is likely to be recovered from the seafloor.
Source: U.S. Bureau of Mines, 1984.

mineral commodities: aluminum, antimony, asbestos, barium, bismuth, cadmium, cesium, chromium, cobalt, copper, diamonds, fluorine, germanium, gold, graphite, indium, lead, magnesium, mercury, mica, nickel, niobium (columbium), platinum, quartz crystal, sand and gravel, silver, sulfur, tantalum, tin, tungsten, and uranium (Table 18.8).

The report further said that:

> the development of marine resources is important to the maintenance of the international economic and political balance and to support the standard of living in the United States. While it is probably not feasible or desirable for the United States to become self-sufficient for the basic mineral commodities, the panel considers it prudent to develop adequate alternate sources of supply from the sea.
>
> Estimates of apparent marine mineral resources have been developed by M. Cruickshank for dissolved, unconsolidated, and consolidated deposits. *With the exception of asbestos, graphite, and quartz crystals, where data are available and deficiencies have been predicted, alternative marine sources for the minerals exist and may exceed existing land resources.* While few of these reserves have been positively identified at the present time, certain specific commodities have been found along the outer continental shelf and on the deep sea bed. As marine mining and extractive technology are developed, it is believed that these apparent resources will become viable mineral sources.

Desalination

Of great significance to nations the world over is the recent development of desalination methods for treating seawater. Industrial nations use huge quantities of fresh water, and many developing nations lack suffi-

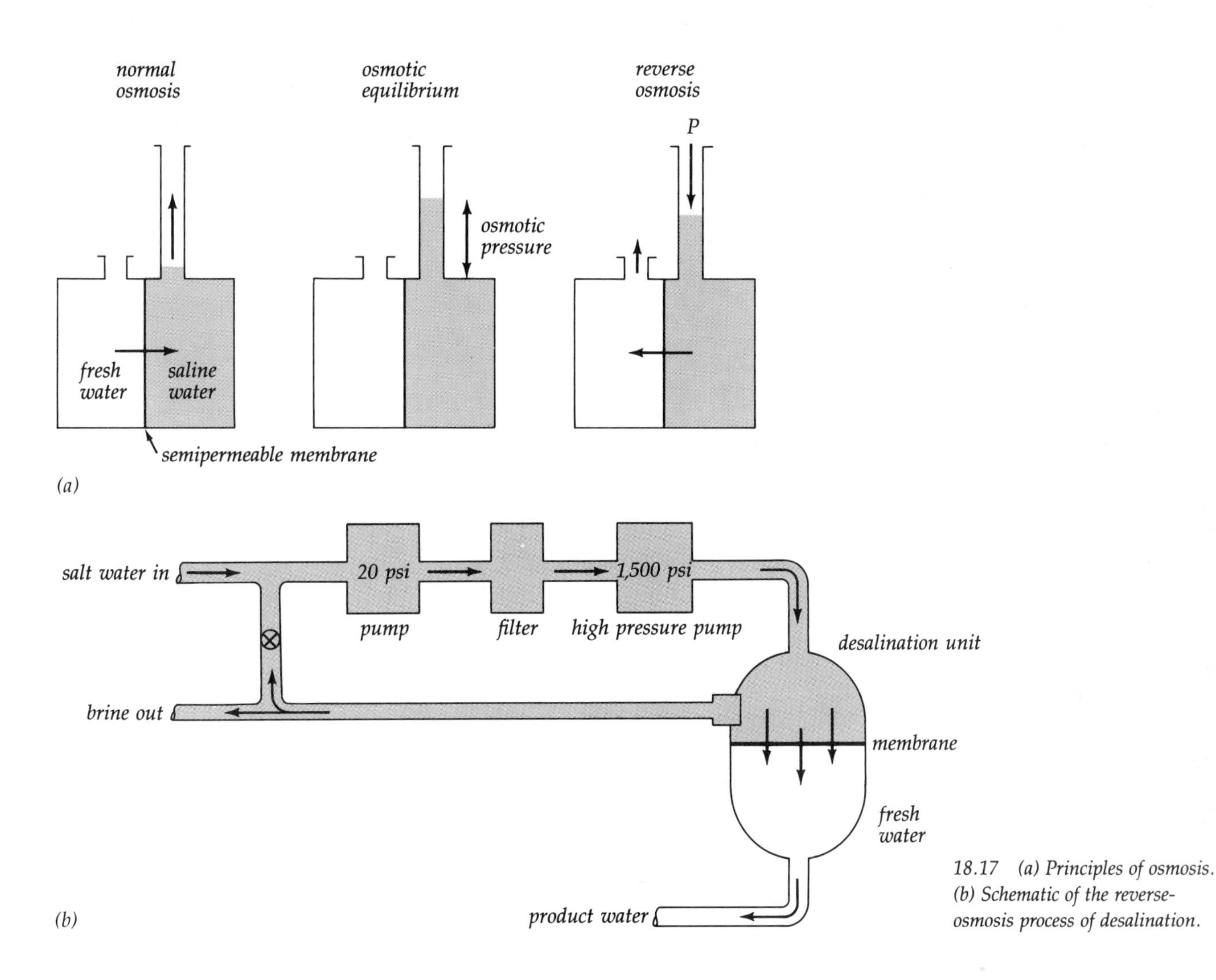

18.17 (a) Principles of osmosis. (b) Schematic of the reverse-osmosis process of desalination.

cient water to grow crops to feed their people. Nations situated near the ocean, such as Ethiopia, Tanzania, and Peru, will especially benefit from desalination technology.

Six methods of desalination are presently in use: distillation, humidification, freezing, reverse osmosis, electrodialysis, and salt absorption. In simple distillation, saline water is boiled and the water vapor is channeled through tubes, where it is cooled. The fresh water that condenses is then collected. The salt precipitates in the container in which the saline water was boiled. This process, though simple, requires an input of nonrenewable energy. Variations on this process have been designed to minimize the energy consumed and maximize the fresh water produced.

A desalination method that does not depend on nonrenewable energy is solar humidification, which employs evaporation and condensation under controlled conditions. Salt water is placed inside a container. Solar radiation then heats the water until it evaporates. As water vapor comes into contact with the cover of the container, it condenses and trickles down to a freshwater catching trough. Experimental, large-scale versions of this process have been used in Israel, West Africa, and Peru to water crops with desalinated ocean water.

When seawater freezes under natural conditions, most of the salt stays in solution in the liquid that remains. The high salinity of the remaining solution gives it high density as well. When ocean water freezes, this dense solution sinks, and the ice left floating on the surface has a density even lower than that of pure water. This natural process has been employed to freeze seawater and produce fresh water.

The reverse-osmosis method of desalination is the most promising for large-scale use. Osmosis can be used to cause less salty water to migrate to more salty water through a semipermeable membrane until equilibrium is reached. In reverse osmosis, water is made to flow from a salty solution to a less salty solution (Figure 18.17). In this process ocean water is separated from fresh water by a semipermeable membrane. Pressure is applied to force the ocean-water solution into the freshwater solution. The water passes through

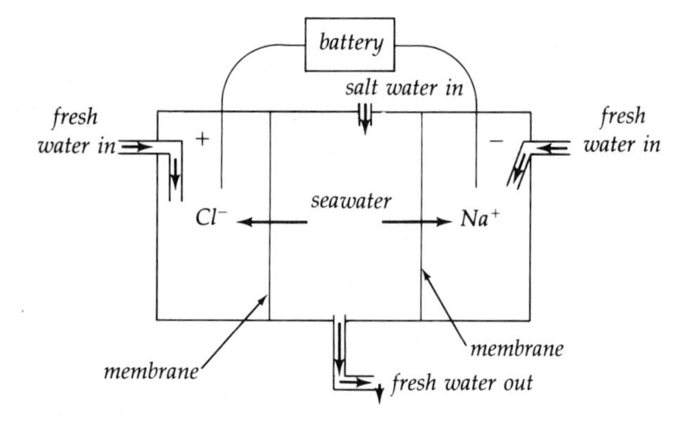

18.18 *Schematic of the electrodialysis process.*

Table 18.9 *Estimated Power of Marine Energy Resources*

Energy Source	Estimated Power (MW × 10^6)
Waves	2.5
Tides	2.7
Currents	5.0
Salinity	1,400
Thermal gradient	40,000

Note: Estimated world power needs by the year 2000 are 33 × 10^6 MW.

the membrane but the salts cannot. This process has also been used to recover fresh water from sewage water and industrial waste water. The semipermeable membranes currently in use are fragile, however, and must be replaced frequently.

In electrodialysis, two semipermeable membranes permit the salt ions to move but not the water. The membranes are positioned to separate volumes of fresh water from a volume of seawater placed between them. An electrode is placed in each container of fresh water. When a current passes through the solutions, positive ions (such as sodium ions) are attracted to the negative electrode, and negative ions (such as chloride ions) are attracted to the positive electrode (Figure 18.18). In this manner the seawater becomes progressively less salty until all the salt ions have been removed. This process is not currently used for large-scale desalination.

The U.S. Navy uses yet another method of desalination in its survival kits. This method absorbs salt by ion exchange. A small cake of activated charcoal and resin placed in a plastic bag containing ocean water absorbs the salts from the water.

Since the per capita consumption of fresh water is rising, and since freshwater pollution is increasing, the need for improved desalination technology is urgent. The costs of desalination have already dropped enough to make its use practicable in many parts of the world, from Key West, Florida, to Israel. The evaporative methods are the least costly, especially when preheated water can be used. Reverse osmosis is being used in Yuma, Arizona, to remove salts from the water of the Colorado River.

ENERGY FROM THE SEA

In the ever-increasing literature on alternative energy resources, energy from the sea receives prominent attention. First, there is a huge energy potential embodied in the sea (Table 18.9). Second, there are some practical methods of harnessing some of that energy to produce electrical power.

Tides

As we mentioned earlier, differences in the levels of high tides and low tides can be exploited to produce electricity. Huge turbines convert the mechanical energy of the tides into electrical energy. In bays and estuaries, dams can be built to retard the tide until high tide. Then the locks are opened to let the water flow over the turbines. A minimal tidal variation of about 3 m (9.8 ft) is required to make such an installation commercially practical. Differences of at least 3 m are found in many coastal regions, including Alaska, the Gulf of California, and the New England coast.

The main advantage of tidal power plants is that the energy needed to run them is free. Once a generating plant has been built, the only cost is maintenance. However, such plants do disrupt the ecology of the estuary.

Waves

The periodic motion of a wind-driven wave as it passes over the surface of the sea is analogous to the rise and fall of a piston. The energy contained in surface waves is tremendous. The problem is how to harness it for commercial use. Many ingenious wave generators have been built, but none has produced more than about 500 W.

Several research programs have used wave-powered generators to pump nutrient-rich water up from the ocean depths to fertilize mariculture projects. This system has proved to be very efficient. On a large scale, it would be of great value throughout the Caribbean Sea and Micronesia, providing additional sources of protein and even enabling the islanders to produce commodities for export, such as shrimp.

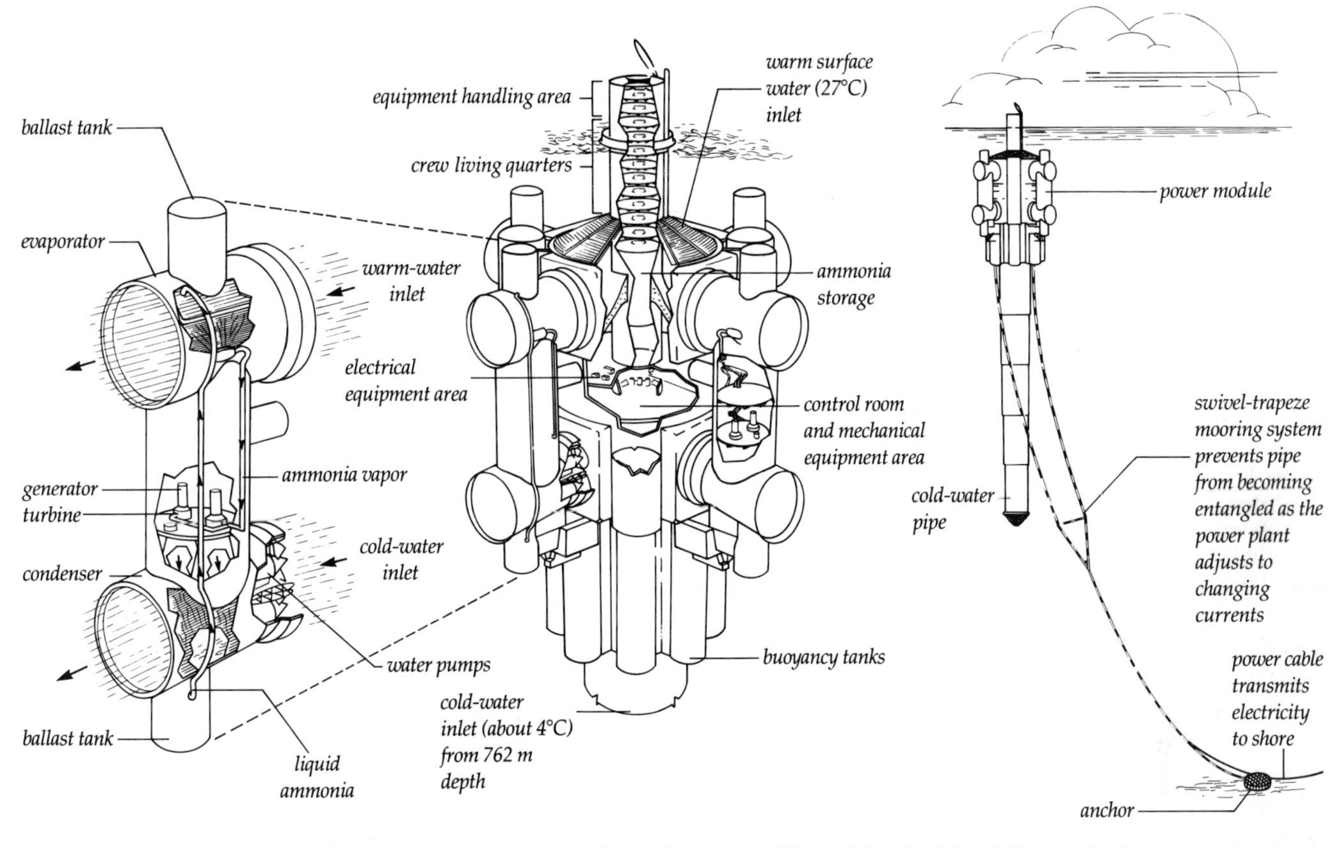

18.19 Cutaway figure of the Lockheed Ocean Thermal Energy Conversion system. The total length of the platform is about 181 m (594 ft), and its diameter is 75 m (246 ft). Each of the power modules is 93 m (305 ft) high and 22 m (72 ft) in diameter. The cold-water pipe, which is used to collect the cold water, is 304.8 m (1,000 ft) long (in five sections). (Courtesy of Lockheed Missiles and Space Company, Inc.)

Thermal Differences

In 1881, the French scientist Jacques d'Arsonval proposed using the temperature gradient between surface waters and deeper waters as an energy source. The temperature gradient between the surface waters in the tropics and the deep water is about 20°C (36°F). A system that could continuously circulate enough water or some other liquid would produce 400 MW or more of power—enough to supply the energy needs of a city the size of San Diego, California, and its suburbs.

The technology for such a facility actually exists. One proposed design calls for a closed system, in which the liquid would be ammonia, propane, or Freon. The vaporization temperature of these liquids suits the temperature differences of the water. Pipes containing the liquid would be immersed in warm surface water, and heated liquid would be vaporized and forced under high pressure through turbine generators to produce electrical power. The vapor would then move on to a condenser in deep, cold water. There the vapor would be condensed into a liquid and passed through a pressurizer back to the warm surface water, completing the cycle. A facility of this type could satisfy all the energy needs of Miami, Jacksonville, or Honolulu, all of which have access to warm surface waters. It could also furnish supplementary power when used with existing conventional generators that produce large amounts of superheated water.

In Hawaii, a privately financed test project called Mini-OTEC (Figure 18.19) has produced 50 kW of power, and a larger OTEC-I is being built to produce 1 MW of power. Within the next few decades, several plants powered by the thermal gradient of ocean waters will almost certainly each be producing at least 250 kW of power.

FISHERIES

During the past 40 years, the world's marine and nonmarine fishery catch has increased steadily, particularly the marine catch. However, Figure 18.20 shows that the upward trend has recently been halted despite improved techniques for locating and capturing fish. Several explanations are possible, including pollution, the degradation of habitats, overfishing, and shifts in the currents that provide nutrients. It is likely that

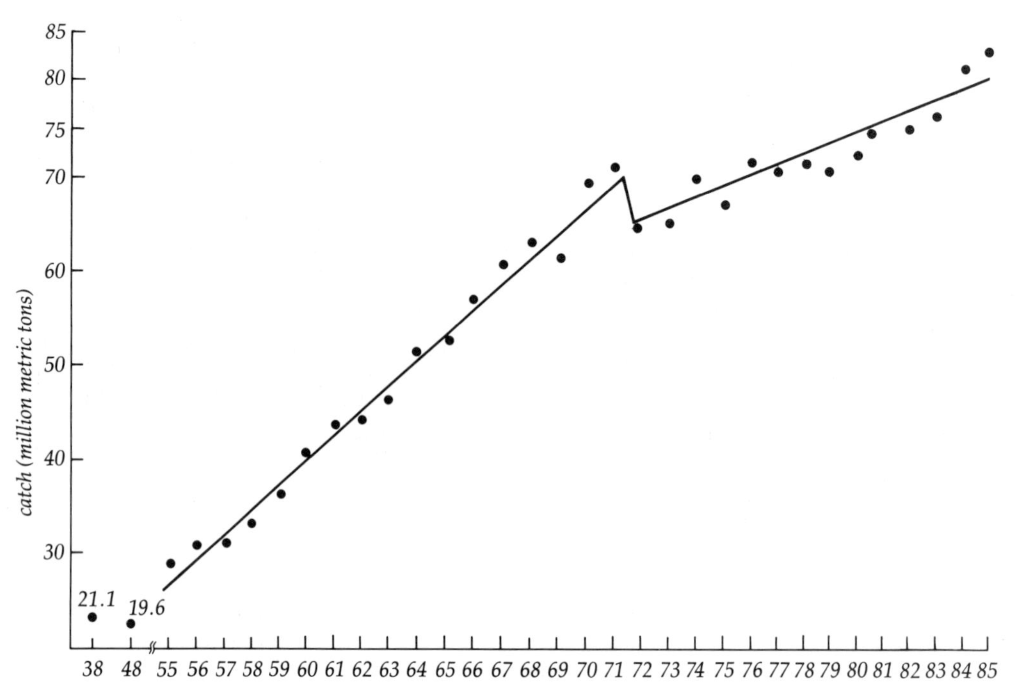

18.20 *World fishery catch of marine and nonmarine organisms.*

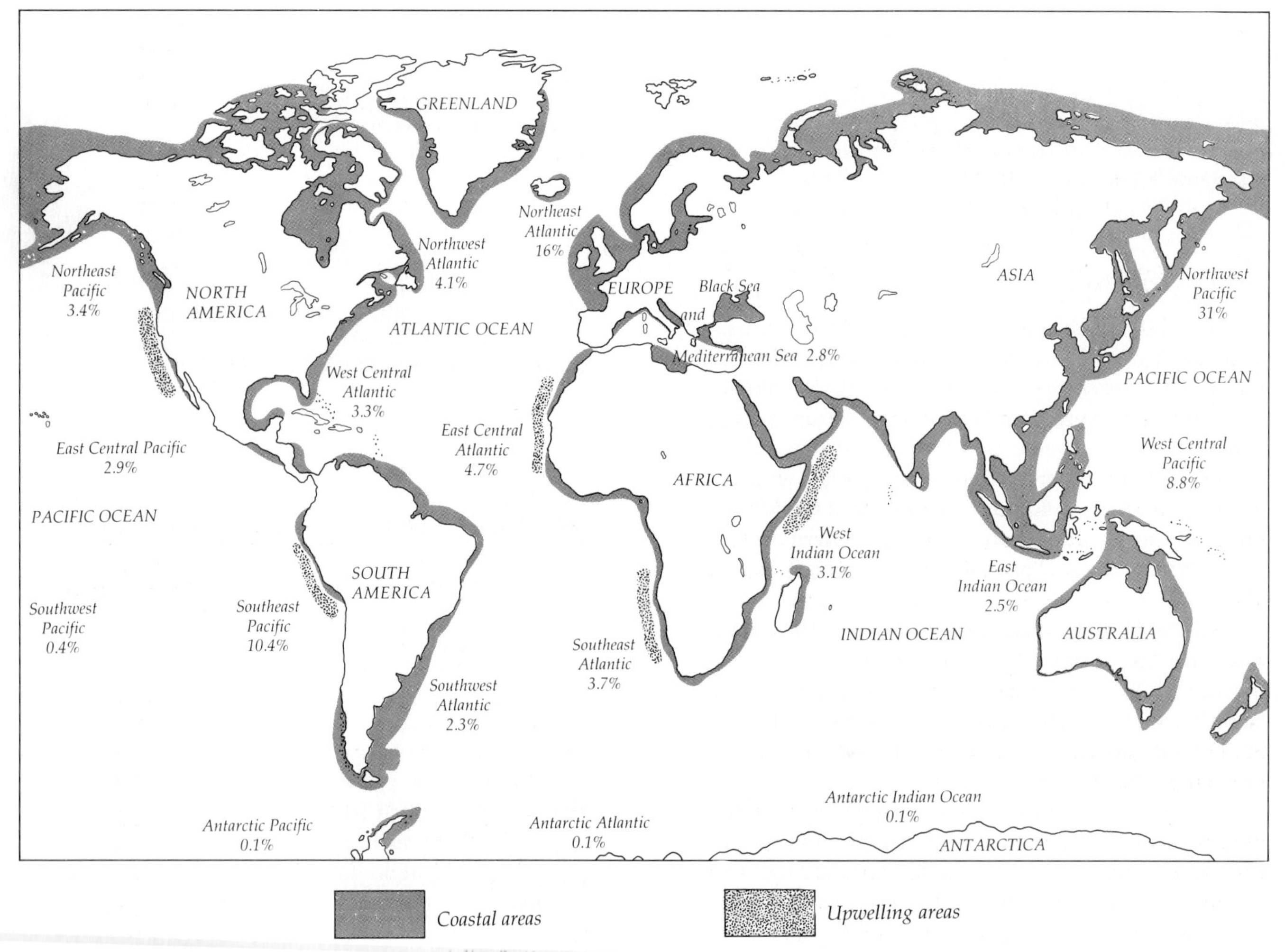

18.21 *Location of the world's major fisheries and distribution of the annual catch. (Data from U.N. Food and Agriculture Organization, 1985)*

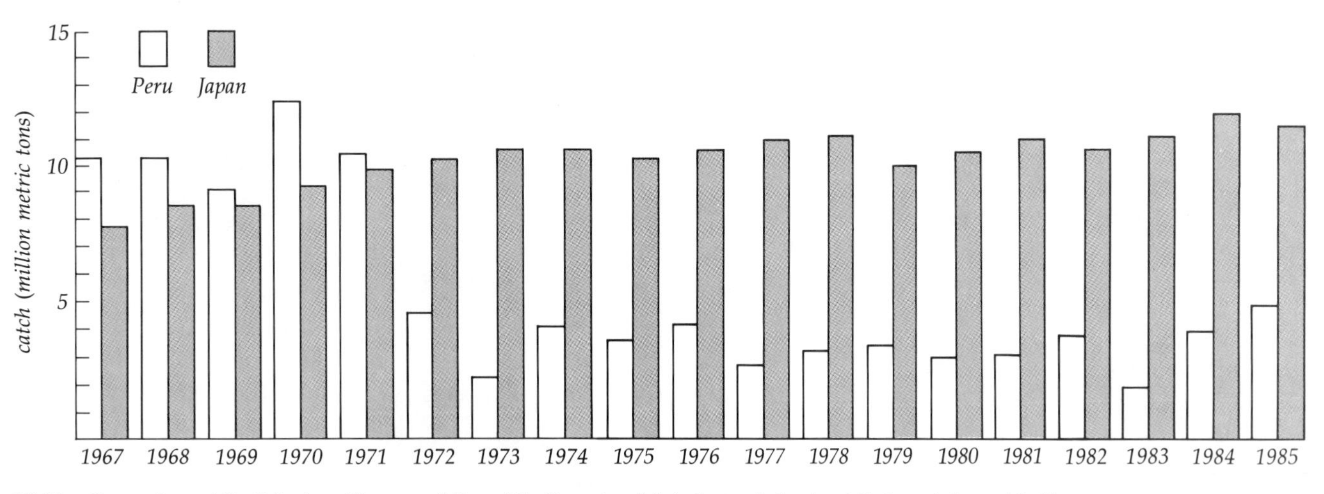

18.22 Comparison of the fisheries of Japan and Peru. The Peruvian fish industry is local, while Japan's is worldwide.

different explanations are appropriate for different regions.

The world fishery produces 84.9 million metric tons a year. About 8 million metric tons of the total are from fresh water. About 74% of the total goes to human consumption, and the rest goes to pet foods, fertilizers, and other products. These proportions have remained constant over the last decade.

The largest fishery in the world is the northwestern Pacific Ocean, from Kamchatka south to the Philippines (Figure 18.21). The southeastern Pacific, along the western coast of South America, has become less productive, probably because of overfishing and El Niños. Almost no commercial fishery exists in the Antarctic region, although Japan and the Soviet Union are currently investigating the fishery potential of krill in that region. From an annual production of 400,000 metric tons in 1979, the potential harvest of krill may ultimately match the total production of the world fishery.

Japan, a small nation about the size of California but with a much larger population, depends largely on resources outside its boundaries for its survival. It leads the world in the total tonnage of fish caught (Table 18.10), and has done so since 1972, although its catch has not increased appreciably (Figure 18.22). Peru, however, has slipped from number one to number four since 1970 as a result of the decline in the anchovy population on which it was dependent. The United States, fifth among fishing nations in 1985, has a production of about 4.8 million metric tons.

The world fishery is dominated by just a few types of fish (Table 18.11 and Figure 18.23). For normal fishery years, herrings, sardines (Figure 18.24), and anchovies top the list, but in 1973 only these fish were displaced by cods, hakes, and haddocks. Again, the decline in the southeastern Pacific fishery, which relied mainly on anchovies, was responsible. The tonnage

Table 18.10 *Countries with the Largest Fishing Industries in 1985*

Country	Catch (million metric tons)
Japan	11.44
Soviet Union	10.52
China	6.78
Chile	4.80
USA	4.77
Peru	4.17
India	2.81
Republic of Korea	2.65
Thailand	2.12
Norway	2.11
World	84.9

Source: Data from U.S. Department of Commerce.

Table 18.11 *World Catch by Fish Type in 1985*

Fish Type	Catch (million metric tons)
Herrings, sardines, anchovies	21.2
Flounders, halibuts, soles	1.4
Cods, hakes, haddocks	12.4
Salmons, trouts, smelts	1.1
Miscellaneous marine fishes	8.9
Jacks, mullets, sauries	8.0
Miscellaneous freshwater fishes	8.4
Redfishes, basses, congers	5.3
Mackerels, snooks, cutlass fishes	3.7
Tunas, bonitos, billfishes	3.2
Crustaceans	3.4
Molluscs	6.1

Source: Data from U.S. Department of Commerce.

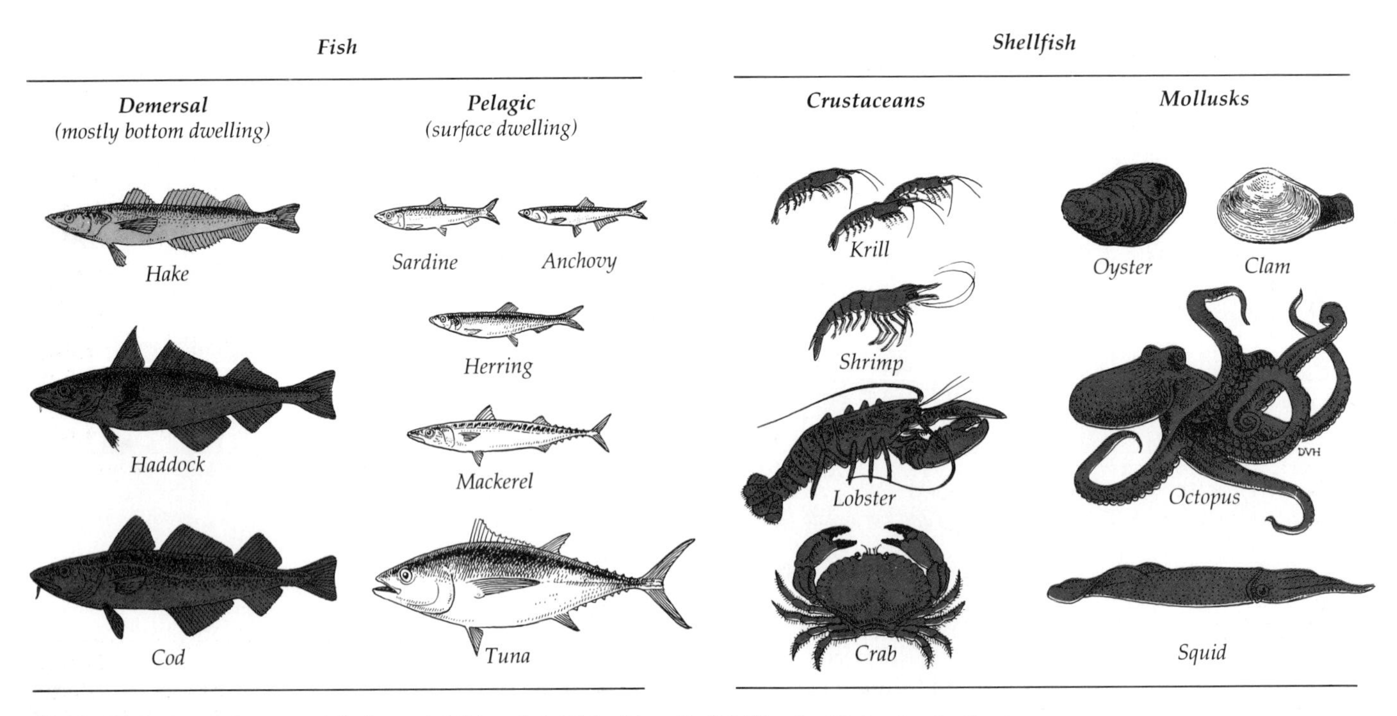

18.23 Major types of commercially harvested fish and shellfish. (From G. T. Miller, Jr., Living in the Environment, *5th ed., Belmont, Calif.: Wadsworth, 1988)*

18.24 Pacific sardines. (U.S. Department of the Interior, Fish and Wildlife Service)

of crustaceans and molluscs brought to the world market is also small.

How much food can be taken from the sea without depleting its populations? This question is especially pertinent given our present population explosion and the diminishing amount of arable land. It is a difficult question to answer, however, because marine populations are hard to assess. We cannot, for instance, count the number of anchovies in a school. Computer analysis based on careful sampling has helped, but the behavior patterns of fish confound statisticians. Most marine biologists feel that they can assess marine populations with no better than 15% accuracy. Given this limitation, the **maximum sustainable yield** (MSY) per year—that is, the maximum amount of fish that can be caught without impairing future fish populations—is estimated to be between 100 and 200 million metric tons. The current MSY seems to have reached that level, which means that yields can be increased only by exploiting regions such as the Antarctic or by using fish species that are currently being discarded or underutilized.

The recent decline in the Peruvian fishery suggests that the MSY for anchovies has already been surpassed in that area of the Pacific. Understandably, Peru and Ecuador are now sensitive to other countries intruding into the southeastern Pacific.

The commercial harvesting of marine mammals is included in fishery statistics, even though mammals are not fish. The Marine Mammal Protection Act protects most marine mammals from commercial exploitation, although it permits native Alaskans to harvest whales for subsistence. However, the act does permit the harvesting of the northern fur seal to continue. In the Aleutians and along the coasts of other Alaskan islands, 25,000 fur seals were taken yearly until 1986. There is no commercial whaling industry in the United States, although some porpoises, seals, and sea lions are taken unintentionally when they become entangled in gill nets or purse seines. A decade ago, more than 250,000 marine mammals were lost in those nets.

Today, the United States allows a limit of 20,500 incidental porpoise deaths in commercial tuna fishing. The regulatory agency, the National Marine Fishery

Table 18.12 *Worldwide Kill of Various Whale Species*

Year	Blue	Fin	Humpback	Sei	Bryde's	Sperm	Others	Total of Whales
1973–74	—	2,142	16	6,239	1,882	21,421	205	31,905
1974–75	—	1,552	17	5,001	1,864	21,338	189	29,961
1975–76	1	741	15	1,870	1,781	17,422	219	22,049
1976–77	2	310	20	2,021	1,412	12,329	215	16,309
1977–78	3	711	35	695	928	11,064	202	13,638
1978–79	—	730	19	163	892	8,655	209	10,668
1979–80	—	472	18	102	522	2,211	217	3,542
1980–81	—	410	12	100	648	1,595	163	2,928
1981–82	—	356	12	71	802	621	188	2,050
1982–83	—	277	15	100	688	414	189	1,683

Source: UN/FAO Fishery Statistics Yearbook, 1984

Table 18.13 *Whaling Results for the Various Countries*

Years	All countries	Australia	Brazil	Canada	Chile	Iceland	Japan	Bahamas	Somali Rep.
1973–74	31,905	1,080	32	—	161	365	10,095	—	451
1974–75	29,961	1,172	57	1	106	420	9,450	—	276
1975–76	22,049	995	12	—	87	389	6,227	—	23
1976–77	16,309	624	30	—	55	386	5,274	—	—
1977–78	13,638	679	24	—	198	391	3,191	—	—
1978–79	10,668	—	27	—	99	440	1,867	—	—
1979–80	3,542	—	30	—	94	437	1,499	—	—
1980–81	2,928	—	—	—	64	397	1,354	—	—
1981–82	2,050	—	—	—	—	352	921	—	—
1982–83	1,683	—	—	—	4	244	929	—	—

Years	Denmark	Norway	Peru	Portugal	Spain	South Africa	United States of America	USSR	Others
1973–74	14	—	1,812	234	497	1,817	21	15,268	58
1974–75	13	—	1,343	386	539	1,707	15	14,455	21
1975–76	18	—	1,923	195	515	—	56	11,560	49
1976–77	30	—	1,192	222	248	—	29	8,189	30
1977–78	38	—	1,070	266	596	—	18	7,119	48
1978–79	33	—	1,042	298	547	—	31	6,261	23
1979–80	28	—	665	330	234	—	36	178	11
1980–81	22	—	387	394	146	—	28	135	1
1981–82	24	—	320	95	150	—	23	165	—
1982–83	27	—	149	21	120	—	20	169	—

Source: UN/FAO Fishery Statistics Yearbook, 1984

Service, reports that incidental deaths have averaged about 28% of the limit since 1981 and that 99.5% of the porpoises encircled by nets are released unharmed. A problem still remains with tuna boats registered in other countries. The Inter-American Tropical Tuna Commission in 1987 estimated that about 100,000 porpoise deaths are attributed annually to fishing activities by other countries.

Internationally, the harvesting of both seals and whales has been a major commercial industry. In the 1980s, Japan and the Soviet Union harvested such large whales as the sperm, Bryde's, finback, sei, and humpback. Until the 1950s, more than 100,000 large whales were taken annually, a figure that had fallen to about 50,000 by the mid-1960s, and to less than 2,000 in 1983 (Tables 18.12 and 18.13 and Figure 18.25).

18.25 *The large-scale, factorylike slaughter of the great whales. Included in the group of dead whales to be rendered by the factory ship are a number of blue whales. (Ray Gilmore)*

Until recently, many countries harvested smaller whales, such as the beluga, bottlenose, grampus, killer, minke, and pilot. The number of small whales taken has decreased, however, from about 20,000 annually in the 1960s to about 9,000 annually during the 1980s. These whales were used for human, livestock, and pet consumption and for various commercial products, especially cosmetics.

The seal industry is much larger than the whaling industry (Table 18.14). In 1982, some 422,477 seals were taken, mainly for their fur. In 1982, Canada took 142,501 harp seals and Norway took 51,386. The United States and the Soviet Union harvested 32,816 northern fur seals in that year; South Africa took 91,425 Cape fur seals, and Uruguay took 1,375 South American fur seals. Other seals of commercial value include the harbor, ringed, hooded, and gray, as well as the South American sea lion. The United States terminated the Alaskan fur seal commercial fishery in 1986, although native Americans maintain a subsistence catch.

There is a small commercial fishery of marine turtles, mainly green sea turtles, hawksbills, and loggerheads. Mexico, Ecuador, Cuba, and Indonesia accounted for most of the 1982 commercial harvest of 6,024.

EXCLUSIVE ECONOMIC ZONE

In 1983, President Reagan proclaimed the ocean area from 3 miles off the coast of the United States and its territories out to 200 nautical miles to be the Exclusive Economic Zone (EEZ) of the United States. This unilateral action, which would have gone into effect anyway if the United Nations' Law of the Sea Treaty had been adopted, expands United States territorial jurisdiction an additional 16 billion hectares (3.9 billion acres) (Figure 18.26). By comparison, the total land area of the United States and its territories is 9.2 billion hectares (2.3 billion acres). Clearly, this provides an

Table 18.14 *Annual Kill of Seals and Sea Lions*

Species	Year 1965	1970	1975	1979	1982
Harp seal	277,112	340,555	114,202	199,871	206,558
Hooded seal	45,065	42,044	10,226	41,303	28,706
Ringed seal	278	16	5	97,628	53,647
Gray seal	1,496	936	3,336	1,565	1,407
Bearded seal	905	596	2	784	1,278
Harbor seal	1,460	2,148	377	388	61
Miscellaneous seals	—	—	—	—	—
Northern fur seal	70,039	58,485	33,049	31,500	32,816
Cape fur seal	63,625	85,776	75,731	75,470	91,425
South American fur seal	7,616	11,940	12,686	10,868	1,375
South American sea lion	2,216	2,950	3,142	8,755	—
Total	471,070	548,315	334,071	471,494	422,477

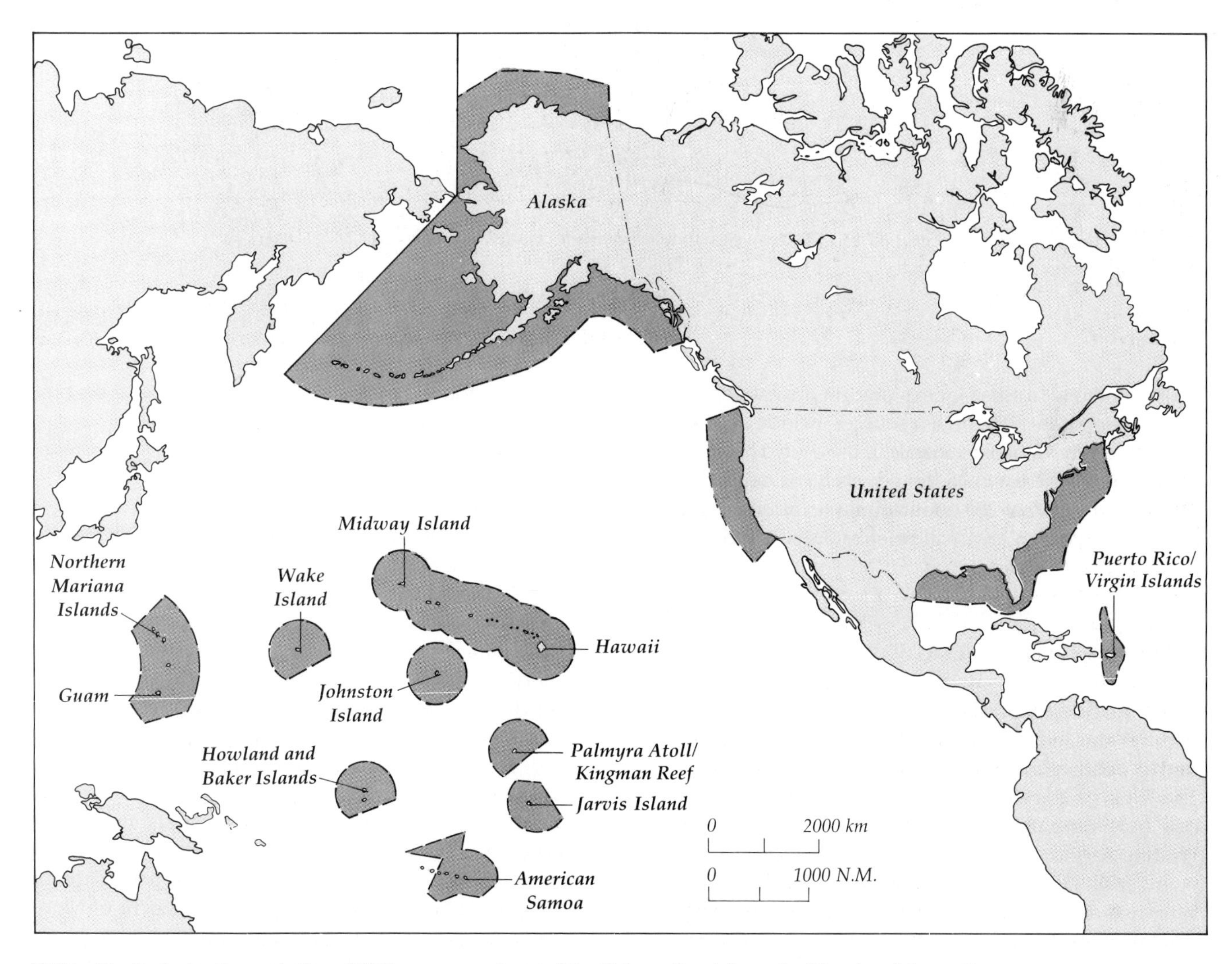

18.26 The Exclusive Economic Zone (EEZ) covers a vast area of the U.S. continental margin. The edge of the continent as it extends out under the ocean contains a wealth of resources. (Dept. of Interior)

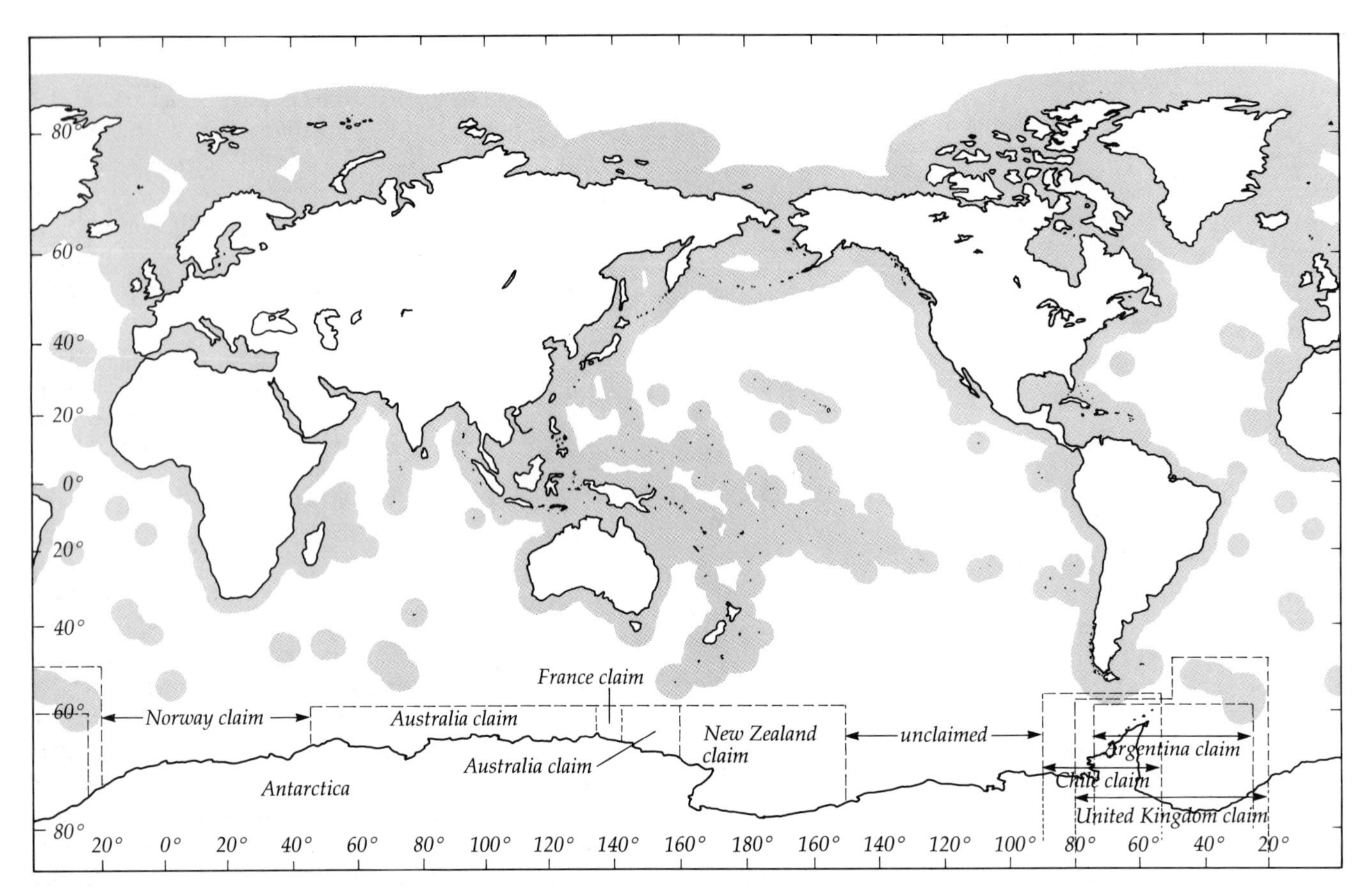

18.27 This map defines the 200 nautical mi (370 km) economic zone boundaries that would apply if the U.N. Law of the Sea Treaty is ratified. (Ross and Landry)

enormous new frontier for exploration, investigation, and development. Potential resources include oil, gas, fish, and many valuable minerals. If the United Nations' Law of the Sea Treaty is adopted, each coastal nation will have an EEZ of 200 nautical miles from its coast, and the free ocean area will be substantially reduced (Figure 18.27).

ENDANGERED LIVING RESOURCES

The Endangered Species Conservation Act of 1969 requires the secretary of the Department of the Interior to publish an annual list of endangered species. The list is prepared by the Office of Endangered Species and International Activities in the U.S. Fish and Wildlife Service, and it must give the number of surviving individuals of each species and describe threats to its natural habitat. Many nongovernmental organizations also study and report on endangered species. For example, the International Union for Conservation of Nature and Natural Resources, based in Switzerland, publishes a "red data book" on rare animals each year.

Overfishing

Overfishing is the harvesting of a species beyond its capability to reestablish its natural population. In other words, overfishing means that the MSY for that species has been exceeded and that its population will grow smaller each year until it is no longer of commercial value. When the population is so reduced that it is in danger of extinction, the species is regarded as endangered.

The Pacific sardine (*Sardinops sagax caeruleus*) is an example of an overfished species. John Steinbeck described the California sardine fishery in *Cannery Row* and *Sweet Thursday*. By the 1950s, fishers had already reached or exceeded the MSY for these fish. Then in 1958 an El Niño event along the coast north of Monterey, California, interfered with the sardine's reproduction and the population was sharply reduced. Nevertheless, fishers continued to take these fish until the commercial fishery collapsed. Recently the Pacific sardine has slowly been returning to the area, but the population has not yet reached its former level. Although this species was overfished and essentially eliminated from a particular area, it was not in danger of extinction. The Peruvian anchovy fishery has experienced a similar series of events.

Demand determines the price of fish. As demand increases, the price goes up and fishers seek more of the species in demand. If overfishing ensues, demand for the scarce fish increases, leading to higher prices, which, in turn, encourage the fishers to keep fishing. Eventually, fishers must spend more time and fuel to catch the fish, or else the prices rise to the point where demand is reduced. Fishing pressure is then reduced, and the overfished population may recover. Overfishing has never driven a species of marine fish to extinction.

None of the marine fish that live exclusively in ocean waters are on the endangered list, and of the estuarine species only the Atlantic shortnose sturgeon is listed as endangered. The Atlantic salmon, though not in danger of extinction, has certainly been overfished. Its habitat in certain parts of Europe has been destroyed by dams, pollution, and the silting up of the gravel beds where it lays its eggs. However, since the species still spawns in Great Britain, Norway, Iceland, Canada, and the United States, it is not yet regarded as endangered.

The Fishery Conservation and Management Act of 1976 has helped reduce the problem of overfishing, and the new U.N. Law of the Sea Treaty will encourage nations to manage their commercial fisheries sensibly. Yet a great need remains for research in this area, and more must be known about the natural population and the reproductive cycles of various species.

Invertebrates

No marine invertebrates appear on the list of endangered species. Still, oyster habitats have been destroyed in Long Island Sound, and pollution has rendered oysters inedible in Raritan Bay, New Jersey. The commercial oyster industry in Chesapeake Bay is also diminishing. Oysters in San Francisco Bay have been rendered inedible by pollution, but uncontaminated oysters continue to be harvested in the Gulf of Mexico, Puget Sound, and numerous Pacific coastal bays.

In California, the abalone has been overfished by both commercial and recreational fishers, and its price has reached more than $25 a pound. The California Fish and Game Department has launched a program to reseed former abalone habitats, such as Palos Verdes, with young abalone raised in hatcheries.

Reptiles

Marine reptiles are in trouble. All 12 living species of crocodiles are endangered, as is the Atlantic salt-marsh snake. In addition, all of the following turtles are endangered: the green sea turtle, the hawksbill sea turtle, the Atlantic ridley sea turtle, the leatherback sea turtle, the loggerhead sea turtle, and the Olive ridley sea turtle. Crocodiles are hunted for sport and by people who perceive them as a threat. Turtles are considered a delicacy in many cultures and are hunted mainly for food. The United States forbids the killing of marine reptiles native to U.S. territorial seas and forbids the importation of marine reptile meat.

Birds

Some 31 species of birds predominantly associated with the marine environment are on the endangered list. Some have been the victims of pollution, such as the peregrine falcons, which were almost wiped out by DDT. The habitats of some have been destroyed by development, as with the Hawaiian goose, stilt, gallinule, and dark-rumped petrel. The coastal nesting grounds of others have been destroyed or impaired by recreational activities, as with the California least tern and the California clapper rail. The United States has implemented conservation measures to protect these species.

Mammals

Many marine mammals are also endangered. For most of them, commercial harvesting has been discontinued, although some subsistence hunting still takes place.

Three species of the order Sirenia are endangered: the Amazon manatee, the West Indies manatee, and the dugong, which formerly ranged from Africa through tropical Asia to the Solomon Islands. Subsistence hunters consider the meat of these herbivorous mammals a delicacy, and their gentle, slow-moving behavior makes them easy prey. In estuaries and lagoons, many of these mammals are killed by careless speedboaters.

Two subspecies of sea otters, the northern and southern, are endangered as a result of the fur trade. Also endangered are two other members of the order Carnivora, the Hawaiian monk seal and the Mediterranean monk seal.

In the order Cetacea, the following species are endangered: great blue whale, bowhead whale, finback whale, gray whale, humpback whale, right whale, sei whale, and sperm whale. There is now a morato-

Table 18.15 *Production of By-Products in the Antarctic and Other Whaling Grounds*

	Whale Meal, Bone Meal	Whale Liver	Liver Oil Mill. Units Vitamin A	Whale Meat, Blubber, Ventral Grooves	Whale Solubles	Other Products
	Long tons*	Long tons		Long tons	Long tons	Long tons
1973–74	37,907	1,826	6,118,000	134,514	5,721	4,494
1974–75	35,459	1,700	5,795,000	128,988	3,917	1,719
1975–76	24,704	1,737	3,819,000	90,895	2,508	1,537
1976–77	20,958	1,302	148 l.t.	89,506	1,726	1,289
1977–78	16,103	1,432	820 l.t.	67,690	1,675	1,836
1978–79	13,286	1,360	114 l.t.	66,016	—	472
1979–80	11,593	193	85 l.t.	47,801	—	97
1980–81	6,438	221	—	43,244	—	88
1981–82	4,693	231	—	47,813	—	152
1982–83	3,393	233	—	46,400	—	180

*1.046 long tons = 1 metric ton
Source: UN/FAO Fishery Statistics Yearbook, 1984

rium on all commercial whaling. The gray whale, after being protected for almost 50 years, has made a spectacular comeback and has been removed from the endangered list.

As far back as 1851, Herman Melville, in *Moby Dick*, questioned whether the whaling practices of his day might not lead to the extinction of these magnificent mammals. In the days of sail-driven ships, whalers took only certain species, avoiding those that sank after being killed. But with steam-driven ships, explosive harpoon guns, and compressed air to keep the dead animals afloat, virtually all species of large whales began to be taken by the early 1900s. Even the fast-swimming blue and the finback were now at the mercy of the whalers.

For many years almost no attempts were made to limit whale catches. In 1900, when the blue whale population of the North Atlantic began to decline, deep-sea whalers simply moved to the Antarctic. In 1931, British, Dutch, and Scandinavian whaling vessels took almost 30,000 blue whales in the Antarctic. In 1934, Japanese whalers joined the fleet. Shortly after the war, eighteen nations joined to form the International Whaling Commission (IWC), whose purpose is to conserve whale resources. The IWC has no power to seize or inspect whaling vessels, however, and neither can it enforce its mandates. It can only suggest the maximum number of each species of whale that should be taken in a year. The IWC uses a unit, the blue whale unit (BWU), which equates 1 blue whale to various numbers of other species. One BWU, for example, equals 2 finback whales or 6 sei whales. Unfortunately, even the modest recommendations of the IWC were violated at will—no one has ever gone to jail for violating whaling regulations.

With the surge in whaling activity that occurred after World War II, approximately 7,000 blue whales were caught each year until the number declined around 1950. In 1952, the catch had dropped to 5,000; in 1956, to less than 2,000; and in 1962, to slightly over 1,000. Yet the sharp decline in the blue whale population did not stop the whalers. Finally, when the 1963–64 whaling season yielded only 110 blue whales, the IWC began to take notice. The hunting of blue whales was officially terminated in 1966, when only 70 were caught.

During those years Britain, Japan, the Netherlands, Norway, and the Soviet Union were all engaged in whaling. In 1963, Britain abandoned whaling as an industry. In 1964, Holland decided that whaling was no longer profitable and sold its fleet to Japan. More recently many Scandinavian interests have decided that whaling is no longer profitable and have withdrawn. Most of the whaling in the 1980s was done by Japan and the Soviet Union, although many other countries did some whaling, primarily from shore stations.

It is estimated that fewer than 1,000 blue whales are left in the ocean. Experts predict that if the number falls below 1,000, the statistical chance of a male finding a female in the vast expanses of the Antarctic Ocean will be too slight to balance the species' normal mortality rate. The blue whale may already be functionally extinct.

With the sharp decline in the number of blue whales, whalers hunted other species more intensively. At the 1980 rate of kills, the finback, humpback, sei, and sperm whales would all become extinct soon. The number of kills of all the great whales was decreasing steadily since the early 1960s, and hunting pressure was increased on the smaller species, such as minke whales.

Whaling was a relatively small industry. In 1967, the world's entire whaling fleet numbered fewer than 300 ships—about 250 catchers and 16 floating processing factories. The industry itself was moving toward self-imposed extinction. The Japanese and Russians use whale oil to make soap and margarine and use the meat as food and fertilizer (Table 18.15). In 1967, the Japanese consumed more than 177,000 metric tons of whale meat.

The United States no longer takes whales, although it permits native Alaskans to harvest a few for subsistence. In 1971, the secretary of the Department of Commerce banned commercial whaling by American firms and by foreign firms using American ports.

The International Whaling Commission (IWC) voted in 1985 for a moratorium on commercial whaling, which began in 1986. By 1987, all commercial whaling ceased. There continues to be a harvest "for scientific purposes," which is allowable by special permit from the IWC. Iceland and Japan have permits that allowed some 600 whales to be taken in 1987. The moratorium will be in effect until 1990, when a review of the condition of whale populations and requests from interested countries takes place.

The U.S. Marine Mammal Protection Act eliminated the commercial harvesting of all marine mammals except for the northern fur seal. The United States, along with many other countries, continues to apply pressure through the IWC on countries that are still commercially harvesting marine mammals. The United States terminated its commercial fur seal industry in 1986. The harvesting of the harp seal has received a great deal of publicity in recent years. Actually, the taking of harp seals is carefully managed and the species is not endangered. Consequently, the United States has adopted a hands-off policy toward this practice.

AQUACULTURE

One of the most promising developments in the management of marine resources is **aquaculture**, a technique for growing plants and animals in a water environment under controlled conditions. Fresh-water carp and tilapia have been raised in ponds or pens in China and Egypt for over 4,000 years. Today, large numbers of trout and catfish are grown under controlled conditions in the United States.

Aquaculture now produces about 13 million metric tons of fish annually, about 15% of the world's total fishery catch. Ten years ago aquaculture produced only about 6%. With the total production of the world fishery slowing to a growth rate of about 2% annually, aquaculture, at an 8% growth rate, promises to become increasingly important in the years ahead.

The term **mariculture** refers to aquaculture conducted in a marine environment. Currently, many species of fish (plaice, salmon, milkfish, and tilapia), invertebrates (shrimp, abalone, mussels, and oysters), and seaweeds *(Porphyra* and *Undaria)* are being grown commercially (Table 18.16). Many countries have active mariculture industries, but the world leader is Japan, which produces prawns, oysters, mussels, salmon, and several varieties of seaweeds.

Aquaculture in the United States is a $150-million-a-year industry. If we exclude salmon hatcheries, mariculture accounts for about $15 million of the total. Most salmon hatcheries are noncommercial enterprises operated by governmental agencies, although many of the salmon are released and caught at sea by commercial fishers. Some commercial salmon ranching has been initiated in Oregon.

Oysters are the main product of the U.S. mariculture industry. Mussels are of minor importance, because the market demand is small in the United States. In the Mediterranean Sea and in Japan, however, demand is high.

Mariculture production can be very efficient, especially for mussels. It requires an adequate supply of seawater that has suitable temperature and salinity, is rich in nutrients, and is low in pollutants. It also requires protected locations or storm-resistant structures, along with suitable characteristics and currents to supply nutrients and remove wastes. Leasing of a suitable site is also a major consideration. In the United States, where coastal waters are in the public domain, leases must be obtained from local, state, or federal agencies.

Mariculture employs a wide range of techniques. Organisms may be transplanted from a less desirable location to a better one to reduce mortality during the early stages of growth. In "pen culture," fish and invertebrates are herded into enclosures to protect them from predators until they reach market size, and their diet is supplemented to accelerate growth rates. Since many species grow more rapidly in warm water, some mariculturists use warm effluents from power plants. Others construct ponds in mangrove swamps to take advantage of the tidal currents.

Finally, raft culture has proven successful for growing such invertebrates as mussels. Raft culture is a highly efficient method of producing protein that eliminates the cost of feed and of disposing of waste. Unfortunately, however, mussels, which are best suited to raft culture, are not popular in many diets around the world. Cultural factors frequently impede the use of readily available sources of protein, even at a time when the world is experiencing a severe food crisis.

Table 18.16 *Some Aquacultural Yields in Increasing Culture Method Control*

Culture Method	Species (Location)	Yield (kg/hectare/year) or Economic Gain
Transplantation	Plaice (Denmark, 1919–57)	Cost:benefit 1:1.1–1.3 in best years (other social benefits)
	Pacific salmon (U.S.)	Cost:benefit, based on return of hatchery fish in commercial catch 1:2.3–5.1
Release of reared young into natural environment	Pacific salmon (Japan)	Cost:benefit 1:14–20, based on return of hatchery fish in commercial catch
	Shrimp, abalone, puffer fish (Japan)	Not assessed; reputed to increase income of fishermen
	Brown trout (Denmark, 1961–63)	Maximum net profit/100 planted fish: 163%
Retention in enclosures of young or juveniles from wild populations with no fertilization or feeding	Mullet	150–300
	Eel, miscellaneous fish (Italy)	
	Shrimp (Singapore)	1,250
Stocking and rearing fertilized enclosures with no feeding	Milkfish (Taiwan)	1,000
	Carp and related species (Israel, SE Asia)	125–700 400–1,200
	Tilapia (Africa)	62,500–125,000
	Carp (Sewage streams in Java; ¼–½ of water area used)	
Stocking and rearing with fertilization and feeding	Channel catfish (U.S.)	3,000
	Carp, mullet (Israel)	2,100
	Tilapia (Cambodia)	8,000–12,000
	Carp and related species (in polyculture) (China, Hong Kong, Malaysia)	3,000–5,000
	Clarias (Thailand)	
		97,000
Intensive cultivation in running water with feeding	Rainbow trout (U.S.)	2,000,000 (170 kg/l·s)
	Carp (Japan)	1,000,000–4,000,000 (about) 100 kg/l·s)
	Shrimp (Japan)	6,000
Intensive cultivation of sessile organisms, molluscs, and algae	Oysters (Japan, Inland Sea)[a]	20,000
	Oysters (U.S.)	5,000 (best yields)
	Mussels (Spain)[a]	300,000
	Porphyra (Japan)[a]	7,500
	Undaria (Japan)[a]	47,500

[a]Raft culture calculations based on an area 25% covered by rafts.

Mariculture can help alleviate the growing food crisis. The industry is growing at a faster rate than the world population, and scientists are improving techniques and developing methods for raising such popular species as tilapia. Knowledge of mariculture is being disseminated by such agencies as Sea Grant, the Department of Agriculture, the National Marine Fisheries Service, the Peace Corps, and the U.N. Food and Agriculture Organization. Such countries as Jamaica, Costa Rica, Mexico, the Philippines, and Indonesia, which are experiencing rapid population growth, are situated in regions well suited to mariculture.

In the United States, mariculture will probably be limited to producing such popular species as abalone, lobster, shrimp, oysters, tuna, and salmon. Since these species generally command high market prices, high profits will help offset the high production costs involved.

MARINE NATURAL PRODUCTS

The phrase *marine natural products* refers to chemicals derived from marine organisms. Such products have been used for medicinal purposes at least since the time of Hippocrates. Currently, most of the natural products used by the pharmaceutical industry are

Table 18.17 *Number of Establishments in Business Categories Related to Marine Recreational Fishing*

Product or Service	Number of Establishments	Category Description
Fishing Tackle		
Manufacturing	260	Manufacturers of freshwater and saltwater tackle in U.S.
Wholesale Trade	175	Establishments that distribute freshwater and saltwater tackle in U.S.
Retail Trade	6,350+	Retailers selling tackle throughout U.S. not including department stores.
Boats		
Manufacturing	361	Manufacturers primarily engaged in producing outboard, inboard and inboard/outdrive boats.
Retail Trade	6,500	Retail boat dealers throughout the U.S.
Motors		
Manufacturing	5	Manufacturers of outboard motors in U.S.
Retail Trade	6,500	Retail boat dealers throughout the U.S.
Boat Trailers		
Manufacturing	100	Manufacturers of trailers throughout the U.S.
Retail Trade	6,500	Retail boat dealers throughout the U.S.
Marinas	2,880	Coastal marinas and boat yards.
Commercial Sportfishing Vessels	3,952	Saltwater head (party) and charter boats.
Boat Insurance	130	Number of insurance carriers selling insurance for recreational boats.
Bait	3,675	Establishments that sell bait for use in salt water as a primary activity.

Source: Sport Fishing Institute, 1983.

derived from terrestrial organisms. Marine products are still underutilized and poorly understood. The Sea Grant program of the National Oceanic and Atmospheric Administration and the Roche Research Institute for Marine Pharmacology have been leaders in supporting research in this area. Through their efforts, a number of compounds have been identified that show promise in the manufacture of drugs and pesticides. For example, sponges from tropical reefs have been identified as valuable sources of antibacterial chemicals. Of 107 species of sponges tested, more than half inhibited bacterial growth. Chemical extracts from seaweeds have long been used in the manufacture of foods and paints, and mariculture is leading to an expansion of their use for such purposes.

SUNKEN TREASURE

Although we do not ordinarily regard sunken treasure as a marine resource, recent recoveries are impressive. Entrepreneur Mel Fisher has recovered more than $400 million in gold and archaeological treasure from sunken Spanish galleons in the Florida Keys. In 1981, a British salvage company recovered $80 million in gold ingots from a British warship sunk in 1942 off Murmansk. Other treasure sites are known and may prove equally exciting and valuable.

MARINE RECREATION

Marine recreation activities represent approximately a $23-billion industry. The largest segments of this industry are boating and fishing. The National Marine Fisheries Service estimated that 17 million anglers caught 400 million marine fish weighing about 317,000 metric tons in 1986. Fishing for fun in the ocean represents about $7.5 billion annually to the economy of the United States (Table 18.17). The most commonly caught fish are mackerel, sea bass, croaker, bluefish, porgies, and flounders (Figure 18.28).

Boating contributes more money to the economy than fishing (as any boat owner will tell you). There are about 14 million registered recreational boats in the United States, according to the U.S. Bureau of Census (U.S. Department of Commerce). The annual expenditure related to these boats in 1986 was $13.3 billion. For those people contemplating marine-related careers, a business related to boating and fishing offers

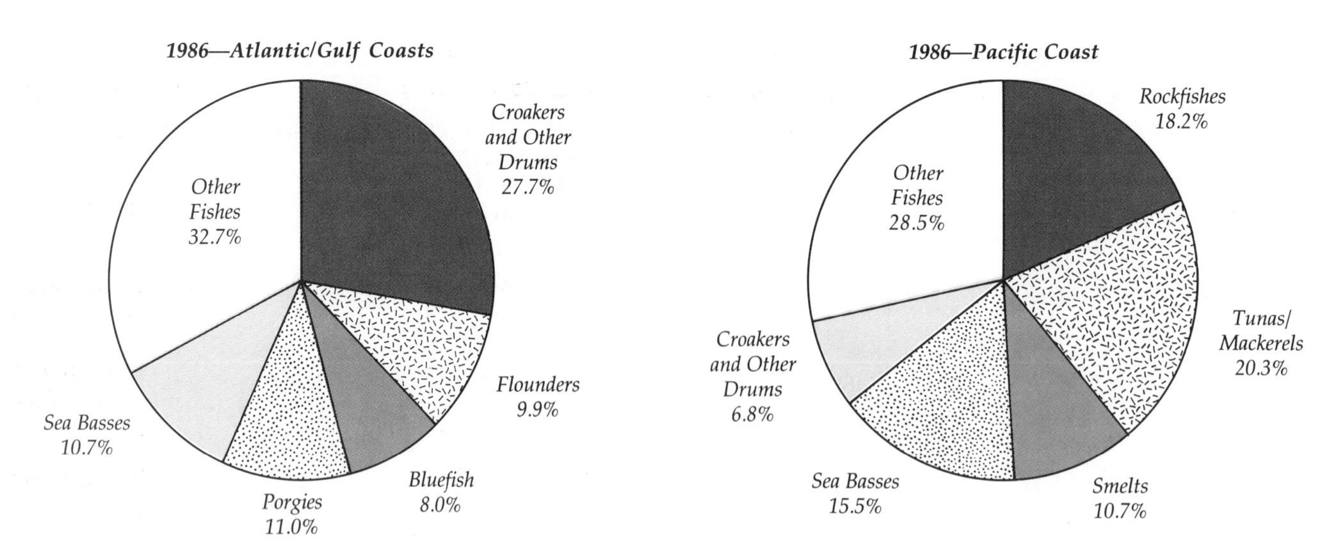

18.28 The five species most commonly caught by marine recreational anglers in 1986 were winter black sea bass, Atlantic croaker, bluefish, scup, and summer flounder. The estimated total marine recreational finfish catch on the Atlantic and Gulf coasts was 410.8 million fish. The total catch on the Pacific coast was estimated to be 55.3 million fish, exclusive of salmon, which historically has been about 2% of the total Pacific marine recreational finfish catch. (National Marine Fisheries Service)

far better financial opportunities than a career in marine biology. There are fewer than 2,500 employed marine biologists in the United States. In San Diego, California, about 20,000 people are employed in recreational boating and fishing.

OVERVIEW

Industries based on marine resources show great potential for expansion. In the United States they currently do a business of $60 billion a year, comparable to that of the agriculture and construction industries. Unlike agriculture, which maintains a strong lobby at all levels of government, ocean industries are poorly organized. Consequently, they will probably develop piecemeal, one segment at a time, often with strong competition between segments. The most rapid advances in the development of marine resources will undoubtedly occur in the less-developed countries.

FURTHER READINGS

Bardach, J. E., J. R. Ryther, and W. O. McLarney. *Aquaculture: The Farming and Husbandry of Freshwater and Marine Organisms.* New York: John Wiley, 1972.

Bascom, W. "Technology and the Ocean." *Scientific American,* September 1969, pp. 198–217.

Borgese, E. M., and N. Ginsberg (eds.). *Ocean Yearbook 4.* Chicago: University of Chicago Press, 1983.

Broadus, J. M. "Seabed Materials." *Science,* 1987, *235,* 853–860.

Brown, G. M., and J. A. Crutchfield. *Economics of Ocean Resources.* Seattle: University of Washington Press, 1981.

Browning, R. J. *Fisheries of the North Pacific.* Anchorage: Alaska NW Publishing, 1980.

Cronan, D. S. "Composition of Atlantic Manganese Nodules." *Nature,* 1972, *235,* 171–72.

Cushing, D. H. *Fisheries Biology: A Study in Population Dynamics.* Madison: University of Wisconsin Press, 1968.

Degens, E. T., and D. A. Ross (eds.). *Hot Brines and Recent Heavy Metal Deposits in the Red Sea.* New York: Springer-Verlag, 1969.

Firth, F. E. (ed.). *Encyclopedia of Marine Resources.* New York: Van Nostrand Reinhold, 1969.

Holt, S. J. "Food Resources of the Ocean." *Scientific American,* September 1969, pp. 93–106.

Iverson, E. S. *Farming the Edge of the Sea.* London: Fishing News Books, 1968.

Ladd, R. S. *A Descriptive List of Treasure Maps and Charts.* Washington, D.C.: U.S. Government Printing Office, 1964.

Laevaster, T., and H. A. Larkins. *Marine Fisheries Ecosystem.* Surrey, England: Fishing News Book Ltd., 1981.

Levine, S. N. *Desalination and Ocean Technology.* New York: Dover, 1968.

Mann, K. H. *Ecology of Coastal Waters.* Berkeley: University of California Press, 1982.

Marine Technology Society. *Remotely Operated Vehicles.* San Diego: Marine Technology Society, 1983.

Mero, J. L. *The Mineral Resources of the Sea.* New York: Elsevier, 1965.

Padelford, N. J. *Public Policy and the Use of the Seas,* 2d ed. Cambridge, Mass.: MIT Press, 1970.

Panel on Operational Safety in Marine Mining. *Mining in the Outer Continental Shelf and in the Deep Ocean.* Washington, D.C.: National Academy of Sciences, 1975, p. 12.

Ross, D. A. *Opportunities and Uses of the Ocean.* New York: Springer-Verlag, 1978.

Rothschild, B. J. *Global Fisheries.* New York: Springer-Verlag, 1983.

Ryther, J. H. "How Much Protein and for Whom?" *Oceanus,* 1975, *18,* 10–22.

Stroud, R. H. *Marine Recreational Fisheries.* Savannah, Ga.: National Coalition for Marine Conservation, 1984.

United Nations, Food and Agriculture Organization. *Yearbook of Fishing Statistics.* Paris: United Nations, Food and Agriculture Organization, published annually.

Wenk, E., Jr. "The Physical Resources of the Ocean." *Scientific American,* September 1969, pp. 166–77.

The sea was wet as wet could be,
The sands were dry as dry.
You could not see a cloud, because
No cloud was in the sky:
No birds were flying overhead—
There were no birds to fly.

Lewis Carroll

NINETEEN

Ocean Pollution and Management

None of us can claim ignorance of urban and industrial sprawl, of overcrowding and congestion, or of air, water, and land pollution. None of us can be unaware of the beaches closed because of contamination, the noxious fumes of rush-hour traffic, or the wastes left by dogs on every city block. But few of us wonder about where the electrical energy comes from to run the stereo, the television, and the dish-washer; where the gasoline comes from to run our automobiles; or where the food in the supermarket comes from. Few of us care to know about the very real limits on environmental resources that we must confront in our overpopulated age. Most people live without complaint in an artificial environment, insulated from the natural world and indifferent to its deterioration. They give little thought to what the world was once like or what it has become.

Most of the United States is still beautiful, although it was even more beautiful 350 years ago. When European settlers first came to North America's eastern coast, they found the land covered with vast, dense, deciduous forests. They did not launch their attack on the forests at the very outset; that would come later. They set up their farms, villages, hamlets, and towns along the coasts, where the salt marshes, estuaries, and riverbanks were easy to settle. Such places were sheltered from the ocean waves and could be drained with less effort than it took to clear the forests. Present-day cities such as Boston were built on marshes that have long since been filled (Figure 19.1). Because roads were difficult to build through the forests, most transportation in the colonies was by boat along the coast and rivers.

The bounty of the marshes, estuaries, and bays seemed inexhaustible. Farmers in seventeenth-century Boston drove pigs into the oyster beds at low tide, where the animals gorged themselves on oysters. Oysters have not grown naturally in Boston harbor for the past century. At low tide along the coastal reaches of Maine, eighteenth-century farmers raked in thousands of pounds of lobster from the shallow waters and plowed them into the fields as fertilizer. While lobsters remain a commercial resource, their abundance is now greatly reduced.

The early settlers saw the forests as an inexhaustible resource that yielded countless board feet of lumber. In the eighteenth century, roads called corduroy sections were made up entirely of logs; some stretched for 100 km. The big oaks and maples used for those roads are all gone now. Only a fraction of the trees that originally grew across the country still stand, and they survive largely because the federal government finally regulated the use of timberland by creating national parks and national forests. Most of the country's forest land has been logged off, often many times. Fewer than 15% of the nation's original redwoods remain.

It has often been said that what the United States does best is to grow things, but U.S. farmers do not really possess thumbs greener than those of the farmers of other nations. The reason for the nation's productivity is the land itself. More than 50% of the land is arable; in many countries, less than 10% is. Some of the richest soil in the world lies in the Great Plains of the American Midwest, in California's central valley, and in Oregon's Willamette Valley. Moreover, the United States is in the temperate zone, with good rainfall, moderate temperature, and a long growing season. On a global scale, prime agricultural land is rare and is becoming more so. Yet as the network of urban sprawl and highways across the country expands, acres of good soil are effectively removed from cultivation. Highways in the United States now cover a land area equal to that of West Virginia.

Only a few rivers flow unimpeded; most are choked with dams. Houses and parking lots cover areas where sand dunes once existed. Factories and shopping centers cover the river floodplains. Even the high mountain trails are littered with tin cans and pull-top tabs. The water we drink and the air we breathe are debased by contaminants.

Millions of years ago, our savanna-dwelling ancestors returned each night to the safety of trees and ledges, letting their wastes drop to the ground below. Later, hunter-gatherers let wastes accumulate at campsites until the group moved on, following the seasons and the game herds. With the coming of agriculture, human groups settled permanently in one place, and so did their wastes. The accumulating wastes became harder to dispose of, but they were natural wastes, and sooner or later nature took care of them. By medieval times, garbage and sewage had begun to collect in city streets, ignored by the population.

Pollution, then, is not a new problem. All organisms, including people, give off by-products of their existence. Over billions of years, in the great cycles of nature, the waste products of one kind of organism have served as the raw materials for metabolic reactions in other kinds. The oxygen "pollutants" of photosynthesis came to be used in cellular respiration, and the carbon "pollutants" of respiration came to be used in photosynthesis. Many substances go through a succession of cycles—not only carbon and oxygen but nitrogen, phosphorus, and other nutrients.

In the beginning, people were closely linked to those cycles, and in many ways they have tried to remain so. The so-called primary waste-water treatment is one such attempt. The primary treatment of waste water and sewage is meant to decompose organic wastes, converting them into substances that are less harmful to people or other living things. But with overloaded treatment plants, the process is not always effective, and primary treatment does nothing to rid the water of such dangerous inorganic wastes as mercury and lead. Increasingly, environmental overload is the rule rather than the exception, and nature can no longer accommodate the overload. The human population has grown so large and the amount of waste so great that nature can no longer do the job (Figure 19.2).

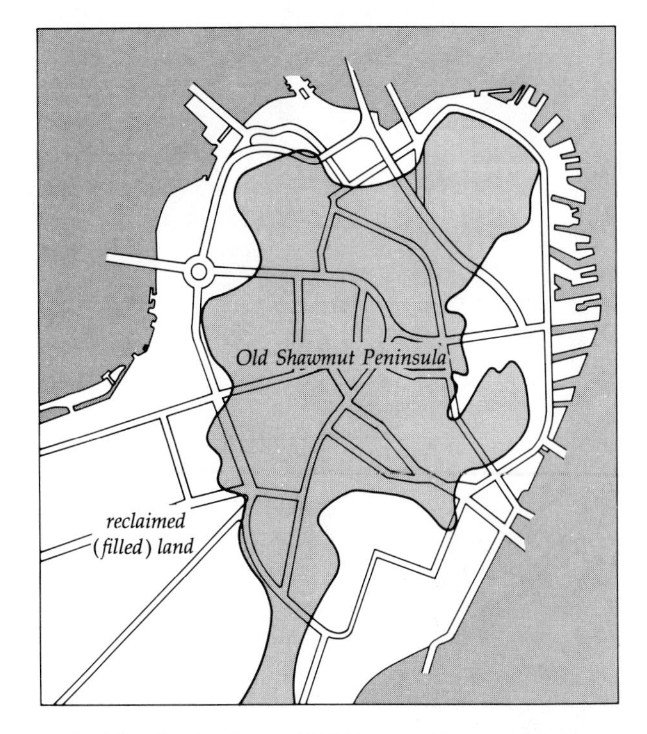

19.1 Map showing Boston (Old Shawmut Peninsula) during the Revolutionary period (shaded area) and today. (John Hancock Life Insurance Company)

Consider, for example, a story that began in the early 1800s, when a network of canals was built in the midwestern and northeastern United States to link rivers with lakes in a vast inland transportation system. Lake Erie was a vital link in the canal traffic. People of the time would have probably thought it impossible that Lake Erie would one day become so squalid that fish would vanish and most of the life in the lake would die off. A single stream might become polluted, but surely not a body of water as great as Lake Erie. Until heavy industry arose in the United States in the twentieth century, that lake remained relatively unpolluted.

But the amazing, continuing saga of Lake Erie is a positive example of the success of international envi-

19.2 *A common scene in most of the world's harbors. (W. J. Wallace)*

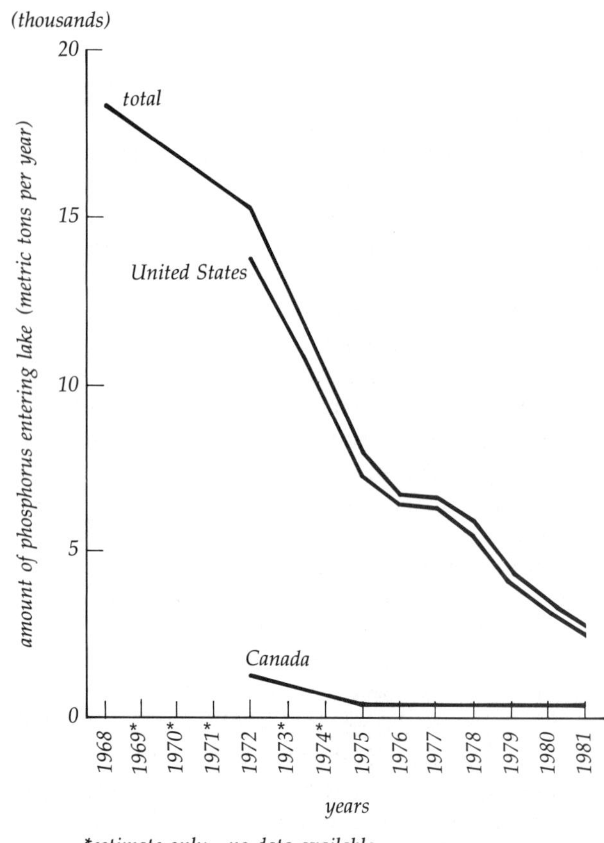

19.3 *Reduction in phosphorus entering Lake Erie from municipal sources. (From Ministry of the Environment, Ontario, Canada)*

ronmental recovery programs. The International Joint Commission for the Great Lakes made recommendations to Canada and the United States that resulted in the Water Quality Agreement of 1972 and a revised, broader agreement in 1978. One of the principal pollutants, phosphorus, has been dramatically reduced (Figure 19.3). Other problems, such as polychlorinated biphenyls (PCBs), exist but are being addressed. Meanwhile, Lake Erie has recovered substantially, and improved water quality has resulted in reduced blue-green algal blooms and increased yellow perch and walleye pike fisheries.

Assuming that the capacity of a lake to absorb wastes was infinite, people at the beginning of the industrial revolution dumped industrial wastes and sewage into the rivers and lakes of the land. As the population grew and the demand for material goods increased, the amount of waste that had to be disposed of also grew. Most of the industrial by-products were inorganic materials that do not occur naturally and that therefore require an extremely long time to be degraded. By the early 1900s, few rivers in the eastern United States near population centers were unpolluted. With few exceptions, increasing pollution has continued into the present. Only recently have attempts been made to rectify or curtail the dumping of wastes into the environment. Pollution has, of course, spread well beyond the eastern states. In the summer of 1970, the Illinois Department of Public Health declared that not a single river or stream in the entire state was safe to swim in. Pollution of rivers has an impact on the ocean, since pollution products transported to the ocean literally make the ocean a dump.

This leads us to the topics of this chapter—pollution of the ocean and the management of ocean resources. What effect do efforts to develop marine resources have on the oceans? What are the consequences of land-based activities? What is being done to resolve the problems of ocean pollution, and the claims and disagreements between nations over the use of the oceans? Could the ocean ever become overloaded with pollutants? Do restoration projects work? Can nations work successfully together to solve pollution problems?

WHAT IS POLLUTION?

There are many ways to define **pollution.** One might say, for example, that the presence of sewage in the ocean is pollution. But sewage consists of organic waste products that serve as nutrients for many organisms. It has been said that anchovies in the waters off southern California produce more waste products than people. Is anchovy waste therefore polluting the ocean? Just the opposite. It contains nutrients that enhance the growth of photosynthetic organisms.

Does pollution depend on the concentration of the waste? If human waste is dumped in a limited area with poor circulation of water, marine plants die off, bacterial populations increase, and the sludge that accumulates on the ocean bottom suffocates organisms living in or on the bottom. Such a situation surely constitutes pollution. If the waste includes a large amount of industrial or agricultural by-products, such as lead and pesticides, the pollution is intensified. Apparently, waste in itself does not constitute pollution unless its concentration reaches a critical level.

What about the natural seepage of oil in, for instance, the Santa Barbara Channel? Oil has been seeping there for millions of years. Paleo-Indians collected tar on the beaches of the area, and bacteria consumed (and still consume) the oil. But oil released by human negligence is pollution. Oil on the surface of the water causes the deaths of cormorants, pelicans, and other diving birds. To the birds, it makes no difference whether the oil is released by natural causes like an oil seep or by careless people.

When oil drifts into Buzzards Bay, Massachusetts, after an oil barge accident, many organisms die in the coastal marshes. Oil is not a naturally occurring substance in Buzzards Bay. Its presence there constitutes pollution, especially if the oil has been refined and lead has been added to it.

Recently, a hurricane hit Jamaica while scientists were monitoring the growth of the coral reef along the Jamaican coast. The high wave energy and the influx of fresh water resulting from the hurricane destroyed more than 90% of the living coral reef. In 1972 the influx of fresh water and mud caused by a hurricane in Chesapeake Bay seriously damaged the oyster beds there. Are we to regard hurricanes as a source of pollution?

What, then, is an acceptable definition of pollution? According to a 1971 United Nations report, "The Sea: Prevention and Control of Marine Pollution," pollution is "the introduction by man, directly or indirectly, of substances or energy into the marine environment (including estuaries) resulting in such deleterious effects as harm to living resources, hazards to human health, hindrance to marine activities including fishing, impairment of quality for use of seawater and reduction of amenities." To assess or predict the environmental impacts of pollution, we must have reliable estimates of natural populations and reliable information on natural habitats. Current research indicates that marine organisms are surprisingly resistant to the effects of pollution.

THE SEA AS A DUMP

Sewage

In the nearshore waters of bays and estuaries, pollution from sewage has closed down thousands of acres of oyster and clam beds (Figure 19.4). The water in many of this nation's estuaries is so foul that people cannot swim in it for fear of getting a disease, and the possibility of an epidemic is ever present. We have almost replicated the conditions of medieval cities, where raw sewage ran down the gutters of the streets.

Many large urban centers on the coast, such as Los Angeles, pump huge volumes of raw or minimally treated sewage directly into the ocean (Figure 19.5). One massive complex of sewage outfalls is located at Palos Verdes, near Los Angeles. When this outfall was built in 1928, an estimated 7.8 km^2 (3 mi^2) around Palos Verdes was ringed by giant kelp (*Macrocystis*). By 1934, the outfall was discharging 64 million L (17 million gal) of sewage a day. A second pipe was added in 1947, increasing the volume to 227 million L (60 million gal) a day. By 1955, the figure had reached 681 million L (180 million gal) a day, and the kelp forests had dwindled considerably. In 1955, a third outfall pipe was added, bringing the discharge capability to 738 million L (195 million gal) a day. Two years later the temperature of the water rose significantly due to an El Niño, and the kelp, a cold-water plant, suffered another setback, this one natural. A new outfall pipe was added in 1966, discharging at a depth of about 55 m (180 ft) at a distance of 3 km (1.9 mi) off the coast. The discharge capacity rose to 1.2 billion L (318 million gal) a day, and within a few months all the kelp was gone.

There is some question about what caused the killing off of the kelp. The sewage sludge at the bottom may not have affected the kelp's holdfasts directly. Large numbers of sea urchins, which eat kelp, live in kelp beds. Recent evidence shows that sea urchins can directly absorb dissolved organic matter, which is abundant near outfalls. This caused an increase in sea urchin numbers and subsequently increased predation on the kelp. And 95% of the kelp once grew in areas subsequently covered by sludge. Moreover, the sludge contains such industrial wastes as cadmium, copper, chromium, lead, and zinc (all of which can be

(a)

(b)

19.4 *(a) Along most of the continental U.S. coastline, clamming is banned for several months of the year. (W. J. Wallace) (b) Toxic chemicals from industry, municipalities, and agriculture cause several hundred fish kills in the United States each year. (Environmental Protection Agency)*

19.5 *A sewer outfall at Point Pinos, Pacific Grove, California. Here, sewage that has undergone only primary treatment is dumped directly into the surf zone. The dark mass on the rocks in the background is the normal seaweed or algae cover in this area. The prevailing current is toward the viewer. Notice that the sewage effluent, containing chlorine, has bleached the rocks in the foreground free of algae. (W. J. Wallace)*

toxic to marine life), in addition to such agricultural chemicals as insecticides.

The Los Angeles story is not over. Even though the discharge of the sewage facility at Palos Verdes now stands at 1.9 billion L (500 million gal) a day, the kelp beds are returning. New outfalls are farther offshore in deeper waters and are designed to distribute the sludge over a wide area. The sludge is treated more thoroughly at the sewage facility before being released, and industries are being encouraged to recover and recycle chemicals before feeding wastes into the system. With the help of scientists and governmental agencies, kelp has been transplanted into the area, where it is thriving. Other organisms found in the kelp community, such as fish and shellfish, are returning or are successfully being reintroduced.

Approximately 90% of sewage consists of water. In an area such as Los Angeles, which is chronically water poor, it makes little sense to dump about 2 billion L of this potentially reclaimable water into the sea every day. It would make far more sense to treat the sewage further and remove the organic material (Figure 19.6). Water reclamation techniques have been available for some time, but they are expensive. Still, analyses show that if such techniques had been implemented in southern California 20 years ago, by now

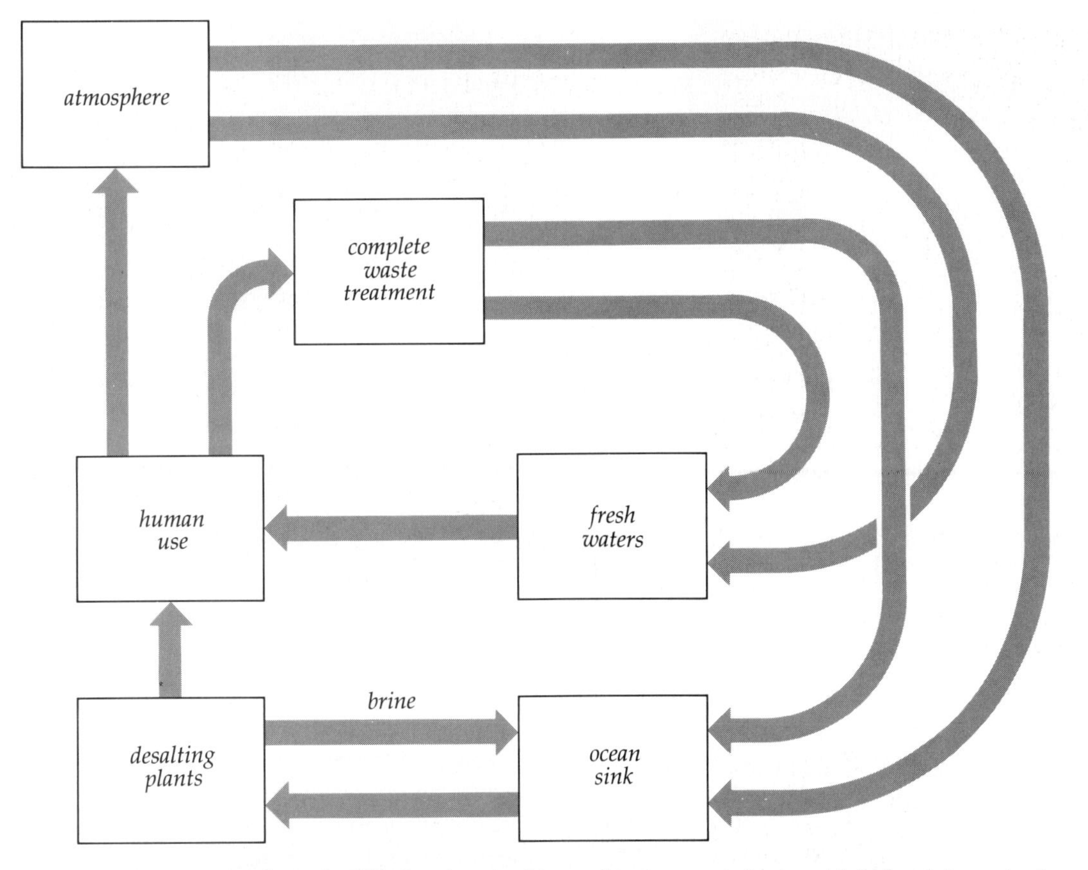

19.6 Our future water system should be based on desalting and tertiary waste treatment (which reclaims water for reuse), especially in heavily populated coastal regions. (From G. T. Miller, Jr., Living in the Environment, *5th ed., Belmont, Calif.: Wadsworth, 1988)*

they would have paid for themselves several times over.

Such technology would be equally useful throughout the urban Southwest. Nature recycles water free of charge, but it costs a lot for people to move it around. It has cost billions of dollars to supply the water-poor areas of California with the precious liquid, and the dams, aqueducts, canals, channels, and pipes that are already despoiling much of the West will do so on an ever-greater scale in the future if corrective measures are not taken. Surely it is shortsighted to throw all this water away like a used Styrofoam cup.

San Diego now has a test facility using aquaculture techniques to recover water from sewage. Bacteria break down the particulate matter, and the sludge that settles is used for landfill. The liquid is transferred to solar greenhouses in which hyacinths are grown in tanks. The hyacinths remove nutrients from the water and are then harvested, processed, and fed to livestock. Fish and crayfish are also raised at this facility. The final product is water that is cleaner than the water of the Colorado River, from which the water originally came. The water is used to irrigate farmland.

Heavy Metals

Hundreds of deaths resulting from the contamination of ocean water by heavy metals have been reported around the world. A classic example is the so-called Minamata disease. Between 1953 and 1960, more than a hundred people living in Minamata on Yatsushiro Bay (Kyūshū Island, Japan) died of this strange affliction. The symptoms varied, but in each case mental and neuromuscular deterioration culminated in death. Some victims survived but became insane. Investigations showed that only people who ate seafood were afflicted by the disease. Shellfish taken from the bay and fed to laboratory animals produced similar symptoms.

The malady was ultimately traced to a large polyvinyl chloride factory located on the headlands of the bay. The factory was pumping waste water that contained mercuric chloride into the bay. The mercury became concentrated in the tissues of shellfish, and people who ate the shellfish suffered from highly toxic mercury poisoning. Mercury disrupts and destroys segments of the central nervous system (Figure 19.7). Although the factory has stopped dumping its waste

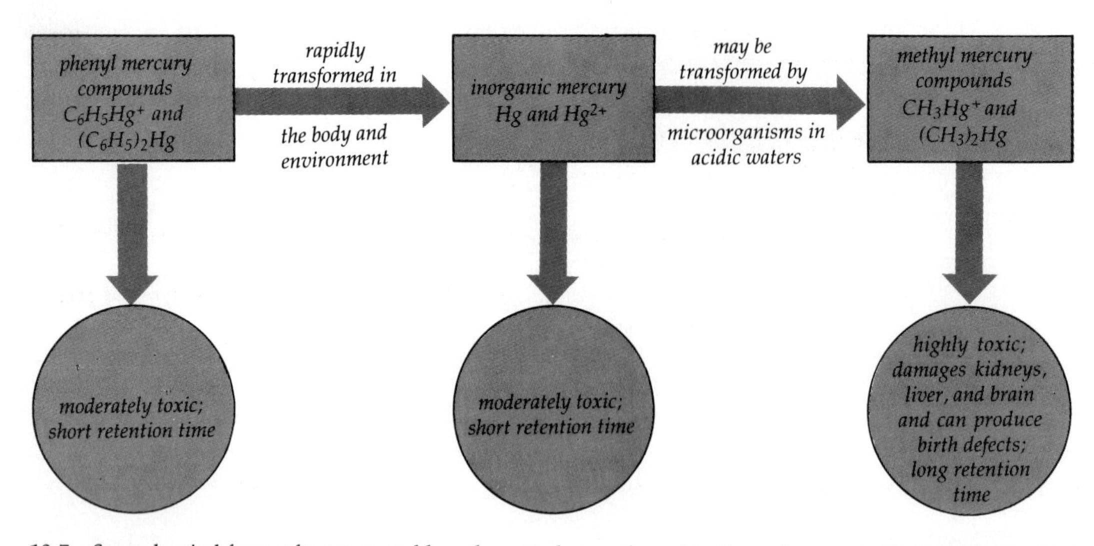

19.7 *Some chemical forms of mercury and how they may be transformed in the environment. (G. T. Miller, Jr.,* Living in the Environment, *5th ed.)*

Table 19.1 *World Heavy Metal Production and Potential Ocean Inputs (million tons/yr)*

Element	Mining Production	Transport by Rivers to Oceans	Atmospheric Washout
Lead	3	0.1	0.3
Copper	6	0.25	0.2
Vanadium	0.02	0.03	0.02
Nickel	0.5	0.01	0.03
Chromium	2	0.04	0.02
Tin	0.2	0.002	0.03
Cadmium	0.01	0.0005	0.01
Arsenic	0.06	0.07	—
Mercury	0.009	0.003	0.08
Zinc	5	0.7	—

Source: D. A. Ross, *Opportunities and Uses of the Sea,* Englewood Cliffs, N.J.: Prentice-Hall, 1978, p. 204.

waters into the bay, the sediment is still heavily contaminated, and fishers and clammers will not be able to use the bay again for years. Similar instances of mercury poisoning on a smaller scale have occurred throughout the world. Unfortunately, mercury is only one of the poisonous heavy metals that have been dumped into waterways (Tables 19.1 and 19.2).

Another heavy metal that produces toxic effects is lead. The amount of lead in the environment is abnormally high and is probably increasing. The prime source is probably tetraethyl lead, which is used to raise octane ratings of gasoline.

One indicator of lead buildup in the environment is the lead content of the ice cover at the poles. Because the use and subsequent release of lead into the environment have been greatest in the Northern Hemisphere, any change in the lead content of polar ice should be greatest at the North Pole. Such is apparently the case. Although the proportions of other materials over an equivalent time span have shown little variation, the lead content in the Greenland ice cap between 1750 and 1940 increased by about 400%. Between 1940 and 1967, it increased an additional 300% (Figure 19.8).

Paint, solders used in tin cans, toothpaste tubes, piping, and certain kinds of crystal glassware and ceramics also release lead through abrasion and wear. In 1968, the lead content in the blood of an average American was about 0.25 parts per million. For city dwellers, especially those in close contact with automobiles (service station attendants, mechanics, and traffic officers), the figure was double the average. Lead is a cumulative poison, and efforts to remove it from fuels are imperative. However, substitutes for tetraethyl lead, especially if they contain heavy metals, may also create serious problems. Fortunately, late-model cars are required by law to burn lead-free gasoline, which contains no heavy metals.

Chemical poisons are particularly hazardous when they are discharged in high concentrations. For example, in 1965 a large amount of copper sulfate was released off Holland. Based on the copper content of the North Sea as a whole, 3 μg/L, the amount released was projected to have a minimal effect. However, the copper sulfate did not disperse rapidly, and concentrations of copper in the coastal waters reached 500 μg/L. More than 100,000 fish were killed, and the commercial mussel beds were decimated.

In the 1970s, the International Mussel Watch began to monitor metals in coastal waters. The various spe-

Table 19.2 *Sources and Effects of Some Widely Used Toxic Metals*

Metal	Major Sources	Major Health Effects
Arsenic (As)	Burning of coal and oil; smelting of nonferrous ores; additive to glass; pesticides; mine tailings	Cumulative poison at high levels; carcinogen
Beryllium (Be)	Burning of coal and oil; cement plants; alloys; ceramics; rocket propellants	Skin lesions; ulcers; respiratory disease (berylliosis); carcinogen
Cadmium (Cd)	Burning of coal and oil; zinc mining and processing; batteries; incineration; fertilizer processing and application	Carcinogen; teratogen; high blood pressure; heart disease; liver, kidney, and lung disease
Lead (Pb)	Auto exhaust (leaded gasoline); lead batteries; lead-based paint; smelting of nonferrous metals	Brain damage; behavioral disorders; hearing damage to children; death
Mercury (Hg) and methyl mercury (CH_3Hg^+)	Burning of coal; many industrial uses; seed fungicides; antifouling paint	Nerve damage; kidney damage; birth defects; death

Source: G. T. Miller, Jr., *Living in the Environment*, 5th ed., Belmont, Calif.: Wadsworth, 1988.

cies of the common mussel, *Mytilus*, occur worldwide, effectively taking up metals from seawater and concentrating the metals so that assays can be made easily and cheaply. The International Mussel Watch monitors metals such as lead, zinc, nickel, cadmium, copper, plutonium, and silver. The program has shown that the use of unleaded fuel has resulted in a reduction of lead in coastal waters. Mussel Watch in the United States is administered by the Environmental Protection Agency (EPA).

In the 1980s, a new heavy-metal problem has been recognized. Organotin compounds have been added to paints for boats and ships as a potent antifouling agent. The organotin compounds have been leaching into coastal waters and sediments where boating use is heavy. Very low concentrations, 1 or 2 parts per billion, are toxic to many marine organisms, especially larvae. This level is toxic, for example, to larvae of the bay mussel, *Mytilus edulis*. The EPA has recently begun to address this new problem.

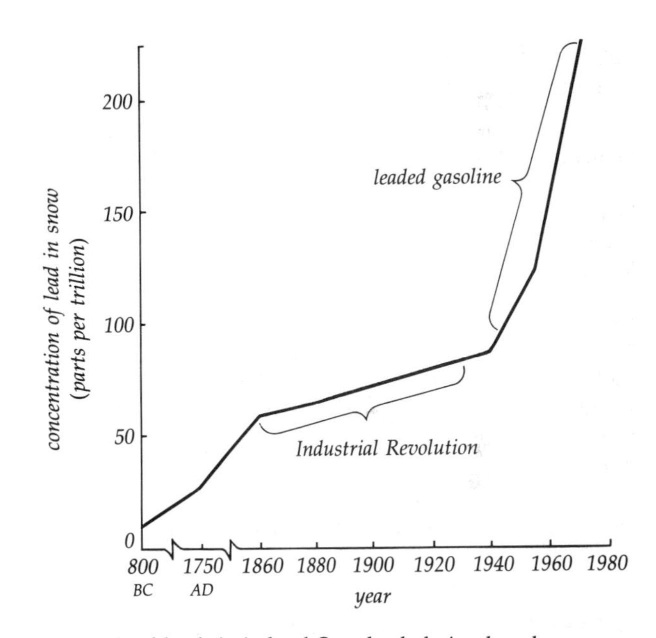

19.8 Lead levels in isolated Greenland glaciers have been increasing since 800 B.C. The actual data are a series of scattered points; the curve represents the best average line through these points. In addition, there is considerable seasonal variation in lead levels. The trend is significant, however. (Data from Murozumi and others, "Chemical Concentrations of Pollutant Lead Aerosols, Terrestrial Dusts, and Sea Salts in Greenland and Antarctic Snow Strata," Geochimica Cosmochimica Acta, *1969, 23, 1247–94.)*

Synthetic Organic Compounds

World production of organic solvents such as benzene and trichloroethylene, organic chemicals used in the plastics industry such as vinyl chloride, and pesticides such as DDT and 2,4-D currently exceeds 91 million metric tons (100 million tons) annually (Figure 19.9 and Table 19.3). Of these, the chlorinated and halogenated hydrocarbons (containing bromine, fluorine, or iodine) are widely used in pesticides, solvents, flame retardants, and other products. These compounds block enzyme action in all cells, many are carcinogenic, and some cause side effects ranging from liver

CH_3Cl
Methyl chloride, or chloromethane (IUC) (refrigerant)

CH_2Cl_2
Methylene chloride, or dichloromethane (IUC) (solvent, local anesthetic by cooling)

$HCCl_3$
Chloroform or trichloromethane (IUC) (anesthetic)

CCl_4
Carbon tetrachloride, or tetrachloromethane (IUC) (solvent, cleaning fluid)

$BrCH_2CH_2CH_2Br$
1,3-Dibromopropane (IUC) (vermicide)

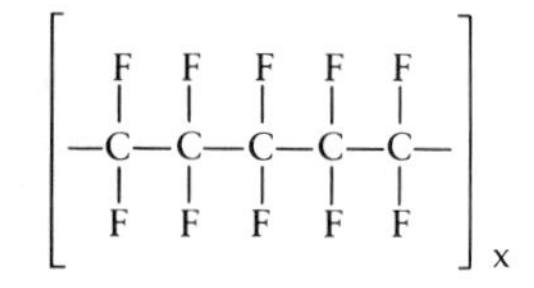

Teflon polymer

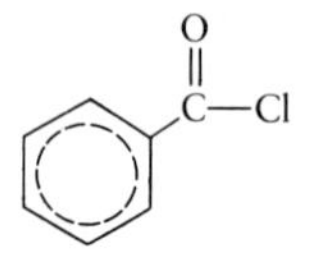

Vinyl chloride, or chloroethane (IUC) (polymers)

Cl Cl
Cl Cl
Cl Cl
Hexachlor, $C_6H_6Cl_6$ (insecticide)

O
C—Cl

Benzoyl chloride (tear gas)

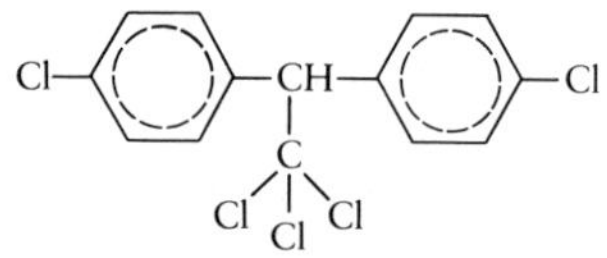

2,4-Dichlorophenoxyacetic acid, or (2,4-D) (weed killer)

H Cl
C=C
Cl Cl
Trichloroethylene (dry cleaning solvent)

Cl CH Cl
C
Cl Cl Cl
2,2-*Bis*-4-chlorophenyl-1,1,1-trichloroethane (DDT) (insecticide)

19.9 Some common synthetic organic compounds. (J. A. Campbell, Chemical Systems, *San Francisco: Freeman, 1970)*

Table 19.3 *Shipping of Synthetic Organic Chemicals by Ocean Tankers in 1970*

Compound	Total Amount Shipped (tons)	Amount Left On Board (tons)
Trichloroethane	4,618	5
Dichloropropane	6,685	unknown
Perchloroethylene	39,986	36
Ethylene dichloride	241,356	80
Benzene	279,852	241
Toluene	366,441	234
Acrylonitrile	93,453	41

Source: D. A. Ross, *Opportunities and Uses of the Sea,* Englewood Cliffs, N.J.: Prentice-Hall, 1978, p. 199.

damage to reduced rates of photosynthesis. DDT, dieldren, aldrin, and hexachlor are effective insecticides, but they do not readily degrade in the environment and are transported to the ocean. Since they are fat (lipid) soluble, they become concentrated in living systems.

The classic example of insecticide pollution is DDT. Its use grew dramatically from after World War II until the early 1970s, when scientists found that homogenized organs of fish-eating seabirds had DDT concentrations of 20 parts per million (ppm), while the ocean concentration was only about 1 part per trillion (Figure 19.10). The DDT concentration in the lipids of these birds was over 1,000 ppm, and these birds were producing eggs with thin shells, apparently because the DDT was retarding the production of calcium carbonate. When the parents sat on the eggs to incubate them, the eggs broke. This phenomenon was observed in peregrine falcons, osprey, bald eagles, brown pelicans, and other seabirds. High levels of DDT were even observed in Antarctic penguins, an indication that the problem was worldwide.

DDT production in the United States has been greatly reduced, and the affected bird populations are recovering. But the production and use of DDT in other countries, such as Brazil, and the increased use of other halogenated hydrocarbons mean that the danger of ocean pollution by these compounds remains.

The National Benthic Surveillance Project was initiated in 1984 as part of the NOAA National Status

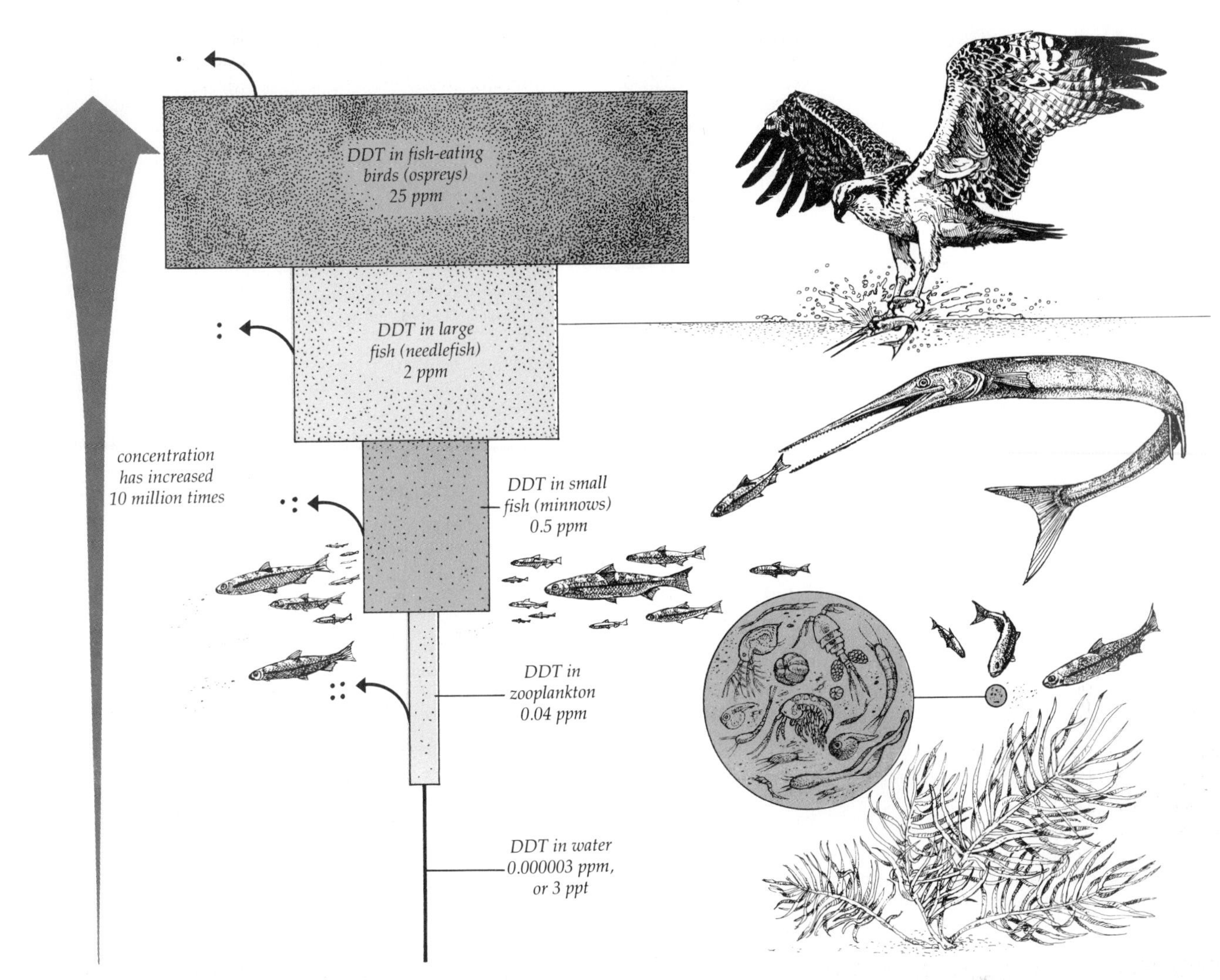

19.10 The concentration of DDT in the fatty tissues of organisms was biologically amplified approximately 10 million times in this food chain of an estuary adjacent to Long Island Sound, near New York City. Dots represent DDT, and arrows show small losses of DDT through respiration and excretion. (G. T. Miller, Living in the Environment, *5th ed.)*

and Trends Program (Table 19.4). Bottom fish and sediments are analyzed for the presence of toxic chemicals, and the fish are also checked for related diseases. The main problem identified to date is liver disease in bottom fish; for example, English sole and white croaker in some bays. The diseases appear to correspond to high concentrations of aromatic hydrocarbons and polychlorinated biphenyls in a number of coastal waters. DDT is still found in high concentrations outside San Pedro Bay near Los Angeles. It must be noted that there is no clear indication of a public-health hazard at any of the sites of the Mussel Watch Program or the Benthic Surveillance Project. The elevated levels of heavy metals and chemical compounds constitute a public health concern. At the same time, there is no documented evidence that eating an English sole that has a diseased liver will affect a person. But don't eat its liver!

Oil

If you look carefully at the water anywhere in the northern shipping lanes, you will probably see fragments of crude oil in the form of small black spheres. The colder temperature of the northern waters tends to make the oil form into tar balls, ranging from the size of a marble to that of a tennis ball. Because oil and water are chemically and structurally different and have little affinity for one another, seawater does not readily dissolve oil. Also, the colder the water, the slower the rate of solution. Even in warmer waters, tar balls may form in choppy seas.

The open ocean is no longer "pure." Many fish, scallops, and oysters taken from the North Atlantic taste of oil because they have ingested tar balls. Thor Heyerdahl, during his *Kon-Tiki* expedition of 1947, observed tar balls in areas of the Pacific remote from shipping lanes. Similarly, marine scientists the world

Table 19.4 *Concentrations of Selected Chemicals in Sediments from West Coast Sampling Sites*

Site Locations	Mean Concentrations ± SD[a] AHs (ng/g)[b]	PCBs (ng/g)	DDTs (ng/g)[c]	Cu (μg/g)[d]	Cr (μg/g)[e]	Pb (μg/g)[f]
San Diego Bay	5,000 ± 4,000	420 ± 100	8 ± 2	219 ± 23	178 ± 19	51 ± 5
Outside San Diego Bay	< 20	7 ± 5	< 0.4	8 ± 0.4	50 ± 9	12 ± 2
Dana Point	23 ± 40	7 ± 4	0.5 ± 0.1	10 ± 4	40 ± 11	19 ± 7
Outside San Pedro Bay	530 ± 220	160 ± 56	620 ± 270	31 ± 8	107 ± 31	17 ± 3
Near Seal Beach	260 ± 230	47 ± 23	27 ± 26	26 ± 10	108 ± 19	27 ± 7
Santa Monica Bay	47 ± 80	15 ± 4	4 ± 1	11 ± 6	54 ± 4	33 ± 2
Hunters Point	2,900 ± 960	39 ± 12	3 ± 0.5	46 ± 23	355 ± 61	12 ± 8
Near Oakland	1,700 ± 360	61 ± 12	6 ± 2	72 ± 5	196 ± 4	44 ± 2
Southampton Shoal	1,300 ± 1,600	12 ± 8	0.3 ± 0.5	13 ± 5	267 ± 63	6 ± 1
San Pablo Bay	116 ± 70	9 ± 2	0.6 ± 0.3	31 ± 9	649 ± 104	5 ± 0.2
Bodega Bay	15 ± 14	4 ± 3	< 0.3	< 0.1	246 ± 156	2 ± 1
Coos Bay	230 ± 400	3 ± 3	< 0.2	1 ± 3	110 ± 95	5 ± 2
Columbia River	150 ± 10	9 ± 8	< 0.5	17 ± 4	30 ± 4	16 ± 5
Nisqually Reach	< 20	4 ± 3	< 0.5	13 ± 3	118 ± 100	25 ± 2
Commencement Bay	1,200 ± 0	6 ± 3	< 0.5	51 ± 4	70 ± 2	35 ± 7
Elliott Bay	4,700 ± 1,200	330 ± 110	5 ± 2	96 ± 13	114 ± 58	20 ± 7
Lutak Inlet	< 20	7 ± 8	< 0.8	27 ± 6	58 ± 23	16 ± 5
Nahku Bay	97 ± 120	7 ± 5	< 0.4	10 ± 0.4	23 ± 2	43 ± 9

[a]SD = Standard Deviation. In cases where concentrations were below the limits of detection in some but not all the samples from a site, these values were considered to be zero.
[b]AH = aromated hydrocarbon; ng/g = nanograms/gram
[c]DDT = dichlorodiphenyltrichloroethane
[d]Cu = copper; μg/g = micrograms/gram
[e]Cr = chromium
[f]Pb = lead
Source: E. D. Goldberg in Wong et al., *Trace Metals in Sea Water.*

over have found their plankton nets fouled in regions far removed from shipping lanes.

Oil is a natural product. In some areas, oil seeps out from deposits beneath the ocean floor. But seepage accounts for only 1 to 10% of the oil in the sea and along the coastlines. Where does the rest of it come from (Figures 19.11 and 19.12 and Table 19.5)?

Each year, the world uses about 4.6 billion metric tons of crude oil. A third of it is shipped across the oceans. Along the way, about 3 million metric tons are lost into the environment, primarily through handling in harbors but also by leakage at sea. Consequently, an ever-present oil slick plagues virtually every harbor in the world. In addition to oil leaking from tankers, considerable quantities of oil enter the sea when shippers drain their bilges with seawater. This common practice is illegal but is difficult to prevent.

Seepage at sea depends on the quality of the particular tanker. Ship captains in the employ of most oil companies are conscientious about trying to prevent seepage, and the seepage from most oil tankers built in recent years is very small. Some companies, however, register their ships under the flag of countries with lax regulations.

The figure of 3.2 million metric tons of oil lost per year includes such calamities as the wreck of the tanker *Torrey Canyon* in 1967 off the English coast of Cornwall. After the wreck, 100,000 metric tons of crude oil spread over the beaches and killed a large number of marine life, especially birds. Such disasters, however, helped focus world attention on the problem and made it evident that the ocean cannot easily absorb massive oil spills and that such accidents cause great ecological damage.

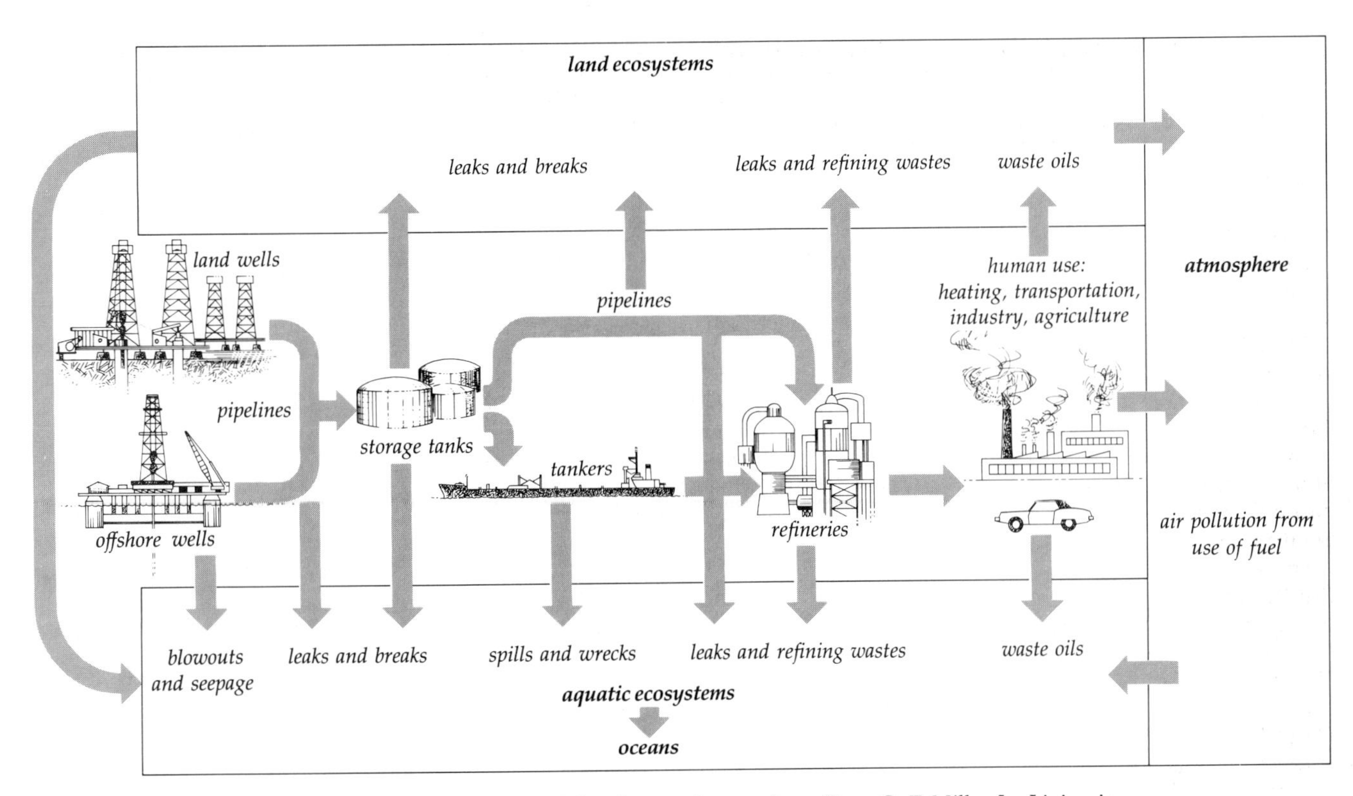

19.11 *Major ways in which oil pollutes the hydrosphere, lithosphere, and atmosphere. (From G. T. Miller, Jr.,* Living in the Environment, *5th ed., Belmont, Calif.: Wadsworth, 1988)*

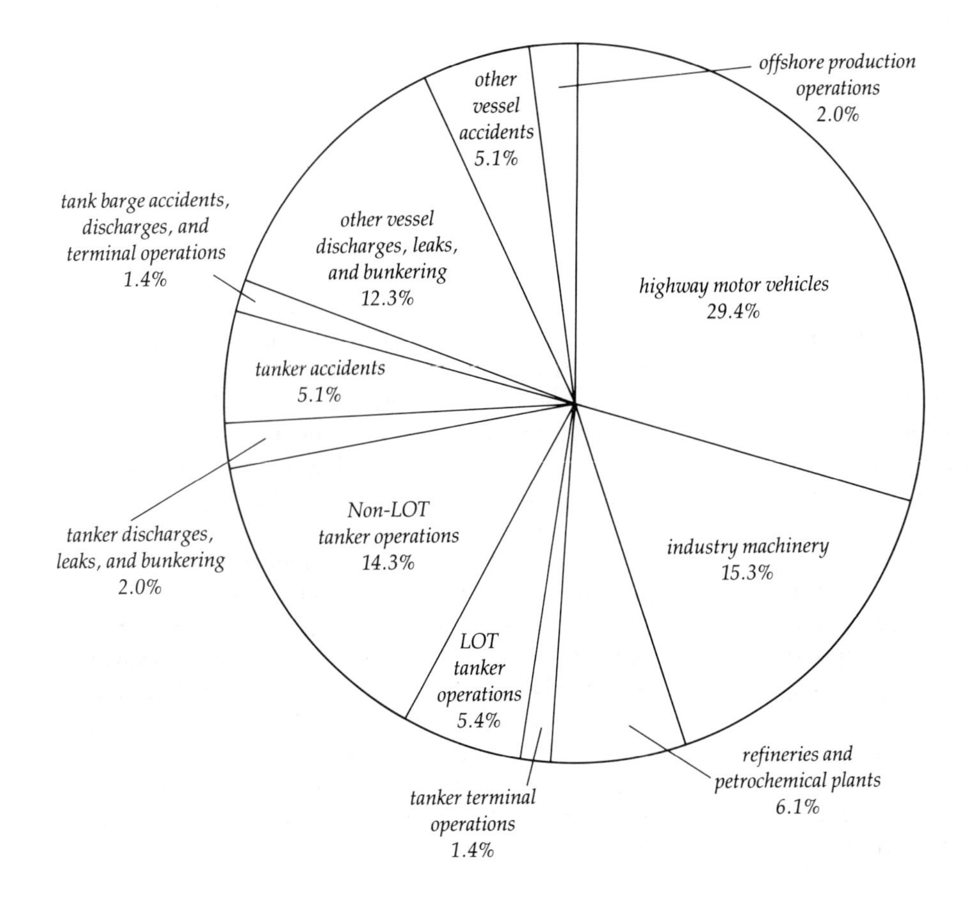

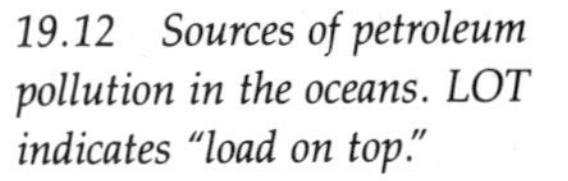

19.12 *Sources of petroleum pollution in the oceans. LOT indicates "load on top."*

Table 19.5 *Input of Petroleum Hydrocarbons into the Marine Environment (mta)*

Source	Probable Range	Best Estimate[a]
Natural sources		
Marine seeps	0.02–2.0	0.2
Sediment erosion	0.005–0.5	0.05
(Total natural sources)	(0.025)–(2.5)	(0.25)
Offshore production	0.04–0.06	0.05
Transportation		
Tanker operations	0.4–1.5	0.7
Dry-docking	0.02–0.05	0.03
Marine terminals	0.01–0.03	0.02
Bilge and fuel oils	0.2–0.6	0.3
Tanker accidents	0.3–0.4	0.4
Nontanker accidents	0.02–0.04	0.02
(Total transportation)	(0.95)–(2.62)	(1.47)
Atmosphere	0.05–0.5	0.3
Municipal and industrial wastes and runoff		
Municipal wastes	0.4–1.5	0.7
Refineries	0.06–0.6	0.1
Nonrefining industrial wastes	0.1–0.3	0.2
Urban runoff	0.01–0.2	0.12
River runoff	0.01–0.5	0.04
Ocean dumping	0.005–0.02	0.02
(Total wastes and runoff)	(0.585)–(3.12)	(1.18)
Total	1.7–8.8	3.2

[a]The total best estimate, 3.2 million metric tons per annum (mta), is a sum of the individual best estimates. A value of 0.3 was used for the atmospheric inputs to obtain the total, although we well realize that this best estimate is only a center point between the range limits and cannot be supported rigorously by the data and calculations used for estimation of this input.
Source: Oil in the Sea, National Academy of Sciences, 1985.

Oil spills have been one of the biggest problems affecting beaches and their wildlife since the 1940s, when millions of tons of oil were spilled along the U.S. coast by torpedoed ships (Figures 19.13 and 19.14). The frontage of many shorelines around the United States, especially near harbors, is constantly covered with a film of oil. The Santa Barbara Channel and *Torrey Canyon* incidents were not unusual; large oil spills occur frequently. Newspapers commonly report local spills, but their reports rarely attract general attention.

When birds land on or plunge into the water, they cannot tell whether it is covered with oil. If it is, the oil penetrates their feathers, which insulate them from the cold and buoy them up in the water. The birds may either freeze to death or drown. Moreover, once their feathers have been matted together, the birds can no longer fly. Most birds that have become coated with oil will die (Figure 19.15). Of the thousands of birds treated by volunteers after an oil spill in San Francisco Bay in 1971, only a few lived.

The largest oil spill from a tanker accident was the *Amoco Cadiz* accident in 1978 off the coast of Brittany, France. A review of the event gives an idea of the impact such a spill can have. Several factors combined to increase its impact. The accident occurred only 13 km from shore in an area of strong currents and wave action. The ship was carrying 220,000 metric tons of crude and Bunker fuel. In the initial rupture, about 50,000 metric tons were spilled and pushed ashore by winds. In the weeks that followed, oil continued to be released and winds shifted, pushing the oil along the north shore of Brittany. Eventually 300 km of coast were affected. Oil persisted in the water column for 6 months and in the sediment for 3 years.

Biological and economic impacts of *Amoco Cadiz* were extensive. More than 4,500 oiled birds were recovered from the water, with an additional 3,200 or more dead birds on the beaches. Most were diving birds such as cormorants. Decreases in the plankton biomass were observed for weeks. In a survey done 70 days after the accident, the biomass showed no signs of recovery. There were minor casualties of fish within 10 km of the wreck. Sizes of commercial flatfish, plaice, and sole from the 1978 catch were below those of other years. Benthic animals that experienced massive casualties were sea urchins, razor clams, and amphipods. Oysters, a major commercial industry in the area, were heavily contaminated but managed to

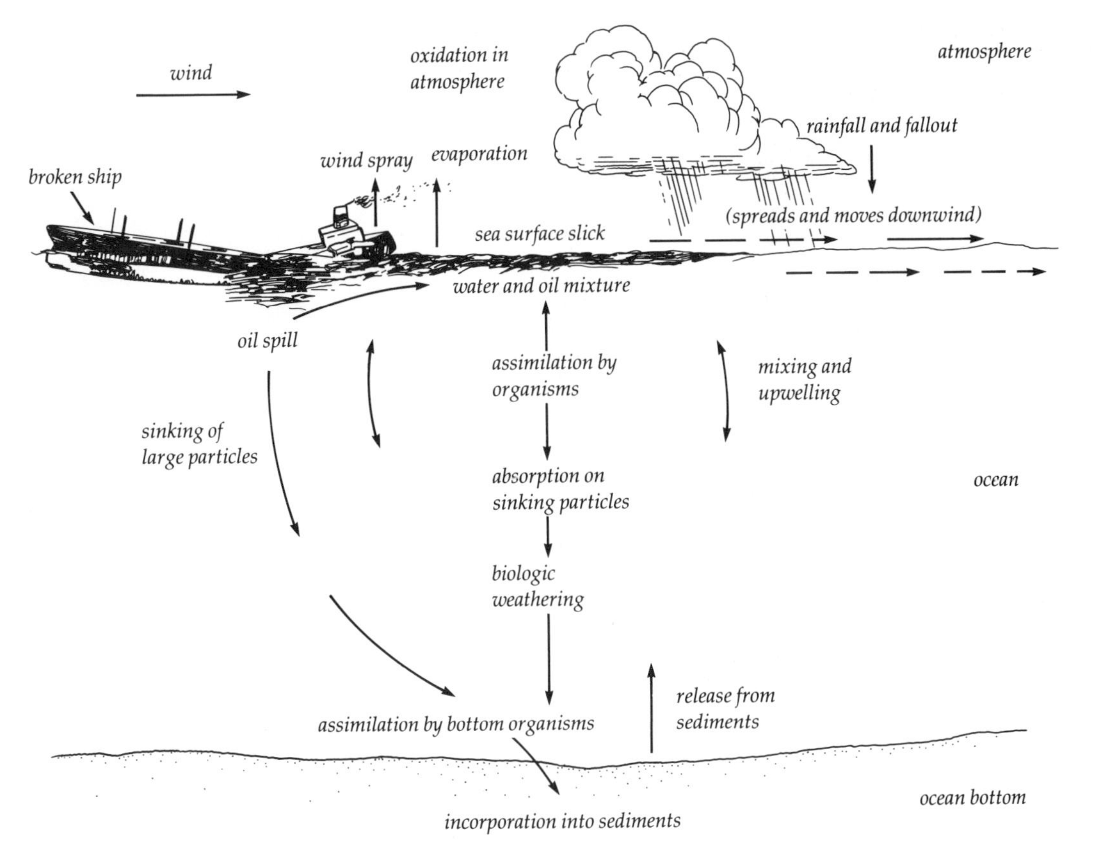

19.13 Some pathways by which an oil spill may travel.

19.14 The oil spill caused by the collision of two tankers on January 18, 1971, in San Francisco Bay. The foreground is outside the bay. Alcatraz is at the top right. (Western Aerial Photos, Inc., Redwood City, California)

19.15 A few of the many thousands of ducks and other waterbirds that succumbed when the Greek tanker Delian Appollon *ran aground. (Carson Baldwin, Jr., Sarasota, Florida. Courtesy of the Outdoor Photographers League, San Diego, California)*

19.16 Crude oil covering rock and creatures in a tide pool at Point Loma, near San Diego. Most rocky areas now show this oil coverage to some degree. (W. J. Wallace)

survive. Salt-marsh vegetation in intertidal mudflats was heavily damaged, as were associated fauna.

Subsequently, scientists have restored the salt marshes, and currents have diluted the oil concentrations. However, the mudflats experienced increased erosion with the loss of vegetation, and the restoration may take more than 10 years, by some estimates. Oil persists in the marsh muds, slowing the recovery of the plants.

The average incidence of oil spills has been reduced by 70%, from the 1970s to the 1980s, according to the International Tanker Owners Pollution Federation Ltd. and the U.S. National Academy of Sciences. During this time period, the amount of oil produced from offshore wells and transported by ship has increased about 15%. The annual average number of spills larger than 5,000 barrels was 26 (range 20–37) from 1974–79, while for 1980–85, the average was 8 (range 3–13). The oil industry has made major improvements in controlling oil pollution. Loss of oil means loss of income, and cleanup costs are expensive.

It is clear that the industrial nations of the world will continue to need oil for some time to come. The best we can hope for is more careful management and reinforcement. Transfer procedures in harbors should be improved to minimize leakage in these areas, which are often adjacent to or part of an estuarine system. International pressure should be placed on countries with lax shipping regulations. Methods must be developed for detecting oil film caused by the flushing of tanks at sea; the U.S. Coast Guard is working on that problem. Simply stopping the flushing of tanks would reduce the presence of oil film on beaches (Figure 19.16).

Radioactivity

A small percentage of the atoms of many elements are naturally radioactive, and radioactivity is continuously being introduced into the environment by high-energy electromagnetic radiation from outer space (Figure 19.17). The large-scale testing of fission and fusion weapons for more than a decade after World War II noticeably increased background radiation counts. The nations of the world have now agreed to cease atmospheric nuclear testing. But the proliferation of nuclear reactors and nuclear power plants is producing large amounts of radioactive waste products, including radioactive gases, most of which are ultimately flushed out of the air by rain.

Before 1961, the Atomic Energy Commission (AEC), now called the Nuclear Regulatory Commission (NRC), usually disposed of radioactive wastes by dumping them into the sea. The waters of the deep sea are not as tranquil as the AEC once thought, however. Fifty-five-gallon drums of radioactive wastes have

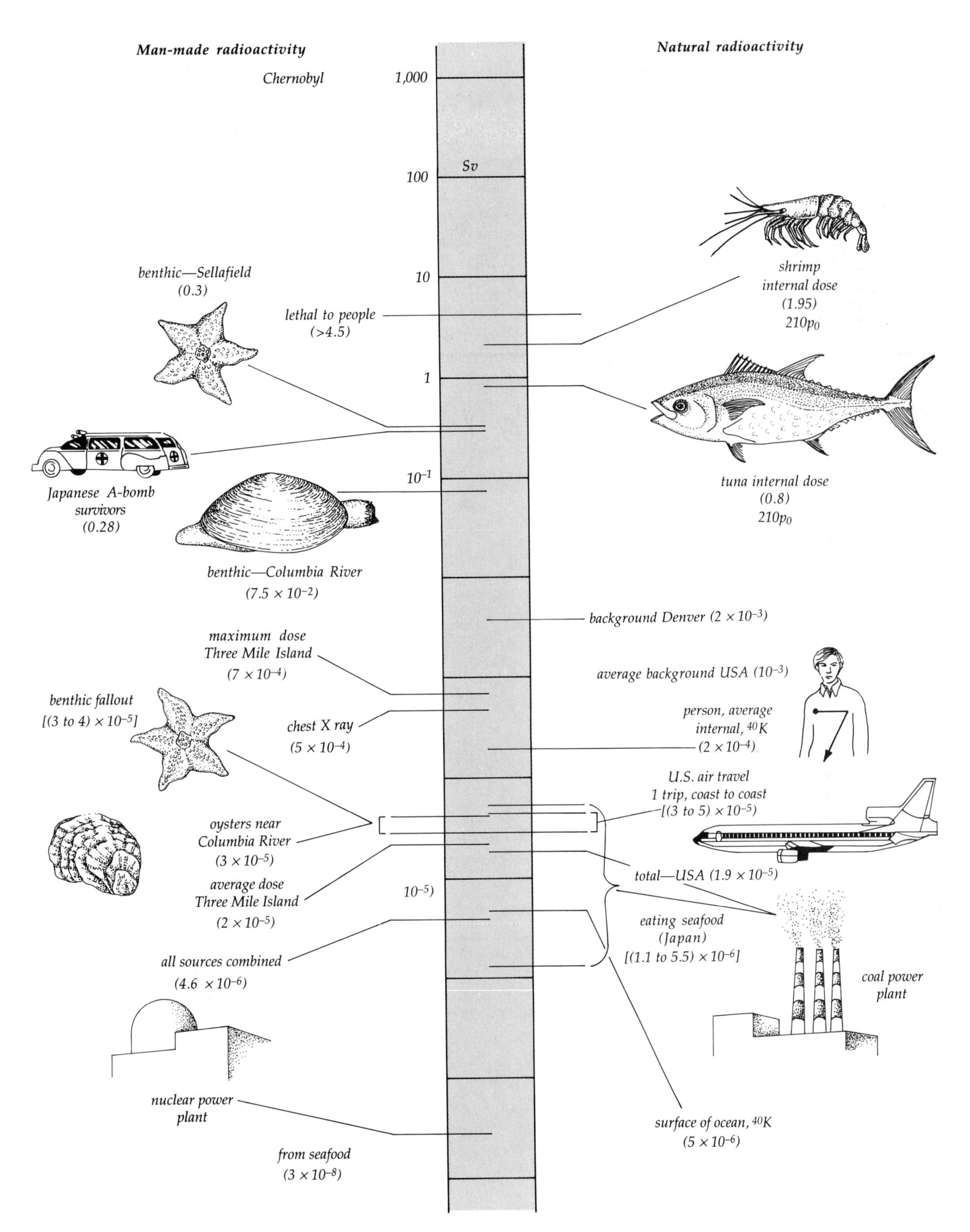

19.17 *Radiation dose (Sv) from man-made and from natural sources.*

washed ashore in Oregon, and North Atlantic fishers have brought up drums in trawl nets far from the restricted dumping regions. Other disturbing signs, such as radioactive crabs, tuna, and clams in the Marshall Islands, have prompted a hard second look at the idea of using the sea as a nuclear-waste dump site.

The 1958 U.N. Convention on the High Seas, Article 25 states:

> Every State shall take measures to prevent pollution of the seas from the dumping of radioactive waste, taking into account any standards and regulations which may be formulated by the competent international organizations. All States shall cooperate with the competent international organizations in taking measures for the prevention of pollution of the seas or air space above, resulting from any activities with radioactive materials or other harmful agents.

The United States and other countries continued dumping some nuclear wastes into the ocean until 1972. In 1972, several laws were enacted to address the general problem of ocean dumping, including radioactive wastes. In the United States, the Federal Water Pollution Control Act and the Marine Protection, Research, and Sanctuaries Act required permits for dumping and transporting wastes at sea and prohibited dumping of high-level radioactive wastes. Also in 1972, an international agreement, the Convention on the Prevention of Marine Pollution by Dumping of Wastes and Other Matter, prohibited dumping high-level radioactive wastes and products developed for chemical and biological warfare. Man-made radioactivity in the ocean is no longer a problem.

The pathway of radioactive particles into the human body is either directly from the air or via food webs. Only about 0.2% of the radioactive particles we receive get to us via marine food webs. This is partly explained by the lack of marine fish in our diet. Also, the ocean disperses particles, and clay sediments on the ocean bottom absorb radioactive particles.

The danger of introducing radioactive materials into the environment is clear, and the hazard of organisms ingesting and perhaps concentrating them is present and very real. Additionally, there is the potential hazard of accidents occurring during waste transport, discharging such extremely poisonous isotopes as plutonium.

The ocean floor may yet turn out to be the safest place to store nuclear wastes. Large areas of the ocean floor lie in the aseismic regions of tectonic plates. Drilling into these stable sediments and sinking nuclear wastes deep below the surface may isolate the wastes for millions of years.

Heat

The United Nations' definition of pollution includes thermal pollution because of the potential problems anticipated with nuclear power plants. Heat is an unusual pollutant because it does not accumulate in the ocean water or sediments and is not concentrated in food webs. Temperatures in coastal waters have not been affected by any power plants except in the immediate vicinity of outfalls. The volume of the ocean so greatly exceeds the volume of hot water effluent that there is no pollution problem. This is not the case for rivers or lakes. Some coastal power plants have aquaculture facilities in which warm effluent actually encourages the growth rates of the fish and shellfish. An example is the Encina plant in San Diego County.

There is, however, a global effect of thermal pollution that could be a problem in the future. W. S. von Arx (in *Wastes in the Ocean*) has pointed out that our civilization uses about 10^{13} W annually that dissipate as heat into the atmosphere. This adds about 0.01% to the heat balance of the atmosphere. If civilization increases its power use by a factor of 10 or 100, the total atmospheric heat balance would have an excess of 0.1% or 1%, respectively. This could substantially warm the earth and alter climates. Present trends suggest that in 50 years, thermal pollution could become a global concern. A shift to the use of solar cells or water to generate power would not affect the heat balance, whereas any power generated by burning coal or hydrocarbons or by nuclear reactions does affect the heat balance.

Solid Refuse

The variety of solid refuse dumped into the ocean is astonishing. Many chemical companies regularly send barges to sea to drop drums of spent or dangerous chemicals. Over the years, the U.S. Defense Department and related industries have dumped tremendous amounts of obsolete hardware into the ocean. It is fairly common for large cities, such as Los Angeles, to dump confiscated weapons into the ocean, and federal and state agents around the country sometimes dispose of narcotics hauls in the sea. A number of California motion picture studios have even thrown old movie sets into the ocean.

Almost any National Oceanic and Atmospheric Administration chart for offshore areas, especially those for some of the larger harbors, shows an area where caution is advised because it is used for bomb disposal. In addition, the U.S. Navy regularly uses certain stretches of the ocean as gunnery ranges, and it is impossible to be sure that all the ammunition has been detonated. Old bombs and shells and radioactive material have often been disposed of in the ocean.

Many people want to throw the rusting carcasses of discarded automobiles into the ocean. This might be beneficial, because the wrecks would shelter a number of small fish and might encourage a sport fishing industry. But they might also snag lines and nets.

These practices are followed not because the ocean is regarded as the ideal dumping place; it is simply that most cities have few other places for dumping their wastes. The amount of debris that a modern industrial society produces each year is staggering. Wastes dumped offshore from the United States amount to more than 9 million tons annually. In the United States alone, more than 7 million cars, 50 million cans, 25 million glass containers, and 18 million metric tons of paper are discarded every year. Add to this the vast amounts of other material that people throw out, and the tonnage becomes awesome. Consider the billions of paper bags, wrappers, and cartons—not to mention the styrene cups and containers—that the nation's leading hamburger and fried chicken chains dispense each year. Many cities have run out of convenient dumping grounds and must ship the stuff farther and farther away. Long trains filled with trash and garbage may one day wind their way out of our cities.

All dumps are potential health hazards, although some are called "sanitary landfills." Historically, most dump sites have been located in the most accessible areas: rivers, flood plains, marshes, estuaries, and canyons. The worst sites imaginable are those next to water. Many of the country's streams, lakes, and estuaries have been poisoned by the proliferation of dumps, and the rivers, along with the barges, carry the wastes to the sea.

In only a few short years, the lowly Styrofoam cup has become ubiquitous in the oceans of the world. When the cup disintegrates, the little nondegradable beads that compose it continue to ride the waves. Marine scientists have collected these beads as well as cellulose fibers from toilet paper in plankton nets thousands of kilometers out at sea.

By weight the principal material dumped at sea is dredged material (Figure 19.18). This consists mainly of rock, sediments, sand, and mud. It is generally not toxic. Its environmental effects are threefold. First, the nature and depth of the bottom may be altered. This may be beneficial if, for example, mud is replaced by rock. Second, some dredged materials such as mud may be reduced chemically. When dumped into the ocean, the mud oxidizes, depleting the dissolved oxygen content of the water. This, of course, affects the biota. Third, any sediment in suspension reduces the compensation depth.

A number of beneficial and innovative uses for solid refuse are being applied now, and their use could be increased. In many places, notably in Florida, old vehicles and ships are sunk and serve as artificial reefs. This encourages the fish populations and enhances sport fishing and diving. Obviously, dredged material could be used for landfill. A third use of solid refuse is to produce ceramic bricks useful for construction purposes.

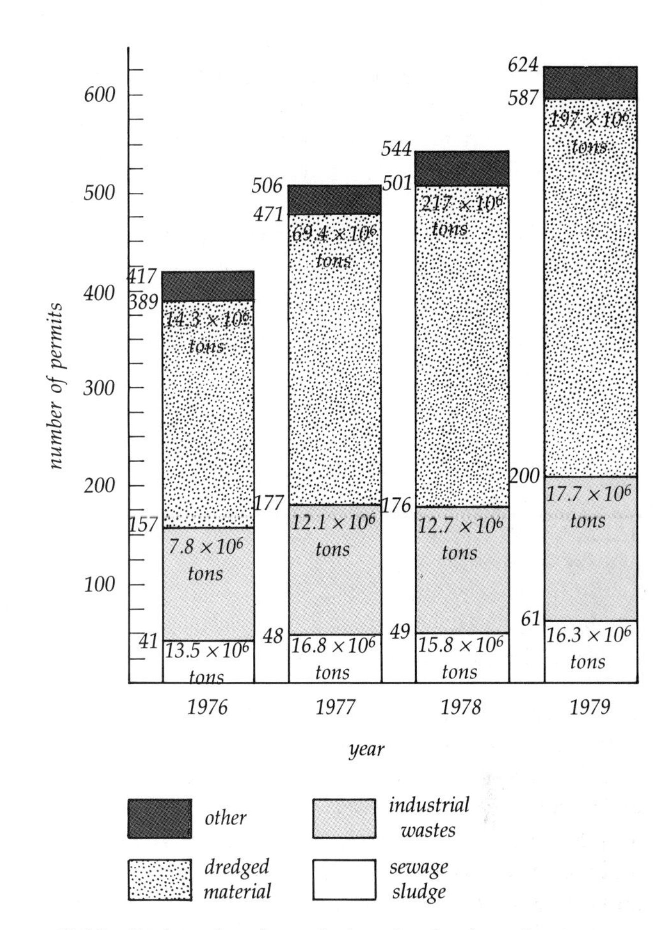

19.18 Total number of permits issued and estimated tonnages (metric tons), on a global basis, for the disposal of wastes in the sea. (From Duedall et al.)

It would be economically sound to recycle most of the debris that we now discard. During World War II, Americans separated their garbage and trash into piles for collection. True, materials were in short supply then. Appearances to the contrary, they still are.

THE CARBON DIOXIDE PROBLEM

Carbon dioxide seems to be an innocuous enough substance. Animals exhale it. Plants absorb it. It is a common constituent of volcanic gas. It is not toxic. On the earth, it is ubiquitous (Table 19.6). Carbon dioxide became less innocuous, however, with the

Table 19.6 *Carbon Contents of the Earth*

	Total Amount of Carbon (10^{16} mol)	Carbon Concentration (mol/m^3)	Percentage of Earth's Carbon
Atmosphere	5	140	1.5
Ocean[a]	325	9,050	94.5
Warm surface	5	140	1.5
Main thermocline	80	2,250	23.3
Deep ocean	240	6,660	69.7
Biosphere	14	370	4.0
Living organics	3	70	0.8
Soil organics	11	300	3.2
Total	343	9,570	100.0

Note: The annual production of carbon dioxide is 2×10^{16} molecules (mol), and this amount is increasing 4.5% annually.
[a]Area of ocean surface is 3.6×10^{14} m^2.

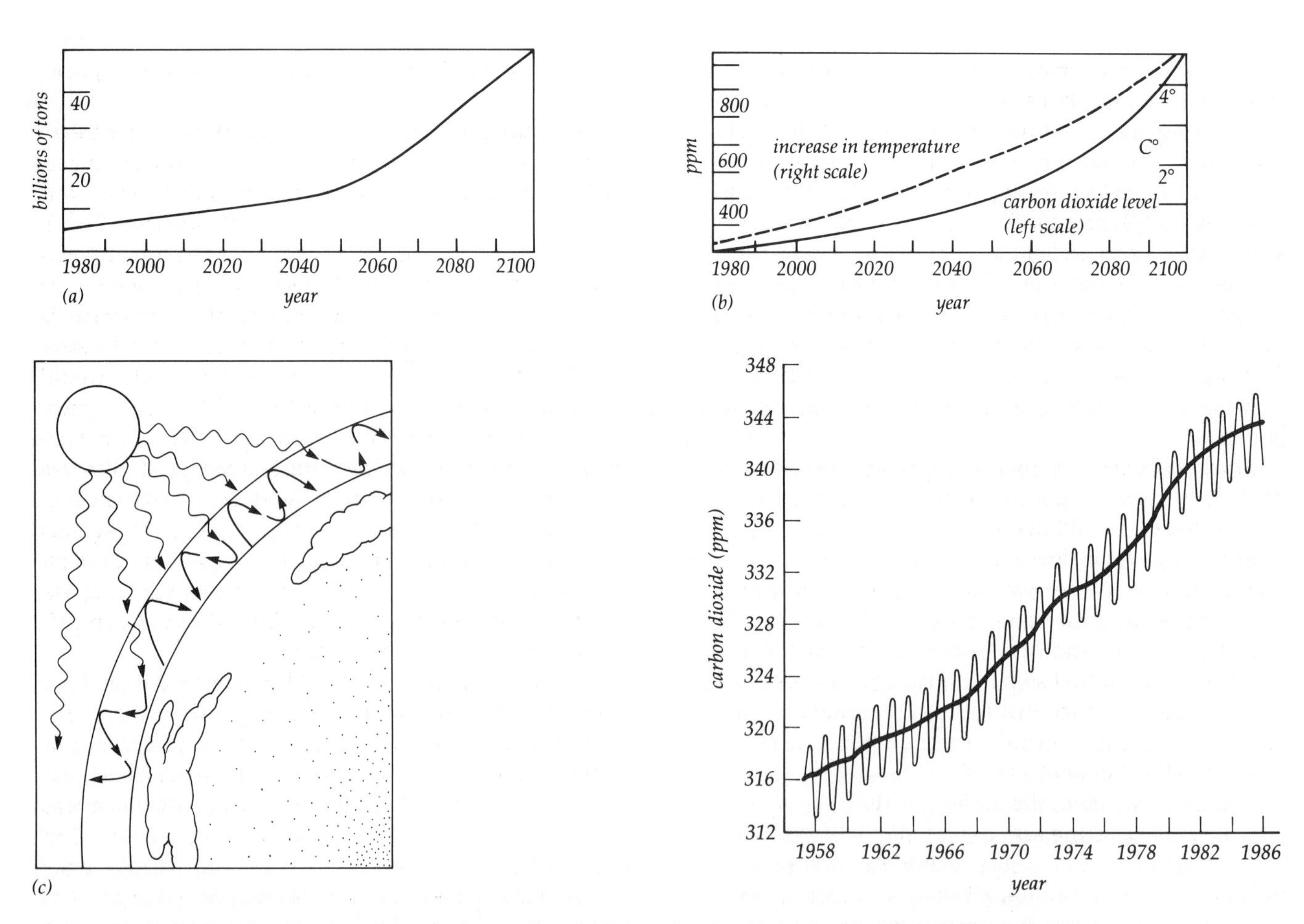

19.19 Atmospheric carbon dioxide and the greenhouse effect. (a) Projected emission of carbon dioxide. (b) Effect of carbon-dioxide level on atmospheric temperature. (c) Carbon dioxide, mostly from fuel burning, traps heat radiation from the earth and atmosphere, preventing dissipation into space. (d) Rising concentration of carbon dioxide in the atmosphere as recorded at Mauna Loa Observatory in Hawaii. Seasonal variations occur because carbon dioxide is removed from the air by plants during the summer growing season and is returned by the decay of fallen leaves in winter. (G. T. Miller, Jr., Living in the Environment, *5th ed.)* (Source: *Environmental Protection Agency, 1985)*

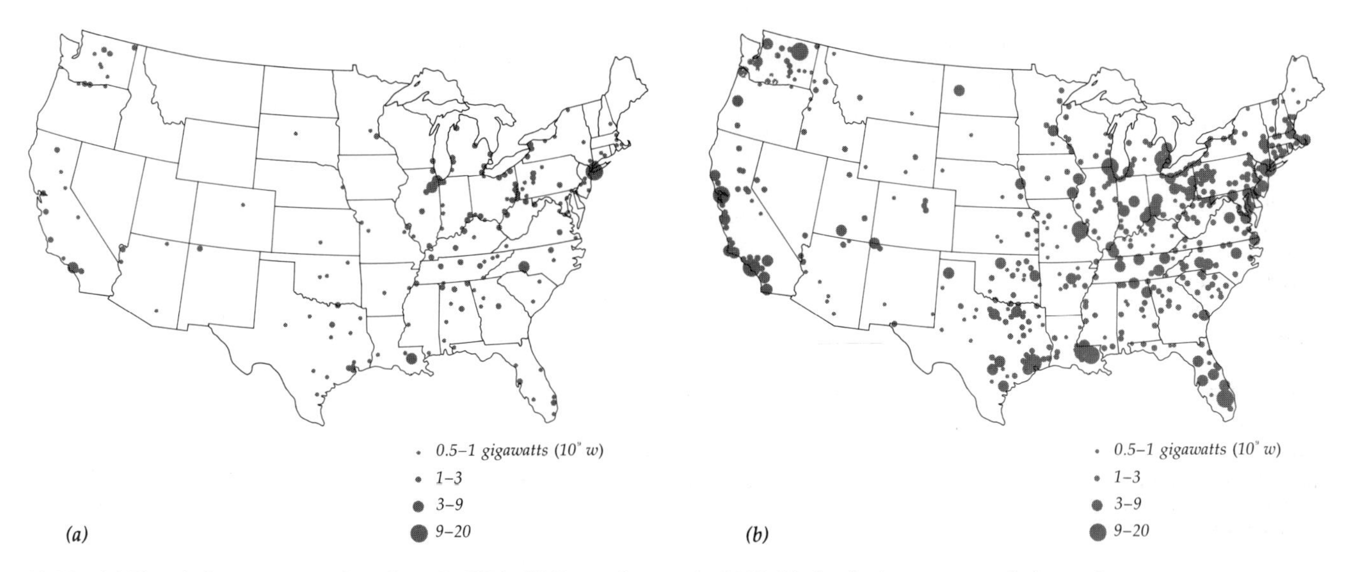

19.20 (a) Electrical power generating plants in 1970. (b) Proposal status by 2000. Notice the increase not only in numbers but also in generating capacity. (The dots are to scale.) (Courtesy of the Federal Power Commission)

advent of the industrial revolution, which has been fueled by hydrocarbons. During the last 100 years, human beings have burned 10^{11} tons of coal, 10^{11} barrels of oil, and 10^{12} m^3 of natural gas. These are all hydrocarbons, which release carbon dioxide and water when burned (Figure 19.19). Carbon dioxide and water vapor absorb infrared light, and their abundance above prehistoric levels in the atmosphere has raised the temperature of the earth's surface about 10°C (18°F) above what it would otherwise be. Further increases of these gases in the atmosphere will raise the average temperature, resulting in the so-called greenhouse effect.

One effect of this global warming would be to melt the polar ice caps. In the Arctic, the disappearance of sea ice would open the Arctic Ocean to shipping and economic development. The increase in the surface area of the sea would also increase evaporation and precipitation along the Arctic coasts of North America, Europe, and Asia. According to one theory, that would be the first step in initiating a new ice age. The melting of the ice that covers the Antarctic continent and Greenland would open those land areas to economic development as well.

At the same time, the meltwater flowing into the ocean would raise the sea level by 50 to 70 m (164 to 230 ft). Many coastal areas would be inundated, including most of Florida, Cape Cod, Holland, the Bahamas, and the Yucatán peninsula. An inland sea would form in the Mississippi River valley, dividing the continental United States in two.

Meteorologists cannot predict exactly how the world climate would change. That it would change is certain, however. The distribution and amount of precipitation would change worldwide, as would growing seasons.

In the ocean, the most serious effect of an increase in carbon dioxide would be the reaction of carbon dioxide with water to produce bicarbonate ions (HCO_3^-), carbonate ions (CO_3^{-2}), calcium ions (Ca^{+2}), and hydrogen ions (H^+). This effect is already evident. The production of hydrogen ions would be especially damaging. The present pH of seawater is about 8. If the concentration of carbon dioxide were to increase, the pH would drop and ocean water would become acidic. It has been estimated that at the present 4.5% annual increase in the release of carbon dioxide, ocean water will be acidic enough to dissolve calcium carbonate in 100 to 200 years. If this happens, corals will become extinct, as will all marine organisms that form calcium carbonate shells (such as oysters, mussels, and foraminifera). Since most marine organisms are in some measure sensitive to changes in pH, the potential effect is catastrophic.

We must, however, consider all factors here. First, mechanisms to buffer pH exist in the ocean. For instance, silicate clays on the seafloor absorb excess hydrogen ions, which would somewhat offset the effects of excess carbon dioxide. Second, other sources of energy may become predominant in the next fifty years. Solar cells, ocean thermal energy, and nuclear fusion may largely replace coal, natural gas, and oil as sources of energy. These are viable alternatives that must be actively pursued now. Petroleum will still be used in the plastics industry, but the carbon-dioxide residue from that industry is minor compared with that of other industries, especially electrical power generation (Figure 19.20).

19.21 Many estuarine areas have been drained, dredged, or filled for development projects, such as this summer cottage colony near Atlantic City, New Jersey. (U.S. Department of Interior, Bureau of Sport Fisheries and Wildlife)

COASTAL LAND USE

Landfill

In the name of progress, developers are draining or filling in bays, marshes, and estuaries to provide more land near the water for marinas and housing (Figure 19.21). The size of most U.S. harbors and bays, including San Francisco and San Diego, has been decreasing over the past 50 years. If this trend continues at its present rate, even realtors and city planners may come to realize that soon there will be no waterfront left for the marinas. San Francisco Bay originally covered 1,131 km^2 (437 mi^2). Slightly more than 463 km^2 (179 mi^2) remain; the rest has been filled in.

It has been said that without its bay San Francisco would be just another city. However, it is more than a question of aesthetics. Several cities around the bay (including Oakland and San Francisco itself), many companies (such as Dow Chemical), and the U.S. Navy all pump large amounts of sewage and industrial slop into the water. As the fill increases, the total amount of tide water moving in and out decreases. As a result, flushing of the bay by the tides grows less effective, and pollution increases. These problems are aggravated by increased diversion of fresh water from the delta to southern California.

Bays and estuaries are regions of biological productivity second to none. An estuary is more than twenty times as productive as the same area of open ocean, and several times as productive as the same acreage of filled land. As we continue to encroach on estuaries by polluting them and filling them in, the rich life they support nearshore and offshore will dwindle.

At present, no federal law controls the filling in of tidal and estuarine waters. Control is left to state and local governments. Unfortunately, these governments have made little attempt at preservation. Politicians and vested interests have proved themselves woefully inadequate in solving such problems as landfill, which of course is related to development and zoning.

With luck, San Francisco may be able to save what is left of its bay. Concerned citizens formed the San Francisco Bay Conservation and Development Commission (SFBCDC) in 1965, which has been able to establish a coordinated, balanced approach to conservation and development around the bay. The model pioneered by SFBCDC is being applied by many communities and regions now.

Unfortunately in other areas, longtime residents have actually promoted the filling in of estuaries, believing in the myth that marinas bring progress, progress brings people, and people bring prosperity.

Even the fill being used in many places may be damaging. It is common to use anything available and to use it as often as desired. In some places, as at Fiesta Island in San Diego's Mission Bay, dry sewage is used. The rich nutrients contained in this material may temporarily upset the ecological balance of the whole region.

A few states have taken a mature, responsible approach to developing estuarine regions. Hawaii, with its statewide zoning, has managed to control development in most areas. Massachusetts has made it illegal to fill any marsh or estuarine region, even those that are privately owned. Recently, a Massachusetts contractor filled in some marshland around his housing development in order to increase the number of units he could build. A wealthy matron touring the area in her chauffeured car happened to see the filling in progress. She called her state representative. Within 2 days, the contractor had most of his work force armed with shovels, laboriously excavating the fill from the marsh. Such examples are heartening; as public awareness grows, they may become more common.

Coastal Development

People seldom have a master plan for their use of land, much less a land ethic. The land is there, there's a lot of it, and people use it as they wish. The unplanned sprawl of most of the world's cities attests to this disregard for the character of the land. Most cities just happened, and they show it. Now, with ever-growing populations and intensifying competition for use of whatever undeveloped land remains, planning is essential.

A classic example of land mismanagement is the treatment of sand dunes. More than a century ago, Henry David Thoreau wrote of mismanagement and restoration of dunes in *Cape Cod* (Boston: Little, Brown and Co., 1985, pp. 172–174):

> In the "Description of the Eastern Coast," which I have already referred to, it is said: "Beach-grass during the spring and summer grows about two feet and a half. If surrounded by naked beach, the storms of autumn and winter heap up the sand on all sides, and cause it to rise nearly to the top of the plant. In the ensuing spring the grass mounts anew; is again covered with sand in the winter; and thus a hill or ridge continues to ascend as long as there is a sufficient base to support it, or till the circumscribing sand, being also covered with beach-grass, will no longer yield to the force of the winds." Sand-hills formed in this way are sometimes one hundred feet high and of every variety of form, like snow-drifts, or Arab tents, and are continually shifting. The grass roots itself very firmly. When I endeavored to pull it up, it usually broke off ten inches or a foot below the surface, at what had been the surface the year before, as appeared by the numerous offshoots there, it being a straight, hard, round shoot, showing by its length how much the sand had accumulated the last year; and sometimes the dead stubs of a previous season were pulled up with it from still deeper in the sand, with their own more decayed shoot attached—so that the age of a sand-hill, and its rate of increase for several years, is pretty accurately recorded in this way. . . .
>
> The inhabitants of Truro were formerly regularly warned under the authority of law in the month of April yearly, to plant beach-grass, as elsewhere they are warned to repair the highways. They dug up the grass in bunches, which were afterward divided into several smaller ones, and set about three feet apart, in rows, so arranged as to break joints and obstruct the passage of the wind. It spread itself rapidly, the weight of the seeds when ripe bending the heads of the grass, and so dropping directly by its side and vegetating there. In this way, for instance, they built up again that part of the Cape between Truro and Provincetown where the sea broke over in the last century. . . .
>
> The attention of the general government was first attracted to the danger which threatened Cape Cod Harbor from the inroads of the sand, about thirty years ago, and commissioners were at that time appointed by Massachusetts, to examine the premises. They reported in June, 1825, that, owing to "the trees and brush having been cut down, and the beach-grass destroyed on the seaward side of the Cape, opposite the Harbor," the original surface of the ground had been broken up and removed by the wind toward the Harbor—during the previous fourteen years—over an extent of "one half a mile in breadth, and about four and a half miles in length."—"The space where a few years since were some of the highest lands on the Cape, covered with trees and bushes," presenting "an extensive waste of undulating sand";— and that, during the previous twelve months, the sand "had approached the Harbor an average distance of fifty rods, for an extent of four and a half miles!" and unless some measures were adopted to check its progress, it would in a few years destroy both the harbor and the town. They therefore recommended that beach-grass be set out on a curving line over a space ten rods wide and four and a half miles long, and that cattle, horses, and sheep be prohibited from going abroad, and the inhabitants from cutting the brush. . . .
>
> Thus Cape Cod is anchored to the heavens, as it were, by a myriad little cables of beach-grass, and, if they should fail, would become a total wreck, and erelong go to the bottom. Formerly, the cows were permitted to go at large, and they ate many strands of the cable by which the Cape is moored, and well-nigh set it adrift, as the bull did the boat which was moored with a grass rope; but now they are not permitted to wander.

19.22 The hedge to the right has been severely trimmed on orders from local authorities. In this New England community, a yacht club had allowed its shrubbery to obscure the view of some of the residents. (W. J. Wallace)

The New Jersey coast, by contrast, is an example of dune mismanagement. Today there is very little that is natural about New Jersey's coastal regions. Large-scale development began in the late nineteenth century, when it became fashionable to vacation at coastal resorts. This was the time when immense wooden resort hotels sprang up around the country. Most are now gone. To build the resorts as close to the water as possible, the dunes were removed by steam shovel. At the same time, the increased mobility afforded by the automobile created enormous pressures on the coastal areas. Land development continued rampant through the 1950s until almost all of the sand dunes had been removed and the land covered with houses, roads, and parking lots; the shoreline was continuously interrupted by groins.

For millions of years before people arrived, the New Jersey shore had been washed by intense fall and winter storms and by an occasional hurricane. The shape of an area might change, but the region remained stable. Indeed, the character of the coast had been shaped by processes and plants that had evolved over many millions of years. What happened when the coast became covered with houses and the dunes destroyed? In the violent storm of March 5, 1962, the 96- to 113-km/h (60- to 70-mph) winds, high waves, spring tides, and an incoming storm surge combined to destroy 2,400 homes and damage 8,300 others. Roads and utilities were ripped out, lives were lost, and damage amounted to $80 million. The area was rebuilt, but when the next violent storm came, the destruction happened all over again.

What the residents of that area did not understand is that the dunes had protected the coast. These dunes were the barrier, the bulwark, between the land and the sea that protected the land and stabilized the coast. Once the dunes had been destroyed, nothing stood in the way of the incoming sea.

In many areas the dunes have simply dwindled away or vanished. The grasses that hold dunes together are hardy plants that can withstand large variations of temperature, moisture, sunlight, and salinity. Their root system forms a dense mat that stabilizes the sand; as the grass grows, so does the dune. A cross section of an average dune complex is shown in Figure 13.14a. However, dunes are intolerant of traffic. Hardy as it is, the dune grass cannot survive trampling. Thus many towns that built nothing on the dunes have lost them to the ravages of motorcycles and dune buggies.

The dunes are too delicate to be opened to public use. Even boardwalks and breaches in the dunes to allow for access to the beach threaten their stability.

19.23 Marblehead, Massachusetts. The center of the photograph shows a public right-of-way, but there is little to indicate its existence. Neighboring landowners have encroached on these public lands. (W. J. Wallace)

Had the residents of the New Jersey shore understood the nature of dunes and built well back from the dune area, their land today would be more attractive, extensive, and valuable—and their insurance cheaper and coverage more comprehensive.

Coastal property is highly prized and is ever-increasing in value. Because the coast offers such rich opportunities for recreation, one state, Oregon, has decided not to trust its shoreline to private developers. Instead, it has purchased most of its coast to ensure the future of that irreplaceable asset—a remarkable feat for a state with some 6,506 km (4,034 mi) of coast and only 2½ million inhabitants!

Other places have taken a quieter approach, permitting public access to the sea to be slowly but relentlessly curtailed. In several eastern states, such as Massachusetts (Figure 19.22), the right-of-way laws are very old and very clear. Once a right-of-way has been established by public usage, access must not be denied to the public by subsequent owners. Many of these rights-of-way date back hundreds of years. But in some areas landowners have been permitted to park cars, build fences, and pile up barriers of debris on rights-of-way. Because most of these rights-of-way are not marked, they tend to go unused and become forgotten. In effect, the landowner bars the public from his or her locale. Figure 19.23 illustrates such a right-of-way. Notice that the adjacent landowner does not offer access to the water.

In the past few years, local governments have intensified their efforts to make the federal government relinquish federally owned land for development. This is especially true of certain federal installations along the coast—for example, Fort Ord in Monterey, California, North Island Naval Air Station in San Diego, and Camp Pendleton in Oceanside, California. Officials in these areas typically declare that these valuable lands are going to waste and should be developed for the public good. Yet Camp Pendleton provides a buffer of open space between Los Angeles and San Diego, preserving the scenic coast for some 10 million southern Californians. True, many federal lands are not being used extensively, but that is a boon. Unused land retains its intrinsic natural value.

Obviously, we need to use the land intelligently and with careful planning. The need is especially acute in coastal regions. A well-balanced master plan for a coastal community must be designed with nature in mind (Figure 19.24). Some areas can remain natural with limited access. Others can be opened for recreation. Planning and development should accommodate the interests of all the inhabitants of an area.

19.24 Plum Island, Massachusetts. The twin jetties at the mouth of the Merrimac River. For the past 20 years, the U.S. Army Corps of Engineers has been battling erosion (largely without success). Millions of dollars have been spent on this problem, but the work has been based on virtually no laboratory model of field data. Notice the large-scale erosion on the right. Efforts thus far have resulted in the destruction of four U.S. Coast Guard buildings at this site. (W. J. Wallace)

OCEAN MANAGEMENT

The extensive development of marine resources in recent years and the pollution that has accompanied that development have triggered conflicts between nations and between communities. In response to those conflicts, interested parties are seeking to establish controls for managing marine resources. Speaking of the multiple uses of marine resources and the need for conservation and preservation, Athelstan Spilhaus, who first conceived the idea of a National Sea Grant program, remarks:

> The problem is to understand the interrelationships of all these different kinds of human activities with each other and with the natural milieu on which they are imposed. The challenge is to transfer that understanding to the Nation as a whole and to devise and execute planning and management schemes to provide the greatest benefit to the greatest number of people in both the present and the future. This requires a fine balance between exploitation and use, on the one hand, and conservation and preservation, on the other hand. This requires management and regulatory strategies and institutions which recognize the needs, expectations, and equities of the present without abrogating responsibilities to the future. It requires continuous and intimate two-way interaction with people and economic entities in ways that are responsive to needs, yet are neither abrasive nor divisive. It requires levels of knowledge and awareness among both managers and the general public that are without precedent. To accomplish these things in the least costly, most effective manner, to balance the do's with the don't's, and to resolve conflicts without creating new ones demands a very special approach in areas of great ecologic, economic, cultural, and political sensitivity. (Source: Hull, E.W.S., *The First Ten Years*, Office of Sea Grant, 1979)

History

Initially, the law of the sea was a matter of naval power. Even so, mariners whose camaraderie transcended their patriotism formulated and abided by ethical standards governing their behavior on the seas. Fishermen did not encroach on one another's territory or foul one another's nets. Ships in distress were assisted. Free port zones and cities were created.

In 1609 the Dutch legal authority and judge Hugo Grotius published *Mare Liberum*—the first legal enunciation of the concept of freedom of the seas. Grotius proclaimed the freedom of all nations to use and pass over the high seas and the right of coastal nations to a sovereign territorial sea. In 1703 the limit of the territorial sea was set as the distance a cannonball could be fired from a ship, which was about 5 km (3 mi).

That 5-km limit was widely accepted until 1945. At the end of World War II, the United States, in critical need of oil, was perfecting techniques for drilling for oil in shallow waters. Prospectors discovered oil on the continental shelf off Louisiana beyond the 5-km limit. President Harry Truman promptly issued a proclamation annexing the resources of the continental shelf contiguous to the United States. In the Gulf of Mexico and east of Cape Cod, the continental shelf extends far beyond the 5-km limit. Truman also declared that the United States had the right to regulate and conserve the fishery resources of the waters over the continental shelf. Unwittingly, perhaps, Truman had opened a Pandora's box.

As development of these offshore oil and gas resources proceeded, their extent proved to be greater than anyone had anticipated. Texas and Louisiana observed the large amount of money that was flowing to Washington, D.C., as a result of leases and royalties. In 1953, at the behest of legislative delegations from those states, Congress passed the Submerged Lands Act and the Outer Continental Shelf Lands Act, which gave the states title to the continental shelf out to 5 km offshore. The federal jurisdiction applied only to the outer continental shelf, beyond 5 km out. Thereafter, the states administered the leasing and collected the royalties derived from the oil, gas, and other resources out to 5 km.

U.N. Law of the Sea

After the Truman proclamation on the continental shelf, other nations hastened to expand their own jurisdictions. Observers noted that pollution was an international problem, since oil spills were affecting many nations. Although manganese nodules had been discovered during the voyage of HMS *Challenger,* the extent of that resource was not appreciated until the 1950s. As a result of these realizations, a series of U.N. conferences and conventions was launched (Table 19.7). The culmination was the Third U.N. Law of the Sea Conference, which first met in 1973 and, after 9 years of negotiations, released a treaty for ratification in 1982. The treaty was sent to the U.N. member nations for ratification after the conference voted 130 for the treaty to 4 against (the United States, Turkey, Venezuela, and Israel), with 17 members abstaining (mostly European nations).

The treaty contains a pledge that the ocean environment and its resources are, in the words of Arvid Pardo (then U.N. representative of Malta), the "common heritage of mankind." Although the treaty is too lengthy to be included here, we shall summarize its main points.

First, a law-of-the-sea tribunal is to be created to enforce the laws embodied in the treaty. Second, the extent of territorial seas is to be established as 22.2 km (12 mi); the extent of a coastal nation's exclusive economic zone is to be 370 km (200 nautical mi; Figure 18.27), and a coastal nation is to have exclusive rights to seabed resources out to a distance of 648 km (350 nautical mi). Third, free passage is to be allowed beyond the 22.2 km (12 nautical mi) territorial sea. Fourth, free passage and overflight are to be allowed through and over straits even though they are less than 44.4 km (24 nautical mi) wide (as is the Strait of Hormuz in the Persian Gulf) or less than 22.2 km (12 mi) when both shores are under the control of the same nation (as are the Dardanelles between the Black and Aegean seas). Fifth, the treaty sets goals for the conservation and management of living marine resources but grants coastal nations full jurisdiction over those resources within their own exclusive economic zones. Sixth, each state is to establish national laws and regulations to prevent and control pollution that are consistent with the rights of adjacent states. Seventh, freedom for scientific research is to prevail on the high seas, and each nation is to control research conducted within its exclusive economic zone.

Eighth—the provision that has prompted the greatest controversy in the United States—an International Seabed Authority (ISA) is to be created to administer the development of seabed resources, most notably manganese nodules. The ISA would grant at no fee priority contracts to corporations that had already carried out research and development activities. New enterprises would have to file an application and pay a fee. The ISA is to allocate areas of the seabed to interested parties and fix the level of production and the price. Participating corporations are to pay a royalty to the ISA and share any technological advances with it. The ISA will have the right to mine the seabed itself and sell the minerals produced, using the money received to provide economic assistance to less-developed countries.

Ninth, the treaty may be altered after 20 years by a three-fourths vote of the signatories. Ratification of the treaty will take place as soon as 60 nations have signed.

Table 19.7 *Chronology of the Law of the Sea Conferences*

Year	Event
1958	First U.N. Conference on Law of the Sea (UNCLOS)—New York Adoption of the four underlying conventions
1960	Second UNCLOS—New York Failed to agree on width of territorial sea
1970	U.N. General Assembly Declaration of the common heritage of humankind Resolutions to start UNCLOS III
1973	UNCLOS III
1973	First session—New York Organizational meeting, list of issues and subjects to be negotiated
1974	Second session—Caracas Transition from U.N. Seabed Committee to a 150-member conference Organization of proposals into working papers
1975	Third session—Geneva Distribution of informal Single Negotiating Texts covering all subjects before the conference
1976	Fourth session—New York Revised Single Negotiating Text issued Deep seabed mining, area of chief disagreement
1976	Fifth session—New York Issuance of revised text on dispute settlement Concentration of deep seabed mining agreement and economic zone delineation
1977	Sixth session—New York Production of Informal Composite Negotiating Text (ICNT)
1978	Seventh session—Geneva Establishment of seven negotiating groups to deal with "hard core" issues
1978	Resumed seventh session—New York Frustration with pace of progress; concern over possibility of unilateral legislation
1979	Eighth session—Geneva Revision of ICNT Remaining concern over deep seabed mining, marine scientific research, continental shelf, and delimitation of offshore boundaries between adjacent or opposite nations
1979	Resumed eighth session—New York Setting of deadlines for adoption of convention
1980	Ninth session—New York Continuing consultations
1980	Resumed ninth session—Geneva Development of draft convention
1981	Tenth session—New York U.S. review
1981	Resumed tenth session—Geneva
1982	Eleventh session—New York
1987	Fourteen of necessary 60 nations have signed

In 1987, only 14 nations have ratified the U.N. Law of the Sea Treaty. There are many disagreements emerging regarding a host of issues. Developed countries (such as the United States, England, and the Federal Republic of Germany) disagree with the proposal for the International Seabed Authority. Some coastal countries disagree with the definition of "innocent passage" of ships through exclusive economic zones. Some countries bordering narrow straits disagree with the concept of "free passage." Others disagree with the formula for establishing the 22.2-km (12-mi) territorial limit (for instance, Libya in the Gulf of Sidra).

The United States, as discussed in Chapter 18, has unilaterally adopted a 22.2-km (12-mi) territorial sea and a 370-km (200-nautical-mi) exclusive economic zone. In addition, the United States is proceeding with bilateral agreements on border and fishing issues with Mexico, Canada, and other countries.

What is the future of the U.N. Law of the Sea Treaty? Much depends on the policy of the next president of the United States. If the next president opposes ratification and does not assist in resolving the growing number of disagreements, the treaty may well die by default. But even without a treaty, many positive results for conservation, planned use of ocean resources, and international relations will have come out of the process.

Coastal Zone Management

The year 1980 was designated as the Year of the Coast by President Carter; the 1970s could well be called the decade of the coast. The National Coastal Zone Management Act (NCZMA) and the Marine Protection, Research, and Sanctuaries Act were passed in 1972, and the Fishery Conservation and Management Act (FCMA) was passed in 1976. The NCZMA encouraged individual coastal and Great Lakes states to develop coastal management plans, and it established an Office of Coastal Zone Management in the National Oceanic and Atmospheric Administration, Department of Commerce, to assist the states and to make federal funds available to them. The FCMA set up eight regional councils to develop management plans for fishery resources. At the same time, the FCMA extended the U.S. territorial sea for fisheries to 370 km (200 mi) and excluded from that area any foreign fishery without a special permit. The regional plans were to prevent overfishing, achieve optimum yield from each species, make consistent overall plans for migratory species, and promote the wise use and development of fishery resources.

The states felt a commensurate sensitivity to the need for coastal zone management. The California Coastal Zone Conservation Act (CZCA), passed in 1972, committed California to "preserve, protect and where possible, to restore the resources of the coastal zone for the enjoyment of current and succeeding generations." The California Coastal Act of 1976 established a State Coastal Commission and mandated the submission of local coastal management plans. The coastal commission has authority to ensure that development in the coastal zone is consonant with local plans and state laws. In 1976, California also passed a Conservancy Act, to provide state funds in support of the CZCA of 1972, and a Coastal Park Bond Act to purchase coastal property for the expansion of state parks and the creation of new parks.

These national and local management activities, coupled with existing laws on pollution control, will help curtail pollution and ensure continuation of a wise and balanced coastal zone multiple-use policy that will preserve some areas and foster the wise development of others.

FURTHER READINGS

Bascom, W. "The Disposal of Waste in the Ocean." *Scientific American,* August 1974, pp. 16–25.

Borgese, E. M., and N. Ginsberg. *Ocean Yearbook 4.* Chicago: University of Chicago Press, 1983.

Cormack, D. *Response to Oil and Chemical Marine Pollution.* New York: Applied Science Publishers, 1983.

Duedall, I. W., B. H. Ketchum, P. K. Park, and D. R. Kester (eds.). *Wastes in the Ocean.* 6 Vol. New York: John Wiley, 1983.

Environmental Protection Agency. *Oil Spills and Spills of Hazardous Substances.* Washington, D.C.: Government Printing Office, 1975.

Hood, D. W. (ed.). *Impingement of Man on the Oceans.* New York: John Wiley, 1971.

Inman, D. L., and B. M. Brush. "The Coastal Challenge." *Science,* 1973, *181,* 20–32.

McHarg, I. L. *Design with Nature.* Garden City, N.Y.: Doubleday, 1969.

"Marine Policy for the 1980s and Beyond." *Oceanus,* Winter 1982/83.

National Academy of Sciences. *Assessing Potential Ocean Pollutants.* Washington, D.C., 1975.

———. *Oil in the Sea.* Washington, D.C., 1985.

Odell, R. *The Saving of San Francisco Bay.* Washington, D. C.: Conservation Foundation, 1972.

Padelford, N. J. *Public Policy and the Use of the Seas,* 2d ed. Cambridge, Mass.: MIT Press, 1970.

Pontecorvo, G. *The New Order of the Oceans.* New York: Columbia University Press, 1986.

Revelle, R. "The Oceans and the Carbon Dioxide Problem." *Oceanus,* 1983, *26*(2), 3–9.

Ringold, P. L., and J. Clark. *The Coastal Almanac for 1980.* Washington, D.C.: The Conservation Foundation, 1980.

Ross, D. A. *Opportunities and Uses of the Ocean.* New York: Springer-Verlag, 1978.

Ruivo, M. (ed.). *Marine Pollution and Sea Life.* London: Fishing News Books, 1972.

"Special Issue: Future Directions in U.S. Marine Policy in the 1980s." *Marine Technology Society Journal,* 1982, *16,* 4.

Wong, C. S., E. Boyle, K. W. Burland, J. D. Burton, and E. D. Goldberg (eds.). *Trace Metals in Seawater.* New York: Plenum Press, 1983.

Appendixes

I Latitude and Longitude

Like streets in a city set up perpendicularly to one another (for example, streets named A, B, C, . . . at right angles to streets numbered 1, 2, 3, . . .), the earth's surface is divided by parallels of latitude perpendicular to meridians of longitude. Because the earth's surface is essentially two-dimensional, it might seem that, like the street grid, the latitude-longitude system is simply the result of linear measurements of the exterior. Such is not the case. Actually, the system is produced by angles measured from the center of the earth.

The rim of the equatorial plane that divides the earth into two hemispheres is the equator, which is 0°N and 0°S latitude. In degrees of 1 to 90 north (the North Pole) and 1 to 90 south (the South Pole) measured from the equatorial plane and the earth's center are the degrees of north and south latitude (Figure I.1). Salem, Oregon, for example, is on the circle formed by the angle of 45°N—its latitude is 45°N.

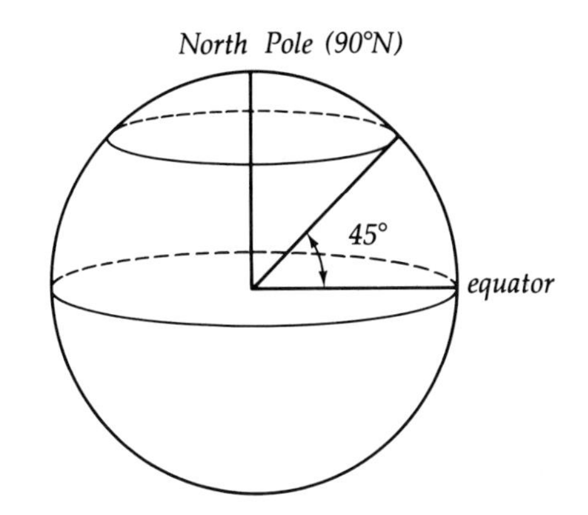

Figure I.1 Determining latitude.

Longitude is expressed in degrees at the north-south axis as the angle (θ), formed by the intersection of two lines drawn from two points (one a reference) to the earth's center (Figure I.2). Greenwich, England, is the reference point and is assigned the meridian of 0°, which runs from pole to pole. A 360° reference could have been used, as the reference system for satellites does, but it has been common to express longitude east and west of Greenwich, up to, of course, 180°E and 180°W.

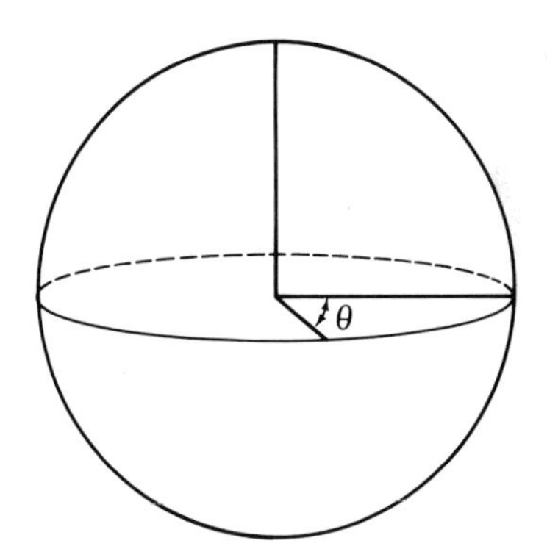

Figure I.2 Determining longitude.

Each degree of latitude and longitude is divided into 60 minutes, and each minute into 60 seconds. A nautical mile (1,852 m) is approximately equal to 1 minute of latitude. Since the earth is not a perfect sphere, a minute of latitude varies somewhat. One degree of latitude at 0° measures 59.701 nautical miles; at 30° it measures 59.853 nautical miles; and at 60° it measures 60.159 nautical miles. The latitude and longitude of Washington, D.C., are 38°53′N and 77°0′W; of San Francisco, 37°47′N and 122°26′W.

A great circle (or geodesic) is a plane that passes through the earth's center. On the earth's surface a great circle is the shortest distance between any two points. Each meridian of longitude is a great circle, but of the circles of latitude, only the equator is a great circle.

The meridian on the opposite side of the planet from Greenwich, corresponding to 180°E and 180°W, is also called the international date line.

II Time and the Date Line

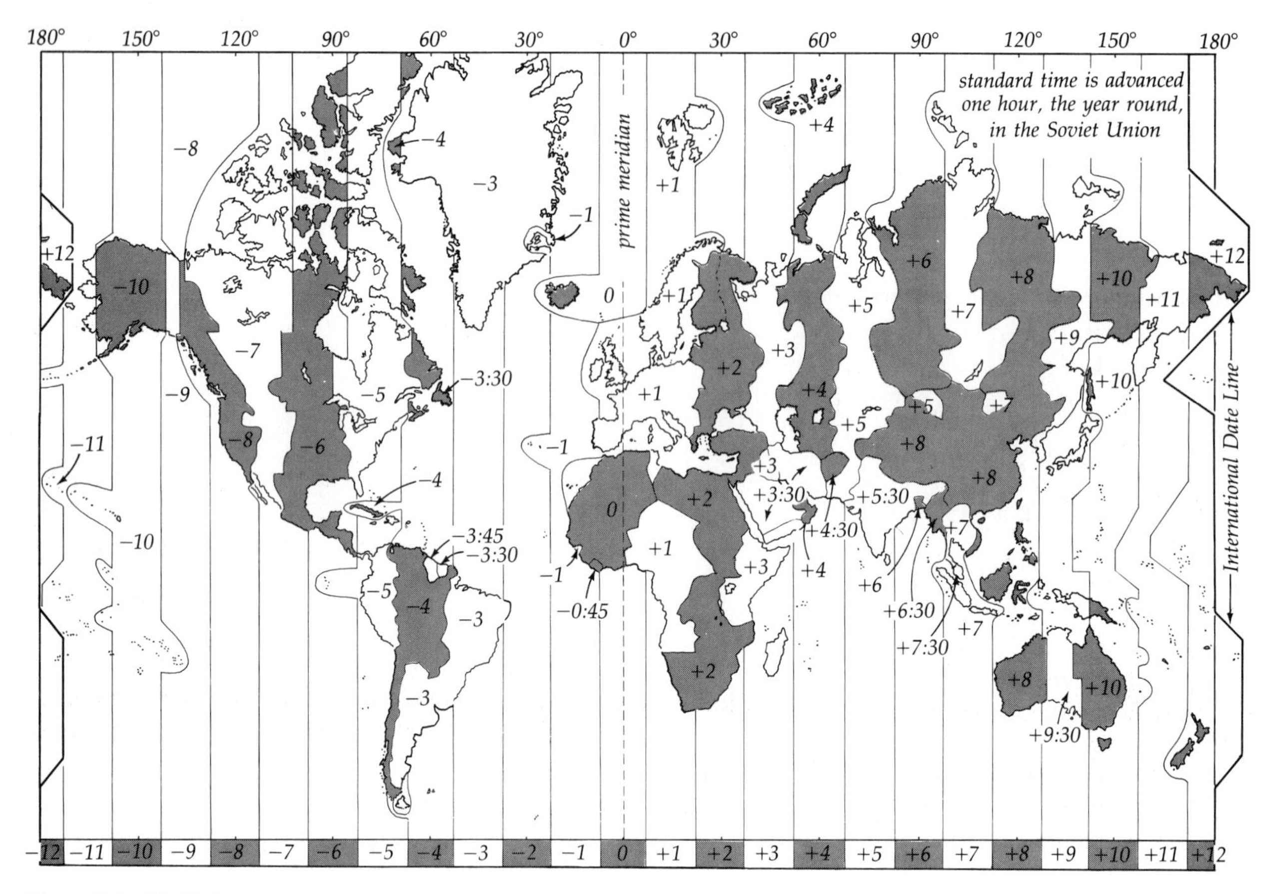

Figure II.1 World time zones.

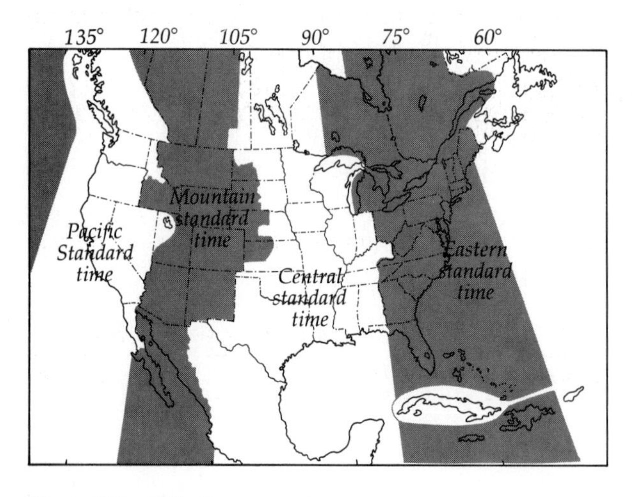

Figure II.2 U.S. time zones.

Most people know that the time varies from place to place. When it is noon in San Diego, for example, it is 3 P.M. in Boston. It might seem simpler and more desirable to have the same time prevailing throughout the world, so that when it was 2 A.M. at one point, it would be 2 A.M. everywhere. But the time zones, though arbitrarily determined, are almost a necessary consequence of the earth's rotation.

Noon at any place occurs when the sun is directly overhead. Thus only one meridian of longitude can have noon at any one time. Indeed, because of the earth's rotation, concepts like noon and relative time in general would have little meaning, especially for a person traveling from place to place, if there were not different time zones.

The earth rotates 360° about once every 24 hours relative to the sun, or 15° each hour. Consequently, the earth is divided into twenty-four time zones according to each 15° of longitude. The prime meridian, passing through Greenwich, is the reference point for time, and time there is called Greenwich Mean Time (GMT). Moving from zone to zone westward, time gets earlier, since the earth rotates counterclockwise (Figure II.1). Because any one zone has the same time throughout, and the earth doesn't rotate in twenty-four equal clicks per day, noon will be off by up to 30 minutes earlier and later at the western and eastern edges of a zone, respectively.

When it is noon at Greenwich, it is 00:00 on the west longitude side of the international date line and 24:00 on the east side. Here at longitude 180°E and 180°W is where the day begins. It is one day earlier east of the date line than west.

As the charts of the world and U.S. time zones (Figure II.2) indicate, the 15° time meridians are not perfectly straight lines. So that time zones bisect as few populated regions, states, and countries as possible, the dividing lines have been reworked slightly to avoid the confusion that would result if the ends of Main Street were in different time zones.

III Scientific Notation

Scientists have used many large and small numbers in describing the world. For example, the average distance from the sun to the earth is 150,000,000 km; the speed of light is 297,800,000 m/s; the age of the earth is 4,700,000,000 yr; the number of atoms or molecules in a molecular weight—Avogadro's number—is 602,300,000,000,000,000,000,000; the rest mass of an electron is 0.0000000000000000000000000000009109 kg; and the distance light travels in a year is 9,460,550,000,000,000 m.

Obviously, these numbers are cumbersome to work with. To shorten the numbers and therefore make them easier to use, a type of shorthand notation is generally used. Avogadro's number becomes 6.023×10^{23}; the rest mass of an electron, 9.109×10^{-31} kg.

The number 1,000 can be written 1×10^3, or 10×10^2, or 100×10^1, or 100×10. However, by convention only one integer is placed to the left of the decimal. So

$$1.4 \times 10^4 = 14{,}000$$

Scientific notation takes the general form $M \times 10^n$, M equaling a number with the decimal placed so that one nonzero integer is to the left of the decimal point and the exponent ^{n} equaling the number of places the decimal must be moved to yield the original number.

If the exponent is positive, the decimal is moved to the left:

$$2.74 \times 10^5 = 274{,}000$$

If the exponent is negative, the decimal is moved to the right:

$$9.26 \times 10^{-6} = 0.00000926$$

Thus the expression 2.74×10^5 means that the number 2.74 is to be multiplied by 10 five times to give 274,000. Numbers expressed in scientific notation are especially valuable because they can easily be used in computations without changing their forms.

Multiplication and division are performed as follows: To multiply $(2 \times 10^4) \times (3 \times 10^3)$, multiply the M factors and add the exponents:

$$2 \times 3 = 6 \qquad 10^4 \times 10^3 = 10^7$$

The answer is 6×10^7.

As a second example, multiply $(4.2 \times 10^5) \times (6.0 \times 10^{-2})$:

$$4.2 \times 6.0 = 25.2 \qquad 10^5 \times 10^{-2} = 10^3$$

The answer is 25.2×10^3, or preferably 2.52×10^4.

For division, such as $(8 \times 10^3) \div (2 \times 10^{-5})$, divide the M factors and subtract the exponents:

$$8 \div 2 = 4 \qquad 10^3 \div 10^{-5} = 10^8$$

So the answer is 4×10^8.

Numbers expressed in scientific notation are not often added or subtracted. When they are, the exponents must be the same:

$$\begin{array}{r} 4.5 \times 10^6 \\ +\underline{5.0 \times 10^6} \\ 9.5 \times 10^6 \end{array}$$

4.5×10^4 cannot be added to (or subtracted from) 5×10^3 unless the exponents agree:

$$\begin{array}{l} 4.5 \times 10^4 \\ +\underline{0.5 \times 10^4\ (= 5 \times 10^3)} \\ 5.0 \times 10^4 \end{array}$$

Table III.1 lists the common prefixes used to refer to these large and small numbers.

Table III.1 *Prefixes Used in the Metric System*

Multiples	Prefixes	Symbols
10^{12}	tera-	T
10^9	giga-	G
10^6	mega-	M
10^3	kilo-	k
10^2	hecto-	h
10	deka-	da
10^{-1}	deci-	d
10^{-2}	centi-	c
10^{-3}	milli-	m
10^{-6}	micro-	μ
10^{-9}	nano-	n
10^{-12}	pico-	p
10^{-15}	femto-	f
10^{-18}	atto	a

IV Constants and Equations

Length

1 micrometer (μm) = 0.0000394 inch
1 millimeter (mm) = 0.0394 inch
1 centimeter (cm) = 0.394 inch
1 meter (m) = 39.4 inches = 3.28 feet
1 kilometer (km) = 0.621 mile = 0.540 nautical mile
1 fathom = 6 feet
1 nautical mile = 1.85 kilometers ≈ 1 minute of latitude

Volume

1 milliliter (mL) = 1 cubic centimeter (cm^3)
= 0.061 cubic inch
1 liter (L) = 1,000 milliliters = 1.0567 U.S. quarts

Mass

1 kilogram (kg) = 2.2 pounds

Speed

1 kilometer per hour = 27.8 centimeters per second
1 knot = 51.5 centimeters per second
= 1 nautical mile per hour

Pressure

1 atmosphere = 1.013 bar
= 14.7 pounds per square inch
= 760 millimeters of mercury at 0°C
= 1.013×10^5 Pa
= 1.013×10^5 N/m^2

Temperature

°C	0	10	20	30	40	100
°F	32	50	68	86	104	212

Heats of Fusion and Vaporization for Water

Heat of fusion (0°C) = 79.71 cal/g

Heat of vaporization (100°C) = 593.55 cal/g

Pressure at Sea Level

Pressure = force ÷ area
$P = F/A$

1 atmosphere = pressure at bottom of a column of mercury 760mm, or 29.921 in, high at 0°C (32°F)
= 14.696 lb/in^2
= 33.899 ft of water at 39.1°F
= 2,116.2 lb/ft^2
= 1,033.2 g/cm^2
= 1.0332×10^4 kg/m^2
= 1.013×10^5 Pa
= 1.013×10^5 N/m^2

Archimedes' Principle

Buoyant force = density × acceleration of gravity × volume of submerged body

$$F = dgV$$

d expressed in grams per cubic centimeter, g = 980 cm/s^2, V expressed in cubic centimeters or d in kg/m^3, g = 9.8 m/s^2 and V in m^3.

Newton's Law of Gravitation

Attractive force = gravitational constant × (mass of object$_1$ × mass of object$_2$) ÷ (distance between objects)^2

$$F = G\frac{M_1M_2}{R^2}$$

In the earth-sun system:

$M_{sun} = 1.971 \times 10^{29}$ kg
$R_{earth\ to\ sun} = 1.495 \times 10^{11}$ m
$F = 5.885 \times 10^{-4}$ m/kg

Table IV.1 *Acceleration of Gravity at Different Locales*

Place	Elevation (m)	g(m/s^2)
New Orleans	2	9.79324
Colorado Springs	1,841	9.79490
Denver	1,638	9.79609
San Francisco	114	9.79965
St. Louis	154	9.80001
New York	38	9.80267
Chicago	182	9.80278
Portland, Oregon	8	9.80646

In the earth-moon system:

$M_{moon} = 7.347 \times 10^{22}$ kg
$R_{earth\ to\ moon} = 3.844 \times 10^8$ m
$F = 3.318 \times 10^{-5}$ m/kg

Force

Force = mass × acceleration
$F = ma$

Since objects tend to move in a straight line, an equation for an object moving in a curved path (circle) must include a force directed toward the center of the circle. In this case:

$$\text{Acceleration} = \text{velocity}^2 \div \text{distance to center}$$

$$a = \frac{v^2}{R}$$

$$\text{therefore } F = \frac{mv^2}{R}$$

Table IV.2 *Coriolis Versus Centrifugal Forces in Tropical Cyclones*

		Centrifugal Force					
		Distance from Center (km)					
v (m/s)	Coriolis Force[a]	20	40	60	80	100	150
10	5	50	25	17	13	10	7
20	10	200	100	67	50	40	27
30	15	450	225	150	112	90	60
40	20	800	400	267	200	160	107
50	25	1,250	625	417	314	250	167

[a]Equals 10^4 m/s^2 at 20°N and 20°S.

Centrifugal Force at Any Latitude per Unit Mass

Force = mass × (velocity)^2 × radius at that latitude × cosine of latitude

$$F = Mv^2R\phi(\cos \phi)$$

Centrifugal acceleration is $FM = v^2R\phi(\cos \phi)$, which is 0 at the North and South poles and a maximum at the equator.

Table IV.3 *Angular Radius of Halos and Rainbows*

Phenomenon	Angle
Corona (caused by small water droplets in air)	1°–10°
Small halo (caused by 60° angles of ice crystals)	22°
Large halo (caused by 90° angles of ice crystals)	46°
Rainbow, primary	41°21′
Rainbow, secondary	52°15′

Source: From *CRC Handbook of Chemistry and Physics*, 37th ed. (Cleveland: Chemical Rubber Publishing Co., 1956).

V Supplementary Topics to Chapter 8

Archimedes' Principle

Does a stone weigh the same in water as it does in air? A scuba diver would know the answer to this question—tanks seem much lighter in the water than in air.

A stone block with density D is submerged to a depth Z in water (Figure V.1). The total force on the top of the block (force down, F_d) is a function of pressure P times the area A, or $F_d = PA$. Furthermore, the pressure is a function of the density of the water, the depth (or height) of the water above the block, and the acceleration of gravity (g), so $P = DgZ$, and the force down becomes $F_d = DgZA$.

The force up (F_u) is the same but with the height of the block (h) increasing the depth (Z) factor. Thus the force up becomes $F_u = Dg(Z + h)A$.

The resultant force of the water on the block becomes $F_u - F_d = DghA$. The quantity hA is the volume of the displaced water, and multiplying it times the density and g yields the mass of the displaced water. In other words, the stone experiences a net upward force equal to the weight of the displaced water.

For example, assume that a granite stone block of 1 m^3 weighs 2,660 kg in air. A cubic meter of water weighs 1,000 kg. Therefore, the block in water would weigh 1,660 kg.

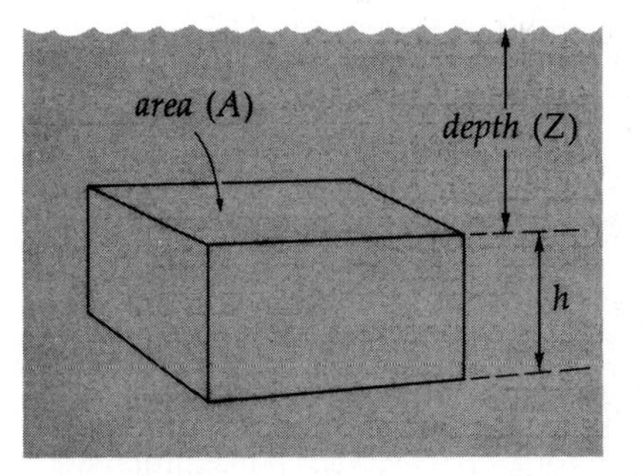

Figure V.1 A block of volume submerged in water with a density of D.

For a body to float in water, then, it must be capable of displacing its own weight of water.

Pressure

Pressure is expressed in dimensions of force per unit area. If the SI system is used, then:

$$P = \text{density}\left(\text{in kg/m}^3\right) \times \text{depth}$$
$$(\text{in m}) \times \text{acceleration}$$
$$\text{due to gravity}\left(\text{in m/s}^2\right)$$

$$P = \frac{\text{kg}}{\text{m}^3} \times \text{m} \times \frac{\text{m}}{\text{s}^2}$$

$$= \frac{\text{kg/m}}{\text{s}^2} = \text{kgm}^{-1}\text{s}^{-2}\,(\text{or N/m}^2)$$

In meteorology, the pressure unit is the millibar (abbreviated "mb"), which equals 10^2 N/m^2 or 10^2 pascals (1 pascal = 1N/m^2). Atmospheric pressure at sea level is equal to 1,013 mb. The pressure unit used in oceanography is the decibar (10^4 N/m^2). (One bar, then, is equal to 10^5 N/m^2.)

VI The Coriolis Effect

Newton's first law states that an object in motion will naturally move in a straight line, but an object at rest on the earth's surface is moving in a curved path. Since the motion is not straight, it must be experiencing some acceleration, which may be written as v^2/R (R equals the distance between the earth's center and that of the object; it is, for all practical purposes, the earth's radius). The force is written $F = v^2/R$. This centrifugal force varies with latitude and is written, per unit mass, as:

$$\text{Force} = \text{mass} \times (\text{velocity})^2 \times \text{radius at that latitude} \times \text{cosine of latitude}$$
$$F = Mv^2R\phi(\cos\phi)$$

Thus the centrifugal acceleration is

$$F/M = v^2R\phi(\cos\phi)$$

The Coriolis effect (F_c) (per unit mass) is a function of an object's velocity across the earth's surface (horizontal velocity v) and the vertical component of the earth's rotational velocity. It is mathematically stated as $F_{CF} = 2wv \sin\phi$, where w is the earth's angular velocity and ϕ is the latitude. This deflecting force is always at right angles to the velocity and, of course, performs no work. The Coriolis effect is greatest at the poles (90° latitude) and zero at the equator (0° latitude). Mathematically, this is due to the fact that the sine is a maximum (1.00) at 90° and zero at 0°. In physical terms, however, the Coriolis effect vanishes at the equator because there the vertical component of the earth's rotation vanishes; the vertical component is greatest at the poles. When an object is at rest, its centrifugal force and deflecting force are equal; but when the object has a horizontal motion across the earth's surface, it has, depending on the direction, an excess or deficit of centrifugal force, which is termed *the Coriolis effect.*

It is by no means a foregone conclusion that the deflection due to the Coriolis effect on a moving object will be greater at higher latitudes, since the velocity of the object can vary. One would expect a large deflection if such an object moves very rapidly. The Coriolis effect would indeed be large, and the distance traversed by the object would also be large per unit time, but in that time period the earth would rotate only a small angle distance, and the object's actual deflection would be small. In the same period of time, a slow-moving object would cover a short distance along its path of travel (velocity = distance/time, so distance = velocity × time), producing a large deflection but a small Coriolis acceleration.

While the deflection force is not a physical force, it is necessarily real to an observer within a rotating system and must be taken into consideration when describing motion, especially for movements of fluids such as water or of air, which move at relatively low velocities and experience appreciable deflection over long distances.

In addition, the equation $C_F = 2wv \sin\phi$ actually expresses the magnitude of the Coriolis acceleration and would, for a unit mass, numerically equal the Coriolis effect. The product of the object's mass and the Coriolis acceleration is the Coriolis effect, $C_F = ma$ (Newton's second law). For objects in vertical motion, the Coriolis acceleration is greatest at the equator and least at the poles.

VII Geologic Time Scale

ERA	PERIOD	EPOCH	AGE IN MILLIONS OF YEARS AGO	GENERALIZED BIOLOGIC CONTINUUM	GENERALIZED GEOLOGIC EVENTS (Mountain building for North America only)
CENOZOIC	QUATERNARY	Recent Pleistocene	2	Paleoindians migrating into North America.	Glaciation on continental scale begins.
	TERTIARY	Pliocene Miocene Oligocene Eocene Paleocene	25 65	Humans first use tools. First humanlike primates. Placental mammals common.	Folding and major uplifting of Rocky Mountain ranges.
MESOZOIC	CRETACEOUS		135	Major extinctions (dinosaurs, and so on). First flowering plants.	
	JURASSIC		190	First birds.	Pangaea breaking up, plates bearing continents toward modern locations. Modern Atlantic forming.
	TRIASSIC		225	First mammals. First dinosaurs.	Ancestral Atlantic closes, folding and uplifting of Appalachian Mountains; Pangaea together.
PALEOZOIC	PERMIAN		280	Major extinctions (trilobites, and so on). First mammal-like reptiles.	Glaciation on continental scale (Gondwanaland).
	PENNSYLVANIAN		325	First reptiles.	Pangaea forming.
	MISSISSIPPIAN		350	First reptile-like amphibians.	
	DEVONIAN		400	First land animals (amphibians). First amphibian-like fish. First rooted land plants.	Acadian mountains rising in New England–Maritime Provinces—subduction zone?
	SILURIAN		430		Glaciation on continental scale.
	ORDOVICIAN		500	First vertebrates (jawless fish). Nearly all major invertebrate types present.	Taconic mountains rising in New York–Quebec region, extensive volcanism—subduction zone? Ancestral Atlantic closing.
	CAMBRIAN		600	Fossils become common as invertebrates develop hard exoskeletons (trilobites, brachiopods).	
PRECAMBRIAN	LATE PRECAMBRIAN		670	Approximate age of earliest known animals (soft bodied worms, and so on); complexity of forms indicates long prehistory.	Glaciation on continental scale.
			1000	Earliest known advanced cell type (distinct nucleus), green algae.	Ancestral Atlantic opening. Proto-Pangaea breaking up; rifting widespread. Incipient rifting of central North America?
	MIDDLE PRECAMBRIAN		2000	Approximate age of oldest well-preserved fossil cells, photosynthetic blue-green algae. Biologically produced oxygen building atmospheric level toward modern atmosphere.	Major iron ore deposits of world forming. Glaciation on continental scale. Widespread magmatic activity, major additions of granitic continental crust. Mountain building, widespread volcanism, formation of greenstone belts-island arcs.
	EARLY PRECAMBRIAN		3000	Approximate age of evidence of first life, algal structures (stromatolites) in rock. Possible bacteria and blue-green algae.? First organisms evolving? Oceans with buildup of nonbiologically produced complex organic molecules (amino acids, nucleic acids, and so on).	Oldest known earth rocks. Oceans and oxygen-free atmosphere forming.
			3800		Initial crust-ocean basins. Earth begins differentiating into core-mantle as radiogenic heat builds up.
			4600	Probable age of Earth and other planets. Evidence: radiometric ages of meteorites and lunar rocks, amounts and ratios of terrestrial lead isotopes.	

VIII The U.S. Coastline

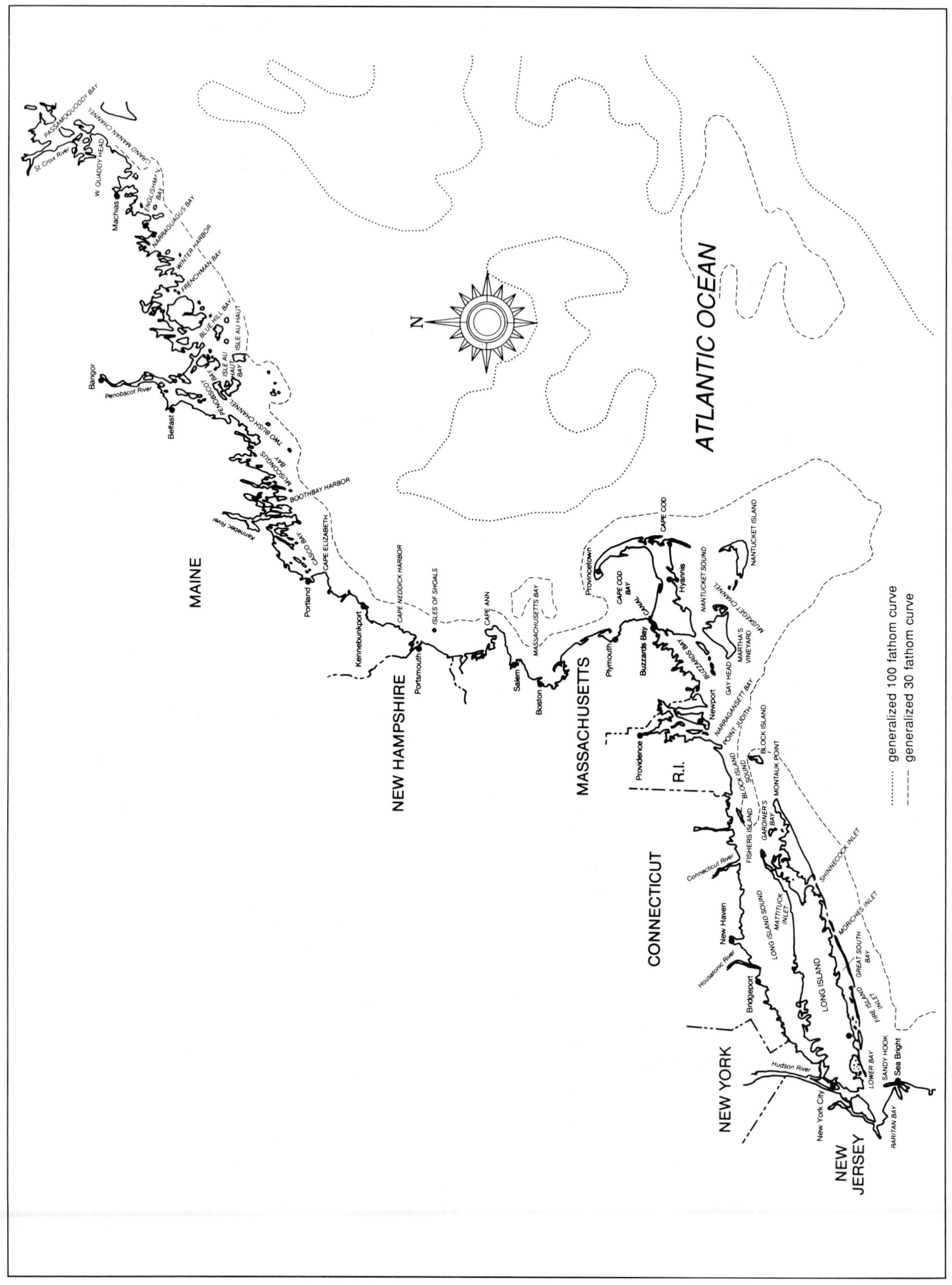

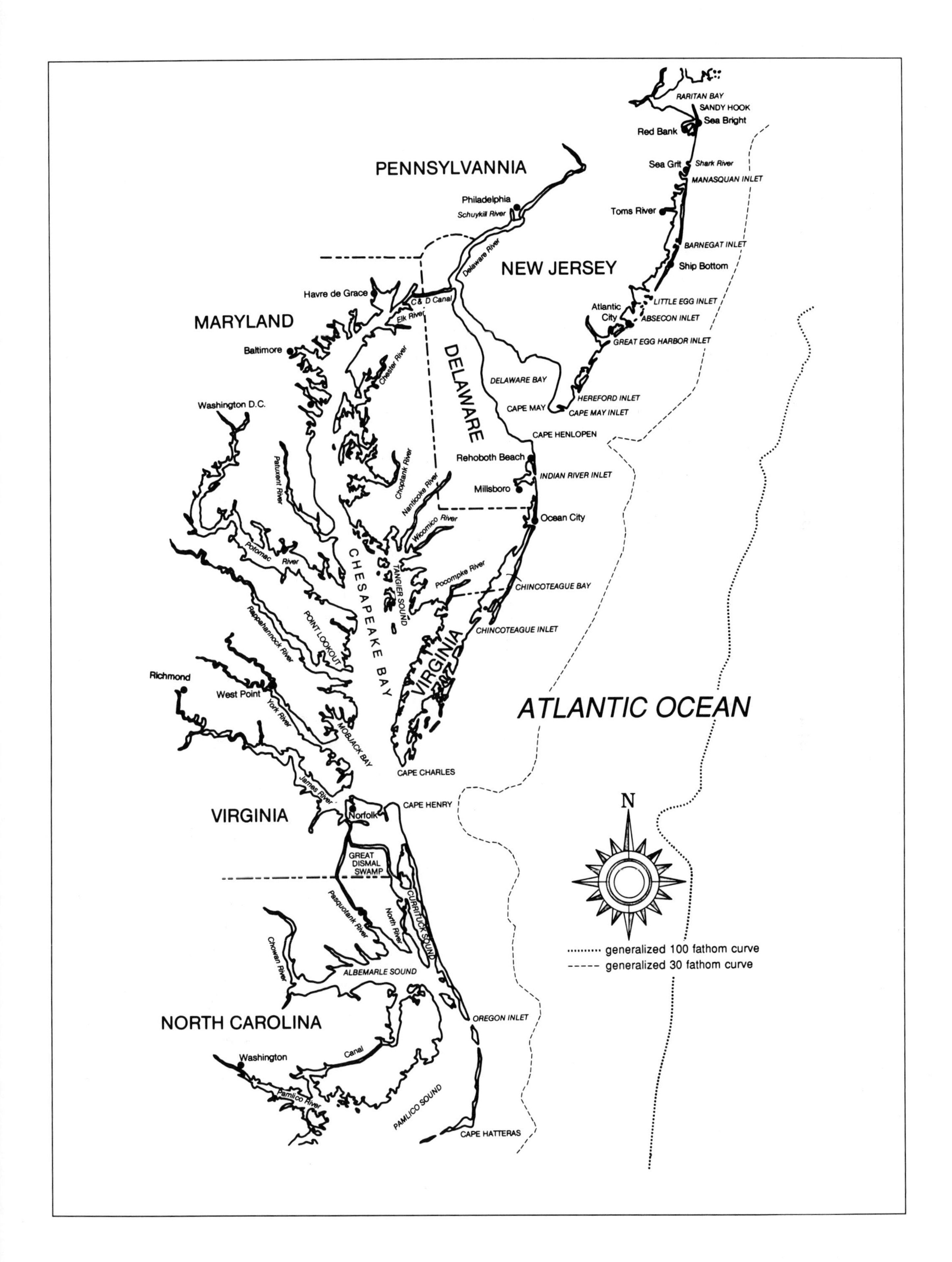
RARITAN BAY
SANDY HOOK
Sea Bright
Red Bank
PENNSYLVANNIA
Sea Grit
Shark River
MANASQUAN INLET
Philadelphia
Schuylkill River
Toms River
BARNEGAT INLET
Delaware River
NEW JERSEY
Ship Bottom
Havre de Grace
C& D Canal
LITTLE EGG INLET
MARYLAND
Elk River
Atlantic City
ABSECON INLET
Baltimore
GREAT EGG HARBOR INLET
Chester River
DELAWARE
DELAWARE BAY
HEREFORD INLET
Washington D.C.
CAPE MAY
CAPE MAY INLET
CAPE HENLOPEN
Rehoboth Beach
Patuxent River
Choptank River
INDIAN RIVER INLET
Nanticoke River
Millsboro
Ocean City
Wicomico River
Potomac River
CHESAPEAKE BAY
TANGIER SOUND
Pocomoke River
CHINCOTEAGUE BAY
Rappahannock River
POINT LOOKOUT
CHINCOTEAGUE INLET
VIRGINIA BAY
Richmond
West Point
York River
ATLANTIC OCEAN
MOBJACK BAY
CAPE CHARLES
James River
N
CAPE HENRY
VIRGINIA
Norfolk
GREAT DISMAL SWAMP
Pasquotank River
North River
CURRITUCK SOUND
Chowan River
generalized 100 fathom curve
generalized 30 fathom curve
ALBEMARLE SOUND
NORTH CAROLINA
OREGON INLET
Washington
Canal
Pamlico River
PAMLICO SOUND
CAPE HATTERAS

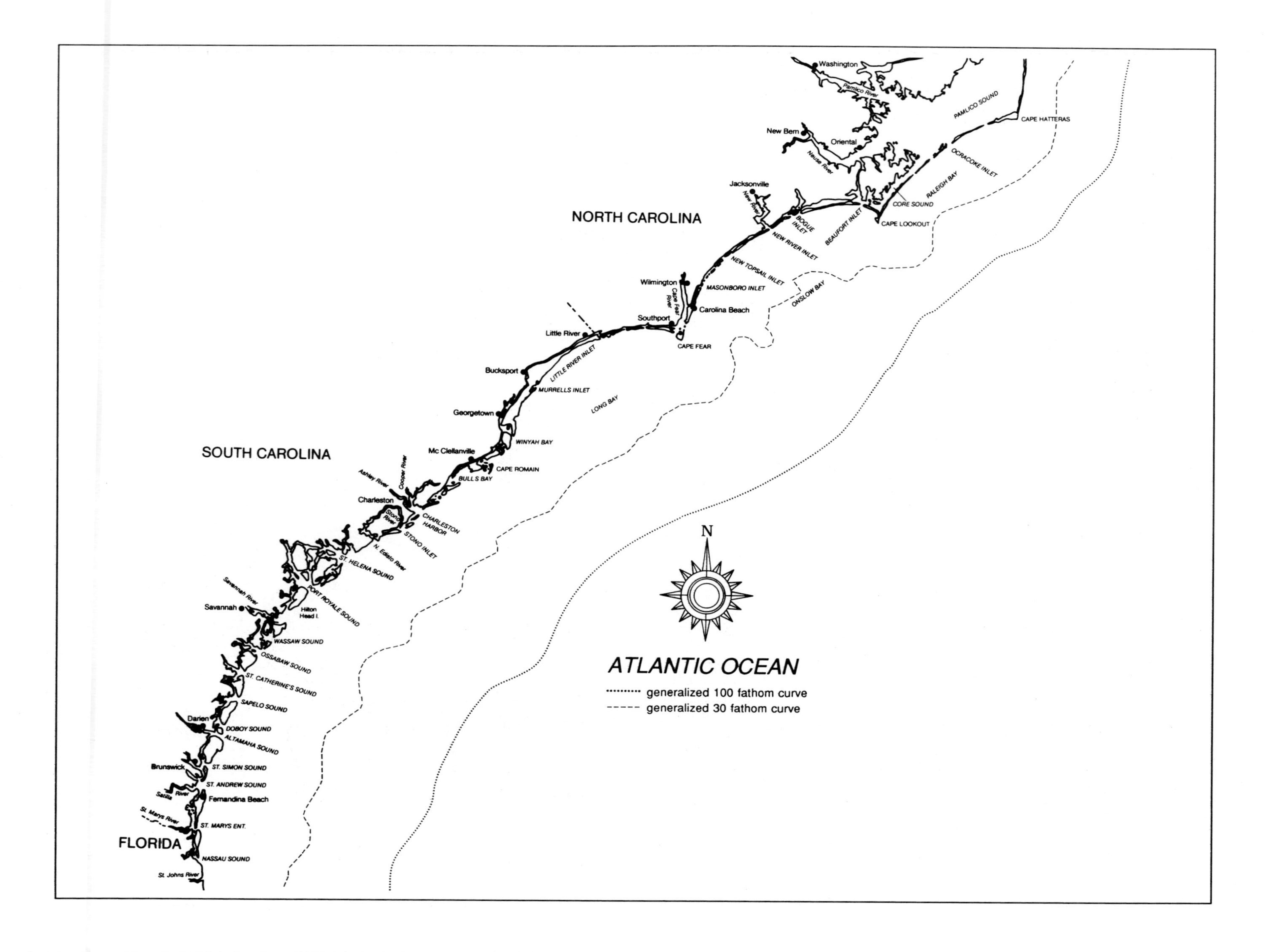
Washington
Pamlico River
PAMLICO SOUND
CAPE HATTERAS
New Bern
Oriental
Neuse River
OCRACOKE INLET
RALEIGH BAY
CORE SOUND
CAPE LOOKOUT
Jacksonville
New River
NORTH CAROLINA
BOGUE INLET
BEAUFORT INLET
NEW RIVER INLET
NEW TOPSAIL INLET
Wilmington
Cape Fear River
MASONBORO INLET
ONSLOW BAY
Carolina Beach
Southport
Little River
CAPE FEAR
Bucksport
LITTLE RIVER INLET
MURRELLS INLET
LONG BAY
Georgetown
WINYAH BAY
SOUTH CAROLINA
Mc Clellanville
CAPE ROMAIN
BULLS BAY
Ashley River
Cooper River
Charleston
Stono River
CHARLESTON HARBOR
STONO INLET
N. Edisto River
ST. HELENA SOUND
Savannah River
PORT ROYALE SOUND
Savannah
Hilton Head I.
WASSAW SOUND
OSSABAW SOUND
ST. CATHERINE'S SOUND
SAPELO SOUND
Darien
DOBOY SOUND
ALTAMAHA SOUND
Brunswick
ST. SIMON SOUND
ST. ANDREW SOUND
Satilla River
Fernandina Beach
St. Marys River
ST. MARYS ENT.
FLORIDA
NASSAU SOUND
St. Johns River
N
ATLANTIC OCEAN
generalized 100 fathom curve
generalized 30 fathom curve

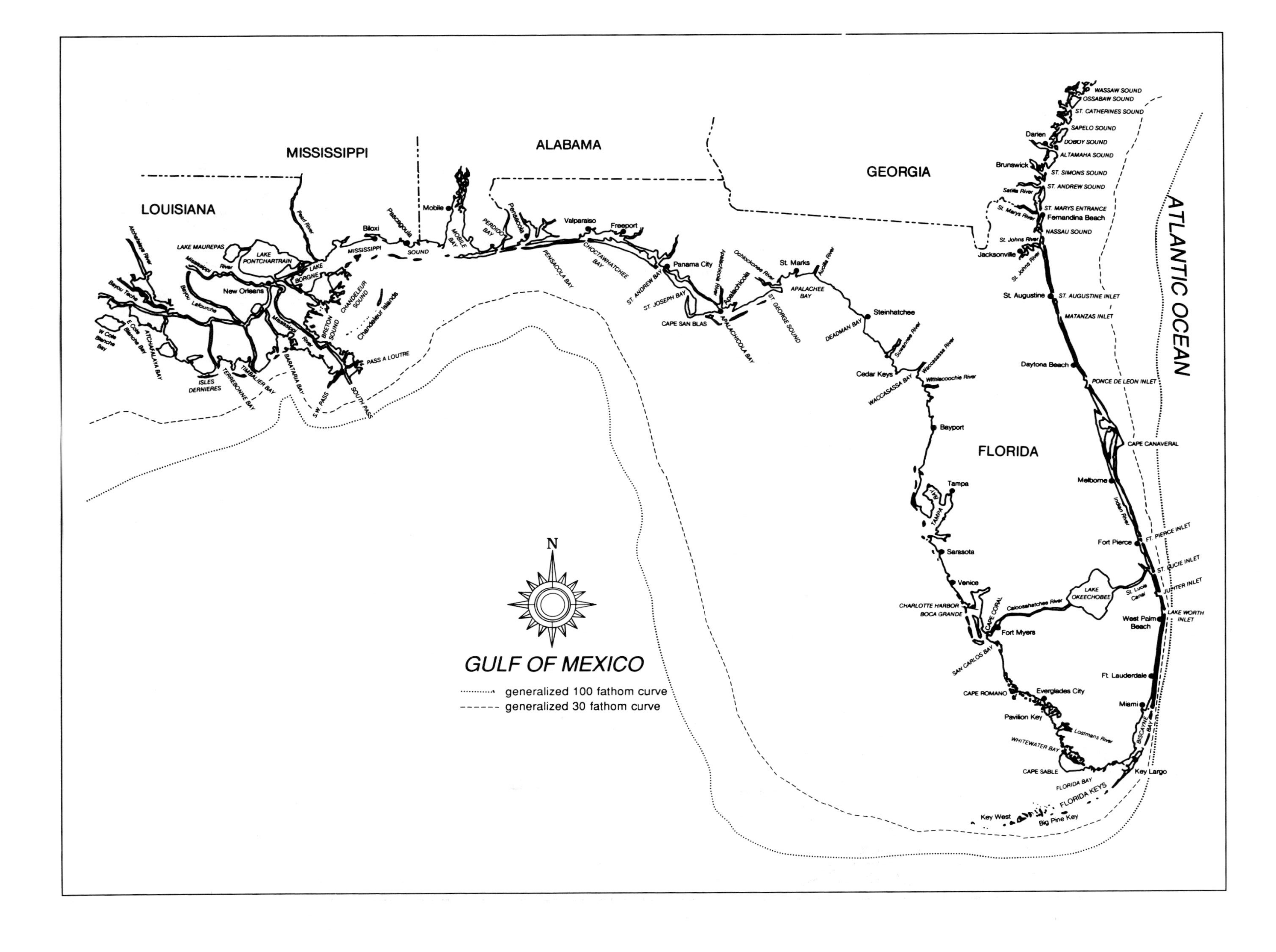
ATLANTIC OCEAN
GULF OF MEXICO
generalized 100 fathom curve
generalized 30 fathom curve
N
LOUISIANA
MISSISSIPPI
ALABAMA
GEORGIA
FLORIDA
WASSAW SOUND
OSSABAW SOUND
ST. CATHERINES SOUND
SAPELO SOUND
DOBOY SOUND
ALTAMAHA SOUND
ST. SIMONS SOUND
ST. ANDREW SOUND
ST. MARYS ENTRANCE
NASSAU SOUND
Darien
Brunswick
Satilla River
St. Marys River
Fernandina Beach
St. Johns River
Jacksonville
St. Augustine
ST. AUGUSTINE INLET
MATANZAS INLET
Daytona Beach
PONCE DE LEON INLET
CAPE CANAVERAL
Melborne
Indian River
Fort Pierce
FT. PIERCE INLET
ST. LUCIE INLET
JUPITER INLET
LAKE WORTH INLET
St. Lucie Canal
LAKE OKEECHOBEE
West Palm Beach
Ft. Lauderdale
Miami
BISCAYNE BAY
Key Largo
FLORIDA KEYS
FLORIDA BAY
Big Pine Key
Key West
CAPE SABLE
WHITEWATER BAY
Lostmans River
Pavilion Key
Everglades City
CAPE ROMANO
SAN CARLOS BAY
Fort Myers
Caloosahatchee River
CAPE CORAL
CHARLOTTE HARBOR
BOCA GRANDE
Venice
Sarasota
Tampa
TAMPA BAY
Bayport
Withlacoochee River
Waccasassa River
WACCASASSA BAY
Cedar Keys
Suwannee River
Steinhatchee
DEADMAN BAY
Aucilla River
St. Marks
APALACHEE BAY
Ochlockonee River
ST. GEORGE SOUND
Apalachicola
Apalachicola River
APALACHICOLA BAY
CAPE SAN BLAS
ST. JOSEPH BAY
ST. ANDREW BAY
Panama City
Freeport
Valparaiso
CHOCTAWHATCHEE BAY
PENSACOLA BAY
Pensacola
PERDIDO BAY
MOBILE BAY
Mobile
SOUND
Pascagoula
Biloxi
MISSISSIPPI
Pearl River
LAKE BORGNE
LAKE PONTCHARTRAIN
LAKE MAUREPAS
Mississippi River
New Orleans
CHANDELEUR SOUND
Chandeleur Islands
BRETON SOUND
PASS A LOUTRE
SOUTH PASS
S.W. PASS
BARATARIA BAY
TIMBALIER BAY
TERREBONNE BAY
ISLES DERNIERES
Bayou Lafourche
Bayou Teche
Atchafalaya River
ATCHAFALAYA BAY
E Cote Blanche Bay
W Cote Blanche Bay

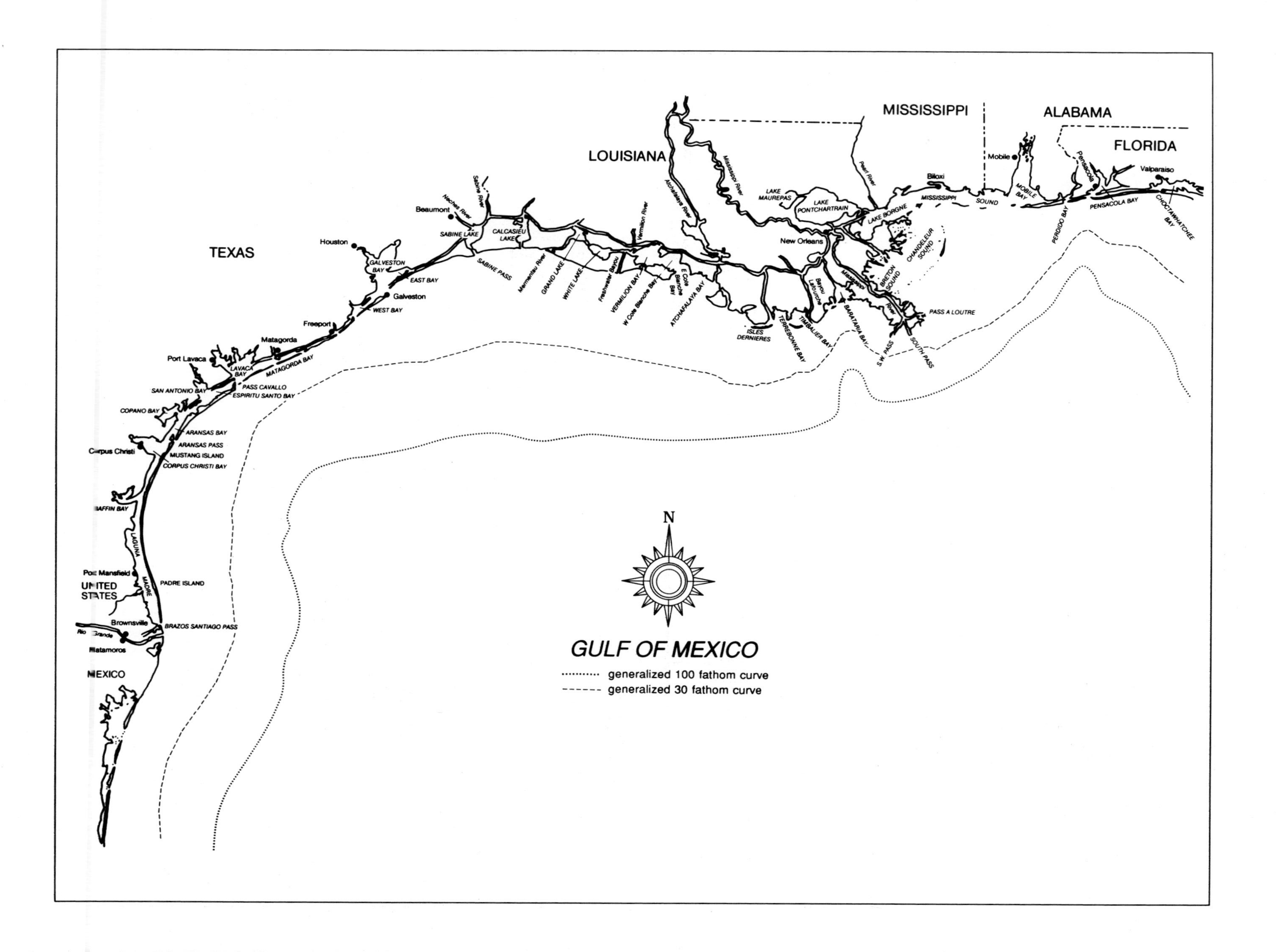
MISSISSIPPI
ALABAMA
FLORIDA
LOUISIANA
TEXAS
UNITED STATES
MEXICO
GULF OF MEXICO
generalized 100 fathom curve
generalized 30 fathom curve
N
Valparaiso
CHOCTAWHATCHEE BAY
PENSACOLA BAY
Pensacola
PERDIDO BAY
Mobile
MOBILE BAY
SOUND
MISSISSIPPI
Biloxi
Pearl River
Mississippi River
LAKE MAUREPAS
LAKE PONTCHARTRAIN
LAKE BORGNE
CHANDELEUR SOUND
BRETON SOUND
New Orleans
Mississippi River
PASS A LOUTRE
SOUTH PASS
S.W. PASS
BARATARIA BAY
Bayou Lafourche
TIMBALIER BAY
TERREBONNE BAY
ISLES DERNIERES
Atchafalaya River
ATCHAFALAYA BAY
E Cote Blanche Bay
W Cote Blanche Bay
VERMILION BAY
Vermilion River
Freshwater Bayou
WHITE LAKE
GRAND LAKE
Mermentau River
CALCASIEU LAKE
SABINE PASS
Sabine River
Neches River
SABINE LAKE
Beaumont
Houston
GALVESTON BAY
EAST BAY
Galveston
WEST BAY
Freeport
Matagorda
MATAGORDA BAY
Port Lavaca
LAVACA BAY
PASS CAVALLO
ESPIRITU SANTO BAY
SAN ANTONIO BAY
COPANO BAY
ARANSAS BAY
ARANSAS PASS
MUSTANG ISLAND
CORPUS CHRISTI BAY
Corpus Christi
LAGUNA MADRE
PADRE ISLAND
Port Mansfield
Brownsville
BRAZOS SANTIAGO PASS
Rio Grande
Matamoros

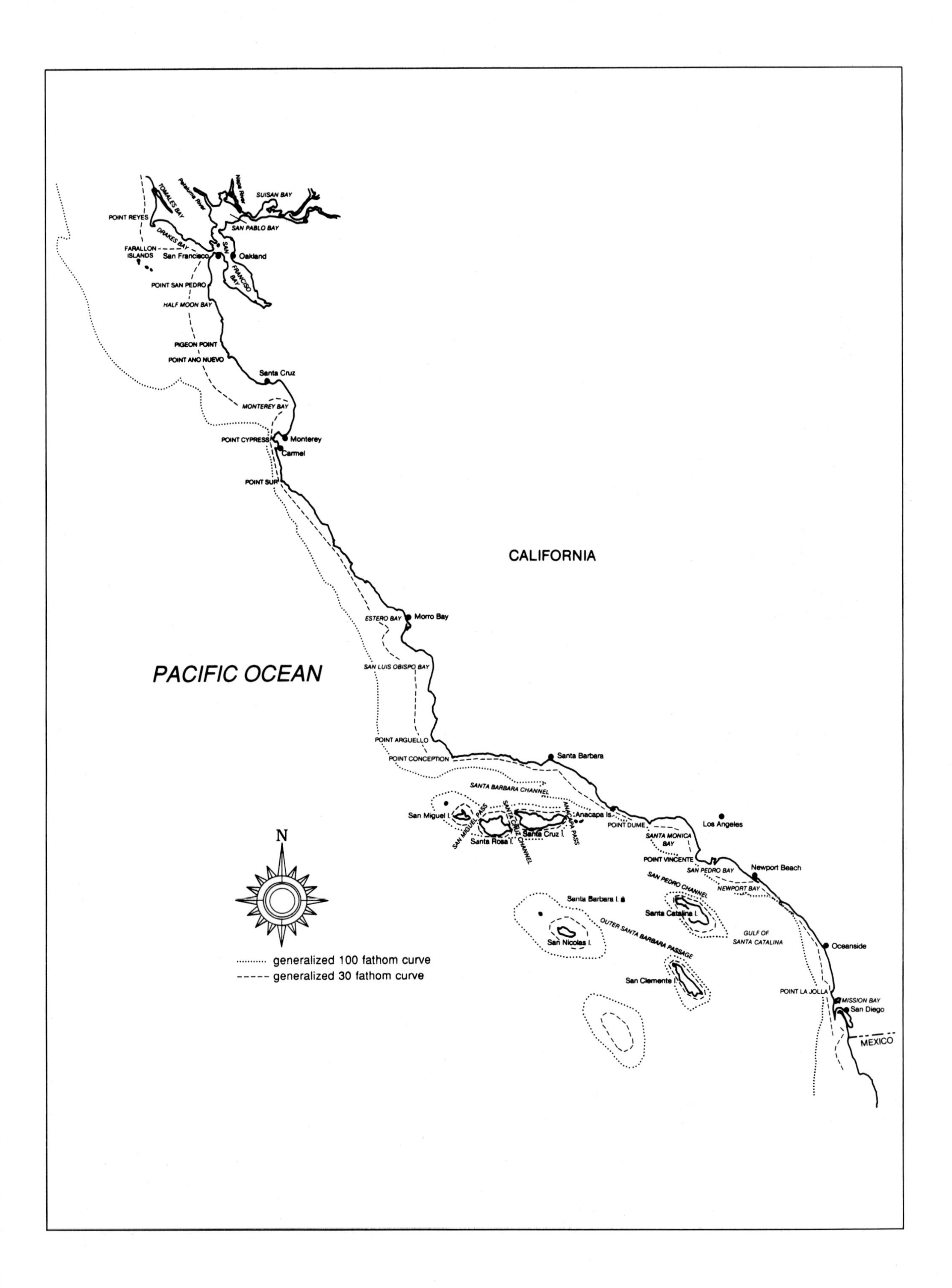

POINT REYES
TOMALES BAY
Petaluma River
Napa River
SUISAN BAY
SAN PABLO BAY
DRAKES BAY
FARALLON ISLANDS
San Francisco
Oakland
SAN FRANCISCO BAY
POINT SAN PEDRO
HALF MOON BAY
PIGEON POINT
POINT ANO NUEVO
Santa Cruz
MONTEREY BAY
POINT CYPRESS
Monterey
Carmel
POINT SUR
CALIFORNIA
ESTERO BAY
Morro Bay
PACIFIC OCEAN
SAN LUIS OBISPO BAY
POINT ARGUELLO
POINT CONCEPTION
Santa Barbara
SANTA BARBARA CHANNEL
San Miguel I.
SAN MIGUEL PASS
SANTA CRUZ CHANNEL
ANACAPA PASS
Anacapa Is.
Santa Rosa I.
Santa Cruz I.
POINT DUME
Los Angeles
SANTA MONICA BAY
POINT VINCENTE
SAN PEDRO BAY
Newport Beach
NEWPORT BAY
SAN PEDRO CHANNEL
N
Santa Barbara I.
Santa Catalina I.
OUTER SANTA BARBARA PASSAGE
San Nicolas I.
GULF OF SANTA CATALINA
Oceanside
San Clemente I.
POINT LA JOLLA
MISSION BAY
San Diego
MEXICO
generalized 100 fathom curve
generalized 30 fathom curve

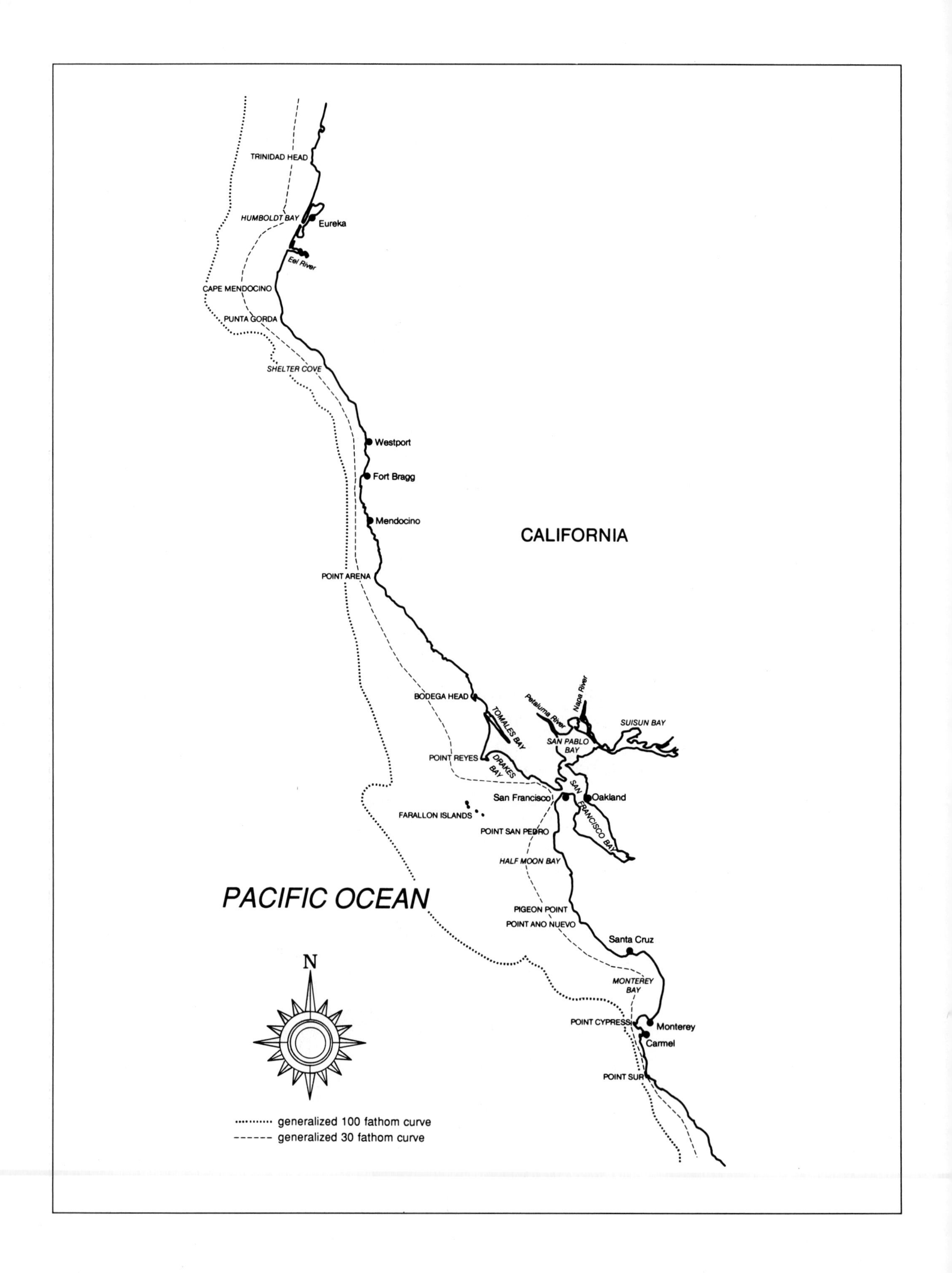
TRINIDAD HEAD
HUMBOLDT BAY
Eureka
Eel River
CAPE MENDOCINO
PUNTA GORDA
SHELTER COVE
Westport
Fort Bragg
Mendocino
CALIFORNIA
POINT ARENA
BODEGA HEAD
TOMALES BAY
Petaluma River
Napa River
SUISUN BAY
SAN PABLO BAY
POINT REYES
DRAKES BAY
SAN FRANCISCO BAY
San Francisco
Oakland
FARALLON ISLANDS
POINT SAN PEDRO
HALF MOON BAY
PACIFIC OCEAN
PIGEON POINT
POINT ANO NUEVO
Santa Cruz
N
MONTEREY BAY
POINT CYPRESS
Monterey
Carmel
POINT SUR
generalized 100 fathom curve
generalized 30 fathom curve

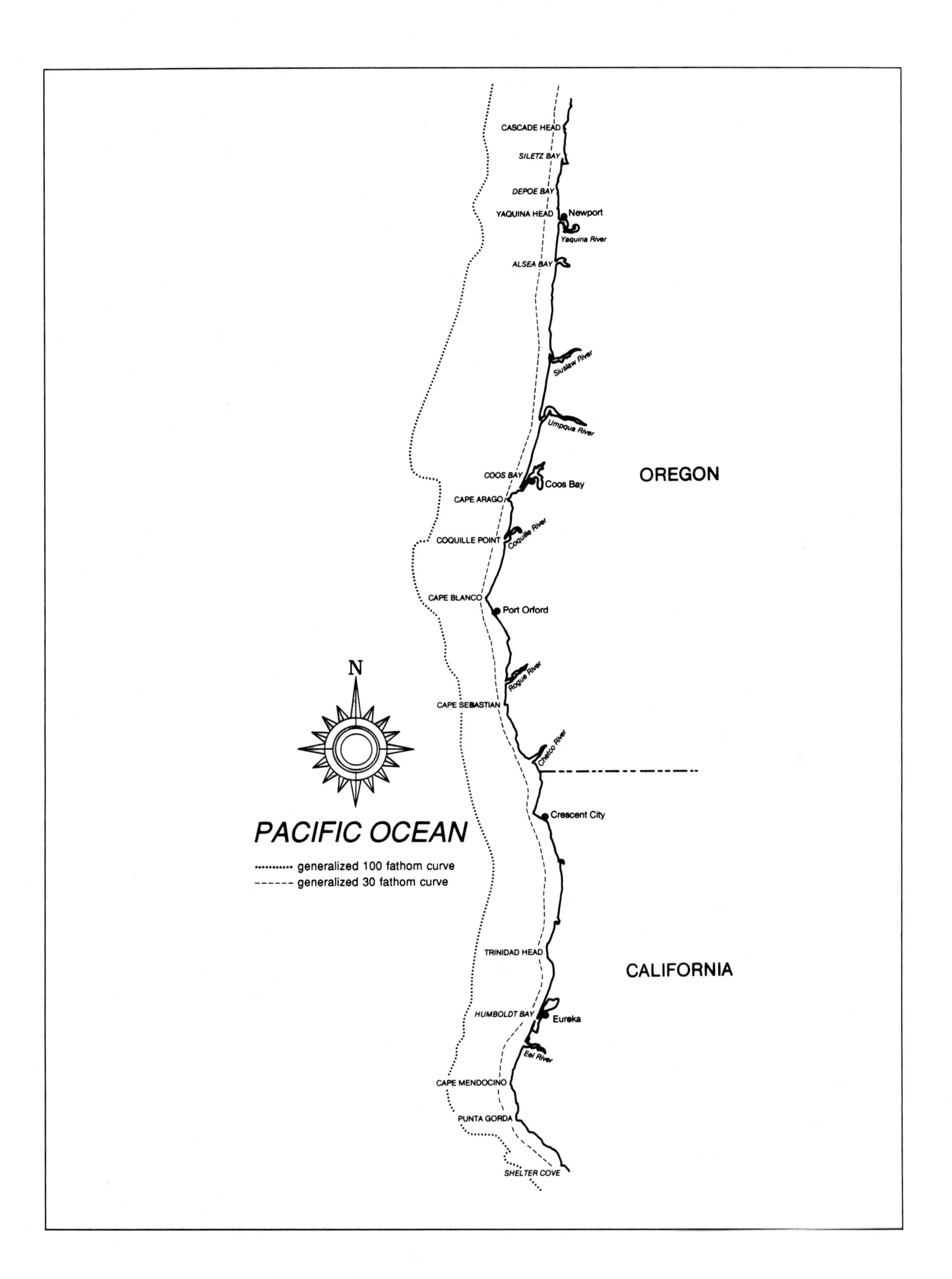
CASCADE HEAD
SILETZ BAY
DEPOE BAY
YAQUINA HEAD
Newport
Yaquina River
ALSEA BAY
Siuslaw River
Umpqua River
COOS BAY
Coos Bay
OREGON
CAPE ARAGO
COQUILLE POINT
Coquille River
CAPE BLANCO
Port Orford
N
Rogue River
CAPE SEBASTIAN
Chetco River
Crescent City
PACIFIC OCEAN
generalized 100 fathom curve
generalized 30 fathom curve
TRINIDAD HEAD
CALIFORNIA
HUMBOLDT BAY
Eureka
Eel River
CAPE MENDOCINO
PUNTA GORDA
SHELTER COVE

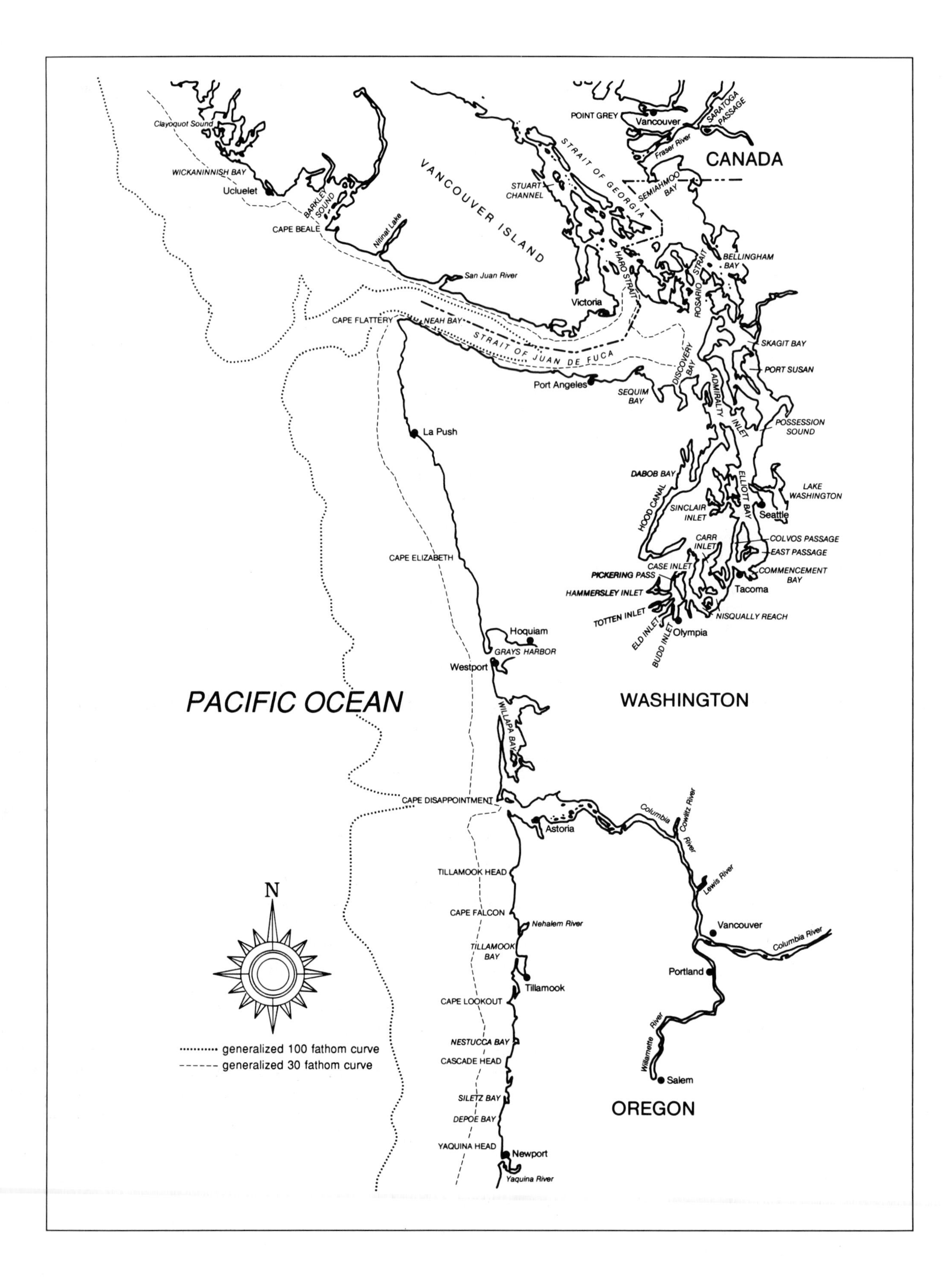
Clayoquot Sound
WICKANINNISH BAY
Ucluelet
BARKLEY SOUND
CAPE BEALE
Nitinat Lake
San Juan River
VANCOUVER ISLAND
STRAIT OF GEORGIA
STUART CHANNEL
POINT GREY
Vancouver
Fraser River
SARATOGA PASSAGE
CANADA
SEMIAHMOO BAY
HARO STRAIT
ROSARIO STRAIT
BELLINGHAM BAY
Victoria
CAPE FLATTERY
NEAH BAY
STRAIT OF JUAN DE FUCA
SKAGIT BAY
PORT SUSAN
DISCOVERY BAY
Port Angeles
SEQUIM BAY
ADMIRALTY INLET
POSSESSION SOUND
La Push
DABOB BAY
HOOD CANAL
ELLIOTT BAY
LAKE WASHINGTON
SINCLAIR INLET
Seattle
CARR INLET
COLVOS PASSAGE
EAST PASSAGE
CAPE ELIZABETH
CASE INLET
PICKERING PASS
COMMENCEMENT BAY
HAMMERSLEY INLET
Tacoma
TOTTEN INLET
NISQUALLY REACH
ELD INLET
BUDD INLET
Olympia
Hoquiam
GRAYS HARBOR
Westport
PACIFIC OCEAN
WASHINGTON
WILLAPA BAY
CAPE DISAPPOINTMENT
Columbia River
Cowlitz River
Astoria
TILLAMOOK HEAD
Lewis River
N
CAPE FALCON
Nehalem River
Vancouver
Columbia River
TILLAMOOK BAY
Portland
Tillamook
CAPE LOOKOUT
Willamette River
NESTUCCA BAY
generalized 100 fathom curve
generalized 30 fathom curve
CASCADE HEAD
Salem
SILETZ BAY
OREGON
DEPOE BAY
YAQUINA HEAD
Newport
Yaquina River

IX A Classification of Living Organisms

The emphasis is on living marine forms, with most extinct phyla and classes omitted.

Kingdom Monera

Procaryotic single-celled forms. Bacteria and "blue-green algae."

Sixteen phyla, distinguished by morphology and biochemical abilities.

Kingdom Protista

Eukaryotic single-celled forms, and marine multicellular forms that are derived from them. Protists and seaweeds.

Phylum Chrysophyta: Diatoms, coccolithophores, silicoflagellates

Phylum Pyrrophyta: Dinoflagellates, zooxanthellae

Phylum Cryptophyta: Some "microflagellates." Cryptomonads

Phylum Euglenophyta: A few "microflagellates." Mostly in fresh water

Phylum Chlorophyta: Green algae. Multicellular and unicellular

Phylum Rhodophyta: Red algae, crustose and coralline algae

Phylum Phaeophyta: Brown algae, kelps

Phylum Charophyta: (Confined to fresh water)

Phylum Zoomastigina: Nonphotosynthesizing flagellated protozoans

Phylum Sarcodina: Amebas and relatives
- Class Rhizopodea: Order Foraminiferida. Foraminiferans
- Class Actinopodea: Subclass Radiolaria. Radiolarians

Phylum Ciliophora: Ciliate protozoans

Phylum Sporozoa: Parasitic protists; mostly terrestrial

Kingdom Fungi

Funguses; mushrooms, molds, bracket fungi, lichens.
Five phyla, mostly land, freshwater, or highest supratidal organisms.

Kingdom Plantae

Division Anthophyta: Flowering plants, or angiosperms. Most species are freshwater or terrestrial. Marine eel grass, manatee grass, turtle grass, mangroves, salt-marsh grasses

Ten other Divisions: (Ferns, mosses, horsetails, liverworts, conifers, cycads, ginkgos, three with no common names. All terrestrial)

Kingdom Animalia

Phylum Placozoa: Ameba-like multicellular animals

Phylum Porifera: Sponges

Phylum Cnidaria: Jellyfish and kin. All are equipped with stinging cells
- Class Hydrozoa: Polyp-like animals that (often) have a medusa-like stage in the life cycle. Includes man-of-wars, fire corals, *Vellela*
- Class Scyphozoa: Jellyfish with no (or reduced) polyp stage in life cycle
- Class Anthozoa: Sea anemones, coral polyps

Phylum Ctenophora: "Sea gooseberries," comb jellies. Most lack stinging cells. Round, gelatinous, predatory, common

Phylum Mesozoa: Wormlike parasites of cephalopods

Phylum Platyhelminthes: Flatworms, tapeworms, flukes. Many free-living predatory forms, many parasites

Phylum Nemertea: Unsegmented worms with extensible "harpoons"

Phylum Gnathostomulida: Microscopic, wormlike, live between grains in marine sediments

Phylum Gastrotricha: Microscopic, ciliated, multicellular, live between grains in marine sediments

Phylum Rotifera: Ciliated, multicellular. Common in fresh water, plankton, and attached to benthic objects

Phylum Kinoryncha: Small, spiny, segmented, wormlike; live between grains in sediments. All marine

Phylum Acanthocephala: Spiny-headed worms. All parasites in vertebrate intestines. Marine, freshwater, terrestrial

Phylum Entoprocta: Polyplike, small, benthic suspension feeders

Phylum Nematoda: Roundworms. Common, free-living and parasitic

Phylum Nematomorpha: Hairworms. Parasitic; most are freshwater or terrestrial

Phylum Ectoprocta: "Bryozoans." Common encrusting colonial marine forms, a few in fresh water. Individuals are polyp-like, suspension feeders, equipped with a crown of tentacles known as a "lophophore"

Phylum Phoronida: Wormlike suspension feeders with lophophores. A few centimeters tall. All marine

Phylum Brachiopoda: Bivalved animals, superficially like clams but with lophophore feeding apparatus. Once widespread in the sea, now scarce, mainly in deep water

Phylum Mollusca:

- Class Gastropoda:
  - Order Prosobranchia: Snails, limpets, abalones
  - Order Opisthobranchia: Sea slugs, pteropods
  - Order Pulmonata: Land snails and slugs; a few have returned to the sea
- Class Cephalopoda: Octopuses, squids, cuttlefish, nautilus, extinct ammonoids
- Bivalvia: Clams, oysters, scallops, mussels, "shipworms"
- Polyplacophora: Chitons
- Three other small classes; tusk shells, wormlike, and limpetlike organisms

Phylum Crustacea:

- Class Copepoda: Copepods
- Class Cirripedia: Barnacles
- Class Ostracoda: Ostracods—small crustaceans with bivalved carapaces enclosing the body
- Class Malacostraca:
  - Order Euphausiacea: Krill
  - Order Isopoda: Isopods
  - Order Amphipoda: Amphipods
  - Order Stomatopoda: Mantis shrimps
  - Order Decapoda: Crabs, shrimps, lobsters, hermit crabs, ghost shrimp, stone crabs, many others
  - Order Mysidacea: "Mysids." Shrimplike animals with loosely-attached carapaces
  - Nine other orders in this class, all small and obscure, marine
  - Five additional classes in this phylum, mainly small crustaceans. One found mainly in nonmarine water

Phylum Chelicerata: Horseshoe crabs, sea spiders

Phylum Uniramia: Insects, centipedes, millipedes, others. Few in the sea

Phylum Trilobita: Extinct multilegged marine arthropods

Phylum Priapulida: Small, rare, subtidal, wormlike. Once very common in the sea. All marine

Phylum Sipunculida: "Peanut worms." Extensible worms. All marine

Phylum Echiura: "Spoon worms," the "fat innkeeper." All marine

Phylum Annelida: Segmented worms

- Class Polychaeta: Feather-duster worms, clam worms, bloodworms
- Class Oligochaeta: Earthworms and kin; mainly freshwater or terrestrial
- Class Hirudinea: Leeches and kin; mainly freshwater or terrestrial

Phylum Tardigrada: "Water bears." Tiny, 8-legged, interstitial animals

Phylum Pentastoma: "Tongue worms." Parasites of vertebrates

Phylum Onychophora: Multilegged worms, intermediate between annelids and arthropods. The only all-terrestrial phylum, but marine fossils are known

Phylum Pogonophora: "Beard worms." No digestive system. Deep-water, all marine

Phylum Echinodermata:
- Class Stelleroidea:
  - Subclass Asteroidea: Sea stars
  - Subclass Ophiuroidea: Brittle stars, basket stars
- Class Echinoidea: Sea urchins, sea biscuits, sand dollars, heart urchins
- Class Holothuroidea: Sea cucumbers
- Class Crinoidea: Sea lilies. (Mostly deepwater, more abundant in prehistoric seas)
- Class Concentricycloidea: Sea daisies. One deepwater species, discovered in 1986

Phylum Chaetognatha: Arrowworms. Stiff-bodied, planktonic, predaceous, common. All marine

Phylum Hemichordata: "Acorn worms." Unsegmented burrowers, fair sized. All marine

Phylum Chordata:
- Subphylum Urochordata:
  - Class Ascidiacea: "Sea squirts," tunicates
  - Class Thaliacea: Salps
  - Class Larvacea: Larvaceans
- Subphylum Cephalochordata: Lancelets, "amphioxus."
- Subphylum Vertebrata:
  - Class Agnatha: Lampreys, hagfish
  - Class Chondrichthyes: Sharks, rays, skates, sawfish, chimaeras
  - Class Osteichthyes: Fish
  - Class Amphibia: Frogs, toads, salamanders
  - Class Reptilia: Reptiles
  - Class Aves: Birds
  - Class Mammalia: Mammals

Glossary

abyssal Refers to organisms or phenomena at depths between 4,000 and 6,000 m, on the sea bottom.

abyssal hills *See* abyssal knoll.

abyssal knoll An elevation rising less than 1,000 m from the seafloor and of limited extent across the summit.

abyssalpelagic zone Pertaining to the portion of the ocean between 2,000 and 4,000 m.

abyssal plain Flat, gently sloping or nearly level region of the seafloor.

abyssopelagic Refers to organisms or phenomena at depths between 4,000 and 6,000 meters, in midwater.

acceleration Rate of change with time of the velocity of a particle.

accretion Addition of sediment by natural processes to form new land as in deltas and bays.

acid Any of a class of chemical compounds capable of transferring a hydrogen ion in solution.

active margin Continental margin that is seismically active; continental margin that is at a global plate boundary.

advection Horizontal or vertical flow of seawater as a current. In meteorology, the predominantly horizontal, large-scale motions of the atmosphere.

advection fog Type of fog caused by the advection of moist air over a cold surface and the consequent cooling of that air to below its dew point. A very common advection fog is that caused by moist air in transport over a cold body of water (sea fog).

aerobe Organism that can live and grow only in the presence of oxygen.

Agulhas Current (sometimes called Agulhas Stream). A fast current flowing southwestward along the southeast coast of Africa.

air mass Often defined as a widespread body of air that is approximately homogeneous in its horizontal extent, particularly with reference to temperature and moisture distribution; in addition, the vertical temperature and moisture variations are approximately the same over its horizontal extent.

Alaska Current Current that flows northwestward and westward along the coasts of Canada and Alaska to the Aleutian Islands. It contains water from the North Pacific Current and has the character of a warm current.

albedo Ratio of the amount of electromagnetic radiation reflected by a body to the amount incident upon it, commonly expressed as a percentage. The albedo is to be distinguished from the reflectivity, which refers to one specific wavelength (monochromatic radiation).

Aleutian low Low-pressure center located near the Aleutian Islands on charts of mean sea-level pressure. It represents one of the main centers of action in the atmospheric circulation of the Northern Hemisphere.

alga(e) Simple plant, without a true stem, leaves, or roots, having a one-celled sex organ and possessing chlorophyll; includes almost all seaweeds.

alluvial fan *See* fan.

alluvium Detrital deposits eroded, transported, and deposited by moving water or ice.

amphidromic point No-tide or nodal point on a chart of cotidal lines from which the cotidal lines radiate.

amphidromic region Area surrounding an amphidromic point in which the cotidal lines radiate from the no-tide point and progress through all hours of the tide cycle.

amplitude For an ocean-surface wave, the vertical distance from still-water level to wave crest, that is, one-half the wave height.

anadromous Migrating from the sea up a river to spawn, as salmon do.

anaerobe Organism requiring a complete or nearly complete absence of oxygen.

anaerobic sediment Highly organic sediment rich in H_2S formed in the absence of free oxygen where little or no circulation or mixing of the bottom water occurs.

andesite Volcanic rock consisting largely of plagioclase feldspar (silicates of calcium and sodium), quartz, and biotite. Generally called intermediate in composition between granite and basalt.

angular momentum Moment of the linear momentum of a particle about a point. If m is the mass of a particle, V the velocity, and r the position vector from the given point O to the particle, the angular momentum M about O is given by $M = rmV$. The angular momentum of a particle about an axis is defined as that

Most of the definitions in this glossary are quoted or adapted from B. B. Baker, Jr., W. R. Deebel, and R. D. Geisenderfer, eds., *Glossary of Oceanographic Terms* (Washington, D.C.: U.S. Naval Oceanographic Office, 1966).

For more complete compilations of terms in the field of oceanography see: R. W. Fairbridge, ed., *The Encyclopedia of Oceanography* (New York: Reinhold Publishing, 1966); M. Gary, R. McAfee, Jr., and C. L. Wolf, eds., *Glossary of Geology* (Washington, D.C.: American Geological Institute, 1972); and L. M. Hunt and D. G. Groves, eds., *A Glossary of Ocean Science and Undersea Technology Terms* (Arlington, Va.: Compass Publications, 1965).

component, along the axis, of the angular momentum of the particle about any point on the axis.

annelid A segmented marine worm. A member of the phylum Annelida.

anomalistic month Average period of about 27½ days, measured from perigee to perigee, during which the moon completes one revolution around the earth.

anomalous Not encompassed by rules governing the majority of cases; distinguished from abnormal by implying a difference of kind rather than a difference merely of degree.

Antarctic air A type of air whose characteristics are developed in an antarctic region. Antarctic air appears to be colder at the surface in all seasons, and at all levels in autumn and winter, than arctic air.

Antarctic Circumpolar Current The current that flows from west to east around the Antarctic continent and therefore through the southern extremities of the Atlantic, Pacific, and Indian oceans; the largest and swiftest ocean current with an approximate volume transport of 110×10^6 m^3/sec (3.9×10^9 ft^3/sec).

Antarctic Convergence Convergence line circling the South Pole. It is the best-defined convergence line in the oceans, being recognized by a relatively rapid northward increase in the surface temperature. *See* convergence.

Antarctic Divergence Those regions near the shores of the Antarctic continent where water rises to the surface from intermediate depths as extensive upwelling replacing surface waters that are drifting in a northern and eastern direction in the West Wind Drift.

Antarctic Ocean Name commonly applied to those portions of the Atlantic, Pacific, and Indian oceans that reach Antarctica on the south and are bounded on the north by the Subtropical Convergence. It is not a recognized ocean body.

anticyclone Atmospheric anticyclonic circulation that is closed. With respect to the relative direction of its rotation, it is the opposite of a cyclone. Because anticyclonic circulation and relative high atmospheric pressure usually coexist, the terms *anticyclone* and *high* are used interchangeably in common practice.

Antilles Current Current formed by part of the North Equatorial Current that flows along the northern side of the Greater Antilles.

aphotic zone Portion of the ocean waters where light is insufficient for plants to carry on photosynthesis. *See* euphotic zone.

apogee Point on the orbit of a satellite farthest from the main body; opposed to perigee.

aquaculture Raising organisms in a water environment under controlled conditions for commercial sale.

Archimedes' principle Statement that a new upward or buoyant force equal in magnitude to the weight of the displaced fluid acts on a body either partly or wholly submerged in a fluid at rest under the influence of gravity. This force is known as the Archimedean buoyant force (or buoyancy) and is independent of the shape of the submerged body and does not depend on any special properties of the fluid.

archipelago A sea or part of a sea studded with islands or island groups; often synonymous with island group.

arthropod An animal with jointed legs and an external skeleton; insects and centipedes, crustaceans, extinct trilobites, spiders. Formerly classed in a single phylum, Arthropoda, these animals are now placed in separate phyla.

asthenosphere Upper part of the earth's mantle. Seismic waves lose energy in this region and the region is believed to undergo plastic flow. Magma is thought to form here.

Atlantic Ocean Ocean extending from Antarctica northward to the southern limits of the Greenland and Norwegian seas. It is separated from the Pacific Ocean by the meridian of Cape Horn and from the Indian Ocean by the meridian of Cape Agulhas.

atmosphere This term when used as a pressure term is defined as the pressure exerted per square centimeter by a column of mercury 760 mm high at a temperature of 0°C where the acceleration of gravity is 980.665 cm/s^2. One atmosphere of pressure equals 1.0133×10^6 dyn/cm^2. The actual effect of atmospheric pressure is ignored when considering pressures within the ocean. Surface pressure is considered to be zero.

atoll Ring-shaped organic reef that encloses a lagoon in which there is no preexisting land, and which is surrounded by the open sea.

attenuation 1. Reduction in sound or light intensity caused by the absorption and scattering of sound or light energy in air or water. 2. A lessening of the amplitude of a wave with distance from the origin.

authigenic sediments Sediments that are formed in place; an example is manganese nodule.

autotroph Organism that makes its organic nutrients from inorganic chemicals.

azoic Without life; but most ocean areas described as azoic contain at least a bacterial flora. Abiotic also means the absence of life.

backshore That inner part of the shore, usually dry, reached only by the highest tides and storms, located between the foreshore and the coastline.

bacterium (bacteria) A microscopic single-celled organism of relatively simple internal architecture, lacking a nucleus and other complex internal structures found in more advanced single-celled organisms such as protists.

baleen Horny material growing down from the upper jaw of large plankton-feeding whales, which forms a strainer or filtering organ consisting of numerous plates with fringed edges.

baleen whale Member of the cetacean suborder Mysticeti, which comprises the right whales, gray whales, and rorquals.

bank An elevation of the seafloor located on a continental (or island) shelf and over which the depth of water is relatively shallow but sufficient for safe surface navigation. Shoals or bars on its surface may be dangerous to navigation.

bar Submerged or emerged embankment of sand, gravel, mud, or mollusc shells built on the seafloor in shallow water by waves and currents. *See* barrier beach, longshore bar.

barrier beach Bar essentially parallel to the shore, the crest of which is above high water.

barrier reef Coral reef parallel to and separated from the coast by a lagoon that is too deep for coral growth.

basalt Igneous rock that is low in silica and that forms a worldwide layer under the oceans and the granitic continents.

base Any ionic or molecular chemical entity capable of accepting or receiving a hydrogen ion from another substance; any chemical entity capable of transferring a hydroxyl ion in solution.

bathyal Refers to organisms or phenomena at depths between about 200 and 4,000 m, on the sea bottom. Often coincident with the continental slope.

bathymetry Science of measuring ocean depths to determine seafloor topography.

bathypelagic Refers to organisms or phenomena at depths between 1,000 and 4,000 m, in mid-water.

bathythermograph (abbreviated BT). Device for obtaining a record of temperature against depth (strictly speaking, pressure) in the ocean from a ship underway.

bay Recess in the shore or an inlet of a sea between two capes or headlands; not as large as a gulf but larger than a cove.

beach Zone of unconsolidated material that extends landward from the low-water line to the line of permanent vegetation (usually the effective limit of storm waves). A beach includes foreshore and backshore. *See also* coast, shoreline.

bends (decompression sickness). Condition resulting from the formation of gas bubbles in the blood or tissues of a diver during ascent. Depending on their number, size, and location, these bubbles can cause pain, paralysis, unconsciousness, and even death. The blocking of an artery by a bubble is called an *air embolism*.

benthic (or benthonic). Portion of the marine environment inhabited by marine organisms that live permanently in or on the bottom.

benthic province All submarine bottom terrain, regardless of water depth.

benthos The community of plants and animals that live on the sea bottom.

berm Nearly horizontal portion of a beach or backshore having an abrupt fall and formed by material deposited by wave action. It marks the limit of ordinary high tides.

biogenic Formed by organisms.

biogenic ooze Pelagic sediment consisting of at least 30% skeletal remains of pelagic organisms such as diatoms, foraminifera, or radiolaria.

biological oceanography (marine biology). Study of the ocean's plant and animal life in relation to the marine environment, including the effects of habitat, sedimentation, physical and chemical changes, and other factors on the distribution of marine organisms and the action of organisms on the environment.

bioluminescence (also called phosphorescence). Production of light without sensible heat by living organisms as a result of a chemical reaction either within certain cells or organs or outside the cells in some form of secretion.

biomass Amount of living matter per unit of water surface or volume expressed in weight units.

biosphere Transition zone between earth and atmosphere within which most terrestrial life is found; the outer portion of the geosphere and inner or lower portion of the atmosphere.

biotic factors Biological factors such as availability of food, competition between species, and predator-prey relationships that affect the distribution and abundance of a given species of plant or animal.

blue-green alga Very simple single-celled or filamentous plants (Monera) in which the blue color is imparted by a water-soluble accessory pigment.

bottom water Water mass at the deepest part of the water column. It is the densest water that can occupy that position as determined by regional topography. *See* water mass.

brackish water Water in which salinity values range from approximately 0.50 to 17.00‰.

breaker Wave breaking on the shore or over a reef. Breakers are roughly classified into three kinds: *spilling breakers* break gradually over a considerable distance; *plunging breakers* tend to curl over and break with a crash; *surging breakers* peak, but then instead of spilling or plunging, they surge up on the beach face.

brine Seawater containing a higher concentration of dissolved salt than the ordinary ocean. Brine is produced by evaporating or freezing seawater or by hot groundwater dissolving salt deposits.

brown alga One of a division (Phaeophyta) of greenish yellow to deep brown, filamentous to massively complex plants, in which the color is imparted by the predominance of orangish pigments over the chlorophylls. This group includes the rockweeds, gulfweeds, and the large kelp. Brown algae are most abundant in the cooler waters of the world.

bryozoan One of a phylum (Ectoprocta) of minute, mostly colonial, aquatic animals with body walls often hardened by calcium carbonate and growing attached to aquatic plants, rocks, and other firm surfaces.

buoyancy 1. Property of an object that enables it to float on the surface of a liquid or ascend through and remain freely suspended in a compressible fluid such as the atmosphere. 2. Upward force exerted on a parcel of fluid (or an object within the fluid) in a gravitational field by virtue of the density difference between the parcel (or object) and that of the surrounding fluids (also known as *buoyant force*).

caballing Mixing of two water masses of identical in situ densities but different in situ temperatures and salinities, such that the resulting mixture is denser than its components.

California Current Ocean current that flows southward along the western coast of the United States to northern Baja California. It is formed by parts of the North Pacific Current and is a wide current that moves sluggishly toward the southeast. Off Central America, the California Current turns toward the west and becomes the North Equatorial Current.

Callao Painter Mariners' reference to the catastrophic destruction of marine life that causes the blackening of paint on ships in the harbor of Callao, Peru. Hydrogen sulfide released during the decomposition of the organisms is responsible for the phenomenon. The immediate cause of this phenomenon is the increase in water temperature when warmer oceanic currents turn inshore; organisms normally accustomed to colder water temperatures die because of this abrupt temperature change.

calm Apparent absence of motion of the water surface when there is no wind or swell; water is generally considered calm if the current speed is less than 0.1 knot.

calorie (abbreviated cal). Unit of heat often defined as the amount of heat required to raise the temperature of 1 g of water through 1°C.

calving Breaking apart of parts of a glacier or sea ice from the main mass.

Canary Current Prevailing southward flow along the northwest coast of Africa; it helps to form the North Equatorial Current.

capillary wave Wave whose velocity of propagation is controlled primarily by the surface tension of the liquid in which the wave is traveling. Water waves less than 1.73 cm long are considered capillary waves.

carapace Chitinous or bony shield covering the whole or part of the back of certain animals, such as crustaceans.

carbon dioxide Heavy, colorless gas of chemical formula CO_2. It is the fourth most abundant constituent of dry air. Over 99% of terrestrial CO_2 is found in the oceans.

catadromous Migrating from rivers to the sea to spawn, as with some eels.

Celsius temperature scale Same as centigrade temperature scale. 0°C = 273.16°/K. *See* Kelvin temperature scale.

center of mass Point at which the mass of an entire body or system (like earth/moon) may be regarded as being concentrated.

centrifugal force Force with which a body moving under constraint along a curved path reacts to the constraint. Centrifugal force acts away from the center of curvature of the path of the moving body. As a force caused by the rotation of the earth on its axis, centrifugal force is opposed to gravitation and combines with it to form gravity.

centripetal force Force that tends to move objects toward the center around which they are moving.

cephalopod A squid, octopus, cuttlefish, or nautilus. A member of the class Cephalopoda.

cetacean A whale, dolphin, or porpoise. An animal of the order Cetacea.

chemical oceanography Study of the chemical composition of dissolved solids and gases, material in suspension, and acidity of ocean waters and the variability of these factors both geographically and temporally in relationship to the atmosphere and the ocean bottom.

chemosynthesis The manufacture of carbohydrates by using the energy stored in chemicals, rather than in sunlight. Many bacteria practice chemosynthesis rather than photosynthesis.

chlorinity Measure of the chloride content by mass of seawater (grams per kilogram of seawater, or per mille). Because the proportion of chloride to sodium is reasonably constant, the amount of chlorinity in a seawater sample is generally used to establish the sample's salinity.

chlorophyll Group of green pigments active in photosynthesis.

chordate An animal equipped with gill slits (either as an adult or during embryological development), a dorsal nerve chord or sometimes a "backbone." These are features of the phylum Chordata. The most conspicuous chordates are vertebrates; some invertebrate chordates include sea squirts.

circulation General term describing a water current flow within a large area; usually a closed circular pattern such as the North Atlantic.

clapotis Standing-wave phenomenon caused by the reflection of a wave train from a breakwater, bulkhead, or steep beach.

class In biological classification, a group of closely related orders. Major subdivision of a biological phylum.

clastic Rock composed principally of detritus transported mechanically to its final place of deposition. Sandstones and shales are the most common clastics.

clay As a size term, refers to sediment particles smaller than silt. Mineralogically, clay is a hydrous aluminum silicate material with plastic properties and a crystal structure.

climate Prevalent or characteristic meteorological conditions of a place or region; in contrast to weather, which is the state of the atmosphere at any time.

Cnidaria The phylum of jellyfish, sea anemones, coral polyps, man-of-wars, and others. Pronounced "Ny-dare'-ia." All possess stinging cells, a key characteristic of members of this phylum.

coast General region of indefinite width that extends from the sea inland to the first major change in terrain features.

coastal current Relatively uniform drift usually flowing parallel to the shore in the deeper water adjacent to the surf zone. The current may be related to tides, winds, or distribution of mass.

cobble (boulderet, cobblestone). Rock fragment between 64 and 256 mm in diameter, larger than a pebble and smaller than a boulder, and rounded or otherwise abraded.

Coccolithophore A tiny, flagellated, single-celled phytoplankton organism that is coated with microscopic ornamented calcium carbonate plates. Photosynthetic, often exceedingly numerous, often important contributors to the sea-bottom sediments and to phytoplankton productivity. Division Chrysophyta in some classifications, Haptophyta in others.

cold wall Steep water-temperature gradient between the Gulf Stream and the slope water inshore of the Gulf Stream.

colligative property Any one of four characteristic properties of solutions, namely, the interdependent changes in vapor pressure, freezing point, boiling point, and osmotic pressure, with a change in amount of dissolved matter. If, under a given set of conditions, the value for any one property is known, the others may be computed. In general, with an increase in dissolved matter (for example, salt in seawater), freezing point and vapor pressure decrease, and boiling point and osmotic pressure increase.

collision coasts (or subduction coasts) Those coasts that occur on the collision side of continents and island arcs in the movement of crustal plates.

colloid As a size term, refers to particles smaller than clay size.

commensalism Symbiotic relationship between two species in which one species is benefited and the other is not harmed.

community An integrated, mutually adjusted assemblage of plants and animals inhabiting a natural area. The assemblage may or may not be self-sufficient and is considered to be in a state of dynamic equilibrium. The community is usually characterized as having a definite species composition and may be defined by the habitat it occupies or by the species present.

compensation depth The depth at which a suspended phytoplankton organism receives just enough light, over a 24-hour period, to enable its photosynthesis to produce exactly enough new carbohydrate to supply its own metabolic needs. The depth at which its photosynthetic production matches its respiratory or metabolic consumption.

competition Interaction of two populations in which each inhibits the other.

condensation Physical process by which a vapor becomes a liquid or solid; the opposite of evaporation. When water vapor condenses, heat is released, and the surrounding temperature is raised.

conduction Transfer of energy (usually heat) within and through a conductor by means of internal particle or molecular activity, and without any net external motion.

conservative property Property whose values do not change in the course of a particular series of events.

consolidated sediment Sediment that has been converted into rocks by compaction, cementation, or other physical or chemical change.

continent Large land mass rising abruptly from the deep-ocean floor, including marginal regions that are shallowly submerged. Continents constitute about one-third of the earth's surface.

continental borderland A region adjacent to a continent, normally occupied by or bordering a continental shelf, that is highly irregular, with depth well in excess of those typical of a continental shelf.

continental drift Concept that the continents can drift on the surface of the earth because of the weakness of the suboceanic crust.

continental margin Zone between continental land and deep-sea floor; consists of continental shelf, slope, rise, and borderland.

continental rise Gentle slope with a generally smooth surface rising toward the foot of the continental slope.

continental shelf Zone adjacent to a continent or around an island, usually extending from the low-water line to the depth at which the slope increases markedly.

continental slope Declivity seaward from a shelf edge into greater depth.

convection In general, mass motions within a fluid resulting in transport and mixing of the properties of that fluid. Convection, along with conduction and radiation, is a principal means of energy transfer.

convergence Situation in which waters of different origins come together at a point or, more commonly, along a line known as a convergence line. Along such a line the denser water from one side sinks under the lighter water from the other side. The recognized convergence lines in the oceans are the polar, subtropical, tropical, and equatorial convergence lines.

convergent boundary Seismically active location at edge of two global plates that are colliding or sliding by one another.

copepod Crustaceans of the class Copepoda. Small (.5– 10 mm long) and numerous, these are the most abundant and important herbivorous animals in the marine zooplankton.

coral reef Ridge or mass of limestone built up of detritus deposited around the skeletal remains of molluscs, colonial coral, and massive calcareous algae. Coral may constitute less than half of the reef material.

core 1. A vertical, cylindrical sample of the bottom sediments, from which the nature and stratification of the bottom may be determined. 2. Central zone of the earth.

Coriolis effect Apparent force on moving particles resulting from the earth's rotation. It causes moving particles to be deflected to the right of motion in the Northern Hemisphere and to the left in the Southern Hemisphere; the force is proportional to the speed and latitude of the moving particle and cannot change the speed of the particle.

cosmogenous sediments Composed of extra-terrestrial material. Tektites are an example.

cotidal line Line on a chart passing through all points where high water occurs at the same time. The lines show the lapse of time, usually in lunar-hour intervals, between the moon's transit over a reference meridian (usually Greenwich) and the occurrence of high water for any point lying along the line.

cove Small bay or baylike recess in the coast, usually affording anchorage and shelter to small craft.

critical depth The greatest depth to which a phytoplankton cell can circulate while still obtaining enough sunlight to produce as much new carbohydrate as its metabolic needs require. Cells that circulate deeper receive too little light, while near the surface, to compensate for the time they spend in darkness.

Cromwell Undercurrent (Pacific Equatorial Undercurrent). Eastward-setting subsurface current that extends about 1 1/2° north and south of the equator, and from about 150°E to 92°W. It is 300 km wide and 0.2 km thick; at its core the speed is 100 to 150 cm/s.

crust Outer shell of the solid earth. The crust varies in thickness from approximately 5–7 km under the ocean basins to 35 km under the continents.

crustacean One of a class (Crustacea) of arthropods that breathe through gills or branchiae and have a body commonly covered by a hard shell or crust. The group includes barnacles, crabs, shrimp, and lobsters.

crystalline Term applied to rocks containing grains of regular polyhedral form bounded by planar surfaces and having an orderly molecular structure. Usually applied to igneous and metamorphic rocks but not to sedimentary rocks.

current Horizontal movement of water.

cusp One of a series of low mounds of beach material separated by crescent A-shaped troughs spaced at more or less regular intervals along the beach.

cuspate foreland Crescent-shaped bar uniting with the shore at each end. It may be formed by a single spit growing from shore turning back to again meet the shore or by two spits growing from shore uniting to form a bar of sharply cuspate form.

cyanobacteria Bacteria capable of performing aerobic photosynthesis. Characterized by blue-green pigments; formerly called blue-green algae. Many are nitrogen fixers.

Davidson Current (Davidson Inshore Current). Coastal countercurrent setting north inshore of the California Current along the western coast of the United States (from northern California to Washington to at least 48°N) during the winter months.

dead water Phenomenon that occurs when a ship of low propulsive power negotiates water in which a thin layer of fresher water is over a layer of more saline water. As the ship moves, part of its energy generates an internal wave, which causes a noticeable drop in efficiency of propulsion.

decay of waves Change that waves undergo after they leave a generating area (fetch) and pass through a calm or region of lighter winds. In the process of decay, the significant wave height decreases and the significant wavelength increases.

deep ocean circulation (or thermohaline) The movements of the ocean water, usually permanent thermocline, caused by changes in the water's density resulting from changes in temperature and salinity.

deep scattering layer (DSL). Stratified population(s) of organisms in most oceanic waters that scatter sound. Such layers are generally found during the day at depths from 200 to 1,000 m and may be from 50 to 200 m thick.

deep-water wave Surface wave with a length less than twice the depth of the water; its velocity is independent of the depth of the water. *See* shallow-water wave.

delta Deposit of alluvium (unconsolidated sediment) at the mouth of a river. The name implies a triangular shape, which is commonly found, as in the Mississippi River delta.

density 1. Ratio of the mass of any substance to its volume; the reciprocal of specific volume. 2. In oceanology, density is equivalent to specific gravity and represents the ratio, at atmospheric pressure, of the weight of a given volume of seawater to that of an equal volume of distilled water at 4.0°C (39.2°F). It is thus dimensionless.

density layer Layer of water in which density increases with depth enough to increase the buoyancy of a submarine.

deposit feeder An animal that eats detritus or sediments rich in organic materials. A detritus feeder. Example: many sea cucumbers.

depth of frictional resistance (or depth of frictional influence). Depth at which wind-induced current direction is 180° from that of the wind.

detritus Dead organic material produced by the breakup of organisms, usually in small particles suspended in the water or resting on the bottom. Edible organic sediment or drifting dead matter.

detritus feeder *See* deposit feeder.

diatom One of a class (Chrysophyta) of microscopic plankton organisms possessing a wall of overlapping halves impregnated with silica. Diatoms are one of the most abundant groups of organisms in the sea and the most important primary food source of marine animals.

diffraction Bending of wave front by an obstacle in the medium. For example, when a portion of a train of waves is interrupted by a barrier such as a breakwater, diffracted waves are propagated in the sheltered region within the barrier's geometric shadow.

diffusion Spread of particles such as atoms, molecules, and ions throughout a medium (water or air) so as to produce an even distribution of the particles in the medium.

dinoflagellate One of a class (Pyrrophyta) of single-celled microscopic or minute organisms. Dinoflagellates may possess both plant (chlorophyll and cellulose plates) and animal (ingestion of food) characteristics.

discontinuity Abrupt change or jump of a variable at a line or surface. *See* interface.

dispersion Sorting out of wave trains generated by storms or earthquakes due to the variation of velocity with wavelength; the velocity is proportional to the wavelength.

diurnal 1. Daily, especially pertaining to actions completed within 24 hours and that recur every 24 hours. 2. Having a period or cycle of approximately one lunar day (24.84 solar hours).

divergence Horizontal flow of water in different directions from a common center or zone; often associated with upwelling.

divergent boundary Edge of two global plates that are separating. An example is an ocean ridge.

division In biological classification, a broad subdivision of the plant or protist kingdoms, equivalent to a phylum in the animal kingdom.

doldrums Nautical term for the equatorial trough, with special reference to the light and variable nature of the winds.

dolphin 1. Member of the cetacean suborder Odontoceti. Name is used interchangeably with *porpoise* by some. More properly, it is given generally to the beaked members of the family Delphinidae, except the larger members, which have been given the name *whale*, such as the killer whale and pilot whale. 2. A pelagic fish of the genus *Coryphaena* noted for its brilliant colors.

dredge A simple cylindrical or rectangular device for collecting samples of bottom sediment and life. It is generally made of heavy gauge steel plate or pipe and depends on a scooping action to obtain the sample.

drowned coast Land submerged beneath water either through a rise in the level of the water or by sinking of the land.

drumlin Asymmetrical elongated gravel hill with its long axis parallel to the direction of the glacier that formed it.

dugong Aquatic herbivorous mammal, sometimes referred to as a sea cow, of the order Sirenia, with a bilobate tail (two lobes) like that of a whale.

dysphotic zone Twilight zone in the ocean, between euphotic and aphotic zone. One percent of sunlight reaches this zone.

earthquake A sudden, transient motion or trembling of the earth's crust resulting from waves in the earth caused by faulting of the rocks or by volcanic activity.

earthquake focus Precise location of earth movement generating shock (seismic) waves.

East Wind Drift Westerly current close to the Antarctic continent, caused by the polar easterlies.

ebb current Tidal current associated with the decrease in the height of a tide. Ebb currents generally flow seaward or in an opposite direction to the tide progression. Erroneously called ebb tide or simply ebb.

echinoderm A sea star, sea cucumber, brittle star, sea urchin, or other organism of the phylum Echinodermata. Spiny skins, 5-ray symmetry, and internal water vascular systems are characteristics of the phylum.

echo sounding Determining the depth of water by measuring the time interval between emission of a sonic or ultrasonic signal and the return of its echo from the bottom.

eddy Circular movement of water usually formed where currents pass obstructions, between two adjacent currents flowing counter to each other or along the edge of a permanent current.

Ekman spiral Theory that a wind blowing steadily over an ocean of unlimited depth and extent and of uniform viscosity causes the surface layer to drift at an angle of 45° to the right of the wind direction in the Northern Hemisphere. Water at successive depths drifts in directions more to the right until at some depth it moves in the direction opposite to the wind. Velocity decreases with depth throughout the spiral. The depth at which this reversal occurs is about 100 m (328 ft). Net water transport is 90° to the right of the direction of the wind in the Northern Hemisphere.

Ekman transport Net water transport in an Ekman spiral condition 90° to the right of the wind direction in the Northern Hemisphere.

elasmobranch A general term for sharks, skates, rays, and sawfishes.

electrical conductivity Facility with which a substance conducts electricity; it is an intrinsic property of seawater and varies with temperature, salinity, and pressure.

El Niño Warm current flowing south along the coast of Ecuador. It generally develops just after Christmas concurrently with a southerly shift in the tropical rain belt. In exceptional years, plankton and fish are killed in the coastal waters.

embayment Indentation in a shoreline forming an open bay.

environment Sum total of all the external conditions that affect an organism, community, material, or energy.

epicenter Point on the earth's surface directly above the focus of an earthquake.

epipelagic zone Portion of the oceanic province extending from the surface to a depth of about 200 m (656 ft).

Equatorial Convergence Zone along which waters from the Northern and Southern hemispheres converge. This zone generally lies in the Northern Hemisphere except in the Indian Ocean.

equilibrium theory Hypothesis that an ideal earth has no continental barriers and is uniformly covered with water of considerable depth. It also assumes that the water responds instantly to the tide-producing forces of the moon and sun to form a surface in equilibrium and moves around the earth without viscosity or friction.

estuary A bay whose salinity is affected by river runoff.

euphotic zone Layer of a body of water that receives ample sunlight for the photosynthetic processes of the plants. Depth of this layer varies with the angle of incidence of the sunlight, length of day, and cloudiness, but it is usually 80 m or more. Ninety-nine percent of sunlight is absorbed.

eutrophic Pertaining to bodies of water containing abundant nutrient matter.

evaporation Physical process by which a liquid or solid is transformed to a gas; the opposite of condensation.

expendable bathythermograph (XBT) Electronic device used to chart a temperature profile of the top 1,500 feet of the ocean, after which the weight falls to the sea floor; low cost and used only once.

Fahrenheit temperature scale Temperature scale with the freezing point of water at 32° and the boiling point at 212° at standard atmospheric pressure.

family In biological classification, a group of closely related genera. Major subdivision of a biological order.

fan Gently sloping, fan-shaped feature normally located near the lower end of a canyon.

fast ice Sea ice that generally remains in the position where originally formed; may attain considerable thickness. It is formed along coasts where it is attached to the shore or over shoals.

fathom Common unit of depth in the ocean for countries using the English system of units, equal to 6 feet (1.83 m). It is also sometimes used in expressing horizontal distances, in which case 120 fathoms make one cable, or very nearly 0.1 nautical mile.

fault Fracture or fracture zone in rock along which one side has been displaced relative to the other side. The intersection of the fault surface with a surface, such as the sea bottom, is called a fault line. If a fault is not a single, clean fracture but a wide zone with small interlacing faults and filled with breccia, it is called a fault zone.

fault block A rock body bounded on at least two opposite sides by faults. It is usually longer than it is wide; when it is depressed relative to the adjacent regions, it is called a graben or rift valley; when it is elevated, it is called a horst.

faulting Earth movement that produces relative displacement of adjacent rock masses along a fracture.

fauna Animal population of a particular location, region, or period.

fetch 1. (Also called generating area). Area of the sea surface over which seas are generated by a wind having a constant direction and speed. 2. Length of the fetch area, measured in the direction of the wind in which the seas are generated.

filter feeder An animal that uses a sophisticated filter system to strain living and nonliving organic particles from the water. A suspension feeder that uses a filter. Example: clam.

fjord (also spelled fiord). Narrow, deep, steep-walled inlet of the sea formed either by the submergence of a mountainous coast or by an entrance of the sea into a deeply excavated glacial trough after the glacier melts away. A fjord may be several hundred meters deep and often has a relatively shallow entrance sill of rock or gravel.

flagellum (-a) A whiplike thread attached to some cells and used by the cell to move itself or to move water currents.

flood current Tidal current associated with the increase in the height of a tide.

Florida Current Fast current with speeds of 2 to 5 knots that sets through the Straits of Florida to a point north of Grand Bahama Island where it joins the Antilles Current to form the Gulf Stream. The Florida Current is traced to the Yucatán Channel, where the greater part of the water flowing through that channel turns clockwise into the Straits of Florida.

flushing time *See* residence time.

fog A visible aggregate of minute water droplets suspended in the atmosphere near the earth's surface that reduces visibility to below 1 km.

food chain A sequence of organisms that are associated with each other by a simple feeding relationship in nature. In a food chain, each organism eats others of only one species, and it, in turn, is eaten by organisms of only one other species in the sequence.

food web A group of organisms associated with each other by their feeding relationships. In a food web, each organism may eat others of many other species, and it can be eaten by organisms of many other species.

foraminifer (foraminiferan, foram). One of an order (Foraminifera) of benthic and planktonic protozoa possessing variously formed shells of calcium carbonate, silica, chitin, or an agglomerate of materials.

foreshore Zone that lies between the ordinary high- and low-water marks and is daily traversed by the oscillating water line as the tides rise and fall.

fracture zone Extensive linear zone of unusually irregular topography of the seafloor characterized by large seamounts, steep-sided or asymmetrical ridges, troughs, or scarps.

freezing point Temperature at which a liquid solidifies under any given set of conditions. Pure water under atmospheric pressure freezes at 0°C (32°F). The freezing point of water decreases with increasing salinity.

frequency (wave). Number of waves passing a specified point in a given unit of time.

fringing reef A reef attached directly to the shore of an island or continental land mass. Its outer margin is submerged and often consists of algal limestone, coral rock, and living coral.

gabbro Igneous rock with such minerals as calcium plagioclase (feldspar), pyroxene, and olivine; sometimes used to refer to any dark igneous rock with crystals visible to the naked eye.

gal Unit of acceleration equal to 1 cm/s^2. The term was invented to honor the memory of Galileo.

Galofalo Whirlpool in the Strait of Messina. Formerly called Charybdis.

gastropod A snail or sea slug. A member of the class Gastropoda.

genus (genera) In biological classification, a group of closely related species. Subdivision of a biological family.

geological oceanography Study of the floors and margins of the oceans, including description of submarine relief features, chemical and physical composition of bottom materials, interaction of sediments and rocks with air and seawater, and action of various forms of wave energy in the submarine crust of the earth.

geomagnetism (or terrestrial magnetism). 1. Magnetic phenomenon exhibited by the earth and its atmosphere. 2. Study of the magnetic field of the earth.

geophysics The physics or nature of the earth. It deals with the composition and physical phenomena of the earth and its liquid and gaseous envelopes; it embraces the study of terrestrial magnetism, atmospheric electricity, and gravity, and includes seismology, volcanology, oceanology, meteorology, and related sciences.

geostrophic flow Horizontal flow resulting when there is a balance between horizontal pressure forces and the Coriolis effect.

geosyncline A large, generally linear subsident trough in which many thousands of feet of sediments are accumulating or have accumulated. Deep oceanic trenches paralleling island arcs are considered to be developing geosynclines.

giant kelp One of a genus *(Macrocystis)* of large, vinelike brown algae, which grow attached to the sea bottom by a massive holdfast and reach lengths to 50 m. Members of this genus are the largest algae in existence.

glacial epoch The Pleistocene epoch, the earlier of the two divisions of geologic time included in the Quaternary period; characterized by continental glaciers that covered extensive land areas now free from ice.

glacial valley A U-shaped valley excavated by a glacier either on land or on the sea bottom.

glaciation Alteration of the earth's solid surface by erosion and deposition due to glacier ice.

glacier Moving mass of ice originating from the compacting of snow by pressure.

graben *See* fault block.

graded bedding A type of stratification in which each stratum displays a downward gradation in grain size from fine to coarse.

gradient flow Current defined by assuming that the horizontal pressure gradient in the sea is balanced by the sum of the Coriolis effect and frictional forces.

gram A metric unit of mass; originally defined as the mass of 1 cm^3 of water at 4°C; now one-thousandth of the standard kilogram.

granite Crystalline rock consisting essentially of alkali feldspar and quartz. *Granitic* is a textural term applied to coarse and medium-grained granular igneous rocks.

gravimeter Weighing instrument of sufficient sensitivity to register variations in the weight of a constant mass when the mass is moved from place to place on the earth and thereby subjected to the influence of gravity at those places.

gravitation In general, the mutual attraction between masses of matter (bodies). Gravitation is the component of gravity that acts toward the earth.

gravity anomaly Measured variation in the force of gravity from a calculated value based on the assumption of an even distribution of mass throughout the earth.

gravity wave Wave whose velocity of propagation is controlled primarily by gravity. Water waves more than 5 cm long are considered gravity waves.

green alga One of a division (Chlorophyta) of grass-green, single-celled filamentous, membranous, or branching plants in which the color, imparted by chlorophylls *a* and *b*, is not masked by accessory pigments.

greenhouse effect In the ocean where a layer of low-salinity water overlies a layer of denser water, the short-wavelength radiation of the sun is absorbed in the deeper layers. The radiation given off by the water is in the far infrared region, and since this cannot radiate through the low-salinity layer, a temperature rise results in the deeper layers. In the atmosphere the same effect is produced by a layer of clouds, and the long-wave radiation is trapped between the clouds and the earth.

groin Low, artificial wall of durable material extending from the land to seaward to protect the coast or to force a current to scour a channel.

gross primary productivity The amount of new carbohydrate material produced each day (or longer time interval) by the plants of an ecosystem. Some of this is consumed by their own metabolic processes; the excess material is the net primary productivity.

group velocity Velocity of a wave disturbance as a whole, that is, of an entire group of component simple harmonic waves. For water-surface waves, the group velocity of deep-water waves is equal to one-half the velocity of individual waves in the group; for shallow-water waves, it is the same as their velocity.

Gulf Stream Warm, well-defined, swift, and relatively narrow ocean current that originates north of Grand Bahama Island where the Florida Current and the Antilles Current meet. The Gulf Stream extends to the Grand Banks at about 40°N, 50°W, where it meets the cold Labrador Current, and the two flow eastward as the North Atlantic Current. The Florida Current, Gulf Stream, and North Atlantic Current together form the Gulf Stream system. Sometimes the entire system is referred to as the Gulf Stream.

guyot (tablemount). Seamount having a relatively smooth, flat top.

gyre Closed circulatory system larger than a whirlpool or eddy.

habitat Place with a particular kind of environment inhabited by organisms.

hadal Refers to organisms or phenomena at depths greater than 6,000 m, living on the sea bottom.

hadalpelagic Refers to organisms or phenomena at depths greater than 6,000 m, living in mid-water.

halocline Steep gradient of salinity, creating a boundary between water levels.

harbor Area of water affording natural or artificial protection for ships.

headland (head, naze, ness, promontory). High, steep-faced promontory extending into the sea. Usually called head when coupled with a specific name.

heat Form of energy transferred between systems through a difference in temperature and existing only in the process of energy transformation. By the first law of thermodynamics, the heat absorbed by a system may be used by the system to do work or to raise its internal energy.

heat budget Accounting for the total amount of the sun's heat received on the earth during any one year as being exactly equal to the total amount lost from the earth by reflection and radiation into space. The portion reflected by the atmosphere does not affect the earth's heat budget. The portion absorbed must balance the long-range radiation into space from the earth's entire system. The portion absorbed into the oceans causes the surface warming critical to the phenomenon of layer depth. Transport by currents further distributes the heat.

heat capacity Ratio of the heat absorbed (or released) by a system to the corresponding temperature rise (or fall).

heat conduction Transfer of heat from one part of a body to another or from one body to another in physical contact with it without displacing the particles of the body.

heat of vaporization The quantity of energy, usually given as heat, required to evaporate a unit mass of a liquid, at constant pressure and temperature; for pure water, the commonly expressed value is 540 calories per gram.

heat probes Instruments that can penetrate sediment to measure the gradient of heat emanating from the sediment.

heat transport Process by which heat is carried past a fixed point or across a fixed plane; thus, a warm current such as the Gulf Stream represents a poleward flux of heat.

heterotrophs Organisms that obtain organic and inorganic nutrients from other organisms, usually by eating the other organisms.

high-energy environment Region with considerable wave and current action that prevents the settling and accumulation of fine-grained sediment smaller than sand size.

higher high water (HHW). Higher of two high waters occurring during a tidal day where the tide exhibits mixed characteristics.

high water (HW; also called high tide). Highest limit of the surface-water level reached by the rising tide. High water is caused by the astronomical tide-producing forces and meteorological conditions.

hook Spit or narrow cape of sand or gravel whose outer end bends sharply landward.

horse latitudes Belts of latitude over the oceans at approximately 30° to 35°N and S where winds are predominantly calm or very light and weather is hot and dry. In the North Atlantic Ocean, these are the latitudes of the Sargasso Sea.

hurricane Severe tropical cyclone in the North Atlantic Ocean, Caribbean Sea, Gulf of Mexico, or the eastern North Pacific off the western coast of Mexico.

hydrocast Survey station measuring temperature and salinity at standard depths in order to calculate densities and map ocean water masses or currents.

hydrogen bond Weak bond between an ion or molecule containing hydrogen and some other ion or molecule; created by slight polarization of the hydrogen atom, which leaves positive edge outward.

hydrography Science dealing with the measurement and description of the physical features of the oceans, seas, lakes, rivers, and their adjoining coastal areas, with particular reference to their use for navigational purposes.

hydrology Scientific study of the waters of the earth, especially the effects of precipitation and evaporation on the water in streams, lakes, or below the land surface.

hydrosphere All the waters of the earth found in oceans, lakes, rivers, and underground; sometimes including water in the atmosphere.

hypothermia State of reduced body core temperature and metabolic processes.

hypsographic chart Chart or part of a chart showing land or submarine bottom relief in terms of height above datum. Hypsography is the science of measuring or describing elevations above datum, usually sea level.

iceberg Large mass of detached land ice floating in the sea or stranded in shallow water. Irregular icebergs generally calve from glaciers, whereas tabular icebergs and ice islands are usually formed from shelf ice. An iceberg is usually defined as being the size of a ship or larger, although any piece of glacier ice more than 5 m high is often called an iceberg.

ice cap Perennial cover of ice and snow over an extensive portion of the earth's land surface. The most important of the existing ice caps are those on Antarctica and Greenland (the latter often called inland ice).

ice floe A single piece of sea ice other than fast ice that may be large or small and is described as "light" or "heavy"according to thickness, if possible. Designations are: vast, over 10 km across; big, 1 to 10 km across; medium, 200 to 1,000 m across; small, 10 to 200 m across.

ice island Large tabular fragment of shelf ice found in the Arctic Ocean. Most such fragments appear to have calved from the Ward Hunt ice shelf off the northern coast of Ellesmere Island. Ice islands are smaller than the largest tabular icebergs of the Antarctic, the largest one known being about 700 km^2 (300 mi^2) in area and about 50 m (164 ft) thick.

igneous rock Rock formed by solidification of molten material or magma.

index of refraction Ratio of the velocity of light in a vacuum to the velocity of light moving through a substance such as water or a crystal.

Indian Ocean Ocean bounded on the north by the southern limits of the Arabian Sea, Laccadive Sea, Bay of Bengal, the limits of the East Indian archipelago and the Great Australian Bight; on the east from the meridian of Southeast Cape, Tasmania; and on the west from Cape Agulhas southward. The limits of the Indian Ocean exclude the seas lying within it.

infrared radiation (abbreviated IR). Electromagnetic radiation lying in the wavelength interval from about 0.8 μm to an indefinite upper boundary, sometimes arbitrarily set at 1,000 μm (0.01 cm). The infrared radiation spectrum is bounded by visible radiation on the lower end and by microwave radiation on the upper end.

inlet A short, narrow waterway connecting a bay or lagoon with the sea.

inshore In beach terminology, the zone of variable width between the shoreface and the seaward limit of the breaker zone.

in situ In place; in the natural or original position.

interface Surface separating two media across which there is a discontinuity of some property, such as density or velocity.

internal wave Wave that occurs within a fluid whose density changes with depth, either abruptly at an interface or gradually. Its amplitude is greatest at the density discontinuity or, in the case of a gradual density change, somewhere inside the fluid. A relatively small amount of energy is required to set up and maintain an internal wave. Wave heights, periods, and lengths are usually large compared with those of surface waves.

intertidal zone (littoral zone). Generally considered to be the zone between mean high-water and mean low-water levels.

Intertropical Convergence (abbreviated ITC). Shallow trough of low pressure where the north and south tradewinds meet. Generally situated at or near the equator lying in the belt of the doldrums over the oceans.

inverse estuary An estuary in which evaporation exceeds land drainage plus precipitation, with a resulting mixture of high salinity estuarine water and sea water.

invertebrate Any animal without a backbone or spinal column.

ion An electrically charged negative or positive atom or group of atoms. The dissolved salts in seawater are ions.

island Body of land surrounded by water; relatively smaller than a continent.

island arc Group of islands usually having a curving, archlike pattern, generally convex toward the open ocean, with a deep trench or trough on the convex side and usually enclosing a deep-sea basin on the concave side.

isobar Line on a chart (or weather map) connecting all points of equal or constant pressure.

isostasy Theoretical balance of all large portions of the earth's crust as though they were floating on a denser underlying layer.

isostatic balance *See* isostasy.

jellyfish (medusa). Any of various free-swimming cnidarians having a disc- or bell-shaped body of jellylike consistency. Many have long tentacles with stinging cells. Some are luminescent.

jetty A structure, such as a wharf, pier, or breakwater, located to influence current or protect the entrance to a harbor or river. *See also* groin.

juvenile water Water that enters for the first time into the hydrologic cycle. It is released from igneous rocks through volcanic activity at a rate probably not exceeding 0.1 km^3 per year.

kelp One of an order (Laminariales) of usually large, blade-shaped, or vinelike brown algae. *See* giant kelp.

Kelvin temperature scale An absolute-temperature scale independent of the thermometric properties of the working substance. For convenience, the Kelvin degree is identified with the Celsius degree (0° K = −273.16°C).

key (cay). A low, flat island or mound of sand built up on a reef flat slightly above high tide; the sand may be mixed with coral or shell fragments.

killer whale Largest member (*Orcinus orca*) of the dolphin family (Delphinidae), having worldwide distribution. Although this animal has been implicated in several attacks on boats containing people and in bumping sea ice bearing people, no documented fatality is known.

kilometer (abbreviated km). Unit of distance measurement in the metric system equal to 0.62 statute miles or 0.54 nautical miles. A statute mile equals 1.61 km; a nautical mile equals 1.85 km.

kinetic energy Energy that a body possesses as a consequence of its motion, defined as one-half the product of its mass and the square of its speed, $1/2mv^2$.

kingdom In biological classification, a broad division of organisms according to fundamental body type.

knot Speed unit of one nautical mile (6,076.12 ft, 1,852 m) per hour. It is equivalent to a speed of 1,688 ft/s or 51.4 cm/s.

krill Term used by whalers and fishers for euphausiids; a type of crustacean.

Kuroshio Current (Japan Current). Fast ocean current (2 to 4 knots) flowing northeastward from Taiwan to Ryūkyū Islands and close to the coast of Japan to about 150°E. The Kuroshio originates from the greater part of the North Equatorial Current, which divides east of the Philippines. Beyond 150°E it widens to form the slower-moving North Pacific Current.

Kuroshio Extension General term for the warm, eastward-transitional flow that connects the Kuroshio and the North Pacific Current.

Kuroshio system System of ocean currents that includes part of the North Equatorial Current, the Tsushima Current, the Kuroshio Current, and the Kuroshio Extension.

Labrador Current Current that flows southward from Baffin Bay, through the Davis Strait, and southeastward along the Labrador and Newfoundland coasts. East of the Grand Banks of Newfoundland, the Labrador Current meets the Gulf Stream, and the two flow eastward as the North Atlantic Current.

lagoon Shallow sound, pond, or lake generally separated from the open sea.

landlocked Pertaining to a body of water enclosed or nearly enclosed by land, thus protected from the sea. San Francisco Bay is a classic example.

larva Embryo that becomes self-sustaining and independent before it has assumed the characteristic features of its parents.

latent heat of evaporation That amount of heat required to change 1 g of water into water vapor without a change in temperature. For example, 536 cal are required to change 1 g of water to water vapor at 100°C at standard atmospheric pressure.

lee Shelter, or the part or side of something (like a ship or island) that is sheltered or turned away from the wind or waves.

leeward Pertaining to the direction toward which the wind is blowing or the direction toward which waves are traveling.

limestone General term for a class of rocks that are at least 80% carbonates of calcium or magnesium. Varieties of limestone take their names from the source material, for example, algal limestone.

limnology Physics and chemistry of freshwater bodies and the classification, biology, and ecology of the organisms living in them.

lithogenous sediment Sediment derived from erosion of other rocks.

lithology Study and description of rocks based on macroscopic and microscopic examination of samples.

lithosphere Solid part of the earth; sometimes used synonymously with the earth's crust.

littoral zone (intertidal zone). Benthic zone between high- and low-water marks; according to some authorities, between the shore and water depths of approximately 200 m.

load Quantity of sediment transported by a current. It includes the suspended load of small particles floating through the whole body of the current and the traction load, bottom load, or bed load of large particles that move along the bottom by traction, that is, saltation, rolling, and sliding.

longshore bar Bar generally parallel to the shore and submerged at high tide.

longshore current Current running roughly parallel to the shoreline and produced by waves being deflected at an angle by the shore. The amount of material a longshore current carries depends on its velocity and particle size of the material; however, any obstruction cutting across the path of the current will cause loss of velocity and consequently loss of carrying power.

longshore trough A long, wide, shallow depression on the seafloor parallel to the shore.

low-energy environment Region characterized by a general lack of wave or current motion, permitting the settling and accumulation of very fine grained sediment (silt and clay).

lower low water (LLW). Lower of two low waters of any tidal day where the tide exhibits mixed characteristics. *See* mixed tide.

low water (LW; also called low tide). Lowest of the surface water level reached by the lowering tide. Low water is caused by the astronomical tide-producing forces and meteorological conditions.

lunar day (tidal day). Interval between two successive upper transits of the moon over a local meridian. The period of the mean lunar day, approximately 24.84 solar hours, is derived from the rotation of the earth on its axis relative to the movement of the moon about the earth.

maelstrom A confused and often destructive current usually caused by the combined effects of high, wind-generated waves and a strong opposing tidal current; the rapid flows may follow eddying patterns or circular paths with whirlpool characteristics. Named after the frequently cited phenomenon along the south shore of Moskenesoy Island in the Lofoten Islands off the Norway coast; here, the maelstrom reaches its strength when the tidal current ebbs westward with speeds up to 9 knots at spring tides during a strong opposing westerly wind. Similar phenomena occur in Pentland Firth, Scotland, and off Cape de la Hague, Normandy.

magma Mobile rock material generated within the earth from which igneous rock is derived by solidification. When extruded it is called lava.

magnetic anomaly Distortion of the regular pattern of the earth's magnetic field due to local concentrations of ferromagnetic minerals.

magnetic field Region in which magnetic forces will cause alignment of magnetized substances or alterations in an electric current.

magnetometer Instrument for measuring the intensity and direction of the earth's magnetic field.

manatee Any of the three species of sea cow that constitute the genus *Trichechus*. All three are confined to shallow tropical marine waters, estuaries, and rivers on both sides of the Atlantic Ocean. The tail is broad and rounded, not whalelike as in the dugong.

manganese nodules Pebble-like rocks on the seafloor; size commonly 1 to 10 cm; contains mainly manganese oxide and iron oxide with other minor components of nickel, cobalt, copper, and other elements.

mangrove One of several genera of tropical trees or shrubs that produce many prop roots and grow along low-lying coasts into shallow water.

mantle Region between the crust and core of the earth.

marginal coasts Those coasts that occur on the protected side of island arcs (that is, away from the collision side) in the movement of crustal plates.

marginal sea Semienclosed sea adjacent to a continent usually underlain by submerged continental rocks, such as the Black Sea.

mariculture Farming of marine organisms in seawater under controlled conditions for commercial purposes.

marsh Area of soft, wet land.

mass transport Transfer of water from one region to another due to the orbital motion of waves.

maximum sustainable yield (MSY). Number or amount of species that can be harvested from a natural population each year without resulting in a subsequent long-term decline in the population; the remaining individuals are capable of replacing the harvested ones by natural reproduction.

mean current Current speed and direction for a specified area; determined by the average of a total number of observations.

mean high water (abbreviated MHW). Average height of all the high waters recorded over a 19-year period or a computed equivalent period.

mean lower low water (abbreviated MLLW). Average height of all the lower low waters (tides) over a 19-year period. It is usually associated with a tide exhibiting mixed characteristics.

mean low water (abbreviated MLW). Average height of all the low waters recorded over a 19-year period or a computed equivalent period.

mean sea level (abbreviated MSL). Mean surface water level determined by averaging heights at all stages of the tide over a 19-year period. Mean sea level is usually determined from hourly height readings measured from a fixed, predetermined reference level.

mean water level (abbreviated MWL). Mean surface level determined by averaging the height of the water at equal intervals of time, usually at hourly intervals, over a considerable period of time.

mediterranean Large body of salt water or inland sea surrounded by land, which may have one or more narrow openings to the ocean or another sea.

mesopelagic Refers to organisms or phenomena at depths between 200 and 1,000 m, in mid-water.

messenger Cylindrical metal weight approximately 7.5 cm long and 2.5 cm in diameter that is attached to an oceanographic wire and sent down to trip devices such as Nansen bottles and current meters after they have been lowered to the desired depth.

metamorphic rock Rock that has undergone structural and mineralogical changes, such as recrystallization, in response to marked changes of temperature, pressure, or chemical environment.

meter Basic unit of length of the metric system, equal to 1,650,763.73 wavelengths of ^{86}Kr orange-red radiation.

metric system (system international, or SI, as it is officially known). System of physical measurements in which the fundamental units of length, mass, and time are the centimeter, gram, and second, respectively. Also called the cgs system or mks system, for meter, kilogram, and second.

microflagellate A tiny, flagellated, single-celled phytoplankton organism. Microflagellates, although difficult to collect on account of their small size, can be more numerous and productive than the larger diatoms or dinoflagellates of the phytoplankton. They are members of many plant divisions; coccolithophores (division Chrysophyta, in the classification used in this text) are the most familiar.

microlayering or microlayer The thin zone(s) beneath the ocean surface within which physical processes are modified by proximity to the air-sea boundary.

mid-oceanic ridge Great median arch or sea-bottom rise extending the length of an ocean basin and roughly paralleling the continental margins.

mid-oceanic rift Deep, narrow-notched cleft, valley, or graben, which is reportedly found almost continuously along the crest of a ridge.

mixed layer Layer of water that is mixed through wave action or thermohaline convection.

mixed tide Type of tide in which a diurnal wave produces large inequalities in heights or durations of successive high and low waters. This term applies to the tides between the predominantly semidiurnal and the predominantly diurnal.

mollusc (mollusk). A snail, bivalve, chiton, octopus, squid, or other organism of the phylum Mollusca. A characteristic gill, a shell, and a radular gnawing apparatus are among the key features of most animals of this group.

monsoon Name for seasonal winds (from Arabic *mawsim*, a season). It was first applied to the winds in the Arabian Sea, which blow for 6 months from the northeast and for 6 months from the southwest, but it has been extended to similar winds in other parts of the world.

moraine Rock debris deposited chiefly by glaciers. Where glaciers float on or discharge into the sea or where glaciated regions are drowned by the sea, moraines form marine deposits.

mudflats Shore where wave action is very mild and particles are clay-sized.

mutualism Symbiotic relationship between two species in which both are benefited but neither requires the interaction.

nannoplankton Planktonic plant cells that are between 2 and 20 μm in length.

Nansen bottle Device used by oceanologists to obtain subsurface samples of seawater.

nauplius The earliest stage in the larval development of many crustaceans.

nautical mile In general, a unit used in marine navigation equal to a minute of arc of a great circle on a sphere. One international nautical mile equals 1,852 m.

neap tide Tide of decreased range that occurs about every 2 weeks when the moon is in quadrature.

nearshore current system Current system caused by wave action in and near the surf zone. The nearshore current system consists of four parts: the shoreward mass transport of water, longshore currents, rip currents, and longshore movement of expanding heads of rip currents. Sometimes called inshore currents.

nearshore zone Zone extending seaward from the shore to an indefinite distance beyond the surf zone.

negative estuary *See* inverse estuary.

negatively charged ion Atom or group of atoms with a negative charge; existing usually in solution.

nekton Pelagic animals that are active swimmers, such as most of the adult squids, fishes, and marine mammals.

neritic division Portion of the pelagic province extending from low-water level to the approximate edge of a continental shelf. (Some writers have used this term to describe bottom organisms of a continental shelf.)

net primary productivity The amount of new carbohydrate produced by plants, per unit time, in excess of their own metabolic needs. *See* gross primary productivity.

neuston The community of organisms that occur right at the sea surface; usually floating or resting on the surface.

neutral Any water solution when it is neither acidic nor basic, that is, its pH is 7 (such as pure water).

neutral estuary An estuary in which neither fresh water inflow or evaporation dominates.

nitrogen cycle The cyclic sequence of changes that nitrogen undergoes when it is converted from atmospheric form to forms useful to living organisms, then back to atmospheric form.

nodal line In a tide area, the line about which the tide oscillates and where there is little or no rise and fall of the tide.

node Part of a standing wave or clapotis where the vertical motion is least and the horizontal velocities are greatest. Nodes are associated with clapotis and with seiche action resulting from resonant wave reflections or oscillations in a harbor or bay.

North Atlantic Current Wide, slow-moving continuation of the Gulf Stream originating in the region east of the Grand Banks of Newfoundland at about 40°N and 50°W. The North Atlantic Current is often masked by shallow and variable wind-driven surface movements.

North Equatorial Current Ocean currents driven by the northeast trade winds blowing over the tropical oceans of the Northern Hemisphere. In the Atlantic Ocean it is known as the Atlantic North Equatorial Current and flows west between the Atlantic Equatorial Countercurrent and the Sargasso Sea. In the Pacific Ocean it is known as the Pacific North Equatorial Current and flows westward between 10°N and 20°N. East of the Philippines it divides, part turning south to join the Equatorial Countercurrent and part going north to form the Kuroshio Current. In the north Indian Ocean there is no equatorial current; monsoon drifts dominate.

North Pacific Current Warm branch of the Kuroshio Extension flowing eastward across the Pacific Ocean.

nutrient In the ocean, any one of a number of inorganic or organic compounds or ions used primarily in the nutrition of primary producers. Nitrogen and phosphorus compounds are essential nutrients. Silicates are essential to diatoms. Vitamins such as B_{12} are essential to many algae.

ocean The vast body of salt water occupying the depressions of the earth's surface, or one of its major primary subdivisions, bounded by continents, the equator, and other imaginary lines.

ocean basin Part of the ocean floor lying, in general, at a depth of more than 2,000 m.

ocean current Regular movement of ocean water, either cyclic or more commonly in a continuous stream flowing along a definable path. Three general classes, by cause, may be distinguished: (1) currents related to seawater density gradients, comprising the various types of gradient currents; (2) wind-driven currents, which are those directly produced by the stress exerted by the wind on the surface; and (3) currents produced by long-wave motions, principally the tidal currents, but including internal wave, tsunami, and seiche currents. The major ocean currents are continuous flowing streams and are of first-order importance in maintaining the earth's thermodynamic balance.

oceanic Pertaining to the portion of the pelagic province seaward from the edge of a continental shelf.

oceanic crust Mass of simatic material approximately 5–7 km thick that lies under the ocean bottom and may be more or less continuous beneath the continental crust.

oceanic division Marine environment not in close proximity to continental land masses.

oceanic ridges Elongate ridges in the ocean that are seismically active; divergent boundaries between global plates; plate separation zone.

oceanology Study of the sea embracing and integrating all knowledge pertaining to the sea's physical boundaries, the chemistry and physics of seawater, and marine biology. In strict usage, oceanography is the description of the marine environment, whereas oceanology is the study of the oceans and related sciences.

ocean slick Area of the sea surface, variable in size and markedly different in color and oiliness from surrounding water; may be caused by internal waves.

offshore Comparatively flat zone of variable width that extends from the breaker line to the edge of the continental shelf.

oligotrophic Refers to water that contains low levels of nutrients, little plant growth.

ooze Fine-grained pelagic sediment containing undissolved sand- or silt-sized, calcareous or siliceous skeletal remains of small marine organisms in proportion of 30% or more, the remainder being amorphous clay-sized material. There are diatomaceous, foraminiferal, pteropod, and radiolarian oozes.

orbit In water waves, the path of a water particle affected by wave motion. In deep-water waves the orbit is nearly circular, and in shallow-water waves the orbit is nearly elliptical. In general, the orbits are slightly open in the direction of wave motion, giving rise to mass transport.

order In biological classification, a group of closely related families. Major subdivision of a biological class.

organic reef Sedimentary rock aggregate composed of living and dead colonial organisms such as algae, coral, crinoids, and bryozoa. When it is covered by more than 12 m of water, it is an organic bank.

osmosis The spontaneous seepage of water through a membrane, from the side that is fresher to the side that is saltier.

oxygen minimum depth Subsurface layer in which the dissolved oxygen content is very low or nil.

Oyashio Current Cold current flowing from the Bering Sea southwest along the coast of Kamchatka, past the Kuril Islands, continuing close to the northeast coast of Japan and reaching nearly 35°N. The Oyashio turns east into the Kuroshio Extension.

Pacific Ocean Ocean area bounded on the east by the western limits of the coastal waters of southwestern Alaska and British Columbia, the southern limits of the Gulf of California, and from the Atlantic Ocean by the meridian of Cape Horn; on the north by the southern limits of Bering Strait and the Gulf of Alaska; on the west by the eastern limits of the Sea of Okhotsk, Sea of Japan, Philippine Sea, the East Indian archipelago from Luzon Island to New Guinea, Bismark Sea, Solomon Sea, Coral Sea, Tasman Sea, and from the Indian Ocean by the meridian of Southeast Cape, Tasmania.

pack ice Term used to denote any area of sea ice other than fast ice, no matter what form it takes or how disposed.

pancake ice Pieces of newly formed sea ice that are usually approximately circular, from 30 cm (12 in) to 3 m (10 ft) in diameter.

parasitism Symbiotic relationship between two species in which one lives on or in the body of its host and obtains food from its tissues, often benefiting at the expense of its host.

passive margin Continental margin that is not seismically active; usually located far from the edge of a global plate.

pelagic Occurring in the water, away from the bottom.

pelagic province Primary division of the sea that includes the whole mass of water. The division is made up of the neritic division (water shallower than 200 m) and the oceanic division (water deeper than 200 m).

pelagic sediments Sediments deposited on the seafloor that have little or no coarse-grained material derived from land.

peridotite Rock composed mainly of olivine and other iron and manganese-containing minerals; main probable constituent of the upper mantle (asthenosphere) of the earth.

perigee Point in the moon's orbit (or any other earth satellite) nearest to the earth.

period (wave). Time interval (in seconds for wind waves) for a wave to pass a point; usually measured from crest to successive crest.

per mille (symbol ‰). Per thousand or 10^{-3}; used in the same way as percent (%, per hundred or 10^{-2}). Per mille by weight is commonly used in oceanology as a measurement of salinity and chlorinity; for example, a salinity of 0.03452 (or 3.452%) is commonly stated as 34.52‰.

Peru Current (Humboldt Current). Cold ocean current flowing north along the coasts of Chile and Peru. The Peru Current originates from the West Wind Drift in the subantarctic Pacific Ocean. The northern limit of the current can be placed a little south of the equator, where the flow turns toward the west, joining the South Equatorial Current.

pH Measure of the hydrogen ion concentration in a solution. A pH of 7 is neutral; higher is basic; lower is acidic. pH 8 means 10^{-8} grams of hydrogen ions per liter of solution.

phosphorus cycle The cyclic sequence of changes that phosphorus experiences when it is converted from mineral phosphate form to organic forms in organisms, then back to mineral form.

photosynthesis Manufacture of carbohydrate food from carbon dioxide and water in the presence of chlorophyll by using light energy and releasing oxygen.

phylum In biological classification, a group of closely related classes. Major subdivision of a biological kingdom.

physical oceanography Study of the physical aspects of the ocean, such as its density, temperature, ability to transmit light and sound, and sea ice; the movements of the sea, such as tides, currents, and waves; and the variability of these factors both geographically and temporally in relationship to the adjoining domains, namely, the atmosphere and the ocean bottom.

physical properties Physical characteristics of seawater; for example, temperature, salinity, density, velocity, sound, electrical conductivity, and transparency.

phytoplankton Planktonic plants, usually single-celled. Example: diatoms.

pilot whale Any of several species of a genus (*Globicephala*) of large dolphins having worldwide distribution. These animals commonly travel in schools, and many strandings of them have been reported.

pinger Battery-powered acoustic device equipped with a transducer that transmits sound waves. When the pinger is attached to a wire and lowered into the water, the direct and bottom-reflected sound can be monitored with a listening device. The difference between the arrival time of the direct and reflected waves is used to compute the distance of the pinger from the ocean bottom.

pinniped Marine mammal of the order Pinnipedia, which comprises the seals, sea lions, and walruses.

plankton Passively drifting or weakly swimming organisms in marine and fresh waters. Members of this group range in size from microscopic plants to jellyfish measuring up to 2 m across the bell and includes the eggs and larval stages of the nekton and benthos. *See* phytoplankton, zooplankton.

plankton bloom Enormous concentration of plankton (usually phytoplankton) in an area, caused by either an explosive or a gradual multiplication of organisms (sometimes of a single species) and usually producing an obvious change in the physical appearance of the sea surface, such as discoloration. Blooms consisting of millions of cells per liter often have been reported. *See* red tide.

plankton net Net for collecting plankton. A great variety of plankton nets have been constructed in attempts to fulfill specific requirements. Typically, the nets are cone shaped, but several modifications of this shape as well as completely different shapes exist.

plate tectonics Theory that describes and explains the movement of large plates of the earth's crust; sometimes called continental drift.

plungers *See* wave.

polar ice cap *See* ice cap.

pollution Introduction by humans, directly or indirectly, of substances or energy into the marine environment (including estuaries) resulting in such deleterious effects as harm to living resources; hazards to human health; hindrance to marine activities, including fishing; impairment of quality for use of seawater; and reduction of amenities.

polyna (or clearing, or ice clearing). Small or thin temporary clearing in salt pack ice.

polyp Individual sessile cnidarian, such as a coral.

population Group of organisms of the same species inhabiting a geographic locality.

porpoise Small to moderate-sized, usually nonbeaked member of the cetacean suborder Odontoceti. *See* dolphin.

positive estuary An estuary in which there is a measurable dilution of sea water by land drainage

positively charged ion Atom or group of atoms possessing a positive charge; existing usually in solution.

predation Interaction in which a population of animals benefits by eating a prey population.

primary coast Coasts whose configuration is due to the action of nonmarine processes.

primary productivity Amount of organic matter synthesized by organisms from inorganic substances in unit time in a unit volume of water or in a column of water of unit area cross section and extending from the surface to the bottom.

prime meridian The meridian of longitude 0°, used as the origin for measurements of longitude. The meridian of Greenwich, England, is the internationally accepted prime meridian.

probe Measuring device or sensor inserted into the environment to be measured. As applied to oceanology, the term is used for devices lowered into the sea for in situ measurements.

productivity Amount of organic matter synthesized by organisms; measured in units such as grams per liter per year.

progressive wave Wave that is manifested by the forward movement of the wave form. *See* standing wave.

propagation Transmission of energy through a medium.

propagation loss Transmission loss associated with any given length of sound path in water.

propagation rate Velocity that waves travel.

pteropod A small swimming snail. Pteropod shells litter the sea bottom in some areas where they are abundant, forming sediments called pteropod oozes. Pronounced "tare'-o-pod."

pycnocline High rate of change of density (sharp gradient) with depth.

quadrature Position in the phase cycle when the two principal tide-producing bodies (moon and sun) are nearly at a right angle to the earth; the moon is in quadrature in its first quarter and last quarter.

race Very fast current flowing through a relatively narrow channel.

radiation Emission and propagation of energy through space or through a material medium in the form of waves; for instance, the emission and propagation of electromagnetic waves.

radiolarian One of an order (Radiolaria) of single-celled planktonic protozoa possessing a skeleton of siliceous spicules and radiating threadlike pseudopodia. Most members are pelagic, and many are luminescent.

ram Underwater ice projection present on many icebergs.

reach 1. An arm of the sea extending into the land. 2. A straight section of restricted waterway of considerable extent; may be similar to a narrows except much longer in extent.

red alga One of a division (Rhodophyta) of reddish, filamentous, membranous, encrusting, or complexly branched plants in which the color is imparted by the predominance of phycoerythrin over the chlorophylls and other pigments. Red algae are distributed worldwide, being more abundant in temperate waters and ranging to greater depths than other algae.

red tide Red or reddish-brown discoloration of surface waters, most frequent in coastal regions, caused by concentrations of microscopic organisms, particularly dinoflagellates. Toxins produced by the dinoflagellates can cause mass kills of fish and other marine animals. In some regions, notably off the western coast of Florida, red tide appears to follow increased rainwater runoff from the land; in this way, it is believed, one or more scarce nutrient elements flow into the sea, permitting dinoflagellates to multiply rapidly.

reef Offshore, consolidated-rock hazard to navigation with a minimum depth of 20 m or less. *See* shoal. (For many years, a depth of 12 m has been considered critical for navigational safety. Because of the increased drafts of modern ships, a depth of 20 m is now considered critical.)

reflection Return of a wave seaward when the wave impinges on a very steep beach, barrier, or other nearly vertical surface.

refraction Process by which the direction of a wave moving in shallow water at an angle to the contours is changed. That part of the wave advancing in shallower water moves more slowly than the part still advancing in deeper water, causing the wave crest to bend toward alignment with the underwater contours.

remote sensing Collection of earth/ocean data from satellites by electromagnetic devices, utilizing various segments of the electromagnetic spectrum such as infrared and microwave.

residence time Length of time a particle of an element remains in seawater.

respiration As used in productivity studies, the amount of food material needed by a plant or animal during a certain period of time. Its metabolic demand for food; often estimated by measuring the respiratory oxygen consumption of the organism.

ridge Long, narrow elevation of the seafloor with steep sides and irregular topography.

rift valley *See* fault block.

ring Large, independent current loop or eddy broken off of a strong, well-defined ocean current.

rip current Return flow of water piled up on shore by incoming waves and wind; a strong, narrow surface current flowing away from the shore.

ripple marks Undulating surface features of various shapes produced in unconsolidated sediments by wave or current action. Compound ripples are characterized by systematically offset crests and are produced by simultaneous interference of wave oscillation with current action. Metaripples are asymmetrical sand ripples. As size increases, ripples grade into sand waves, sand ridges, sand dunes, and migratory sandbanks or shoals.

rotary current Tidal current that flows continually with the direction of flow changing through all points of the compass during a tide cycle. Rotary currents usually occur offshore where the direction of flow is not restricted; unless modified by local conditions, the change in direction is generally clockwise in the Northern Hemisphere and counterclockwise in the Southern Hemisphere. The speed of the current usually varies throughout the tide cycle, passing through two maximums in approximately opposite directions and two minimums where the direction of the current is approximately 90° from the direction at time of maximum speed.

salinity Measure of the quantity of dissolved salts in seawater. It is formally defined as the total amount of dissolved solids in seawater in parts per thousand (‰) by weight when all the carbonate has been converted to oxide, the bromide and iodide to chloride, and all organic matter is completely oxidized. In practice, salinity is not determined directly but is computed from chlorinity, electrical conductivity, refractive index, or some other property whose relationship to salinity is well established. The equation presently used for determining salinity from chlorinity is $S = 1.80655\ Cl$.

salinometer One of a number of electrically operated meters that determine the salinity of seawater in situ by measuring its conductivity.

salt Any substance that yields ions other than hydrogen or hydroxyl ions. A salt is obtained by displacing the hydrogen of an acid by a metal.

salt marsh Flat, poorly drained coastal swamps that are flooded by most high tides.

saltwater wedge Intrusion in a tidal estuary of seawater in the form of a wedge characterized by a pronounced increase in salinity from surface to bottom.

sand Loose material that consists of grains ranging between 0.0625 and 2.000 mm in diameter.

Sargasso Sea Region of the North Atlantic Ocean to the east and south of the Gulf Stream system. This is a region of convergence of the surface waters and is characterized by clear, warm water, a deep-blue color, and large quantities of floating *Sargassum* or gulfweed.

scarp Elongated, comparatively steep slope of the seafloor separating flat or gently sloping areas.

scattering Random dispersal of sound energy after it is reflected from the sea surface or sea bottom or off the surface of solid, liquid, or gaseous particles suspended in the water.

sea Subdivision of an ocean. All seas except inland seas are physically interconnected parts of the earth's total saltwater system. Two types are distinguished, mediterranean and adjacent. Mediterraneans are seas separated from the major body of water. Adjacent seas are those connected individually to the larger body.

sea cow An aquatic, herbivorous mammal of the order Sirenia, which includes the dugong, the manatee, and the extinct Steller sea cow.

seafloor Bottom of the ocean where there is generally a smooth, gentle gradient. In many uses depth is disregarded, and the term may be used to designate areas in basins or plains or on the continental shelf.

seafloor spreading Idea that new ocean crust is forming at mid-oceanic ridges, moving toward continents and diving back into the earth's mantle at the edges of continents.

sea ice Ice formed by the freezing of seawater; opposed, principally, to land ice. In brief, it forms first as small crystals, thickens into sludge, and coagulates into sheet ice, pancake ice, or ice floes of various shapes and sizes. Thereafter, sea ice may develop into pack ice or become a form of pressure ice.

sea level Height of the surface of the sea at any time.

seamount Elevation rising 1,000 m or more from the seafloor and of limited extent across the summit.

sea smoke Fog formed when water vapor is added to air that is much colder than the vapor's source, most commonly when very cold air drifts across relatively warm water.

sea state Numerical or written description of ocean-surface roughness. For more precise usage, sea state may be defined as the average height of the highest one-third of the waves observed in a wave train, referred to a numerical code.

seawater Water of the seas, distinguished from fresh water by its appreciable salinity. Commonly, seawater is used as the antithesis of specific types of fresh water, as river water, lake water, or rainwater, whereas salt water is merely the antithesis of fresh water in general.

seaweed Any macroscopic marine alga.

secondary coasts Coasts shaped primarily by marine agencies or marine organisms.

second law of thermodynamics A fundamental property of energy; the second law says that energy cannot be transformed from one useful form to another without some loss as heat.

sediment Particles of organic and inorganic matter that accumulate in a loose, unconsolidated form.

sedimentary rocks Rocks formed by the accumulation of sediment in water or from the air. The sediment may consist of rock fragments, the remains or products of animals or plants, the products of chemical action or evaporation, or mixtures of these materials.

seiche Standing-wave oscillation of an enclosed or semienclosed water body that continues like a pendulum, after the cessation of the originating force, which may have been either seismic, atmospheric, or wave-induced.

seismic profile Data from a single series of observations made at one geographic location with a linear arrangement of seismometers.

seismic seawave Tsunami; ocean wave of long period generated by large earth movements and displacement of the sea floor.

seismic wave *See* tsunami.

semidiurnal tide Type of tide having two high waters and two low waters each tidal day, with small variations between successive high and successive low water heights and durations.

sessile Permanently attached; not free to move about.

shadow zone Region into which very little sound energy penetrates.

shallow water Commonly, water of such a depth that surface waves are noticeably affected by bottom topography. It is customary to consider water of depths less than half the surface wavelength as shallow water.

shallow-water wave Progressive gravity wave that is in water less than one-half the wavelength in depth.

shark Any of approximately 250 species of cartilage-skeletoned, fishlike vertebrates (order: Selachii) and including the large plankton-feeding basking sharks and predaceous sharks.

shoal Submerged ridge, bank, or bar consisting of or covered by unconsolidated sediments (mud, sand, gravel) that is near enough to the water surface to be a hazard to navigation. If composed of coral or rock, it is called a reef.

shoreface Narrow zone seaward from the low tide shoreline permanently covered by water over which beach sands and gravels actively oscillate with changing wave conditions.

shoreline Boundary line between a body of water and the land at high tide (usually mean high water). *See* coast.

sial (sialic rocks). *See* granite.

sill Lower part of the ridge or rise separating ocean basins from one another or from the adjacent seafloor.

silt Unconsolidated sediment whose particles are between clay and sand sizes.

sima (simatic rocks). *See* basalt.

sinking (downwelling). Downward movement of surface water generally caused by converging currents or when a water mass becomes more dense than the surrounding water. *See* upwelling.

sirenian A sea cow. An animal of the order Sirenia.

slack water Interval when the speed of the tidal current is very weak or zero; usually refers to the period of reversal between ebb and flood currents. In most places slack water occurs near times of high and low water; in other localities, slack water may occur midway between high and low water.

slick Smooth region of an otherwise rippled water surface; usually due to the presence of a thin layer of organic matter such as oil or the movement of a ship or other large mass through the water.

slough Shallow estuary in which large areas of the sandy or muddy bottom are exposed during low tides.

slump deposit Sediment produced when material slides down a slope en masse and comes to rest in a new position at the base of the slope.

SOFAR Acronym derived from the expression *SO*und *F*ixing *A*nd *R*anging.

solution State in which a substance, or solute, is homogeneously mixed with a liquid called the solvent. Thus pure water is a solvent and seawater is a solution of many substances.

sonar Acronym derived from the expression *so*und *na*vigation *r*anging. It refers to method or equipment for determining by underwater sound techniques the presence, location, or nature of objects in the sea.

sound Periodic variation in pressure, particle displacement, or particle velocity in an elastic medium. *See* sound velocity.

sounding Measurement of the depth of water beneath a ship.

sound velocity Rate at which sound energy moves through a medium. Velocity of sound in seawater is a function of temperature, salinity, and the changes in pressure associated with changes in depth. Increasing any of these factors tends to increase the velocity. Sound is propagated at a speed of 1,447 m at 0°C (4,742 ft/s, 32°F), one atmosphere pressure, and a salinity of 35‰.

South Atlantic Current Eastward-flowing current of the South Atlantic Ocean that is continuous with the northern edge of the West Wind Drift.

South Equatorial Current Any of several ocean currents driven by the southeast trade winds flowing over the tropical oceans of the Southern Hemisphere. In the Atlantic Ocean it is known as the Atlantic South Equatorial Current and flows westward with its axis through 2°N, 25°W. Part flows northwest along the northeast coast of South America (the Guianas) as the Guiana Current. The other part turns below Natal and flows south along the coast of Brazil as the Brazil Current. In the Pacific Ocean, the Pacific South Equatorial Current flows westward between approximately 3°N and 10°S. Much of it turns south in mid-ocean, forming a large anticyclonic whirl. The portion that continues across the ocean divides as it approaches Australia, part moving north toward New Guinea and part turning south along the east coast of Australia as the East Australia Current.

South Indian Current Eastward-flowing current of the southern Indian Ocean that is continuous with the northern edge of the West Wind Drift.

South Pacific Current Eastward-flowing current of the South Pacific Ocean that is continuous with the northern edge of the West Wind Drift.

spat Spawn or young of bivalve molluscs.

species In biological classification, a "kind" of organism. Organisms of the same species are usually capable of interbreeding to produce fertile young.

specific gravity Ratio of the density of a given substance to that of distilled water usually at 4°C and at a pressure of one atmosphere.

specific heat Heat capacity of a system per unit mass, that is, the ratio of the heat absorbed (or released) by unit mass of the system to the corresponding temperature rise (or fall). The amount of heat required to raise the temperature of 1 g of water by 1°C. The specific heat of water, which for pure water at 17.5°C (63.5°F) is 1 cal/g, decreases with increasing temperature and salinity.

spicule Minute, needlelike calcareous or siliceous body in sponges, radiolarians, primitive chitons, and echinoderms.

spillers *See* wave.

spit Small point of land or narrow shoal projecting into a body of water from the shore.

sponge One of a phylum (Porifera) of solitary or colonial sessile animals of simple construction. Sponges are of many sizes and forms and are varied in color.

spring tide Tide of increased range, which occurs about every 2 weeks when the moon is new or full.

squid One of an order (Decapoda) of cephalopods in which the body is cigar-shaped or globose and bears ten arms, eight of which are of equal length, with suckers along the entire length, and two of which are longer, with suckers only on a broad terminal portion; shell, in most, is embedded in the body or absent. The giant squid is the largest known invertebrate and a food of the sperm whale.

standard depth Depth below the sea surface at which water properties should be measured and reported. The internationally accepted depths (in meters) are: 0, 10, 20, 30, 50, 75, 100, 150, 200, 250, 300, 400, 500, 600, 800, 1,000, 1,200, 1,500, 2,000, 2,500, 3,000, 4,000, 5,000, 6,000, 7,000, 8,000, 9,000, 10,000, to which the National Oceanographic Data Center has added 125, 700, 900, 1,100, 1,300, 1,400, and 1,750.

standing wave Type of wave in which the surface of the water oscillates vertically between fixed points, called nodes, without progression. It may result from two equal progressive wave trains traveling through each other in opposite directions.

starved current Current with very little suspended sediment load that is capable of carrying more sediment.

storm surge Rise above normal water level on the open coast due only to the action of wind stress on the water surface. A storm surge resulting from a hurricane or other intense storm also includes the rise in level due to atmospheric pressure reduction as well as that due to wind stress. A storm surge is more severe when it occurs in conjunction with a high tide.

strait Narrow sea channel that separates two land masses.

subbottom reflection Return of sound energy from a discontinuity in material below the sea-bottom surface.

subduction zone Elongate crustal block that apparently is descending from the crust into the mantle of the earth; characterized by a zone of descending earthquake foci recorded over a period of time and frequent earthquakes.

sublittoral Benthic region extending from mean low water to a depth of about 200 m, or the edge of a continental shelf.

submarine canyon Relatively narrow, deep, undersea depression with steep slopes, the bottom of which grades continuously downward.

subphylum Category of related biological classes within a phylum.

substrate Base on which an organism lives.

subsurface current Current usually flowing below the thermocline, generally at slower speeds and frequently in a different direction from the currents near the surface.

Subtropical Convergence Zone of converging currents generally located in mid-latitudes. It is fairly well defined in the Southern Hemisphere, where it appears as an earth-girding region within which the surface temperature increases equatorward.

supralittoral Pertaining to the shore zone immediately above high-tide level; commonly, the zone kept more or less moist by waves and spray.

surf Collective term for breakers, or the wave activity between the shoreline and the outermost limit of breakers.

surface circulation (or wind-driven) A general term referring to that part of the directly observable ocean surface waters that extends from the surface to 150 m (495 ft) and is caused largely by atmospheric circulation.

surface current Part of a directly observed water movement that, in nearshore areas, does not extend more than 1 to 3 m below the surface; in deep or open ocean areas, surface currents generally are considered to extend from the surface to depths of about 10 m. When surface currents are computed by theoretical methods, the volume of water in the mixed layer (above the thermocline) from the surface to depths of about 50 to 150 m is generally referred to as surface current.

surface duct Zone immediately below the sea surface where sound rays are refracted toward the surface and then reflected. They are refracted because the sound velocity at some depth near the surface is greater than at the surface. The rays are alternately refracted and reflected along the duct to considerable distances from the sound source.

surface tension Phenomenon peculiar to the surface of liquids, caused by a strong attraction toward the interior of the liquid acting on the liquid molecules in or near the surface in such a way as to reduce the surface area. An actual tension results.

surface waters Layer of mixed water above the thermocline.

surface wave Progressive gravity wave in which the disturbance (that is, the particle movement in the fluid mass as well as the surface movement) is confined to the upper limits of a body of water. Strictly speaking, this term applies to progressive gravity waves whose speed depends only on wavelength.

surf beat Irregular oscillations of the nearshore water level with periods of the order of several minutes.

surf zone Area between the outermost breaker and the limit of wave uprush.

surge Name applied to wave motion with a period intermediate between that of the ordinary wind wave and that of the tide, from about ½ to 60 minutes. It is of low height, usually less than 9 cm.

suspension feeder An animal that feeds by collecting living and nonliving organic particles from the water. May use a sophisticated filter system or other means of collecting the particles. Example: sponges.

sverdrup Unit of volume transport equal to 1 million cubic meters per second.

swamp Lowland, marshlike region saturated usually with fresh water rather than seawater.

swash Rush of water up onto the beach following the breaking of a wave.

swash mark Thin, wavy line of fine sand, mica scales, bits of seaweed, and so on, left by the swash when it recedes from its upward limit of movement on the beach face.

swell Ocean waves that have traveled out of their generating area. Swell characteristically exhibits a more regular and longer period and has flatter crests than waves within their fetch.

symbiosis Relationship between two species in which one or both species are benefited and neither is harmed. *See* commensalism, mutualism, parasitism, predation.

synodic month (also called the lunar month). Period of a complete revolution of the moon around the earth. The period between successive new moons.

syzygy Two points in the moon's orbit when the moon is in conjunction or opposition to the sun relative to the earth; time of new or full moon in the cycle of phases.

tabular iceberg Flat-topped iceberg, usually calved from an ice shelf.

taxon (taxa) In biological classification, a category or group of related organisms.

tectonics Study of origin and development of the broad structural features of the earth.

tektite Black to greenish glassy object of rounded shape; unknown origin but thought to be extra-terrestrial (exploded meteor or asteroid).

temperature-salinity (T-S) diagram Plot of temperature versus salinity data of a water column.

tephra Collective term for all clastic volcanic materials that are ejected during an eruption from a crater.

terminal moraine Ridge of poorly sorted sediment that marks the farthest extent of a glacier or ice sheet, such as Long Island or Cape Cod.

terrace Benchlike structure bordering an undersea feature.

terrigenous sediment Sediment derived from the land.

test Shell or supporting structure of many invertebrates.

thermistor Solid state resistor whose action depends on changes in resistance as a function of changes in temperature.

thermistor chain Instrument-carrying chain (up to 400 m long) generally towed astern to get continuous temperature recordings from upper water layers at sea.

thermocline Vertical negative temperature gradient in some layer of a body of water that is appreciably greater than the gradients above and below it; also a layer in which such a gradient occurs. The principal thermoclines in the ocean are either seasonal, due to heating of the surface water in summer, or permanent.

thermohaline Pertaining to both temperature and salinity, for example, thermohaline circulation.

thermohaline circulation Vertical circulation induced by surface cooling, which causes convective overturning and consequent mixing.

thermohaline convection Vertical movement of water observed when seawater, because of its decreasing temperature or increasing salinity, becomes heavier than the water underneath it, and the vertical equilibrium is disturbed.

tidal age Difference in time between the local observed zenith of the moon's daily transit and the accompanying high tide.

tidal bore High, breaking wave of water advancing rapidly up an estuary. Bores can occur at the mouths of shallow rivers if the tide range at the mouth is large. They can also be generated in a river when tsunamis enter shallow coastal water and propagate up the river.

tidal current Alternating horizontal movement of water associated with the rise (flood) and fall (ebb) of the tide caused by the astronomical tide-producing forces.

tidal prism Difference between the mean high-water volume and the mean low-water volume of an estuary. The volume of water that is moved in and out of an estuary, bay, or harbor with the change in tides.

tidal range Difference in height between consecutive high and low waters. Where the tide is diurnal, the mean range is the same as the diurnal range.

tidal wave In popular usage, any unusually high (and therefore destructive) water level along a shore. It usually refers incorrectly to either a storm surge or a tsunami. The term may correctly be applied to the arrival tide, for example, if the tide is treated as a wave, the crest being a high tide.

tide Periodic rising and falling of the earth's oceans and atmosphere. It results from the tide-producing forces of the moon and sun acting on the rotating earth. Sometimes the periodic horizontal movements of the water along coastlines are also called tides, but these are more correctly called tidal currents.

tide force(s) Slight local difference between the gravitational attraction of two astronomical bodies and the centrifugal force that holds them apart. These forces are exactly equal and opposite at the center of gravity of either of the bodies, but since gravitational attraction is inversely proportional to the square of the distance, it varies from point to point on the surface of the bodies. Therefore, gravitational attraction predominates at the surface point nearest to the other body, while centrifugal "repulsion" predominates at the surface point farthest from the other body.

tideland Land that is under water at high tide and uncovered at low tide. Tideland, beach, strand, and seashore have nearly the same meanings. Tideland refers to the land sometimes covered by tidewater.

titration Chemical method for determining the concentration of a substance in solution. This concentration is established in terms of the smallest amount of the substance required to bring about a given effect in reaction with another known solution or substance. The most common titration is that for chlorinity.

topography Configuration of a surface, including its relief. In oceanology, the term is applied to a surface such as the sea bottom or a surface of given characteristics within the water mass.

tracer Foreign substance mixed with (or attached to) a given substance used to determine distribution or location.

trade winds The wind system occupying most of the tropics that blows from the subtropical highs toward the equatorial trough; a major component of the general circulation of the atmosphere. The winds are northeasterly in the Northern Hemisphere and southeasterly in the Southern Hemisphere.

trailing-edge coast Those coasts that occur where coasts of continents and islands are moving away from mid-oceanic ridges.

transform boundary Seismically active boundary between two global plates sliding past one another; example is San Andreas Fault, which is the boundary between Pacific and North American plates; plate boundary along which there is lateral displacement.

transform faults Fractures in the earth revealing large, horizontal displacement in the region of mid-oceanic ridges.

translation margin Continental margin cut by a transform fault system as found in southern California.

trawl Bag- or funnel-shaped net to catch bottom fish by dragging along the bottom. Also applied to research nets designed on the same principles.

trench A long, narrow, and deep depression of the seafloor with relatively steep sides.

trophic level A position in a food chain. Plants occupy the first trophic level, herbivores the second trophic level, and carnivores occupy the third, fourth and higher levels, depending on which animals they eat.

tropical cyclone General term for a cyclone that originates over the tropical oceans (hurricane and typhoon, for example).

tropical year (that is, solar year, equinoctial year, astronomical year). The interval between one vernal equinox and the next; 365 d, 5 h, 48 m, 46 s.

trough 1. Long depression of the seafloor normally wider and shallower than a trench. 2. A shallow, small-scale depression between a beach and its offshore sand bar (longshore bar). Commonly called a longshore trough. *See* wave trough.

tsunami (seismic sea wave) Long-period sea wave produced by a submarine earthquake or volcanic eruption. It may travel unnoticed across the ocean for thousands of miles from its point of origin and build up to great heights over shoal water.

turbidites Sediment or rock deposited by a turbidity current.

turbidity current Highly turbid, relatively dense current carrying large quantities of clay, silt, and sand in suspension, which flows down a submarine slope through less dense seawater.

typhoon (typhon). Severe tropical cyclone in the western Pacific.

ultrasonics Technology of sound at frequencies above the audio range (that is, above 20,000 cycles per second).

ultraviolet radiation Electromagnetic radiation of shorter wavelength than visible radiation but longer than X rays; roughly, radiation in the wavelength interval from 1 to 400 nm.

upwelling Process by which water rises from a lower to a higher depth, usually as a result of divergence and offshore currents. *See* sinking. It influences climate by bringing colder water to the surface. The upwelled water, besides being cooler, is richer in plant nutrients, so that regions of upwelling are also generally regions of rich fisheries.

vertebrate Member of the subphylum Vertebrata (phylum: Chordata). It contains the fish, amphibians, reptiles, birds, and mammals. Vertebrates have a skull that surrounds a well-developed brain and skeleton of cartilage or bone that includes a vertebral column extending longitudinally through the main body axis to provide support.

viscosity Molecular property of a fluid that enables it to support tangential stresses for a finite time and thus to resist deformation.

vorticity Corresponds to the angular velocity in a tiny part of a fluid involved in curved motion.

wake Region of turbulence immediately behind a boat or other solid object in motion on the water.

water age Time elapsed since a water mass was last at the surface and in contact with the atmosphere. Water age helps indicate the rate of overturn, an important factor in the use of the oceans for dumping radioactive wastes and determining the rate of replenishment of nutrients.

water mass Body of water usually identified by its temperature-salinity curve or chemical content, normally consisting of a mixture of two or more water types. *See* temperature-salinity diagram.

waterspout Usually, a tornado occurring over water; rarely, a lesser whirlwind over water, comparable in intensity to a dust devil over land.

water type Seawater of a specific temperature and salinity and hence defined by a single point on a temperature-salinity diagram.

wave 1. Disturbance that moves through or over the surface of a medium with speed dependent on the properties of the medium. 2. Ridge, deformation, or undulation of the surface of a liquid.

wave crest Highest point of a wave.

wave height Vertical distance between a wave crest and the preceding wave trough.

wavelength Distance between corresponding points of two successive periodic waves in the direction of propagation, for which the oscillation has the same phase. Unit of measurement is meters.

wave train Series of waves moving in the same direction.

wave trough Lowest point of a wave form between successive wave crests. Also that part of a wave below still-water level.

westerlies Winds that normally blow from west to east; found between latitudes 35° and 65° in the Northern and Southern hemispheres.

West Wind Drift Ocean current with the largest volume transport (approximately 100–150 x 10^6 m^3/s (100–150 sv); it flows from west to east around the Antarctic continent and is formed partly by the strong westerly wind in this region and partly by density differences.

wetlands Land on which water dominates the soil development, and the types of plant and animal life in the soil and its surface.

whirlpool Water rapidly moving in a circular path; an eddy or vortex of water.

windward Pertaining to the direction from which the wind is blowing.

wind wave Wave generated by wind.

XBT *See* expendable bathythermograph.

zooplankton Animal forms of plankton. They include various crustaceans, such as copepods and euphausiids; jellyfish; certain protozoans; worms; molluscs; and the eggs and larvae of benthic and nektonic animals.

zooxanthella (-ae) Single-celled plants that live within the tissues of coral polyps and other animals. The plants supply their animal hosts with carbohydrates; the hosts supply the plants with CO_2 and nutrients.

Index

Note: Page numbers appearing in **boldface** indicate where definitions of key terms can be found in the text; these terms also appear in the glossary. Page numbers in *italics* indicate illustrations, tables, and figures.

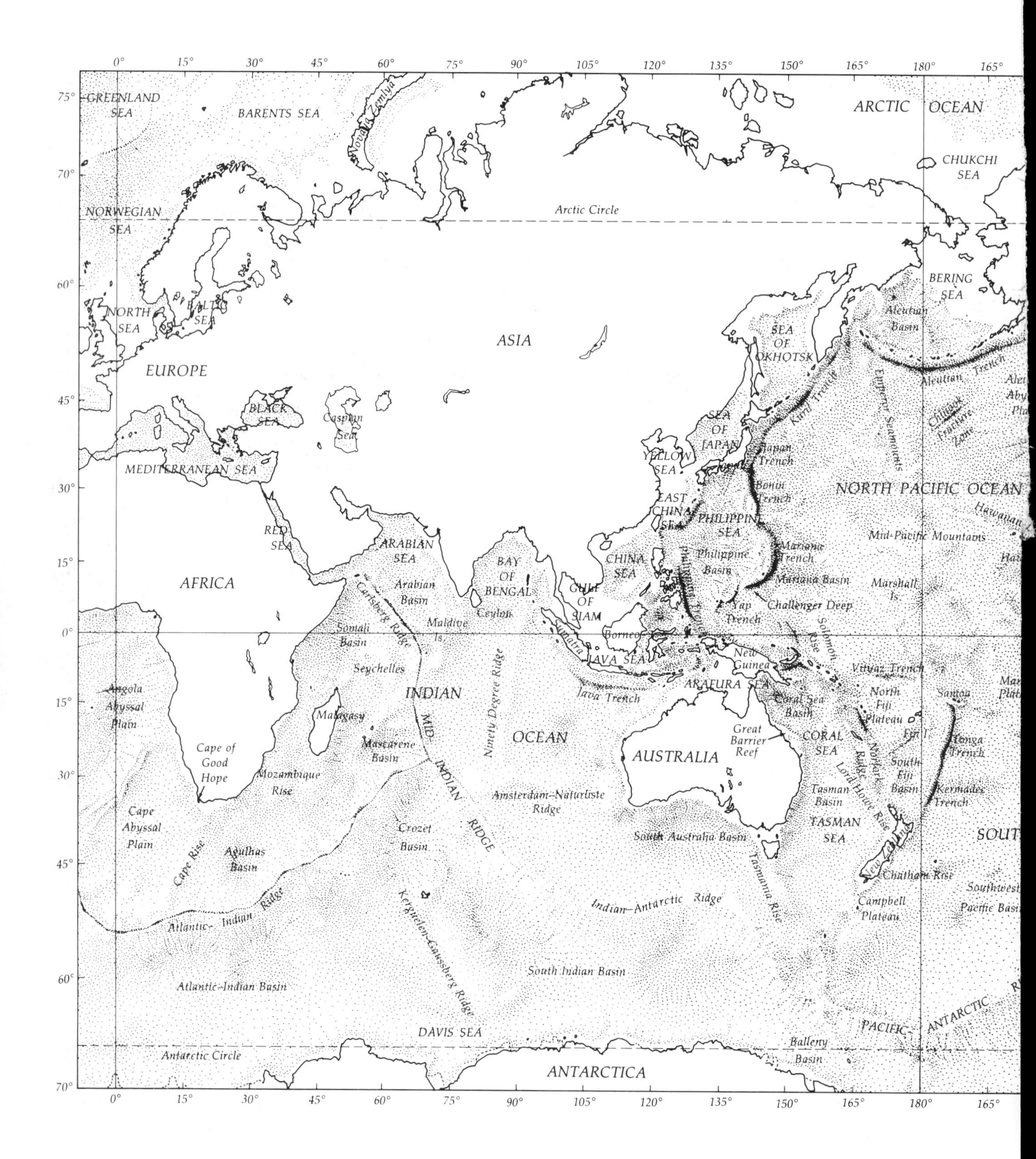

GREENLAND SEA
BARENTS SEA
Novaya Zemlya
ARCTIC OCEAN
CHUKCHI SEA
NORWEGIAN SEA
Arctic Circle
BERING SEA
Aleutian Basin
NORTH SEA
BALTIC SEA
SEA OF OKHOTSK
ASIA
EUROPE
Aleutian Trench
Kuril Trench
Emperor Seamounts
BLACK SEA
Caspian Sea
SEA OF JAPAN
Chinook Fracture Zone
Japan Trench
YELLOW SEA
MEDITERRANEAN SEA
Bonin Trench
NORTH PACIFIC OCEAN
EAST CHINA SEA
PHILIPPINE SEA
RED SEA
ARABIAN SEA
Mid-Pacific Mountains
Philippine Basin
Mariana Trench
CHINA SEA
BAY OF BENGAL
Arabian Basin
AFRICA
Mariana Basin
Marshall Is.
GULF OF SIAM
Carlsberg Ridge
Yap Trench
Challenger Deep
Ceylon
Maldive Is.
Somali Basin
Borneo
Sumatra
JAVA SEA
New Guinea
Seychelles
Vityaz Trench
ARAFURA SEA
Angola Abyssal Plain
INDIAN
Java Trench
North Fiji Plateau
Coral Sea Basin
Ninety Degree Ridge
Malagasy
MID
Great Barrier Reef
CORAL SEA
Fiji I.
Tonga Trench
Mascarene Basin
OCEAN
AUSTRALIA
Cape of Good Hope
Norfolk Ridge
South Fiji Basin
INDIAN
Mozambique Rise
Tasman Basin
Kermadec Trench
Cape Abyssal Plain
Amsterdam–Naturaliste Ridge
Lord Howe Rise
RIDGE
TASMAN SEA
Crozet Basin
South Australia Basin
Cape Rise
Agulhas Basin
Tasmania Rise
New Zealand
Chatham Rise
Southwest Pacific Basin
Indian–Antarctic Ridge
Campbell Plateau
Atlantic–Indian Ridge
Kerguelen–Gaussberg Ridge
South Indian Basin
Atlantic–Indian Basin
PACIFIC–ANTARCTIC
DAVIS SEA
Balleny Basin
Antarctic Circle
ANTARCTICA
0°
15°
30°
45°
60°
75°
90°
105°
120°
135°
150°
165°
180°
70°

135°
120°
105°
90°
75°
60°
45°
30°
15°
0°
15°
30°
45°
60°
75°
70°
60°
45°
30°
15°
0°
15°
30°
45°
60°
70°
BAFFIN BAY
GREENLAND
GREENLAND SEA
BARENTS SEA
Novaya Zemlya
Norwegian Basin
NORWEGIAN SEA
Iceland
Surtsey I.
Faeroe Is.
HUDSON BAY
LABRADOR SEA
Reykjanes Ridge
Rockall Bank
NORTH SEA
BALTIC SEA
Labrador Basin
Ireland
England
NORTH AMERICA
NORTH ATLANTIC OCEAN
EUROPE
Hudson Canyon
Grand Banks
Flemish Cap
Biscay Abyssal Plain
AMERICA
Fracture Zone
Monterey Canyon
Blake Plateau
Kelvin Seamounts
Corner Seamounts
Azores
Azores Rise
BLACK SEA
Caspian Sea
Baja California Seamount Province
Tongue of the Ocean
Hatteras Abyssal Plain
Bermuda Rise
Atlantis F. Z.
MID ATLANTIC RIDGE
Madeira Abyssal Plain
MEDITERRANEAN SEA
Scripps-LaJolla canyons
GULF OF MEXICO
Puerto Rico Trench
Cape Verde Abyssal Plain
RED SEA
Mid America Trench
CARIBBEAN SEA
AFRICA
Clipperton Fracture Zone
Lesser Antilles
Demerara-Ceara Abyssal Plains
Sierra Leone Rise
Guinea Abyssal Plain
Carlsberg Ridge
Cocos Ridge
Galapagos Fracture Zone
Galapagos Is.
Romanche Fracture Zone
Congo Canyon
Seychelles
Marquesas Fracture Zone
Saint Paul's Rocks
Peru Trench
SOUTH AMERICA
Pernambuco Abyssal Plain
Angola Abyssal Plain
INDIAN OCEAN
Tuamoto Archipelago
Galapagos Rise
Nasca Ridge
Malagasy
EAST PACIFIC RISE
Chile Trench
Trindade Seamounts
Easter Fracture Zone
Walvis Ridge
Cape of Good Hope
Rio Grande Rise
Mozambique Rise
SOUTH
Tristan de Cunha
Cape Abyssal Plain
PACIFIC
Chile Rise
Crozet Basin
Argentine Basin
ATLANTIC
Cape Rise
Agulhas Basin
OCEAN
OCEAN
Eltanin Fracture Zone
Falkland Rise
South Georgia Rise
Atlantic-Indian Ridge
Cape Horn
South Sandwich Trench
West Scotia Basin
East Scotia Basin
SCOTIA SEA
Atlantic-Indian Basin
Southeast Pacific Basni
Antarctic Circle
ANTARCTICA
DNH 73